The field of quantum optics has witnessed significant developments in recent years, from the laboratory realization of counter-intuitive concepts such as lasing without inversion and micromasers, to the investigation of fundamental issues in quantum mechanics, such as complementarity and hidden variables. This book provides an in-depth and wide-ranging introduction to the subject of quantum optics, emphasizing throughout the basic principles and their applications.

The book begins by developing the basic tools of quantum optics, and goes on to show the application of these tools in a variety of quantum optical systems, including resonance fluorescence, lasers, micromasers, squeezed states, and atom optics. The final four chapters are devoted to a discussion of quantum optical tests of the foundations of quantum mechanics, and particular aspects of measurement theory.

Assuming only a background of standard quantum mechanics and electromagnetic theory, and containing many problems and references, this book will be invaluable to graduate students of quantum optics, as well as to established researchers in this field.

Quantum optics

Marlan O. Scully has received numerous honors as a result of his many pioneering contributions as a researcher and educator on both sides of the Atlantic. These include: the Adolph Lomb Medal of the Optical Society of America, the Elliot Cresson Medal of the Franklin Society, Guggenheim Fellowship, and the Alexander von Humboldt Award. He is an elected member of the Max-Planck Society and a Fellow of the American Physical Society, the Optical Society of America, and the American Association of the Advancement of Science. He has held the faculty positions at Yale, MIT, University of Arizona, University of New Mexico and is currently Burgess Distinguished Professor of Physics, Director of the Center for Theoretical Physics at Texas A&M University, Co-Director of the Texas Laser Lab., and Auswrtiges Wissenschaftliches Mitglied at the Max-Planck Institut für Quantenoptik.

M. Suhail Zubairy received his PhD degree from the University of Rochester in 1978. He held research and faculty positions at the University of Rochester, University of Arizona, and the University of New Mexico before joining the Quaid-i-Azam University, Islamabad in 1984 where he is presently a Professor and founding Chairman of the Department of Electronics. He has held visiting appointments at the Max-Planck Institut für Quantenoptik, Universität Ulm, University of New Mexico, and the University of Campinas. He has been an Associate Member of the International Centre for Theoretical Physics, Trieste. He was awarded the Order of Sitara-e-Imtiaz by the President of Pakistan in 1993. He has received the Salam Prize for Physics and a Gold Medal from the Pakistan Academy of Sciences. He is a Fellow of the Pakistan Academy of Sciences and the Optical Society of America.

Quantum optics

Marlan O. Scully
Texas A&M University and Max-Planck-Institut für Quantenoptik

M. Suhail Zubairy
Quaid-i-Azam University

CAMBRIDGE
UNIVERSITY PRESS

CAMBRIDGE UNIVERSITY PRESS
Cambridge, New York, Melbourne, Madrid, Cape Town, Singapore,
São Paulo, Delhi, Dubai, Tokyo, Mexico City

Cambridge University Press
The Edinburgh Building, Cambridge CB2 8RU, UK

Published in the United States of America by Cambridge University Press, New York

www.cambridge.org
Information on this title: www.cambridge.org/9780521435956

First published 1997
Sixth printing 2008

A catalogue record for this publication is available from the British Library

Library of Congress Cataloguing in Publication Data

Scully, Marlan O. (Marlan Orvil), 1939–
Quantum optics/Marlan O. Scully and M. Suhail Zubairy.
p. cm
Includes bibliographical references and Index.
ISBN 0 521 43458 0 (hardback) - ISBN 0 521 43595 1 (paperback)
1. Quantum optics. I. Zubairy, Muhammad Suhail, 1952–. II. Title.
QC446.2.S4 1996
535-dc20 94-42949 CIP

ISBN 978-0-521-43458-4 Hardback
ISBN 978-0-521-43595-6 Paperback

Contents

To

Thelma T. Scully and Naseem Fatima Zubairy

and to the memory of

Orvil O. Scully and Muhibul Islam Zubairy

Preface

Quantum optics, the union of quantum field theory and physical optics, is undergoing a time of revolutionary change. The subject has evolved from early studies on the coherence properties of radiation like, for example, quantum statistical theories of the laser in the sixties to modern areas of study involving, for example, the role of squeezed states of the radiation field and atomic coherence in quenching quantum noise in interferometry and optical amplifiers. On the one hand, counter intuitive concepts such as lasing without inversion and single atom (micro) masers and lasers are now laboratory realities. Many of these techniques hold promise for new devices whose sensitivity goes well beyond the standard quantum limits. On the other hand, quantum optics provides a powerful new probe for addressing fundamental issues of quantum mechanics such as complementarity, hidden variables, and other aspects central to the foundations of quantum physics and philosophy.

The intent of this book is to present these and many other exciting developments in the field of quantum optics to students and scientists, with an emphasis on fundamental concepts and their applications, so as to enable the students to perform independent research in this field. The book (which has developed from our lectures on the subject at various universities, research institutes, and summer schools) may be used as a textbook for beginning graduate students with some background in standard quantum mechanics and electromagnetic theory. Each chapter is supplemented by problems and general references. Some of the problems rely heavily on the treatment given in a research paper, leading students directly to the scientific literature. The role of the references is to identify original papers, and to refer the reader to review articles and related papers for in-depth study. No attempt is made to give an exhaustive list of references.

The book is divided roughly into three parts. In the first six chapters, we develop the 'tools' of quantum optics. In the next eleven chapters, these 'tools' are applied to various quantum optical systems. In the last four chapters, we consider the application of modern quantum optical physics to testing the foundations of quantum mechanics.

The book opens with the presentation of the quantization of the radiation field by associating each mode of the field with a quantized harmonic oscillator. The strong motivation to quantize the radiation field in many quantum optical systems comes from phenomena such as quantum beats, two-photon interferometry, and the generation of nonclassical states of the radiation field, e.g., Fock states. Some of these phenomena shed new light on our understanding of the elusive concept of the photon. In the first part of the book, we discuss the various states of the radiation field, e.g., coherent and squeezed states, and introduce the distribution functions of the field which form a correspondence between the quantum and the classical theories of radiation. We then develop a quantum theory of coherence in terms of the correlation functions of the field, which provides a framework for discussing the outcome of interferometric experiments. We proceed to develop the semiclassical and quantum theories of the interaction of the radiation field with matter, with an emphasis on formulating a theoretical framework directed toward understanding the many faceted problems of modern quantum optics.

In the second part, we use this theoretical framework to develop theories of atomic and field damping, resonance fluorescence, laser and micromaser operation, and the study of the quantum noise properties of such nonlinear optical processes as parametric amplification and four-wave mixing. Atomic coherence effects in many novel systems are discussed in detail. For example, the role of atomic coherence in suppressing absorption leads to interesting effects such as lasing without inversion and electromagnetically induced transparency. Atomic coherence can also play a role in quenching Schawlow–Townes spontaneous emission noise in lasers, as in the correlated emission laser (CEL). Such CEL systems have potential applications in, e.g., laser gyro physics and 'noise-free' amplification.

In the third part, we move on to the application of modern quantum optical physics to fundamental questions related to the foundation of quantum mechanics. These include Bell's theorem, quantum nondemolition measurements, 'which-path' detectors, and two-photon interferometry.

We have benefited greatly from our interaction with many of our colleagues, friends, and students in the preparation of this book. They are too numerous to be individually acknowledged and we are able to express our gratitude to only a few of them here.

We would especially like to thank Stephen Harris, Willis Lamb, Julian Schwinger, and Herbert Walther, who have strongly influenced our thinking through their profound contributions to physics in general and many fruitful collaborations in particular. Their imprint on this book is evident: but for them, entire chapters would be missing. We are grateful to Peter Knight for providing the encouragement for writing this book. Critical comments from and helpful discussions with him and with Girish Agarwal, Richard Arnowitt, Chris Bednar, Janos Bergou, Leon Cohen, Jonathan Dowling, Joe Eberly, Michael Fleischhauer, Edward Fry, Julio Gea-Banacloche, Roy Glauber, Trung-Dung Ho, Hwang Lee, Lorenzo Narducci, Robert O'Connell, Norman Ramsey, Ulrich Rathe, Wolfgang Schleich, Krzysztof Wodkiewicz, Bernand Yurke, and Shi-Yao Zhu provided invaluable assistance concerning many subtle points. One of us (MSZ) would like to express his gratitude to the Pakistan Atomic Energy Commission for the financial support over the years, and particularly its Chairman, Ishfaq Ahmad, for his deep interest and commitment that played a vital role in the completion of this project. MOS would like to acknowledge the support of the Office of Naval Research, and particularly Herschel Pilloff, whose wisdom and dedication to scientific excellence have resulted in many successful joint projects and conferences which have had a marked impact on this book. The support of the Houston Advanced Research Center (HARC), and the Welch Foundation is also deeply appreciated. We thank Jeanne Williams for the careful typing in TEX, and Jim and Andrey Bailey for the hospitality of the Bailey ranch, where the manuscript was completed.

Finally, we are grateful to our family members, Judith, James, Debra, Robert, Steven, and Jacquelyn, and Parveen, Sarah, Sahar, and Raheel, for their support, and understanding, especially during the extended absences in the course of the last decade, when this book was contemplated, planned, and written.

Marlan O. Scully
M. Suhail Zubairy

CHAPTER 1

Quantum theory of radiation

Light occupies a special position in our attempts to understand nature both classically and quantum mechanically. We recall that Newton, who made so many fundamental contributions to optics, championed a particle description of light and was not favorably disposed to the wave picture of light. However, the beautiful unification of electricity and magnetism achieved by Maxwell clearly showed that light was properly understood as the wave-like undulations of electric and magnetic fields propagating through space.

The central role of light in marking the frontiers of physics continues on into the twentieth century with the ultraviolet catastrophe associated with black-body radiation on the one hand and the photoelectric effect on the other. Indeed, it was here that the era of quantum mechanics was initiated with Planck's introduction of the quantum of action that was necessary to explain the black-body radiation spectrum. The extension of these ideas led Einstein to explain the photoelectric effect, and to introduce the photon concept.

It was, however, left to Dirac* to combine the wave- and particle-like aspects of light so that the radiation field is capable of explaining all interference phenomena and yet shows the excitation of a specific atom located along a wave front absorbing one photon of energy. In this chapter, following Dirac, we associate each mode of the radiation field with a quantized simple harmonic oscillator, this is the essence of the quantum theory of radiation. An interesting consequence of the quantization of radiation is the fluctuations associated with the zero-

* The pioneering papers on the quantum theory of radiation by Dirac [1927] and Fermi [1932] should be read by every student of the subject. Excellent modern treatments are to be found in the textbooks by: Loudon, *The Quantum Theory of Light* [1973], Cohen-Tannoudji, Dupont-Roc, and Grynberg, *Atom–Photon Interactions* [1992], Weinberg, *Theory of Quantum Fields* [1995], and Pike and Sarkar, *Quantum Theory of Radiation* [1995].

point energy or the so-called vacuum fluctuations. These fluctuations have no classical analog and are responsible for many interesting phenomena in quantum optics. As is discussed at length in Chapters 5 and 7, a semiclassical theory of atom–field interaction in which only the atom is quantized and the field is treated classically, can explain many of the phenomena which we observe in modern optics. The quantization of the radiation field is, however, needed to explain effects such as spontaneous emission, the Lamb shift, the laser linewidth, the Casimir effect, and the full photon statistics of the laser. In fact, each of these physical effects can be understood from the point of view of vacuum fluctuations perturbing the atoms, e.g., spontaneous emission is often said to be the result of 'stimulating' the atom by vacuum fluctuations. However, as compelling as these reasons are for quantizing the radiation field, there are other strong reasons and logical arguments for quantizing the radiation field.

For example, the problem of quantum beat phenomena provides us with a simple example in which the results of self-consistent fully quantized calculation differ qualitatively from those obtained via a semiclassical theory with or without vacuum fluctuations. Another experiment wherein a quantized theory of radiation is required for the proper interpretation of the observed results is two-photon interferometry and the production of entangled states associated with such a configuration. This is discussed in detail in Chapter 21. Further support that the electromagnetic field is quantized is provided by the experimental observations of nonclassical states of the radiation field, e.g., squeezed states, sub-Poissonian photon statistics, and photon antibunching.

Following this brief motivation for the quantum theory of radiation, we now turn to the quantization of the free electromagnetic field.

1.1 Quantization of the free electromagnetic field

With the objective of quantizing the electromagnetic field in free space, it is convenient to begin with the classical description of the field based on Maxwell's equations. These equations relate the electric and magnetic field vectors $\mathbf{E}$ and $\mathbf{H}$, respectively, together with the displacement and inductive vectors $\mathbf{D}$ and $\mathbf{B}$, respectively, and have the form (in mks units):

$$\nabla \times \mathbf{H} = \frac{\partial \mathbf{D}}{\partial t}, \tag{1.1.1a}$$

$$\nabla \times \mathbf{E} = -\frac{\partial \mathbf{B}}{\partial t}, \tag{1.1.1b}$$

$$\nabla \cdot \mathbf{B} = 0, \tag{1.1.1c}$$

$$\nabla \cdot \mathbf{D} = 0, \tag{1.1.1d}$$

with the constitutive relations

$$\mathbf{B} = \mu_0 \mathbf{H}, \tag{1.1.2}$$

$$\mathbf{D} = \epsilon_0 \mathbf{E}. \tag{1.1.3}$$

Here ϵ_0 and μ_0 are the free space permittivity and permeability, respectively, and $\mu_0 \epsilon_0 = c^{-2}$ where c is the speed of light in vacuum.

It follows, on taking the curl of Eq. (1.1.1b) and using Eqs. (1.1.1a), (1.1.1d), (1.1.2), and (1.1.3), that $\mathbf{E}(\mathbf{r}, t)$ satisfies the wave equation

$$\nabla^2 \mathbf{E} - \frac{1}{c^2} \frac{\partial^2 \mathbf{E}}{\partial t^2} = 0. \tag{1.1.4}$$

In deriving Eq. (1.1.4) we also used $\nabla \times (\nabla \times \mathbf{E}) = \nabla(\nabla \cdot \mathbf{E}) - \nabla^2 \mathbf{E}$.

1.1.1 Mode expansion of the field

We first consider the electric field to have the spatial dependence appropriate for a cavity resonator of length L (Fig. 1.1). We take the electric field to be linearly polarized in the x-direction and expand in the normal modes of the cavity

$$E_x(z, t) = \sum_j A_j q_j(t) \sin(k_j z), \tag{1.1.5}$$

where q_j is the normal mode amplitude with the dimension of a length, $k_j = j\pi/L$, with $j = 1, 2, 3, \ldots$, and

$$A_j = \left(\frac{2\nu_j^2 m_j}{V \epsilon_0} \right)^{1/2}, \tag{1.1.6}$$

with $\nu_j = j\pi c/L$ being the cavity eigenfrequency, $V = LA$ (A is the transverse area of the optical resonator) is the volume of the resonator and m_j is a constant with the dimension of mass. The constant m_j has been included only to establish the analogy between the dynamical problem of a single mode of the electromagnetic field and that of the simple harmonic oscillator. The equivalent mechanical oscillator will have a mass m_j, and a Cartesian coordinate q_j. The nonvanishing component of the magnetic field H_y in the cavity* is obtained from Eq. (1.1.5):

* In the present treatment of field quantization in vacuum we are focussing on the electric $\mathbf{E}(\mathbf{r}, t)$ and magnetic $\mathbf{H}(\mathbf{r}, t)$ fields. In a material medium it is preferable to work with $\mathbf{D}(\mathbf{r}, t)$ and $\mathbf{B}(\mathbf{r}, t)$; see Bialynicki-Birula and Bialynicka-Birula [1976].

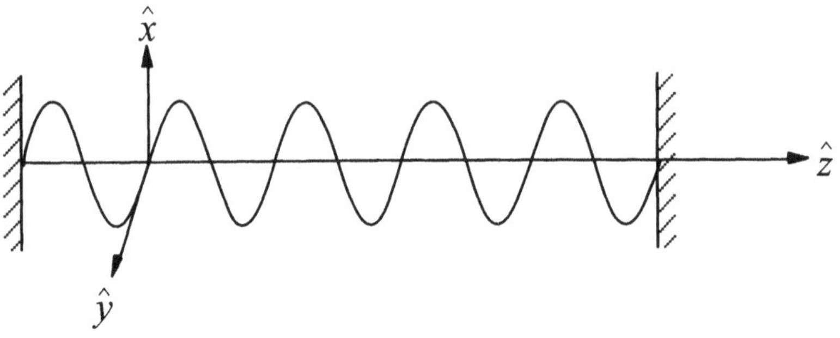

Fig. 1.1
Electromagnetic field of frequency ν inside a cavity. The field is assumed to be transverse with the electric field polarized in the x-direction.

$$H_y = \sum_j A_j \left(\frac{\dot{q}_j \epsilon_0}{k_j} \right) \cos(k_j z). \tag{1.1.7}$$

The classical Hamiltonian for the field is

$$\mathscr{H} = \frac{1}{2} \int_V d\tau (\epsilon_0 E_x^2 + \mu_0 H_y^2), \tag{1.1.8}$$

where the integration is over the volume of the cavity. It follows, on substituting from Eqs. (1.1.5) and (1.1.7) for E_x and H_y, respectively, in Eq. (1.1.8), that

$$\begin{aligned} \mathscr{H} &= \frac{1}{2} \sum_j (m_j \nu_j^2 q_j^2 + m_j \dot{q}_j^2) \\ &= \frac{1}{2} \sum_j \left(m_j \nu_j^2 q_j^2 + \frac{p_j^2}{m_j} \right), \end{aligned} \tag{1.1.9}$$

where $p_j = m_j \dot{q}_j$ is the canonical momentum of the jth mode. Equation (1.1.9) expresses the Hamiltonian of the radiation field as a sum of independent oscillator energies. Each mode of the field is therefore dynamically equivalent to a mechanical harmonic oscillator.

1.1.2 Quantization

The present dynamical problem can be quantized by identifying q_j and p_j as operators which obey the commutation relations

$$[q_j, p_{j'}] = i\hbar \delta_{jj'}, \tag{1.1.10a}$$

$$[q_j, q_{j'}] = [p_j, p_{j'}] = 0. \tag{1.1.10b}$$

It is convenient to make a canonical transformation to operators a_j and $a_j^\dagger$:

$$a_j e^{-i\nu_j t} = \frac{1}{\sqrt{2 m_j \hbar \nu_j}} (m_j \nu_j q_j + i p_j), \tag{1.1.11a}$$

$$a_j^\dagger e^{i\nu_j t} = \frac{1}{\sqrt{2m_j\hbar\nu_j}}(m_j\nu_j q_j - ip_j). \tag{1.1.11b}$$

In terms of a_j and $a_j^\dagger$, the Hamiltonian (1.1.9) becomes

$$\mathscr{H} = \hbar\sum_j \nu_j\left(a_j^\dagger a_j + \frac{1}{2}\right). \tag{1.1.12}$$

The commutation relations between a_j and $a_j^\dagger$ follow from those between q_j and p_j:

$$[a_j, a_{j'}^\dagger] = \delta_{jj'}, \tag{1.1.13}$$

$$[a_j, a_{j'}] = [a_j^\dagger, a_{j'}^\dagger] = 0. \tag{1.1.14}$$

The operators a_j and $a_j^\dagger$ are referred to as the annihilation and the creation operators, respectively. The reason for these names will become clear in the next section. In terms of a_j and $a_j^\dagger$, the electric and magnetic fields (Eqs. (1.1.5) and (1.1.7)) take the form

$$E_x(z,t) = \sum_j \mathscr{E}_j(a_j e^{-i\nu_j t} + a_j^\dagger e^{i\nu_j t})\sin k_j z, \tag{1.1.15}$$

$$H_y(z,t) = -i\epsilon_0 c\sum_j \mathscr{E}_j(a_j e^{-i\nu_j t} - a_j^\dagger e^{i\nu_j t})\cos k_j z, \tag{1.1.16}$$

where the quantity

$$\mathscr{E}_j = \left(\frac{\hbar\nu_j}{\epsilon_0 V}\right)^{1/2} \tag{1.1.17}$$

has the dimensions of an electric field.

So far we have considered the quantization of the radiation field in a finite one-dimensional cavity. We can now quantize the field in unbounded free space as follows.

We consider the field in a large but finite cubic cavity of side L. Here we regard the *cavity* merely as a region of space with no specific boundaries. We consider the running-wave solutions instead of the standing-wave solutions considered above and impose periodic boundary conditions.

The classical electric and magnetic fields can be expanded in terms of the plane waves

$$\mathbf{E}(\mathbf{r},t) = \sum_{\mathbf{k}} \hat{\epsilon}_{\mathbf{k}}\mathscr{E}_{\mathbf{k}}\alpha_{\mathbf{k}}e^{-i\nu_k t + i\mathbf{k}\cdot\mathbf{r}} + \text{c.c.}, \tag{1.1.18}$$

$$\mathbf{H}(\mathbf{r},t) = \frac{1}{\mu_0}\sum_{\mathbf{k}} \frac{\mathbf{k}\times\hat{\epsilon}_{\mathbf{k}}}{\nu_k}\mathscr{E}_{\mathbf{k}}\alpha_{\mathbf{k}}e^{-i\nu_k t + i\mathbf{k}\cdot\mathbf{r}} + \text{c.c.}, \tag{1.1.19}$$

where the summation is taken over an infinite discrete set of values of the wave vector $\mathbf{k} \equiv (k_x, k_y, k_z)$, $\hat{\epsilon}_{\mathbf{k}}$ is a unit polarization vector, $\alpha_{\mathbf{k}}$ is a dimensionless amplitude and

$$\mathscr{E}_{\mathbf{k}} = \left(\frac{\hbar \nu_k}{2\epsilon_0 V} \right)^{1/2}. \tag{1.1.20}$$

In Eqs. (1.1.18) and (1.1.19) c.c. stands for complex conjugate. The periodic boundary conditions require that

$$k_x = \frac{2\pi n_x}{L}, \quad k_y = \frac{2\pi n_y}{L}, \quad k_z = \frac{2\pi n_z}{L}, \tag{1.1.21}$$

where n_x, n_y, n_z are integers $(0, \pm 1, \pm 2, \ldots)$. A set of numbers (n_x, n_y, n_z) defines a mode of the electromagnetic field. Equation (1.1.1d) requires that

$$\mathbf{k} \cdot \hat{\epsilon}_{\mathbf{k}} = 0, \tag{1.1.22}$$

i.e., the fields are purely transverse. There are, therefore, two independent polarization directions of $\hat{\epsilon}_{\mathbf{k}}$ for each $\mathbf{k}$.

The change from a discrete distribution of modes to a continuous distribution can be made by replacing the sum in Eqs. (1.1.18) and (1.1.19) by an integral:

$$\sum_{\mathbf{k}} \rightarrow 2 \left(\frac{L}{2\pi} \right)^3 \int d^3k, \tag{1.1.23}$$

where the factor 2 accounts for two possible states of polarization.

In many problems, we shall be interested in the density of modes between the frequencies ν and $\nu + d\nu$. This can be obtained by transforming from the rectangular components $(k_x,\ k_y,\ k_z)$ to the polar coordinates $(k \sin\theta \cos\phi,\ k \sin\theta \sin\phi,\ k\cos\theta)$, so that the volume element in $\mathbf{k}$ space is

$$d^3k = k^2 dk \sin\theta d\theta d\phi = \frac{\nu^2}{c^3} d\nu \sin\theta d\theta d\phi. \tag{1.1.24}$$

The total number of modes in volume L^3 in the range between ν and $\nu + d\nu$ is given by

$$d\mathscr{N} = 2 \left(\frac{L}{2\pi} \right)^3 \frac{\nu^2 d\nu}{c^3} \int_0^{\pi} d\theta \sin\theta \int_0^{2\pi} d\phi = \frac{L^3 \nu^2}{\pi^2 c^3} d\nu. \tag{1.1.25}$$

Therefore the number of modes with frequencies in the range ν to $\nu + d\nu$ is

$$D(\nu) d\nu = \frac{L^3 \nu^2}{\pi^2 c^3} d\nu, \tag{1.1.26}$$

where $D(\nu)$ is called the mode density.

As before, the radiation field is quantized by identifying $\alpha_{\mathbf{k}}$ and $\alpha_{\mathbf{k}}^*$ with the harmonic oscillator operators $a_{\mathbf{k}}$ and $a_{\mathbf{k}}^\dagger$, respectively, which satisfy the commutation relation $[a_{\mathbf{k}}, a_{\mathbf{k}}^\dagger] = 1$. The quantized electric and magnetic fields take the form

$$\mathbf{E}(\mathbf{r}, t) = \sum_{\mathbf{k}} \hat{\epsilon}_{\mathbf{k}} \mathscr{E}_{\mathbf{k}} a_{\mathbf{k}} e^{-i\nu_k t + i\mathbf{k}\cdot\mathbf{r}} + \text{H.c.}, \tag{1.1.27}$$

$$\mathbf{H}(\mathbf{r}, t) = \frac{1}{\mu_0} \sum_{\mathbf{k}} \frac{\mathbf{k} \times \hat{\epsilon}_{\mathbf{k}}}{\nu_k} \mathscr{E}_{\mathbf{k}} a_{\mathbf{k}} e^{-i\nu_k t + i\mathbf{k}\cdot\mathbf{r}} + \text{H.c.}, \tag{1.1.28}$$

where H.c. stands for Hermitian conjugate. Usually the positive and negative frequency parts of these field operators are written separately. For example, the electric field operator $\mathbf{E}(\mathbf{r}, t)$ is written as

$$\mathbf{E}(\mathbf{r}, t) = \mathbf{E}^{(+)}(\mathbf{r}, t) + \mathbf{E}^{(-)}(\mathbf{r}, t), \tag{1.1.29}$$

where

$$\mathbf{E}^{(+)}(\mathbf{r}, t) = \sum_{\mathbf{k}} \hat{\epsilon}_{\mathbf{k}} \mathscr{E}_{\mathbf{k}} a_{\mathbf{k}} e^{-i\nu_k t + i\mathbf{k}\cdot\mathbf{r}}, \tag{1.1.30}$$

$$\mathbf{E}^{(-)}(\mathbf{r}, t) = \sum_{\mathbf{k}} \hat{\epsilon}_{\mathbf{k}} \mathscr{E}_{\mathbf{k}} a_{\mathbf{k}}^\dagger e^{i\nu_k t - i\mathbf{k}\cdot\mathbf{r}}. \tag{1.1.31}$$

Here $\mathbf{E}^{(+)}(\mathbf{r}, t)$ contains only the annihilation operators and its adjoint $\mathbf{E}^{(-)}(\mathbf{r}, t)$ contains only the creation operators.

1.1.3 Commutation relations between electric and magnetic field components

An important consequence of imposing the quantum conditions (1.1.13) and (1.1.14) is that as the electric and magnetic field strengths do not commute they are thus not measurable simultaneously. In order to show this we rewrite the quantized mode expansions (1.1.27) and (1.1.28) for $\mathbf{E}(\mathbf{r}, t)$ and $\mathbf{H}(\mathbf{r}, t)$, respectively by including explicitly the two states of polarization denoted by the symbol λ:

$$\mathbf{E}(\mathbf{r}, t) = \sum_{\mathbf{k},\lambda} \hat{\epsilon}_{\mathbf{k}}^{(\lambda)} \mathscr{E}_{\mathbf{k}} a_{\mathbf{k},\lambda} e^{-i\nu_k t + i\mathbf{k}\cdot\mathbf{r}} + \text{H.c.}, \tag{1.1.32}$$

$$\mathbf{H}(\mathbf{r}, t) = \frac{1}{\mu_0} \sum_{\mathbf{k},\lambda} \frac{\mathbf{k} \times \hat{\epsilon}_{\mathbf{k}}^{(\lambda)}}{\nu_k} \mathscr{E}_{\mathbf{k}} a_{\mathbf{k},\lambda} e^{-i\nu_k t + i\mathbf{k}\cdot\mathbf{r}} + \text{H.c.} \tag{1.1.33}$$

The corresponding commutation relations between the operators $a_{\mathbf{k},\lambda}$ and $a_{\mathbf{k},\lambda}^\dagger$ are

$$\begin{aligned} &[a_{\mathbf{k},\lambda}, a_{\mathbf{k}',\lambda'}] = [a_{\mathbf{k},\lambda}^\dagger, a_{\mathbf{k}',\lambda'}^\dagger] = 0, \\ &[a_{\mathbf{k},\lambda}, a_{\mathbf{k}',\lambda'}^\dagger] = \delta_{\mathbf{k}\mathbf{k}'} \delta_{\lambda\lambda'}. \end{aligned} \tag{1.1.34}$$

It then follows that the equal time commutator between the field components is given by

$$[E_x(\mathbf{r},t), H_y(\mathbf{r}',t)] = \frac{\hbar c^2}{2V}\sum_{\mathbf{k},\lambda} \epsilon_{\mathbf{k}x}^{(\lambda)}\left[\epsilon_{\mathbf{k}x}^{(\lambda)}k_z - \epsilon_{\mathbf{k}z}^{(\lambda)}k_x\right] \times \left[e^{i\mathbf{k}\cdot(\mathbf{r}-\mathbf{r}')} - e^{-i\mathbf{k}\cdot(\mathbf{r}-\mathbf{r}')}\right], \quad (1.1.35)$$

where $\epsilon_{\mathbf{k}i}^{(\lambda)}$ $(i = x, y, z)$ is the ith component of $\hat{\epsilon}_{\mathbf{k}}^{(\lambda)}$. We proceed by using the operator identity of Problem 1.9 to write

$$\hat{\epsilon}_{\mathbf{k}}^{(1)}\hat{\epsilon}_{\mathbf{k}}^{(1)} + \hat{\epsilon}_{\mathbf{k}}^{(2)}\hat{\epsilon}_{\mathbf{k}}^{(2)} + \frac{\mathbf{kk}}{k^2} = \mathbf{1}, \quad (1.1.36)$$

where $\hat{\epsilon}_{\mathbf{k}}^{(1)}\hat{\epsilon}_{\mathbf{k}}^{(1)}$, $\hat{\epsilon}_{\mathbf{k}}^{(2)}\hat{\epsilon}_{\mathbf{k}}^{(2)}$, and $\mathbf{kk}$ denote dyadic products. One can verify that taking the inner product of (1.1.36) with the Cartesian unit vector $\hat{\mathbf{e}}_i$ from the left and $\hat{\mathbf{e}}_j$ from the right yields

$$\epsilon_{\mathbf{k}i}^{(1)}\epsilon_{\mathbf{k}j}^{(1)} + \epsilon_{\mathbf{k}i}^{(2)}\epsilon_{\mathbf{k}j}^{(2)} = \delta_{ij} - \frac{k_i k_j}{k^2}. \quad (1.1.37)$$

The summation over the polarization states in Eq. (1.1.35) can now be carried out using (1.1.37). The resulting expression for the commutator is

$$[E_x(\mathbf{r},t), H_y(\mathbf{r}',t)] = \frac{\hbar c^2}{2V}\sum_{\mathbf{k}} k_z \left[e^{i\mathbf{k}\cdot(\mathbf{r}-\mathbf{r}')} - e^{-i\mathbf{k}\cdot(\mathbf{r}-\mathbf{r}')}\right]. \quad (1.1.38)$$

We now replace the summation by an integral via

$$\sum_{\mathbf{k}} \to \frac{V}{(2\pi)^3}\int d^3k. \quad (1.1.39)$$

The factor of 2 has not been included as was done in Eq. (1.1.23) because, in the present case, we have summed over two polarization states explicitly. We obtain

$$[E_x(\mathbf{r},t), H_y(\mathbf{r}',t)] = -i\hbar c^2 \frac{\partial}{\partial z}\delta^{(3)}(\mathbf{r}-\mathbf{r}'). \quad (1.1.40)$$

In general

$$[E_j(\mathbf{r},t), H_j(\mathbf{r}',t)] = 0 \quad (j = x, y, z), \quad (1.1.41)$$

$$[E_j(\mathbf{r},t), H_k(\mathbf{r}',t)] = -i\hbar c^2 \frac{\partial}{\partial \ell}\delta^{(3)}(\mathbf{r}-\mathbf{r}'), \quad (1.1.42)$$

where j, k, and ℓ form a cyclic permutation of x, y, and z.

We, therefore, conclude that the parallel components of **E** and **H** may be measured simultaneously whereas the perpendicular components cannot.

1.2 Fock or number states

In this section we first restrict ourselves to a single mode of the field of frequency ν having creation and annihilation operators $a^\dagger$ and a, respectively. Let $|n\rangle$ be the energy eigenstate corresponding to the energy eigenvalue E_n, i.e.,

$$\mathscr{H}|n\rangle = \hbar\nu\left(a^\dagger a + \frac{1}{2}\right)|n\rangle = E_n|n\rangle. \tag{1.2.1}$$

If we apply the operator a from the left, we obtain after using the commutation relation $[a, a^\dagger] = 1$ and some rearrangement

$$\mathscr{H}a|n\rangle = (E_n - \hbar\nu)a|n\rangle. \tag{1.2.2}$$

This means that the state

$$|n-1\rangle = \frac{a}{\alpha_n}|n\rangle, \tag{1.2.3}$$

is also an energy eigenstate but with the reduced eigenvalue

$$E_{n-1} = E_n - \hbar\nu. \tag{1.2.4}$$

In Eq. (1.2.3), α_n is a constant which will be determined from the normalization condition

$$\langle n-1|n-1\rangle = 1. \tag{1.2.5}$$

If we repeat this procedure n times we move down the energy ladder in steps of $\hbar\nu$ until we obtain

$$\mathscr{H}a|0\rangle = (E_0 - \hbar\nu)a|0\rangle. \tag{1.2.6}$$

Here E_0 is the ground state energy such that $(E_0 - \hbar\nu)$ would correspond to an energy eigenvalue smaller than E_0. Since we do not allow energies lower than E_0 for the oscillator, we must conclude

$$a|0\rangle = 0. \tag{1.2.7}$$

The state $|0\rangle$ is referred to as the vacuum state. Using this relation we can find the value of E_0 from the eigenvalue equation

$$\mathscr{H}|0\rangle = \frac{1}{2}\hbar\nu|0\rangle = E_0|0\rangle. \tag{1.2.8}$$

This gives

$$E_0 = \frac{1}{2}\hbar\nu. \tag{1.2.9}$$

It then follows from Eq. (1.2.4) that

$$E_n = \left(n + \frac{1}{2}\right)\hbar\nu. \tag{1.2.10}$$

From Eq. (1.2.1), we obtain

$$a^\dagger a|n\rangle = n|n\rangle, \tag{1.2.11}$$

i.e., the energy eigenstate $|n\rangle$ is also an eigenstate of the 'number' operator

$$n = a^\dagger a. \tag{1.2.12}$$

The normalization constant α_n in Eq. (1.2.3) can now be determined.

$$\langle n-1|n-1\rangle = \frac{1}{|\alpha_n|^2}\langle n|a^\dagger a|n\rangle = \frac{n}{|\alpha_n|^2}\langle n|n\rangle = \frac{n}{|\alpha_n|^2} = 1. \tag{1.2.13}$$

If we take the phase of the normalization constant α_n to be zero then $\alpha_n = \sqrt{n}$. Equation (1.2.3) then becomes

$$a|n\rangle = \sqrt{n}|n-1\rangle. \tag{1.2.14}$$

We can proceed along the same lines with the operator $a^\dagger$. The resulting equation is

$$a^\dagger|n\rangle = \sqrt{n+1}\,|n+1\rangle. \tag{1.2.15}$$

A repeated use of this equation gives

$$|n\rangle = \frac{(a^\dagger)^n}{\sqrt{n!}}|0\rangle. \tag{1.2.16}$$

It is useful to interpret the energy eigenvalues (1.2.10) as corresponding to the presence of n quanta or *photons* of energy $\hbar\nu$. The eigenstates $|n\rangle$ are called Fock states or photon number states. They form a complete set of states, i.e.,

$$\sum_{n=0}^{\infty} |n\rangle\langle n| = 1. \tag{1.2.17}$$

The energy eigenvalues are discrete, in contrast to classical electromagnetic theory where energy can have any value. The energy expectation value can however take on any value, for the state vector is, in general, an arbitrary superposition of energy eigenstates, i.e.,

$$|\psi\rangle = \sum_n c_n|n\rangle \tag{1.2.18}$$

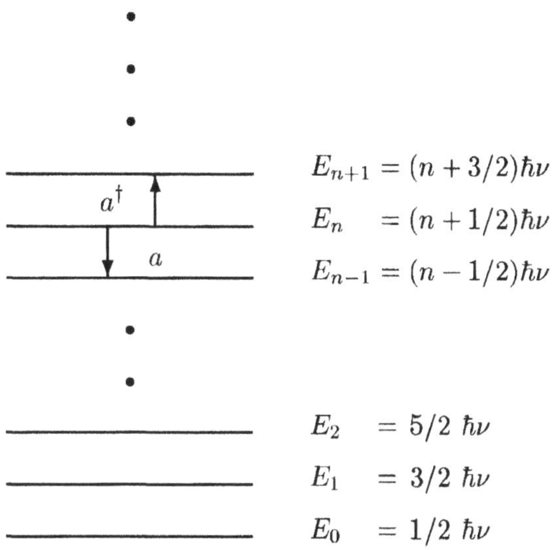

Fig. 1.2
Energy levels for the quantum mechanical harmonic oscillators associated with the electromagnetic field. The creation operator $a^\dagger$ adds a quantum of energy $\hbar\nu$, whereas the destruction operator a subtracts the same amount of energy.

where c_n are complex coefficients. The residual energy $\hbar\nu/2$ corresponding to E_0 is called the zero-point energy. In Fig. 1.2, the energy levels for the quantum mechanical oscillations associated with the electromagnetic field are given.

An important property of the number state $|n\rangle$ is that the corresponding expectation value of the single-mode linearly polarized field operator

$$E(\mathbf{r}, t) = \mathscr{E} a e^{-i\nu t + i\mathbf{k}\cdot\mathbf{r}} + \text{H.c.} \tag{1.2.19}$$

vanishes, i.e.,

$$\langle n|E|n\rangle = 0. \tag{1.2.20}$$

However, the expectation value of the intensity operator E^2 is given by

$$\langle n|E^2|n\rangle = 2|\mathscr{E}|^2 \left(n + \frac{1}{2}\right), \tag{1.2.21}$$

i.e., there are fluctuations in the field about its zero ensemble average. It is interesting to note that there are nonzero fluctuations even for a *vacuum* state $|0\rangle$. These vacuum fluctuations are responsible for many interesting phenomena in quantum optics as discussed earlier. For example, it may be considered that they *stimulate* an excited atom to emit spontaneously. They also account for the Lamb shift of $2P_{1/2} \to 2S_{1/2}$ energy levels of atomic hydrogen. In particular in Section 1.3, we shall see how the vacuum fluctuations of the electromagnetic field are responsible for the Lamb shift.

The operators a and $a^\dagger$ annihilate and create photons, respectively, for, as seen in Eqs. (1.2.14) and (1.2.15), they change a state with n photons into one with $n-1$ or $n+1$ photons. The operators a and $a^\dagger$ are therefore referred to as annihilation (or destruction) and creation operators, respectively. These operators are not themselves Hermitian ($a \neq a^\dagger$) and do not represent observable quantities such as the electric and magnetic field amplitudes. However, some combinations of the operators are Hermitian such as $a_1 = (a + a^\dagger)/2$ and $a_2 = (a - a^\dagger)/2i$.

So far we have considered a single-mode field and have found that, in general, the wave function can be written as a linear superposition of photon number states $|n\rangle$. We now extend this formalism to deal with multi-mode fields.

We can rewrite the Hamiltonian $\mathscr{H}$ in Eq. (1.1.12) as

$$\mathscr{H} = \sum_{\mathbf{k}} \mathscr{H}_{\mathbf{k}} \tag{1.2.22}$$

where

$$\mathscr{H}_{\mathbf{k}} = \hbar\nu_k \left(a_{\mathbf{k}}^\dagger a_{\mathbf{k}} + \frac{1}{2} \right). \tag{1.2.23}$$

The energy eigenstate $|n_{\mathbf{k}}\rangle$ of $\mathscr{H}_{\mathbf{k}}$ is defined in a manner similar to the single-mode field via the energy eigenvalue equation

$$\mathscr{H}_{\mathbf{k}}|n_{\mathbf{k}}\rangle = \hbar\nu_k \left(n_{\mathbf{k}} + \frac{1}{2} \right) |n_{\mathbf{k}}\rangle. \tag{1.2.24}$$

The general eigenstate of $\mathscr{H}$ can therefore have $n_{\mathbf{k}_1}$ photons in the first mode, $n_{\mathbf{k}_2}$ in the second, $n_{\mathbf{k}_\ell}$ in the ℓth and so forth, and can be written as $|n_{\mathbf{k}_1}\rangle|n_{\mathbf{k}_2}\rangle \ldots |n_{\mathbf{k}_\ell}\rangle \ldots$ or more conveniently

$$|n_{\mathbf{k}_1}, n_{\mathbf{k}_2}, \ldots, n_{\mathbf{k}_\ell}, \ldots\rangle \equiv |\{n_{\mathbf{k}}\}\rangle. \tag{1.2.25}$$

The annihilation and creation operators $a_{\mathbf{k}_\ell}$ and $a_{\mathbf{k}_\ell}^\dagger$ lower and raise the ℓth entry alone, i.e.,

$$a_{\mathbf{k}_\ell}|n_{\mathbf{k}_1}, n_{\mathbf{k}_2}, \ldots, n_{\mathbf{k}_\ell}, \ldots\rangle = \sqrt{n_{\mathbf{k}_\ell}}|n_{\mathbf{k}_1}, n_{\mathbf{k}_2}, \ldots, n_{\mathbf{k}_\ell} - 1, \ldots\rangle, \tag{1.2.26}$$

$$a_{\mathbf{k}_\ell}^\dagger|n_{\mathbf{k}_1}, n_{\mathbf{k}_2}, \ldots, n_{\mathbf{k}_\ell}, \ldots\rangle = \sqrt{n_{\mathbf{k}_\ell} + 1}|n_{\mathbf{k}_1}, n_{\mathbf{k}_2}, \ldots, n_{\mathbf{k}_\ell} + 1, \ldots\rangle. \tag{1.2.27}$$

The general state vector for the field is a linear superposition of these eigenstates:

$$\begin{aligned} |\psi\rangle &= \sum_{n_{\mathbf{k}_1}} \sum_{n_{\mathbf{k}_2}} \cdots \sum_{n_{\mathbf{k}_\ell}} \cdots c_{n_{\mathbf{k}_1}, n_{\mathbf{k}_2}, \ldots, n_{\mathbf{k}_\ell}, \ldots} |n_{\mathbf{k}_1}, n_{\mathbf{k}_2}, \ldots, n_{\mathbf{k}_\ell}, \ldots\rangle \\ &\equiv \sum_{\{n_{\mathbf{k}}\}} c_{\{n_{\mathbf{k}}\}} |\{n_{\mathbf{k}}\}\rangle. \end{aligned} \tag{1.2.28}$$

This is a more general superposition than

$$|\psi\rangle = |\psi_{\mathbf{k}_1}\rangle|\psi_{\mathbf{k}_2}\rangle \dots |\psi_{\mathbf{k}_\ell}\rangle \dots, \tag{1.2.29}$$

where $|\psi_{\mathbf{k}_\ell}\rangle$ are state vectors for individual modes. Equation (1.2.28) includes state vectors of the type (1.2.29) as well as more general states having correlations between the field modes which can result from interaction of the various field modes with a common system.

1.3 Lamb shift

The precision observation of the Lamb shift, between the $2S_{1/2}$ and $2P_{1/2}$ levels in hydrogen, was in a real sense the stimulus for modern quantum electrodynamics (QED). According to Dirac theory, the $2S_{1/2}$ and $2P_{1/2}$ levels should have equal energies. However, radiative corrections due to the interaction between the atomic electron and the vacuum, shift the $2S_{1/2}$ level higher in energy by around 1057 MHz relative to the $2P_{1/2}$ level.

Early attempts to calculate such 'vacuum induced' radiative corrections were frustrated in that they predicted infinite level shifts. However, the beautiful measurement of Lamb and Retherford provided the stimulus for renormalization theory which has been so successful in handling these divergences.

On the occasion of Lamb's sixty-fifth birthday, Freeman Dyson* wrote:

> Your work on the hydrogen fine structure led directly to the wave of progress in quantum electrodynamics on which I took a ride to fame and fortune. You did the hard, tedious, exploratory work. Once you had started the wave rolling, the ride for us theorists was easy. And after we had zoomed ashore with our fine, fancy formalisms, you still stayed with your stubborn experiment. For many years thereafter you were at work, carefully coaxing the hydrogen atom to give us the accurate numbers which provided the solid foundations for all our speculations...
>
> Those years, when the Lamb shift was the central theme of physics, were golden years for all the physicists of my generation. You were the first to see that that tiny shift, so elusive and hard to measure, would clarify in a fundamental way our thinking about particles and fields.

* Dyson [1978].

Shortly after the experimental results were announced, Bethe produced a simple nonrelativistic calculation which was in good qualitative agreement with theory, by using the suggestion of Kramers, Schwinger, and Weisskopf for 'subtracting off' infinities. This was extended to a full relativistic theory in quantitative agreement with experiments by Kroll and Lamb, and French and Weisskopf; and was the harbinger of modern QED as developed by Schwinger, Feynman, and Dyson.

The excellent agreement between the full quantum theory of radiation and matter, and experiment, e.g., the Lamb shift, provides strong support for the quantization of the radiation field. However, a detailed calculation of the Lamb shift would take us too far from mainstream quantum optics. Therefore, we will present here a heuristic derivation of the electromagnetic level shift following Welton.

The effect of the fluctuations in the electric and magnetic fields associated with the vacuum is a perturbation of the electron in a hydrogen atom from the standard orbits of the Coulomb potential $-e^2/4\pi\epsilon_o r$ due to the proton; so the electron radius $r \to r + \delta r$, where δr is the fluctuation in the position of the electron due to the fluctuating fields. The change in potential energy, and thus the associated level shift, is given by

$$\begin{aligned}\Delta V &= V(\mathbf{r}+\delta\mathbf{r}) - V(\mathbf{r})\\ &= \delta\mathbf{r}\cdot\nabla V + \frac{1}{2}(\delta\mathbf{r}\cdot\nabla)^2 V(\mathbf{r}) + \ldots\end{aligned} \tag{1.3.1}$$

Since the fluctuations are isotropic, $\langle\delta\mathbf{r}\rangle_{\text{vac}} = 0$, the first term can be neglected. Moreover,

$$\langle(\delta\mathbf{r}\cdot\nabla)^2\rangle_{\text{vac}} = \frac{1}{3}\langle(\delta\mathbf{r})^2\rangle_{\text{vac}}\nabla^2, \tag{1.3.2}$$

again due to the isotropy of the fluctuations. We therefore obtain

$$\langle\Delta V\rangle = \frac{1}{6}\langle(\delta\mathbf{r})^2\rangle_{\text{vac}}\left\langle\nabla^2\left(\frac{-e^2}{4\pi\epsilon_0 r}\right)\right\rangle_{\text{at}}, \tag{1.3.3}$$

where $\langle\ldots\rangle_{\text{at}}$ represents the quantum average with respect to the atomic states.

For the $2S$ state of hydrogen

$$\begin{aligned}\left\langle\nabla^2\left(\frac{-e^2}{4\pi\epsilon_0 r}\right)\right\rangle_{\text{at}} &= \frac{-e^2}{4\pi\epsilon_0}\int d\mathbf{r}\,\psi_{2S}^*(\mathbf{r})\nabla^2\left(\frac{1}{r}\right)\psi_{2S}(\mathbf{r})\\ &= \frac{e^2}{\epsilon_0}|\psi_{2S}(0)|^2\\ &= \frac{e^2}{8\pi\epsilon_0 a_0^3},\end{aligned} \tag{1.3.4}$$

where $a_0 = 4\pi\epsilon_0\hbar^2/me^2$ (m is the mass of the electron) is the Bohr radius and we use

$$\nabla^2\left(\frac{1}{r}\right) = -4\pi\delta(\mathbf{r}), \tag{1.3.5}$$

and

$$\psi_{2S}(0) = \frac{1}{(8\pi a_0^3)^{1/2}}. \tag{1.3.6}$$

For P-states, the nonrelativistic wave function vanishes at the origin and hence so does the energy shift.

Next we consider the contribution $\langle(\delta\mathbf{r})^2\rangle_{\text{vac}}$ due to the vacuum fluctuations in Eq. (1.3.3). The classical equation of motion for the electron displacement $(\delta r)_{\mathbf{k}}$ induced by a single mode of the field of wave vector $\mathbf{k}$ and frequency v is

$$m\frac{d^2}{dt^2}(\delta r)_{\mathbf{k}} = -eE_{\mathbf{k}}. \tag{1.3.7}$$

This is valid if the field frequency v is greater than the frequency v_0 in the Bohr orbit, i.e., if $v > \pi c/a_0$. For the field oscillating at frequency v,

$$\delta r(t) \cong \delta r(0)e^{-ivt} + \text{c.c.} \tag{1.3.8}$$

We thus have

$$(\delta r)_{\mathbf{k}} \cong \frac{e}{mc^2k^2}E_{\mathbf{k}}, \tag{1.3.9}$$

where, from Eq. (1.1.27),

$$E_{\mathbf{k}} = \mathscr{E}_{\mathbf{k}}(a_{\mathbf{k}}e^{-ivt+i\mathbf{k}\cdot\mathbf{r}} + \text{H.c.}). \tag{1.3.10}$$

After summing over all modes, we obtain

$$\begin{aligned}\langle(\delta\mathbf{r})^2\rangle_{\text{vac}} &= \sum_{\mathbf{k}}\left(\frac{e}{mc^2k^2}\right)^2\langle 0|(E_{\mathbf{k}})^2|0\rangle \\ &= \sum_{\mathbf{k}}\left(\frac{e}{mc^2k^2}\right)^2\left(\frac{\hbar ck}{2\epsilon_0 V}\right),\end{aligned} \tag{1.3.11}$$

where we have made the substitution $\mathscr{E}_{\mathbf{k}} = (\hbar ck/2\epsilon_0 V)^{1/2}$. For the continuous mode distribution, the summation in Eq. (1.3.11) is changed to an integral (Eq. (1.1.23)). We then obtain after carrying out the angular integrations

$$\begin{aligned}\langle(\delta\mathbf{r})^2\rangle_{\text{vac}} &= 2\frac{V}{(2\pi)^3}4\pi\int dk k^2\left(\frac{e}{mc^2k^2}\right)^2\left(\frac{\hbar ck}{2\epsilon_0 V}\right) \\ &= \frac{1}{2\epsilon_0\pi^2}\left(\frac{e^2}{\hbar c}\right)\left(\frac{\hbar}{mc}\right)^2\int\frac{dk}{k}.\end{aligned} \tag{1.3.12}$$

This gives a divergent result. However as noted before, the present method is only valid for $\nu > \pi c/a_0$, or equivalently $k > \pi/a_0$. It is also valid only for wavelengths longer than the Compton wavelength, i.e., $k < mc/\hbar$, because of magnetic effects on the motion which begin when $v/c = p/mc = \hbar k/mc \lesssim 1$. The present method is invalid if the electron is relativistic. We can therefore choose the lower and upper limits for the integral in Eq. (1.3.12) to be π/a_0 and $mc/\hbar$, respectively. We then obtain

$$\langle(\delta\mathbf{r})^2\rangle_{\text{vac}} \cong \frac{1}{2\epsilon_0\pi^2}\left(\frac{e^2}{\hbar c}\right)\left(\frac{\hbar}{mc}\right)^2 \ln\left(\frac{4\epsilon_0\hbar c}{e^2}\right). \tag{1.3.13}$$

On substituting Eqs. (1.3.4) and (1.3.13) into Eq. (1.3.3), we obtain the following expression for the Lamb shift

$$\langle\Delta V\rangle = \frac{4}{3}\frac{e^2}{4\pi\epsilon_0}\frac{e^2}{4\pi\epsilon_0\hbar c}\left(\frac{\hbar}{mc}\right)^2\frac{1}{8\pi a_0^3}\ln\left(\frac{4\epsilon_0\hbar c}{e^2}\right). \tag{1.3.14}$$

This shift is about 1 GHz in good agreement with the observed shift, considering the crude approximations made in the calculation.

Finally, we note the exciting developments in Lamb shift physics made possible by modern quantum optical techniques, namely the measurement of the radiation shift of the $1S$ state via precise measurements of the two-photon $1S$–$2S$ transition first performed by Hänsch and co-workers.

1.4 Quantum beats

Over the past decades several alternative theories to quantum electrodynamics (QED) have been proposed. One such theory is based on stochastic electrodynamics. In this theory, matter is treated quantum mechanically while radiation is described according to Maxwell's equations, to which one adds *vacuum fluctuations*. In this picture, it would seem that almost all quantum phenomena, such as spontaneous emission, Lamb shift, and the laser linewidth, can be understood in a semiquantitative fashion.

Quantum beat* phenomena however provide us with a simple example of a case in which the results of a self-consistent fully quantized calculation differ substantially from those obtained via a semiclassical theory (SCT) even when augmented by the notion of vacuum fluctuations. This is a good example of a problem which cannot be explained, let alone calculated, by semiclassical-type arguments.

* Svanberg [1991].

In later chapters, we shall present elaborate theories of atom–field interaction based on semiclassical and fully quantum mechanical treatments. In this section, however, we discuss quantum beats via QED and SCT in three-level atomic systems using simple arguments. We consider two different types of three-level atoms in the so-called V and Λ type configurations which are prepared in a coherent superposition of all three states. Both systems are first treated semiclassically and then by QED methods in order to compare the results of both approaches.

As depicted in Fig. 1.3, an ensemble of atoms prepared in a coherent superposition of states is described by a state vector,

$$\begin{aligned}|\psi(t)\rangle = {} & c_a \exp(-i\omega_a t)|a\rangle + c_b \exp(-i\omega_b t)|b\rangle \\ & + c_c \exp(-i\omega_c t)|c\rangle,\end{aligned} \tag{1.4.1}$$

where c_a, c_b, and c_c are probability amplitudes for the atom to be in levels $|a\rangle$, $|b\rangle$, and $|c\rangle$, respectively. Furthermore, if the nonvanishing dipole matrix elements are denoted by

$$\begin{array}{ll} V \text{ type atoms} & \Lambda \text{ type atoms} \\ & \\ \mathscr{P}_{ac} = e\langle a|r|c\rangle & \mathscr{P}_{ac} = e\langle a|r|c\rangle \\ \mathscr{P}_{bc} = e\langle b|r|c\rangle & \mathscr{P}_{ab} = e\langle a|r|b\rangle, \end{array} \tag{1.4.2}$$

where the designations V and Λ are explained in Fig. 1.3, then the state (1.4.1) implies that each atom contains two microscopic oscillating dipoles, that is,

$$\begin{array}{ll} V \text{ type atoms} & \Lambda \text{ type atoms} \\ & \\ P(t) = \mathscr{P}_{ac}(c_a^* c_c)\exp(i\nu_1 t) & P(t) = \mathscr{P}_{ac}(c_a^* c_c)\exp(i\nu_2 t) \\ \quad + \mathscr{P}_{bc}(c_b^* c_c)\exp(i\nu_2 t) & \quad + \mathscr{P}_{ab}(c_a^* c_b)\exp(i\nu_1 t) \\ \quad + \text{c.c.} & \quad + \text{c.c.}, \end{array} \tag{1.4.3}$$

where $\nu_1 = \omega_a - \omega_c$, $\nu_2 = \omega_b - \omega_c$ for V type atoms and $\nu_1 = \omega_a - \omega_b$, $\nu_2 = \omega_a - \omega_c$ for Λ type atoms. From a semiclassical perspective, the field radiated will then be a sum of two terms

$$E^{(+)} = \mathscr{E}_1 \exp(-i\nu_1 t) + \mathscr{E}_2 \exp(-i\nu_2 t), \tag{1.4.4}$$

in an obvious notation. Hence it is clear that a square law detector contains an interference or *beat note* term

$$|E^{(+)}|^2 = |\mathscr{E}_1|^2 + |\mathscr{E}_2|^2 + \{\mathscr{E}_1^* \mathscr{E}_2 \exp[i(\nu_1 - \nu_2)t] + \text{c.c.}\}. \tag{1.4.5}$$

Fig. 1.3
Three-level atomic structures for (a) V type and (b) Λ type quantum beats.

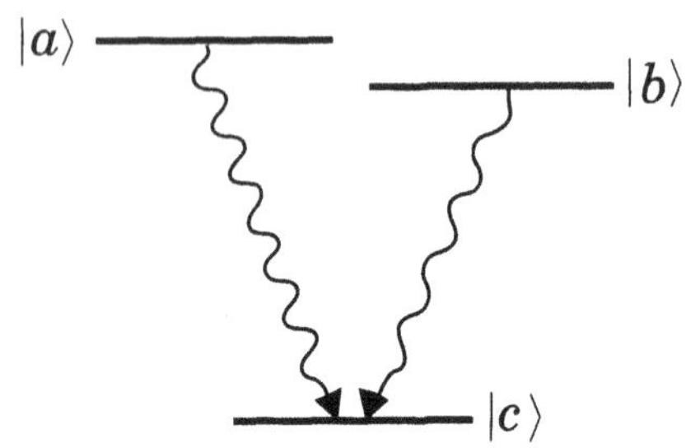

(a) V-type atom

$$|\psi_V(t)\rangle = \sum_{i=a,b,c} c_i\,|i,0\rangle + c_1\,|c,1_{\nu_1}\rangle + c_2\,|c,1_{\nu_2}\rangle$$

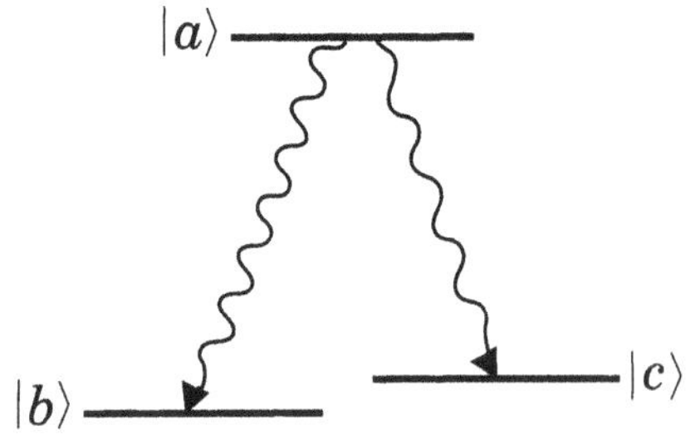

(b) Λ-type atom

$$|\psi_\Lambda(t)\rangle = \sum_{i=a,b,c} c'_i\,|i,0\rangle + c'_1\,|b,1_{\nu_1}\rangle + c'_2\,|c,1_{\nu_2}\rangle$$

Such a beat note is frequently observed in beam–foil spectroscopy experiments.

Finally we note, and this is the central point, that such an interference term is predicted by SCT for atoms of both types V and Λ.

Let us now consider the same problem as viewed from a QED perspective. For an atom of the V type we now calculate a beat note

$$\langle\psi_V(t)|E_1^{(-)}(t)E_2^{(+)}(t)|\psi_V(t)\rangle, \tag{1.4.6}$$

where $E_1^{(-)}(t)$ and $E_2^{(+)}(t)$ are proportional to the creation and annihilation operator expressions $a_1^\dagger\exp(i\nu_1 t)$ and $a_2\exp(-i\nu_2 t)$, respectively. In view of $|\psi_V(t)\rangle$, as given in Fig. 1.3(a), Eq. (1.4.6) reduces to

$$\kappa\langle 1_{\nu_1}0_{\nu_2}|a_1^\dagger a_2|0_{\nu_1}1_{\nu_2}\rangle\exp[i(\nu_1-\nu_2)t]\langle c|c\rangle, \tag{1.4.7}$$

where κ is a constant. Hence, the beat note calculated via QED is given by

$$\kappa \exp[i(\nu_1 - \nu_2)t] \underbrace{\langle c|c\rangle}_{=1} . \qquad (1.4.8)$$

On the other hand for Λ type atoms we have

$$\langle \psi_\Lambda(t)|E_1^{(-)}(t)E_2^{(+)}(t)|\psi_\Lambda(t)\rangle, \qquad (1.4.9)$$

and taking $|\psi_\Lambda(t)\rangle$ from Fig. 1.3(b) this becomes

$$\begin{aligned} &\kappa'\langle 1_{\nu_1}0_{\nu_2}|a_1^\dagger a_2|0_{\nu_1}1_{\nu_2}\rangle \exp[i(\nu_1 - \nu_2)t]\langle c|b\rangle \\ &= \kappa' \exp[i(\nu_1 - \nu_2)t] \underbrace{\langle c|b\rangle}_{=0} . \end{aligned} \qquad (1.4.10)$$

Summarizing these QED considerations,

$$\begin{aligned} &V \text{ type atoms} : \langle \psi_V(t)|E_1^{(-)}(t)E_2^{(+)}(t)|\psi_V(t)\rangle = \kappa \exp[i(\nu_1 - \nu_2)t], \\ &\Lambda \text{ type atoms} : \langle \psi_\Lambda(t)|E_1^{(-)}(t)E_2^{(+)}(t)|\psi_\Lambda(t)\rangle = 0 \end{aligned} \qquad (1.4.11)$$

whereas in the SCT calculations one finds the beat note amplitude to be nonvanishing for both V type and Λ type atoms.

The following argument based on the *quantum theory of measurement* provides some physical insight concerning the *missing* beats. A V type atom when coherently excited will decay via the emission of a photon with frequency ν_1 or ν_2. Since both transitions lead to the same final atomic state, one cannot determine along *which path*, ν_1 or ν_2, the atom decayed. Analogous to Young's double-slit problem, this uncertainty in atomic trajectory leads to an interference between photons with frequencies ν_1 and ν_2, giving rise to quantum beats. The complementary nature of *which-path* information and the appearance of quantum beats will be discussed in detail in Chapter 19. A coherently excited Λ type atom will also decay via the emission of a photon with frequency ν_1 or ν_2. However, after the emission is long past, an observation of the atom would now tell us which decay channel (1 or 2) was taken (atom in $|c\rangle$ or $|b\rangle$). Consequently, we expect no beats in this case.

The clear conclusion is that a QED calculation is consistent with our most fundamental notions of quantum theory, while SCT applied to this problem is not.

1.5 What is light? – The photon concept

The quantum theory of radiation provides a complete description of radiation–matter interactions (when supplemented by certain renormalization presumptions). It is however tempting to argue that the conceptual underpinnings of the quantum theory of radiation and the concept of a photon can be best thought of as involving a classical electromagnetic field plus the fluctuations associated with vacuum. However, advances in quantum optics have brought forward new arguments for quantizing the electromagnetic field, and with them, deeper insight into the conceptual nature of photons. With such examples as quantum beat phenomena, the quantum eraser, and certain two-photon interference phenomena, as discussed later in this book, it becomes necessary to think of the photon as a quantum mechanical entity whose basic physics is much deeper than the semiclassical theory plus vacuum fluctuation logic. We also note that there are deep questions associated with the question of metric in a quantized field theory, and that, in one of his last papers, Feynman* makes interesting comments connecting the possibility of a deeper understanding of renormalization theory by combining negative probability concepts with indefinite-metric physics. Some of these ideas and the extensions of our conceptual understanding of the photon are the subject of this concluding section of this chapter.

1.5.1 Vacuum fluctuations and the photon concept

While the formal quantum theory of radiation and quantum electrodynamics has had amazing success in explaining the interaction of electromagnetic radiation with matter, there are certain conceptual problems. For example, the various infinities associated with the calculations of quantities, such as the Lamb shift, the anomalous magnetic moment.

On the other hand, as we shall see in later chapters of this book, there are many processes associated with the radiation–matter interaction which can be well explained by a semiclassical theory in which the field is treated classically and the matter is treated quantum mechanically. Examples of physical phenomena which can be explained either totally or largely by semiclassical theory include the photoelectric effect which was first explained semiclassically by Wentzel in 1927. Stimulated emission, resonance fluorescence, and many other effects

* In: Negative Probabilities in Quantum Mechanics, ed. B. Hiley and F. Peat (Routledge, London, 1978).

do not require the full machinery of the quantum theory of radiation for their explanation; they can rather be explained by a semiclassical analysis.

In the same spirit, it is interesting to note that the two clouds on the horizon of physics at the beginning of the twentieth century both involved electromagnetic radiation. As the reader will no doubt recall, it was stated that the only two issues that were not completely understood in physics at that time were the null result of the Michelson–Morley experiment and the Rayleigh–Jeans catastrophe associated with black-body radiation. The Michelson–Morley experiment, of course, led to special relativity, which was the logical capstone of classical mechanics and electrodynamics, and the Planck solution to the Rayleigh–Jeans catastrophe was the beginning of quantum mechanics.

It is, however, interesting and important to realize that neither of these phenomena involved the concept of a photon. In the first instance, Einstein was thinking essentially of transformations involving Maxwell's equations and in the second instance, Planck was thinking in terms of quantizing the energies of the oscillators in the walls of his cavity, not quantizing the radiation field. Up to this point, neither the quantum theory of radiation nor the ideal concept of the photon had been conceived.

The first introduction of the photon concept was Einstein's utilization of the idea to explain the photoelectric effect. It is again interesting to note, as we alluded to earlier, that most of the photoelectric effect can be understood semiclassically. We recall for the reader that there are three issues associated with the photoelectric effect that any theory needs to explain. First, when light of frequency ν falls on a photoemissive surface, the energy of the ejected electrons T_e obeys the expression

$$\hbar\nu = \Phi + T_e, \tag{1.5.1}$$

where Φ is a work function and is a parameter characterizing the particular material under discussion. Second, the rate of electron ejection is proportional to the square of the electric field of the incident light. Third, there is no time delay between the time in which the field begins falling on the photoactive surface and the instance of photoelectron emission. The first two of these phenomena can, in contrast to what we read in most textbooks, be explained fully by simply quantizing the atoms associated with the photodetector. However, the third point, namely, the lack of a delay is a bit more subtle. It may be reasonably argued that quantum mechanics teaches us that the rate of ejection is finite even for small times, i.e., times involving a few optical cycles of the radiation field. Nevertheless, one may argue that the concept of the photon is really explicit here in the sense that conservation of energy is at stake. That is, if we have only a short period of time τ elapsing between the instants that the radiation field begins to interact with the photoemitting atoms and the emission of the photoelectron, the amount of energy which has fallen on the surface would be governed by $\epsilon_0 E^2 A\tau$, where A is the cross-section of the incident beam. For sufficiently short times, the energy which has fallen on the photodetector may not exceed Φ. This clearly shows that we are not able to conserve energy if we take a semiclassical point of view. However, the photon concept in which the ejection of the photoelectron implies that a photon is annihilated gets around this problem completely. This is one of the triumphs of the quantum field theory.

In any case, it is a tribute to Einstein's deep understanding of physics that he was able to introduce the photon concept from such limited, and in some ways, misleading information. Having listed some of the virtues of the semiclassical theory, we now turn to the question of where it breaks down. In many arguments of this type, one hears the statement that it is the lack of the back-action of the field on the atom that is missing in semiclassical theory. This is, of course, not the case, as this back-action is contained by forcing the theory to be self-consistent as shown in Fig. 1.4. There we see that the existence of a field enters into the Schrödinger equation in such a way as to induce a dipole in an otherwise unperturbed atom. This dipole then radiates and is the source of absorption, stimulated emission, resonance fluorescence, etc. Now, the radiation which is emitted by the dipole is itself a source of perturbation of the atomic wave function (i.e., back-action) in a self-consistent analysis, as indicated in Fig. 1.4. However, the success of the semiclassical theory can only go so far and we now turn to the problems in which it breaks down and indicate how these examples can be understood by supplementing semiclassical theory with fluctuations due to the vacuum.

1.5.2 Vacuum fluctuations

Perhaps the most important example of a situation which is not covered by the semiclassical theory of Fig. 1.4 is the spontaneous emission of light. We note that an atom, which is initially in the excited state, will remain in the excited state since there is no dipole associated with an atom in any pure quantum state and therefore the atom never starts radiating. The situation is that of unstable

Quantum Mechanics

$$i\hbar \frac{\partial}{\partial t} \left|\psi^i\right\rangle = \left[\mathcal{H}_0 - e\mathbf{r} \cdot \mathbf{E}'(\mathbf{r}_0, t)\right] \left|\psi^i\right\rangle$$

Statistical summation

$$\mathbf{P}(\mathbf{r}_0, t) = e \sum_i \left\langle \psi^i \right| \mathbf{r} \left| \psi^i \right\rangle \delta(\mathbf{r}_{0i} - \mathbf{r}_0)$$

Electrodynamics

$$\nabla \times [\nabla \times \mathbf{E}(\mathbf{r}_0, t)] + \frac{1}{c^2}\frac{\partial^2}{\partial t^2}\mathbf{E}(\mathbf{r}_0, t) = -\mu_0 \frac{\partial^2}{\partial t^2}\mathbf{P}(\mathbf{r}_0, t)$$

Self-consistent field: $\mathbf{E}' = \mathbf{E}$

Fig. 1.4 Self-consistent equations demonstrating that an assumed field $\mathbf{E}'(\mathbf{r}_0, t)$ perturbs the ith atom according to the laws of quantum mechanics and induces an electric dipole expectation value. Values for atoms localized at $\mathbf{z}_0$ are added to yield the macroscopic polarization, $\mathbf{P}(\mathbf{r}_0, t)$. This polarization acts as a source in Maxwell's equations for a field $\mathbf{E}(\mathbf{r}_0, t)$. The loop is completed by the self-consistency requirement that the field assumed, $\mathbf{E}'$, is equal to the field produced, $\mathbf{E}$.

equilibrium and the atom remains in the excited state for a long, potentially infinite, time if there are no fluctuations to get things started. Furthermore, the Lamb shift is a good example of a physical situation which is only understood with the introduction of the vacuum into the problem. As we recall, the Dirac solution of the hydrogen atom shows a complete degeneracy between the $2^2S_{1/2}$ and the $2^2P_{1/2}$ levels of the hydrogen atom. However, when vacuum fluctuations are included, as in Section 1.3, we see that the Lamb shift is qualitatively accounted for and conceptually understood. Other phenomena, such as the Planck distribution of black-body radiation and the linewidth of the laser, can be understood by such semiclassical plus vacuum fluctuation arguments.

The general feeling in the early 1970s then, was that vacuum fluctuations play a very important role in our understanding of the photon concept and that perhaps the best paradigm to apply to such problems was the notion of a classical field plus a vacuum fluctuation noise or uncertainty. The discussions of squeezing as a redistribution of this uncertainty (as discussed in Chapter 2) and other related physical arguments tend to support this perspective. However, we soon realize that this concept of a 'photon", while useful, is incomplete and we now turn to a deeper and more compelling argument for quantizing the radiation field.

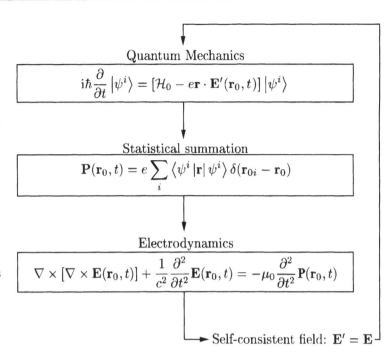

1.5.3 Quantum beats, the quantum eraser, Bell's theorem, and more

As we discussed in Section 1.4, the existence of quantum beats in an upper-state V type doublet ensemble in contrast to the absence of quantum beats associated with a lower-doublet in a Λ type atomic configuration forms the basis for an alternative argument for quantizing the radiation field which has nothing to do with the previous vacuum fluctuations. The quantum beat argument provides an example of the insufficiency of semiclassical theory plus vacuum fluctuations to understand the physics of the phenomenon. From this early example sprang concepts such as the quantum eraser and the two-photon correlation interference phenomena. This eventually showed that the early arguments and statements to the effect 'photons interfere only with themselves' were to be understood only within the context of Young's double-slit type experiments, and should not be pushed beyond that limit. We here have a great example of the importance of photon entangled states. Such entangled states are used in optical tests of Bell's inequalities and it could therefore be argued that they provide a deeper insight into the photon concept and indeed all quantum mechanics. As discussed in the last chapter of this book, we have a deeper appreciation of the nature of the quantum theory of light as a result of recent quantum optical studies.

1.5.4 'Wave function for photons'

The heading of this section is put in quotes for two reasons. First, it is the heading of a section in Power's classic book on QED. Second, the quotes serve to alert the reader to the fact that there is, strictly speaking, no such a thing as a 'photon wave function'.

For example, Power and also Kramers make the point that one may not think* of the 'photon' in the same sense as a massive (nonrelativistic) particle. On the other hand, some physicists argue that a single photon in free space is analogous to the meson if we let the meson mass go to zero. It is therefore interesting to consider the evidence and arguments for and against the concept of a 'photon wave function'.

The 'wave–particle duality' of light was the philosophical notion which led De Broglie to suggest that electrons might display wave-like behavior. However from the perspective of modern quantum optics, the wave mechanical, Maxwell–Schrödinger, treatment makes a clear distinction between light and matter waves. The interference

* See also Bialynicki-Birula [1994].

and diffraction of matter waves are the essence of *quantum* mechanics. However the corresponding behavior in light is described by the *classical* Maxwell equations.

But the question naturally rises: can we think of the electric field of light as a kind of 'wave function for the photon'? Specifically in his book on quantum mechanics Kramers asks in the section entitled 'The Photon Wave Function: Motivation and Definition',

> How far and how exactly can one consistently compare the radiation field with an ensemble of independent particles?
>
> When in 1924 De Broglie suggested that material particles should show wave phenomena ... such a comparison was of great heuristic importance. Now that wave mechanics has become a consistent formalism one could ask whether it is possible to consider the Maxwell equations to be a kind of Schrödinger equation for light particles, instead of considering them, as we have done up to now, to be classical equations of motion which formally look like a wave equation, and which are quantized only later on; or are both ideas equivalent?

At the end of the section Kramers answers the question as follows:

> The answer to the question put at the beginning of this section is thus that *one can not speak of particles in a radiation field in the same sense as in the (non-relativistic) quantum mechanics of systems of point particles.*

Kramers' reason for this conclusion is the same as that clearly stated by Power who says (in Section 5.1 entitled 'Wave Function For Photons')

> Thus it is natural to ask what are the ϕ's for photons? Strictly speaking there are no such wave functions! One may not speak of particles in a radiation field in the same sense as in the elementary quantum mechanics of systems of particles as used in the last chapter. The reason is that the wave equation ... solutions of Schrödinger's time-dependent wave function corresponding to an energy E_λ have a circular frequency $\omega_\lambda = +E_\lambda/\hbar$, while the monochromatic solutions of the wave equation have both $\pm\omega_\lambda$. The E and B fields satisfying the Maxwell equations in free space, and therefore satisfying the wave equation too, are real and are not eigenfunctions of $i\hbar\partial/\partial t$. A Schrödinger wave of given energy must be complex.

That is, the *real* electric wave (Eq. (1.1.27)) has both $\exp(-i\nu_k t)$ and $\exp(i\nu_k t)$ parts while the matter wave has only $\exp(-i\nu_p t)$ type terms. We shall return to this point later, but let us first recall the arguments of Bohm in his classic *Quantum Theory* book on the subject. On page 98 he notes that

> The probability that an electron can be found with position between x and $x + dx$ is
>
> $$P(x) = \psi^*(x)\psi(x)dx.$$

He then compares this with the situation for light and goes on to say:

> There is, strictly speaking, no function that represents the probability of finding a light quantum at a given point. If we choose a region large compared with a wavelength, we obtain approximately
>
> $$P(x) \cong \frac{\mathscr{E}^2(x) + \mathscr{H}^2(x)}{8\pi h\nu(x)},$$
>
> but if this region is defined too well, $\nu(x)$ has no meaning.

Later on Bohm makes the statement that for matter

> There is a probability current
>
> $$S = \frac{\hbar}{2mi}(\psi^*\Delta\psi - \psi\Delta\psi^*)$$
>
> which satisfies the relation
>
> $$\frac{\partial P}{\partial t} + \text{div} S = 0,$$

but he notes that

> There is no corresponding quantity for light.

We agree with the conclusion of Kramers and Bohm, namely that the concept of a photon wave function must be used with care and can be very misleading. However as we shall see, each of the above objections to the concept can be overcome.

We begin by noting that, from the perspective of a semiclassical theory, we are dealing with a wave description of the (classical) radiation, and (quantum) matter systems. Only when we proceed to quantize the radiation field are the radiation–matter equations treated on the same footing. In this fully quantized theory, it is instructive to consider matter from a second quantized vantage. Recall the quantization procedure of Section 1.1 in which we replaced the Fourier amplitudes of the field by operators. Consider the classical complex field $E(\mathbf{r}, t)$ for polarized light. Since the light is polarized, we can ignore the vector character of the field. In passing from the classical to the quantum description of the field we replace the coefficients of the field eigenfunctions, $U_{\mathbf{k}}(\mathbf{r})$, by operators, i.e.,

$$E^{(+)}(\mathbf{r},t) = \sum_{\mathbf{k}} \mathscr{E}_{\mathbf{k}} \alpha_{\mathbf{k}} e^{-i\nu_k t} U_{\mathbf{k}}(\mathbf{r}), \tag{1.5.2}$$

where $\alpha_{\mathbf{k}}$ are classical field amplitudes, is replaced by

$$E^{(+)}(\mathbf{r},t) = \sum_{\mathbf{k}} \mathscr{E}_{\mathbf{k}} a_{\mathbf{k}} e^{-i\nu_k t} U_{\mathbf{k}}(\mathbf{r}), \tag{1.5.3}$$

where $a_{\mathbf{k}}$ are quantum field operators.

Now, a corresponding quantization procedure can be, and is, applied to matter. For example, the wave function of a massive system (atom, electron, meson, etc.) is described by the superposition of states

$$\psi(\mathbf{r},t) = \sum_{\mathbf{p}} c_{\mathbf{p}} e^{-i\nu_p t} \phi_{\mathbf{p}}(\mathbf{r}), \tag{1.5.4}$$

where $\nu_p = E_p/\hbar$ and $c_{\mathbf{p}}$ is the probability amplitude for a particle being in state $\phi_{\mathbf{p}}(\mathbf{r})$, e.g., for a particle of momentum $\mathbf{p}$ we have

$$\phi_{\mathbf{p}}(\mathbf{r}) = \frac{1}{\sqrt{V}} e^{i\mathbf{p}\cdot\mathbf{r}}. \tag{1.5.5}$$

The (*second*) quantization procedure now is to turn each probability amplitude $c_{\mathbf{p}}$ into an annihilation operator $\hat{c}_{\mathbf{p}}$ obeying Fermi–Dirac or Bose–Einstein commutation relations, etc. In such a case, the wave function becomes an operator

$$\hat{\psi}(\mathbf{r},t) = \sum_{\mathbf{p}} \hat{c}_{\mathbf{p}} e^{-i\nu_p t} \phi_{\mathbf{p}}(\mathbf{r}), \tag{1.5.6}$$

which annihilates a particle at $\mathbf{r}$ and the state of the system is described by a state vector $|\psi\rangle$. At this level both the matter and photons are described by quantized fields and the state of the photon and/or meson *field* is described by a state vector $|\psi\rangle$. The logic of semiclassical and fully second quantized treatments of the radiation–matter system is summarized in Fig. 1.5.

Notice that the terminology 'second' quantization is appropriate for the matter field, since we are introducing operators for the second time; i.e., we first set $p_x \to (\hbar/i)\partial/\partial x$, etc., and second we replace probability amplitudes by operators $c_{\mathbf{p}}(t) \to \hat{c}_{\mathbf{p}}(t)$. However, this does not appear to be the case for the photon since $\hbar$ appears only once. In this sense, the quantization of the radiation field can be argued to be a 'first' quantization procedure.

We now turn the picture around and pretend that we first learn of photons and mesons, etc., from a fully quantized field perspective. The particle wave function is obtained from the state vector by taking the inner product between the position eigenstate $|\mathbf{r}\rangle$ and the state vector $|\psi(t)\rangle$

	Light	Matter
Semiclassical	$\mathbf{E}(\mathbf{r},t)$ $\square^2\mathbf{E} = -\mu_0\mathbf{P}$ Maxwell	$\psi(\mathbf{r},t)$ $\dot{\psi}(\mathbf{r},t) = -\frac{\mathrm{i}}{\hbar}\mathcal{H}\psi(\mathbf{r},t)$ Schrödinger
Quantum field	$\lvert\dot{\psi}_f\rangle = -\frac{\mathrm{i}}{\hbar}\mathcal{H}_f\lvert\psi_f\rangle$ $\mathbf{E}(\mathbf{r},t) = \sum_{\mathbf{k}} \alpha_{\mathbf{k}}(t)U_{\mathbf{k}}(\mathbf{r})$ Dirac	$\lvert\dot{\psi}_m\rangle = -\frac{\mathrm{i}}{\hbar}\mathcal{H}_m\lvert\psi_m\rangle$ $\hat{\psi}(\mathbf{r},t) = \sum_{\mathbf{p}} \hat{c}_{\mathbf{p}}(t)\phi_{\mathbf{p}}(\mathbf{r})$ Schwinger

Fig. 1.5
The semiclassical theory of the radiation and matter 'fields' are treated according to the Maxwell and Schrödinger equations. Both fields display wave-like behavior but $\hbar$ appears only in the matter equation. Applying the full quantum field theory of, e.g., Dirac and Schwinger, the radiation and matter are treated on the same footing.

$$\Psi(\mathbf{r},t) = \langle\mathbf{r}|\psi(t)\rangle. \tag{1.5.7}$$

We recall that the state $|\mathbf{r}\rangle$ can be written as

$$|\mathbf{r}\rangle = \hat{\psi}^\dagger(\mathbf{r})|0\rangle, \tag{1.5.8}$$

that is, the creation operator

$$\hat{\psi}^\dagger(\mathbf{r}) = \sum_{\mathbf{p}} \hat{c}^\dagger_{\mathbf{p}} e^{i\nu_p t}\phi^*_{\mathbf{p}}(\mathbf{r}), \tag{1.5.9}$$

acting on the vacuum creates a particle at $\mathbf{r}$. So from Eqs. (1.5.7) and (1.5.8), we have the usual result for the matter wave function

$$\Psi(\mathbf{r},t) = \langle 0|\hat{\psi}(\mathbf{r})|\psi(t)\rangle. \tag{1.5.10}$$

Now it is natural to ask: can we write something like Eq. (1.5.7) for the photon? The answer is, strictly speaking, 'no'; because there is no $|\mathbf{r}\rangle$ state for the photon.

With that in mind, let us push on and ask the operational question: what is the probability that a single-photon state of the radiation field, that is

$$|\psi\rangle = \sum_{\{n\}} c_{\{n\}}(t)|\{n\}\rangle, \tag{1.5.11}$$

where $\{n\}$ stands for the set of states with one (and only one) photon in each mode $\mathbf{k}$, will lead to the ejection of a photoelectron by a detector (atom) placed at point $\mathbf{r}$?

For example, the state Eq. (1.5.11) might be produced by an excited atom decaying to a ground state, an example we will return to later. In any case, we have in mind a wave packet representing a single photon propagating through space and the probability amplitudes $c_{\{n\}}$ contain the information normally associated with the Fourier coefficients for the single-photon pulse.

Now, as we will discuss in Section 4.2, the probability of exciting an atom (a detector atom) at $\mathbf{r}$ is governed by

$$P_\psi(\mathbf{r},t) \propto \langle\psi|E^{(-)}(\mathbf{r},t)E^{(+)}(\mathbf{r},t)|\psi\rangle, \tag{1.5.12}$$

where the annihilation operator $E^{(+)}(\mathbf{r},t)$ is given by

$$E^{(+)}(\mathbf{r},t) = \sum_{\mathbf{k}} \mathscr{E}_{\mathbf{k}} a_{\mathbf{k}} e^{-i\nu_k t} U_{\mathbf{k}}(\mathbf{r}), \tag{1.5.13}$$

and the creation operator $E^{(-)}(\mathbf{r},t)$ is just the adjoint of Eq. (1.5.13). We insert a sum over a complete set of states, $\sum_{\{n'\}} |\{n'\}\rangle\langle\{n'\}| = 1$ in Eq. (1.5.12) and write

$$P_\psi(\mathbf{r},t) \propto \sum_{\{n'\}} \langle\psi|E^{(-)}(\mathbf{r},t)|\{n'\}\rangle\langle\{n'\}|E^{(+)}(\mathbf{r},t)|\psi\rangle. \tag{1.5.14}$$

But since there is only one photon in ψ and $E^{(+)}(\mathbf{r},t)$ annihilates it, only the vacuum term $|0\rangle\langle 0|$ will contribute to Eq. (1.5.14). Hence we have

$$P_\psi(\mathbf{r},t) \propto \langle\psi|E^{(-)}(\mathbf{r},t)|0\rangle\langle 0|E^{(+)}(\mathbf{r},t)|\psi\rangle, \tag{1.5.15}$$

and we are therefore led to define the 'electric field' associated with the single photon state $|\psi_\gamma\rangle$ as

$$\Psi_{\mathscr{E}}(\mathbf{r},t) = \langle 0|E^{(+)}(\mathbf{r},t)|\psi_\gamma\rangle. \tag{1.5.16}$$

Now for the state $|\psi_\gamma\rangle$ prepared by atomic decay, Eq. (6.3.24), we find

$$\Psi_{\mathscr{E}}(\mathbf{r},t) = \frac{\mathscr{E}_0}{r}\Theta\left(t - \frac{r}{c}\right) e^{-i(t-r/c)(\omega - i\Gamma/2)}, \tag{1.5.17}$$

where $\mathscr{E}_0$ is a constant, r is the distance from the atom to the detector, $\Theta(x)$ is the usual step function and Γ is the atomic decay rate. We note that the wave packet (1.5.17) is sharply peaked about the atomic transition frequency ω. This will be the case in all the packets we consider in this section.

Let us write Eq. (1.5.16) more explicitly using the positive frequency part in Eq. (1.1.32) for the electric field annihilation operator, that is

$$\begin{aligned}\boldsymbol{\Psi}_{\mathscr{E}}(\mathbf{r},t) &= \langle 0|\mathbf{E}^{(+)}(\mathbf{r},t)|\psi_\gamma\rangle \\ &= \langle 0|\sum_{\mathbf{k},\lambda} \hat{\epsilon}_{\mathbf{k}}^{(\lambda)} \sqrt{\frac{\hbar\nu_k}{2\epsilon_0 V}} a_{\mathbf{k},\lambda} e^{-i\nu_k t + i\mathbf{k}\cdot\mathbf{r}}|\psi_\gamma\rangle.\end{aligned} \tag{1.5.18}$$

As discussed in the previous paragraph, the field is sharply peaked about the frequency ω so that we may replace the slowly varying frequency ν_k as it appears in the square-root factor by ω and write

$$\mathbf{\Psi}_{\mathscr{E}}(\mathbf{r},t) = \sqrt{\frac{\hbar\omega}{2\epsilon_0 V}}\langle 0|\sum_{\mathbf{k},\lambda}\hat{\epsilon}_{\mathbf{k}}^{(\lambda)}a_{\mathbf{k},\lambda}e^{-i\nu_k t+i\mathbf{k}\cdot\mathbf{r}}|\psi_\gamma\rangle. \tag{1.5.19}$$

Comparing (1.5.19) with the wave function (1.5.4) we are led to define the photodetection probability amplitude as

$$\varphi_\gamma(\mathbf{r},t) = \sum_{\mathbf{k},\lambda}\hat{\epsilon}_{\mathbf{k}}^{(\lambda)}\langle 0|a_{\mathbf{k},\lambda}\frac{e^{-i\nu_k t+i\mathbf{k}\cdot\mathbf{r}}}{\sqrt{V}}|\psi_\gamma\rangle, \tag{1.5.20}$$

which is to say

$$\mathbf{\Psi}_{\mathscr{E}}(\mathbf{r},t) = \sqrt{\frac{\hbar\omega}{2\epsilon_0}}\varphi_\gamma(\mathbf{r},t). \tag{1.5.21}$$

We may write an equation of motion for $\varphi_\gamma(\mathbf{r},t)$ by using Maxwell's equations, which couple together the electric field (1.5.16) with the magnetic field

$$\mathbf{\Psi}_{\mathscr{H}}(\mathbf{r},t) = \langle 0|\mathbf{H}^{(+)}(\mathbf{r},t)|\psi_\gamma\rangle, \tag{1.5.22}$$

where $\mathbf{H}^{(+)}(\mathbf{r},t)$ is the positive frequency part of the magnetic field operator (1.1.33), which we here write in the form

$$\mathbf{H}^{(+)}(\mathbf{r},t) = \sum_{\mathbf{k},\lambda}\frac{\mathbf{k}}{k}\times\hat{\epsilon}_{\mathbf{k}}^{(\lambda)}\sqrt{\frac{\hbar\nu_k}{2\mu_0}}a_{\mathbf{k},\lambda}\frac{e^{-i\nu_k t+i\mathbf{k}\cdot\mathbf{r}}}{\sqrt{V}}. \tag{1.5.23}$$

Using Eqs. (1.5.22) and (1.5.23) and proceeding as in the case of $\Psi_{\mathscr{E}}(\mathbf{r},t)$ we find

$$\begin{aligned}\mathbf{\Psi}_{\mathscr{H}}(\mathbf{r},t) &= \sqrt{\frac{\hbar\omega}{2\mu_0}}\langle 0|\sum_{\mathbf{k},\lambda}\frac{\mathbf{k}}{k}\times\hat{\epsilon}_{\mathbf{k}}^{(\lambda)}a_{\mathbf{k},\lambda}\frac{e^{-i\nu_k t+i\mathbf{k}\cdot\mathbf{r}}}{\sqrt{V}}|\psi_\gamma\rangle \\ &= \sqrt{\frac{\hbar\omega}{2\mu_0}}\boldsymbol{\chi}_\gamma(\mathbf{r},t).\end{aligned} \tag{1.5.24}$$

Now we may write Maxwell's equations (1.1.1) in terms of $\boldsymbol{\varphi}_\gamma$ (Eq. (1.5.21)) and $\boldsymbol{\chi}_\gamma$ (Eq. (1.5.24)) as

$$\nabla\times\boldsymbol{\chi}_\gamma = \frac{1}{c}\frac{\partial\boldsymbol{\varphi}_\gamma}{\partial t}, \tag{1.5.25a}$$

$$\nabla\times\boldsymbol{\varphi}_\gamma = -\frac{1}{c}\frac{\partial\boldsymbol{\chi}_\gamma}{\partial t}, \tag{1.5.25b}$$

$$\nabla\cdot\boldsymbol{\chi}_\gamma = 0, \tag{1.5.25c}$$

$$\nabla\cdot\boldsymbol{\varphi}_\gamma = 0. \tag{1.5.25d}$$

We proceed to express Eqs. (1.5.25a-1.5.25d) in an aesthetically pleasing matrix form by writing $\boldsymbol{\varphi}$ and $\boldsymbol{\chi}$ as 1×3 column matrices

$$\boldsymbol{\varphi}_\gamma = \begin{bmatrix} \varphi_x \\ \varphi_y \\ \varphi_z \end{bmatrix}; \qquad \boldsymbol{\chi}_\gamma = \begin{bmatrix} \chi_x \\ \chi_y \\ \chi_z \end{bmatrix} \tag{1.5.26}$$

in terms of which, see Problem 1.7, Maxwell's equations (1.5.25a-1.5.25d) may be written as

$$i\hbar \frac{\partial}{\partial t} \begin{bmatrix} \boldsymbol{\varphi}_\gamma \\ \boldsymbol{\chi}_\gamma \end{bmatrix} = \begin{bmatrix} 0 & -c\mathbf{s} \cdot \mathbf{p} \\ c\mathbf{s} \cdot \mathbf{p} & 0 \end{bmatrix} \begin{bmatrix} \boldsymbol{\varphi}_\gamma \\ \boldsymbol{\chi}_\gamma \end{bmatrix}, \tag{1.5.27a}$$

and

$$\nabla \cdot \begin{bmatrix} \boldsymbol{\varphi}_\gamma \\ \boldsymbol{\chi}_\gamma \end{bmatrix} = 0, \tag{1.5.27b}$$

where s_x, s_y, and s_z are the 3×3 matrices given in Problem 1.7, and $\mathbf{p}$ is the usual momentum operator $(\hbar/i)\boldsymbol{\nabla}$.

It is interesting to compare Maxwell's equations in the form (1.5.27a, 1.5.27b) to the Dirac equations* for the neutrino

$$i\hbar \frac{\partial}{\partial t} \begin{bmatrix} \boldsymbol{\varphi}_\eta \\ \boldsymbol{\chi}_\eta \end{bmatrix} = \begin{bmatrix} 0 & c\boldsymbol{\sigma} \cdot \mathbf{p} \\ c\boldsymbol{\sigma} \cdot \mathbf{p} & 0 \end{bmatrix} \begin{bmatrix} \boldsymbol{\varphi}_\eta \\ \boldsymbol{\chi}_\eta \end{bmatrix}, \tag{1.5.28}$$

where the two-component spinors $\boldsymbol{\varphi}_\eta$ and $\boldsymbol{\chi}_\eta$ make up the Dirac wave function for the neutrino

$$\boldsymbol{\Psi}_\eta = \begin{bmatrix} \boldsymbol{\varphi}_\eta \\ \boldsymbol{\chi}_\eta \end{bmatrix}. \tag{1.5.29}$$

With equations of motion (1.5.27a,1.5.27b) in hand we easily derive the equation of continuity

$$\frac{\partial}{\partial t} \boldsymbol{\Psi}_\gamma^\dagger \boldsymbol{\Psi}_\gamma = -\nabla \cdot \mathbf{j}, \tag{1.5.30}$$

where

$$\boldsymbol{\Psi}_\gamma = \begin{bmatrix} \boldsymbol{\varphi}_\gamma \\ \boldsymbol{\chi}_\gamma \end{bmatrix}, \tag{1.5.31}$$

and the current density $\mathbf{j}$ is found, see Problem 1.8, to be

$$\mathbf{j} = \boldsymbol{\Psi}_\gamma^\dagger \mathbf{v} \boldsymbol{\Psi}_\gamma \tag{1.5.32}$$

* The correspondence between the Maxwell and Dirac equations is well known. See, for example, Bialynicki-Birula, [1994].

	Photon	Neutrino
Quantum field theory	$\vert\dot{\psi}_\gamma\rangle = -\frac{i}{\hbar}\mathcal{H}_\gamma \vert\psi_\gamma\rangle$	$\vert\dot{\psi}_\eta\rangle = -\frac{i}{\hbar}\mathcal{H}_\eta \vert\psi_\eta\rangle$
"Wave" mechanics	$\Psi_\gamma = \begin{bmatrix} \phi_\gamma \\ \chi_\gamma \end{bmatrix}$ $\dot{\Psi}_\gamma = -\frac{i}{\hbar}\begin{bmatrix} 0 & -c\mathbf{s}\cdot\mathbf{p} \\ c\mathbf{s}\cdot\mathbf{p} & 0 \end{bmatrix}\Psi_\gamma$	$\Psi_\eta = \begin{bmatrix} \phi_\eta \\ \chi_\eta \end{bmatrix}$ $\dot{\Psi}_\eta = -\frac{i}{\hbar}\begin{bmatrix} 0 & -c\sigma\cdot\mathbf{p} \\ c\sigma\cdot\mathbf{p} & 0 \end{bmatrix}\Psi_\eta$
Classical limit: Eikonal physics	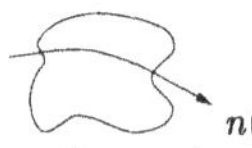 $n(\mathbf{r})$ Ray optics $\delta\int n\,ds = 0$ Fermat's principle	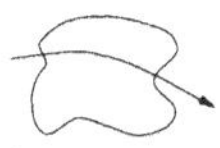 $V(\mathbf{r})$ Classical mechanics $\delta\int L\,dt = 0$ Hamilton's principle

Fig. 1.6
A symmetric description for a photon and a neutrino. In the classical limit (Last row), light is described by ray optics whereas matter is described by the analogous classical Hamilton's principle. The quantum field theoretical description of a photon and a neutrino in the first row is also quite symmetric. The 'wave' mechanics row indicates the equations of motion for $\Psi_\gamma(\mathbf{r}, t)$ and $\Psi_\eta(\mathbf{r}, t)$.

with the 'velocity' operator given by

$$\mathbf{v} = c\begin{bmatrix} 0 & -\mathbf{s} \\ \mathbf{s} & 0 \end{bmatrix}. \tag{1.5.33}$$

The comparison between $\boldsymbol{\Psi}_\gamma$ and $\boldsymbol{\Psi}_\eta$ is summarized in Fig. 1.6.

While it is amusing to note the similarities between the photon and the neutrino equations of motion, important and basic differences must be noted. For example, if we consider the electronic cousin to (1.5.28) in the nonrelativistic limit we have, for example, a plane wave relation of the form

$$\varphi_{\text{electron}}(\mathbf{r}, t) = \frac{1}{\sqrt{V}} e^{i(k_z z - \omega_k t)}, \tag{1.5.34}$$

where $k_z = p_z/\hbar$ and $\omega_k = p_z^2/2m\hbar$. Now to give the electron a momentum kick in the $\hat{x}$-direction we need only apply the boost operation $\exp(ik_x x)$ where now $x = \partial/\partial k_x$. Thus the new momentum is given by

$$\mathbf{k} = \hat{\mathbf{e}}_x k_x + \hat{\mathbf{e}}_z k_z. \tag{1.5.35}$$

But now consider the same sort of operation applied to (1.5.20). That is if we initially write $\varphi(\mathbf{r}, t)$ for a plane wave propagating in the $\hat{z}$-direction with polarization in the $\hat{x}$-direction, that is

$$\varphi(\mathbf{r}, t) = \hat{\mathbf{e}}_x \frac{e^{i(k_z z - \omega_k t)}}{\sqrt{V}}, \tag{1.5.36}$$

and we then apply a boost operation as before, we might think we could write the new function

$$\tilde{\varphi}(\mathbf{r},t) = \hat{\mathbf{e}}_x \frac{e^{i(k_z z + k_x x - \omega t)}}{\sqrt{V}}. \tag{1.5.37}$$

But now the Maxwell equation (1.5.27b) is no longer satisfied

$$\nabla \cdot \tilde{\varphi} = \frac{\partial}{\partial x}\left[\frac{e^{i(k_z z + k_x x - \omega t)}}{\sqrt{V}}\right] \neq 0. \tag{1.5.38}$$

This is just one example of how the photon 'wave function' is different from that of a nonrelativistic massive particle, and a 'photon-as-a-particle' picture can be misleading.

As another, even more dramatic example, we turn to the question of two-photon events. Specifically we have in mind two photon emission and detection as in Fig. 1.7. As discussed in detail in Chapter 4, the probability of two photoelectrons being counted at detectors D_1 and D_2 is governed by the two-photon correlation function. To that end, we calculate the two-photon correlation function

$$\begin{aligned} &G^{(2)}(\mathbf{r}_1,t_1;\mathbf{r}_2,t_2) \\ &\quad = \langle\psi|E^{(-)}(\mathbf{r}_1,t_1)E^{(-)}(\mathbf{r}_2,t_2)E^{(+)}(\mathbf{r}_2,t_2)E^{(+)}(\mathbf{r}_1,t_1)|\psi\rangle, \end{aligned} \tag{1.5.39}$$

corresponding to two detectors at points $\mathbf{r}_1$ and $\mathbf{r}_2$ and where the interaction with the photon field, described by $|\psi\rangle$, is switched on at times t_1 and t_2, respectively as in Fig. 1.7.

We note that for the radiation from a single atom only two photons are involved so that

$$\begin{aligned} \langle\psi|E_1^{(-)}E_2^{(-)}E_2^{(+)}E_1^{(+)}|\psi\rangle &= \sum_{\{n\}}\langle\psi|E_1^{(-)}E_2^{(-)}|\{n\}\rangle\langle\{n\}|E_2^{(+)}E_1^{(+)}|\psi\rangle \\ &= \langle\psi|E_1^{(-)}E_2^{(-)}|0\rangle\langle 0|E_2^{(+)}E_1^{(+)}|\psi\rangle, \end{aligned} \tag{1.5.40}$$

and therefore it is the two-photon detection amplitude

$$\Psi^{(2)}(\mathbf{r}_1,t_1;\mathbf{r}_2,t_2) \equiv \langle 0|E^{(+)}(\mathbf{r}_2,t_2)E^{(+)}(\mathbf{r}_1,t_1)|\psi\rangle \tag{1.5.41}$$

which we must now consider.

First consider the case in which the atomic decay rates from level $|a\rangle$ to level $|b\rangle$, γ_a, and from level $|b\rangle$ to level $|c\rangle$, γ_b, are such that $\gamma_a \gg \gamma_b$, that is the atom decays very quickly to level $|b\rangle$ and then after some time decays to level $|c\rangle$. In such a case we find, see Section 21.4.1, that

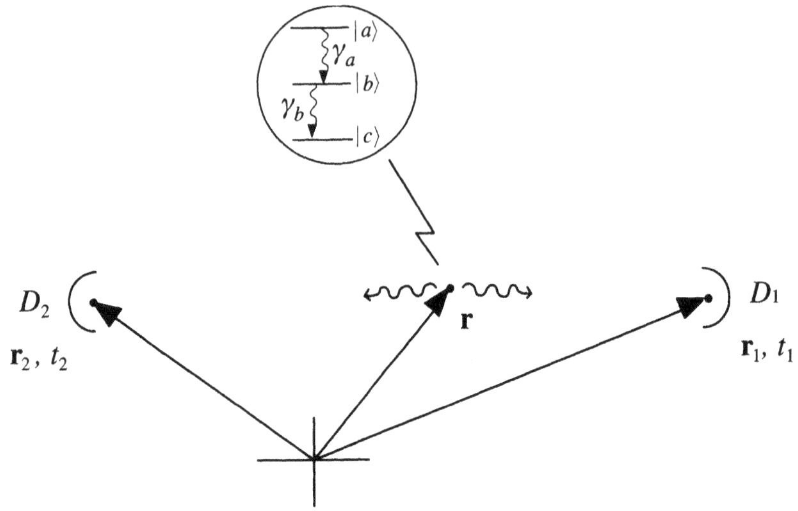

Fig. 1.7
Three-level atom located at $\mathbf{r}$ decays from $|a\rangle \to |b\rangle$ with rate γ_a and $|b\rangle \to |c\rangle$ with rate γ_b. Detectors D_1 and D_2 at $\mathbf{r}_1$ and $\mathbf{r}_2$ are switched on at times t_1 and t_2.

$$\Psi^{(2)}(\mathbf{r}_1, t_1; \mathbf{r}_2, t_2) = \Psi_\alpha(\mathbf{r}_1, t_1)\Psi_\beta(\mathbf{r}_2, t_2) + \Psi_\beta(\mathbf{r}_1, t_1)\Psi_\alpha(\mathbf{r}_2, t_2), \tag{1.5.42}$$

where

$$\Psi_\alpha(\mathbf{r}_i, t_i) = \frac{\mathscr{E}_a}{\Delta r_i}\Theta\left(t_i - \frac{\Delta r_i}{c}\right) e^{-\gamma_a\left(t_i - \frac{\Delta r_i}{c}\right)} e^{-i\omega_{ab}\left(t_i - \frac{\Delta r_i}{c}\right)}, \tag{1.5.43a}$$

and

$$\Psi_\beta(\mathbf{r}_i, t_i) = \frac{\mathscr{E}_b}{\Delta r_i}\Theta\left(t_i - \frac{\Delta r_i}{c}\right) e^{-\gamma_b\left(t_i - \frac{\Delta r_i}{c}\right)} e^{-i\omega_{bc}\left(t_i - \frac{\Delta r_i}{c}\right)}, \tag{1.5.43b}$$

in which $i = 1, 2$, ω_{ab} and ω_{bc} are the atomic frequencies for the $|a\rangle \to |b\rangle$ and $|b\rangle \to |c\rangle$ transitions, Δr_i is the distance from the atom to the ith detector and $\mathscr{E}_a$ and $\mathscr{E}_b$ are uninteresting constants. The immediate comparison between Eqs. (1.5.43a) and (1.5.43b) and the single photoelectron detection amplitude (1.5.17) is apparent. We clearly have here a Bose–Einstein expression of the type we might write for two helium atoms.

But things are very different when we make the simple change to the case $\gamma_b \gg \gamma_a$. That is when the atoms which decay at some time to level $|b\rangle$ rapidly decay to level $|c\rangle$. Then Section 21.4.1, we find a two-photon detection amplitude of the form

$$\begin{aligned}
&\Psi^{(2)}(\mathbf{r}_1, t_1; \mathbf{r}_2; t_2) \\
&= \frac{-\kappa}{\Delta r_1 \Delta r_2} \exp\left[-(i\omega_{ac} + \gamma_a)\left(t_1 - \frac{\Delta r_1}{c}\right)\right]\Theta\left(t_1 - \frac{\Delta r_1}{c}\right) \\
&\quad \times \exp\left\{-(i\omega_{bc} + \gamma_b)\left[\left(t_2 - \frac{\Delta r_2}{c}\right) - \left(t_1 - \frac{\Delta r_1}{c}\right)\right]\right\} \\
&\quad \times \Theta\left[\left(t_2 - \frac{\Delta r_2}{c}\right) - \left(t_1 - \frac{\Delta r_1)}{c}\right)\right] + (1 \leftrightarrow 2).
\end{aligned} \tag{1.5.44}$$

The message is clear. When $\gamma_a \gg \gamma_b$ we have essentially independent photons emitted. But when $\gamma_a \ll \gamma_b$ the two events are strongly correlated and the 'photon-as-a-particle' picture is very misleading.

In conclusion, we can say that while we have perhaps overcome the main objection of Kramers (the probability amplitude $\langle 0|E|\psi\rangle \sim e^{-i\omega t}$ only) and partially overcome that of Bohm (photodetection events are indeed localized* to distances smaller than the wavelength), naively visualizing $\varphi(\mathbf{r}, t)$ as a literal particle-like wave function can be misleading. "Photon" physics is very different from that of Schrödinger particles.

The proper operational "photon" philosophy is well summarized by Willis Lamb who says:

> What do we do next? We can, and should, use the Quantum Theory of Radiation. Fermi showed how to do this for the case of Lippmann fringes. The idea is simple, but the details are somewhat messy. A good notation and lots of practice makes it easier. Begin by deciding how much of the universe needs to be brought into the discussion. Decide what normal modes are needed for an adequate treatment. Decide how to model the light sources and work out how they drive the system.

This is what we will be doing in the next 20 chapters.

1.A Equivalence between a many-particle Bose gas and a set of quantized harmonic oscillators

In Section 1.1 we quantized the radiation field by associating each mode of the field with a quantized simple harmonic oscillator. This procedure led to the introduction of the Fock or number state of the field containing, for each oscillator, n photons and the associated operators a and $a^\dagger$ which annihilate and create photons, respectively. In this section, we argue that a set of harmonic oscillators is dynamically equivalent to a many-particle Bose gas.

Consider a Bose gas of N particles inside a volume V. The N-particle wave function can be written by symmetrizing the product of

* We note however that the localization of the photon (as opposed to the localizability of the photon-detection probability amplitude) is qualitatively different from the localization of a massive particle e.g., an electron. For an electron it is possible to 'fit' the electron into a small box greater than or equal to the Compton wavelength. For the photon, however, it is not possible to 'fit' or 'force' the photon into a box smaller than its wavelength. See Deutsch and Garrison [1991].

the single-particle wave functions $\psi_{\mathbf{s}}(\mathbf{r})$:

$$\Psi^N_{n_{\mathbf{p}},n_{\mathbf{q}},\ldots,n_{\mathbf{k}},\ldots}(\mathbf{r}_1,\mathbf{r}_2,\ldots\mathbf{r}_N) = \left[\frac{n_{\mathbf{p}}!n_{\mathbf{q}}!\ldots n_{\mathbf{k}}!\ldots}{N!}\right]^{1/2}$$
$$\times\sum_{\mathbf{P}}\begin{bmatrix} \psi_{\mathbf{p}}(\mathbf{r}_1)\psi_{\mathbf{p}}(\mathbf{r}_2)\ldots\psi_{\mathbf{p}}(\mathbf{r}_{n_{\mathbf{p}}}) \\ \times\psi_{\mathbf{q}}(\mathbf{r}_{n_{\mathbf{p}}+1})\psi_{\mathbf{q}}(\mathbf{r}_{n_{\mathbf{p}}+2})\ldots\psi_{\mathbf{q}}(\mathbf{r}_{n_{\mathbf{p}}+n_{\mathbf{q}}}) \\ \vdots \\ \times\psi_{\mathbf{k}}(\mathbf{r}_{\sigma+1})\psi_{\mathbf{k}}(\mathbf{r}_{\sigma+2})\ldots\psi_{\mathbf{k}}(\mathbf{r}_{\sigma+n_{\mathbf{k}}}) \\ \vdots \end{bmatrix} \tag{1.A.1}$$

where $\mathbf{P}$ denotes the permutation on N objects. The integers $n_{\mathbf{s}}$ ($\mathbf{s} = \mathbf{p},\mathbf{q},\ldots,\mathbf{k},\ldots$) are the occupation numbers of the single-particle wave functions $\psi_{\mathbf{s}}(\mathbf{r}_j)$ such that

$$\sum_{\mathbf{s}} n_{\mathbf{s}} = N, \tag{1.A.2}$$

and $n_{\mathbf{s}}$ can take the values $0, 1, 2, \ldots, N$. The single-particle wave function for a free particle is given by

$$\psi_{\mathbf{s}}(\mathbf{r}) = \frac{1}{\sqrt{V}}e^{i\mathbf{s}\cdot\mathbf{r}}. \tag{1.A.3}$$

Here $\hbar\mathbf{s}$ is the momentum of the particle.

Let the N particles interact with each other via a potential

$$\mathscr{V} = \sum_{j=1}^{N} v(\mathbf{r}_j). \tag{1.A.4}$$

A particle in the state $\psi_{\mathbf{p}}(\mathbf{r}_j)$ can go to the state $\psi_{\mathbf{k}}(\mathbf{r}_j)$ by interacting with the potential. The transition amplitude for this process is proportional to

$$v_{\mathbf{kp}} = \int d\mathbf{r}_j\psi^*_{\mathbf{k}}(\mathbf{r}_j)v(\mathbf{r}_j)\psi_{\mathbf{p}}(\mathbf{r}_j). \tag{1.A.5}$$

As an example, if a free particle with momentum $\hbar\mathbf{k}$ scatters with a phonon wave with wave vector $\tilde{\mathbf{k}}$, i.e.,

$$v(\mathbf{r}_j) = v_0 e^{i\tilde{\mathbf{k}}\cdot\mathbf{r}_j}, \tag{1.A.6}$$

to a state with momentum $\hbar\mathbf{p}$, we have

$$v_{\mathbf{kp}} = v_0\delta(\mathbf{p} + \tilde{\mathbf{k}} - \mathbf{k}). \tag{1.A.7}$$

We now consider the many-particle analysis of the problem.

Before considering the general case of a Bose gas of N particles inside a volume V, we will consider the simple case of a three-boson system. The wave function for a three-particle system initially having $n_{\mathbf{p}} = 2, n_{\mathbf{k}} = 1$ is given by

$$\begin{aligned}\psi^3_{n_{\mathbf{p}}=2,n_{\mathbf{k}}=1}(\mathbf{r}_1, \mathbf{r}_2, \mathbf{r}_3) = \frac{1}{\sqrt{3}}[&\psi_{\mathbf{p}}(\mathbf{r}_1)\psi_{\mathbf{p}}(\mathbf{r}_2)\psi_{\mathbf{k}}(\mathbf{r}_3) \\ &+\psi_{\mathbf{p}}(\mathbf{r}_3)\psi_{\mathbf{p}}(\mathbf{r}_1)\psi_{\mathbf{k}}(\mathbf{r}_2) \\ &+\psi_{\mathbf{p}}(\mathbf{r}_2)\psi_{\mathbf{p}}(\mathbf{r}_3)\psi_{\mathbf{k}}(\mathbf{r}_1)]. \end{aligned} \tag{1.A.8}$$

An interaction between the particles via a potential (Eq. (1.A.4)) with $N = 3$ transforms one particle in state $\mathbf{p}$ to state $\mathbf{k}$, i.e.,

$$\begin{aligned}\psi^3_{n_{\mathbf{p}}=1,n_{\mathbf{k}}=2}(\mathbf{r}_1, \mathbf{r}_2, \mathbf{r}_3) = \frac{1}{\sqrt{3}}[&\psi_{\mathbf{p}}(\mathbf{r}_1)\psi_{\mathbf{k}}(\mathbf{r}_2)\psi_{\mathbf{k}}(\mathbf{r}_3) \\ &+\psi_{\mathbf{p}}(\mathbf{r}_3)\psi_{\mathbf{k}}(\mathbf{r}_1)\psi_{\mathbf{k}}(\mathbf{r}_2) \\ &+\psi_{\mathbf{p}}(\mathbf{r}_2)\psi_{\mathbf{k}}(\mathbf{r}_3)\psi_{\mathbf{k}}(\mathbf{r}_1)]. \end{aligned} \tag{1.A.9}$$

The three-particle matrix element for the process is then

$$\begin{aligned}\mathcal{M}_3 = &\int\int\int d\mathbf{r}_1 d\mathbf{r}_2 d\mathbf{r}_3 \psi^{3*}_{n_{\mathbf{p}}=1,n_{\mathbf{k}}=2}(\mathbf{r}_1, \mathbf{r}_2, \mathbf{r}_3) \\ &\times \sum_{i=1}^{3} v(\mathbf{r}_i)\psi^3_{n_{\mathbf{p}}=2,n_{\mathbf{k}}=1}(\mathbf{r}_1, \mathbf{r}_2, \mathbf{r}_3). \end{aligned} \tag{1.A.10}$$

Now each particle in the sum $\sum_{i=1}^{3} v(\mathbf{r}_i)$ contributes equally so that we may simply choose a particle, say particle 1, and replace $\sum_{i=1}^{3}$ by the factor 3. Then we have

$$\begin{aligned}\mathcal{M}_3 = 3&\int\int\int d\mathbf{r}_1 d\mathbf{r}_2 d\mathbf{r}_3 \frac{1}{\sqrt{3}}[\psi^*_{\mathbf{k}}(\mathbf{r}_2)\psi^*_{\mathbf{p}}(\mathbf{r}_3) + \psi^*_{\mathbf{k}}(\mathbf{r}_3)\psi^*_{\mathbf{p}}(\mathbf{r}_2)] \\ &\times \psi^*_{\mathbf{k}}(\mathbf{r}_1)v(\mathbf{r}_1)\psi_{\mathbf{p}}(\mathbf{r}_1)\frac{1}{\sqrt{3}}[\psi_{\mathbf{k}}(\mathbf{r}_2)\psi_{\mathbf{p}}(\mathbf{r}_3) + \psi_{\mathbf{k}}(\mathbf{r}_3)\psi_{\mathbf{p}}(\mathbf{r}_2)]. \end{aligned} \tag{1.A.11}$$

If we multiply and divide by $\sqrt{2}$ each of the expressions in square brackets and use the definition of $\Psi^2_{n_{\mathbf{p}}=1,n_{\mathbf{k}}=1}(\mathbf{r}_2, \mathbf{r}_3)$ from Eq. (1.A.1), we obtain

$$\begin{aligned}\mathcal{M}_3 = \sqrt{2}\sqrt{2}&\int d\mathbf{r}_1 \psi^*_{\mathbf{k}}(\mathbf{r}_1)v(\mathbf{r}_1)\psi_{\mathbf{p}}(\mathbf{r}_1) \\ &\times \int\int d\mathbf{r}_2 d\mathbf{r}_3 |\Psi^2_{n_{\mathbf{p}}=1,n_{\mathbf{k}}=1}(\mathbf{r}_2, \mathbf{r}_3)|^2. \end{aligned} \tag{1.A.12}$$

Since the two-particle wave function is normalized we have

$$\mathcal{M}_3 = \sqrt{2}\sqrt{2}\, v_{\mathbf{kp}}, \tag{1.A.13}$$

where $v_{\mathbf{kp}}$ is defined by Eq. (1.A.5).

Consider next the same process for scattering a single particle from the initial state (Eq. (1.A.1)) via the interaction (Eq. (1.A.4)) to the final state

$$\Psi^{N}_{n_{\mathbf{p}}-1,n_{\mathbf{q}},\ldots,n_{\mathbf{k}}+1,\ldots}(\mathbf{r}_1,\mathbf{r}_2,\ldots,\mathbf{r}_N)=\left[\frac{(n_{\mathbf{p}}-1)!n_{\mathbf{q}}!\ldots(n_{\mathbf{k}}+1)!\ldots}{N!}\right]^{1/2}$$

$$\times\sum_{\mathbf{P}}\begin{bmatrix}\psi_{\mathbf{p}}(\mathbf{r}_1)\psi_{\mathbf{p}}(\mathbf{r}_2)\ldots\psi_{\mathbf{p}}(\mathbf{r}_{n_{\mathbf{p}}-1})\\ \times\psi_{\mathbf{q}}(\mathbf{r}_{n_{\mathbf{p}}})\psi_{\mathbf{q}}(\mathbf{r}_{n_{\mathbf{p}}+1})\ldots\psi_{\mathbf{q}}(\mathbf{r}_{n_{\mathbf{p}}+n_{\mathbf{q}}-1})\\ \vdots\\ \times\psi_{\mathbf{k}}(\mathbf{r}_{\sigma})\psi_{\mathbf{k}}(\mathbf{r}_{\sigma+1})\ldots\psi_{\mathbf{k}}(\mathbf{r}_{\sigma+n_{\mathbf{k}}})\\ \vdots\end{bmatrix}. \tag{1.A.14}$$

Now, as in the three-particle case, we want to evaluate the matrix element

$$\mathcal{M}_N=\int\ldots\int d\mathbf{r}_1\ldots d\mathbf{r}_N\Psi^{N*}_{n_{\mathbf{p}}-1,n_{\mathbf{q}},\ldots,n_{\mathbf{k}}+1,\ldots}(\mathbf{r}_1,\ldots,\mathbf{r}_N)$$
$$\times\sum_j v(\mathbf{r}_j)\Psi^{N}_{n_{\mathbf{p}},n_{\mathbf{q}},\ldots,n_{\mathbf{k}},\ldots}(\mathbf{r}_1,\ldots,\mathbf{r}_N). \tag{1.A.15}$$

Again, as in the three-particle case, we recognize that all permutations are identical and replace $\sum_{i=1}^{N} v(\mathbf{r}_i)$ by $Nv(\mathbf{r}_1)$. Equation (1.A.15) can then be rewritten in terms of $(N-1)$-particle wave functions as

$$\mathcal{M}_N=\int\ldots\int d\mathbf{r}_1\ldots d\mathbf{r}_N\sqrt{\frac{n_{\mathbf{k}}+1}{N}}\psi^*_{\mathbf{k}}(\mathbf{r}_1)\Psi^{N-1\,*}_{n_{\mathbf{p}}-1,n_{\mathbf{q}},\ldots,n_{\mathbf{k}},\ldots}(\mathbf{r}_2,\ldots,\mathbf{r}_N)$$
$$\times Nv(\mathbf{r}_1)\sqrt{\frac{n_{\mathbf{p}}}{N}}\psi_{\mathbf{p}}(\mathbf{r}_1)\Psi^{N-1}_{n_{\mathbf{p}}-1,n_{\mathbf{q}},\ldots,n_{\mathbf{k}},\ldots}(\mathbf{r}_2,\ldots,\mathbf{r}_N)$$
$$=\int d\mathbf{r}_1\sqrt{\frac{n_{\mathbf{k}}+1}{N}}\psi^*_{\mathbf{k}}(\mathbf{r}_1)Nv(\mathbf{r}_1)\psi_{\mathbf{p}}(\mathbf{r}_1)\sqrt{\frac{n_{\mathbf{p}}}{N}}. \tag{1.A.16}$$

Thus we see that the multi-particle character of the problem is contained in the $\sqrt{n_{\mathbf{p}}}$ and $\sqrt{n_{\mathbf{k}}+1}$ factors, associated with the removal (annihilation) of a particle in state $\psi_{\mathbf{p}}$ and addition (creation) of a particle in state $\psi_{\mathbf{k}}$.

It is natural (and much easier!) to introduce a multi-particle state vector

$$|n_{\mathbf{p}},n_{\mathbf{q}},\ldots,n_{\mathbf{k}},\ldots\rangle, \tag{1.A.17}$$

and operators which transform state vectors into one another by changing the numbers of particles in the various states. To this end, we introduce annihilation (or destruction or absorption) operators for our boson system as

$$a_{\mathbf{p}}|n_{\mathbf{p}},n_{\mathbf{q}},\ldots,n_{\mathbf{k}},\ldots\rangle=\sqrt{n_{\mathbf{p}}}|n_{\mathbf{p}}-1,n_{\mathbf{q}},\ldots,n_{\mathbf{k}},\ldots\rangle, \tag{1.A.18}$$

and the corresponding creation operators

$$a^\dagger_{\mathbf{p}}|n_{\mathbf{p}}, n_{\mathbf{q}}, \ldots, n_{\mathbf{k}}, \ldots\rangle = \sqrt{n_{\mathbf{p}}+1}|n_{\mathbf{p}}+1, n_{\mathbf{q}}, \ldots, n_{\mathbf{k}}, \ldots\rangle. \quad (1.A.19)$$

From the definitions it is clear that we have the commutation relations

$$[a_{\mathbf{p}}, a^\dagger_{\mathbf{p}'}] = \delta_{\mathbf{p},\mathbf{p}'}, \quad (1.A.20)$$

and

$$[a_{\mathbf{p}}, a_{\mathbf{p}'}] = [a^\dagger_{\mathbf{p}}, a^\dagger_{\mathbf{p}'}] = 0, \quad (1.A.21)$$

as is apparent from the action of such ordered operations on the state vectors. In order to regain the results of our matrix element calculation we are thus led to introduce the interaction Hamiltonian

$$\mathscr{V} = \sum_{\mathbf{k},\mathbf{p}} a^\dagger_{\mathbf{p}+\mathbf{k}} a_{\mathbf{p}} v_{\mathbf{k}} \quad (1.A.22)$$

and the free particle Hamiltonian

$$\mathscr{H} = \sum_{p} \frac{p^2}{2m} a^\dagger_{\mathbf{p}} a_{\mathbf{p}}. \quad (1.A.23)$$

To summarize: the physics is in the occupancy of the number states where information is contained in the states $|n_{\mathbf{p}}\rangle$ and the matrix elements $v_{\mathbf{kp}} = \int d\mathbf{r}\psi^*_{\mathbf{k}}(\mathbf{r})v(\mathbf{r})\psi_{\mathbf{p}}(\mathbf{r})$. That is, we never have to worry about complicated combinations, the operator formalism takes care of all that in a very neat way.

The main point of this section, however, is not the convenience of the operator approach but rather the deep connection between many-boson quantum mechanics and that of quantized harmonic oscillators. In the words of Dirac

> The dynamical system consisting of an ensemble of similar bosons is equivalent to the dynamical system consisting of a set of oscillators – the two systems are just the same system looked at from two different points of view. There is one oscillator associated with each independent boson state. We have here one of the most fundamental results of quantum mechanics, which enables a unification of the wave and corpuscular theories of light.

However, as compelling as the 'boson' ↔ 'oscillator set' comparison is, there are fundamental differences. For example, in the oscillator problem we end up with a vacuum fluctuation contribution that does not appear in the boson collection argument. In Section 1.3 we used this vacuum state of the electromagnetic field to obtain the Lamb shift.

Problems

1.1 The radiation field in an empty cubic cavity of side L satisfies the wave equation

$$\nabla^2 \mathbf{A} - \frac{1}{c^2}\frac{\partial^2 \mathbf{A}}{\partial t^2} = 0,$$

together with the Coulomb gauge condition $\nabla \cdot \mathbf{A} = 0$. Show that the solution that satisfies the boundary conditions has components

$$A_x(\mathbf{r}, t) = A_x(t)\cos(k_x x)\sin(k_y y)\sin(k_z z),$$
$$A_y(\mathbf{r}, t) = A_y(t)\sin(k_x x)\cos(k_y y)\sin(k_z z),$$
$$A_z(\mathbf{r}, t) = A_z(t)\sin(k_x x)\sin(k_y y)\cos(k_z z),$$

where $\mathbf{A}(t)$ is independent of position and the wave vector $\mathbf{k}$ has components given by Eq. (1.1.21). Hence show that the integers n_x, n_y, n_z in Eq. (1.1.21) are restricted in that only one of them can be zero at a time.

1.2 If A and B are two noncommuting operators that satisfy the conditions

$$[[A, B], A] = [[A, B], B] = 0,$$

then show that

$$e^{A+B} = e^{-\frac{1}{2}[A,B]}e^A e^B,$$
$$= e^{+\frac{1}{2}[A,B]}e^B e^A.$$

This is a special case of the so-called Baker–Hausdorff theorem of group theory.

1.3 If A and B are two noncommuting operators and α is a parameter, then show that

$$e^{-\alpha A}Be^{\alpha A} = B - \alpha[A, B] + \frac{\alpha^2}{2!}[A, [A, B]] + \ldots$$

1.4 If $f(a, a^\dagger)$ is a function which can be expanded in a power series of a and $a^\dagger$, then show that

(a) $[a, f(a, a^\dagger)] = \frac{\partial f}{\partial a^\dagger}$,

(b) $[a^\dagger, f(a, a^\dagger)] = -\frac{\partial f}{\partial a}$,

(c) $e^{-\alpha a^\dagger a}f(a, a^\dagger)e^{\alpha a^\dagger a} = f(ae^{\alpha}, a^\dagger e^{-\alpha})$,

where α is a parameter.

1.5 Show that

$$[a, e^{-\alpha a^\dagger a}] = (e^{-\alpha} - 1)e^{-\alpha a^\dagger a}a,$$
$$[a^\dagger, e^{-\alpha a^\dagger a}] = (e^{\alpha} - 1)e^{-\alpha a^\dagger a}a^\dagger,$$

where α is a parameter.

1.6 Show that the free-field Hamiltonian

$$\mathscr{H} = \hbar\nu\left(a^\dagger a + \frac{1}{2}\right)$$

can be written in terms of the number states as

$$\mathscr{H} = \sum_n E_n|n\rangle\langle n|,$$

and hence

$$e^{i\mathscr{H}t/\hbar} = \sum_n e^{iE_nt/\hbar}|n\rangle\langle n|.$$

1.7 Show that Maxwell's equations in free space may be written in the form of Eqs. (1.5.27a) and (1.5.27b) by first showing that

$$\frac{1}{c}\frac{\partial \tilde{\mathbf{E}}}{\partial t} = \nabla \times \tilde{\mathbf{H}}, \qquad \nabla \cdot \tilde{\mathbf{E}} = 0,$$

$$-\frac{1}{c}\frac{\partial \tilde{\mathbf{H}}}{\partial t} = \nabla \times \tilde{\mathbf{E}}, \qquad \nabla \cdot \tilde{\mathbf{H}} = 0,$$

where $\tilde{\mathbf{E}} = \sqrt{\epsilon_0}\,\mathbf{E}$ and $\tilde{\mathbf{H}} = \sqrt{\mu_0}\,\mathbf{H}$. Then prove that

$$\mathbf{s} \cdot \nabla\mathbf{V} = \nabla \times \mathbf{V},$$

$$s_x = \begin{bmatrix} 0 & 0 & 0 \\ 0 & 0 & -1 \\ 0 & 1 & 0 \end{bmatrix}, \quad s_y = \begin{bmatrix} 0 & 0 & 1 \\ 0 & 0 & 0 \\ -1 & 0 & 0 \end{bmatrix},$$

$$s_z = \begin{bmatrix} 0 & -1 & 0 \\ 1 & 0 & 0 \\ 0 & 0 & 0 \end{bmatrix},$$

where **s** and **V** on the left-hand side are regarded as 1×3 column vectors. Use this identity to obtain Eqs. (1.5.27a) and (1.5.27b).

1.8 Derive the current density (1.5.32) by writing the equations of motion for $\boldsymbol{\varphi}_\gamma$ and $\boldsymbol{\chi}_\gamma$ in the form

$$\dot{\boldsymbol{\varphi}}_\gamma = c\mathbf{s}\cdot\nabla\boldsymbol{\chi}_\gamma,$$
$$\dot{\boldsymbol{\chi}}_\gamma = -c\mathbf{s}\cdot\nabla\boldsymbol{\varphi}_\gamma, \qquad \dot{\boldsymbol{\varphi}}_\gamma^\dagger = c\nabla\boldsymbol{\chi}_\gamma^\dagger\cdot\mathbf{s}^\dagger,$$
$$\dot{\boldsymbol{\chi}}_\gamma^\dagger = -c\nabla\boldsymbol{\varphi}_\gamma^\dagger\cdot\mathbf{s}^\dagger,$$

and noting that $\mathbf{s}^\dagger = -\mathbf{s}$.

1.9 Verify that $\sum_i \hat{\mathbf{e}}_i\hat{\mathbf{e}}_i = \mathbf{1}$ by taking the dot product with any vector $\mathbf{v}$. Thus if $\hat{\mathbf{e}}_1 = \hat{\epsilon}_{\mathbf{k}}^{(1)}$, $\hat{\mathbf{e}}_2 = \hat{\epsilon}_{\mathbf{k}}^{(2)}$, and $\hat{\mathbf{e}}_3 = \mathbf{k}/k$ we have equation (1.1.36). It is also possible to prove (1.1.37) by letting k, θ, ϕ be the polar coordinates of the wave vector $\mathbf{k}$, so that

$$\mathbf{k} \equiv k(\sin\theta\cos\phi, \sin\theta\sin\phi, \cos\theta).$$

The two transverse polarization vectors can then be represented by

$$\hat{\epsilon}_{\mathbf{k}}^{(1)} \equiv (\sin\phi, -\cos\phi, 0),$$
$$\hat{\epsilon}_{\mathbf{k}}^{(2)} \equiv (\cos\theta\cos\phi, \cos\theta\sin\phi, -\sin\theta),$$

and it can be verified that

$$\epsilon_{\mathbf{k}i}^{(1)}\epsilon_{\mathbf{k}j}^{(1)} + \epsilon_{\mathbf{k}i}^{(2)}\epsilon_{\mathbf{k}j}^{(2)} = \delta_{ij} - \frac{k_i k_j}{k^2},$$

where i, j represent the Cartesian components. Demonstrate this by direct substitution.

References and bibliography

Quantization of the radiation field

P. A. M. Dirac, *Proc. Roy. Soc. A* **114**, 243 (1927).

E. Fermi *Rev. Mod. Phys.* **4**, 87 (1932).

Some useful books dealing with the quantum theory of radiation and quantum optics

W. Heitler, *The Quantum Theory of Radiation*, (Oxford University Press, New York 1954).

H. A. Kramers, *Quantum Mechanics*, (North-Holland, Amsterdam 1958).

E. A. Power, *Introductory Quantum Electrodynamics*, (Longman, London 1964).

J. R. Klauder and E. C. G. Sudarshan, *Fundamentals of Quantum Optics*, (W. A. Benjamin, New York 1970).

J. Peřina, *Coherence of Light*, (Van Nostrand, London 1972).

F. Haake, *Statistical Treatment of Open Systems by Generalised Master Equations*, (Springer Tracts in Modern Physics, Vol. 66), (Springer-Verlag, Berlin 1973).

R. Loudon, *The Quantum Theory of Light*, (Oxford University Press, New York 1973).

W. H. Louisell, *Quantum Statistical Properties of Radiation*, (John Wiley, New York 1973).

G. S. Agarwal, *Quantum Statistical Theories of Spontaneous Emission*, (Springer Tracts in Modern Physics, Vol. 70), (Springer-Verlag, Berlin 1974).

H. M. Nussenzveig, *Introduction to Quantum Optics*, (Gordon and Breach, New York 1974).

L. Allen and J. H. Eberly, *Optical Resonance and Two-Level Atoms*, (John Wiley, New York 1975).

I. Bialynicki-Birula and Z. Bialynicka-Birula, *Quantum Electrodynamics*, (Pergamon Press, Oxford 1976).

M. Sargent III, M. O. Scully, and W. E. Lamb, Jr., *Laser Physics*, (Addison-Wesley, Mass. 1974).

C. Itzykson and J. B. Zuber, *Quantum Field Theory*, (McGraw Hill, New York 1980).

H. Haken, *Light*, Vols. I and II, (North-Holland, Amsterdam 1981).

P. L. Knight and L. Allen, *Concepts of Quantum Optics*, (Pergamon Press, Oxford 1983).

J. Peřina, *Quantum Statistics of Linear and Nonlinear Optical Phenomena*, (D. Reidel, Dordrecht 1984).

C. Cohen-Tannoudji, J. Dupont-Roc, and G. Grynberg, *Photons and Atoms, Introduction to Quantum Electrodynamics*, (Wiley, New York 1989).

P. Meystre and M. Sargent III, *Elements of Quantum Optics*, (Springer-Verlag, Berlin 1990).

B. W. Shore, *The Theory of Coherent Atomic Excitation*, Vols. 1 and 2, (John Wiley, New York 1990).

C. W. Gardiner, *Quantum Noise*, (Springer-Verlag, Berlin 1991).

C. Cohen-Tannoudji, J. Dupont-Roc, and G. Grynberg, *Atom–Photon Interactions*, (Wiley, New York 1992).
H. Carmichael, *An Open Systems Approach to Quantum Optics*, (Springer-Verlag, Berlin 1993).
P. W. Milonni, *The Quantum Vacuum: An Introduction to Quantum Electrodynamics*, (Academic, New York 1994).
W. Vogel and D.-G. Welsch, *Lectures on Quantum Optics*, (Akademie Verlag, Berlin 1994).
J. Pĕrina, Z. Hradil, and B. Jurčo, *Quantum Optics and Fundamentals of Physics*, (Kluwer, Dordrecht 1994).
D. F. Walls and G. J. Milburn, *Quantum Optics*, (Springer-Verlag, Berlin 1994).
L. Mandel and E. Wolf, *Optical Coherence and Quantum Optics*, (Cambridge, London 1995).
S. Weinberg, *Theory of Quantum Fields*, (Cambridge, London 1995).
E. R. Pike and S. Sarkar, *Quantum Theory of Radiation*, (Cambridge, London 1995).
U. Leonhardt, *Measuring the Quantum State of Light*, (Cambridge, London 1997).

Lamb shift

W. E. Lamb, Jr. and R. C. Retherford, *Phys. Rev.* **72**, 241 (1947).
H. Bethe, *Phys. Rev.* **72**, 339 (1947).
T. A. Welton, *Phys. Rev.* **74**, 1157 (1948). We have followed the simple, though not rigorous, approach of this paper to calculate the Lamb shift.
N. H. Kroll and W. E. Lamb, Jr., *Phys. Rev.* **75**, 388 (1949).
J. B. French and V. F. Weisskopf, *Phys. Rev.* **75**, 1240 (1949).
F. Dyson's letter, in *W. E. Lamb, Jr., A Festschrift on the Occasion of his 65th Birthday*, ed. D. ter Haar and M. O. Scully (North-Holland, Amsterdam 1978), p. XXXVI.
E. A. Hildum, U. Boest, D. H. McIntyre, R. G. Beausoleil, and T. W. Hänsch, *Phys. Rev. Lett.* **56**, 576 (1986). This paper reports the first precise measurement of the $1S$–$2S$ energy interval in atomic hydrogen.

Quantum beats

A. Corney and G. W. Series, *Proc. Phys. Soc.* **83**, 207 (1964).
W. W. Chow, M. O. Scully, and J. O. Stoner, Jr., *Phys. Rev. A* **11**, 1380 (1975).
R. M. Herman, H. Groth, R. Kornblith, and J. H. Eberly, *Phys. Rev. A* **11**, 1389 (1975).
I. C. Khoo and J. H. Eberly, *Phys. Rev. A* **14**, 2174 (1976).
E. T. Jaynes, in *Foundations of Radiation Theory and Quantum Electrodynamics*, ed. A. O. Barut (Plenum, New York 1980).
M. O. Scully, in *Foundations of Radiation Theory and Quantum Electrodynamics*, ed. A. O. Barut (Plenum, New York 1980).
S. Svanberg, *Atomic and Molecular Spectroscopy*, (Springer-Verlag, Berlin, 1991).

Localizability of photons and electrons and the photon wave function

T. D. Newton and E. P. Wigner, *Rev. Mod. Phys.* **21**, 400 (1949).
L. Mandel, *Phys. Rev.* **144**, 1071 (1961).
J. M. Jauch and C. Piron, *Helv. Phys. Acta* **40**, 559 (1967).
W. O. Amrein, *Helv. Phys. Acta* **42**, 149 (1969).
R. J. Cook, *Phys. Rev. A* **25**, 2164 (1982); *ibid* **26**, 2754 (1982).
E. R. Pike and S. Sarkar, *Phys. Rev. A* **35**, 926 (1987).
I. H. Deutsch and J. C. Garrison *Phys. Rev.* **43**, 2498 (1991).
I. Bialynicki-Birula, *Acta Phys. Polonica A* **86**, 97 (1994).
W. E. Lamb, Jr., *Appl. Phys. B* **60**, 77 (1995).

CHAPTER 2

Coherent and squeezed states of the radiation field

Following the development of the quantum theory of radiation and with the advent of the laser, the states of the field that most nearly describe a classical electromagnetic field were widely studied. In order to realize such 'classical' states, we will consider the field generated by a classical monochromatic current, and find that the quantum state thus generated has many interesting properties and deserves to be called a *coherent state*.* An important consequence of the quantization of the radiation field is the associated uncertainty relation for the conjugate field variables. It therefore appears reasonable to propose that the wave function which corresponds most closely to the classical field must have *minimum* uncertainty for all times subject to the appropriate simple harmonic potential.

In this chapter we show that a displaced simple harmonic oscillator ground state wave function satisfies this property and the wave packet oscillates sinusoidally in the oscillator potential without changing shape as shown in Fig. 2.1. This *coherent* wave packet always has minimum uncertainty, and resembles the classical field as nearly as quantum mechanics permits. The corresponding state vector is the *coherent state* $|\alpha\rangle$, which is the eigenstate of the positive frequency part of the electric field operator, or, equivalently, the eigenstate of the destruction operator of the field.

Classically an electromagnetic field consists of waves with well-defined amplitude and phase. Such is not the case when we treat the field quantum mechanically. There are fluctuations associated with both the amplitude and phase of the field. An electromagnetic field in a number state $|n\rangle$ has a well-defined amplitude but completely

* The coherent state concept was introduced by Schrödinger [1926]. For an excellent treatment of the subject see the Les Houches lectures of Glauber [1965].

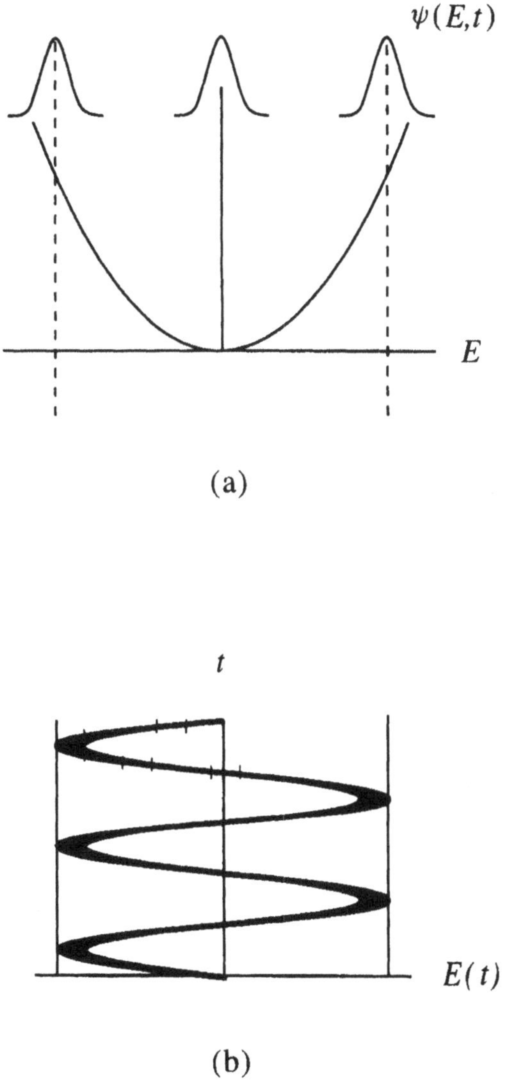

Fig. 2.1
(a) Minimum-uncertainty wave packet at different times in a harmonic oscillator potential.
(b) Corresponding electric field.

uncertain phase, whereas a field in a coherent state has equal amount of uncertainties in the two variables. Equivalently, we can describe the field in terms of the two conjugate quadrature components. The uncertainties in the two conjugate variables satisfy the Heisenberg uncertainty principle such that the product of the uncertainties in the two variables is equal to or greater than half the magnitude of the expectation value of the commutator of the variables (see Eq. (2.6.2) below). A field in a coherent state is a minimum-uncertainty state with equal uncertainties in the two quadrature components.

After developing the coherent states of the radiation field, we turn to the so-called *squeezed states*. In principle, it is possible to generate states in which fluctuations are reduced below the symmetric quantum limit in one quadrature component. This is accomplished at the expense of enhanced fluctuations in the canonically conjugate quadrature, such that the Heisenberg uncertainty principle is not violated. Such states of the radiation field are called *squeezed states*. A quadrature of electromagnetic field with reduced fluctuations below the standard quantum limit, has attractive applications in optical communication, photon detection techniques, gravitational wave detection, and noise-free amplification. In this chapter, we physically motivate and present the definition and properties of the squeezed states, with special reference to the so-called squeezed coherent states. These states result from applying the 'squeeze operator' to the coherent state.

2.1 Radiation from a classical current

In this section, we define the coherent state and show that the radiation emitted by a classical current distribution is such a state. By *classical* we mean that the current can be described by a prescribed vector $\mathbf{J}(\mathbf{r}, t)$ which is not an operator. We consider coupling of this current to the vector potential operator (cf. Eq. (1.1.27) and Section 5.1)

$$\mathbf{A}(\mathbf{r}, t) = -i \sum_{\mathbf{k}} \frac{1}{\nu_k} \hat{\epsilon}_{\mathbf{k}} \mathscr{E}_{\mathbf{k}} a_{\mathbf{k}} e^{-i\nu_k t + i\mathbf{k}\cdot\mathbf{r}} + \text{H.c.} \tag{2.1.1}$$

The Hamiltonian that describes the interaction between the field and the current is then given by

$$\mathscr{V}(t) = \int \mathbf{J}(\mathbf{r}, t) \cdot \mathbf{A}(\mathbf{r}, t) d^3r \tag{2.1.2}$$

and the state vector $|\psi(t)\rangle$ for the combined system obeys the interaction picture Schrödinger equation

$$\frac{d}{dt}|\psi(t)\rangle = -\frac{i}{\hbar}\mathscr{V}|\psi(t)\rangle. \tag{2.1.3}$$

The vector function $\mathbf{J}(\mathbf{r}, t)$ commutes with itself at different times, but the operator $\mathbf{A}(\mathbf{r}, t)$ does not. Hence the interaction energy $\mathscr{V}(t)$ does not either, and ordinarily the Schrödinger equation cannot be integrated as

$$|\psi(t)\rangle = \exp\left[-\frac{i}{\hbar}\int_0^t dt' \mathscr{V}(t')\right]|\psi(0)\rangle. \tag{2.1.4}$$

However, the various commutators introduced in obtaining the correct integration yield (2.1.4) multiplied by an overall phase factor which we discard. With (2.1.1) and (2.1.2), the exponential in (2.1.4) becomes

$$\exp\left[-\frac{i}{\hbar}\int_0^t dt'\mathscr{V}(t')\right] = \prod_{\mathbf{k}} \exp(\alpha_{\mathbf{k}} a_{\mathbf{k}}^{\dagger} - \alpha_{\mathbf{k}}^{*} a_{\mathbf{k}}), \tag{2.1.5}$$

where the complex time-dependent amplitude $\alpha_{\mathbf{k}}$ is

$$\alpha_{\mathbf{k}} = \frac{1}{\hbar \nu_k}\mathscr{E}_{\mathbf{k}} \int_0^t dt' \int d\mathbf{r}\hat{\epsilon}_{\mathbf{k}} \cdot \mathbf{J}_{\nu}(\mathbf{r},t)e^{i\nu_k t' - i\mathbf{k}\cdot\mathbf{r}}. \tag{2.1.6}$$

In Eq. (2.1.6) the dipole current $\mathbf{J}_{\nu}(\mathbf{r},t)$ is given the subscript ν to denote the fact that it is a monochromatic dipole oscillating at frequency $\nu = ck$. We choose the initial state $|\psi(0)\rangle$ to be the vacuum $|0\rangle$, and the state vector (2.1.4) then becomes

$$|\psi(t)\rangle = \prod_{\mathbf{k}} \exp(\alpha_{\mathbf{k}} a_{\mathbf{k}}^{\dagger} - \alpha_{\mathbf{k}}^{*} a_{\mathbf{k}})|0\rangle_{\mathbf{k}}. \tag{2.1.7}$$

This state of the radiation field is called a coherent state and is denoted as $|\{\alpha_{\mathbf{k}}\}\rangle$. It is apparent that the multi-mode coherent state in Eq. (2.1.7) can be expressed as a product of single-mode coherent states $|\alpha_{\mathbf{k}}\rangle$:

$$|\{\alpha_{\mathbf{k}}\}\rangle = \prod_{\mathbf{k}} |\alpha_{\mathbf{k}}\rangle, \tag{2.1.8}$$

where

$$|\alpha_{\mathbf{k}}\rangle = \exp(\alpha_{\mathbf{k}} a_{\mathbf{k}}^{\dagger} - \alpha_{\mathbf{k}}^{*} a_{\mathbf{k}})\,|0\rangle_{\mathbf{k}}. \tag{2.1.9}$$

In the remainder of this chapter, we shall be mostly concerned with a single-mode coherent state. We shall therefore remove the index **k** from our definition in Eq. (2.1.9) and write

$$|\alpha\rangle = \exp(\alpha a^{\dagger} - \alpha^{*} a)\,|0\rangle. \tag{2.1.10}$$

In the following, we present alternative approaches to the coherent state.

2.2 The coherent state as an eigenstate of the annihilation operator and as a displaced harmonic oscillator state

Expression (2.1.10) was obtained by defining the coherent state of the radiation field $|\alpha\rangle$ as a state of the field which is generated by a classically oscillating current distribution. The same expression for $|\alpha\rangle$ can be obtained by defining it as an eigenstate of the annihilation operator a with an eigenvalue α, i.e.,

$$a|\alpha\rangle = \alpha|\alpha\rangle. \tag{2.2.1}$$

An expression of $|\alpha\rangle$ in terms of the number state $|n\rangle$ is given by

$$|\alpha\rangle = e^{-|\alpha|^2/2} \sum_{n=0}^{\infty} \frac{\alpha^n}{\sqrt{n!}} |n\rangle, \tag{2.2.2}$$

and since $|n\rangle = [(a^\dagger)^n/\sqrt{n!}]|0\rangle$ this can be written as

$$|\alpha\rangle = e^{\alpha a^\dagger}|0\rangle e^{-|\alpha|^2/2}. \tag{2.2.3}$$

Next we note that since $\exp(-\alpha^* a)|0\rangle = |0\rangle$, Eq. (2.2.3) can be rewritten as

$$|\alpha\rangle = D(\alpha)|0\rangle, \tag{2.2.4}$$

where

$$D(\alpha) = e^{-|\alpha|^2/2} e^{\alpha a^\dagger} e^{-\alpha^* a}. \tag{2.2.5}$$

Now, in view of the Baker–Hausdorff formula, if A and B are any two operators such that

$$[[A, B], A] = [[A, B], B] = 0, \tag{2.2.6}$$

then

$$e^{A+B} = e^{-[A,B]/2} e^A e^B. \tag{2.2.7}$$

If we write $A = \alpha a^\dagger, B = -\alpha^* a$, it follows that

$$D(\alpha) = e^{\alpha a^\dagger - \alpha^* a}, \tag{2.2.8}$$

in agreement with Eq. (2.1.10). Another equivalent antinormal form of $D(\alpha)$ is

$$D(\alpha) = e^{|\alpha|^2/2} e^{-\alpha^* a} e^{\alpha a^\dagger}. \tag{2.2.9}$$

The operator $D(\alpha)$ is a unitary operator, i.e.,

$$D^\dagger(\alpha) = D(-\alpha) = [D(\alpha)]^{-1}. \tag{2.2.10}$$

It acts as a displacement operator upon the amplitudes a and $a^\dagger$, i.e.,

$$D^{-1}(\alpha)aD(\alpha) = a + \alpha, \tag{2.2.11}$$

$$D^{-1}(\alpha)a^\dagger D(\alpha) = a^\dagger + \alpha^*. \tag{2.2.12}$$

The displacement property can be proved by writing

$$D^{-1}(\alpha)aD(\alpha) = e^{\alpha^* a}e^{-\alpha a^\dagger} a e^{\alpha a^\dagger} e^{-\alpha^* a}, \tag{2.2.13}$$

where we have used the form (2.2.9) for $D^{-1}(\alpha)$ and the form (2.2.5) for $D(\alpha)$. For any operators A and B

$$e^{-\alpha A}Be^{\alpha A} = B - \alpha[A, B] + \frac{\alpha^2}{2!}[A, [A, B]] + \ldots \tag{2.2.14}$$

For $A = a^\dagger, B = a$, this becomes

$$e^{-\alpha a^\dagger} a e^{\alpha a^\dagger} = a + \alpha. \tag{2.2.15}$$

Use of this result in Eq. (2.2.13) gives the displacement property (2.2.11) for $D(\alpha)$. The displacement property (2.2.12) can be proved in a similar way.

According to Eq. (2.2.4), a coherent state is obtained by applying the displacement operator on the vacuum state. The coherent state is therefore the displaced form of the harmonic oscillator ground state.

2.3 What is so coherent about coherent states?

To answer this question it is instructive to consider the coordinate representation of the oscillator number state $|n\rangle$. The coordinate representation of $|n\rangle$ is given by

$$\phi_n(q) = \langle q|n\rangle. \tag{2.3.1}$$

It follows from Eqs. (1.1.11) that

$$a = \frac{1}{\sqrt{2\hbar\nu}}\left(\nu q + \hbar\frac{\partial}{\partial q}\right), \quad a^\dagger = \frac{1}{\sqrt{2\hbar\nu}}\left(\nu q - \hbar\frac{\partial}{\partial q}\right), \tag{2.3.2}$$

where we have used $p = -i\hbar\partial/\partial q$. Equation (1.2.7) then leads to

$$\left(\nu q + \hbar\frac{\partial}{\partial q}\right)\phi_0(q) = 0. \tag{2.3.3}$$

A normalized solution of this equation is

$$\phi_0(q) = \left(\frac{\nu}{\pi\hbar}\right)^{1/4}\exp\left(-\frac{\nu q^2}{2\hbar}\right). \tag{2.3.4}$$

Higher order eigenfunctions in the coordinate representation can be obtained from Eqs. (1.2.16), (2.3.1), and (2.3.2):

$$\phi_n(q) = \frac{(a^\dagger)^n}{\sqrt{n!}}\phi_0(q) = \frac{1}{\sqrt{n!}}\frac{1}{(2\hbar\nu)^{n/2}}\left(\nu q - \hbar\frac{\partial}{\partial q}\right)^n \phi_0(q)$$
$$= \frac{1}{(2^n n!)^{1/2}} H_n\left(\sqrt{\frac{\nu}{\hbar}}q\right)\phi_0(q), \tag{2.3.5}$$

where H_n are the Hermite polynomials. These are the well-known eigenfunctions of the harmonic oscillator. It can be verified that these wave functions satisfy the orthonormality condition

$$\int_{-\infty}^{\infty} \phi_n^*(q)\phi_m(q)dq = \delta_{nm}. \tag{2.3.6}$$

It follows from the definition of the harmonic oscillator wave functions $\phi_n(q)$ that

$$\langle q\rangle = \int_{-\infty}^{\infty} \phi_n^*(q)q\phi_n(q)dq = 0. \tag{2.3.7}$$

Similarly

$$\langle p\rangle = 0, \tag{2.3.8}$$
$$\langle p^2\rangle = \hbar\nu\left(n+\frac{1}{2}\right), \tag{2.3.9}$$
$$\langle q^2\rangle = \frac{\hbar}{\nu}\left(n+\frac{1}{2}\right). \tag{2.3.10}$$

The uncertainties in the generalized momentum and coordinate variables are therefore given by

$$(\Delta p)^2 = \langle p^2\rangle - \langle p\rangle^2$$
$$= \hbar\nu\left(n+\frac{1}{2}\right), \tag{2.3.11}$$
$$(\Delta q)^2 = \frac{\hbar}{\nu}\left(n+\frac{1}{2}\right). \tag{2.3.12}$$

The uncertainty product is

$$\Delta p\Delta q = \left(n+\frac{1}{2}\right)\hbar. \tag{2.3.13}$$

This has minimum possible value of $\hbar/2$ for the ground state wave function $\phi_0(q)$.

It is of special interest to find a wave packet which maintains the same variance Δq while undergoing simple harmonic motion. Such a wave function would correspond most closely to a classical field. In order to investigate this possibility we assume that, at time $t = 0$, the wave function $\psi(q,t)$ is of the form (2.3.4) of the minimum-uncertainty wave packet except that it is displaced in the positive q direction by an amount q_0. We then have

$$\psi(q,0) = \left(\frac{\nu}{\pi\hbar}\right)^{1/4} \exp\left[-\frac{\nu}{2\hbar}(q-q_0)^2\right]. \tag{2.3.14}$$

The time evolution of this wave packet is derived in Problem 2.3, where it is shown that the initial packet given by Eq. (2.3.14) implies that the probability density later in time is

$$|\psi(q,t)|^2 = \left(\frac{\nu}{\pi\hbar}\right)^{1/2} \exp\left[-\frac{\nu}{\hbar}(q-q_0\cos\nu t)^2\right]. \tag{2.3.15}$$

We note that the wave packet (2.3.14) oscillates back and forth in a simple harmonic oscillator potential without changing its shape, i.e., it sticks together or *coheres*. This is to be contrasted with the wave packet which is a delta function at $t = 0$, goes to a plane wave at $\nu t = \pi/2$, and is again a delta function at $\nu t = \pi$, see Section 2.5 for more details. Although the delta function packet returns to its original shape at the end of a period, it has a variance which is a strong function of time, i.e., it does not *cohere*.

The packet ψ has the minimum-uncertainty product allowed by quantum mechanics, namely $\Delta p \Delta q = \hbar/2$. These states therefore provide the closest quantum mechanical analog to a free classical single-mode field.

The minimum-uncertainty wave packet (2.3.14) which *coheres* in a simple harmonic oscillator potential is given by (Problem 2.4)

$$\psi(q,0) = e^{-|\alpha|^2/2} \sum_{n=0}^{\infty} \frac{\alpha^n}{\sqrt{n!}} \langle q|n\rangle, \tag{2.3.16}$$

with $\alpha = (\nu/2\hbar)^{1/2} q_0$, where we use $\phi(q) = \langle q|n\rangle$. The state $|\alpha\rangle$ associated with $\psi(q,0)$ therefore has an expansion in number states identical to that for a coherent state, as given by Eq. (2.2.2). The minimum-uncertainty wave packet $\psi(q,0)$ is therefore the coordinate representation of the coherent state.

2.4 Some properties of coherent states

In this section, we list some important properties of the coherent states of the radiation field.

(a) The mean number of photons in the coherent state $|\alpha\rangle$ is given by

$$\langle\alpha|a^\dagger a|\alpha\rangle = |\alpha|^2. \tag{2.4.1}$$

The probability of finding n photons in $|\alpha\rangle$ is given by a Poisson distribution, i.e.,

$$p(n) = \langle n|\alpha\rangle\langle\alpha|n\rangle = \frac{|\alpha|^{2n}e^{-\langle n\rangle}}{n!} = \frac{\langle n\rangle^n e^{-\langle n\rangle}}{n!}. \tag{2.4.2}$$

where $\langle n\rangle = |\alpha|^2$. As we shall see in Chapter 11, the photon distribution for the laser approaches this distribution for sufficiently high excitations. In Fig. 2.2 we have plotted $p(n)$ versus n for different values of $|\alpha|^2$. It is seen that, for $|\alpha|^2 \leq 1$, $p(n)$ is maximum at $n = 0$, whereas, for $|\alpha|^2 > 1$, $p(n)$ has a peak at $n = |\alpha|^2$.

(b) As discussed earlier, the coherent state is a minimum-uncertainty state so that

$$\Delta p \Delta q = \frac{\hbar}{2}. \tag{2.4.3}$$

(c) The set of all coherent states $|\alpha\rangle$ is a complete set. To show this, we first consider the integral identity (with $\alpha = |\alpha|e^{i\theta}$)

$$\begin{aligned}\int (\alpha^*)^n \alpha^m e^{-|\alpha|^2} d^2\alpha &= \int_0^\infty |\alpha|^{n+m+1} e^{-|\alpha|^2} d|\alpha| \int_0^{2\pi} e^{i(m-n)\theta} d\theta \\ &= \pi n! \delta_{nm},\end{aligned} \tag{2.4.4}$$

in which the integration is carried out over the entire area of the complex plane. With the help of this identity it follows, on using the expansion (2.2.2) for the coherent states, that

$$\int |\alpha\rangle\langle\alpha| d^2\alpha = \pi \sum_n |n\rangle\langle n|. \tag{2.4.5}$$

Since the Fock states $|n\rangle$ form a complete orthonormal set, the sum over n is simply the unit operator. We thus have

$$\frac{1}{\pi}\int |\alpha\rangle\langle\alpha| d^2\alpha = 1, \tag{2.4.6}$$

which is the completeness relation for the coherent states.

Fig. 2.2
The photon distribution $p(n)$ for a coherent state with
(a) $|\alpha|^2 = 0.1$,
(b) $|\alpha|^2 = 1$, and
(c) $|\alpha|^2 = 10$.

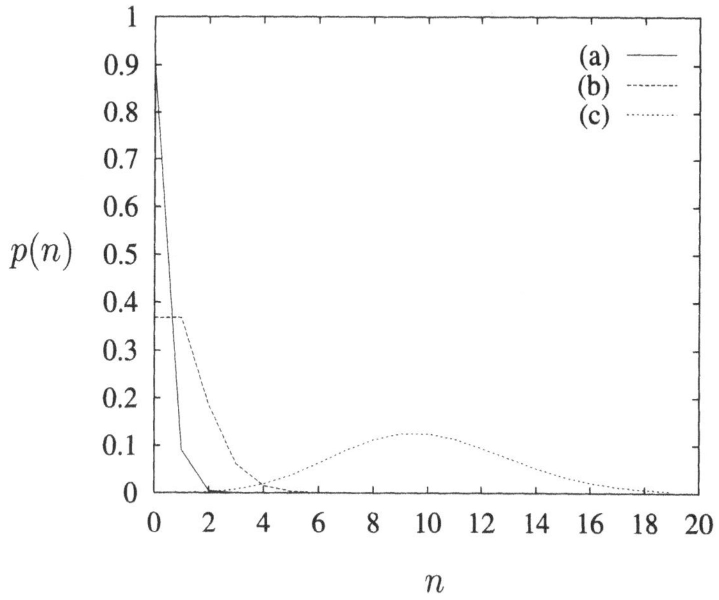

(d) Two coherent states corresponding to different eigenstates α and α' are not orthogonal, i.e.,

$$\langle\alpha|\alpha'\rangle = \exp\left(-\frac{1}{2}|\alpha|^2 + \alpha'\alpha^* - \frac{1}{2}|\alpha'|^2\right), \tag{2.4.7}$$

and

$$|\langle\alpha|\alpha'\rangle|^2 = \exp(-|\alpha - \alpha'|^2). \tag{2.4.8}$$

Here we see that, if the magnitude of $\alpha - \alpha'$ is much greater than unity, the states $|\alpha\rangle$ and $|\alpha'\rangle$ are nearly orthogonal to one another. The degree to which these wave functions overlap determines the size of the inner product $\langle\alpha|\alpha'\rangle$. A consequence of Eq. (2.4.7) is the fact that any coherent state can be expanded in terms of the other states:

$$\begin{aligned}|\alpha\rangle &= \frac{1}{\pi}\int d^2\alpha'|\alpha'\rangle\langle\alpha'|\alpha\rangle \\ &= \frac{1}{\pi}\int d^2\alpha'|\alpha'\rangle \exp\left(-\frac{1}{2}|\alpha|^2 + \alpha'^*\alpha - \frac{1}{2}|\alpha'|^2\right).\end{aligned} \tag{2.4.9}$$

This indicates that the coherent states are *overcomplete.*

2.5 Squeezed state physics

Natural philosophy, the union of experimental and theoretical science, abounds with wonderful examples of the fruitful interplay between experimental and theoretical thought. The 'ultraviolet catastrophe' observed in black-body radiation led Planck to introduce the notion of the quantum. These considerations led Einstein to the concept of 'stimulated emission' which was the key to understanding the differences between the radiation distributions of Planck and Wien. Stimulated emission is, of course, the basis for the laser which ushered in the modern era of quantum optics.

Squeezed states of the radiation field provide another, near term, example of the rich interplay between experiment and theory. By itself, the squeezing of states of the field is of limited interest. For example, the number state consisting of n photons clearly exists, but how to make it and who cares if we do?

One answer to the 'who cares?' question comes from the search for gravitational radiation. As is further discussed in Chapter 4, the acceleration of distant matter, e.g., the explosion of a supernova, leads to tiny forces on laboratory instruments. For example, an oscillating gravity wave can drive a mechanical oscillator which thus serves as a gravity wave detector.

But the amplitudes of oscillation generated by many sources of gravitational radiation are anticipated to be much smaller than the width of the ground state wave function. This prompted people to think about squeezing the ground state wave function (zero-point noise) of quantum mechanical oscillators.

That such 'squeezing' is possible in principle is made clear by considering the elementary quantum mechanics of the simple harmonic oscillator (SHO). As is depicted in Fig. 2.3, a wave packet which is sharply peaked (i.e., squeezed) initially will spread out and return to its initial state periodically. A little review of the SHO time evolution makes this clear.* Recall that the wave function at time t is related to that at $t = 0$ by the expression

$$\psi(x,t) = \int dx' G(x,x',t)\psi(x',0), \tag{2.5.1}$$

where the well known SHO propagator, as given in quantum mechanics texts, is

$$G(x,x',t) = \sqrt{\frac{m\nu}{2\pi\hbar|\sin\nu t|}} \exp\left\{\frac{im\nu}{2\hbar\sin\nu t}[(x^2+x'^2)\cos\nu t - 2xx']\right\}, \tag{2.5.2}$$

* See, for example, Sargent, Scully, and Lamb, *Laser Physics* [1974] Appendix H.

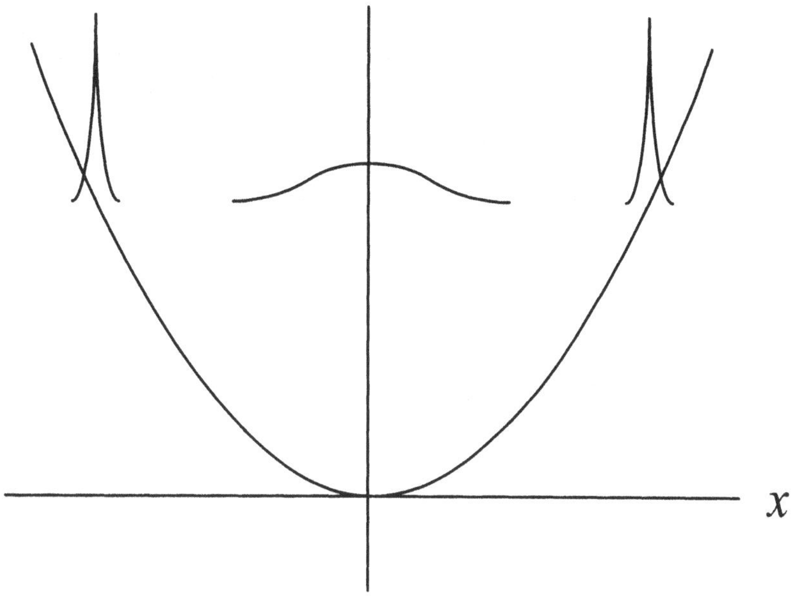

Fig. 2.3
Evolution of a squeezed state of a simple harmonic oscillator.

with m and v being the mass and frequency of the oscillator.

Now if we begin at $t = 0$ with a δ-function wave packet $\psi(x', 0) = \delta(x' - x_0)$ then at a time $t = \pi/2v$ later the wave function will be a plane wave; that is, our squeezed state evolves as

$$\psi(x, t = 0) = \delta(x - x_0), \tag{2.5.3a}$$

$$\psi(x, t = \pi/2v) = \sqrt{\frac{mv}{2\pi\hbar}} \exp\left[i\left(\frac{mvx_0}{\hbar}\right)x\right], \tag{2.5.3b}$$

$$\psi(x, t = \pi/v) = \delta(x + x_0). \tag{2.5.3c}$$

Thus, from Fig. 2.3 and Eqs. (2.5.3), we see that if we start with a sharp or squeezed state we will return to a sharp state every half period. In this sense we have the possibility of a kind of 'stroboscopic' measurement, in which we look at our oscillator at $t = 0, \pi/v, 2\pi/v, \ldots$, so that we are not limited by the width of the ground state wave function.

Having motivated and illustrated squeezed states, let us proceed to a better understanding of these states by considering a *gedanken* experiment illustrating how we might prepare such states. To this end, let us return briefly to the question of how we might prepare a coherent state.

In classical mechanics we can excite a SHO into motion by, e.g., stretching the spring of Fig. 2.4 to a new equilibrium position and releasing it to produce oscillation. In quantum mechanics a similar procedure can be followed but we must be more specific about how

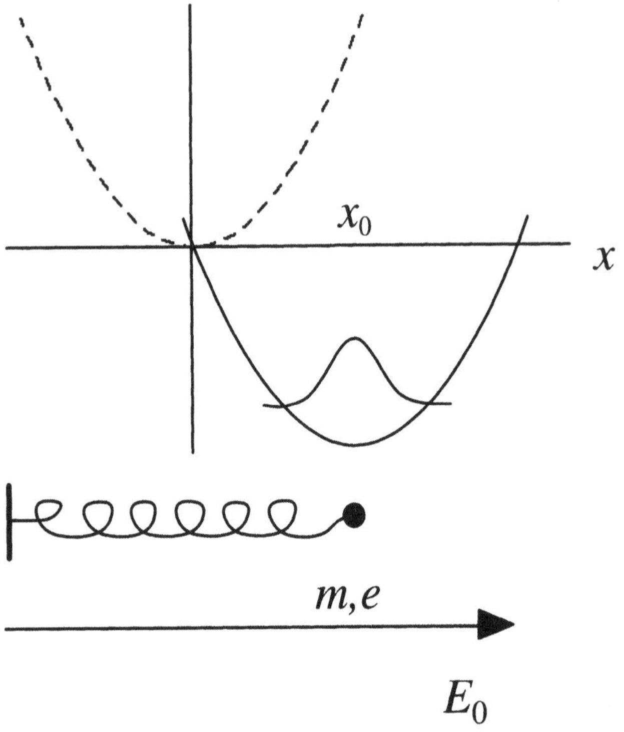

Fig. 2.4
Dashed potential applies for a spring-type SHO and causes a particle of mass m and charge e to oscillate about $x = 0$. Applying a dc electric field stretches the spring to a new equilibrium position x_0 about which the point charge particle now oscillates.

we prepare the initial state of the SHO. Let us envision a SHO characterized by mass m and charge e in a field E_0, as in Fig. 2.4; then the Hamiltonian is

$$\mathscr{H} = \frac{p^2}{2m} + \frac{1}{2}kx^2 - eE_0x, \tag{2.5.4a}$$

which we may write as

$$\mathscr{H} = \frac{p^2}{2m} + \frac{1}{2}k\left(x - \frac{eE_0}{k}\right)^2 - \frac{1}{2}k\left(\frac{eE_0}{k}\right)^2. \tag{2.5.4b}$$

We have in (2.5.4b) the well-known fact that applying a linear potential to a SHO just shifts its equilibrium point. Clearly the same solutions obtain. We have thus prepared a displaced ground state as in Fig. 2.4. And upon turning off the dc field, i.e., setting $E_0 = 0$, we will have a coherent state $|\alpha\rangle$ which oscillates without changing its shape.

It is to be noted that applying the dc field to the SHO is mathematically equivalent to applying the displacement operator (2.2.8) to the state $|0\rangle$. This is summarized in Fig. 2.4.

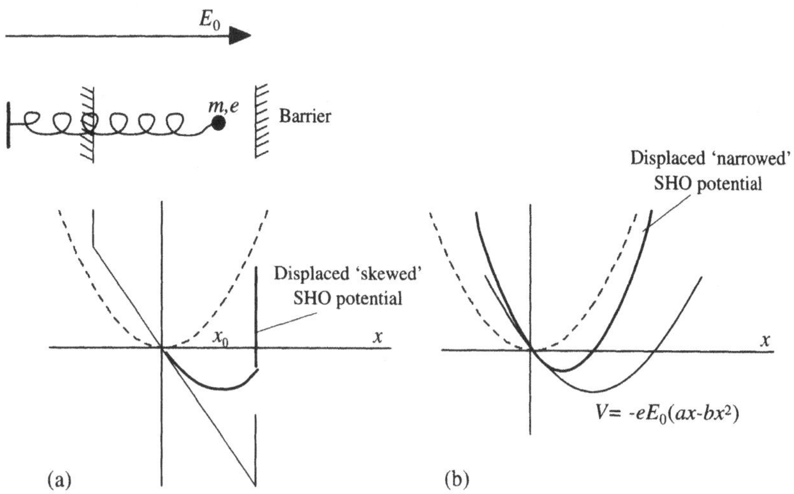

Fig. 2.5
(a) The SHO potential is first displaced by a dc elecric field and then 'skewed' by barriers which limit the charge oscillation to a finite region.
(b) The SHO potential is displaced and 'narrowed' by a quadratic displacement potential.

Next, let us consider how we might prepare a squeezed state. Suppose we again apply a dc field but this time with a 'wall' which limits the SHO to a finite region as in Fig. 2.5(a).

In such a case, it would be expected that the wave packet would be deformed or 'squeezed' when it is pushed against the barrier. Similarly the quadratic displacement potential of Fig. 2.5(b) would be expected to produce a squeezed wave packet. To see that this is indeed the case, consider the Hamiltonian for the SHO in the presence of the quadratic potential

$$\mathscr{H} = \frac{p^2}{2m} + \frac{1}{2}kx^2 - eE_0(ax - bx^2), \tag{2.5.5a}$$

where the ax term will displace the oscillator and the bx^2 is added in order to give us a barrier to 'squeeze the packet against'. We rewrite (2.5.5a) as

$$\mathscr{H} = \frac{p^2}{2m} + \frac{1}{2}(k + 2ebE_0)x^2 - eaE_0x. \tag{2.5.5b}$$

From Eq. (2.5.5b) it is clear that we again have a displaced ground state, but this time with the larger effective spring constant $k' = k + 2ebE_0$. This, of course, means that we have a *squeezed* displaced wave packet as depicted in Fig. 2.6. This is the desired result.

In conclusion we note that, just as it is the creation operator part of the linear displacement potential which is most important in preparing a coherent state; we shall find that it is the two-photon $a^{\dagger 2}$ and a^2 contributions, contained within the bx^2 term in Eqs. (2.5.5), that are most important in preparing a squeezed coherent state.

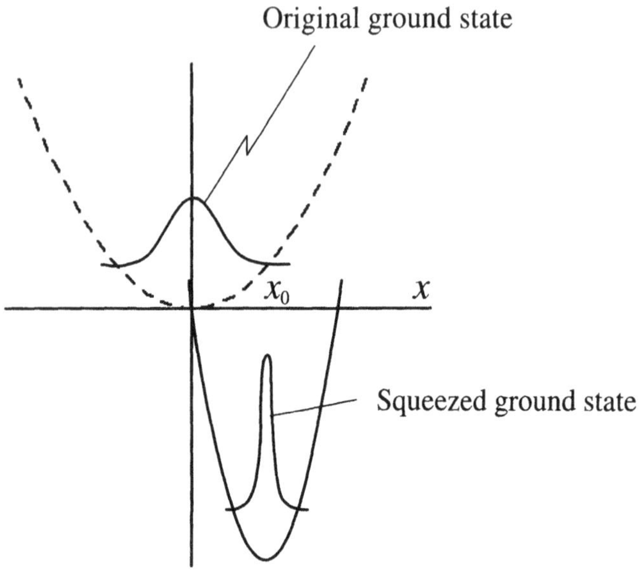

Fig. 2.6
The displaced 'narrowed' SHO potential squeezes the wave packet.

2.6 Squeezed states and the uncertainty relation

Having motivated the study and nature of squeezed states, let us consider what other properties we might expect from them. Consider two Hermitian operators A and B which satisfy the commutation relation

$$[A, B] = iC. \tag{2.6.1}$$

According to the Heisenberg uncertainty relation, the product of the uncertainties in determining the expectation values of two variables A and B is given by

$$\Delta A \Delta B \geq \frac{1}{2}|\langle C \rangle|. \tag{2.6.2}$$

A state of the system is called a squeezed state if the uncertainty in one of the observables (say A) satisfies the relation

$$(\Delta A)^2 < \frac{1}{2}|\langle C \rangle|. \tag{2.6.3}$$

If, in addition to the condition (2.6.3), the variances satisfy the minimum-uncertainty relation, i.e.,

$$\Delta A \Delta B = \frac{1}{2}|\langle C \rangle|, \tag{2.6.4}$$

then the state is called an ideal squeezed state.

In a squeezed state, therefore, the quantum fluctuations in one variable are reduced below their value in a symmetric minimum-uncertainty state ($(\Delta A)^2 = (\Delta B)^2 = |\langle C\rangle|/2$) at the expense of the corresponding increased fluctuations in the conjugate variable such that the uncertainty relation is not violated.

As an illustration, we consider a quantized single-mode electric field of frequency ν:

$$\mathbf{E}(t) = \mathscr{E}\hat{\epsilon}(ae^{-i\nu t} + a^\dagger e^{i\nu t}), \tag{2.6.5}$$

where a and $a^\dagger$ obey the commutation relation

$$[a, a^\dagger] = 1. \tag{2.6.6}$$

We introduce the Hermitian amplitude operators

$$X_1 = \frac{1}{2}(a + a^\dagger), \tag{2.6.7}$$

$$X_2 = \frac{1}{2i}(a - a^\dagger). \tag{2.6.8}$$

It is, of course, clear that X_1 and X_2 are essentially dimensionless position and momentum operators

$$x = \frac{\sqrt{2\hbar/m\nu}}{2}(a + a^\dagger),$$

$$p = \frac{\sqrt{2m\hbar\nu}}{2i}(a - a^\dagger).$$

It follows from the commutation relation (2.6.6) that X_1 and X_2 satisfy

$$[X_1, X_2] = \frac{i}{2}. \tag{2.6.9}$$

In terms of these operators, Eq. (2.6.5) can be rewritten as

$$\mathbf{E}(t) = 2\mathscr{E}\hat{\epsilon}(X_1 \cos \nu t + X_2 \sin \nu t). \tag{2.6.10}$$

The Hermitian operators X_1 and X_2 are now readily seen to be the amplitudes of the two quadratures of the field having a phase difference $\pi/2$. From Eq. (2.6.9), the uncertainty relation for the two amplitudes is

$$\Delta X_1 \Delta X_2 \geq \frac{1}{4}. \tag{2.6.11}$$

A squeezed state of the radiation field is obtained if

$$(\Delta X_i)^2 < \frac{1}{4} \qquad (i = 1 \text{ or } 2). \tag{2.6.12}$$

An ideal squeezed state is obtained if in addition to Eq. (2.6.12), the relation

$$\Delta X_1 \Delta X_2 = \frac{1}{4} \tag{2.6.13}$$

also holds.

In the next section we will consider the two-photon coherent state which is an example of an ideal squeezed state. Here we mention that the coherent state $|\alpha\rangle$ and the Fock state $|n\rangle$ are not squeezed states. It follows from Eq. (2.6.7) that, in a coherent state,

$$\begin{aligned}(\Delta X_1)^2 &= \langle\alpha|X_1^2|\alpha\rangle - (\langle\alpha|X_1|\alpha\rangle)^2 \\ &= \frac{1}{4}\langle\alpha|[a^2 + aa^\dagger + a^\dagger a + (a^\dagger)^2]|\alpha\rangle - \frac{1}{4}[\langle\alpha|(a + a^\dagger)|\alpha\rangle]^2 \\ &= \frac{1}{4}.\end{aligned} \tag{2.6.14}$$

Similarly

$$(\Delta X_2)^2 = \frac{1}{4}. \tag{2.6.15}$$

In a similar manner, in a Fock state,

$$\begin{aligned}(\Delta X_1)^2 &= \langle n|X_1^2|n\rangle - (\langle n|X_1|n\rangle)^2, \\ &= \frac{1}{4}(2n+1),\end{aligned} \tag{2.6.16}$$

$$(\Delta X_2)^2 = \frac{1}{4}(2n+1). \tag{2.6.17}$$

In Fig. 2.7 error contours of the uncertainties in X_1 and X_2, along with the corresponding graphs of the electric field versus time are shown for a coherent state, a squeezed state with reduced noise in X_1, and a squeezed state with reduced noise in X_2. Each point in the error contour for various states corresponds to a wave with a certain amplitude and a certain phase. A summation of all such waves in the error contours thus leads to the uncertainties of the electric field represented by the shaded region. A coherent state (Fig. 2.7(a)), having identical uncertainties in both X_1 and X_2, has a constant value for the variance of the electric field. A squeezed state with reduced noise in X_1 (Fig. 2.7(b)) has reduced uncertainty in the amplitude at the expense of large uncertainty in the phase of the electric field whereas the situation is reversed for a squeezed state with reduced noise in X_2 (Fig. 2.7(c)).

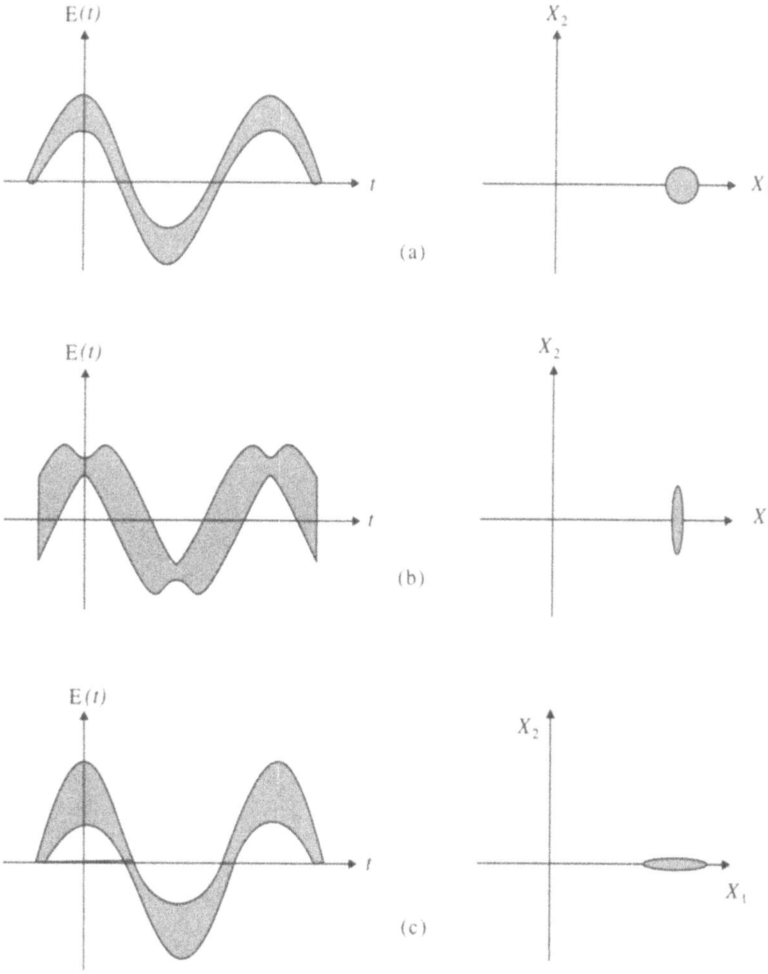

Fig. 2.7
Error contours and the corresponding graphs of electric field versus time for (a) a coherent state, (b) a squeezed state with reduced nosie in X_1, and (c) a squeezed state with reduced noise in X_2. (From C. Caves, *Phys. Rev. D* **23**, 1693 (1981).)

2.7 The squeeze operator and the squeezed coherent states

In Section 2.5 we found that quadratic terms in x, i.e., terms of the form $(a + a^\dagger)^2$, were important in the preparation of squeezed states. With that thought in mind, we are naturally motivated to consider degenerate parametric processes in connection with the generation of such states of the radiation field. In fact, much of squeezed state physics is nicely illustrated by the degenerate parametric process, as discussed in Chapter 16. The associated two-photon Hamiltonian can be written as

$$\mathscr{H} = i\hbar\left(ga^{\dagger 2} - g^* a^2\right), \tag{2.7.1}$$

where g is a coupling constant. Hence the state of the field generated by this expression is

$$|\psi(t)\rangle = e^{\left(g a^{\dagger 2} - g^* a^2\right)t}\,|0\rangle \tag{2.7.2}$$

and this leads us to define the unitary squeeze operator

$$S(\xi) = \exp\left(\frac{1}{2}\xi^* a^2 - \frac{1}{2}\xi a^{\dagger 2}\right), \tag{2.7.3}$$

where $\xi = r\exp(i\theta)$ is an arbitrary complex number. It is easy to see that

$$S^\dagger(\xi) = S^{-1}(\xi) = S(-\xi). \tag{2.7.4}$$

A straightforward application of the formula

$$e^A B e^{-A} = B + [A, B] + \frac{1}{2!}[A, [A, B]] + \ldots, \tag{2.7.5}$$

leads to the following useful unitary transformation properties of the squeeze operator

$$S^\dagger(\xi) a S(\xi) = a\cosh r - a^\dagger e^{i\theta}\sinh r, \tag{2.7.6}$$

$$S^\dagger(\xi) a^\dagger S(\xi) = a^\dagger \cosh r - a e^{-i\theta}\sinh r. \tag{2.7.7}$$

If we define a rotated complex amplitude at an angle $\theta/2$

$$Y_1 + iY_2 = (X_1 + iX_2)e^{-i\theta/2}, \tag{2.7.8}$$

it follows from Eq. (2.7.6) that

$$S^\dagger(\xi)(Y_1 + iY_2)S(\xi) = Y_1 e^{-r} + iY_2 e^r. \tag{2.7.9}$$

A squeezed coherent state $|\alpha, \xi\rangle$ is obtained by first acting with the displacement operator $D(\alpha)$ on the vacuum followed by the squeeze operator $S(\xi)$, i.e.,

$$|\alpha, \xi\rangle = S(\xi)D(\alpha)|0\rangle, \tag{2.7.10}$$

with $\alpha = |\alpha|\exp(i\varphi)$. As discussed earlier, whereas a coherent state is generated by linear terms in a and $a^\dagger$ in the exponent, the squeezed coherent state requires quadratic terms.

In the following we discuss some properties of the squeezed coherent state since it is a canonical example of a squeezed state.

2.7.1 Quadrature variance

The operator expectation values of the state $|\alpha, \xi\rangle$ can be determined from the definition (2.7.10) by making use of the transformation properties of the displacement and squeezing operators (Eq. (2.7.3)). It then follows that

$$\begin{aligned}
\langle a\rangle &= \langle \alpha, \xi|a|\alpha, \xi\rangle \\
&= \langle 0|D^{\dagger}(\alpha)S^{\dagger}(\xi)aS(\xi)D(\alpha)|0\rangle \\
&= \langle \alpha|(a\cosh r - a^{\dagger}e^{i\theta}\sinh r)|\alpha\rangle \\
&= \alpha\cosh r - \alpha^{*}e^{i\theta}\sinh r, \qquad (2.7.11)\\
\langle a^2\rangle &= \langle (a^{\dagger})^2\rangle^{*} \\
&= \langle 0|D^{\dagger}(\alpha)S^{\dagger}(\xi)a^2S(\xi)D(\alpha)|0\rangle \\
&= \langle \alpha|S^{\dagger}(\xi)aS(\xi)S^{\dagger}(\xi)aS(\xi)|\alpha\rangle \\
&= \alpha^2\cosh^2 r + (\alpha^{*})^2e^{2i\theta}\sinh^2 r - 2|\alpha|^2e^{i\theta}\sinh r\cosh r \\
&\quad -e^{i\theta}\cosh r\sinh r, \qquad (2.7.12)\\
\langle a^{\dagger}a\rangle &= |\alpha|^2(\cosh^2 r + \sinh^2 r) - (\alpha^{*})^2e^{i\theta}\sinh r\cosh r \\
&\quad -\alpha^2e^{-i\theta}\sinh r\cosh r + \sinh^2 r. \qquad (2.7.13)
\end{aligned}$$

The variances of the rotated amplitudes Y_1 and Y_2 can be determined from these expectation values. On substituting for X_1 and X_2 from Eqs. (2.6.7) and (2.6.8) into Eq. (2.7.8) we obtain

$$Y_1 + iY_2 = a\exp(-i\theta/2), \qquad (2.7.14)$$

so that

$$\begin{aligned}
(\Delta Y_1)^2 &= \langle Y_1^2\rangle - \langle Y_1\rangle^2 \\
&= \frac{1}{4}\langle (ae^{-i\theta/2} + a^{\dagger}e^{i\theta/2})^2\rangle - \frac{1}{4}(\langle ae^{-i\theta/2} + a^{\dagger}e^{i\theta/2}\rangle)^2 \\
&= \frac{1}{4}\langle a^2e^{-i\theta} + a^{\dagger 2}e^{i\theta} + aa^{\dagger} + a^{\dagger}a\rangle \\
&\quad -\frac{1}{4}(\langle ae^{-i\theta/2} + a^{\dagger}e^{i\theta/2}\rangle)^2 = \frac{1}{4}e^{-2r}, \qquad (2.7.15)\\
(\Delta Y_2)^2 &= \frac{1}{4}e^{2r}, \qquad (2.7.16)\\
\Delta Y_1\Delta Y_2 &= \frac{1}{4}. \qquad (2.7.17)
\end{aligned}$$

A squeezed coherent state is therefore an ideal squeezed state. As shown in Fig. 2.8, in the complex amplitude plane the coherent state error circle is *squeezed* into an *error ellipse* of the same area. The principal axes of the ellipse lie along Y_1 and Y_2 rotated at an angle $\theta/2$ from X_1 and X_2, respectively. The degree of squeezing is determined by $r = |\xi|$ which is therefore called the squeeze parameter.

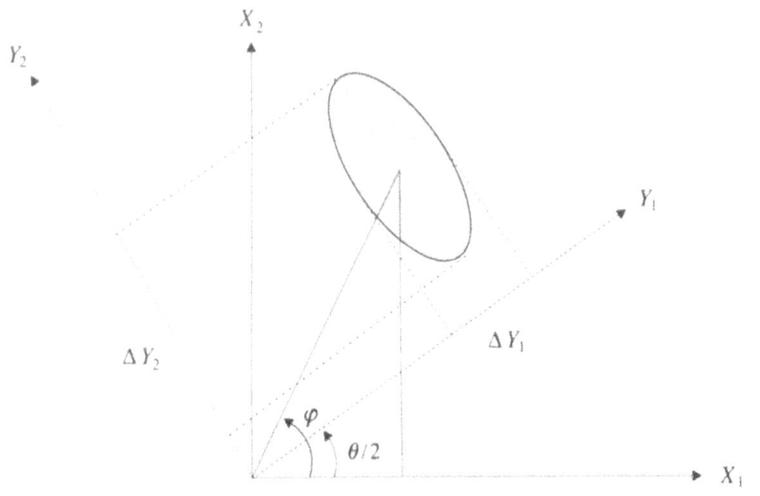

Fig. 2.8
Error contour for a squeezed coherent state.

2.8 Multi-mode squeezing

The single-mode two-photon coherent state can be generalized to a multi-mode squeezed state by using a generator which incorporates the product of annihilation (and creation) operators for correlated pairs of modes symmetrically placed around a mode of frequency, say, ν. First, we discuss the simple case of two-mode squeezing and then generalize it to the multi-mode case. The two-mode squeezed state is obtained by the action of the unitary operator

$$S(\xi) = e^{\xi^* a_{\nu+\nu'} a_{\nu-\nu'} - \xi a^\dagger_{\nu+\nu'} a^\dagger_{\nu-\nu'}}, \tag{2.8.1}$$

on the two-mode vacuum.

To show that the operators spanning the two modes exhibit squeezing, we define collective creation and destruction operators

$$b^\dagger = \frac{1}{\sqrt{2}} \left[a^\dagger_{\nu+\nu'} + e^{i\delta} a^\dagger_{\nu-\nu'} \right], \tag{2.8.2}$$

$$b = \frac{1}{\sqrt{2}} \left[a_{\nu+\nu'} + e^{-i\delta} a_{\nu-\nu'} \right]. \tag{2.8.3}$$

The in-phase and in-quadrature components are given by

$$b_1 = \frac{1}{2}(b + b^\dagger), \tag{2.8.4}$$

$$b_2 = \frac{1}{2i}(b - b^\dagger). \tag{2.8.5}$$

The corresponding uncertainty relation is

$$\Delta b_1 \Delta b_2 \geq \frac{1}{4}. \tag{2.8.6}$$

The variances in the two components in the two-mode squeezed vacuum are

$$(\Delta b_1)^2 = \frac{1}{4}\left[\exp(-2r)\cos^2\left(\frac{\delta}{2}-\frac{\theta}{2}\right) + \exp(2r)\sin^2\left(\frac{\delta}{2}-\frac{\theta}{2}\right)\right], \tag{2.8.7}$$

$$(\Delta b_2)^2 = \frac{1}{4}\left[\exp(2r)\cos^2\left(\frac{\delta}{2}-\frac{\theta}{2}\right) + \exp(-2r)\sin^2\left(\frac{\delta}{2}-\frac{\theta}{2}\right)\right]. \tag{2.8.8}$$

For the particular choices of the phase $\delta - \theta = 0$ and π, it is an ideal squeezed state with reduced fluctuations in b_1 and b_2, respectively.

In a similar manner, a large number of modes of the vacuum can be squeezed. The multi-mode squeeze operator is defined as

$$S[\xi(\nu)] = \int \frac{d\nu'}{2\pi} \exp\left[\xi^*(\nu')a_{\nu+\nu'}a_{\nu-\nu'} - \xi(\nu')a^\dagger_{\nu+\nu'}a^\dagger_{\nu-\nu'}\right]. \tag{2.8.9}$$

Here the integration is over the positive half-band of frequencies and $\xi(\nu) = r(\nu)\exp[i\theta(\nu)]$. A multi-mode squeezed coherent state is obtained, as in definition (2.7.10), by first displacing the vacuum and then squeezing it through a multi-mode displacement operator

$$|\alpha(\nu), \xi(\nu)\rangle \equiv S[\xi(\nu)]D[\alpha(\nu)]|\widetilde{0}\rangle, \tag{2.8.10}$$

where $|\widetilde{0}\rangle$ is a multi-mode vacuum state.

Problems

2.1 Show that

$$a^\dagger|\alpha\rangle\langle\alpha| = \left(\alpha^* + \frac{\partial}{\partial\alpha}\right)|\alpha\rangle\langle\alpha|,$$

and

$$|\alpha\rangle\langle\alpha|a = \left(\alpha + \frac{\partial}{\partial\alpha^*}\right)|\alpha\rangle\langle\alpha|.$$

2.2 Show that the expectation value of the displacement operator $D(\alpha)$ for a thermal field is given by

$$\langle D(\alpha)\rangle = \exp\left[-|\alpha|^2\left(\langle n\rangle + \frac{1}{2}\right)\right],$$

where $\langle n\rangle$ is the mean number of photons in the field.

2.3 The time evolution of the wave packet (2.3.14) is determined by the Schrödinger equation for the harmonic oscillator

$$i\hbar\frac{\partial\psi}{\partial t} = \left(-\frac{\hbar^2}{2}\frac{\partial^2}{\partial q^2} + \frac{v^2q^2}{2}\right)\psi.$$

A general solution of this equation can be given in terms of the stationary wave functions

$$\psi(q,t) = \sum_{n=0}^{\infty} a_n\phi_n(q)e^{-iE_nt/\hbar},$$

where $E_n = (n+1/2)\hbar v$ and a_n are arbitrary coefficients. Using the orthonormality conditions on the wave functions $\phi_n(q)$, find a_n and hence prove Eq. (2.3.15).

2.4 Derive Eq. (2.3.16).

2.5 An alternate definition of a squeezed coherent state is

$$|\alpha,\xi\rangle = D(\alpha)S(\xi)|0\rangle,$$

where $\xi = r\exp(i\theta)$. Show that the variances in the quadrature components Y_1 and Y_2, such that

$$Y_1 + iY_2 = ae^{-i\theta/2},$$

are given by

$$(\Delta Y_1)^2 = \frac{1}{4}e^{-2r},$$
$$(\Delta Y_2)^2 = \frac{1}{4}e^{2r}.$$

2.6 Consider a two-mode squeezed state defined by

$$|\alpha_1,\alpha_2,\xi\rangle = D_1(\alpha_1)D_2(\alpha_2)S_{12}(\xi)|0\rangle,$$

where

$$D_i(\alpha_i) = \exp(\alpha_i a_i^\dagger - \alpha_i^* a_i) \quad (i = 1,2),$$

is the coherent displacement operator for the two modes described by destruction and creation operators a_i and $a_i^\dagger$, respectively,

$$S_{12}(\xi) = \exp(\xi^* a_1 a_2 + \xi a_1^\dagger a_2^\dagger)$$

is the two-mode squeeze operator, and $|0\rangle$ is the two-mode vacuum state. Show that there is no squeezing in the two individual modes. (Hint: see S. M. Barnett and P. L. Knight, *J. Opt. Soc. Am. B* **2**, 467 (1985).)

2.7 A state is said to be squeezed in the Nth order if $\langle(\Delta X_i)^N\rangle$ ($i = 1$ or 2) is lower than its corresponding coherent state value. Here

$$X_1 = \frac{1}{2}(a + a^\dagger),$$
$$X_2 = \frac{1}{2i}(a - a^\dagger).$$

Show that the condition of the Nth-order squeezing is

$$q^N < 0,$$

where

$$q^N = (\Delta X_i)^N - \left(\frac{1}{4}\right)^{N-2} (N-1)!!$$

(Hint: see C. K. Hong and L. Mandel, *Phys. Rev. Lett.* **54**, 323 (1985).)

2.8 Consider the Hermitian operators corresponding to the real and imaginary parts of the square of the complex amplitude of the field

$$X_1 = \frac{1}{2}(a^2 + a^{\dagger 2}),$$
$$X_2 = \frac{1}{2i}(a^2 - a^{\dagger 2}).$$

Show that the squeezing condition is

$$\langle \Delta X_i^2 \rangle < \langle a^\dagger a \rangle + \frac{1}{2} \quad (i = 1 \text{ or } 2).$$

This type of squeezing is called amplitude-squared squeezing. Show that the amplitude-squared squeezing is a nonclassical effect. (Hint: see M. Hillery, *Phys. Rev. A* **36**, 3796 (1987).)

References and bibliography

Coherent states of the radiation field

E. Schrödinger, *Naturwissenschaften* **14**, 664 (1926).
E. H. Kennard, *Z. Physik* **44**, 326 (1927).
R. J. Glauber, *Phys. Rev.* **130**, 2529 (1963).
R. J. Glauber, in *Quantum Optics and Electronics*, Les Houches, ed. C. DeWitt, A. Blandin, and C. Cohen-Tannoudji (Gordon and Breach, New York 1965).
L. Mandel and E. Wolf, *Rev. Mod. Phys.* **37**, 231 (1965).
J. R. Klauder and B. S. Skagerstam, *Coherent States*, (World Scientific, Singapore 1985). This book contains an excellent collection of papers on coherent states.

Review of squeezed states

D. F. Walls, *Nature* **306**, 141 (1983).
G. Leuchs, in *Non-equilibrium Quantum Statistical Physics*, ed. G. T. Moore and M. O. Scully (Plenum, New York 1985), p. 329.
R. Loudon and P. L. Knight, *J. Mod. Opt.* **34**, 709 (1987).
K. Zaheer and M. S. Zubairy, in *Advances in Atomic, Molecular and Optical Physics* Vol. 28, ed. D. R. Bates and B. Bederson (Academic Press, New York 1990), p. 143.

Squeezed coherent state

D. Stoler, *Phys. Rev. D* **1**, 3217 (1970).
C. Y. E. Lu, *Lett. Nuovo Cim.* **2**, 1241 (1971).
M. Sargent III, M. O. Scully, and W. E. Lamb, Jr., *Laser Physics*, (Addison-Wesley, Mass. 1974).
H. P. Yuen, *Phys. Rev. A* **13**, 2226 (1976).
J. N. Hollenhorst, *Phys. Rev. D* **19**, 1669 (1979).
W. Schleich and J. A. Wheeler, *Nature* **326**, 574 (1987); *J. Opt. Soc. Am. B* **4**, 1715 (1987).

Applications of squeezed states to interferometry

C. M. Caves, *Phys. Rev. D* **23**, 1693 (1981).
M. Xiao, L. A. Wu, and H. J. Kimble, *Phys. Rev. Lett.* **59**, 278 (1987).
J. Gea-Banacloche and G. Leuchs *J. Opt. Soc. Am. B* **4**, 1667 (1987).

Multi-mode squeezing

C. M. Caves, *Phys. Rev. D* **26**, 1817 (1982).
S. M. Barnett and P. L. Knight, *J. Opt. Soc. Am. B* **2**, 467 (1985).
C. M. Caves and B. L. Schumaker, *Phys. Rev. A* **31**, 3068 (1985).
B. L. Schumaker and C. M. Caves, *Phys. Rev. A* **31**, 3093 (1985).

Higher order squeezing

R. A. Fisher, M. M. Nieto, and V. D. Sandberg, *Phys. Rev. D* **29**, 1107 (1984).
M. Hillery, M. S. Zubairy, and K. Wodkiewicz, *Phys. Lett.* **103A**, 259 (1984).
G. D'Ariano, M. Rasetti, and N. Vadacchino, *Phys. Rev. D* **32**, 1034 (1985).
C. K. Hong and L. Mandel, *Phys. Rev. Lett.* **54**, 323 (1985).
S. L. Braunstein and R. I. McLachlam, *Phys. Rev. A* **35**, 1659 (1987).
M. Hillery, *Phys. Rev. A* **36**, 3796 (1987); *Opt. Commun.* **62**, 135 (1987).

CHAPTER 3

Quantum distribution theory and partially coherent radiation

As we have seen in the previous chapters, there are quantum fluctuations associated with the states corresponding to classically well-defined electromagnetic fields. The general description of fluctuation phenomena requires the density operator. However, it is possible to give an alternative but equivalent description in terms of distribution functions. In the present chapter, we extend our treatment of quantum statistical phenomena by developing the theory of quasi-classical distributions. This is of interest for several reasons.

First of all, the extension of the quantum theory of radiation to involve nonquantum stochastic effects such as thermal fluctuations is needed. This is an important ingredient in the theory of partial coherence. Furthermore, the interface between classical and quantum physics is elucidated by the use of such distributions. The arch type example being the Wigner distribution.*

In this chapter, we introduce various distribution functions. These include the coherent state representation or the Glauber–Sudarshan P-representation. The P-representation is used to evaluate the normally ordered correlation functions of the field operators. As we shall see in the next chapter, the P-representation forms a correspondence between the quantum and the classical coherence theory. This distribution function does not have all of the properties of the classical distribution functions for certain states of the field, e.g., it can be negative. We also discuss the so-called Q-representation associated with the antinormally

* The first quasiclassical distribution, Wigner [1932], was written from a wave function perspective. The later work of Moyal [1949] introduced the characteristic function approach to obtaining the Wigner distribution. For reviews of the subject see Hillery, O'Connell, Scully, and Wigner [1984], and Reichl, chapter 7 [1980]. The very readable textbooks by Louisell [1974], Walls and Milburn [1994], and Cohen [1995] extend the quasiclassical distribution concept and are recommended reading.

ordered correlation functions. Other distribution functions and their properties are also presented.

3.1 Coherent state representation

The study of the interface between quantum and classical physics is a fascinating subject. Nowhere is this better illustrated than in quantum optics, where we are often faced with the problem of characterizing fields which are nearly classical but have important quantum features. The coherent states are well suited to such studies. In order to see why this is the case, let us recall that for a fluctuating *classical* field we are generally dealing with a probability distribution $P(\mathscr{E})$ for the complex field amplitude $\mathscr{E} = |\mathscr{E}|e^{i\phi}$ as indicated in Fig. 3.1.

Now in quantum mechanical problems, a probability distribution for the system comes from the statistical or density operator which is defined as follows. Suppose we know that the system is in state $|\psi\rangle$, then an operator O has the expectation value

$$\langle O\rangle_{\mathrm{QM}} = \langle\psi|O|\psi\rangle, \tag{3.1.1}$$

but we typically do not know that we are in $|\psi\rangle$. We only have a probability P_ψ for being in this state so we must perform an ensemble average as well

$$\langle\langle O\rangle_{\mathrm{QM}}\rangle_{\mathrm{ensemble}} = \sum_\psi P_\psi\langle\psi|O|\psi\rangle. \tag{3.1.2}$$

Now using completeness $\sum_n |n\rangle\langle n| = 1$

$$\begin{aligned}\langle\langle O\rangle\rangle &= \sum_n\sum_\psi P_\psi\langle\psi|O|n\rangle\langle n|\psi\rangle\\ &= \sum_n\sum_\psi P_\psi\langle n|\psi\rangle\langle\psi|O|n\rangle\\ &= \sum_n\langle n|\rho O|n\rangle.\end{aligned} \tag{3.1.3}$$

Thus the radiation field is, in general, described by the density operator

$$\rho = \sum_\psi P_\psi|\psi\rangle\langle\psi|, \tag{3.1.4}$$

where P_ψ is the probability of being in the state $|\psi\rangle$. The expectation value of any field operator O is then given by

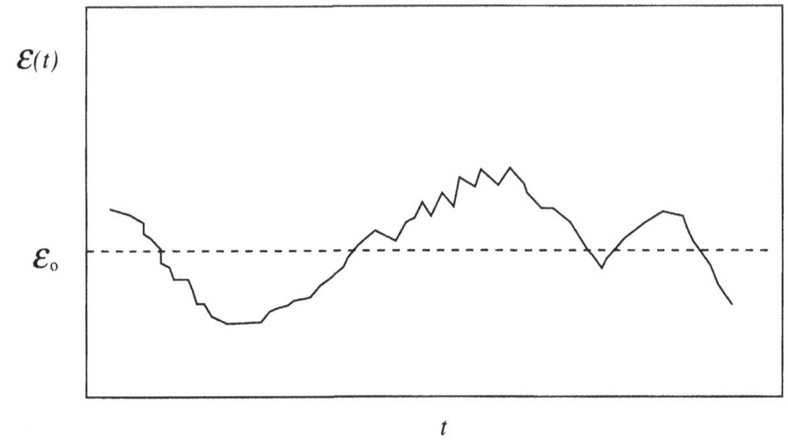

(a)

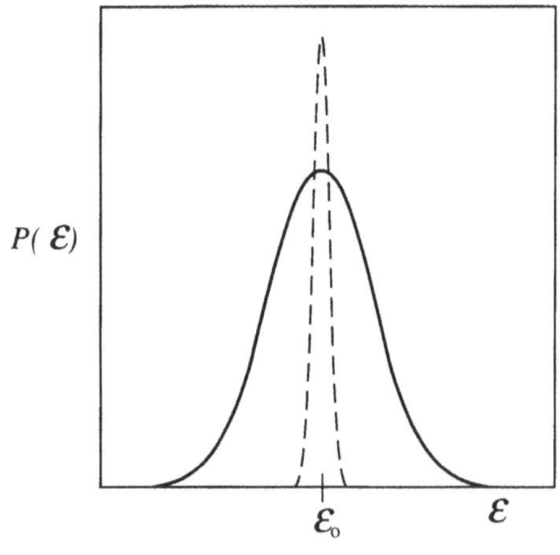

(b)

Fig. 3.1
(a) The fluctuating classical field as a function of time for a field with large fluctuations (solid line) and a well stabilized field (dashed line), and (b) associated probability distributions.

$$\langle O \rangle = \mathrm{Tr}(O\rho), \tag{3.1.5}$$

where Tr stands for trace. Now the density operator ρ can be expanded in terms of the photon occupation number states:

$$\rho = \sum_n \sum_m |n\rangle\langle n|\rho|m\rangle\langle m| = \sum_n \sum_m \rho_{nm}|n\rangle\langle m|. \tag{3.1.6}$$

Likewise the expansion may be made in terms of coherent states as

$$\rho = \int\int \frac{d^2\alpha}{\pi}\frac{d^2\beta}{\pi}|\alpha\rangle\langle\alpha|\rho|\beta\rangle\langle\beta|. \tag{3.1.7}$$

Following Glauber's convention we define the R-representation as

$$R(\alpha^*, \beta) = \langle\alpha|\rho|\beta\rangle e^{\frac{1}{2}(|\alpha|^2+|\beta|^2)}, \tag{3.1.8}$$

so that the density matrix may be written as

$$\rho = \int\int \frac{d^2\alpha}{\pi}\frac{d^2\beta}{\pi} |\alpha\rangle\langle\beta|\, R(\alpha^*, \beta) e^{-\frac{1}{2}(|\alpha|^2+|\beta|^2)}. \tag{3.1.9}$$

We thus have used two indices n and m or α and β in order to specify the density matrix.

We next make contact with $P(\mathscr{E})$, as discussed earlier, by developing a *diagonal* coherent state representation. That is, we express the density operator ρ in terms of the diagonal pair $|\alpha\rangle\langle\alpha|$ in the following.

3.1.1 Definition of the coherent state representation

Consider an operator $O_N(a, a^\dagger)$, which is a function of a and $a^\dagger$ in the normal order (all the creation operators $a^\dagger$ on the left-hand side and all the annihilation operators a on the right-hand side), i.e.,

$$O_N(a, a^\dagger) = \sum_n \sum_m c_{nm}(a^\dagger)^n a^m. \tag{3.1.10}$$

It may be noted that any operator involving a and $a^\dagger$ can be converted into a normal ordered form by using the commutation relation $[a, a^\dagger] = 1$. For example $a^2 a^\dagger = a^\dagger a^2 + 2a$. The expectation value of the operator $O_N(a, a^\dagger)$ can then be written as

$$\begin{aligned}\langle O_N(a, a^\dagger)\rangle &= \mathrm{Tr}[\rho O_N(a, a^\dagger)] \\ &= \sum_n \sum_m c_{nm} \mathrm{Tr}[\rho(a^\dagger)^n a^m]. \end{aligned} \tag{3.1.11}$$

As discussed in Appendix 3.A, we define the operator

$$\begin{aligned}&\delta(\alpha^* - a^\dagger)\delta(\alpha - a) \\ &= \frac{1}{\pi^2}\int \exp[-\beta(\alpha^* - a^\dagger)] \exp[\beta^*(\alpha - a)] d^2\beta, \end{aligned} \tag{3.1.12a}$$

or, in an equivalent form

$$\begin{aligned}&\delta(\alpha^* - a^\dagger)\delta(\alpha - a) \\ &= \frac{1}{\pi^2}\int \exp[-i\beta(\alpha^* - a^\dagger)] \exp[-i\beta^*(\alpha - a)] d^2\beta. \end{aligned} \tag{3.1.12b}$$

We will use (3.1.12a) and (3.1.12b) interchangeably in the text. Equation (3.1.11) can then be rewritten as

$$\begin{aligned}\langle O_N(a, a^\dagger)\rangle &= \int d^2\alpha \sum_n \sum_m c_{nm} \mathrm{Tr}[\rho\delta(\alpha^* - a^\dagger)\delta(\alpha - a)](\alpha^*)^n \alpha^m \\ &= \int d^2\alpha P(\alpha, \alpha^*) O_N(\alpha, \alpha^*), \end{aligned} \tag{3.1.13}$$

where

$$P(\alpha, \alpha^*) = \text{Tr}[\rho\delta(\alpha^* - a^\dagger)\delta(\alpha - a)]. \tag{3.1.14}$$

It is seen from Eq. (3.1.13) that the function $P(\alpha, \alpha^*)$ can be used to evaluate the expectation values of any normal ordered function of a and $a^\dagger$ using the methods of classical statistical mechanics. Due to the Hermiticity of the density operator ρ, the distribution function $P(\alpha, \alpha^*)$ is real. Moreover, since $\text{Tr}(\rho) = 1$, $P(\alpha, \alpha^*)$ is normalized to unity, i.e.,

$$\int P(\alpha, \alpha^*) d^2\alpha = 1. \tag{3.1.15}$$

The function $P(\alpha, \alpha^*)$ is referred to as the P-representation or the coherent state representation. The name *coherent state representation* is due to the following representation of the density operator ρ by means of a diagonal representation in terms of the coherent states:

$$\rho = \int P(\alpha, \alpha^*)|\alpha\rangle\langle\alpha| d^2\alpha. \tag{3.1.16}$$

The equivalence of the definitions of $P(\alpha, \alpha^*)$ as given by Eqs. (3.1.14) and (3.1.16) can be seen simply by substituting for ρ from Eq. (3.1.16) into Eq. (3.1.14). As we shall see in the next chapter, $P(\alpha, \alpha^*)$ forms a connection between the classical and quantum coherence theory.

Before considering some examples of the P-representation, we give a simple procedure to find $P(\alpha, \alpha^*)$ from a knowledge of ρ. Let $|\beta\rangle$ and $|-\beta\rangle$ be the coherent states with β and $-\beta$ being the eigenvalues of a, respectively. Then, using Eq. (2.4.7),

$$\begin{aligned} \langle -\beta|\rho|\beta\rangle &= \int P(\alpha, \alpha^*)\langle -\beta|\alpha\rangle\langle\alpha|\beta\rangle d^2\alpha \\ &= e^{-|\beta|^2} \int [P(\alpha, \alpha^*)e^{-|\alpha|^2}] e^{\beta\alpha^* - \beta^*\alpha} d^2\alpha. \end{aligned} \tag{3.1.17}$$

At this point we note that if $\alpha = x_\alpha + iy_\alpha$ and $\beta = x_\beta + iy_\beta$, then $d^2\alpha = dx_\alpha dy_\alpha$ and $\beta\alpha^* - \beta^*\alpha = 2i(y_\beta x_\alpha - x_\beta y_\alpha)$, and Eq. (3.1.17) becomes

$$\begin{aligned} &\langle -\beta|\rho|\beta\rangle e^{|\beta|^2} \\ &= \int\int [P(x_\alpha, y_\alpha)e^{-(x_\alpha^2 + y_\alpha^2)}] e^{2i(y_\beta x_\alpha - x_\beta y_\alpha)} dx_\alpha dy_\alpha. \end{aligned} \tag{3.1.18}$$

Thus, $\langle -\beta|\rho|\beta\rangle e^{|\beta|^2}$ is the two-dimensional Fourier transform of $P(\alpha, \alpha^*)e^{-|\alpha|^2}$. This shows the utility of considering the matrix element $\langle -\beta|\rho|\beta\rangle$, since the inverse Fourier transform readily gives $P(\alpha, \alpha^*)$ in terms of the density operator ρ. On taking the Fourier inverse of Eq. (3.1.17), we obtain

$$P(\alpha,\alpha^*) = \frac{e^{(x_\alpha^2+y_\alpha^2)}}{\pi^2}\int\int\langle-\beta|\rho|\beta\rangle e^{(x_\beta^2+y_\beta^2)}e^{2i(y_\alpha x_\beta - x_\alpha y_\beta)}dx_\beta dy_\beta$$
$$= \frac{e^{|\alpha|^2}}{\pi^2}\int\langle-\beta|\rho|\beta\rangle e^{|\beta|^2}e^{-\beta\alpha^*+\beta^*\alpha}d^2\beta. \quad (3.1.19)$$

This is the required expression.

3.1.2 Examples of the coherent state representation

As a first example, we calculate $P(\alpha,\alpha^*)$ for the thermal field. A field emitted by a source in thermal equilibrium at temperature T is described by a canonical ensemble

$$\rho = \frac{\exp(-\mathscr{H}/k_B T)}{\mathrm{Tr}[\exp(-\mathscr{H}/k_B T)]}, \quad (3.1.20)$$

where k_B is the Boltzmann constant and $\mathscr{H}$ is the free-field Hamiltonian, $\mathscr{H} = \hbar\nu(a^\dagger a + 1/2)$. For simplicity, we restrict ourselves to a single mode of the field. On substituting this form of $\mathscr{H}$ into Eq. (3.1.20) we obtain

$$\rho = \sum_n \left[1 - \exp\left(-\frac{\hbar\nu}{k_B T}\right)\right]\exp\left(-\frac{n\hbar\nu}{k_B T}\right)|n\rangle\langle n|. \quad (3.1.21)$$

Correspondingly

$$\langle n\rangle = \mathrm{Tr}(a^\dagger a\rho) = \left[\exp\left(\frac{\hbar\nu}{k_B T}\right) - 1\right]^{-1}. \quad (3.1.22)$$

Equation (3.1.21) can therefore be rewritten in terms of $\langle n\rangle$ as

$$\rho = \sum_n \frac{\langle n\rangle^n}{(1+\langle n\rangle)^{n+1}}|n\rangle\langle n|. \quad (3.1.23)$$

This leads to the well-known result that the photon distribution in a thermal field is described by the Bose–Einstein distribution, i.e.,

$$\rho_{nn} = \langle n|\rho|n\rangle$$
$$= \frac{\langle n\rangle^n}{(1+\langle n\rangle)^{n+1}}. \quad (3.1.24)$$

Next we substitute for ρ from Eq. (3.1.23) into Eq. (3.1.19). We note that

$$\begin{aligned}\langle -\beta|\rho|\beta\rangle &= \sum_n \frac{\langle n\rangle^n}{(1+\langle n\rangle)^{n+1}}\langle -\beta|n\rangle\langle n|\beta\rangle \\ &= \frac{e^{-|\beta|^2}}{1+\langle n\rangle}\sum_{n=0}^{\infty}\frac{(-|\beta|^2)^n}{n!}\left(\frac{\langle n\rangle}{1+\langle n\rangle}\right)^n \\ &= \frac{e^{-|\beta|^2}}{1+\langle n\rangle}\exp\left[-|\beta|^2\Big/\left(1+\frac{1}{\langle n\rangle}\right)\right], \end{aligned} \tag{3.1.25}$$

so that

$$\begin{aligned}P(\alpha,\alpha^*) &= \frac{e^{|\alpha|^2}}{\pi^2(1+\langle n\rangle)}\int e^{-|\beta|^2/\left(1+\frac{1}{\langle n\rangle}\right)}e^{-\beta\alpha^*+\alpha\beta^*}d^2\beta \\ &= \frac{1}{\pi\langle n\rangle}e^{-|\alpha|^2/\langle n\rangle}, \end{aligned} \tag{3.1.26}$$

i.e., the P-representation of the thermal distribution is given by a Gaussian distribution.

As another example, we consider the P-representation of a coherent state $|\alpha_0\rangle$. Here $\rho = |\alpha_0\rangle\langle\alpha_0|$ so that

$$\begin{aligned}\langle -\beta|\rho|\beta\rangle &= \langle -\beta|\alpha_0\rangle\langle\alpha_0|\beta\rangle \\ &= \exp(-|\alpha_0|^2 - |\beta|^2 - \alpha_0\beta^* + \beta\alpha_0^*). \end{aligned} \tag{3.1.27}$$

It then follows from Eq. (3.1.19) that

$$\begin{aligned}P(\alpha,\alpha^*) &= \frac{1}{\pi^2}e^{|\alpha|^2-|\alpha_0|^2}\int e^{-\beta(\alpha^*-\alpha_0^*)+\beta^*(\alpha-\alpha_0)}d^2\beta \\ &= \delta^{(2)}(\alpha-\alpha_0), \end{aligned} \tag{3.1.28}$$

i.e., the P-representation of a coherent state is a two-dimensional delta function.

Even though the P-representation allows us to evaluate the normally ordered correlation functions of the field operators a and $a^\dagger$, it is not nonnegative definite and as such cannot be described as a distribution function for certain field states. This can be readily seen by evaluating the P-representation of a number state $|n\rangle$, for which $\rho = |n\rangle\langle n|$ and

$$\begin{aligned}\langle -\beta|\rho|\beta\rangle &= \langle -\beta|n\rangle\langle n|\beta\rangle \\ &= \exp(-|\beta|^2)\frac{(-1)^n|\beta|^{2n}}{n!}. \end{aligned} \tag{3.1.29}$$

The corresponding P-representation is, therefore, given by

$$P(\alpha, \alpha^*) = \frac{(-1)^n e^{|\alpha|^2}}{\pi^2 n!} \int |\beta|^{2n} e^{-\beta\alpha^* + \beta^*\alpha} d^2\beta$$
$$= \frac{e^{|\alpha|^2}}{\pi^2 n!} \frac{\partial^{2n}}{\partial\alpha^n \partial\alpha^{*n}} \int e^{-\beta\alpha^* + \beta^*\alpha} d^2\beta$$
$$= \frac{e^{|\alpha|^2}}{n!} \frac{\partial^{2n}}{\partial\alpha^n \partial\alpha^{*n}} \delta^{(2)}(\alpha). \quad (3.1.30)$$

For $n > 0$, this is clearly not a nonnegative definite function and, therefore, a number state does not have a well-defined P-representation.

As we will discuss in the next chapter, whenever the photon distribution ρ_{nn} is narrower than the Poisson distribution, as in the case of number state $|n\rangle$, $P(\alpha, \alpha^*)$ becomes *badly* behaved. This is the price we pay for forcing quantum physics into a classical format, i.e., for using $P(\alpha, \alpha^*)$ instead of say, $R(\alpha, \beta^*)$.

3.2 Q-representation

Just as the P-representation is associated with the evaluation of normally ordered correlation functions of the field operators a and $a^\dagger$, we may define other distribution functions which may be associated with different orderings of a and $a^\dagger$. The distribution function which helps in determining the antinormally ordered correlation functions is the so-called Q-representation. It is defined as

$$Q(\alpha, \alpha^*) = \mathrm{Tr}[\rho\delta(\alpha - a)\delta(\alpha^* - a^\dagger)]. \quad (3.2.1)$$

It follows, on inserting the representation (2.4.6) for unity between $\delta(\alpha - a)$ and $\delta(\alpha^* - a^\dagger)$ and using (2.2.1) that

$$Q(\alpha, \alpha^*) = \frac{1}{\pi} \mathrm{Tr} \int d^2\alpha' [\rho\delta(\alpha - a)|\alpha'\rangle\langle\alpha'|\delta(\alpha^* - a^\dagger)]$$
$$= \frac{1}{\pi} \mathrm{Tr} \int d^2\alpha' \{\rho\delta(\alpha - \alpha')|\alpha'\rangle\langle\alpha'|\delta[\alpha^* - (\alpha')^*]\}$$
$$= \frac{1}{\pi} \mathrm{Tr}(\rho|\alpha\rangle\langle\alpha|)$$
$$= \frac{1}{\pi} \langle\alpha|\rho|\alpha\rangle, \quad (3.2.2)$$

i.e., $Q(\alpha, \alpha^*)$ is proportional to the diagonal element of the density operator in the coherent state representation. It follows from the completeness of the coherent states $|\alpha\rangle$ (Eq. (2.4.6)) and the condition $\mathrm{Tr}(\rho) = 1$ that $Q(\alpha, \alpha^*)$ is normalized to unity, i.e.,

$$\int Q(\alpha, \alpha^*) d^2\alpha = 1. \tag{3.2.3}$$

In order to see how the antinormally ordered correlation functions of a and $a^\dagger$ are evaluated using the Q-representation, we first define a function $O_A(a, a^\dagger)$ in antinormal order, i.e.,

$$O_A(a, a^\dagger) = \sum_n \sum_m d_{nm} a^n (a^\dagger)^m. \tag{3.2.4}$$

It then follows that

$$\begin{aligned}
\langle O_A(a, a^\dagger) \rangle &= \mathrm{Tr}[O_A(a, a^\dagger)\rho] \\
&= \sum_n \sum_m d_{nm} \mathrm{Tr}[a^n (a^\dagger)^m \rho] \\
&= \sum_n \sum_m d_{nm} \mathrm{Tr} \left[\frac{1}{\pi} \int a^n |\alpha\rangle\langle\alpha| (a^\dagger)^m \rho d^2\alpha \right] \\
&= \sum_n \sum_m d_{nm} \frac{1}{\pi} \int \alpha^n (\alpha^*)^m \langle\alpha|\rho|\alpha\rangle d^2\alpha \\
&= \int Q(\alpha, \alpha^*) O_A(\alpha, \alpha^*) d^2\alpha,
\end{aligned} \tag{3.2.5}$$

where, in the third line, we inserted

$$\frac{1}{\pi} \int |\alpha\rangle\langle\alpha| d^2\alpha = 1. \tag{3.2.6}$$

Unlike the P-representation, $Q(\alpha, \alpha^*)$ is nonnegative definite and bounded. This can be seen by substituting for ρ from Eq. (3.1.4) into Eq. (3.2.2). We then obtain

$$Q(\alpha, \alpha^*) = \frac{1}{\pi} \sum_\psi P_\psi |\langle\psi|\alpha\rangle|^2. \tag{3.2.7}$$

Since $|\langle\psi|\alpha\rangle|^2 \leq 1$, we have

$$Q(\alpha, \alpha^*) \leq \frac{1}{\pi}. \tag{3.2.8}$$

The Q-representation may be related to the P-representation by taking the coherent state diagonal element of ρ in Eq. (3.1.16). The resulting equation is

$$Q(\alpha, \alpha^*) = \frac{1}{\pi} \int P(\alpha', \alpha'^*) e^{-|\alpha-\alpha'|^2} d^2\alpha'. \tag{3.2.9}$$

As an example, $Q(\alpha, \alpha^*)$ for a number state $|n\rangle$ is given by

$$Q(\alpha, \alpha^*) = \frac{1}{\pi} |\langle n|\alpha\rangle|^2 = \frac{e^{-|\alpha|^2} |\alpha|^{2n}}{\pi n!}, \tag{3.2.10}$$

which is a well-behaved function. The Q-representation of a squeezed state is given in Section 3.5.

3.3 The Wigner–Weyl distribution

So far we have discussed various distribution functions, namely P- and Q-representations associated with the normal and the antinormal orderings, respectively, of the operators a and $a^\dagger$. We can similarly derive distribution functions associated with other orderings.

To summarize, we have introduced

$$P(\alpha,\alpha^*) = \mathrm{Tr}[\delta(\alpha^* - a^\dagger)\delta(\alpha - a)\rho], \tag{3.3.1a}$$

$$Q(\alpha,\alpha^*) = \mathrm{Tr}[\delta(\alpha - a)\delta(\alpha^* - a^\dagger)\rho], \tag{3.3.1b}$$

which we can write in terms of the so-called characteristic functions. For example, inserting (3.1.12b) into (3.3.1a) we have

$$P(\alpha,\alpha^*) = \frac{1}{\pi^2}\int d^2\beta e^{-i\beta\alpha^*-i\beta^*\alpha}C^{(n)}(\beta,\beta^*), \tag{3.3.2}$$

where the characteristic function $C^{(n)}(\beta,\beta^*)$ is defined as

$$C^{(n)}(\beta,\beta^*) = \mathrm{Tr}\left(e^{i\beta a^\dagger}e^{i\beta^* a}\rho\right). \tag{3.3.3}$$

Likewise, we may write (3.3.1b) as

$$Q(\alpha,\alpha^*) = \frac{1}{\pi^2}\int d^2\beta e^{-i\beta\alpha^*-i\beta^*\alpha}C^{(a)}(\beta,\beta^*), \tag{3.3.4}$$

with the characteristic function

$$C^{(a)}(\beta,\beta^*) = \mathrm{Tr}\left(e^{i\beta^* a}e^{i\beta a^\dagger}\rho\right). \tag{3.3.5}$$

Another useful distribution, due to Wigner and Weyl, is defined as

$$W(\alpha,\alpha^*) = \frac{1}{\pi^2}\int d^2\beta e^{-i\beta\alpha^*-i\beta^*\alpha}C^{(s)}(\beta,\beta^*), \tag{3.3.6}$$

where the characteristic function $C^{(s)}(\beta,\beta^*)$ is given by

$$C^{(s)}(\beta,\beta^*) = \mathrm{Tr}\left(e^{i\beta a^\dagger+i\beta^* a}\rho\right). \tag{3.3.7}$$

This distribution function $W(\alpha,\alpha^*)$ is associated with symmetric ordering. It can be used to evaluate expectation values of any symmetrically ordered functions of a and $a^\dagger$ in a classical fashion. For example,

$$\frac{1}{2}\langle aa^\dagger + a^\dagger a\rangle = \int W(\alpha,\alpha^*)\alpha\alpha^* d^2\alpha. \tag{3.3.8}$$

In Appendix 3.B, we give a procedure to find the c-number function $O_S(\alpha,\alpha^*)$ corresponding to the symmetrically ordered form of an operator $O(a,a^\dagger)$.

Historically, the $W(\alpha,\alpha^*)$ distribution was introduced in terms of the position $\hat{q}$ and momentum $\hat{p}$ operators in a form equivalent to

$$W(p,q) = \frac{1}{(2\pi)^2}\int d\sigma \int d\tau e^{i(\tau p+\sigma q)} \mathrm{Tr}\left[e^{-i(\tau\hat{p}+\sigma\hat{q})}\rho\right]. \tag{3.3.9}$$

To cast this into the form first introduced by Wigner we use the operator identity

$$e^{A+B} = e^A e^B e^{-[A,B]/2},$$

which holds when the commutator $[A,B]$ commutes with A and B, to write (3.3.9) as

$$W(p,q) = \frac{1}{(2\pi)^2}\int d\sigma \int d\tau e^{i(\tau p+\sigma q)} \mathrm{Tr}\left(e^{-i\tau\hat{p}} e^{-i\sigma\hat{q}} e^{-i\hbar\sigma\tau/2}\rho\right), \tag{3.3.10}$$

which by cyclic invariance under the trace may be written as

$$W(p,q) = \frac{1}{(2\pi)^2}\int d\sigma \int d\tau e^{i(\tau p+\sigma q)} \mathrm{Tr}\left(e^{-i\tau\hat{p}/2} e^{-i\sigma\hat{q}} \rho e^{-i\tau\hat{p}/2}\right) e^{-i\hbar\sigma\tau/2}. \tag{3.3.11}$$

Writing the trace in the coordinate representation this becomes

$$W(p,q) = \frac{1}{(2\pi)^2}\int d\sigma \int d\tau e^{i(\tau p+\sigma q)} \int dq' \langle q'| e^{-i\tau\hat{p}/2} e^{-i\sigma\hat{q}} \rho e^{-i\tau\hat{p}/2} |q'\rangle e^{-i\hbar\sigma\tau/2}, \tag{3.3.12}$$

and noting that $\exp(-i\tau\hat{p}/2)|q'\rangle = |q' - \hbar\tau/2\rangle$ etc., we find

$$W(p,q) = \frac{1}{(2\pi)^2}\int d\sigma \int d\tau \int dq' e^{i\sigma(q-q')} \langle q' + \hbar\tau/2|\rho|q' - \hbar\tau/2\rangle e^{i\tau p}. \tag{3.3.13}$$

Finally, we carry out the σ-integration to obtain a delta function $\delta(q - q')$, which allows us to carry out the q'-integration, and introducing the notation $y = -\hbar\tau/2$, we write $W(p,q)$ in the usual form

$$W(p,q) = \frac{1}{\pi\hbar}\int dy e^{-i2yp/\hbar} \langle q - y|\rho|q + y\rangle. \tag{3.3.14}$$

The Wigner function in the form (3.3.14) has been widely used in a host of problems; and we further elaborate on its connection with the P- and Q-distributions in the next section.

3.4 Generalized representation of the density operator and connection between the P-, Q-, and W-distributions

In the following, we present a generalized representation of the density operator originally due to Cohen and applied to quantum optics by Agarwal and Wolf. The P-, Q-, and W-representations can be derived as special cases of this generalized representation.

A generalized representation $F^{(\Omega)}(\alpha, \alpha^*)$ of the density operator is given by

$$\rho = \pi \int F^{(\Omega)}(\alpha, \alpha^*) \Delta^{(\Omega)}(\alpha - a, \alpha^* - a^\dagger) d^2\alpha, \tag{3.4.1}$$

where

$$\Delta^{(\Omega)}(\alpha - a, \alpha^* - a^\dagger) = \frac{1}{\pi^2} \int \exp[\Omega(\beta, \beta^*)] \times \exp[-\beta(\alpha^* - a^\dagger) + \beta^*(\alpha - a)] d^2\beta. \tag{3.4.2}$$

Here $\Omega(\beta, \beta^*)$ (such that $\Omega(0,0) = 0$) is a function which characterizes different orderings. For example, when $\Omega(\beta, \beta^*) = -|\beta|^2/2$ we have $F^{(\Omega)}(\alpha, \alpha^*) \equiv P(\alpha, \alpha^*)$ and when $\Omega(\beta, \beta^*) = |\beta|^2/2$ we have $F^{(\Omega)}(\alpha, \alpha^*) \equiv Q(\alpha, \alpha^*)$.

To see these results explicitly, we first consider

$$\Omega(\beta, \beta^*) = -\frac{|\beta|^2}{2}. \tag{3.4.3}$$

It follows from Eqs. (2.2.6) and (2.2.7) that

$$\exp\left(-\frac{|\beta|^2}{2} + \beta a^\dagger - \beta^* a\right) = \exp(-\beta^* a) \exp(\beta a^\dagger), \tag{3.4.4}$$

and we obtain

$$\begin{aligned}
\Delta^{(\Omega)}(\alpha - a, \alpha^* - a^\dagger) &= \frac{1}{\pi^2} \int e^{\beta^*(\alpha - a)} e^{-\beta(\alpha^* - a^\dagger)} d^2\beta \\
&= \frac{1}{\pi^3} \int\int e^{\beta^*(\alpha - a)} |\alpha_1\rangle\langle\alpha_1| e^{-\beta(\alpha^* - a^\dagger)} d^2\beta d^2\alpha_1 \\
&= \frac{1}{\pi^3} \int\int e^{\beta^*(\alpha - \alpha_1) - \beta(\alpha^* - \alpha_1^*)} |\alpha_1\rangle\langle\alpha_1| d^2\beta d^2\alpha_1 \\
&= \frac{1}{\pi} |\alpha\rangle\langle\alpha|.
\end{aligned} \tag{3.4.5}$$

On substituting this expression for $\Delta^{(\Omega)}(\alpha - a, \alpha^* - a^\dagger)$ into Eq. (3.4.1) we recover the definition of the P-representation (Eq. (3.1.16)) with $F^{(\Omega)}(\alpha, \alpha^*) \equiv P(\alpha, \alpha^*)$.

On the other hand, if we choose $\Omega(\beta,\beta^*) = |\beta|^2/2$,

$$\Delta^{(\Omega)}(\alpha - a, \alpha^* - a^\dagger) = \frac{1}{\pi^2}\int e^{-\beta(\alpha^*-a^\dagger)}e^{\beta^*(\alpha-a)}d^2\beta. \tag{3.4.6}$$

It follows from Eq. (3.4.1) that

$$\frac{1}{\pi}\langle\alpha'|\rho|\alpha'\rangle = \int F^{(\Omega)}(\alpha,\alpha^*)\langle\alpha'|\Delta^{(\Omega)}(\alpha - a, \alpha^* - a^\dagger)|\alpha'\rangle d^2\alpha. \tag{3.4.7}$$

However, from Eq. (3.4.6),

$$\begin{aligned}\langle\alpha'|\Delta^{(\Omega)}(\alpha - a, \alpha^* - a^\dagger)|\alpha'\rangle &= \frac{1}{\pi^2}\int \langle\alpha'|e^{-\beta(\alpha^*-a^\dagger)}e^{\beta^*(\alpha-a)}|\alpha'\rangle d^2\beta \\ &= \frac{1}{\pi^2}\int e^{-\beta(\alpha^*-\alpha'^*)+\beta^*(\alpha-\alpha')}d^2\beta \\ &= \delta^{(2)}(\alpha-\alpha'). \end{aligned}\tag{3.4.8}$$

On carrying out the α-integration in Eq. (3.4.7) we recover Eq. (3.2.2) with $Q(\alpha,\alpha^*) \equiv F^{(\Omega)}(\alpha,\alpha^*)$.

Another distribution, the Wigner–Weyl distribution, is recovered for the proper choice of Ω, namely, $\Omega(\alpha,\alpha^*) = 0$. To that end, we invert Eq. (3.4.1) by using the function

$$\begin{aligned}\bar{\Delta}^{(\Omega)}(\alpha - a, \alpha^* - a^\dagger) = \frac{1}{\pi^2}\int &\exp[-\Omega(\beta,\beta^*)] \\ &\times \exp[\beta(\alpha^* - a^\dagger) - \beta^*(\alpha - a)]d^2\beta.\end{aligned}\tag{3.4.9}$$

Now, it can be shown that (see Problem 3.3)

$$\begin{aligned}&\mathrm{Tr}\left[\Delta^{(\Omega)}(\alpha - a, \alpha^* - a^\dagger)\bar{\Delta}^{(\Omega)}(\alpha' - a, \alpha'^* - a^\dagger)\right] \\ &= \frac{1}{\pi}\delta^{(2)}(\alpha-\alpha').\end{aligned}\tag{3.4.10}$$

It then follows from Eq. (3.4.1) that

$$F^{(\Omega)}(\alpha,\alpha^*) = \mathrm{Tr}\left[\rho\bar{\Delta}^{(\Omega)}(\alpha - a, \alpha^* - a^\dagger)\right]. \tag{3.4.11}$$

From Eqs. (3.4.9) and (3.4.11), we obtain

$$\begin{aligned}&W(\alpha,\alpha^*) \\ &= \frac{1}{\pi^2}\int \mathrm{Tr}[\rho\exp(-\beta a^\dagger + \beta^* a)]\exp(\beta\alpha^* - \beta^*\alpha)d^2\beta,\end{aligned}\tag{3.4.12}$$

which, as expected, is the same as Eq. (3.3.6) with β replaced by $-i\beta$ and β^* by $i\beta^*$. Equations (3.1.14) for the P-representation and (3.2.2) for the Q-representation can be recovered from expression (3.4.11) for $\Omega(\beta,\beta^*) = -|\beta|^2/2$ and $\Omega(\beta,\beta^*) = |\beta|^2/2$, respectively.

In the following we derive an explicit expression for the Wigner–Weyl distribution $W(\alpha,\alpha^*)$. First we mention that $W(\alpha,\alpha^*)$ is the Fourier transform of the function $\mathrm{Tr}[\rho\exp(-\beta a^\dagger + \beta^* a)]/\pi^2$. We also note that $\exp(-2|\alpha|^2)$ is the Fourier transform of $\exp(-|\beta|^2/2)/2\pi$, i.e.,

$$\exp(-2|\alpha|^2) = \frac{1}{2\pi}\int \exp\left(-\frac{1}{2}|\beta|^2\right)\exp(\beta\alpha^* - \beta^*\alpha)d^2\beta. \quad (3.4.13)$$

It then follows from the convolution theorem that

$$W(\alpha,\alpha^*)\exp(-2|\alpha|^2) = \int C(\beta,\beta^*)\exp(\beta\alpha^* - \beta^*\alpha)d^2\beta, \quad (3.4.14)$$

where $C(\beta,\beta^*)$ is the convolution product

$$C(\beta,\beta^*) = \frac{1}{2\pi^3}\int \mathrm{Tr}\{\rho\exp[-(\beta-\beta_1)a^\dagger + (\beta^* - \beta_1^*)a]\} \exp\left(-\frac{1}{2}|\beta_1|^2\right)d^2\beta_1. \quad (3.4.15)$$

An explicit expression for $C(\beta,\beta^*)$ can be obtained by using the identity (2.2.7) and inserting the resolution of the identity in terms of coherent states (Eq. (2.4.6)) as follows:

$$\begin{aligned}
C(\beta,\beta^*) &= \frac{1}{2\pi^5}\int\int\int \mathrm{Tr}\{\rho|\beta_2\rangle\langle\beta_2|\exp[-(\beta-\beta_1)a^\dagger] \\
&\quad \times\exp[(\beta^*-\beta_1^*)a]|\beta_3\rangle\langle\beta_3|\} \\
&\quad \times\exp\left(-\frac{1}{2}|\beta-\beta_1|^2 - \frac{1}{2}|\beta_1|^2\right)d^2\beta_1 d^2\beta_2 d^2\beta_3 \\
&= \frac{1}{2\pi^5}\int\int\int \langle\beta_3|\rho|\beta_2\rangle\langle\beta_2|\beta_3\rangle \\
&\quad \times\exp\left[-(\beta-\beta_1)\beta_2^* + (\beta^*-\beta_1^*)\beta_3 \right. \\
&\quad \left. -\frac{1}{2}|\beta-\beta_1|^2 - \frac{1}{2}|\beta_1|^2\right] \\
&\quad \times d^2\beta_1 d^2\beta_2 d^2\beta_3. \qquad (3.4.16)
\end{aligned}$$

On carrying out the integrations over β_1, Eq. (3.4.16) reduces to

$$\begin{aligned}
C(\beta,\beta^*) &= \frac{1}{2\pi^4}\int\int \langle\beta_3|\rho|\beta_2\rangle\langle\beta/2|\beta_3\rangle\langle\beta_2|-\beta/2\rangle d^2\beta_2 d^2\beta_3 \\
&= \frac{1}{2\pi^2}\langle\beta/2|\rho|-\beta/2\rangle. \qquad (3.4.17)
\end{aligned}$$

Finally, on substituting for $C(\beta,\beta^*)$ from Eq. (3.4.17) into Eq. (3.4.14) and changing the variables of integration from β,β^* to $-2\beta,-2\beta^*$, we obtain

$$\begin{aligned}
&W(\alpha,\alpha^*) \\
&= \frac{2}{\pi^2}\exp(2|\alpha|^2)\int \langle-\beta|\rho|\beta\rangle\exp[-2(\beta\alpha^* - \beta^*\alpha)]d^2\beta. \qquad (3.4.18)
\end{aligned}$$

This expression, which is very similar to the corresponding expression for P-representation (Eq. (3.1.19)), can be used to evaluate the Wigner–Weyl distribution for the given density operator of the field.

3.5 Q-representation for a squeezed coherent state

In this section, we derive the Q-representation for the squeezed coherent state $|\beta, \xi\rangle$. According to Eq. (3.2.2)

$$Q(\alpha, \alpha^*) = \frac{1}{\pi}\langle\alpha|\rho|\alpha\rangle = \frac{1}{\pi}|\langle\alpha|\beta, \xi\rangle|^2. \tag{3.5.1}$$

Now

$$\langle\alpha|\beta, \xi\rangle = \langle\alpha|S(\xi)D(\beta)|0\rangle = \langle\alpha|S(\xi)|\beta\rangle. \tag{3.5.2}$$

We therefore need to calculate the function $\langle\alpha|S(\xi)|\beta\rangle$.

It follows, on using the properties of the coherent state and the transformation property (2.7.7) of $S(\xi)$ that

$$\begin{aligned}\langle\alpha|S(\xi)|\beta\rangle &= \frac{1}{\alpha^*}\langle\alpha|a^\dagger S(\xi)|\beta\rangle \\ &= \frac{1}{\alpha^*}\langle\alpha|S(\xi)S^\dagger(\xi)a^\dagger S(\xi)|\beta\rangle \\ &= \frac{1}{\alpha^*}\langle\alpha|S(\xi)(a^\dagger\cosh r - ae^{-i\theta}\sinh r)|\beta\rangle \\ &= \frac{1}{\alpha^*}\left[\cosh r\left(\frac{\partial}{\partial\beta} + \frac{1}{2}\beta^*\right) - e^{-i\theta}\beta\sinh r\right]\langle\alpha|S(\xi)|\beta\rangle.\end{aligned} \tag{3.5.3}$$

The function $\langle\alpha|S(\xi)|\beta\rangle$ therefore satisfies the following differential equation

$$\left[\cosh r\frac{\partial}{\partial\beta} - \beta e^{-i\theta}\sinh r + \left(\frac{1}{2}\beta^*\cosh r - \alpha^*\right)\right]\langle\alpha|S(\xi)|\beta\rangle = 0. \tag{3.5.4}$$

The solution of this equation is

$$\begin{aligned}&\langle\alpha|S(\xi)|\beta\rangle \\ &= K\exp\left(-\frac{1}{2}|\beta|^2 + \alpha^*\beta\,\mathrm{sech}\, r + \frac{1}{2}e^{-i\theta}\beta^2\tanh r\right).\end{aligned} \tag{3.5.5}$$

The form of K, which may depend upon $\alpha, \alpha^*, \beta^*, r$, and θ, can be determined using the unitarity of $S(\xi)$. It follows from

$$\langle\alpha|S(\xi)|\beta\rangle^* = \langle\beta|S^\dagger(\xi)|\alpha\rangle = \langle\beta|S(-\xi)|\alpha\rangle, \tag{3.5.6}$$

that

$$K^*(\alpha,\alpha^*,\beta^*,r,\theta)\exp\left[-\frac{1}{2}|\beta|^2+\frac{1}{2}e^{i\theta}(\beta^*)^2\tanh r\right]$$
$$=K(\beta,\beta^*,\alpha^*,r,\theta+\pi)\exp\left(-\frac{1}{2}|\alpha|^2-\frac{1}{2}e^{-i\theta}\alpha^2\tanh r\right). \quad (3.5.7)$$

The form of K is therefore

$$K(\alpha,\alpha^*,\beta^*,r,\theta)$$
$$=(\text{sech}\, r)^{1/2}\exp\left[-\frac{1}{2}|\alpha|^2-\frac{1}{2}e^{i\theta}(\alpha^*)^2\tanh r\right]. \quad (3.5.8)$$

The coefficient $(\text{sech}\, r)^{1/2}$ is chosen so that the normalization condition

$$\frac{1}{\pi}\int|\langle\alpha|S(\xi)|\beta\rangle|^2 d^2\alpha=1 \quad (3.5.9)$$

is satisfied.

On substituting this expression for K in Eq. (3.5.5) we obtain

$$\langle\alpha|S(\xi)|\beta\rangle=(\text{sech}\, r)^{1/2}\exp\left\{-\frac{1}{2}(|\alpha|^2+|\beta|^2)+\alpha^*\beta\,\text{sech}\, r\right.$$
$$\left.-\frac{1}{2}\left[e^{i\theta}(\alpha^*)^2-e^{-i\theta}\beta^2\right]\tanh r\right\}. \quad (3.5.10)$$

The Q-representation for the state $|\beta,\xi\rangle$ is therefore

$$Q(\alpha,\alpha^*)=\frac{\text{sech}\, r}{\pi}\exp\left\{-(|\alpha|^2+|\beta|^2)+(\alpha^*\beta+\beta^*\alpha)\text{sech}\, r\right.$$
$$\left.-\frac{1}{2}[e^{i\theta}(\alpha^{*2}-\beta^{*2})+e^{-i\theta}(\alpha^2-\beta^2)]\tanh r\right\}. \quad (3.5.11)$$

In Fig. 3.2, $Q(\alpha,\alpha^*)\equiv Q(X_1,X_2)$ ($X_1=(\alpha+\alpha^*)/2$, $X_2=(\alpha-\alpha^*)/2i$) is plotted as a function of the amplitudes X_1, X_2 of the two quadratures. We clearly see the unequal variances in X_1 and X_2 in the state $|\alpha,\xi\rangle$. We can employ expression (3.5.10) for $\langle\alpha|S(\xi)|\beta\rangle$ to calculate the photon distribution function of a squeezed coherent state.

The photon distribution function $p(n)$ for the field in state $|\beta,\xi\rangle$ is given by

$$p(n)=|\langle n|\beta,\xi\rangle|^2. \quad (3.5.12)$$

The quantity $\langle n|\beta,\xi\rangle$ can be determined by writing

$$\langle\alpha|\beta,\xi\rangle=\sum_{n=0}^{\infty}\langle\alpha|n\rangle\langle n|\beta,\xi\rangle=e^{-\frac{1}{2}|\alpha|^2}\sum_{n=0}^{\infty}\frac{(\alpha^*)^n}{\sqrt{n!}}\langle n|\beta,\xi\rangle, \quad (3.5.13)$$

and expanding the right-hand side of Eq. (3.5.10) in powers of α^* by means of the generating function for the Hermite polynomials $H_n(z)$:

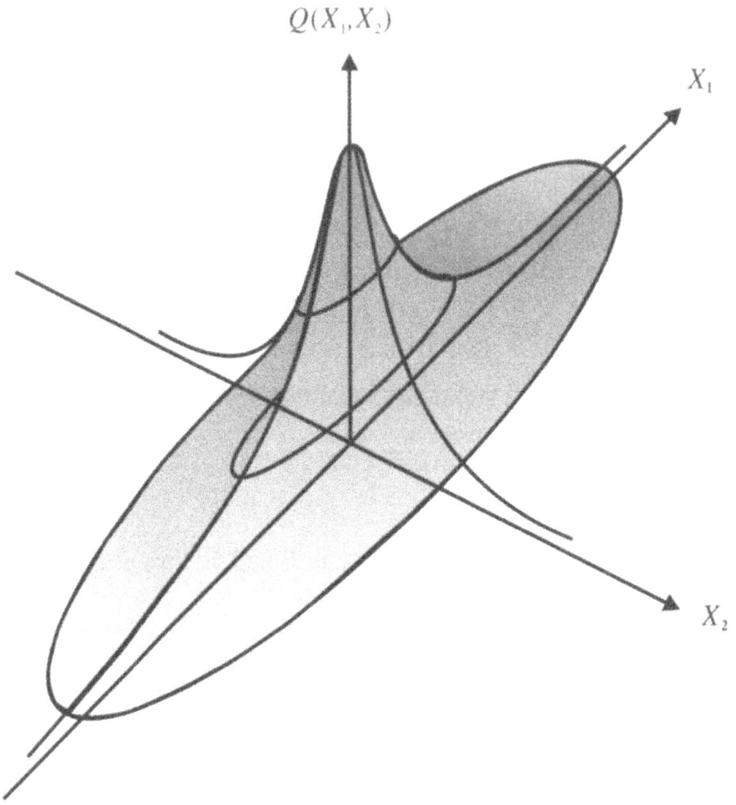

Fig. 3.2
A plot of $Q(\alpha, \alpha^*) \equiv Q(X_1, X_2)$ as a function of the amplitudes X_1 and X_2 in a squeezed coherent state. (From H. P. Yuen, *Phys. Rev. A* **13**, 2226 (1976).)

$$\exp(2zt - t^2) = \sum_{n=0}^{\infty} \frac{H_n(z)t^n}{n!}. \tag{3.5.14}$$

On comparing the resulting expansion with the expansion in Eq. (3.5.13), it follows that

$$\langle n|\beta, \xi\rangle = \frac{(e^{i\theta}\tanh r)^{n/2}}{2^{n/2}(n!\cosh r)^{1/2}} \exp\left[-\frac{1}{2}(|\beta|^2 - e^{-i\theta}\beta^2 \tanh r)\right] \times H_n\left(\frac{\beta e^{-i\theta/2}}{\sqrt{2\cosh r \sinh r}}\right). \tag{3.5.15}$$

The photon distribution function $p(n)$ for an ideal squeezed state is therefore given by

$$p(n) = \frac{(\tanh r)^n}{2^n n! \cosh r} \exp\left\{-|\beta|^2 + \frac{1}{2}[e^{-i\theta}\beta^2 + e^{i\theta}(\beta^*)^2]\tanh r\right\} \times \left|H_n\left(\frac{\beta e^{-i\theta/2}}{\sqrt{2\cosh r \sinh r}}\right)\right|^2. \tag{3.5.16}$$

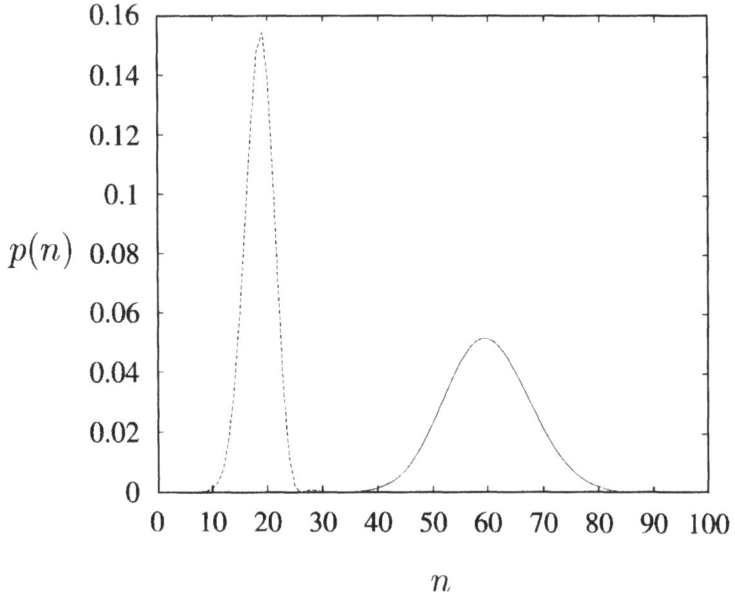

Fig. 3.3 Comparison of photon distribution function for a coherent state $|\alpha\rangle$ with $|\alpha|^2 = 60$ (solid line) with the squeezed coherent state $|\beta, \xi\rangle$ ($\beta = |\beta| \exp(i\phi)$, $\xi = r \exp(i\theta)$) with $|\beta|^2 = 60$, $r = 0.6$, and $\phi = \theta/2$ (dashed line).

Generally, sources of squeezing produce a radiation field in a squeezed vacuum state $|0, \xi\rangle$. The detection schemes, however, add a coherent component to it. The detected state is therefore described by the distribution (3.5.16). The fluctuations in the mean number of photons can be found either from Eq. (3.5.16), by using

$$\langle n^r \rangle = \sum_{n=0}^{\infty} n^r p(n), \tag{3.5.17}$$

or through the use of the unitary transformation properties of the squeeze operator (2.7.6) and (2.7.7). We obtain

$$(\Delta n)^2 = |\beta|^2[\cosh 4r - \cos(\theta - 2\phi) \sinh 4r] + 2 \sinh^2 r \cosh^2 r. \tag{3.5.18}$$

In the following, we discuss three cases of interest. First, when $|\beta|^2 \gg \sinh^2 r$, the coherent component is larger than the squeeze component. Figure 3.3 compares the probability distribution for a squeezed state with a coherent state. If the squeezing is along the coherent amplitude, the state has sub-Poissonian photon statistics. In the second case (Fig. 3.4) when the squeeze component is larger than the coherent component and squeezing is along the coherent amplitude, the squeezed state exhibits oscillations. The main peak as well as the subsequent peaks are narrower than the corresponding $\sqrt{n}$ value. But the overall distribution shows super-Poissonian statistics.

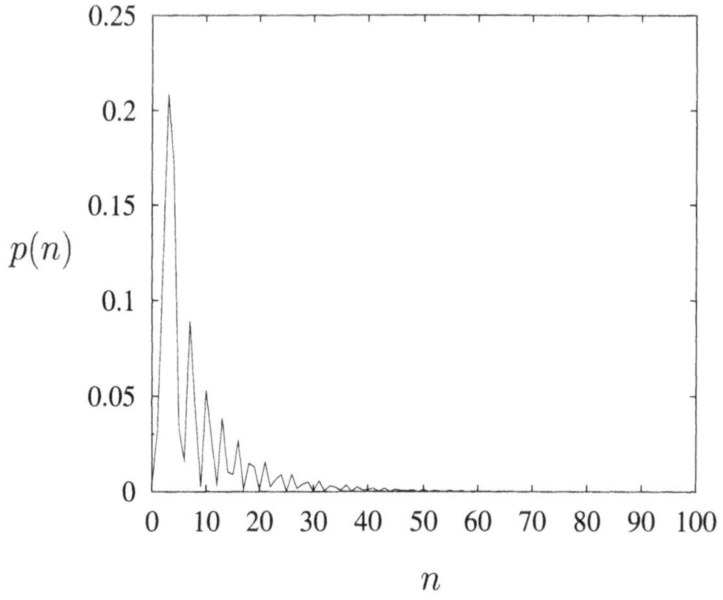

Fig. 3.4
Photon distribution function $p(n)$ for a two-photon coherent state (Eq.(3.5.16)) for $|\beta|^2 = 60$, $r = 1.6$. The squeeze component is larger than the coherent component and squeezing is along the coherent amplitude.

Finally, for zero displacement, i.e., for the squeezed vacuum state, the distribution function (3.5.16) reduces to

$$p(2n) = (\cosh r)^{-1}\frac{(2n)!}{(n!)^2}\left(\frac{1}{2}\tanh r\right)^{2n},$$
$$p(2n+1) = 0. \tag{3.5.19}$$

In the above equations, a nonzero value for even terms arises due to squeezing of the vacuum and clearly shows the 'two-photon' nature of the field. Figure 3.5 shows a plot of the probability distribution (3.5.19). The distribution peaks sharply at $n = 0$ and has a very long tail similar to a thermal distribution.

3.A Verifying equations (3.1.12a, 3.1.12b)

It can be verified that the two-dimensional delta function has the form (3.1.12a)

$$\delta(\alpha^* - a^\dagger)\delta(\alpha - a) = \frac{1}{\pi^2}\int \exp[-\beta(\alpha^* - a^\dagger)]\exp[\beta^*(\alpha - a)]d^2\beta \tag{3.A.1}$$

by taking the expectation values in a coherent state $|\gamma\rangle$ of both sides of Eq. (3.A.1). Indeed, on doing so and utilizing the fact that $|\gamma\rangle$ is an

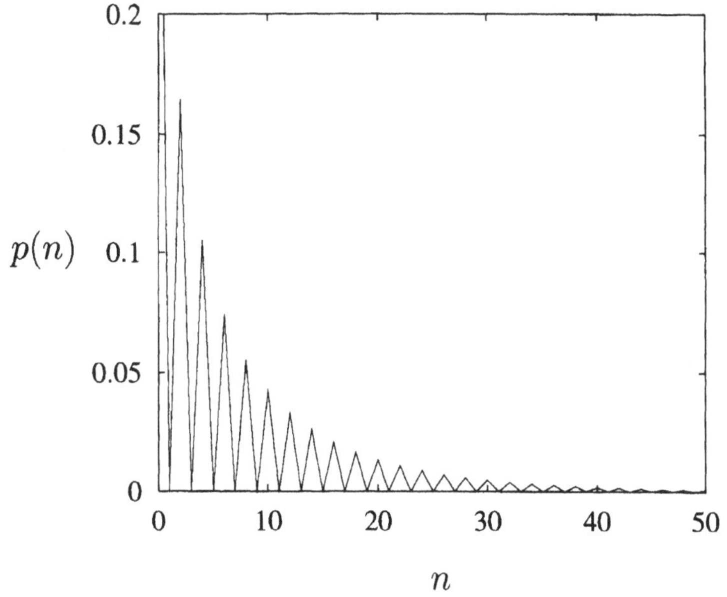

Fig. 3.5
Photon distribution function $p(n)$ for a squeezed vacuum state for $r = 1.6$.

eigenstate of the annihilation operator a with an eigenvalue γ, we get

$$\delta(\alpha^* - \gamma^*)\delta(\alpha - \gamma) = \frac{1}{\pi^2}\int \exp[-\beta(\alpha^* - \gamma^*)]\exp[\beta^*(\alpha - \gamma)]d^2\beta. \tag{3.A.2}$$

If we write $\alpha = x_\alpha + iy_\alpha$, $\beta = x_\beta + iy_\beta$, and $\gamma = x_\gamma + iy_\gamma$, then $d^2\beta = dx_\beta dy_\beta$ and the right-hand side of Eq. (3.A.2) becomes

$$\begin{aligned}
&\frac{1}{\pi^2}\int \exp[-\beta(\alpha^* - \gamma^*)]\exp[\beta^*(\alpha - \gamma)]d^2\beta \\
&= \frac{1}{\pi^2}\int \exp\left\{2i[x_\beta(y_\alpha - y_\gamma) - y_\beta(x_\alpha - x_\gamma)]\right\}dx_\beta dy_\beta \\
&= \left(\frac{1}{2\pi}\right)^2 \int\int \exp\left\{i[x_\beta(y_\alpha - y_\gamma) - y_\beta(x_\alpha - x_\gamma)]\right\}dx_\beta dy_\beta \\
&= \delta[\mathrm{Im}(\alpha - \gamma)]\delta[\mathrm{Re}(\alpha - \gamma)] \\
&\equiv \delta(\alpha - \gamma)\delta(\alpha^* - \gamma^*),
\end{aligned} \tag{3.A.3}$$

where we have replaced $2x_\beta$ and $2y_\beta$ by x_β and y_β, respectively, in the second line and used the following expression for the delta function

$$\delta(x) = \frac{1}{2\pi}\int_{-\infty}^{\infty} e^{ikx}dk. \tag{3.A.4}$$

Equation (3.1.12b)

$$\delta(\alpha^* - a^\dagger)\delta(\alpha - a)$$
$$= \frac{1}{\pi^2} \int \exp[-i\beta(\alpha^* - a^\dagger)] \exp[-i\beta^*(\alpha - a)] d^2\beta \qquad (3.A.5)$$

can be obtained from (3.A.1) simply by changing the variables $\beta \to i\beta$ and $\beta^* \to -i\beta^*$.

Another formula for the antinormally ordered two-dimensional delta function, namely,

$$\delta(\alpha - a)\delta(\alpha^* - a^\dagger)$$
$$= \frac{1}{\pi^2} \int \exp[\beta^*(\alpha - a)] \exp[-\beta(\alpha^* - a^\dagger)] d^2\beta, \qquad (3.A.6)$$

which has been used to define the Q-representation (Eq. (3.2.1)), can be proven by inserting

$$\frac{1}{\pi} \int |\gamma\rangle\langle\gamma| d^2\gamma = 1, \qquad (3.A.7)$$

as follows:

$$\frac{1}{\pi^2} \int \exp[\beta^*(\alpha - a)] \exp[-\beta(\alpha^* - a^\dagger)] d^2\beta$$
$$= \frac{1}{\pi^3} \int\int e^{\beta^*(\alpha - a)} |\gamma\rangle\langle\gamma| e^{-\beta(\alpha^* - a^\dagger)} d^2\beta d^2\gamma$$
$$= \frac{1}{\pi^3} \int\int e^{\beta^*(\alpha - \gamma)} |\gamma\rangle\langle\gamma| e^{-\beta(\alpha^* - \gamma^*)} d^2\beta d^2\gamma$$
$$= \frac{1}{\pi} \int \delta(\alpha - \gamma) |\gamma\rangle\langle\gamma| \delta(\alpha^* - \gamma^*) d^2\gamma$$
$$= \delta(\alpha - a) \left(\frac{1}{\pi} \int |\gamma\rangle\langle\gamma| d^2\gamma \right) \delta(\alpha^* - a^\dagger)$$
$$= \delta(\alpha - a)\delta(\alpha^* - a^\dagger). \qquad (3.A.8)$$

3.B c-number function correspondence for the Wigner–Weyl distribution

Given an operator $O(a, a^\dagger)$ and the Wigner–Weyl distribution $W(\alpha, \alpha^*)$, we calculate the c-number function $O_S(\alpha, \alpha^*)$ such that

$$\langle O(a, a^\dagger) \rangle = \mathrm{Tr}(O\rho) = \int d^2\alpha O_S(\alpha, \alpha^*) W(\alpha, \alpha^*). \qquad (3.B.1)$$

Recall that the Wigner–Weyl distribution is defined as (Eqs. (3.3.6) and (3.3.7))

$$W(\alpha, \alpha^*) = \frac{1}{\pi^2} \int d^2\beta \mathrm{Tr}\left(e^{i\beta^* a + i\beta a^\dagger} \rho\right) e^{-i\beta^* \alpha - i\beta \alpha^*} \tag{3.B.2}$$

with the characteristic function

$$\begin{aligned} C^{(s)}(\beta, \beta^*) &= \int e^{i\beta^* \alpha + i\beta \alpha^*} W(\alpha, \alpha^*) d^2\alpha \\ &= \mathrm{Tr}\left(e^{i\beta^* a + i\beta a^\dagger} \rho\right) \\ &= \mathrm{Tr}\left(e^{i\beta a^\dagger} e^{i\beta^* a} e^{-|\beta|^2/2} \rho\right). \end{aligned} \tag{3.B.3}$$

where, in the last line, we use the Baker–Hausdorff formula (2.2.7). Now for any normally ordered operator $O(a, a^\dagger)$, one can write

$$O(a, a^\dagger) = \sum_{n,m} c_{n,m} a^{\dagger n} a^m. \tag{3.B.4}$$

It can be easily found that

$$\langle a^\dagger \rangle = \left[\frac{\partial}{\partial(i\beta)} + \frac{\beta^*}{2i}\right] C^{(s)}(\beta, \beta^*)\Big|_{\beta^*=\beta=0},$$

and

$$\langle a \rangle = \left[\frac{\partial}{\partial(i\beta^*)} + \frac{\beta}{2i}\right] C^{(s)}(\beta, \beta^*)\Big|_{\beta^*=\beta=0}.$$

Then we have

$$\begin{aligned} &\langle O(a, a^\dagger) \rangle \\ &= \sum_{n,m} c_{n,m} \left[\frac{\partial}{\partial(i\beta)} + \frac{\beta^*}{2i}\right]^n \left[\frac{\partial}{\partial(i\beta^*)} + \frac{\beta}{2i}\right]^m C^{(s)}(\beta, \beta^*)\Big|_{\beta^*=\beta=0} \\ &= \int d^2\alpha \sum_{n,m} c_{n,m} \left[\frac{\partial}{\partial(i\beta)} + \frac{\beta^*}{2i}\right]^n \\ &\qquad \times \left[\frac{\partial}{\partial(i\beta^*)} + \frac{\beta}{2i}\right]^m e^{i\beta^* \alpha + i\beta \alpha^*}\Big|_{\beta^*=\beta=0} W(\alpha, \alpha^*) \\ &\equiv \int d^2\alpha O_S(\alpha, \alpha^*) W(\alpha, \alpha^*), \end{aligned} \tag{3.B.5}$$

which yields

$$\begin{aligned} &O_S(\alpha, \alpha^*) \\ &= \sum_{n,m} c_{n,m} \left[\frac{\partial}{\partial(i\beta)} + \frac{\beta^*}{2i}\right]^n \left[\frac{\partial}{\partial(i\beta^*)} + \frac{\beta}{2i}\right]^m e^{i\beta^* \alpha + i\beta \alpha^*}\Big|_{\beta^*=\beta=0}. \end{aligned} \tag{3.B.6}$$

Equation (3.B.6) is our desired result. Consider some examples:

(a) $O(a, a^\dagger) = a^\dagger a$

$$
\begin{aligned}
&O_S(\alpha, \alpha^*) \\
&= \left[\frac{\partial}{\partial(i\beta)} + \frac{\beta^*}{2i}\right]\left[\frac{\partial}{\partial(i\beta^*)} + \frac{\beta}{2i}\right] e^{i\beta^*\alpha + i\beta\alpha^*}\Big|_{\beta^*=\beta=0} \\
&= \left[\frac{\partial}{\partial(i\beta)} + \frac{\beta^*}{2i}\right]\left(\alpha + \frac{\beta}{2i}\right) e^{i\beta^*\alpha + i\beta\alpha^*}\Big|_{\beta^*=\beta=0} \\
&= \alpha^*\alpha - \frac{1}{2}.
\end{aligned}
\tag{3.B.7}
$$

(b) $O(a, a^\dagger) = a^{\dagger 2} a$

$$
\begin{aligned}
&O_S(\alpha, \alpha^*) \\
&= \left(\frac{\partial}{\partial(i\beta)} + \frac{\beta^*}{2i}\right)^2 \left(\frac{\partial}{\partial(i\beta^*)} + \frac{\beta}{2i}\right) e^{i\beta^*\alpha + i\beta\alpha^*}\Big|_{\beta^*=\beta=0} \\
&= \left(\frac{\partial}{\partial(i\beta)} + \frac{\beta^*}{2i}\right)^2 \left(\alpha + \frac{\beta}{2i}\right) e^{i\beta^*\alpha + i\beta\alpha^*}\Big|_{\beta^*=\beta=0} \\
&= \left(\frac{\partial}{\partial(i\beta)} + \frac{\beta^*}{2i}\right) \\
&\quad \left[\left(\alpha^* + \frac{\beta^*}{2i}\right)\left(\alpha + \frac{\beta}{2i}\right) - \frac{1}{2}\right] e^{i\beta^*\alpha + i\beta\alpha^*}\Big|_{\beta^*=\beta=0} \\
&= \left\{-\frac{1}{2}\left(\alpha^* + \frac{\beta^*}{2i}\right) + \left(\alpha^* + \frac{\beta^*}{2i}\right)\right. \\
&\quad \left.\left[\left(\alpha^* + \frac{\beta^*}{2i}\right)\left(\alpha + \frac{\beta}{2i}\right) - \frac{1}{2}\right]\right\} e^{i\beta^*\alpha + i\beta\alpha^*}\Big|_{\beta^*=\beta=0} \\
&= -\frac{\alpha^*}{2} + \alpha^*\left(\alpha^*\alpha - \frac{1}{2}\right) \\
&= \alpha^{*2}\alpha - \alpha^*.
\end{aligned}
\tag{3.B.8}
$$

The operator corresponding to the Wigner distribution function in the coordinate-momentum representation is given by Cohen (1986).

Problems

3.1 Show that

$$
\frac{1}{2}\langle aa^\dagger + a^\dagger a\rangle = \int W(\alpha, \alpha^*)|\alpha|^2 d^2\alpha,
$$

where $W(\alpha, \alpha^*)$ is the Wigner–Weyl distribution.

3.2 Show that

$$\mathrm{Tr}[D(\alpha)] = \pi\delta^{(2)}(\alpha),$$
$$\mathrm{Tr}[D(\alpha)D^{\dagger}(\alpha')] = \pi\delta^{(2)}(\alpha - \alpha'),$$

where $D(\alpha)$ is the displacement operator. Using these results, show that

$$\mathrm{Tr}[\Delta^{(\Omega)}(\alpha - a, \alpha^* - a^{\dagger})\bar{\Delta}^{(\Omega)}(\alpha' - a, \alpha^{*\prime} - a^{\dagger})]$$
$$= \frac{1}{\pi}\delta^{(2)}(\alpha - \alpha'),$$

The operators $\Delta^{(\Omega)}$ and $\bar{\Delta}^{(\Omega)}$ are defined in Eqs. (3.4.2) and (3.4.9), respectively.

3.3 Show that the Wigner–Weyl distribution $W(\alpha, \alpha^*)$ can be expressed in terms of the P-representation $P(\alpha, \alpha^*)$ via the relation

$$W(\alpha, \alpha^*) = \frac{2}{\pi}\int P(\beta, \beta^*)\exp(-2|\alpha - \beta|^2)d^2\beta.$$

3.4 Determine $Q(\alpha, \alpha^*)$ and $W(\alpha, \alpha^*)$ for a coherent state and a thermal state.

References and bibliography

P-representation

R. J. Glauber, *Phys. Rev.* **131**, 2766 (1963).
E. C. G. Sudarshan, *Phys. Rev. Lett.* **10**, 277 (1963).
D. F. Walls and G. J. Milburn, *Quantum Optics*, (Springer-Verlag, Berlin 1994).

Q-, Wigner–Weyl and other distributions

E. Wigner, *Phys. Rev.* **40**, 749 (1932).
J. Moyal, *Proc. Cambridge Phil. Soc.* **45**, 99 (1949).
Y. Kano, *J. Math. Phys.* **6**, 1913 (1965).
J. R. Klauder, J. McKenna, and D. G. Currie, *J. Math. Phys.* **6**, 734 (1965).
C. L. Mehta and E. C. G. Sudarshan, *Phys. Rev.* **138**, B274 (1965).
L. Cohen, *J. Math. Phys.* **7**, 781 (1966).
M. Lax, *1968 Brandis Lectures on Fluctuation and Coherence Phenomena* (Gordon and Breach, New York 1968).
K. E. Cahill and R. J. Glauber, *Phys. Rev.* **177**, 1857 (1969).
G. S. Agarwal and E. Wolf, *Phys. Rev. D* **2**, 2161, 2187, 2206 (1970).
W. Louisell, *Quantum Statistical Properties of Radiation*, (Wiley, New York 1974).
P. Carruthers and F. Zachariaser, *Phys. Rev. D* **13**, 950 (1976).
L. A. Reichl, *A Modern Course in Statistcal Physics*, (University of Texas Press 1980).
M. Hillery, R. F. O'Connell, M. O. Scully, and E. P. Wigner, *Physics Reports*, **106**, 123 (1984).
L. Cohen, in *Frontiers of Nonequilibrium Statistical Physics*, ed. G. T. Moore and M. O. Scully (Plenum, New York 1986), p. 97.
L. Cohen, *Time Frequency Analysis*, (Prentice Hall, New York 1995).
W. Vogel and D.-G. Welsch, *Lectures on Quantum Optics*, (Akademe Verlag, Berlin 1994).

CHAPTER 4

Field–field and photon–photon interferometry

Optical interferometry was at the heart of the revolution which ushered in the new era of twentieth century physics. For example, the Michelson interferometer was used to show that there is no detectable motion relative to the 'ether'; a key experiment in support of special relativity.

It is a wonderful tribute to Michelson that the same interferometer concept is central to the gravity-wave detectors which promise to provide new insights into general relativity and astrophysics in the twenty-first century. Similar tales can be told about the Sagnac and Mach–Zehnder interferometers as discussed in this chapter. We further note that the intensity correlation stellar interferometer of Hanbury-Brown and Twiss* was a driving force in ushering in the modern era of quantum optics.

We are thus motivated to develop the theory of field (amplitude) and photon (intensity) correlation interferometry. In doing so we will find that the subject provides us with an exquisite probe of the micro and macrocosmos, i.e., quantum mechanics and general relativity.

With these thoughts in mind we here develop a framework to study the quantum statistical correlations of light. We will motivate the quantum correlation functions of the field operators from the standpoint of photodetection theory. Many experimentally observed quantities, such as photoelectron statistics and the spectral distribution of the field, can be related to the appropriate field correlation functions. These correlation functions are essential in the description of Young's double-slit experiment and the notion of the power spectrum of light. The intensity correlation functions are usually associated with the intensity–intensity correlation measurements as required in

* See the pioneering work of Hanbury-Brown and Twiss [1954, 1956]. Excellent pedagogical treatments of the problem are given by Fano [1961], Glauber [1965], and Baym [1969]. For a review of the subject see Hanbury-Brown [1974].

the descriptions of the famous Hanbury-Brown–Twiss effect and other two-photon interference experiments which we discuss in this chapter.

Quantum coherence theory also allows us to examine field states which exhibit certain nonclassical features, i.e., states which cannot be described by a classical statistical theory. Such states can arise when the quantum nature of light is explicitly exhibited. Examples are the number and squeezed states of the radiation field.

In the next two sections we discuss the application of interferometry to astrophysics and general relativity and then turn to a general discussion of photon optics.

4.1 The interferometer as a cosmic probe

The foundations of physics are anchored in the bedrock of curved spacetime. 'Spacetime' in the sense of Minkowski who showed us that physical events (e.g., the emission of a photon) are best viewed as occurring in a (flat) four-dimensional geometry having one time and three spatial coordinates. The adjective 'curved' enters the picture when gravitation is included in the problem. That is, according to Einstein's theory of general relativity, we view gravity as arising from (or described by) the curvature of this four-dimensional space. This curvature itself is produced by the presence of massive bodies in the universe, the earth, sun, Crab Nebula, etc.

Many theorists regard the general theory of relativity to be the most beautiful of all physical theories. However, due to the smallness of the gravitational coupling constant,

$$G = 6.67 \times 10^{-8}\ \text{cm}^3/\text{g s}^2,$$

experimental tests of this theory are very scarce. This fact is underscored by Misner, Thorne, and Wheeler, who observed that: '*For the first half century of its life, general relativity was a theorist's paradise but an experimentalist's hell.*' However, thanks in large part to advances in modern laser optics, new tests of metric gravity (general relativity) have been, and will continue to be, carried out. The optical interferometer is the main tool in these astrophysical and general relativistic studies.

4.1.1 Michelson interferometer and general relativity

As mentioned earlier, the Michelson interferometer was used to search for motion through the ether and was one of the key experiments in formulating special relativity and modern physics.

At present a type of Michelson interferometer is being built to detect gravity waves. As depicted in Fig. 4.1, gravitational radiation acts so as to effectively change the path length for light in one arm of the interferometer and thus introduces a phase shift. How this phase shift comes about can be viewed in two different ways: (1) the gravity wave changes the distance between the mirrors (2) the gravity wave changes or perturbes spacetime and acts much as a dielectric. We here take the first point of view.

In time-independent (Newtonian) gravity, the (scalar) potential Φ (in free space) obeys the Laplace equation

$$\nabla^2 \Phi = 0, \tag{4.1.1}$$

whereas, in time-dependent (Einsteinian) metric gravity, the tensor field* $\Phi_{\mu\nu}(\mathbf{r}, t)$, where the indices μ and ν run from 1 to 4, obeys a wave equation of the form

$$\left(\nabla^2 - \frac{1}{c^2} \frac{\partial^2}{\partial t^2} \right) \Phi_{\mu\nu}(\mathbf{r}, t) = 0. \tag{4.1.2}$$

Thus, the effects of gravity propagate with the speed of light c from their point of origin (binary stars, exploding galaxies, etc.) to our laboratory on earth. This 'gravitational wave' causes points in the laboratory to experience tiny amplitude-relative oscillations.

A scheme to measure the gravitational waves (g-waves) is based on the Michelson interferometer. The effect of gravitational radiation is to stretch or compress a rod of length L which is perpendicular to the direction of propagation. For example, the gravitational wave of frequency ν_g will cause the length L_x between the mirror M_1 and the beam-splitter in Fig. 4.1(a) to vary as

$$L_x = L[1 + h_0 \cos(\nu_g t)], \tag{4.1.3}$$

where L is the length of the interferometer arm in the absence of a gravitational wave and h_0 represents the amplitude of the gravitational wave and is of order $\leq 10^{-21}$ for the envisioned sources.

Therefore, there will be a phase shift between the light traversing the two arms of the interferometer of an amount

$$\begin{aligned} \delta &= k(L_x - L_y) \\ &= kLh_0 \cos(\nu_g t). \end{aligned} \tag{4.1.4}$$

Hence the intensity recorded by the detector in Fig. 4.1(a) will be

* For a discussion of general relativity directed toward the student of modern quantum optics see Schleich and Scully [1984].

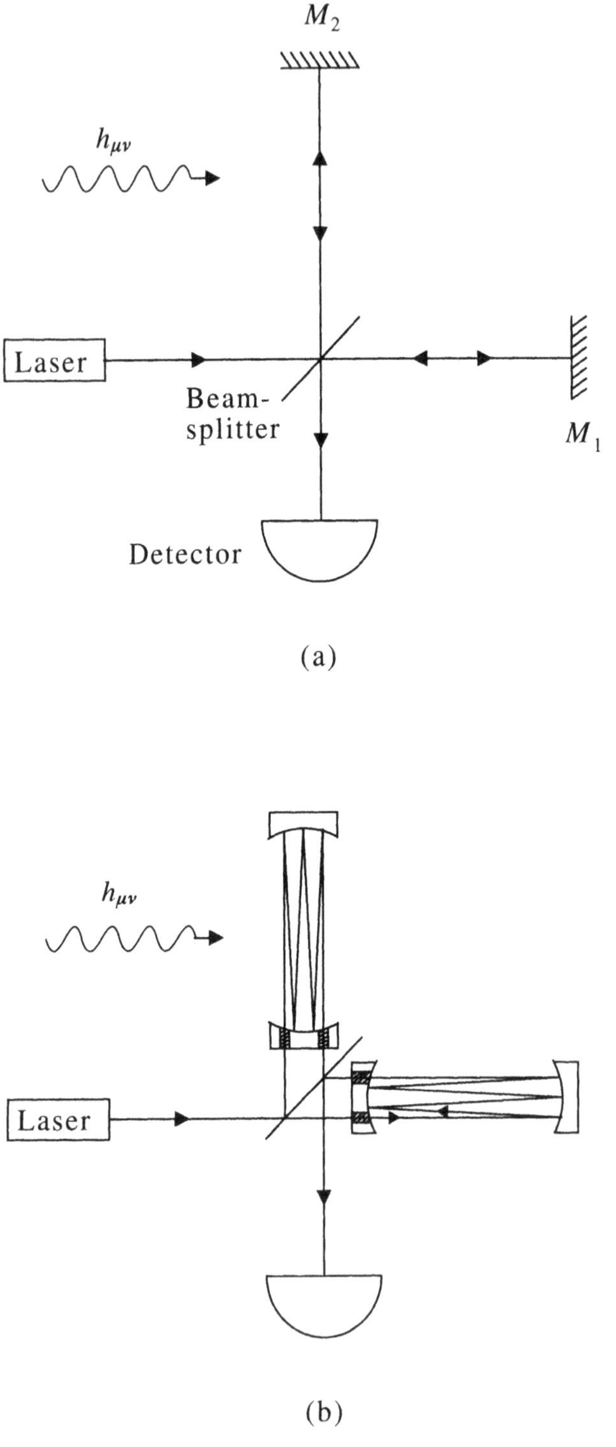

Fig. 4.1
(a) An external laser drives a Michelson interferometer which is indluenced by an incident gravity wave denoted by $h_{\mu\nu}$. (b) A Michelson interferometer with cavities in both arms, 'folds' the light many times, thus lengthening the effective optical path lengths in each arm.

$$I = \frac{1}{2} I_0 (1 + \cos \delta), \tag{4.1.5}$$

where I_0 is the incident intensity.

In the actual experiments, cavities are used in the two arms of the Michelson interferometer as in Fig. 4.1(b). Now the signal due to a gravity wave translates into a time-dependent phase shift obtained from Eq. (4.1.4) by replacing L_x by the effective path length $\tilde{L}$, which is essentially the number of bounces times the length of the arm L. Therefore, for times $t \ll v_g^{-1}$, the g-wave-induced phase shift is given by $\Delta\theta^{(p)} = v\tilde{L}h_0/c$, where v is the frequency of the laser light. In such an experiment the fundamental quantum limit is given by 'photon shot noise'. Denoting the average number of laser photons by $\bar{n}$, the power at the detector by P and assuming unit quantum efficiency for present purposes, one has the phase uncertainty due to shot noise for a measurement of duration t_m

$$\Delta\theta_n \simeq \frac{1}{\sqrt{\bar{n}}} = \sqrt{\frac{\hbar v}{P t_m}}. \tag{4.1.6}$$

Equating $\Delta\theta^{(p)}$ to $\Delta\theta_n$, we find the minimum detectable g-wave amplitude for such a passive system to be

$$h_{\min}^{(p)} \simeq \frac{c}{v\tilde{L}} \sqrt{\frac{\hbar v}{P t_m}} = \frac{\mathscr{C}}{v} \sqrt{\frac{\hbar v}{P t_m}}, \tag{4.1.7}$$

where we have introduced the cavity decay rate $\mathscr{C} = c/\tilde{L}$.

4.1.2 The Sagnac ring interferometer

In 1913 Sagnac considered the use of a ring resonator to search for the 'ether drift' relative to a rotating frame. However, as often happens, his results turned out to be useful in ways that Sagnac himself never dreamt of. As shown in Fig. 4.2, the real physics associated with the Sagnac effect is simply that it takes longer for a short pulse of light to 'get back' to its point of origin if it goes in the direction of rotation and it takes less time if it is moving in a counter-propagating sense.

To quantify this, consider Fig. 4.2. There we see that laser light enters the interferometer at point A and is split into clockwise (CW) and counter-clockwise (CCW) propagating beams by a beam-splitter. If the interferometer is not rotating, the CW and CCW propagating beams recombine at point A after a time given by

$$t = \frac{2\pi b}{c}, \tag{4.1.8}$$

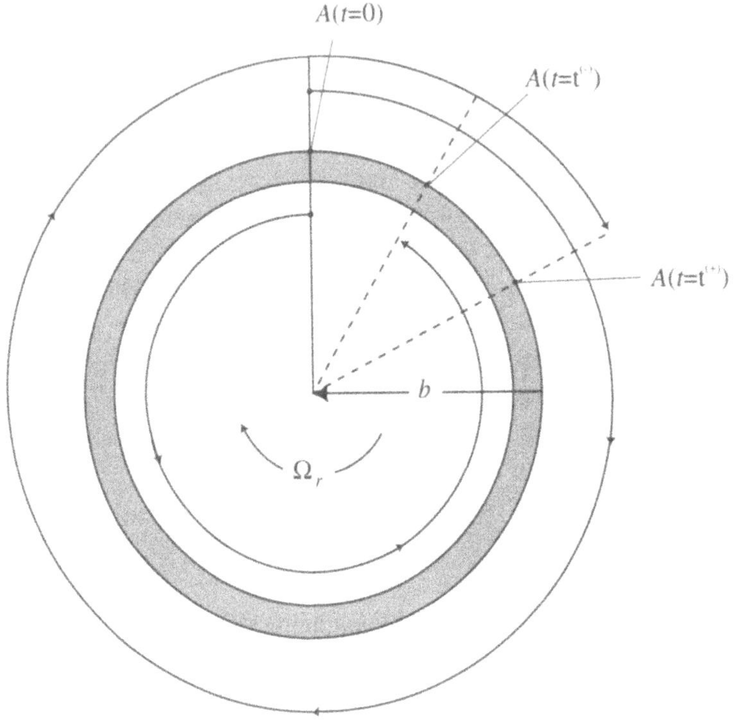

Fig. 4.2 Schematics of a Sagnac ring interferometer.

where b is the radius of the circular beam path. However, if the interferometer is rotating, with angular velocity Ω_r, about an axis through the center and perpendicular to the plane of the interferometer, then the beams reencounter the beam-splitter at different times because the CW (co-directional with Ω_r) propagating beam must traverse a path length of slightly more than $2\pi b$ in order to complete one round trip, since the interferometer rotates through a small angle during the round-trip transit time. Similarly, the CCW propagating beam traverses a path length slightly less than $2\pi b$ during one round trip. If we denote the round-trip transit time of the CW beam by t^+ and that of the CCW beam by t^-, then t^+ is given by

$$\begin{aligned} t^+ &= \frac{2\pi b + b\Omega_r t^+}{c} \\ &= \frac{2\pi b}{c}\left(1 - \frac{b\Omega_r}{c}\right)^{-1}, \end{aligned} \tag{4.1.9a}$$

where, in the first line, $b\Omega_r t^+$ is the arc length the interferometer rotates through before the CW beam arrives back at the beam-splitter.

Similarly,

$$t^- = \frac{2\pi b - b\Omega_r t^-}{c}$$
$$= \frac{2\pi b}{c}\left(1 + \frac{b\Omega_r}{c}\right)^{-1}. \tag{4.1.9b}$$

The difference between t^+ and t^- is given by

$$\Delta t = t^+ - t^- = \frac{4\pi b^2 \Omega_r}{c^2 - b^2\Omega_r^2}. \tag{4.1.10}$$

For reasonable values of b and Ω_r, $(b\Omega_r)^2 \ll c^2$, so that

$$\Delta t \cong \frac{4\pi b^2 \Omega_r}{c^2}, \tag{4.1.11}$$

the round-trip optical path difference, ΔL, is given by

$$\Delta L = c\Delta t = \frac{4\pi b^2 \Omega_r}{c}. \tag{4.1.12}$$

From Eq. (4.1.12) we see the round-trip optical path difference, according to this analysis, is directly proportional to the rotation rate of the interferometer. A more general approach valid for an arbitrary interferometer shape leads to the result

$$\Delta L = \frac{4\mathbf{\Omega}_r \cdot \hat{\mathbf{z}} A}{c}, \tag{4.1.13}$$

where A is the area enclosed by the light path and $\hat{\mathbf{z}}$ is a unit vector normal to the surface of the interferometer.

The effectiveness of the Sagnac interferometer is limited by the fact that the optical path difference given by Eq. (4.1.12) is much less than a wavelength. (For instance, if $b = 1$ m and $\Omega_r = 10$ deg/h, then $\Delta L \cong 4.1 \times 10^{-12}$ m.) At first glance this would seem to make the use of ring laser gyros impractical as rotation sensing devices, since sensitivities of 10^{-3} deg/h or less are desirable. However, there are two different schemes used to greatly increase the sensitivity of ring laser gyros.

The first of these is to increase the total round-trip path length of the light by the use of a kilometer-long optical fiber as the interferometer cavity. To see why this increases the sensitivity of the gyroscope, we shall recast Eq. (4.1.12) into a more general form. From Eq. (4.1.12) we see that the phase difference, $\Delta\theta$, between the counter-propagating beams after one round trip is given by

$$\Delta\theta = \frac{2\pi\Delta L}{\lambda} = \frac{8\pi^2 b^2 \Omega_r}{c\lambda} = \frac{4A\Omega_r}{c\bar{\lambda}}, \tag{4.1.14}$$

where $\bar{\lambda} = \lambda/2\pi$ is the reduced wavelength of the laser light and $A = \pi b^2$ is the area enclosed by the light beams. Equation (4.1.14) is valid for a one loop circular light path. If an optical fiber is used, the light path typically consists of a fiber coil of radius b and many turns. In particular, in such a fiber coil with N turns, Eq. (4.1.14) becomes

$$\Delta\theta = \frac{8\pi^2 b^2 N\Omega_r}{c\lambda}, \tag{4.1.15}$$

or, in terms of the total length, $L = 2\pi bN$, of the optical fiber,

$$\Delta\theta = \frac{4\pi L b\Omega_r}{c\lambda}. \tag{4.1.16}$$

Equation (4.1.16) represents the important result that the phase shift induced by rotation of a Sagnac fiber ring interferometer increases linearly with the total length of the optical fiber.

The second scheme devised to increase the signal from a ring laser gyroscope is the introduction of an active laser medium into the ring cavity. This arrangement is illustrated by Fig. 4.3. For convenience, throughout the rest of this subsection, such an arrangement will be called an active ring laser gyro. Then the CW and CCW ring laser modes have different frequencies because of the difference in effective round-trip optical path lengths caused by the rotation of the cavity. Thus we have only oscillations with frequencies satisfying the resonance condition associated with $L_\pm$ corresponding to the effective cavity lengths seen by the CW and CCW propagating beams, respectively, namely

$$\nu_\pm = \frac{m\pi c}{L_\pm}, \tag{4.1.17}$$

where m is an integer and

$$L_\pm = L\left(1 \pm \frac{b\Omega_r}{c}\right). \tag{4.1.18}$$

Using Eq. (4.1.17) the frequency difference between the CW and CCW propagating beams can be approximated by

$$\Delta\nu = \nu_- - \nu_+ = \frac{m\pi c}{L_-} - \frac{m\pi c}{L_+} \cong \frac{m\pi c\Delta L}{L^2} = \nu\frac{\Delta L}{L}. \tag{4.1.19}$$

The approximation arises out of setting $L_+ L_- \cong L^2$.

Fig. 4.3
Schematics of an active ring laser gyroscope.

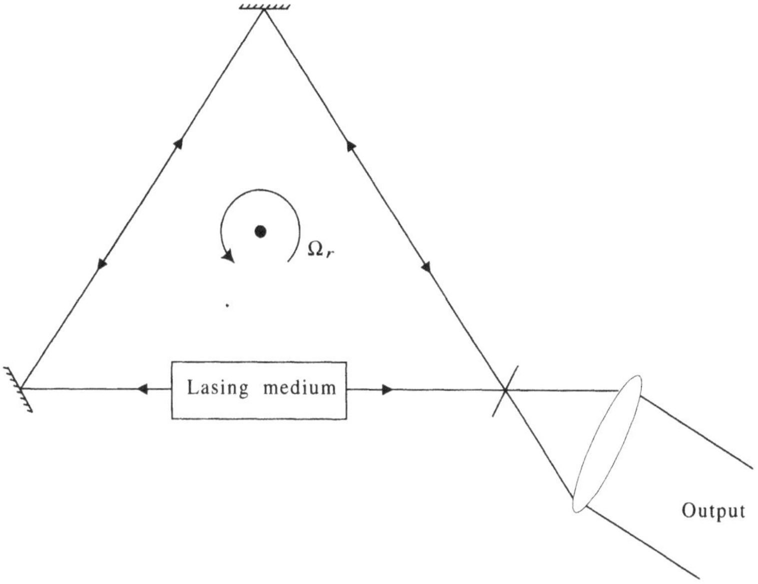

Now, a couple of important points need to be made. The first of these is that when using an active ring laser gyro it is the frequency difference (not the optical path difference) between the counter-propagating beams which is measured. This frequency difference is generally measured by heterodyning the two output beams. Also note that the frequency difference given by Eq. (4.1.19) is a factor of v/L larger than the optical path length difference given by Eq. (4.1.12). This increased scale factor together with the relative experimental ease associated with small frequency difference measurements makes the active ring laser gyro the most common and, currently, the most sensitive interferometer rotation sensor.

Inserting Eq. (4.1.12) into Eq. (4.1.19) gives (for a circular ring)

$$\Delta v = \frac{2vb\Omega_r}{c} = \frac{2b\Omega_r}{\bar{\lambda}}. \tag{4.1.20}$$

Note that Δv does not depend on the total length of the cavity so an increased scale factor is not achieved by using long fiber optic coils in active ring laser gyros. For an arbitrary cavity geometry, Eq. (4.1.20) becomes

$$\Delta v = \frac{4A\Omega_r}{p\bar{\lambda}}, \tag{4.1.21}$$

where A is the area enclosed by the light path and p is the perimeter of the light path. The constant of proportionality, $4A/\bar{\lambda}p$, between Δv

and Ω_r is often called the scale factor, which we will later represent by the symbol S.

4.1.3 Proposed ring laser test of metric gravitation theories

Recent progress in research using ring laser gyroscopic devices indicates that rotation rates as slow as $10^{-10}\,\Omega_\oplus$, where $\Omega_\oplus$ is the earth's rotation rate, are potentially measurable. With this in mind, experiments sensitive to Machian frame-dragging (Lense–Thirring effect), the presence of a preferred frame in the universe (preferred frame cosmology), and the curvature of local spacetime can now be envisioned.

Since Einstein formulated the general theory of relativity, there have been many other alternative theories of gravitation, e.g., due to Brans–Dicke and Ni. These theories, which have been motivated by different considerations, lead to different predictions for the effects discussed above. The theoretical framework of the parametrized post-Newtonian (PPN) formalism, which provides a means for studying a very wide class of metric theories of gravitation in the weak-field and slow-motion setting of the solar system, has been developed to systematically compare the various theories with experiment.

When an ultrasensitive ring laser is placed on the rotating earth, we expect to have several 'effective rotations' depending on the particular theory of metric gravity and spacetime we choose. These are summarized in Fig. 4.4 and Table 4.1. There, we see that in addition to the rotation of the ring at Ω_o and the earth's rotation $\Omega_\oplus$, we have three other contributions corresponding to Ω_{Mach}, Ω_{Cosmos}, and Ω_{Curve}, respectively.

The first of these effective rotations, Ω_{Mach}, is regarded as a "weak" verification of Mach's principle. That is, our gyro experiences an effective rotation even if it is fixed relative to the fixed stars (i.e., if we step off the earth so that the Ω_o and $\Omega_\oplus$ do not directly affect our ring laser). This effective rotation rate is due solely to the fact that we are near another massive rotating body – the earth. Another way to interpret this is as a kind of magnetic gravity analogous to the magnetic moment associated with a spinning electron.

The second contribution, Ω_{Cosmos}, arises from the presence of a preferred (rest) coordinate system. This 'preferred frame' might be thought to be that implied by the 3 K black-body background. Einstein's theory of general relativity involves no preferred coordinate frame, while in the theory of Ni the universe is at rest in a preferred frame. This effect is especially interesting since it is one of the least well established in gravitation physics.

Table 4.1. *Theories of spacetime and effective rotations.*

Effective rotation	Physical origin	PPN form	Order of magnitude	Einstein	Brans–Dicke	Ni
Ω_{Mach}	Mach principle or magnetic gravity	$\frac{C_1}{8}(7\Delta_1 + \Delta_2)\Omega_{\oplus}$	$5.6 \times 10^{-10} \times \frac{[7\Delta_1 + \Delta_2]}{8}\Omega_{\oplus}$	$\Delta_1 = 1$	$\Delta_1 = \frac{10 + 7\omega}{14 + 7\omega}$	$\Delta_1 = \frac{1}{7}$
		C_1 depends on latitude etc. and is of order 5.6×10^{-10}	$\Delta_2 = 1$	$\Delta_2 = 1$	$\Delta_2 = 1$ ω = number of order 27	$\Delta_2 = 1$
Ω_{Cosmos}	Preferred frame or motion through the 3 K cosmic black-body background	$\frac{C_2}{2}\alpha_1\Omega_{\oplus}$ C_2 depends on velocity of earth through 3 K background and is of order 1.2×10^{-7}	$1.2 \times 10^{-7} \times \alpha_1\Omega_{\oplus}$	$\alpha_1 = 0$	$\alpha_2 = 0$	$\alpha_1 = -8$
Ω_{Curve}	Spacetime curvature	$\frac{C_3}{2}(1 + \gamma)\Omega_{\oplus}$ C_3 depends on latitude etc. and is of order 1.4×10^{-9}	$1.4 \times 10^{-9} \times \left[\frac{\gamma + 1}{2}\right] \Omega_{\oplus}$	$\gamma = 1$	$\gamma = \frac{1 + \omega}{2 + \omega}$	$\gamma = 1$

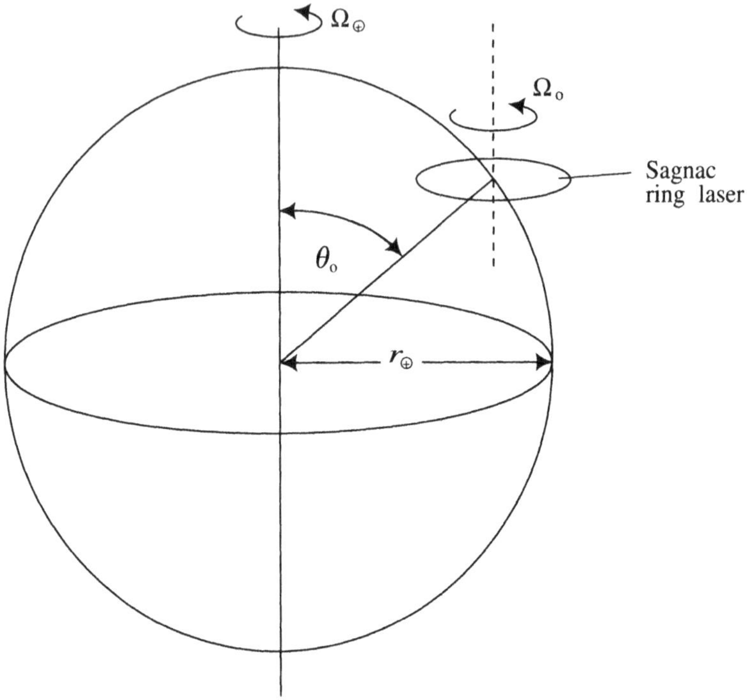

Fig. 4.4
Sagnac ring laser interferometer used to test metric theories of gravity.

The final term, Ω_{Curve}, is due to the fact that we use a curved space metric. Similar 'curved space' physics leads to the bending of starlight and the gravitational red shift, etc.

The application of modern quantum optical tools to problems in gravitational physics calls for heroic and imaginative experimental effort. However, it is clear that such effort will yield rich dividends in both fundamental and applied science.

4.1.4 The Michelson stellar interferometer

Consider the simple double (i.e., double source) interference setup as in Fig. 4.5. In Fig. 4.5(a), we see a binary star 'sending' light to earth with wave vectors $\mathbf{k}$ and $\mathbf{k}'$, and we wish to measure their angular separation, φ.

One way to accomplish this is to collect the light by mirrors M_1 and M_2, as in Fig. 4.5(b), and to beat the light from two stars on the photodetector located at the point P chosen so that the two paths

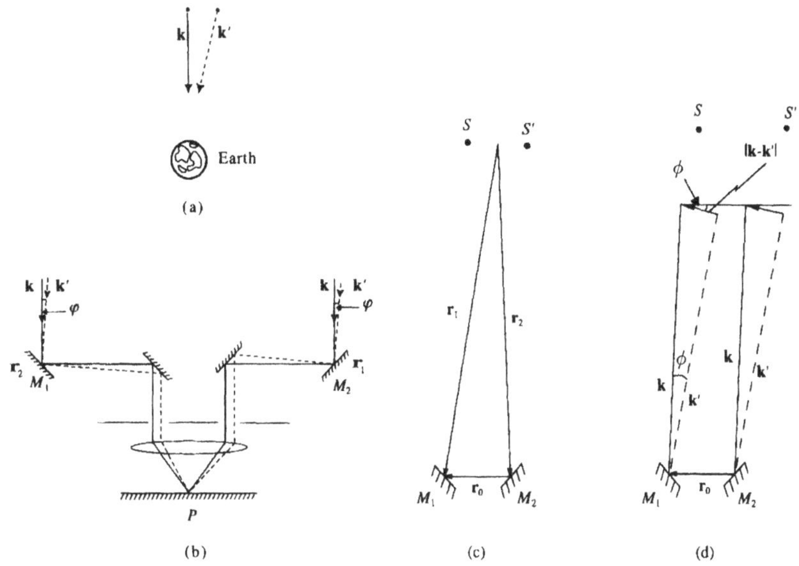

Fig. 4.5
(a) A binary star sending light to earth with wave vectors **k** and **k**′.
(b) Schematics of a Michelson stellar interferometer to measure the angular separation of athe binary star.
(c) Filtered light from star S arrives at mirrors M_1 and M_2 with phase factors $\exp(-i\nu_k t + i\mathbf{k}\cdot\mathbf{r}_1)$ and $\exp(-i\nu_k t + i\mathbf{k}\cdot\mathbf{r}_2)$, respectively, while that from star S'goes as $\exp(-i\nu_{k'} t + i\mathbf{k}'\cdot\mathbf{r}_1)$ and $\exp(-i\nu_{k'} t + i\mathbf{k}'\cdot\mathbf{r}_2)$.
(d) Illustration that for small angls, $(\mathbf{k}-\mathbf{k}')\cdot(\mathbf{r}_1-\mathbf{r}_2) = |\mathbf{k}-\mathbf{k}'|r_0\cos\phi \approx \phi k r_0$, since $|\mathbf{k}-\mathbf{k}'| \simeq k\phi$ and $\cos\phi \simeq 1$.

$\overline{M_1P}$ and $\overline{M_2P}$ are equal. The photocurrent is then given by

$$\begin{aligned}
I &= \kappa\langle E^*E\rangle \\
&= \kappa\langle |E_{\mathbf{k}}(e^{i\mathbf{k}\cdot\mathbf{r}_1} + e^{i\mathbf{k}\cdot\mathbf{r}_2}) + E_{\mathbf{k}'}(e^{i\mathbf{k}'\cdot\mathbf{r}_1} + e^{i\mathbf{k}'\cdot\mathbf{r}_2})|^2\rangle \\
&= \kappa\langle 2(|E_{\mathbf{k}}|^2 + |E_{\mathbf{k}'}|^2) + |E_{\mathbf{k}}|^2[e^{i\mathbf{k}\cdot(\mathbf{r}_1-\mathbf{r}_2)} + \text{c.c.}] \\
&\quad + |E_{\mathbf{k}'}|^2[e^{i\mathbf{k}'\cdot(\mathbf{r}_1-\mathbf{r}_2)} + \text{c.c.}]\rangle, \qquad (4.1.22)
\end{aligned}$$

where we have made the simplifying assumption that the light from the stars has been filtered so that we may take $\nu_{\mathbf{k}} = \nu_{\mathbf{k}'}$ and therefore the temporal factors like $\exp(i\nu_{\mathbf{k}}t)$ and $\exp(i\nu_{\mathbf{k}'}t)$ cancel from Eq. (4.1.22). Furthermore, since the radiation from a star is thermal $\langle E_{\mathbf{k}}\rangle = \langle E_{\mathbf{k}'}\rangle = 0$ and $\langle E^*_{\mathbf{k}}E_{\mathbf{k}'}\rangle = \langle E^*_{\mathbf{k}}\rangle\langle E_{\mathbf{k}'}\rangle = 0$. Finally, we note that κ is an uninteresting constant depending of the characteristics of the photodetector and the distance to the star, etc.

If $\langle |E_{\mathbf{k}}|^2\rangle = \langle |E_{\mathbf{k}'}|^2\rangle = I_0$, we have

$$\begin{aligned}
I &= 2\kappa I_0\left\{2 + \cos[\mathbf{k}\cdot(\mathbf{r}_1-\mathbf{r}_2)] + \cos[\mathbf{k}'\cdot(\mathbf{r}_1-\mathbf{r}_2)]\right\} \\
&= 4\kappa I_0\left\{1 + \cos\left[(\mathbf{k}+\mathbf{k}')\cdot(\mathbf{r}_1-\mathbf{r}_2)/2\right]\right. \\
&\quad \left.\times\cos\left[(\mathbf{k}-\mathbf{k}')\cdot(\mathbf{r}_1-\mathbf{r}_2)/2\right]\right\}. \qquad (4.1.23)
\end{aligned}$$

From Fig. 4.5(d) we see that $(\mathbf{k}-\mathbf{k}')\cdot(\mathbf{r}_1-\mathbf{r}_2) \cong \varphi k r_0$, so that (4.1.23) may be written as

$$I = 4\kappa I_0\left\{1 + \cos\left[(\mathbf{k}+\mathbf{k}')\cdot(\mathbf{r}_1-\mathbf{r}_2)/2\right]\cos\left(\frac{\pi r_0\varphi}{\lambda}\right)\right\}, \qquad (4.1.24)$$

where we have noted $|\mathbf{k}| = |\mathbf{k}'| = 2\pi/\lambda$. Thus, we see that the photocurrent will contain an interference term which is modulated as we vary r_0 and would serve to determine φ varying r_0 until $\pi r_0 \varphi/\lambda = \pi$, etc.

This clever scheme has been applied to several nearby binaries. Unfortunately, atmospheric and instrumental fluctuations enter strongly into the term $\cos\left[(\mathbf{k}+\mathbf{k}')\cdot(\mathbf{r}_1-\mathbf{r}_2)/2\right]$ in Eq. (4.1.24) and limit the utility of the approach. This is where Hanbury-Brown and Twiss make their dramatic entrance.

4.1.5 Hanbury-Brown–Twiss interferometer

The essence of the Hanbury-Brown–Twiss (HB–T) stellar interferometer is to recognize that if we consider two photodetectors at points A_1 and A_2 with position vectors $\mathbf{r}_1$ and $\mathbf{r}_2$, respectively, as in Fig. 4.6, then we have the photocurrents

$$I(\mathbf{r}_i,t)=\kappa\left\{|E_{\mathbf{k}}|^2+|E_{\mathbf{k}'}|^2+\left[E_{\mathbf{k}}E^*_{\mathbf{k}'}e^{i(\mathbf{k}-\mathbf{k}')\cdot\mathbf{r}_i}+\text{c.c.}\right]\right\}\quad(i=1,2), \tag{4.1.25}$$

and there is phase information in the $\exp[i(\mathbf{k}-\mathbf{k}')\cdot\mathbf{r}_i]$ terms.

What if we multiply the currents from two detectors (at A_1 and A_2 in Fig. 4.6)? From Eq. (4.1.25) this will yield

$$\begin{aligned}
&\langle I(\mathbf{r}_1,t)I(\mathbf{r}_2,t)\rangle \\
&=\kappa^2\Big\langle\Big\{|E_{\mathbf{k}}|^2+|E_{\mathbf{k}'}|^2+\left[E_{\mathbf{k}}E^*_{\mathbf{k}'}e^{i(\mathbf{k}-\mathbf{k}')\cdot\mathbf{r}_1}+\text{c.c.}\right]\Big\} \\
&\qquad\times\Big\{|E_{\mathbf{k}}|^2+|E_{\mathbf{k}'}|^2+\left[E_{\mathbf{k}}E^*_{\mathbf{k}'}e^{i(\mathbf{k}-\mathbf{k}')\cdot\mathbf{r}_2}+\text{c.c.}\right]\Big\}\Big\rangle \\
&=\kappa^2\Big\{\left\langle\left(|E_{\mathbf{k}}|^2+|E_{\mathbf{k}'}|^2\right)^2\right\rangle \\
&\qquad+\langle|E_{\mathbf{k}}|^2\rangle\langle|E_{\mathbf{k}'}|^2\rangle\left[e^{i(\mathbf{k}-\mathbf{k}')\cdot(\mathbf{r}_1-\mathbf{r}_2)}+\text{c.c.}\right]\Big\},
\end{aligned} \tag{4.1.26}$$

where we have used the fact that $\langle|E_{\mathbf{k}}|^2E^*_{\mathbf{k}}E_{\mathbf{k}'}\rangle = 0$, etc. Thus we see that the desired low frequency interference term is present; but atmospherically sensitive terms like $\cos[(\mathbf{k}+\mathbf{k}')\cdot(\mathbf{r}_1-\mathbf{r}_2)/2]$ are absent. This is the key insight of Hanbury Brown and Twiss.

It is fair to say, however, that the Hanbury-Brown–Twiss effect created quite a stir when it was first announced. Many questions were voiced, e.g., how can we get phase information by beating photocurrents? Does this not somehow violate quantum mechanics? And what about Dirac's statement that photons only interfere with themselves? The confusion is resolved by considering the quantum theory of photon detection and correlation to which we now turn.

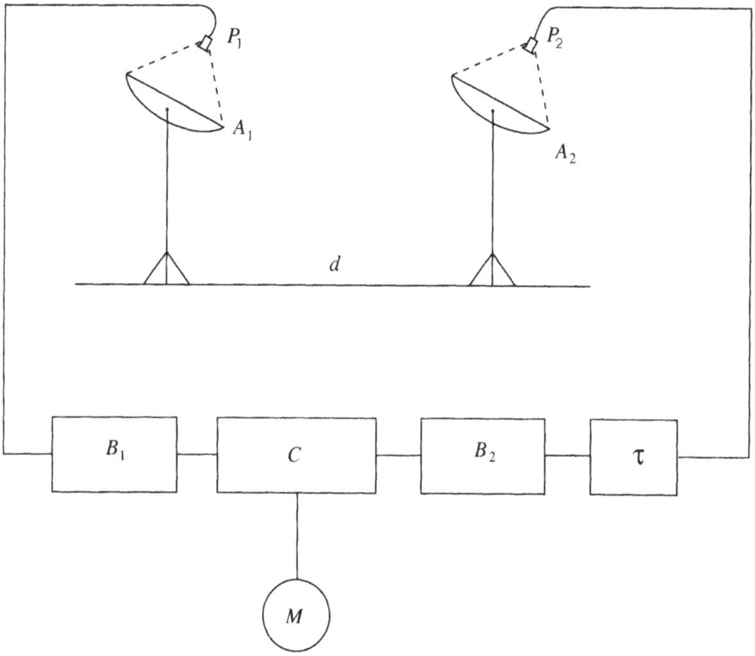

Fig. 4.6
Schematic diagram of the Hanbury Brown-Twiss stellar intensity interferometer. Here P_1 and P_2 are the photodetectors, A_1 and A_2 are the mirrors, B_1 and B_2 are the amplifiers, τ is the delay time, C is a multiplier, and M is the integrator.

4.2 Photon detection and quantum coherence functions

A more complete account of photodetection theory is given in Section 6.5. Here we present a heuristic derivation of photodetection and correlation which is sufficient for the present purposes.

As shown in Chapter 1, the field operator $\mathbf{E}(\mathbf{r},t)$ can be separated into the sum of its positive and negative frequency parts

$$\mathbf{E}(\mathbf{r},t) = \mathbf{E}^{(+)}(\mathbf{r},t) + \mathbf{E}^{(-)}(\mathbf{r},t), \tag{4.2.1}$$

where

$$\mathbf{E}^{(+)}(\mathbf{r},t) = \sum_{\mathbf{k}} \hat{\epsilon}_{\mathbf{k}} \mathscr{E}_{\mathbf{k}} a_{\mathbf{k}} e^{-i\nu_k t + i\mathbf{k}\cdot\mathbf{r}}, \tag{4.2.2}$$

$$\mathbf{E}^{(-)}(\mathbf{r},t) = \sum_{\mathbf{k}} \hat{\epsilon}_{\mathbf{k}} \mathscr{E}_{\mathbf{k}} a_{\mathbf{k}}^{\dagger} e^{i\nu_k t - i\mathbf{k}\cdot\mathbf{r}}. \tag{4.2.3}$$

In the following we shall assume, for simplicity, that the field is linearly polarized so that we deal with the scalar quantities $E^{(+)}(\mathbf{r},t) = \hat{\epsilon}\cdot\mathbf{E}^{(+)}(\mathbf{r},t)$ and $E^{(-)}(\mathbf{r},t) = \hat{\epsilon}\cdot\mathbf{E}^{(-)}(\mathbf{r},t)$.

In the optical region, the detectors usually use the photoelectric effect to make local field measurements. Schematically an atom is placed in the radiation field at position $\mathbf{r}$ in its ground state. The photoelectrons produced by photoionization are then observed. In such absorptive detectors, the measurements are destructive as the photons responsible for producing photoelectrons disappear. In this case, therefore, only the annihilation operator $E^{(+)}$ contributes. The transition probability of the detector atom for absorbing a photon from the field at position $\mathbf{r}$ between times t and $t+dt$ is proportional to $w_1(\mathbf{r},t)dt$, with

$$w_1(\mathbf{r},t) = |\langle f|E^{(+)}(\mathbf{r},t)|i\rangle|^2, \tag{4.2.4}$$

where $|i\rangle$ is the initial state of the field before the detection process and $|f\rangle$ is the final state in which the field could be found after the process. The final state of the field is never measured. We can therefore sum over all the final states

$$\begin{aligned} w_1(\mathbf{r},t) &= \sum_f |\langle f|E^{(+)}(\mathbf{r},t)|i\rangle|^2 \\ &= \sum_f \langle i|E^{(-)}(\mathbf{r},t)|f\rangle\langle f|E^{(+)}(\mathbf{r},t)|i\rangle \\ &= \langle i|E^{(-)}(\mathbf{r},t)E^{(+)}(\mathbf{r},t)|i\rangle, \end{aligned} \tag{4.2.5}$$

where in the last line we use the completeness relation

$$\sum_f |f\rangle\langle f| = 1. \tag{4.2.6}$$

The photon counting rate w_1 is therefore proportional to the expectation value of the positive definite Hermitian operator $E^{(-)}(\mathbf{r},t)E^{(+)}(\mathbf{r},t)$ taken in the initial state of the field $|i\rangle$. In practice, however, we almost never know precisely the state $|i\rangle$. Since the precise knowledge of the field does not usually exist, we resort to a statistical description by averaging over all the possible realizations of the initial field

$$w_1(\mathbf{r},t) = \sum_i P_i\langle i|E^{(-)}(\mathbf{r},t)E^{(+)}(\mathbf{r},t)|i\rangle. \tag{4.2.7}$$

If we introduce the density operator for the field

$$\rho = \sum_i P_i|i\rangle\langle i|, \tag{4.2.8}$$

we can rewrite Eq. (4.2.7) as

$$w_1(\mathbf{r},t) = \mathrm{Tr}[\rho E^{(-)}(\mathbf{r},t)E^{(+)}(\mathbf{r},t)]. \tag{4.2.9}$$

We define the first-order correlation function of the field

$$G^{(1)}(\mathbf{r}_1,\mathbf{r}_2;t_1,t_2) = \mathrm{Tr}[\rho E^{(-)}(\mathbf{r}_1,t_1)E^{(+)}(\mathbf{r}_2,t_2)] \\ = \langle E^{(-)}(\mathbf{r}_1,t_1)E^{(+)}(\mathbf{r}_2,t_2)\rangle. \quad (4.2.10)$$

Usually we deal with statistically stationary fields in optics, i.e., the correlation functions of the field are invariant under displacements of the time variable. The correlation function $G^{(1)}(\mathbf{r}_1,\mathbf{r}_2;t_1,t_2)$ then depends on t_1 and t_2 only through the time difference $\tau = t_2 - t_1$, i.e.,

$$G^{(1)}(\mathbf{r}_1,\mathbf{r}_2;t_1,t_2) \equiv G^{(1)}(\mathbf{r}_1,\mathbf{r}_2;\tau). \quad (4.2.11)$$

In terms of $G^{(1)}$, the counting rate w_1 is given by

$$w_1 = G^{(1)}(\mathbf{r},\mathbf{r};0). \quad (4.2.12)$$

We now consider the joint counting rate at two photodetectors at $\mathbf{r}_1$ and $\mathbf{r}_2$. The joint probability of observing one photoionization at point $\mathbf{r}_2$ between t_2 and $t_2 + dt_2$ and another one at point $\mathbf{r}_1$ between t_1 and $t_1 + dt_1$ with $t_1 \leq t_2$ is proportional to $w_2(\mathbf{r}_1,t_1;\mathbf{r}_2,t_2)dt_1dt_2$, where

$$w_2(\mathbf{r}_1,t_1;\mathbf{r}_2,t_2) = |\langle f|E^{(+)}(\mathbf{r}_2,t_2)E^{(+)}(\mathbf{r}_1,t_1)|i\rangle|^2. \quad (4.2.13)$$

It follows, on summing over all the final states and averaging over all the possible realizations of the initial field as before, that

$$w_2(\mathbf{r}_1,t_1;\mathbf{r}_2,t_2) \\ = \mathrm{Tr}[\rho E^{(-)}(\mathbf{r}_1,t_1)E^{(-)}(\mathbf{r}_2,t_2)E^{(+)}(\mathbf{r}_2,t_2)E^{(+)}(\mathbf{r}_1,t_1)]. \quad (4.2.14)$$

The joint probability of photodetection is thus governed by the second-order quantum mechanical correlation function

$$G^{(2)}(\mathbf{r}_1,\mathbf{r}_2,\mathbf{r}_3,\mathbf{r}_4;t_1,t_2,t_3,t_4) = \mathrm{Tr}[\rho E^{(-)}(\mathbf{r}_1,t_1)E^{(-)}(\mathbf{r}_2,t_2) \\ \times E^{(+)}(\mathbf{r}_3,t_3)E^{(+)}(\mathbf{r}_4,t_4)] \\ = \langle E^{(-)}(\mathbf{r}_1,t_1)E^{(-)}(\mathbf{r}_2,t_2) \\ \times E^{(+)}(\mathbf{r}_3,t_3)E^{(+)}(\mathbf{r}_4,t_4)\rangle. \quad (4.2.15)$$

In general, we can define the nth-order correlation function

$$G^{(n)}(\mathbf{r}_1,\ldots,\mathbf{r}_n,\mathbf{r}_{n+1},\ldots,\mathbf{r}_{2n};t_1,\ldots,t_n,t_{n+1},\ldots,t_{2n}) \\ = \mathrm{Tr}[\rho E^{(-)}(\mathbf{r}_1,t_1)\ldots E^{(-)}(\mathbf{r}_n,t_n)E^{(+)}(\mathbf{r}_{n+1},t_{n+1})\ldots E^{(+)}(\mathbf{r}_{2n},t_{2n})] \\ = \langle E^{(-)}(\mathbf{r}_1,t_1)\ldots E^{(-)}(\mathbf{r}_n,t_n)E^{(+)}(\mathbf{r}_{n+1},t_{n+1})\ldots E^{(+)}(\mathbf{r}_{2n},t_{2n})\rangle. \\ (4.2.16)$$

In this definition of the nth-order correlation function we have included equal numbers of creation and destruction operators because such correlation functions are measured in typical multi-photon counting experiments.

It is apparent from the above discussion that the correlation functions of the field operators which are encountered in any photon detection experiment based on the photoelectric effect are in normal order (that is, with all the destruction operators on the right and all the creation operators on the left). For example, the average light intensity at point **r** at time t is

$$\langle I(\mathbf{r},t)\rangle = \langle E^{(-)}(\mathbf{r},t)E^{(+)}(\mathbf{r},t)\rangle, \tag{4.2.17}$$

and the measured intensity–intensity correlation function is equal to $\langle E^{(-)}(\mathbf{r},t)E^{(-)}(\mathbf{r},t)E^{(+)}(\mathbf{r},t)E^{(+)}(\mathbf{r},t)\rangle$, which is different from $\langle I(\mathbf{r},t)$ $I(\mathbf{r},t)\rangle$.

We can define the quantum mechanical first- and second-order degrees of coherence at the position **r** as

$$g^{(1)}(\mathbf{r},\tau) = \frac{\langle E^{(-)}(\mathbf{r},t)E^{(+)}(\mathbf{r},t+\tau)\rangle}{\sqrt{\langle E^{(-)}(\mathbf{r},t)E^{(+)}(\mathbf{r},t)\rangle\langle E^{(-)}(\mathbf{r},t+\tau)E^{(+)}(\mathbf{r},t+\tau)\rangle}}, \tag{4.2.18}$$

$$g^{(2)}(\mathbf{r},\tau) = \frac{\langle E^{(-)}(\mathbf{r},t)E^{(-)}(\mathbf{r},t+\tau)E^{(+)}(\mathbf{r},t+\tau)E^{(+)}(\mathbf{r},t)\rangle}{\langle E^{(-)}(\mathbf{r},t)E^{(+)}(\mathbf{r},t)\rangle\langle E^{(-)}(\mathbf{r},t+\tau)E^{(+)}(\mathbf{r},t+\tau)\rangle}, \tag{4.2.19}$$

where we have assumed the field to be statistically stationary. In the definition of $g^{(2)}(\mathbf{r},\tau)$, we have chosen not only the normal ordering of the field operators in the numerator but a certain time ordering. This time ordering is a consequence of the way the photoelectron rate is calculated above (note that $t_2 \geq t_1$ in Eq. (4.2.14)). Considerably simpler forms for these quantities are obtained in the special case when the radiation field consists of only a single mode. Then most factors cancel when the mode expansions for $E^{(+)}$ and $E^{(-)}$ are substituted from Eqs. (4.2.2) and (4.2.3) into Eqs. (4.2.18) and (4.2.19), leaving

$$g^{(1)}(\tau) = \frac{\langle a^\dagger(t)a(t+\tau)\rangle}{\langle a^\dagger a\rangle}, \tag{4.2.20}$$

$$g^{(2)}(\tau) = \frac{\langle a^\dagger(t)a^\dagger(t+\tau)a(t+\tau)a(t)\rangle}{\langle a^\dagger a\rangle^2}. \tag{4.2.21}$$

Since only the normally ordered correlation functions are involved in the photodetection processes, the P-representation $P(\alpha,\alpha^*)$ forms a correspondence between classical and quantum coherence theory. This happens because the quantum mechanical expectation values of the normally ordered functions can be calculated from the P-representation just as we would evaluate the corresponding classical coherence function from a classical distribution function. The P-representation, however, does not have all the properties of a classical distribution function. In particular, as discussed in Section 3.1, the P-representation is not nonnegative definite. Light fields for which the P-representation is not a well-behaved distribution will exhibit nonclassical features of light. We will discuss some of them in Section 4.4.

We now derive the normalized correlation function $g^{(2)}(\tau)$ for thermal and coherent fields within the framework of the quantum theory of coherence. The P-representation of a single-mode thermal field is given by a Gaussian distribution (Eq. (3.1.26)):

$$P(\alpha, \alpha^*) = \frac{1}{\pi\langle n\rangle} \exp(-|\alpha|^2/\langle n\rangle). \tag{4.2.22}$$

We then have

$$g^{(2)}(0) = \frac{\int P(\alpha, \alpha^*)|\alpha|^4 d^2\alpha}{[\int P(\alpha, \alpha^*)|\alpha|^2 d^2\alpha]^2} = 2. \tag{4.2.23}$$

However, for a laser operating far above threshold, the field is in a coherent state $|\alpha_0\rangle$, for which (see Eq. (3.1.28))

$$P(\alpha, \alpha^*) = \delta^{(2)}(\alpha - \alpha_0). \tag{4.2.24}$$

The normalized correlation then is

$$g^{(2)}(0) = 1. \tag{4.2.25}$$

4.3 First-order coherence and Young-type double-source experiments

4.3.1 Young's double-slit experiment

One of the classic experiments that exhibits the first-order coherence properties of light is Young's double-slit experiment (see Fig. 4.7). The complex field generated by a quasimonochromatic light source is split at the screen S_1 by placing an opaque screen across the beam with pinholes at points P_1 and P_2. The positive frequency part of the field operator at a point P on the screen S_2 at time t may be approximated by a linear superposition of the field operators present at P_1 and P_2 at earlier times:

$$E^{(+)}(\mathbf{r}, t) = K_1 E^{(+)}(\mathbf{r}_1, t - t_1) + K_2 E^{(+)}(\mathbf{r}_2, t - t_2), \tag{4.3.1}$$

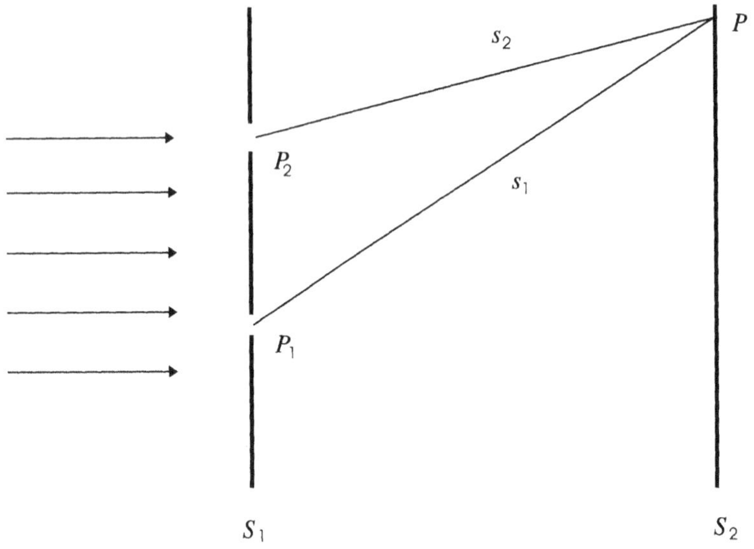

Fig. 4.7
Schematic diagram of an idealized Young's double-slit experiment.

where $t_i = s_i/c$ $(i = 1, 2)$ is the time needed for the light to travel from the pinhole P_i to the point P and $\mathbf{r}_1$ and $\mathbf{r}_2$ are the position vectors at the pinholes P_1 and P_2, respectively. The coefficients K_1 and K_2 depend on the size and geometry of the pinholes. From diffraction theory it follows that K_1 and K_2 are purely imaginary numbers.

A photodetector placed at the point P measures the intensity

$$\begin{aligned}\langle I(\mathbf{r},t)\rangle &= \mathrm{Tr}[\rho E^{(-)}(\mathbf{r},t)E^{(+)}(\mathbf{r},t)] \\ &= |K_1|^2\mathrm{Tr}[\rho E^{(-)}(\mathbf{r}_1,t-t_1)E^{(+)}(\mathbf{r}_1,t-t_1)] \\ &\quad +|K_2|^2\mathrm{Tr}[\rho E^{(-)}(\mathbf{r}_2,t-t_2)E^{(+)}(\mathbf{r}_2,t-t_2)] \\ &\quad +2\mathrm{Re}\{K_1^*K_2\mathrm{Tr}[\rho E^{(-)}(\mathbf{r}_1,t-t_1)E^{(+)}(\mathbf{r}_2,t-t_2)]\}. \end{aligned} \tag{4.3.2}$$

We can rewrite this equation in terms of the first-order correlation function
$G^{(1)}(\mathbf{r}_1, \mathbf{r}_2; t_1, t_2)$ in the following way:

$$\begin{aligned}\langle I(\mathbf{r},t)\rangle &= |K_1|^2G^{(1)}(\mathbf{r}_1,\mathbf{r}_1;t-t_1,t-t_1) \\ &\quad +|K_2|^2G^{(1)}(\mathbf{r}_2,\mathbf{r}_2;t-t_2,t-t_2) \\ &\quad +2\mathrm{Re}[K_1^*K_2G^{(1)}(\mathbf{r}_1,\mathbf{r}_2;t-t_1,t-t_2)]. \end{aligned} \tag{4.3.3}$$

For statistically stationary fields, expression (4.3.3) for the average intensity at the point P becomes

$$\begin{aligned}\langle I(\mathbf{r},t)\rangle &= |K_1|^2G^{(1)}(\mathbf{r}_1,\mathbf{r}_1;0) + |K_2|^2G^{(1)}(\mathbf{r}_2,\mathbf{r}_2;0) \\ &\quad +2\mathrm{Re}[K_1^*K_2G^{(1)}(\mathbf{r}_1,\mathbf{r}_2;\tau)], \end{aligned} \tag{4.3.4}$$

where $\tau = t_1 - t_2$. The average intensity $\langle I(\mathbf{r}, t)\rangle$ is therefore independent of the time t.

The first two terms in Eq. (4.3.4) represent the average intensities at the point P due to the light field at the pinholes P_1 and P_2, respectively. The last term, however, gives a contribution due to fields at both the pinholes and is responsible for the interference. In order to see this clearly we set

$$\langle I^{(i)}(\mathbf{r})\rangle = |K_i|^2 G^{(1)}(\mathbf{r}_i, \mathbf{r}_i; 0) \qquad (i = 1, 2). \tag{4.3.5}$$

We next define the normalized first-order correlation function

$$g^{(1)}(\mathbf{r}_1, \mathbf{r}_2; \tau) = \frac{G^{(1)}(\mathbf{r}_1, \mathbf{r}_2; \tau)}{\sqrt{G^{(1)}(\mathbf{r}_1, \mathbf{r}_1; 0) G^{(1)}(\mathbf{r}_2, \mathbf{r}_2; 0)}}. \tag{4.3.6}$$

In terms of $g^{(1)}(\mathbf{r}_1, \mathbf{r}_2; \tau)$, Eq. (4.3.4) can be rewritten as

$$\begin{aligned}\langle I(\mathbf{r}, t)\rangle &= \langle I^{(1)}(\mathbf{r})\rangle + \langle I^{(2)}(\mathbf{r})\rangle \\ &\quad + 2[\langle I^{(1)}(\mathbf{r})\rangle\langle I^{(2)}(\mathbf{r})\rangle]^{1/2} \mathrm{Re}[g^{(1)}(\mathbf{r}_1, \mathbf{r}_2; \tau)].\end{aligned} \tag{4.3.7}$$

Next we set

$$g^{(1)}(\mathbf{r}_1, \mathbf{r}_2; \tau) = |g^{(1)}(\mathbf{r}_1, \mathbf{r}_2; \tau)| e^{i\alpha(\mathbf{r}_1, \mathbf{r}_2; \tau) - i\nu_0\tau}, \tag{4.3.8}$$

where $\alpha(\mathbf{r}_1, \mathbf{r}_2; \tau) = \arg[g^{(1)}(\mathbf{r}_1, \mathbf{r}_2; \tau)] + \nu_0\tau$ and ν_0 is the field frequency. We then obtain

$$\begin{aligned}\langle I(\mathbf{r}, t)\rangle &= \langle I^{(1)}(\mathbf{r})\rangle + \langle I^{(2)}(\mathbf{r})\rangle + 2[\langle I^{(1)}(\mathbf{r})\rangle\langle I^{(2)}(\mathbf{r})\rangle]^{1/2} \\ &\quad \times |g^{(1)}(\mathbf{r}_1, \mathbf{r}_2; \tau)| \cos[\alpha(\mathbf{r}_1, \mathbf{r}_2; \tau) - \nu_0\tau].\end{aligned} \tag{4.3.9}$$

For a quasimonochromatic source of light, $\langle I^{(1)}(\mathbf{r})\rangle$, $\langle I^{(2)}(\mathbf{r})\rangle$, $|g^{(1)}(\mathbf{r}_1, \mathbf{r}_2; \tau)|$, and $\alpha(\mathbf{r}_1, \mathbf{r}_2; \tau)$ vary slowly with respect to position on the screen. However, the cosine term varies rapidly due to the term $\nu_0\tau = \nu_0(s_1 - s_2)/c$ and will lead to sinusoidal variation of intensity on the screen.

The physical meaning of $g^{(1)}(\mathbf{r}_1, \mathbf{r}_2; \tau)$ can be understood if we consider the visibility of the interference fringes on the screen. The visibility, which is a measure of the sharpness of the interference fringes, is defined as

$$U = \frac{\langle I(\mathbf{r})\rangle_{\max} - \langle I(\mathbf{r})\rangle_{\min}}{\langle I(\mathbf{r})\rangle_{\max} + \langle I(\mathbf{r})\rangle_{\min}}, \tag{4.3.10}$$

where $\langle I(\mathbf{r})\rangle_{\max}$ and $\langle I(\mathbf{r})\rangle_{\min}$ represent the maximum and minimum average intensity, respectively, in the neighborhood of the point P. To a good approximation for $\cos[\alpha(\mathbf{r}_1, \mathbf{r}_2; \tau) - \nu_0\tau]$ they are equal to $+1$ and -1 in Eq. (4.3.9). We then obtain

$$U = \frac{2[\langle I^{(1)}(\mathbf{r})\rangle\langle I^{(2)}(\mathbf{r})\rangle]^{1/2}}{\langle I^{(1)}(\mathbf{r})\rangle + \langle I^{(2)}(\mathbf{r})\rangle}|g^{(1)}(\mathbf{r}_1,\mathbf{r}_2;\tau)|, \tag{4.3.11}$$

i.e., the visibility of the fringes is proportional to the magnitude of $g^{(1)}(\mathbf{r}_1,\mathbf{r}_2;\tau)$, which is called the complex degree of coherence. In particular, when the averaged intensities of the two beams are equal, $\langle I^{(1)}(\mathbf{r})\rangle = \langle I^{(2)}(\mathbf{r})\rangle$, the visibility U is equal to $|g^{(1)}(\mathbf{r}_1,\mathbf{r}_2;\tau)|$. Thus when $g^{(1)}(\mathbf{r}_1,\mathbf{r}_2;\tau) = 0$, no interference fringes are formed in the region around P and it would be implied that the two light beams reaching the point P are mutually incoherent. A maximum visibility of the fringes is obtained around P when $|g^{(1)}(\mathbf{r}_1,\mathbf{r}_2;\tau)| = 1$ and the two light beams reaching P are mutually completely coherent. This happens when

$$\langle E^{(-)}(\mathbf{r}_1,t)E^{(+)}(\mathbf{r}_2,t+\tau)\rangle = \mathscr{E}^*(\mathbf{r}_1,t)\mathscr{E}(\mathbf{r}_2,t+\tau). \tag{4.3.12}$$

The intermediate cases $0 < |g^{(1)}(\mathbf{r}_1,\mathbf{r}_2;\tau)| < 1$ characterize partial coherence.

As an example, the emission from a Doppler-broadened spectral light source, such as that from a thermal lamp, is described by

$$G^{(1)}(\mathbf{r}_1,\mathbf{r}_2;\tau) = \mathscr{E}_0^2 \exp(-i\nu_0\tau - \tau^2/2\tau_c^2), \tag{4.3.13}$$

where τ_c is a constant. It is therefore clear that as the path difference $c\tau$ becomes much larger than $c\tau_c$, $|g^{(1)}(\mathbf{r}_1,\mathbf{r}_2;\tau)| = \exp(-\tau^2/2\tau_c^2)$ goes to zero and the interference fringes disappear. The constant τ_c, which will be related to the light bandwidth (shown below), is thus a measure of the coherence time of the light.

An important property of the first-order correlation function

$$G^{(1)}(\mathbf{r},\mathbf{r};\tau) = \langle E^{(-)}(\mathbf{r},t)E^{(+)}(\mathbf{r},t+\tau)\rangle$$

is that it forms a Fourier transform pair with the power spectrum $S(\mathbf{r},\nu)$ of the statistically stationary field at the position $\mathbf{r}$, i.e.,

$$S(\mathbf{r},\nu) = \frac{1}{\pi}\mathrm{Re}\int_0^\infty d\tau G^{(1)}(\mathbf{r},\mathbf{r};\tau)e^{i\nu\tau}. \tag{4.3.14}$$

We therefore need the first-order correlation function at positive τ to compute the power spectrum.

We consider the example of the Doppler-broadened spectral light source whose first-order correlation function is given by Eq. (4.3.13). The power spectrum for the light source, as computed from Eq. (4.3.14) is therefore equal to

$$S(\mathbf{r},\nu) = \frac{\mathscr{E}_0^2\tau_c}{\sqrt{2\pi}}\exp[-(\nu-\nu_0)^2\tau_c^2/2]. \tag{4.3.15}$$

This is a Gaussian spectrum centered around $\nu = \nu_0$ with a full-width at half-maximum equal to $2\sqrt{2\ln 2}/\tau_c$. Thus $1/\tau_c$, which is the inverse of the coherence time of the light field, is a measure of the light bandwidth.

4.3.2 *Young's experiment with light from two atoms**

Consider the Young-type experiment shown in Fig. 4.8. There we see two atoms at locations S and S'. At $t = 0$ both atoms are allowed to interact with a single photon, designated by $|\phi\rangle$, and one or other of the atoms may be excited. In this way we prepare the state

$$\alpha\left(|a, b'\rangle + |b, a'\rangle\right)|0\rangle + \beta|b, b'\rangle|\phi\rangle, \tag{4.3.16}$$

where $|a\rangle$, $|b\rangle$, and $|a'\rangle$, $|b'\rangle$ denote the excited and ground states of atoms at S and S', and α and β are the probability amplitudes for the states associated with excited and ground state atoms, respectively. Thus, with a probability $|\alpha|^2$ we have prepared the state

$$|\psi(0)\rangle = \frac{1}{\sqrt{2}}(|a, b'\rangle + |b, a'\rangle)|0\rangle \tag{4.3.17}$$

by single-photon absorption. Later in time, this state will decay into the state

$$|\psi(\infty)\rangle = \frac{1}{\sqrt{2}}|b, b'\rangle(|\gamma\rangle + |\gamma'\rangle), \tag{4.3.18}$$

where $|\gamma\rangle$ and $|\gamma'\rangle$ denote the photon states associated with emission from sites S and S'. For present purposes, it will suffice to take $|\gamma\rangle$ and $|\gamma'\rangle$ as plane wave states $|1_{\mathbf{k}}\rangle$ and $|1_{\mathbf{k}'}\rangle$ where $\mathbf{k}/k$ and $\mathbf{k}'/k'$ are the unit vectors from S and S' to the detectors at $\mathbf{r}$, see Fig. 4.8. However, the question of how to most simply choose the states $|\gamma\rangle$ and $|\gamma'\rangle$ while still being faithful to the physics is an important and subtle one, and is treated in Appendix 4.A.*

The correlation function $G^{(1)}(\mathbf{r}, \mathbf{r}; t, t)$ now takes the form

$$G^{(1)}(\mathbf{r}, \mathbf{r}; t, t) = \langle\psi(\infty)|E^{(-)}(\mathbf{r})E^{(+)}(\mathbf{r})|\psi(\infty)\rangle = G^{(1)}(\mathbf{r}, \mathbf{r}; 0), \tag{4.3.19}$$

where we have noted that the time-dependent factors cancel because $\nu_k = \nu_{k'}$. By completeness as in Eq. (1.5.16), this may be written as

$$G^{(1)}(\mathbf{r}, \mathbf{r}; 0) = \Psi_{\mathscr{E}}^*(\mathbf{r})\Psi_{\mathscr{E}}(\mathbf{r}), \tag{4.3.20}$$

* See Scully and Drühl, *Phys. Rev.* A **25**, 2208 (1982).

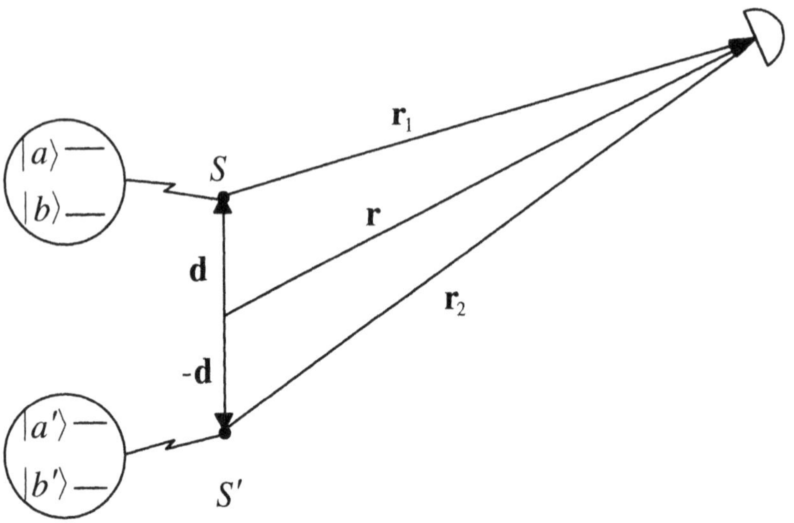

Fig. 4.8 Schematic diagram of an Young-type experiment via light from two atoms.

where

$$\Psi_{\mathscr{E}}(\mathbf{r}) = \langle 0|E^{(+)}(\mathbf{r})|\psi(\infty)\rangle = \frac{\mathscr{E}_{\mathbf{k}}}{\sqrt{2}}\left(e^{i\mathbf{k}\cdot\mathbf{r}} + e^{i\mathbf{k}'\cdot\mathbf{r}}\right). \tag{4.3.21}$$

Thus we see that an interference pattern is obtained which is governed by

$$G^{(1)}(\mathbf{r},\mathbf{r};0) = \mathscr{E}_{\mathbf{k}}^2\{1+\cos[(\mathbf{k}-\mathbf{k}')\cdot\mathbf{r}]\}, \tag{4.3.22}$$

and as is discussed in Appendix 4.A, this can be written as

$$G^{(1)}(\mathbf{r},\mathbf{r};0) = \mathscr{E}_{\mathbf{k}}^2\left[1+\cos\left(\frac{2k}{r}\mathbf{d}\cdot\mathbf{r}\right)\right] = \mathscr{E}_{\mathbf{k}}^2[1+\cos(2kxd/D)], \tag{4.3.23}$$

which is the usual result.

4.4 Second-order coherence

In the previous section we considered the first-order correlation functions and their properties. For fields with identical spectral properties, it is not possible to distinguish the nature of the light source from only the first-order correlation function. For example, a laser beam and the light generated by a conventional thermal source can both have the same first-order coherence properties. The same, however, is not true when we consider the second- and higher-order coherence properties of the light sources. We therefore turn to the applications of the second-order correlation functions of the field.

4.4.1 The physics behind the Hanbury-Brown–Twiss effect

Armed with a theory of photoelectron correlations, we now return to the Hanbury-Brown–Twiss effect. Let us begin by considering the state $|\psi\rangle = |1_\mathbf{k}, 1_{\mathbf{k}'}\rangle$, i.e., the case of two independent photons one having momentum $\mathbf{k}$ and one having momentum $\mathbf{k}'$. Now it is clear that the second-order correlation function may be written as

$$\begin{aligned}&G^{(2)}(\mathbf{r}_1, \mathbf{r}_2; t, t)\\&= \langle 1_\mathbf{k}, 1_{\mathbf{k}'}|E^{(-)}(\mathbf{r}_1, t)E^{(-)}(\mathbf{r}_2, t)E^{(+)}(\mathbf{r}_2, t)E^{(+)}(\mathbf{r}_1, t)|1_\mathbf{k}, 1_{\mathbf{k}'}\rangle, \quad (4.4.1)\end{aligned}$$

and using $\sum_{\{n\}} |\{n\}\rangle\langle\{n\}| = 1$ this becomes

$$\begin{aligned}G^{(2)}(\mathbf{r}_1, \mathbf{r}_2; t, t) = &\sum_{\{n\}} \langle 1_\mathbf{k}, 1_{\mathbf{k}'}|E^{(-)}(\mathbf{r}_1, t)E^{(-)}(\mathbf{r}_2, t)|\{n\}\rangle\\&\times \langle\{n\}|E^{(+)}(\mathbf{r}_2, t)E^{(+)}(\mathbf{r}_1, t)|1_\mathbf{k}, 1_{\mathbf{k}'}\rangle. \quad (4.4.2)\end{aligned}$$

As $|1_\mathbf{k}, 1_{\mathbf{k}'}\rangle$ is a two-photon state which is annihilated by $E^{(+)}(\mathbf{r}_2, t)$ $E^{(+)}(\mathbf{r}_1, t)$, only the $|0\rangle\langle 0|$ term survives.

In view of the above, we see that for the case of two single photons we may write

$$G^{(2)}(\mathbf{r}_1, \mathbf{r}_2; t, t) = \Psi^{(2)^*}(\mathbf{r}_1, \mathbf{r}_2; t, t)\Psi^{(2)}(\mathbf{r}_1, t; \mathbf{r}_2, t), \quad (4.4.3)$$

where

$$\Psi^{(2)}(\mathbf{r}_1, t; \mathbf{r}_2, t) = \langle 0|E^{(+)}(\mathbf{r}_2, t)E^{(+)}(\mathbf{r}_1, t)|1_\mathbf{k}, 1_{\mathbf{k}'}\rangle. \quad (4.4.4)$$

From

$$E^{(+)}(\mathbf{r}_i, t) = \mathscr{E}_\mathbf{k}\left(a_\mathbf{k}e^{-ivt+i\mathbf{k}\cdot\mathbf{r}_i} + a_{\mathbf{k}'}e^{-ivt+i\mathbf{k}'\cdot\mathbf{r}_i}\right) \quad (i = 1, 2), \quad (4.4.5)$$

these become

$$\begin{aligned}\Psi^{(2)}(\mathbf{r}_1, t; \mathbf{r}_2, t) &= \mathscr{E}_\mathbf{k}^2 e^{-2ivt}\langle 0|a_\mathbf{k}e^{i\mathbf{k}\cdot\mathbf{r}_1}a_{\mathbf{k}'}e^{i\mathbf{k}'\cdot\mathbf{r}_2}|1_\mathbf{k}, 1_{\mathbf{k}'}\rangle\\&\quad + \mathscr{E}_\mathbf{k}^2 e^{-2ivt}\langle 0|a_{\mathbf{k}'}e^{i\mathbf{k}'\cdot\mathbf{r}_1}a_\mathbf{k}e^{i\mathbf{k}\cdot\mathbf{r}_2}|1_\mathbf{k}, 1_{\mathbf{k}'}\rangle\\&= \mathscr{E}_\mathbf{k}^2 e^{-2ivt}\left(e^{i\mathbf{k}\cdot\mathbf{r}_1+\mathbf{k}'\cdot\mathbf{r}_2} + e^{i\mathbf{k}'\cdot\mathbf{r}_1+\mathbf{k}\cdot\mathbf{r}_2}\right), \quad (4.4.6)\end{aligned}$$

and

$$G^{(2)}(\mathbf{r}_1, \mathbf{r}_2; t, t) = 2\mathscr{E}_\mathbf{k}^4\left\{1 + \cos[(\mathbf{k} - \mathbf{k}') \cdot (\mathbf{r}_1 - \mathbf{r}_2)]\right\}. \quad (4.4.7)$$

$\Psi^{(2)}(\mathbf{r}_1, \mathbf{r}_2; t, t) =$

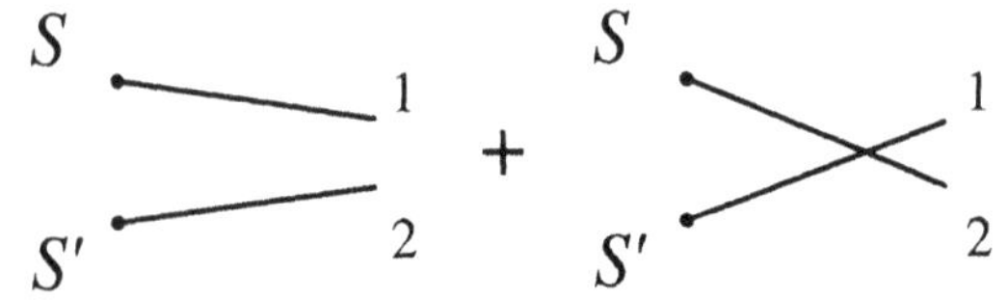

Fig. 4.9 Pictorial representation of terms in Eq. (4.4.6).

Photon-correlation interferometry from two atoms

Consider next the case of two atoms at S and S' as in Fig. 4.9 in which both atoms are initially excited, that is,

$$|\psi(0)\rangle = |a, a'\rangle|0\rangle. \tag{4.4.8}$$

Then after many decay times this goes into

$$|\psi(\infty)\rangle = |b, b'\rangle|\gamma, \gamma'\rangle, \tag{4.4.9}$$

where, as in the previous section, we may take $|\gamma\rangle = |1_{\mathbf{k}}\rangle$, $|\gamma'\rangle = |1_{\mathbf{k}'}\rangle$. The two-photon correlation function is then identical with that given by Eqs. (4.4.1) and (4.4.7).

Next we turn to incoherent atom excitation in order to display the real power of the HB–T effect. Specifically, suppose we excite the atoms at S and S' by electron impact. Then at some instant, call it $t = 0$, we will have a state of the form

$$|\psi(0)\rangle = \Big[|\alpha|e^{i\varphi}|a, a'\rangle + |\beta|\left(e^{i\theta}|a, b'\rangle + e^{i\theta'}|b, a'\rangle\right) + |\gamma||b, b'\rangle\Big] \otimes |0\rangle, \tag{4.4.10}$$

which, see Appendix 4.A for a discussion of the spherical-versus plane-wave description of interference physics, evolves into

$$|\psi(\infty)\rangle = \Big[|\alpha|e^{i\varphi}|1_{\mathbf{k}}, 1_{\mathbf{k}'}\rangle + |\beta|\left(e^{i\theta}|1_{\mathbf{k}}\rangle + e^{i\theta'}|1_{\mathbf{k}'}\rangle\right) + |\gamma||0\rangle\Big] \otimes|b, b'\rangle, \tag{4.4.11}$$

where φ, θ, and θ' are random phases due, for example, to random excitation times of the atoms.

In such a case, the interference terms in the first-order correlation function will be multiplied by a random phase factor, which we must average over, that is

$$\left[G^{(1)}(\mathbf{r}, \mathbf{r}; t)\right]_{\text{interference cross terms}} \longrightarrow \left\langle e^{-i(\theta-\theta')}\right\rangle e^{-i(\mathbf{k}-\mathbf{k}')\cdot\mathbf{r}}. \tag{4.4.12}$$

This vanishes due to the random nature of θ and θ'. Thus one might conclude that atoms described by Eq. (4.4.11) would never yield spatial interference. This is not the case. If we use Eq. (4.4.11) to calculate $G^{(2)}(\mathbf{r}_1, \mathbf{r}_2; t, t)$, we find

$$\begin{aligned} & G^{(2)}(\mathbf{r}_1, \mathbf{r}_2; t, t) \\ & = |\alpha|^2 \langle 1_{\mathbf{k}}, 1_{\mathbf{k}'} | E^{(-)}(\mathbf{r}_1) E^{(-)}(\mathbf{r}_2) E^{(+)}(\mathbf{r}_2) E^{(+)}(\mathbf{r}_1) | 1_{\mathbf{k}}, 1_{\mathbf{k}'} \rangle \\ & = 2|\alpha|^2 \mathscr{E}_{\mathbf{k}}^4 \{1 + \cos[(\mathbf{k} - \mathbf{k}') \cdot (\mathbf{r}_1 - \mathbf{r}_2)]\}. \end{aligned} \tag{4.4.13}$$

Here we see again that the random phases which destroy first-order coherence do not affect second-order HB–T type coherences.

The Hanbury-Brown–Twiss effect for thermal and laser light

We now turn to the case of many-photon states associated with thermal and laser light and calculate the HB–T correlations for two such sources at S and S'.

As before, we look for the rate of coincidences in the photocount rates of detectors at $\mathbf{r}_1$ and $\mathbf{r}_2$ governed by the second-order correlation function

$$G^{(2)}(\mathbf{r}_1, \mathbf{r}_2; t, t) = \langle E^{(-)}(\mathbf{r}_1, t) E^{(-)}(\mathbf{r}_2, t) E^{(+)}(\mathbf{r}_2, t) E^{(+)}(\mathbf{r}_1, t) \rangle, \tag{4.4.14}$$

and consider the case in which the essential terms in the electric field operators $E(\mathbf{r}_i, t)$ $(i = 1, 2)$ are given by

$$E^{(+)}(\mathbf{r}_i, t) = \mathscr{E}_{\mathbf{k}} \left(a_{\mathbf{k}} e^{-i\nu t + i\mathbf{k}\cdot\mathbf{r}_i} + a_{\mathbf{k}'} e^{-i\nu t + i\mathbf{k}'\cdot\mathbf{r}_i} \right), \tag{4.4.15}$$

where $\mathbf{k}$ and $\mathbf{k}'$ are the wave vectors of light from the two sources S and S'. Furthermore, as before, we are considering only equal frequency intervals such that $\nu = c|\mathbf{k}| = c|\mathbf{k}'|$. Noting that only 'pairwise' operator orderings remain for thermal light, phase-diffused laser light, and light from two atoms (see Appendix 4.B), we have

$$\begin{aligned} & G^{(2)}(\mathbf{r}_1, \mathbf{r}_2; t, t) \\ & = \mathscr{E}_{\mathbf{k}}^4 \Big\langle \left(a_{\mathbf{k}}^\dagger e^{-i\mathbf{k}\cdot\mathbf{r}_1} + a_{\mathbf{k}'}^\dagger e^{-i\mathbf{k}'\cdot\mathbf{r}_1} \right) \left(a_{\mathbf{k}}^\dagger e^{-i\mathbf{k}\cdot\mathbf{r}_2} + a_{\mathbf{k}'}^\dagger e^{-i\mathbf{k}'\cdot\mathbf{r}_2} \right) \\ & \quad \times \left(a_{\mathbf{k}} e^{i\mathbf{k}\cdot\mathbf{r}_2} + a_{\mathbf{k}'} e^{i\mathbf{k}'\cdot\mathbf{r}_2} \right) \left(a_{\mathbf{k}} e^{i\mathbf{k}\cdot\mathbf{r}_1} + a_{\mathbf{k}'} e^{i\mathbf{k}'\cdot\mathbf{r}_1} \right) \Big\rangle \\ & = \mathscr{E}_{\mathbf{k}}^4 \Big\langle a_{\mathbf{k}}^\dagger a_{\mathbf{k}}^\dagger a_{\mathbf{k}} a_{\mathbf{k}} + a_{\mathbf{k}'}^\dagger a_{\mathbf{k}'}^\dagger a_{\mathbf{k}'} a_{\mathbf{k}'} \\ & \quad + a_{\mathbf{k}}^\dagger a_{\mathbf{k}'}^\dagger a_{\mathbf{k}} a_{\mathbf{k}'} \left[1 + e^{-i(\mathbf{k}-\mathbf{k}')\cdot(\mathbf{r}_1 - \mathbf{r}_2)} \right] \\ & \quad + a_{\mathbf{k}'}^\dagger a_{\mathbf{k}}^\dagger a_{\mathbf{k}'} a_{\mathbf{k}} \left[1 + e^{i(\mathbf{k}-\mathbf{k}')\cdot(\mathbf{r}_1 - \mathbf{r}_2)} \right] \Big\rangle. \end{aligned} \tag{4.4.16}$$

If we assume $\langle n_{\mathbf{k}}\rangle = \langle n_{\mathbf{k}'}\rangle \equiv \langle n\rangle$ and likewise $\langle n_{\mathbf{k}}^2\rangle = \langle n_{\mathbf{k}'}^2\rangle \equiv \langle n^2\rangle$, we may write Eq. (4.4.16) as

$$G^{(2)}(\mathbf{r}_1,\mathbf{r}_2;t,t)$$
$$= 2\mathscr{E}_{\mathbf{k}}^4\Big(\langle n^2\rangle - \langle n\rangle + \langle n\rangle^2\Big\{1+\cos\big[(\mathbf{k}-\mathbf{k}')\cdot(\mathbf{r}_1-\mathbf{r}_2)\big]\Big\}\Big). \quad (4.4.17)$$

Next we calculate $\langle n^2\rangle$ for the two different cases in question: stars and phase-diffused laser light.

(a) Stars: the light from stars is thermal, therefore

$$\langle n^2\rangle = 2\langle n\rangle^2 + \langle n\rangle, \qquad \langle n\rangle = [\exp(\hbar\nu/k_B T) - 1]^{-1},$$

and Eq. (4.4.17) yields

$$G^{(2)}(\mathbf{r}_1,\mathbf{r}_2;t,t)$$
$$= 2\mathscr{E}_{\mathbf{k}}^4\Big(2\langle n\rangle^2 + \langle n\rangle^2\Big\{1+\cos\big[(\mathbf{k}-\mathbf{k}')\cdot(\mathbf{r}_1-\mathbf{r}_2)\big]\Big\}\Big). \quad (4.4.18)$$

The last term in Eq. (4.4.18) is the Hanbury-Brown–Twiss term which allows us to measure the angle between $\mathbf{k}$ and $\mathbf{k}'$ as in the discussion following Eq. (4.1.24).

(b) Lasers: far above threshold, the photon statistics for the lasers are Poissonian, therefore, $\langle n^2\rangle = \langle n\rangle^2 + \langle n\rangle$, and we have

$$G^{(2)}(\mathbf{r}_1,\mathbf{r}_2;t,t)$$
$$= 2\mathscr{E}_{\mathbf{k}}^4\Big(\langle n\rangle^2 + \langle n\rangle^2\Big\{1+\cos\big[(\mathbf{k}-\mathbf{k}')\cdot(\mathbf{r}_1-\mathbf{r}_2)\big]\Big\}\Big). \quad (4.4.19)$$

So, in both cases, we can measure the angular separation without the troublesome $\cos\big[(\mathbf{k}+\mathbf{k}')\cdot(\mathbf{r}_1-\mathbf{r}_2)/2\big]$-type terms which plague the Michelson stellar interferometer.

The Hanbury-Brown–Twiss spatial interference effect for neutrons

By now, it is clear (contrary to what one frequently hears and reads) that the HB–T interference pattern, i.e., the interference cross terms in $G^{(2)}(\mathbf{r}_1,\mathbf{r}_2)$, has nothing to do with the boson nature of the photons. That is, the HB–T interference cross terms are present for radiation emitted by two independent atoms or lasers as shown in the previous two sections. In both of these cases, 'boson clumping' is absent.

Furthermore, it is clear from Eq. (4.4.4) and Fig. 4.9 that the effect carries over for neutrons as well. In such a case, the photon annihilation operators such as that given by Eq. (4.4.5) are replaced by a fermion operator of the form

$$\hat{\psi}(\mathbf{r}_i, t) = c_{\mathbf{k}} e^{-i\nu t + i\mathbf{k}\cdot\mathbf{r}_i} + c_{\mathbf{k}'} e^{-i\nu t + i\mathbf{k}'\cdot\mathbf{r}_i}, \tag{4.4.20}$$

where the relevant fermion annihilation operators $c_{\mathbf{k}}$ and $c_{\mathbf{k}'}$ now obey the anticommutation relations

$$c_{\mathbf{k}} c_{\mathbf{k}'}^{\dagger} + c_{\mathbf{k}'}^{\dagger} c_{\mathbf{k}} = \delta_{\mathbf{k},\mathbf{k}'}, \tag{4.4.21}$$

$$c_{\mathbf{k}}^{\dagger} c_{\mathbf{k}'}^{\dagger} + c_{\mathbf{k}'}^{\dagger} c_{\mathbf{k}}^{\dagger} = 0, \tag{4.4.22}$$

$$c_{\mathbf{k}} c_{\mathbf{k}'} + c_{\mathbf{k}'} c_{\mathbf{k}} = 0. \tag{4.4.23}$$

Now Eq. (4.4.4) is replaced by the two-fermion wave function

$$\begin{aligned}\Psi^{(2)}(\mathbf{r}_1, t; \mathbf{r}_2, t) &= \langle 0|\hat{\psi}(\mathbf{r}_2, t)\hat{\psi}(\mathbf{r}_1, t)|\psi\rangle \\ &= e^{-2i\nu t}\langle 0|c_{\mathbf{k}} e^{i\mathbf{k}\cdot\mathbf{r}_2} c_{\mathbf{k}'} e^{i\mathbf{k}'\cdot\mathbf{r}_1}|1_{\mathbf{k}}, 1_{\mathbf{k}'}\rangle \\ &\quad + e^{-2i\nu t}\langle 0|c_{\mathbf{k}'} e^{i\mathbf{k}'\cdot\mathbf{r}_2} c_{\mathbf{k}} e^{i\mathbf{k}\cdot\mathbf{r}_1}|1_{\mathbf{k}}, 1_{\mathbf{k}'}\rangle,\end{aligned} \tag{4.4.24}$$

and because

$$\begin{aligned}\langle 0|c_{\mathbf{k}} c_{\mathbf{k}'}|1_{\mathbf{k}}, 1_{\mathbf{k}'}\rangle &= \langle 0|c_{\mathbf{k}} c_{\mathbf{k}'} c_{\mathbf{k}}^{\dagger} c_{\mathbf{k}'}^{\dagger}|0\rangle \\ &= -\langle 0|c_{\mathbf{k}} c_{\mathbf{k}}^{\dagger}|0\rangle\langle 0|c_{\mathbf{k}'} c_{\mathbf{k}'}^{\dagger}|0\rangle \\ &= -1,\end{aligned} \tag{4.4.25}$$

while an equivalent operator algebra for the second term in (4.4.24) yields +1, the fermion–fermion correlation function takes the form

$$G^{(2)}(\mathbf{r}_1, \mathbf{r}_2; t, t) = 2\Big\{1 - \cos[(\mathbf{k} - \mathbf{k}')\cdot(\mathbf{r}_1 - \mathbf{r}_2)]\Big\}. \tag{4.4.26}$$

Thus we see that the Hanbury-Brown–Twiss effect works as well for two radiative point sources, S and S' of Fig. 4.9, emitting neutrons or β particles, as it does for γ rays or α particles. The only difference is the sign of the interference term.

4.4.2 *Detection and measurement of squeezed states via homodyne detection*

As seen earlier, direct photon count experiments, in which light of photon number distribution $p(n)$ falls directly on a photodetector, provide information about the mean photon number and higher-order moments only. Such intensity measurements, therefore, are not particularly sensitive to squeezing but to antibunching and sub- or super-Poissonian statistics, which can also occur for nonsqueezed fields. Detection of squeezed states, on the other hand, requires a phase-sensitive scheme that measures the variance of a quadrature of the field. In this section, we consider the problem of detection of squeezed states of radiation via homodyne detection.

The schematic arrangement for homodyne detection is shown in

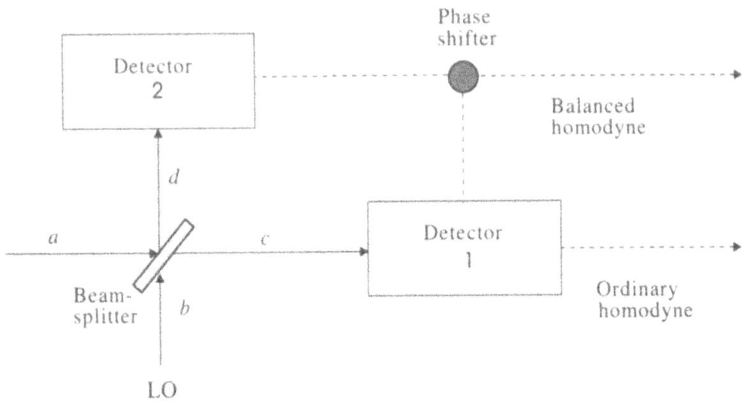

Fig. 4.10
Schematic diagram for homodyne detection.

Fig. 4.10. The input field is superimposed on the field from a local oscillator (LO) at a lossless beam-splitter of transmissivity T and reflectivity R such that $R + T = 1$. The input and the oscillator modes are described by the annihilation operators a and b, respectively. Then denoting the two out-modes reaching photodetectors 1 and 2 by c and d, respectively, we have

$$c = \sqrt{T}\, a + i\sqrt{1-T}\, b, \tag{4.4.27}$$
$$d = i\sqrt{1-T}\, a + \sqrt{T}\, b. \tag{4.4.28}$$

There is a $\pi/2$ phase shift between the reflected and the transmitted waves for a symmetric beam-splitter which we have included by the factor i in Eqs. (4.4.27) and (4.4.28). The signals measured by the two detectors are determined by the operators

$$c^\dagger c = T a^\dagger a + (1-T) b^\dagger b + i\sqrt{T(1-T)}(a^\dagger b - b^\dagger a), \tag{4.4.29}$$
$$d^\dagger d = (1-T) a^\dagger a + T b^\dagger b - i\sqrt{T(1-T)}(a^\dagger b - b^\dagger a). \tag{4.4.30}$$

The frequency of the LO is equal to the input frequency so that the above operators do not have any time dependence. In the following we discuss the ordinary and balanced homodyne detectors.

Ordinary homodyne detection

In ordinary homodyne detection, the transmissivity of the beam-splitter is close to unity, i.e.,

$$T \gg R, \tag{4.4.31}$$

and only the photocurrent from detector 1 is measured. The LO mode is excited into a large amplitude coherent state $|\beta_l\rangle$ with phase ϕ_l. From Eq. (4.4.29) the signal reaching detector 1 is obtained as

$$\langle c^\dagger c\rangle = T\langle a^\dagger a\rangle + (1-T)|\beta_l|^2 - 2\sqrt{T(1-T)}|\beta_l|\langle X(\phi_l+\pi/2)\rangle, \tag{4.4.32}$$

where

$$X(\phi) \equiv X_\phi = \frac{1}{2}(ae^{-i\phi} + a^\dagger e^{i\phi}). \tag{4.4.33}$$

We see that the signal contains the transmitted part of the input photons, reflected LO field, and most importantly, an interference term between the input field and the LO field. It is precisely this interference term that contains a quadrature of the input field depending upon the phase of the LO. In this detection scheme, a strong LO is used so that

$$(1-T)|\beta_l|^2 \gg T\langle a^\dagger a\rangle. \tag{4.4.34}$$

The inequalities (4.4.31) and (4.4.34) together imply that almost all the input field reaches the photodetector but the fraction of the LO field reaching the detector is still dominant. We can, therefore, neglect the first term in Eq. (4.4.32) and the mean number of photons in mode c is

$$\langle n_c\rangle = (1-T)|\beta_l|^2 - 2\sqrt{T(1-T)}|\beta_l|\langle X(\phi_l+\pi/2)\rangle. \tag{4.4.35}$$

The first term constitutes a known constant value which can be subtracted from the signal and the remaining signal contains the quadrature of the input only.

The input and the LO modes are independent, i.e., $\langle ab\rangle = \langle a\rangle\langle b\rangle$. The photon number fluctuations can then be calculated in a straightforward manner using Eqs. (4.4.29) and (4.4.30)

$$(\Delta n_c)^2 = (1-T)|\beta_l|^2\{(1-T) + 4T[\Delta X(\phi_l+\pi/2)]^2\}. \tag{4.4.36}$$

In obtaining Eq. (4.4.36), we have used the inequality (4.4.34) and retained terms of second order in $|\beta_l|$. The signal noise is now seen to contain reflected LO noise (first term) and the transmitted input quadrature noise (second term). When the input is incoherent (or vacuum), $[\Delta X(\phi_l+\pi/2)]^2 = 1/4$, and the remaining term represents the LO shot noise. The squeezing condition for the input is

$$[\Delta X(\phi_l+\pi/2)]^2 < 1/4 \tag{4.4.37}$$

for certain values of the LO phase ϕ_l for which either quadrature X_1 or X_2 is squeezed.

In practice, the input is first blocked to determine the shot-noise level. The input is then allowed to reach the beam-splitter and the variance is determined with reference to the shot-noise level. Squeezing therefore manifests itself in sub-Poissonian statistics in homodyne detection.

Note, however, that intensity measurements in homodyne detection are quite different from those in direct detection, i.e., (a) intensity fluctuations in this case directly measure the fluctuations in a quadrature of the input, and (b) the signal and its variance depend upon the local oscillator phase angle, which is an external parameter.

Balanced homodyne detection

In the discussion following Eq. (4.4.35), we assumed a perfectly coherent LO field and the oscillator excess noise has been neglected. The LO shot noise and the excess noise that enter through the reflectivity of the beam-splitter cannot be suppressed in ordinary homodyne detection because T, in principle, can never be 1. The LO noise can therefore limit ordinary homodyne detection. In particular, the detection is not quantum limited if the transmitted input noise is smaller than the reflected oscillator noise, as may be the case when the input noise is too small.

An alternative scheme is based on two-port homodyne detection which balances the output from the two ports of the beam-splitter. The fact that the noninterference terms at the two ports have the same sign and the interference terms appear with opposite signs (see Eqs. (4.4.29) and (4.4.30)) can be exploited to completely eliminate the noninterference terms. In this scheme, a 50/50 beam-splitter is used and the difference of two photodetector measurements is obtained. The output signal is determined by the operator

$$n_{cd} = c^\dagger c - d^\dagger d = -i(a^\dagger b - b^\dagger a). \tag{4.4.38}$$

The measured signal then is

$$\langle n_{cd} \rangle = -2|\beta_l| \langle X(\phi_l + \pi/2) \rangle. \tag{4.4.39}$$

We see that the LO contribution to the signal has been eliminated and only the interference between the LO mean field and the input quadrature survives. The variance of the output signal can be found to be

$$(\Delta n_{cd})^2 = 4|\beta_l|^2 [\Delta X(\phi_l + \pi/2)]^2. \tag{4.4.40}$$

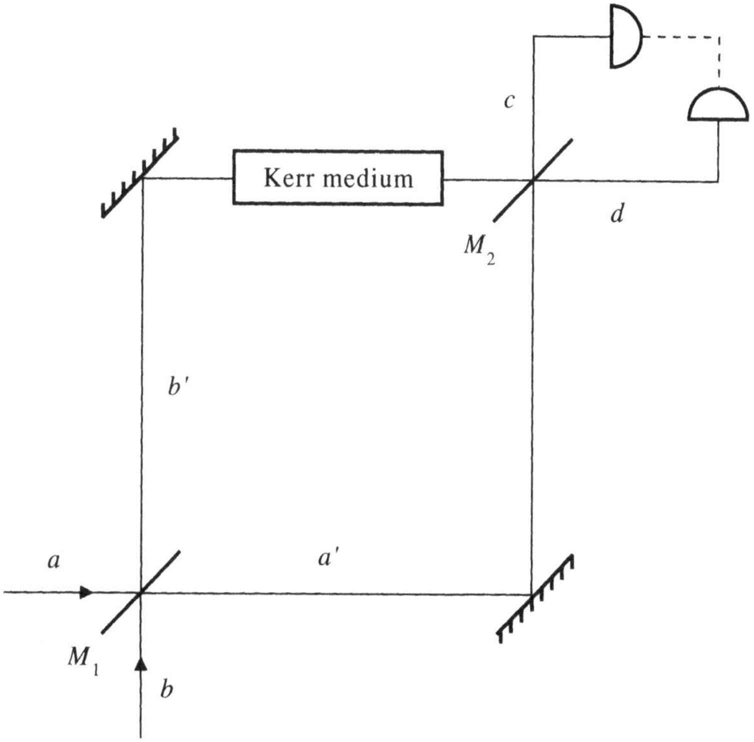

Fig. 4.11
A Mach-Zehnder interferometer with a phase sensitive element in the upper arm operating in the balanced mode. The operators a and b are the annihilation operators for the signal and local oscillator modes.

Here once again we assume a strong LO. The dominant term now is only due to the interference between the input signal noise and the LO power, and the LO noise is eliminated completely. This makes the strong LO condition less stringent in this case.

Measurement of phase uncertainty*

The use of balanced homodyne detection in precision interferometry yields several interesting results. Here we discuss the application of balanced homodyne detection in the measurement of phase uncertainty of optical signals.

The system is depicted in Fig. 4.11, where we see a Mach–Zehnder interferometer with a phase sensitive element in the upper arm operating in the balanced mode. The phase sensitive element introduces a phase shift ϕ_p, e.g., by a Kerr effect medium discussed in Section 19.2.

We assume M_1 and M_2 in Fig. 4.11 to be 50/50 beam-splitters

* This section follows, in part, the unpublished lecture notes of B. Yurke, to whom the authors are indebted.

and assume the two path lengths between them to be equal. The annihilation operators of the various modes in Fig. 4.11 are related to each other via

$$a' = \frac{1}{\sqrt{2}}(a + ib), \tag{4.4.41}$$

$$b' = \frac{1}{\sqrt{2}}(ia + b), \tag{4.4.42}$$

and

$$\begin{aligned} c &= \frac{1}{\sqrt{2}}\left(a' + ib'e^{i\phi_p}\right) \\ &= \frac{1}{2}\left[\left(1 - e^{i\phi_p}\right)a + i\left(1 + e^{i\phi_p}\right)b\right], \end{aligned} \tag{4.4.43}$$

$$\begin{aligned} d &= \frac{1}{\sqrt{2}}\left(ia' + b'e^{i\phi_p}\right) \\ &= \frac{1}{2}\left[i\left(1 + e^{i\phi_p}\right)a - \left(1 - e^{i\phi_p}\right)b\right]. \end{aligned} \tag{4.4.44}$$

Here, as before, we assume a $\pi/2$ phase shift for the reflected field.

The output signal in the balanced homodyne detector is given by the operator

$$\begin{aligned} n_{cd} &= c^\dagger c - d^\dagger d \\ &= (b^\dagger b - a^\dagger a)\cos\phi_p - (a^\dagger b + b^\dagger a)\sin\phi_p. \end{aligned} \tag{4.4.45}$$

If the local oscillator mode is in a large amplitude coherent state $|\beta_l\rangle$ and the signal mode is in a vacuum state $|0\rangle$, the signal is

$$\langle n_{cd}\rangle = n_l \cos\phi_p, \tag{4.4.46}$$

where $n_l = |\beta_l|^2$.

It is interesting to note that, for $\phi_p = \pi/2$,

$$n_{cd} = -(a^\dagger b + b^\dagger a), \tag{4.4.47}$$

i.e., n_{cd} does not depend on the photon number operators and the system in Fig. 4.11 is essentially equivalent to a balanced homodyne detector of Fig. 4.10.

Now, for the signal mode in a vacuum state, the difference operator (4.4.47) has a variance

$$(\Delta n_{cd})^2 = |\beta_l|^2 = n_l. \tag{4.4.48}$$

This can be related to the phase error by noting that, from Eq. (4.4.46), we have

$$\frac{\partial\langle n_{cd}\rangle}{\partial\phi_p} = -n_l \sin\phi_p, \tag{4.4.49}$$

and, since on balance ($\phi_p = \pi/2$),

$$\left|\frac{\partial\langle n_{cd}\rangle}{\partial\phi_p}\right| = n_l. \tag{4.4.50}$$

Hence the phase error is given by

$$\begin{aligned}\Delta\phi &= \frac{\Delta n_{cd}}{|\partial\langle n_{cd}\rangle/\partial\phi_p|} = \frac{\sqrt{n_l}}{n_l} \\ &= \frac{1}{\sqrt{n_l}}.\end{aligned} \tag{4.4.51}$$

If we now take the signal to be a squeezed vacuum state, $|0, \xi\rangle$ with $\xi = r\exp(i\theta)$,

$$\begin{aligned}\langle n_{cd}\rangle &= (n_l + \sinh^2 r)\cos\phi_p \\ &\cong n_l\cos\phi_p.\end{aligned} \tag{4.4.52}$$

On balance, $\phi_p = \pi/2$, $|\partial\langle n_{cd}\rangle/\partial\phi_p| = n_l$, and

$$(\Delta n_{cd})^2 = n_l[\cosh 2r - \cos(\theta - 2\phi_l)\sinh 2r] + \sinh^2 r, \tag{4.4.53}$$

where we have used Eqs. (2.7.11)–(2.7.13). If we take $\theta = 2\phi_l$, Eq. (4.4.53) becomes

$$(\Delta n_{cd})^2 = n_l e^{-2r} + \sinh^2 r, \tag{4.4.54}$$

and, for $n_l \gg 1$, we may neglect the $\sinh^2 r$ term in Eq. (4.4.54) to yield the reduced phase noise

$$\begin{aligned}\Delta\phi &= \frac{\Delta n_{cd}}{|\partial\langle n_{cd}\rangle/\partial\phi_p|} \\ &= \frac{e^{-r}}{\sqrt{n_l}}.\end{aligned} \tag{4.4.55}$$

4.4.3 Interference of two photons

We now describe an experiment in which the joint probability for the detection of two photons at two points is measured as a function of the separation between the points. This two-photon interference experiment is an example of the intensity correlation experiment where the predictions of the quantum theoretical analysis are quite different from the corresponding predictions of the classical coherence theory. The experimental results agree with the predictions of the quantum coherence theory for the choice of parameters where the classical and quantum theories yield different results. The existence of nonclassical effects in two-photon interference is just one example of a large number of related phenomena where the quantum nature of light is exhibited

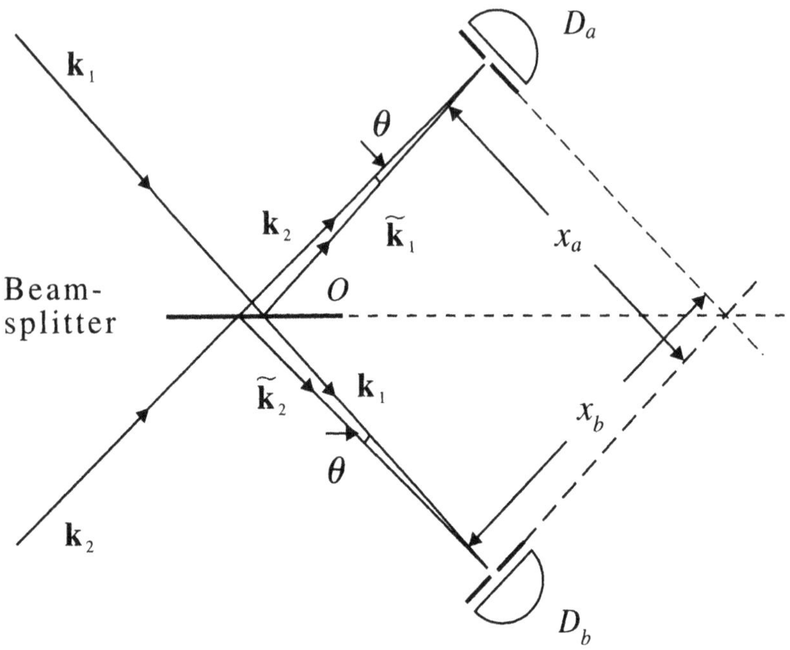

Fig. 4.12 Schematic diagram for the two-photon interference experiment. (From Z. Y. Ou and L. Mandel, *Phys. Rev. Lett.* **62**, 2941 (1989).)

explicitly. Some of these phenomena will be discussed later in this book, particularly in Chapter 21.

In the two-photon interferometer, two randomly phased light waves of narrow bandwidth impinge simultaneously on the surface of a beam-splitter. The reflected and transmitted waves are brought together on the detectors D_a and D_b located at $\mathbf{r}_a$ and $\mathbf{r}_b$, respectively, as shown in Fig. 4.12. The outputs, after amplification, are sent to a correlator. The measured coincidence rate provides a measure of the joint detection probability $P(x_a, x_b)\delta x_a \delta x_b$ of detection at x_a and x_b within δx_a and δx_b, respectively. Here x_a and x_b are the projections of $\mathbf{r}_a$ and $\mathbf{r}_b$ onto the vectors $\mathbf{k}_2 - \tilde{\mathbf{k}}_1$ and $\tilde{\mathbf{k}}_2 - \mathbf{k}_1$, respectively (see Fig. 4.12), where $\tilde{\mathbf{k}}_1$, $\tilde{\mathbf{k}}_2$ are the wave vectors corresponding to $\mathbf{k}_1$, $\mathbf{k}_2$ after reflection at the beam-splitter.

As discussed in Section 4.2, the joint detection probability is governed by w_2. We thus have

$$P(x_a, x_b) = \kappa_a \kappa_b \langle E^{(-)}(x_a) E^{(-)}(x_b) E^{(+)}(x_b) E^{(+)}(x_a) \rangle, \qquad (4.4.56)$$

where κ_a and κ_b are factors which depend on the characteristics of the detectors. We now calculate the joint detection probability for incident correlated photons within the framework of both quantum and classical coherence theories.

If we treat O (see Fig. 4.12) as the origin, we can express the positive frequency part of the fields $E^{(+)}(x_a)$ and $E^{(+)}(x_b)$ in the form

$$E^{(+)}(x_a) = \mathscr{E}\left(i\sqrt{R}a_1 e^{i\tilde{\mathbf{k}}_1\cdot\mathbf{r}_a} + \sqrt{T}a_2 e^{i\mathbf{k}_2\cdot\mathbf{r}_a}\right), \tag{4.4.57}$$

$$E^{(+)}(x_b) = \mathscr{E}\left(\sqrt{T}a_1 e^{i\mathbf{k}_1\cdot\mathbf{r}_b} + i\sqrt{R}a_2 e^{i\tilde{\mathbf{k}}_2\cdot\mathbf{r}_b}\right), \tag{4.4.58}$$

where R and T are the reflectivity and the transmittivity of the beam-splitter, a_1 and a_2 are the destruction operators for the input fields at the beam-splitter, and $\mathscr{E} = (\hbar\nu/2\epsilon_0 V)^{1/2}$. If the beam-splitter is 50/50, then $R = T = 1/2$. Equations (4.4.57) and (4.4.58) then simplify and are given in the form

$$E^{(+)}(x_a) = \frac{\mathscr{E}}{\sqrt{2}}(ia_1 e^{i\tilde{\mathbf{k}}_1\cdot\mathbf{r}_a} + a_2 e^{i\mathbf{k}_2\cdot\mathbf{r}_a}), \tag{4.4.59}$$

$$E^{(+)}(x_b) = \frac{\mathscr{E}}{\sqrt{2}}(a_1 e^{i\mathbf{k}_1\cdot\mathbf{r}_b} + ia_2 e^{i\tilde{\mathbf{k}}_2\cdot\mathbf{r}_b}). \tag{4.4.60}$$

The initial state of the field for single photons is the two-photon Fock state $|1_1, 1_2\rangle$. Such a state can be prepared in the process of degenerate parametric amplification in a nonlinear medium (Chapter 16). The joint detection probability density, Eq. (4.4.56), is therefore given by

$$\begin{aligned} P(x_a, x_b) &= \kappa_a\kappa_b\langle 1_1, 1_2|E^{(-)}(x_a)E^{(-)}(x_b)E^{(+)}(x_b)E^{(+)}(x_a)|1_1, 1_2\rangle \\ &= \frac{1}{2}\kappa_a\kappa_b\mathscr{E}^4 \\ &\quad \{1 - \cos[(\mathbf{k}_2 - \tilde{\mathbf{k}}_1)\cdot\mathbf{r}_a - (\tilde{\mathbf{k}}_2 - \mathbf{k}_1)\cdot\mathbf{r}_b]\} \end{aligned} \tag{4.4.61}$$

where we have substituted for $E^{(+)}(x_a)$ and $E^{(+)}(x_b)$ and their Hermitian conjugates from Eqs. (4.4.59) and (4.4.60). If the angles θ between $\tilde{\mathbf{k}}_1$ and $\mathbf{k}_2$ and between $\tilde{\mathbf{k}}_2$ and $\mathbf{k}_1$ are very small, then the associated interference pattern has a fringe spacing given by

$$L \approx \frac{2\pi}{|\tilde{\mathbf{k}}_1 - \mathbf{k}_2|} = \frac{2\pi}{|\tilde{\mathbf{k}}_2 - \mathbf{k}_1|} \approx \frac{2\pi}{k\theta}, \tag{4.4.62}$$

where $k = |\mathbf{k}_1| = |\mathbf{k}_2|$, and we obtain

$$P(x_a, x_b) = \frac{1}{2}\kappa_a\kappa_b\mathscr{E}^4\{1 - \cos[2\pi(x_a - x_b)/L]\}. \tag{4.4.63}$$

Thus the joint detection probability exhibits a cosine modulation in $x_a - x_b$ with visibility

$$U = \frac{P_{\max} - P_{\min}}{P_{\max} + P_{\min}} = 1. \tag{4.4.64}$$

Therefore, there is an interference between two two-photon amplitudes associated with both photons being reflected and both photons being transmitted.

A unity visibility implies that if a photon is detected at the position x_a then there are certain positions x_b where the other photon cannot be found, and vice versa. This situation is in contrast to classical optics (as seen below) which predicts a nonvanishing optical field at both positions x_a and x_b.

Next we calculate the visibility by treating the incident fields classically. We can replace the operators a_1 and a_2 in Eqs. (4.4.59) and (4.4.60) by the classical c-number amplitudes α_1 and α_2, respectively. We also assume that the fields have random phases. This is a reasonable assumption because the single-photon states have arbitrary phase. The classical ensemble averages of phase-dependent quantities, such as α_1 and $|\alpha_1|^2\alpha_2$, therefore vanish.

The joint detection probability $P(x_a, x_b)$ is now given by Eq. (4.4.56) where $E^{(+)}$ and $E^{(-)}$ are classical c-number fields and the angle brackets indicate the classical ensemble average. It is readily seen that

$$P(x_a, x_b) = \frac{1}{4}\kappa_a\kappa_b\{\langle(I_1 + I_2)^2\rangle - 2\langle I_1 I_2\rangle \cos[2\pi(x_a - x_b)/2]\}, \tag{4.4.65}$$

where $I_1 = \mathscr{E}^2|\alpha_1|^2$ and $I_2 = \mathscr{E}^2|\alpha_2|^2$. The visibility U of the interference is given by

$$U = \frac{2\langle I_1 I_2\rangle}{\langle I_1^2\rangle + \langle I_2^2\rangle + 2\langle I_1 I_2\rangle}. \tag{4.4.66}$$

As $\langle I_1^2\rangle + \langle I_2^2\rangle \geq 2\langle I_1 I_2\rangle$, it follows that

$$U \leq \frac{1}{2}, \tag{4.4.67}$$

which gives a classical limit. This shows that the visibility cannot exceed 50 percent in contradiction to the prediction of the quantum mechanical result.

An observation of a larger than 50 percent visibility therefore corresponds to nonclassical behavior. A visibility of over 75 percent has been observed in the two-photon interference experiments.

4.4.4 *Photon antibunching, Poissonian, and sub-Poissonian light*

In Section 4.2 we showed that a correspondence between the quantum and classical coherence theories can be established via *P*-representation.

However, as we discussed, the P-representation does not have all the properties of a classical distribution function. Thus it is possible that certain inequalities for the correlation functions which implicitly assume a well-defined probability distribution may not be satisfied. A violation of these inequalities for certain radiation fields would therefore provide explicit evidence for the quantum nature of light. In this section we consider some examples of such fields.

In the classical coherence theory, the field operators are replaced by c-number fields. For such classical fields, it follows from the Schwarz inequality, $|\langle a^* b\rangle|^2 \leq \langle|a|^2\rangle\langle|b|^2\rangle$ (with $a = I(\mathbf{r}, t)$ and $b = I(\mathbf{r}, t+\tau)$), that

$$|\langle I(\mathbf{r}, t) I(\mathbf{r}, t+\tau)\rangle|^2 \leq \langle I^2(\mathbf{r}, t)\rangle\langle I^2(\mathbf{r}, t+\tau)\rangle. \tag{4.4.68}$$

The corresponding inequality in the quantum coherence theory is obtained by replacing the product of intensities within the angle brackets by the corresponding normally ordered operators, i.e.,

$$|\langle : I(\mathbf{r}, t) I(\mathbf{r}, t+\tau) :\rangle|^2 \leq \langle : I^2(\mathbf{r}, t) :\rangle\langle : I^2(\mathbf{r}, t+\tau) :\rangle, \tag{4.4.69}$$

where : : represents normal ordering, i.e., the creation operators to the left and the annihilation operators to the right. This inequality is satisfied for fields with a well-defined P-representation. It follows from the definition of $g^{(2)}(\tau)$ (Eq. (4.2.21)), that, for statistically stationary fields, this inequality can be recast in the following simple form

$$g^{(2)}(\tau) \leq g^{(2)}(0). \tag{4.4.70}$$

This inequality was seen to be satisfied by thermal and coherent light. We recall from the definition of $g^{(2)}(\tau)$ that it is a measure of the photon correlations between some time t and a later time $t+\tau$. When the field satisfies the inequality $g^{(2)}(\tau) < g^{(2)}(0)$ for $\tau < \tau_c$, the photons exhibit excess correlations for times less than the correlation time τ_c. This is called *photon bunching* as the photons tend to distribute themselves preferentially in bunches rather than at random. When such a light beam falls on a photodetector, more photon pairs are detected close together than further apart (Fig. 4.13(a)). The thermal field is an example of photon bunching.

In certain quantum optical systems, the inequality (4.4.70) may be violated with the result

$$g^{(2)}(\tau) > g^{(2)}(0). \tag{4.4.71}$$

This would correspond to the phenomenon of *photon antibunching*. This is the opposite effect, in which fewer photon pairs are detected

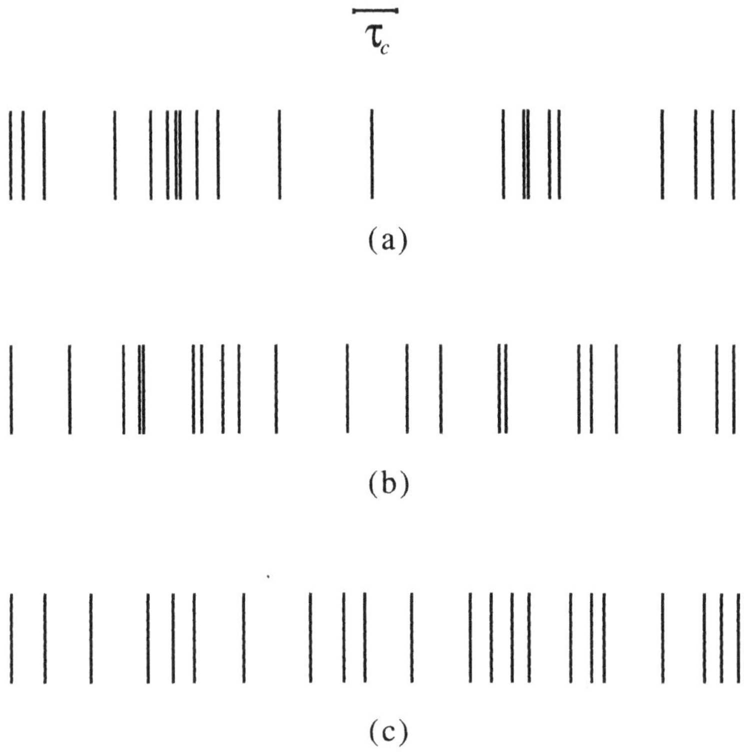

Fig. 4.13
Photon counts as functions of time for light beams which are (a) bunched, (b) random, and (c) antibunched.

close together than further apart (Fig. 4.13(c)). Photon antibunching will be discussed in the process of resonance fluorescence from an atom in Section 10.6.

Another nonclassical inequality is given by

$$g^{(2)}(0) < 1. \tag{4.4.72}$$

This nonclassical inequality is satisfied by fields whose P-representation is not nonnegative definite. To see this explicitly, we first rewrite this inequality, after some rearrangement, in the form (see Eq. (4.2.21))

$$(\langle a^\dagger a^\dagger aa\rangle - \langle a^\dagger a\rangle^2) < 0, \tag{4.4.73}$$

or, in terms of the P-representation,

$$\int P(\alpha, \alpha^*)(|\alpha|^2 - \langle a^\dagger a\rangle)^2 d^2\alpha < 0. \tag{4.4.74}$$

Since $(|\alpha|^2 - \langle a^\dagger a\rangle)^2$ is positive definite for all values of α, the only way this inequality may be satisfied is if $P(\alpha, \alpha^*)$ is negative for at least some values of α. Thus $P(\alpha, \alpha^*)$ does not satisfy the properties of a classical distribution function. The inequality (4.4.72) is satisfied by fields whose photon distribution function is narrower than the Poisson distribution. Such fields are referred to as *sub-Poissonian.* Fields for which $g^{(2)}(0) = 1$ and $g^{(2)}(0) > 1$ are similarly referred to as *Poissonian* and *super-Poissonian*, respectively. For example, a thermal field for which $g^{(2)}(0) = 2$ is super-Poissonian, a field in a coherent state $|\alpha_0\rangle$ for which $g^{(2)}(0) = 1$ is Poissonian, and the field in a number state $|n_0\rangle$ for which $g^{(2)}(0) = 1 - 1/n_0$ is sub-Poissonian.

It is evident from the above discussion that many other field states can be constructed for which the P-representation will not be well-behaved. One such state is the squeezed state of the radiation field. To show this, we express $(\Delta X_i)^2$ $(i = 1, 2)$ as an average with respect to the P-representation:

$$(\Delta X_i)^2 = \frac{1}{4} + (: \Delta X_i :)^2$$
$$= \frac{1}{4}\left\{1 + \int d^2\alpha P(\alpha, \alpha^*)[(\alpha + \alpha^*) - (\langle\alpha\rangle + \langle\alpha^*\rangle)]^2\right\}. \tag{4.4.75}$$

The condition for squeezing $(\Delta X_i)^2 < 1/4$ $(i = 1$ or $2)$ requires that $P(\alpha, \alpha^*)$ is negative for at least some values of α, i.e., it is not "non-negative definite". A squeezed state of the radiation field, therefore, is a nonclassical state.

4.5 Photon counting and photon statistics

In this section we determine the photoelectron counting statistics produced by a fully quantum mechanical field. The problem of obtaining the photocount distribution from the photon statistics can be solved in a completely quantum mechanical fashion. Here we give a simple derivation of this relationship based on a simple probabilistic argument.

Let the probability of having a photoelectron ejected from a detector interacting with a field having just one photon $|1\rangle$ for a certain time be given by η. The *quantum efficiency* η depends on the characteristics of the detector atoms and the interaction time. Now, if the state of the radiation field is $|n\rangle$, the probability of observing m photoelectrons, $P_m^{(n)}$, is proportional to η^m which is to be multiplied by the probability that $(n - m)$ quanta were not absorbed, i.e., $(1 - \eta)^{n-m}$. This gives

$$P_m^{(n)} \propto \eta^m (1 - \eta)^{n-m}. \tag{4.5.1}$$

However, we do not know which m photons of the original number n were absorbed, so we must include a combinatorial factor:

$$P_m^{(n)} = \binom{n}{m} \eta^m (1-\eta)^{n-m}. \tag{4.5.2}$$

This is Bernoulli's distribution for m successful events (counts) and $n-m$ failures, each event having a probability η. Since we have a distribution of n values given by the photon distribution function ρ_{nn}, we must multiply Eq. (4.5.2) by ρ_{nn} and sum over n:

$$P_m = \sum_n P_m^{(n)} \rho_{nn}, \tag{4.5.3}$$

which yields the following expression for the photoelectron counting distribution:

$$P_m = \sum_{n=m}^{\infty} \binom{n}{m} \eta^m (1-\eta)^{n-m} \rho_{nn}. \tag{4.5.4}$$

This expression is valid for all η $(0 \leq \eta \leq 1)$. Clearly, if we wish to obtain the photon statistics by counting photoelectrons, we must require $\eta = 1$. In that case, we obtain from Eq. (4.5.4)

$$P_m = \rho_{mm}. \tag{4.5.5}$$

In all other cases, $\eta < 1$, and the measured photoelectron statistics can be very different from the photon statistics.

Alternatively, we can write P_m in terms of the P-representation, $P(\alpha, \alpha^*)$, of the field by noting that

$$\rho_{nn} = \int d^2\alpha P(\alpha, \alpha^*) \frac{|\alpha|^{2n}}{n!} e^{-|\alpha|^2}, \tag{4.5.6}$$

so that Eq. (4.5.4) becomes

$$P_m = \int d^2\alpha \sum_{n=m}^{\infty} \binom{n}{m} P(\alpha, \alpha^*) \frac{|\alpha|^{2n}}{n!} e^{-|\alpha|^2} \eta^m (1-\eta)^{n-m}. \tag{4.5.7}$$

By changing n to $\ell + m$ and summing over ℓ, we obtain

$$P_m = \int d^2\alpha P(\alpha, \alpha^*) \frac{(\eta|\alpha|^2)^m}{m!} e^{-\eta|\alpha|^2}. \tag{4.5.8}$$

It may be pointed out that this equation can be inverted, i.e., it is possible to derive the P-representation of the field from the knowledge of P_m, given that ρ is diagonal in the n representation.

4.A Classical and quantum descriptions of two-source interference

Classically, the radiation from the two slits in Young's experiment is correctly described by two spherical waves. In the notation of Fig. 4.14, the intensity at the screen then goes as

$$I(\mathbf{r}) = \left| \frac{\mathscr{E} e^{ikr_1}}{r_1} + \frac{\mathscr{E} e^{ikr_2}}{r_2} \right|^2, \tag{4.A.1}$$

and the interference cross term is given by

$$I_{12} = \frac{\mathscr{E}^* \mathscr{E}}{r_1 r_2} e^{ik(r_1 - r_2)} + \text{c.c.} \tag{4.A.2}$$

Noting that $r_{1,2} = \sqrt{D^2 + (x \mp d)^2} \cong D + d^2/(2D) \mp xd/D$, where the '$-$' goes with source 1 and the '$+$' with source 2, we have

$$I_{12} \cong \frac{\mathscr{E}^* \mathscr{E}}{r^2} e^{-2ikxd/D} + \text{c.c.} \tag{4.A.3}$$

However, some texts give a plane-wave treatment of Young's setup, in which it is argued that the radiation at the detector site $\mathbf{r}$ consists of two plane waves. In such a case, we have

$$I(\mathbf{r}) = \left| \mathscr{E}_0 e^{\mathbf{k}_1 \cdot \mathbf{r}} + \mathscr{E}_0 e^{\mathbf{k}_2 \cdot \mathbf{r}} \right|^2, \tag{4.A.4}$$

and the interference cross term is

$$I_{12}(\mathbf{r}) = \mathscr{E}_0^* \mathscr{E}_0 e^{i(\mathbf{k}_1 - \mathbf{k}_2) \cdot \mathbf{r}} + \text{c.c.} \tag{4.A.5}$$

Hence if, in the notation of Fig. 4.14, we write $\mathbf{k}_i = k(\hat{z} \cos\theta_i + \hat{x} \sin\theta_i)$, then $\mathbf{k}_i \cdot \mathbf{r} = k(D \cos\theta_i + x \sin\theta_i) \cong k(D \mp xd/D)$, where the '$\mp$' signs go with 1 and 2, and we find

$$I_{12}(\mathbf{r}) \cong \mathscr{E}_0^* \mathscr{E}_0 e^{-2ikxd/D} + \text{c.c.}, \tag{4.A.6}$$

in agreement with the spherical-wave treatment.

The quantum field theoretic description of Young's experiment is well illustrated by replacing the two slits by two atoms as in Section 4.3.2 and Section 21.1. There the state vector for the photon emitted by the ith atom is given by

$$|\gamma_i\rangle = \sum_{\mathbf{k}} \frac{g_{\mathbf{k}} e^{-i\mathbf{k} \cdot \mathbf{d}_i}}{(\nu_{\mathbf{k}} - \omega) + i\Gamma/2} |1_{\mathbf{k}}\rangle, \tag{4.A.7}$$

where $g_{\mathbf{k}}$ is a constant depending on the strength of atom-field coupling, ω is the atomic frequency between levels $|a\rangle$ and $|b\rangle$, Γ is the

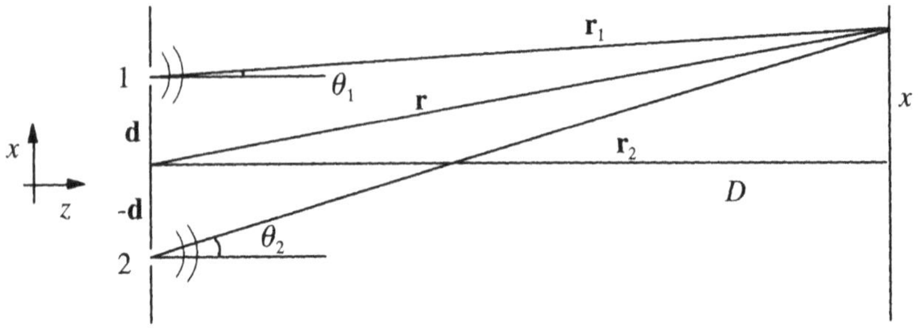

Fig. 4.14 Schematic diagram for a plane-wave treatment of Young's setup.

decay rate for the $|a\rangle \rightarrow |b\rangle$ transition, and $\mathbf{d}_i$ is the position of the ith atom. The correlation function for the scattered field is then found to be

$$G^{(1)}(\mathbf{r},\mathbf{r};0) = \frac{1}{2}\left|\langle 0|E^{(+)}(\mathbf{r},t)|\gamma_1\rangle + \langle 0|E^{(+)}(\mathbf{r},t)|\gamma_2\rangle\right|^2$$
$$= \left|\frac{\tilde{\mathscr{E}}e^{ikr_1}}{r_1} + \frac{\tilde{\mathscr{E}}e^{ikr_2}}{r_2}\right|^2, \tag{4.A.8}$$

where $\tilde{\mathscr{E}}$ is an effective electric field. Thus we have the same result as in the classical spherical-wave problem.

Finally, we note that single photon plane-wave states can be used to demonstrate the two-source interference fringes. At the risk of belaboring the obvious, we note that if we consider the radiation from source 1 to be described by the single photon state $|1_{\mathbf{k}_1}\rangle$, and that from source 2 by $|1_{\mathbf{k}_2}\rangle$, then we have $|\psi\rangle = (|1_{\mathbf{k}_1}\rangle + |1_{\mathbf{k}_2}\rangle)/\sqrt{2}$ and

$$G^{(1)}(\mathbf{r},\mathbf{r};0) = \langle\psi|E^{(-)}(\mathbf{r},t)E^{(+)}(\mathbf{r},t)|\psi\rangle$$
$$= \frac{1}{2}\left|\langle 0|E^{(+)}(\mathbf{r},t)|1_{\mathbf{k}_1}\rangle + \langle 0|E^{(+)}(\mathbf{r},t)|1_{\mathbf{k}_2}\rangle\right|^2$$
$$= \mathscr{E}_0^*\mathscr{E}_0\left|e^{i\mathbf{k}_1\cdot\mathbf{r}} + e^{i\mathbf{k}_2\cdot\mathbf{r}}\right|^2. \tag{4.A.9}$$

Here again, interference is observed as in the classical case, and the utilization of both states $|\gamma_i\rangle$ and $|1_{\mathbf{k}_i}\rangle$ in Young-type experiments is justified.

4.B Calculation of the second-order correlation function

From Eqs. (4.4.14) and (4.4.15), we have (note that some terms are underlined)

$$
\begin{aligned}
&G^{(2)}(\mathbf{r}_1, \mathbf{r}_2; t, t) \\
&= \mathscr{E}_{\mathbf{k}}^4 \langle [a_{\mathbf{k}}^\dagger(1) + a_{\mathbf{k}'}^\dagger(1)][a_{\mathbf{k}}^\dagger(2) + a_{\mathbf{k}'}^\dagger(2)][a_{\mathbf{k}}(2) + a_{\mathbf{k}'}(2)] \\
&\qquad [a_{\mathbf{k}}(1) + a_{\mathbf{k}'}(1)] \rangle \\
&= \mathscr{E}_{\mathbf{k}}^4 \langle \underline{a_{\mathbf{k}}^\dagger(1) a_{\mathbf{k}}^\dagger(2) a_{\mathbf{k}}(2) a_{\mathbf{k}}(1)} \\
&\quad + [a_{\mathbf{k}}^\dagger(1) a_{\mathbf{k}}^\dagger(2) a_{\mathbf{k}}(2) a_{\mathbf{k}'}(1) + a_{\mathbf{k}}^\dagger(1) a_{\mathbf{k}}^\dagger(2) a_{\mathbf{k}'}(2) a_{\mathbf{k}}(1) \\
&\quad + a_{\mathbf{k}}^\dagger(1) a_{\mathbf{k}}^\dagger(2) a_{\mathbf{k}'}(2) a_{\mathbf{k}'}(1)] \\
&\quad + \underline{a_{\mathbf{k}}^\dagger(1) a_{\mathbf{k}'}^\dagger(2) a_{\mathbf{k}}(2) a_{\mathbf{k}'}(1)} + \underline{a_{\mathbf{k}}^\dagger(1) a_{\mathbf{k}'}^\dagger(2) a_{\mathbf{k}'}(2) a_{\mathbf{k}}(1)} \\
&\quad + [a_{\mathbf{k}}^\dagger(1) a_{\mathbf{k}'}^\dagger(2) a_{\mathbf{k}}(2) a_{\mathbf{k}}(1) + a_{\mathbf{k}}^\dagger(1) a_{\mathbf{k}'}^\dagger(2) a_{\mathbf{k}'}(2) a_{\mathbf{k}'}(1)] \\
&\quad + \underline{a_{\mathbf{k}'}^\dagger(1) a_{\mathbf{k}}^\dagger(2) a_{\mathbf{k}}(2) a_{\mathbf{k}'}(1)} + \underline{a_{\mathbf{k}'}^\dagger(1) a_{\mathbf{k}}^\dagger(2) a_{\mathbf{k}'}(2) a_{\mathbf{k}}(1)} \\
&\quad + [a_{\mathbf{k}'}^\dagger(1) a_{\mathbf{k}}^\dagger(2) a_{\mathbf{k}}(2) a_{\mathbf{k}}(1) + a_{\mathbf{k}'}^\dagger(1) a_{\mathbf{k}}^\dagger(2) a_{\mathbf{k}'}(2) a_{\mathbf{k}'}(1)] \\
&\quad + \underline{a_{\mathbf{k}'}^\dagger(1) a_{\mathbf{k}'}^\dagger(2) a_{\mathbf{k}'}(2) a_{\mathbf{k}'}(1)} \\
&\quad + [a_{\mathbf{k}'}^\dagger(1) a_{\mathbf{k}'}^\dagger(2) a_{\mathbf{k}}(2) a_{\mathbf{k}'}(1) + a_{\mathbf{k}'}^\dagger(1) a_{\mathbf{k}'}^\dagger(2) a_{\mathbf{k}'}(2) a_{\mathbf{k}}(1) \\
&\quad + a_{\mathbf{k}'}^\dagger(1) a_{\mathbf{k}'}^\dagger(2) a_{\mathbf{k}}(2) a_{\mathbf{k}}(1)] \rangle,
\end{aligned}
\tag{4.B.1}
$$

where

$$
\begin{aligned}
a_{\mathbf{k}}^\dagger(i) &= a_{\mathbf{k}}^\dagger \, e^{-i\mathbf{k}\cdot\mathbf{r}_i}, \\
a_{\mathbf{k}'}^\dagger(i) &= a_{\mathbf{k}'}^\dagger \, e^{-i\mathbf{k}'\cdot\mathbf{r}_i}.
\end{aligned}
\tag{4.B.2}
$$

Note that all the termsin square brackets for the final equation vanish when averaged for stars, phase-diffused lasers, thermal light, and atoms. Therefore, keeping only the underlined terms, we find

$$
\begin{aligned}
G^{(2)}(\mathbf{r}_1, \mathbf{r}_2; t, t) = \mathscr{E}_{\mathbf{k}}^4 \langle & a_{\mathbf{k}}^\dagger a_{\mathbf{k}}^\dagger a_{\mathbf{k}} a_{\mathbf{k}} + a_{\mathbf{k}'}^\dagger a_{\mathbf{k}'}^\dagger a_{\mathbf{k}}' a_{\mathbf{k}}' \\
& + a_{\mathbf{k}}^\dagger a_{\mathbf{k}'}^\dagger a_{\mathbf{k}} a_{\mathbf{k}'} [1 + e^{-i(\mathbf{k}-\mathbf{k}')\cdot(\mathbf{r}_1 - \mathbf{r}_2)}] \\
& + a_{\mathbf{k}'}^\dagger a_{\mathbf{k}}^\dagger a_{\mathbf{k}'} a_{\mathbf{k}} [1 + e^{i(\mathbf{k}-\mathbf{k}')\cdot(\mathbf{r}_1 - \mathbf{r}_2)}] \rangle.
\end{aligned}
\tag{4.B.3}
$$

Problems

4.1 Show that the radiation field state which is a linear superposition of the vacuum state and a single photon state, i.e.,

$$|\psi\rangle = a_0|0\rangle + a_1|1\rangle,$$

where a_0 and a_1 are complex coefficients, is a nonclassical state.

4.2 Let $m_n = \langle a^{\dagger n} a^n \rangle$ be the nth-order moment of the intensity variable. Consider the matrix defined by

$$\mathscr{M} = \begin{bmatrix} 1 & m_1 & m_2 \\ m_1 & m_2 & m_3 \\ m_2 & m_3 & m_4 \end{bmatrix}.$$

Show that for a classical P-representation det$\mathscr{M}$ must be positive definite. (Hint: see G. S. Agarwal and K. Tara, *Phys. Rev. A* **46**, 485 (1992).)

4.3 Consider a state described by the density operator

$$\rho = \mathscr{N} a^{\dagger m} e^{-\kappa a^\dagger a} a^m,$$

where $\mathscr{N}$ is a normalization constant and $\kappa = \hbar\nu/k_B T$.

(a) Show that it goes over to a Fock state if $\kappa \to \infty$ and to a thermal state if $\kappa \to 0$.

(b) Find $g^{(2)}(0)$ and show that the photon statistics are sub-Poissonian if

$$\bar{n} < \sqrt{\frac{m}{m+1}},$$

where $\bar{n} = [\exp(\kappa) - 1]^{-1}$.

4.4 Find the photoelectron distribution function P_m for the coherent state $|\alpha\rangle$, the number state $|n\rangle$, and the single-mode thermal field at temperature T.

References and bibliography

Review articles on optical coherence theory

R. J. Glauber, in *Quantum Optics and Electronics*, Les Houches, ed. C. DeWitt, A. Blandin, and C. Cohen-Tannoudji (Gordon and Breach, New York 1965).

L. Mandel and E. Wolf, *Rev. Mod. Phys.* **37**, 231 (1965).

L. Mandel and E. Wolf, editors, *Selected Papers on Coherence and Fluctuations of Light*, Vols. 1 and 2 (Dover, New York 1970). These volumes contain an excellent collection of papers on optical coherence theory until 1966.

R. J. Glauber, in *Quantum Optics*, ed. S. Kay and A. Maitland (Academic, New York 1970).

L. Mandel, in *Progress in Optics*, Vol. 13, ed. E. Wolf (North-Holland, Amsterdam 1976).

R. Loudon, *Rep. Prog. Phys.* **43**, 913 (1980).

H. Paul, *Rev. Mod. Phys.* **54**, 1061 (1982).

Sagnac interferometry and ring laser gyroscope

G. Sagnac, *C. R. Acad. Sci.* **157**, 708 (1913).

F. Aronowitz, in *Laser Applications*, ed. M. Ross (Academic, New York 1971), p. 113.

W. W. Chow, J. Gea-Banacloche, L. M. Pedrotti, V. E. Sanders, W. Schleich, and M. O. Scully, *Rev. Mod. Phys.* **57**, 61 (1985).

Applications of ring laser gyroscopes to general relativity

M. O. Scully, M. S. Zubairy, and M. P. Haugan, *Phys. Rev. A* **24**, 2009 (1981).

W. Schleich and M. O. Scully, 'General Relativity and Modern Optics' in *Modern Trends in Atomic and Molecular Physics, Proceedings of Les Houches Summer School, Session* XXXVIII, ed. R. Stora and G. Grynberg (North-Holland, Amsterdam 1984).

Hanbury-Brown–Twiss experiment

H. Hanbury-Brown and R. Q. Twiss, *Phil. Mag.* **45**, 663 (1954); *Nature* **178**, 1046 (1956); *Proc. Roy. Soc.* **A242**, 300 (1957).

A. Forrester, R. Gudmundsen, and P. Johnson, *Phys. Rev.* **99**, 1691 (1955).

E. M. Purcell, *Nature* **178**, 1449 (1956).

U. Fano, *Am J. Phys.* **29**, 539 (1961).

G. Baym, *Lectures on Quantum Mechanics*, (Benjamin Pub., New York 1969), p. 431.

R. Hanbury-Brown, *The Intensity Interferometer* (Taylor and Frances, London 1974).

Two-photon correlation experiment

L. Mandel, *Phys. Rev. A* **28**, 929 (1983).
R. Ghosh and L. Mandel, *Phys. Rev. Lett.* **59**, 1903 (1987).
Z. Y. Ou and L. Mandel, *Phys. Rev. Lett.* **62**, 2941 (1989).

Photon counting and photon statistics

L. Mandel, *Proc. Phys. Soc.* **72**, 1037 (1958).
L. Mandel, in *Progress in Optics*, Vol. 2, ed. E. Wolf (North-Holland, Amsterdam 1963), p. 181.
P. L. Kelley and W. H. Kleiner, *Phys. Rev.* **136**, A316 (1964).
M. O. Scully and W. E. Lamb, Jr., *Phys. Rev.* **179**, 368 (1969).

Power spectrum

J. H. Eberly and K. Wódkiewicz, *J. Opt. Soc. Am.* **67**, 1252 (1977). This paper discusses the power spectrum for nonstationary fields.
J. H. Eberly, C. V. Kunasz, and K. Wódkiewicz, *J. Phys. B* **13**, 217 (1980).
J. D. Cresser, *Phys. Rep.* **94**, 47 (1983). This paper gives an exhaustive review of the subject.

Balanced homodyne detection

H. P. Yuen and V. W. S. Chan, *Opt. Lett.* **8**, 177 (1983).
N. G. Walker and J. E. Carrol, *Opt. Quantum Electron.* **18**, 355 (1986).

CHAPTER 5

Atom–field interaction – semiclassical theory

One of the simplest nontrivial problems involving the atom–field interaction is the coupling of a two-level atom with a single mode of the electromagnetic field. A two-level atom description is valid if the two atomic levels involved are resonant or nearly resonant with the driving field, while all other levels are highly detuned. Under certain realistic approximations, it is possible to reduce this problem to a form which can be solved exactly; allowing essential features of the atom–field interaction to be extracted.

In this chapter we present a semiclassical theory of the interaction of a single two-level atom with a single mode of the field in which the atom is treated as a quantum two-level system and the field is treated classically. A fully quantum mechanical theory will be presented in Chapter 6.

A two-level atom is formally analogous to a spin-1/2 system with two possible states. In the dipole approximation, when the field wavelength is larger than the atomic size, the atom–field interaction problem is mathematically equivalent to a spin-1/2 particle interacting with a time-dependent magnetic field. Just as the spin-1/2 system undergoes the so-called Rabi oscillations between the spin-up and spin-down states under the action of an oscillating magnetic field, the two-level atom also undergoes *optical* Rabi oscillations under the action of the driving electromagnetic field. These oscillations are damped if the atomic levels decay. An understanding of this simple model of the atom–field interaction enables us to consider more complicated problems involving an ensemble of atoms interacting with the field. Perhaps the most important example of such a problem is the laser, which we discuss later in this chapter.*

5.1 Atom–field interaction Hamiltonian

An electron of charge e and mass m interacting with an external electromagnetic field is described by a minimal-coupling Hamiltonian

$$\mathscr{H} = \frac{1}{2m}\left[\mathbf{p} - e\mathbf{A}(\mathbf{r}, t)\right]^2 + eU(\mathbf{r}, t) + V(r), \tag{5.1.1}$$

where $\mathbf{p}$ is the canonical momentum operator, $\mathbf{A}(\mathbf{r}, t)$ and $U(\mathbf{r}, t)$ are the vector and scalar potentials of the external field, respectively, and $V(r)$ is an electrostatic potential that is normally the atomic binding potential. In this section, we first derive this Hamiltonian from a gauge invariance point of view, before reducing it to a simple form suitable for describing the interaction of a two-level atom with the radiation field.

5.1.1 *Local gauge (phase) invariance and minimal-coupling Hamiltonian*

The motion of a free electron is described by the Schrödinger equation

$$\frac{-\hbar^2}{2m}\nabla^2\psi = i\hbar\frac{\partial\psi}{\partial t}, \tag{5.1.2}$$

such that

$$P(\mathbf{r}, t) = |\psi(\mathbf{r}, t)|^2 \tag{5.1.3}$$

gives the probability density of finding an electron at position $\mathbf{r}$ and time t. In Eq. (5.1.2), if $\psi(\mathbf{r}, t)$ is a solution so is $\psi_1(\mathbf{r}, t) = \psi(\mathbf{r}, t)\exp(i\chi)$ where χ is an arbitrary constant phase. The probability density $P(\mathbf{r}, t)$ would also remain unaffected by an arbitrary choice of χ. Thus the choice of the phase of the wave function $\psi(\mathbf{r}, t)$ is completely arbitrary, and two functions differing only by a constant phase factor represent the same physical state.

The situation is different, however, if the phase is allowed to vary *locally*, i.e. to be a function of space and time variables, i.e.,

$$\psi(\mathbf{r}, t) \rightarrow \psi(\mathbf{r}, t)e^{i\chi(\mathbf{r}, t)}. \tag{5.1.4}$$

* The semiclassical theory of laser behavior as developed by the schools of Lamb and Haken (see Lamb [1963,1964] and Haken [1964]) are the pioneering treatments of the problem. Lamb begins from the coupled Maxwell–Schrödinger equations, while Haken and co-workers take a semiclassical (factorized) limit of quantum fields.

The probability $P(\mathbf{r},t)$ remains unaffected by this transformation, but the Schrödinger equation (5.1.2) is no longer satisfied. If we want to satisfy *local* gauge (phase) invariance, then the Schrödinger equation must be modified by adding new terms to Eq. (5.1.2)

$$\left\{-\frac{\hbar^2}{2m}\left[\nabla - i\frac{e}{\hbar}\mathbf{A}(\mathbf{r},t)\right]^2 + eU(\mathbf{r},t)\right\}\psi = i\hbar\frac{\partial\psi}{\partial t}, \tag{5.1.5}$$

where $\mathbf{A}(\mathbf{r},t)$ and $U(\mathbf{r},t)$ are functions which must be inserted into (5.1.2) if we want to be able to make the transformation (5.1.4), and are given by

$$\mathbf{A}(\mathbf{r},t) \to \mathbf{A}(\mathbf{r},t) + \frac{\hbar}{e}\nabla\chi(\mathbf{r},t), \tag{5.1.6}$$

$$U(\mathbf{r},t) \to U(\mathbf{r},t) - \frac{\hbar}{e}\frac{\partial\chi}{\partial t}(\mathbf{r},t). \tag{5.1.7}$$

The functions $\mathbf{A}(\mathbf{r},t)$ and $U(\mathbf{r},t)$ are identified as the vector and scalar potentials of the electromagnetic field, respectively. These are the gauge-dependent potentials. The gauge-independent quantities are the electric and magnetic fields

$$\mathbf{E} = -\nabla U - \frac{\partial\mathbf{A}}{\partial t}, \tag{5.1.8}$$

$$\mathbf{B} = \nabla \times \mathbf{A}. \tag{5.1.9}$$

Equation (5.1.5), which is the logical extension of Eq. (5.1.2) due to the requirement of local gauge (phase) invariance, has the form

$$\mathscr{H}\psi = i\hbar\partial\psi/\partial t, \tag{5.1.10}$$

with $\mathscr{H}$ being the minimal-coupling Hamiltonian (recall $\mathbf{p} = -i\hbar\nabla$) described in Eq. (5.1.1). The Schrödinger equation (5.1.5) represents the interaction of an electron with a given electromagnetic field. The electrons are described by the wave function $\psi(\mathbf{r},t)$ whereas the field is described by the vector and scalar potentials $\mathbf{A}$ and U, respectively.

It is interesting to note that the Hamiltonian (5.1.1) has been 'derived' from a gauge invariance argument and is expressed in terms of the gauge-dependent quantities $\mathbf{A}(\mathbf{r},t)$ and $U(\mathbf{r},t)$. The vector and scalar potentials have therefore a larger physical significance than is usually attributed to them. They are not merely introduced for the sake of mathematical simplicity in problems dealing with 'observable' electric and magnetic fields. Instead, they arise naturally in any gauge (phase) invariance argument as shown above.

We also note that the Schrödinger equation plus the concept of local gauge invariance has led us to the introduction of the electromagnetic field. In this way, we can and do argue that the 'photon' (in our derivation, the classical field limit of the same) has been 'derived' from the Schrödinger equation plus the local gauge invariance arguments.

We have here a taste of one of the most fundamental concepts in modern physics, namely, that of the gauge field theory. Gauge theory, in the hands of Steven Weinberg and Abdus Salam, led to the unification of the weak and the electromagnetic interactions.

5.1.2 Dipole approximation and $\mathbf{r} \cdot \mathbf{E}$ Hamiltonian

We now examine the problem of an electron bound by a potential $V(r)$ to a force center (nucleus) located at $\mathbf{r}_0$. The minimal-coupling Hamiltonian (5.1.1) for an interaction between an atom and the radiation field can be reduced to a simple form by using the dipole approximation. The entire atom is immersed in a plane electromagnetic wave described by a vector potential $\mathbf{A}(\mathbf{r}_0 + \mathbf{r}, t)$. This vector potential may be written in the dipole approximation, $\mathbf{k} \cdot \mathbf{r} \ll 1$, as

$$\begin{aligned}\mathbf{A}(\mathbf{r}_0 + \mathbf{r}, t) &= \mathbf{A}(t) \exp[i\mathbf{k} \cdot (\mathbf{r}_0 + \mathbf{r})] \\ &= \mathbf{A}(t) \exp(i\mathbf{k} \cdot \mathbf{r}_0)(1 + i\mathbf{k} \cdot \mathbf{r} + \ldots) \\ &\simeq \mathbf{A}(t) \exp(i\mathbf{k} \cdot \mathbf{r}_0). \end{aligned} \tag{5.1.11}$$

The Schrödinger equation for this problem (in the dipole approximation) is given by Eq. (5.1.5) with $\mathbf{A}(\mathbf{r}, t) \equiv \mathbf{A}(\mathbf{r}_0, t)$, i.e.,

$$\left\{ -\frac{\hbar^2}{2m} \left[\nabla - \frac{ie}{\hbar} \mathbf{A}(\mathbf{r}_0, t) \right]^2 + V(r) \right\} \psi(\mathbf{r}, t) = i\hbar \frac{\partial \psi(\mathbf{r}, t)}{\partial t}, \tag{5.1.12}$$

where we have added a binding potential $V(r)$. We note that in Eq. (5.1.12), and elsewhere in this book, we are working in the radiation gauge, in which

$$U(\mathbf{r}, t) = 0, \tag{5.1.13}$$

and

$$\nabla \cdot \mathbf{A} = 0. \tag{5.1.14}$$

We have added the term $V(r)$ in the Hamiltonian which arises from the electrostatic potential that binds the electron to the nucleus.

We proceed to simplify Eq. (5.1.12) by defining a new wave function $\phi(\mathbf{r}, t)$ as

$$\psi(\mathbf{r}, t) = \exp \left[\frac{ie}{\hbar} \mathbf{A}(\mathbf{r}_0, t) \cdot \mathbf{r} \right] \phi(\mathbf{r}, t). \tag{5.1.15}$$

Inserting Eq. (5.1.15) into Eq. (5.1.12), we find

$$i\hbar\left[\frac{ie}{\hbar}\dot{\mathbf{A}}\cdot\mathbf{r}\phi(\mathbf{r},t)+\dot{\phi}(\mathbf{r},t)\right]\exp\left(\frac{ie}{\hbar}\mathbf{A}\cdot\mathbf{r}\right)$$
$$=\exp\left(\frac{ie}{\hbar}\mathbf{A}\cdot\mathbf{r}\right)\left[\frac{p^2}{2m}+V(r)\right]\phi(\mathbf{r},t). \tag{5.1.16}$$

This equation, after the cancellation of the exponential factor and some rearrangement, takes the simple form

$$i\hbar\dot{\phi}(\mathbf{r},t)=[\mathscr{H}_0-e\mathbf{r}\cdot\mathbf{E}(\mathbf{r}_0,t)]\phi(\mathbf{r},t), \tag{5.1.17}$$

where

$$\mathscr{H}_0=\frac{p^2}{2m}+V(r), \tag{5.1.18}$$

is the unperturbed Hamiltonian of the electron and we use $\mathbf{E}=-\dot{\mathbf{A}}$. Notice that the total Hamiltonian

$$\mathscr{H}=\mathscr{H}_0+\mathscr{H}_1 \tag{5.1.19}$$

with

$$\mathscr{H}_1=-e\mathbf{r}\cdot\mathbf{E}(\mathbf{r}_0,t), \tag{5.1.20}$$

is given in terms of the gauge-independent field **E**. We shall use this Hamiltonian in our subsequent studies of atom–field interaction. Note also that this Hamiltonian has been obtained from the radiation gauge Hamiltonian (5.1.12) by applying the gauge transformation $\chi(\mathbf{r},t)=-e\mathbf{A}(\mathbf{r}_0,t)\cdot\mathbf{r}/\hbar$.

5.1.3 $\mathbf{p}\cdot\mathbf{A}$ *Hamiltonian*

In many textbooks one finds the atom–field Hamiltonian expressed in terms of the canonical momentum **p** and the vector potential **A** instead of the simple gauge invariant expression (5.1.17). This has resulted in considerable confusion, and we therefore consider the problem in some detail. We again choose a radiation gauge in which $U(\mathbf{r},t)=0$ and $\nabla\cdot\mathbf{A}=0$. The condition $\nabla\cdot\mathbf{A}=0$ implies, in quantum mechanics, that $[\mathbf{p},\mathbf{A}]=0$. The total Hamiltonian (5.1.1) can, therefore, be written as

$$\mathscr{H}'=\mathscr{H}_0+\mathscr{H}_2, \tag{5.1.21}$$

where $\mathscr{H}_0$ is given by Eq. (5.1.18) and, in the dipole approximation (5.1.11),

$$\mathscr{H}_2=-\frac{e}{m}\mathbf{p}\cdot\mathbf{A}(\mathbf{r}_0,t)+\frac{e^2}{2m}A^2(\mathbf{r}_0,t), \tag{5.1.22}$$

and the Schrödinger equation reads

$$\left[\mathscr{H}_0 - \frac{e}{m}\mathbf{p}\cdot\mathbf{A}(\mathbf{r}_0,t) + \frac{e^2}{2m}A^2(\mathbf{r}_0,t)\right]\psi(\mathbf{r},t) = i\hbar\frac{\partial}{\partial t}\psi(\mathbf{r},t). \quad (5.1.23)$$

The A^2 term in Eq. (5.1.23) is ususally small and can be ignored. The wave function $\psi(\mathbf{r},t)$ then obeys the equation of motion

$$i\hbar\frac{\partial}{\partial t}\psi(\mathbf{r},t) = \left[\mathscr{H}_0 - \frac{e}{m}\mathbf{p}\cdot\mathbf{A}(\mathbf{r}_0,t)\right]\psi(\mathbf{r},t), \quad (5.1.24)$$

corresponding to a Hamiltonian

$$\mathscr{H}' = \mathscr{H}_0 - \frac{e}{m}\mathbf{p}\cdot\mathbf{A}(\mathbf{r}_0,t), \quad (5.1.25)$$

and

$$\mathscr{H}_2 = -\frac{e}{m}\mathbf{p}\cdot\mathbf{A}(\mathbf{r}_0,t). \quad (5.1.26)$$

The two different Hamiltonians $\mathscr{H}_1$ and $\mathscr{H}_2$ given in Eqs. (5.1.20) and (5.1.26), respectively, seem to give different physical results since the matrix elements of these Hamiltonians, calculated between the eigenstates of the *unperturbed* Hamiltonian $\mathscr{H}_0$, given by Eq. (5.1.18), are not the same. In order to show this explicitly, we consider a linearly polarized monochromatic plane-wave field interacting with an atom placed at $\mathbf{r}_0 = 0$. The electric field then takes the form

$$\mathbf{E}(0,t) = \mathscr{E}\cos\nu t, \quad (5.1.27)$$

and the corresponding vector potential in the radiation gauge is

$$\mathbf{A}(0,t) = -\frac{1}{\nu}\mathscr{E}\sin\nu t. \quad (5.1.28)$$

Consider now the time-independent amplitudes associated with $\mathscr{H}_1$ and $\mathscr{H}_2$,

$$W_1 = -e\mathbf{r}\cdot\mathscr{E}, \quad (5.1.29a)$$
$$W_2 = \frac{e}{m\nu}\mathbf{p}\cdot\mathscr{E}. \quad (5.1.29b)$$

We may relate W_1 and W_2 by noting that

$$\mathbf{p} = m\mathbf{v} = -m\left(\frac{i}{\hbar}\right)[\mathbf{r}, \mathscr{H}_0]. \quad (5.1.30)$$

We then find for the matrix elements of W_1 and W_2, calculated between an initial eigenstate $|i\rangle$ of $\mathscr{H}_0$ (with $\mathscr{H}_0|i\rangle = \hbar\omega_i|i\rangle$) and a final eigenstate $|f\rangle$ (with $\mathscr{H}_0|f\rangle = \hbar\omega_f|f\rangle$), the ratio

$$\left|\frac{\langle f|W_2|i\rangle}{\langle f|W_1|i\rangle}\right| = \left|-\frac{(e/mv)\langle f|\mathbf{p}|i\rangle \cdot \mathscr{E}}{e\langle f|\mathbf{r}|i\rangle \cdot \mathscr{E}}\right| = \frac{\omega}{v}, \tag{5.1.31}$$

where $\omega = \omega_f - \omega_i$ is the transition frequency. Hence, the matrix elements of the two interaction Hamiltonians $\mathscr{H}_1$ and $\mathscr{H}_2$ differ by the ratio of the transition frequency over the field frequency. As was first pointed out by Lamb, this makes a difference in measurable quantities like transition rates. We present a resolution of this in Appendix 5.A.

5.2 Interaction of a single two-level atom with a single-mode field

5.2.1 *Probability amplitude method*

Consider the interaction of a single-mode radiation field of frequency v with a two-level atom (Fig. 5.1). Let $|a\rangle$ and $|b\rangle$ represent the upper and lower level states of the atom, i.e., they are eigenstates of the unperturbed part of the Hamiltonian $\mathscr{H}_0$ with the eigenvalues $\hbar\omega_a$ and $\hbar\omega_b$, respectively. The wave function of a two-level atom can be written in the form

$$|\psi(t)\rangle = C_a(t)|a\rangle + C_b(t)|b\rangle, \tag{5.2.1}$$

where C_a and C_b are the probability amplitudes of finding the atom in states $|a\rangle$ and $|b\rangle$, respectively. The corresponding Schrödinger equation is

$$|\dot{\psi}(t)\rangle = -\frac{i}{\hbar}\mathscr{H}|\psi(t)\rangle, \tag{5.2.2}$$

with

$$\mathscr{H} = \mathscr{H}_0 + \mathscr{H}_1, \tag{5.2.3}$$

where $\mathscr{H}_0$ and $\mathscr{H}_1$ represent the unperturbed and interaction parts of the Hamiltonian, respectively. By using the completeness relation $|a\rangle\langle a| + |b\rangle\langle b| = \mathbf{1}$, we can write $\mathscr{H}_0$ as

$$\begin{aligned}\mathscr{H}_0 &= (|a\rangle\langle a| + |b\rangle\langle b|)\mathscr{H}_0(|a\rangle\langle a| + |b\rangle\langle b|)\\ &= \hbar\omega_a|a\rangle\langle a| + \hbar\omega_b|b\rangle\langle b|,\end{aligned} \tag{5.2.4}$$

where we use $\mathscr{H}_0|a\rangle = \hbar\omega_a|a\rangle$ and $\mathscr{H}_0|b\rangle = \hbar\omega_b|b\rangle$. Similarly, the part of the Hamiltonian $\mathscr{H}_1$ that represents the interaction of the atom with the radiation field can be written as

$$\begin{aligned}\mathscr{H}_1 &= -exE(t)\\ &= -e(|a\rangle\langle a| + |b\rangle\langle b|)x(|a\rangle\langle a| + |b\rangle\langle b|)E(z,t)\\ &= -(\wp_{ab}|a\rangle\langle b| + \wp_{ba}|b\rangle\langle a|)E(t),\end{aligned} \tag{5.2.5}$$

where $\wp_{ab} = \wp_{ba}^{*} = e\langle a|x|b\rangle$ is the matrix element of the electric dipole moment and $E(t)$ is the field at the atom. Here, we assume that the electric field is linearly polarized along the x-axis. In the dipole approximation, the field can be expressed as

$$E(t) = \mathscr{E}\cos\nu t, \tag{5.2.6}$$

where $\mathscr{E}$ is the amplitude and $\nu = ck$ is the frequency of the field. The equations of motion for the amplitudes C_a and C_b may be written as

$$\dot{C}_a = -i\omega_a C_a + i\Omega_R e^{-i\phi}\cos(\nu t)C_b, \tag{5.2.7}$$

$$\dot{C}_b = -i\omega_b C_b + i\Omega_R e^{i\phi}\cos(\nu t)C_a, \tag{5.2.8}$$

where the Rabi frequency Ω_R is defined as

$$\Omega_R = \frac{|\wp_{ba}|\mathscr{E}}{\hbar}, \tag{5.2.9}$$

and ϕ is the phase of the dipole matrix element $\wp_{ba} = |\wp_{ba}|\exp(i\phi)$. In order to solve for C_a and C_b, we first write the equations of motion for the slowly varying amplitudes:

$$c_a = C_a e^{i\omega_a t}, \tag{5.2.10}$$

$$c_b = C_b e^{i\omega_b t}. \tag{5.2.11}$$

It then follows from Eqs. (5.2.7) and (5.2.8) that

$$\dot{c}_a = i\frac{\Omega_R}{2}e^{-i\phi}c_b e^{i(\omega-\nu)t}, \tag{5.2.12}$$

$$\dot{c}_b = i\frac{\Omega_R}{2}e^{i\phi}c_a e^{-i(\omega-\nu)t}, \tag{5.2.13}$$

where $\omega = \omega_a - \omega_b$ is the atomic transition frequency. In deriving Eqs. (5.2.12) and (5.2.13), we have ignored counter-rotating terms proportional to $\exp[\pm i(\omega+\nu)t]$ on the right-hand side in the *rotating-wave approximation.* This is generally a very good approximation. Furthermore, in some cases the counter-rotating terms never appear (as seen later in section 5.2.3) and Eqs. (5.2.12) and (5.2.13) are exact.

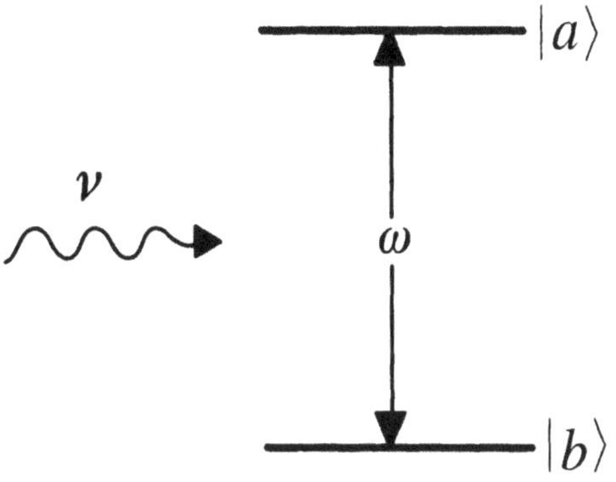

Fig. 5.1
Interaction of a two-level atom with a single-mode field.

The solutions for c_a and c_b can be written as

$$c_a(t) = \left(a_1 e^{i\Omega t/2} + a_2 e^{-i\Omega t/2}\right) e^{i\Delta t/2}, \tag{5.2.14}$$

$$c_b(t) = \left(b_1 e^{i\Omega t/2} + b_2 e^{-i\Omega t/2}\right) e^{-i\Delta t/2}, \tag{5.2.15}$$

where $\Delta = \omega - \nu$,

$$\Omega = \sqrt{\Omega_R^2 + (\omega - \nu)^2}, \tag{5.2.16}$$

and a_1, a_2, b_1, and b_2 are constants of integration which are determined from the initial conditions:

$$a_1 = \frac{1}{2\Omega}\left[(\Omega - \Delta)c_a(0) + \Omega_R e^{-i\phi} c_b(0)\right], \tag{5.2.17}$$

$$a_2 = \frac{1}{2\Omega}\left[(\Omega + \Delta)c_a(0) - \Omega_R e^{-i\phi} c_b(0)\right], \tag{5.2.18}$$

$$b_1 = \frac{1}{2\Omega}\left[(\Omega + \Delta)c_b(0) + \Omega_R e^{i\phi} c_a(0)\right], \tag{5.2.19}$$

$$b_2 = \frac{1}{2\Omega}\left[(\Omega - \Delta)c_b(0) - \Omega_R e^{i\phi} c_a(0)\right]. \tag{5.2.20}$$

We then have

$$c_a(t) = \left\{ c_a(0)\left[\cos\left(\frac{\Omega t}{2}\right) - \frac{i\Delta}{\Omega}\sin\left(\frac{\Omega t}{2}\right)\right] + i\frac{\Omega_R}{\Omega} e^{-i\phi} c_b(0)\sin\left(\frac{\Omega t}{2}\right)\right\} e^{i\Delta t/2}, \tag{5.2.21}$$

$$c_b(t) = \left\{ c_b(0)\left[\cos\left(\frac{\Omega t}{2}\right) + \frac{i\Delta}{\Omega}\sin\left(\frac{\Omega t}{2}\right)\right] + i\frac{\Omega_R}{\Omega} e^{i\phi} c_a(0)\sin\left(\frac{\Omega t}{2}\right)\right\} e^{-i\Delta t/2}. \tag{5.2.22}$$

It is not difficult to verify that

$$|c_a(t)|^2 + |c_b(t)|^2 = 1, \tag{5.2.23}$$

which is a simple statement of the conservation of probability since the atom is in state $|a\rangle$ or $|b\rangle$.

If we assume that the atom is initially in the state $|a\rangle$ then $c_a(0) = 1, c_b(0) = 0$. The probabilities of the atom being in states $|a\rangle$ and $|b\rangle$ at time t are then given by $|c_a(t)|^2$ and $|c_b(t)|^2$. The inversion is given by

$$\begin{aligned} W(t) &= |c_a(t)|^2 - |c_b(t)|^2 \\ &= \left(\frac{\Delta^2 - \Omega_R^2}{\Omega^2}\right) \sin^2\left(\frac{\Omega t}{2}\right) + \cos^2\left(\frac{\Omega t}{2}\right). \end{aligned} \tag{5.2.24}$$

Under the action of the incident field, a dipole moment is induced between the two atomic levels. This induced dipole moment is given by the expectation value of the dipole moment operator

$$P(t) = e\langle\psi(t)|r|\psi(t)\rangle = C_a^* C_b \wp_{ab} + \text{c.c.} = c_a^* c_b \wp_{ab} e^{i\omega t} + \text{c.c.} \tag{5.2.25}$$

On substituting Eqs. (5.2.21) and (5.2.22) into Eq. (5.2.25), we obtain, for an atom initially in the upper level,

$$\begin{aligned} &P(t) \\ &= 2\text{Re}\left\{\frac{i\Omega_R}{\Omega}\wp_{ab}\left[\cos\left(\frac{\Omega t}{2}\right) + \frac{i\Delta}{\Omega}\sin\left(\frac{\Omega t}{2}\right)\right]\sin\left(\frac{\Omega t}{2}\right)e^{i\phi}e^{i\nu t}\right\}. \end{aligned} \tag{5.2.26}$$

The dipole moment therefore oscillates with the frequency of the incident field.

In the special case when the atom is at resonance with the incident field ($\Delta = 0$), we get $\Omega = \Omega_R$ and

$$W(t) = \cos(\Omega_R t). \tag{5.2.27}$$

The inversion oscillates between -1 and 1 at a frequency Ω_R (see Fig. 5.2).

In 1937, Rabi considered the problem of a spin-1/2 magnetic dipole undergoing precessions in a magnetic field and obtained an expression for the probability that a spin-1/2 atom incident on a Stern–Gerlach apparatus would be flipped from the $\binom{1}{0}$ or $\binom{0}{1}$ state to the $\binom{0}{1}$ or $\binom{1}{0}$ state, respectively, by an applied radio-frequency magnetic field. In the present problem, the atom undergoes a Rabi 'flopping' between the upper and lower levels under the action of the electromagnetic field in complete analogy with the spin-1/2 system.

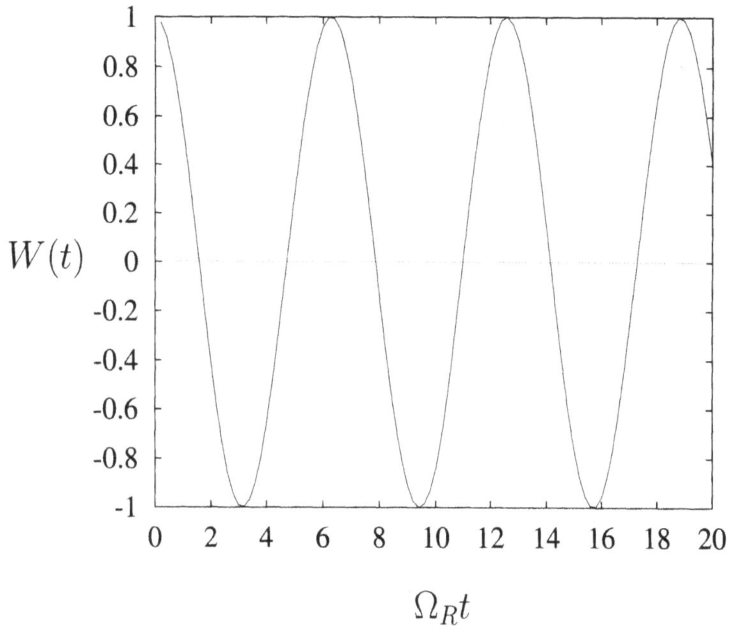

Fig. 5.2
Oscillations of the population inversion $W(t)$ as a function of time.

5.2.2 *Interaction picture*

Consider the Schrödinger equation

$$\frac{\partial}{\partial t}|\psi(t)\rangle = -\frac{i}{\hbar}\mathscr{H}|\psi(t)\rangle. \tag{5.2.28}$$

This equation can be integrated formally to give

$$|\psi(t)\rangle = U(t)|\psi(0)\rangle, \tag{5.2.29}$$

where the unitary time-evolution operator is defined by

$$\dot{U}(t) = -\frac{i}{\hbar}\mathscr{H}U(t), \tag{5.2.30}$$

and $U(0) = 1$.

A useful approach to the atom–field interaction problem exists in the *interaction picture* in which we assign to the state vector the time dependence due only to the interaction energy. This is accomplished by defining the state vector $|\psi_I\rangle$ in the interaction picture via

$$|\psi_I(t)\rangle = U_0^\dagger(t)|\psi(t)\rangle, \tag{5.2.31}$$

where

$$U_0(t) = \exp\left(-\frac{i}{\hbar}\mathscr{H}_0 t\right). \tag{5.2.32}$$

It then follows that

$$\frac{\partial}{\partial t}|\psi_I(t)\rangle = \left[\frac{\partial}{\partial t}U_0^\dagger(t)\right]|\psi(t)\rangle + U_0^\dagger(t)\frac{\partial}{\partial t}|\psi(t)\rangle, \tag{5.2.33}$$

and hence, from Eqs. (5.2.28), (5.2.31), and (5.2.32), we obtain

$$\frac{\partial}{\partial t}|\psi_I(t)\rangle = -\frac{i}{\hbar}\mathscr{V}(t)|\psi_I(t)\rangle. \tag{5.2.34}$$

Here

$$\mathscr{V}(t) = U_0^\dagger(t)\mathscr{H}_1 U_0(t), \tag{5.2.35}$$

is the interaction picture Hamiltonian. An operator O in the Schrödinger picture will accordingly transform as

$$O_I(t) = U_0^\dagger(t) O U_0(t). \tag{5.2.36}$$

This can be seen from the expectation value

$$\begin{aligned}\langle O\rangle &= \langle\psi(t)|O|\psi(t)\rangle\\ &= \langle\psi_I(t)|U_0^\dagger(t) O U_0(t)|\psi_I(t)\rangle\\ &= \langle\psi_I(t)|O_I(t)|\psi_I(t)\rangle.\end{aligned} \tag{5.2.37}$$

A formal solution of Eq. (5.2.34) is

$$|\psi_I(t)\rangle = U_I(t)|\psi_I(0)\rangle, \tag{5.2.38}$$

where

$$U_I(t) = \mathscr{T}\exp\left[-\frac{i}{\hbar}\int_0^t \mathscr{V}(\tau)d\tau\right] \tag{5.2.39}$$

is the time-evolution operator in the interaction picture, and $\mathscr{T}$ is the time-ordering operator, which is a shorthand notation for

$$\begin{aligned}&\mathscr{T}\exp\left[-\frac{i}{\hbar}\int_0^t \mathscr{V}(\tau)d\tau\right]\\ &= 1-\frac{i}{\hbar}\int_0^t dt_1\mathscr{V}(t_1)+\left(-\frac{i}{\hbar}\right)^2\int_0^t dt_1\int_0^{t_1} dt_2\mathscr{V}(t_1)\mathscr{V}(t_2)+\ldots\end{aligned} \tag{5.2.40}$$

In order to demonstrate the usefulness of the above formalism, we consider the interaction of a two-level atom with a monochromatic field of frequency ν. The Hamiltonian for this problem is given by Eqs. (5.2.3), (5.2.4), and (5.2.5). It follows, on using

$$\mathscr{H}_0^n = (\hbar\omega_a)^n|a\rangle\langle a| + (\hbar\omega_b)^n|b\rangle\langle b|, \tag{5.2.41}$$

that

$$U_0(t) = \exp\left(-\frac{i}{\hbar}\mathscr{H}_0 t\right)$$
$$= \exp(-i\omega_a t)|a\rangle\langle a| + \exp(-i\omega_b t)|b\rangle\langle b|. \tag{5.2.42}$$

For an atom at $z = 0$, the interaction picture Hamiltonian is, therefore, given by

$$\mathscr{V}(t) = -\hbar\Omega_R U_0^\dagger(t)(e^{-i\phi}|a\rangle\langle b| + e^{i\phi}|b\rangle\langle a|)U_0(t)\cos\nu t$$
$$= -\frac{\hbar\Omega_R}{2}[e^{-i\phi}|a\rangle\langle b|e^{i\Delta t} + e^{i\phi}|b\rangle\langle a|e^{-i\Delta t}$$
$$+e^{-i\phi}|a\rangle\langle b|e^{i(\omega+\nu)t} + e^{i\phi}|b\rangle\langle a|e^{-i(\omega+\nu)t}], \tag{5.2.43}$$

where $\Delta = \omega - \nu$. The terms proportional to $\exp[\pm i(\omega + \nu)t]$ vary very rapidly and their average over a time scale larger than $1/\nu$ is zero. These terms can therefore be neglected in the *rotating-wave approximation*. The simplified interaction picture Hamiltonian is

$$\mathscr{V}(t) = -\frac{\hbar\Omega_R}{2}\left(e^{-i\phi}|a\rangle\langle b| + e^{i\phi}|b\rangle\langle a|\right), \tag{5.2.44}$$

where we assume resonance, $\Delta = 0$. The time-evolution operator in the interaction picture $U_I(t)$ can be obtained simply from Eq. (5.2.39) by using

$$\mathscr{V}^{2n}(t) = \left(\frac{\hbar\Omega_R}{2}\right)^{2n}(|a\rangle\langle a| + |b\rangle\langle b|)^n,$$
$$\mathscr{V}^{2n+1}(t) = -\left(\frac{\hbar\Omega_R}{2}\right)^{2n+1}\left(e^{-i\phi}|a\rangle\langle b| + e^{i\phi}|b\rangle\langle a|\right). \tag{5.2.45}$$

The resulting expression for $U_I(t)$ is

$$U_I(t) = \cos\left(\frac{\Omega_R t}{2}\right)(|a\rangle\langle a| + |b\rangle\langle b|) + i\sin\left(\frac{\Omega_R t}{2}\right)\left(e^{-i\phi}|a\rangle\langle b|\right.$$
$$\left. + e^{i\phi}|b\rangle\langle a|\right). \tag{5.2.46}$$

If the atom is initially in the excited state ($|\psi(0)\rangle \equiv |a\rangle$),

$$|\psi(t)\rangle = U_I(t)|a\rangle$$
$$= \cos\left(\frac{\Omega_R t}{2}\right)|a\rangle + i\sin\left(\frac{\Omega_R t}{2}\right)e^{i\phi}|b\rangle, \tag{5.2.47}$$

and we obtain the probability amplitudes

$$c_a(t) = \langle a|\psi\rangle = \cos\left(\frac{\Omega_R t}{2}\right), \tag{5.2.48a}$$

$$c_b(t) = \langle b|\psi\rangle = i\sin\left(\frac{\Omega_R t}{2}\right)e^{i\phi}, \tag{5.2.48b}$$

in agreement with Eqs. (5.2.21) and (5.2.22).

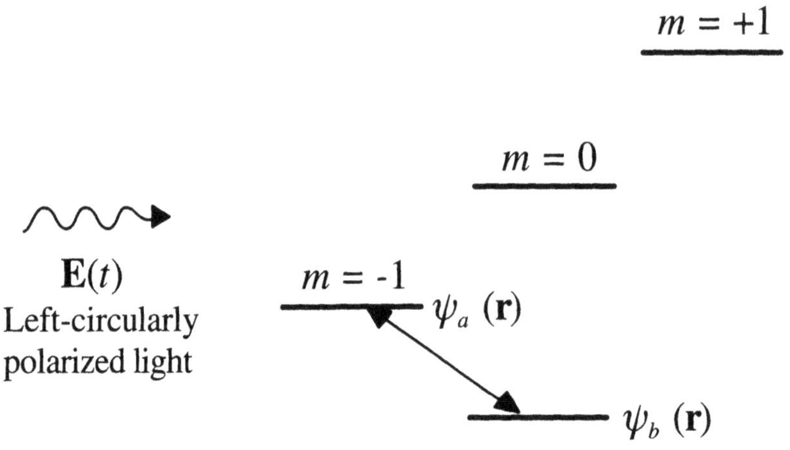

Fig. 5.3
Figure illustrating an incident electric field interacting with a hydrogen (Rydberg) atom such that the energy difference $\epsilon_{m=1} - \epsilon_b$ is much larger than $\epsilon_{m=-1} - \epsilon_b$.

5.2.3 *Beyond the rotating-wave approximation*

In quantum optics, the so-called rotating-wave approximation, as discussed in connection with Eq. (5.2.13), is frequently encountered. Of course, it is a very good approximation and amounts to keeping only energy-conserving terms in the Hamiltonian.

Moreover, as we show here, there are situations in which it is "exact", i.e., for all practical purposes the 'counter-rotating terms' never show up. Consider the case of a hydrogen atom in a strong magnetic field interacting with a monochromatic field as shown in Fig. 5.3. If the levels are sharp and well separated, we may focus on only the two levels, see Problem 5.7, for which

$$\psi_a(\mathbf{r}) = \frac{1}{\sqrt{64\pi a_0^3}} \frac{1}{a_0}(x - iy)\exp(-r/2a_0), \tag{5.2.49a}$$

$$\psi_b(\mathbf{r}) = \frac{1}{\sqrt{\pi a_0^3}} \exp(-r/a_0), \tag{5.2.49b}$$

where a_0 is the Bohr radius.

Using the dipole approximation (see Section 5.1.2) and placing the atom at the origin so that $\mathbf{R} = 0$, we have the interaction picture Hamiltonian

$$\mathscr{V} = -e\mathbf{r}(t) \cdot \mathbf{E}(t), \tag{5.2.50a}$$

where

$$\mathbf{r}(t) = e^{i\mathscr{H}_0 t}\mathbf{r}e^{-i\mathscr{H}_0 t}, \tag{5.2.50b}$$

and therefore

$$\mathscr{V}_{ab}(t) = -e\mathbf{r}_{ab}(t) \cdot \mathbf{E}(t) = -e\mathbf{r}_{ab} \cdot \mathbf{E}(t)e^{i\omega t}, \tag{5.2.51a}$$

$$\mathscr{V}_{ba}(t) = -e\mathbf{r}_{ba}(t) \cdot \mathbf{E}(t) = -e\mathbf{r}_{ba} \cdot \mathbf{E}(t)e^{-i\omega t}, \tag{5.2.51b}$$

where ω is the atomic frequency.

Now, for the case of linear polarization in which

$$\mathbf{E}(t) = \hat{x}\mathscr{E} \cos \nu t, \tag{5.2.52}$$

Eqs. (5.2.51a, 5.2.51b) and (5.2.52) imply

$$\begin{aligned} \mathscr{V}_{ab}(t) &= -ex_{ab}\mathscr{E} \cos \nu t e^{i\omega t} \\ &= -ex_{ab}\frac{\mathscr{E}}{2}\left[e^{i(\nu+\omega)t} + e^{-i(\nu-\omega)t}\right] \\ &\cong -ex_{ab}\frac{\mathscr{E}}{2}e^{-i(\nu-\omega)t}, \end{aligned} \tag{5.2.53}$$

and likewise

$$\begin{aligned} \mathscr{V}_{ba}(t) &= -ex_{ba}\mathscr{E} \cos \nu t e^{-i\omega t} \\ &= -ex_{ba}\frac{\mathscr{E}}{2}\left[e^{i(\nu-\omega)t} + e^{-i(\nu+\omega)t}\right] \\ &\cong -ex_{ba}\frac{\mathscr{E}}{2}e^{i(\nu-\omega)t}. \end{aligned} \tag{5.2.54}$$

Thus we make the rotating-wave approximation in neglecting counter terms that go like $\exp[\pm i(\omega + \nu)t]$.

Now consider the case of left-circular polarization (LCP), which connects $\psi_a(r)$ and $\psi_b(r)$, as given by Eqs. (5.2.49). The electric field is given by

$$\mathbf{E}(t) = \hat{x}\mathscr{E} \cos \nu t - \hat{y}\mathscr{E} \sin \nu t. \tag{5.2.55}$$

Equations (5.2.53) and (5.2.54) now take the form

$$\mathscr{V}_{ab}(t) = -e\mathscr{E}\left(x_{ab} \cos \nu t + y_{ab} \sin \nu t\right)e^{i\omega t} \tag{5.2.56a}$$

$$\mathscr{V}_{ba}(t) = -e\mathscr{E}\left(x_{ba} \cos \nu t + y_{ba} \sin \nu t\right)e^{-i\omega t} \tag{5.2.56b}$$

where, in view of Eqs. (5.2.49a) and (5.2.49b), we can write

$$ex_{ab} = \int \psi_a^*(\mathbf{r})x\psi_b(\mathbf{r})dr = \wp, \tag{5.2.57a}$$

$$ey_{ab} = \int \psi_a^*(\mathbf{r})y\psi_b(\mathbf{r})dr = -i\wp, \tag{5.2.57b}$$

and similarly, $ex_{ba} = \wp$ and $ey_{ba} = i\wp$. Therefore Eqs. (5.2.56a) and (5.2.56b) take the simple form

$$\mathscr{V}_{ab}(t) = -\wp\mathscr{E}\left(\cos \nu t - i \sin \nu t\right)e^{i\omega t} = -\wp\mathscr{E}e^{-i(\nu-\omega)t} \tag{5.2.58a}$$

$$\mathscr{V}_{ba}(t) = -\wp\mathscr{E}\left(\cos \nu t + i \sin \nu t\right)e^{-i\omega t} = -\wp\mathscr{E}e^{i(\nu-\omega)t}, \tag{5.2.58b}$$

and the counter-rotating terms never appear.

Finally, we note that although there are no counter terms of the form $e^{i(\nu+\omega)t}$ associated with the LCP light inducing $\Delta m = -1$ transitions, there are counter terms associated with LCP and transitions to a state $n = 2, l = 1, m_l = +1$, i.e., $\Delta m = +1$. Such transitions are usually said to vanish due to angular momentum selection rules. Here they are seen to 'vanish' since they go as counter rotating terms. That is, they are allowed in the sense of an atom making a transition to an excited state with the emission of a photon. Such terms can be much smaller than the usual counter-rotating terms; see Problem 5.7.

5.3 Density matrix for a two-level atom*

For a given physical system, there exists a state vector $|\psi\rangle$ which contains all possible information about the system. If we want to extract a piece of the system's information, we must calculate the expectation value of the corresponding operator O,

$$\langle O\rangle_{\mathrm{QM}} = \langle\psi|O|\psi\rangle. \tag{5.3.1}$$

In many situations we may not know $|\psi\rangle$; we may only know the probability P_ψ that the system is in the state $|\psi\rangle$. For such a situation, we not only need to take the quantum mechanical average but also the ensemble average over many identical systems that have been similarly prepared. Instead of Eq. (5.3.1), we now have (see Section 3.1)

$$\langle\langle O\rangle_{\mathrm{QM}}\rangle_{\mathrm{ensemble}} = \mathrm{Tr}(O\rho), \tag{5.3.2}$$

where the density operator ρ is defined by

$$\rho = \sum_\psi P_\psi |\psi\rangle\langle\psi|. \tag{5.3.3}$$

It can be seen that

$$\mathrm{Tr}(O\rho) = \mathrm{Tr}(\rho O). \tag{5.3.4}$$

In the particular case where all P_ψ are zero except the one for a state $|\psi_0\rangle$, then

$$\rho = |\psi_0\rangle\langle\psi_0|, \tag{5.3.5}$$

and the state is called a pure state. It follows from the conservation of probability that $\mathrm{Tr}(\rho) = 1$. Also, for a pure state,

$$\mathrm{Tr}(\rho^2) = 1. \tag{5.3.6}$$

* The more complete picture of stimulated emission as developed by Lamb and Scully [1971] is found in Chapter III of Sargent, Scully and Lamb [1974].

5.3.1 Equation of motion for the density matrix

We can obtain the equation of motion for the density matrix from the Schrödinger equation,

$$|\dot{\psi}\rangle = -\frac{i}{\hbar}\mathscr{H}|\psi\rangle. \tag{5.3.7}$$

Taking the time derivative of ρ (Eq. (5.3.3)) we have

$$\dot{\rho} = \sum_{\psi} P_{\psi}(|\dot{\psi}\rangle\langle\psi| + |\psi\rangle\langle\dot{\psi}|), \tag{5.3.8}$$

where P_{ψ} is time independent. Using Eq. (5.3.7) to replace $|\dot{\psi}\rangle$ and $\langle\dot{\psi}|$ in Eq. (5.3.8) we get

$$\dot{\rho} = -\frac{i}{\hbar}[\mathscr{H}, \rho]. \tag{5.3.9}$$

Equation (5.3.9) is often called the Liouville or Von Neumann equation of motion for the density matrix. It is more general than the Schrödinger equation since it uses the density operator instead of a specific state vector and can therefore give statistical as well as quantum mechanical information.

In Eq. (5.3.9), we have not included the decay of the atomic levels due to spontaneous emission. The excited atomic levels can also decay because of collisions and other phenomena. The finite lifetime of the atomic levels can be described very well by adding phenomenological decay terms to the density operator equation (5.3.9) (see also Problem 5.2).

The decay rates can be incorporated in Eq. (5.3.9) by a relaxation matrix Γ, which is defined by the equation

$$\langle n|\Gamma|m\rangle = \gamma_n \delta_{nm}. \tag{5.3.10}$$

With this addition, the density matrix equation of motion becomes

$$\dot{\rho} = -\frac{i}{\hbar}[\mathscr{H}, \rho] - \frac{1}{2}\{\Gamma, \rho\}, \tag{5.3.11}$$

where $\{\Gamma, \rho\} = \Gamma\rho + \rho\Gamma$. In general, the relaxation processes are more complicated.

The ijth matrix element of Eq. (5.3.11) is

$$\dot{\rho}_{ij} = -\frac{i}{\hbar}\sum_k(\mathscr{H}_{ik}\rho_{kj} - \rho_{ik}\mathscr{H}_{kj}) - \frac{1}{2}\sum_k(\Gamma_{ik}\rho_{kj} + \rho_{ik}\Gamma_{kj}). \tag{5.3.12}$$

This formula is useful in the treatment of many-level systems.

5.3.2 Two-level atom

We now consider the two-level atomic system again where the state of the system is a linear combination of states $|a\rangle$ and $|b\rangle$, i.e., $|\psi\rangle = C_a|a\rangle + C_b|b\rangle$. Then the density matrix operator can be written as

$$\begin{aligned}\rho = |\psi\rangle\langle\psi| &= [C_a(t)|a\rangle + C_b(t)|b\rangle]\ [C_a^*(t)\langle a| + C_b^*(t)\langle b|]\\ &= |C_a|^2|a\rangle\langle a| + C_aC_b^*|a\rangle\langle b| + C_bC_a^*|b\rangle\langle a| + |C_b|^2|b\rangle\langle b|.\end{aligned} \tag{5.3.13}$$

Taking the matrix elements, we get

$$\rho_{aa} = \langle a|\rho|a\rangle = |C_a(t)|^2, \tag{5.3.14}$$

$$\rho_{ab} = \langle a|\rho|b\rangle = C_a(t)C_b^*(t), \tag{5.3.15}$$

$$\rho_{ba} = \rho_{ab}^*, \tag{5.3.16}$$

$$\rho_{bb} = \langle b|\rho|b\rangle = |C_b(t)|^2. \tag{5.3.17}$$

The matrix form of the density operator is

$$\rho = \begin{pmatrix} \rho_{aa} & \rho_{ab} \\ \rho_{ba} & \rho_{bb} \end{pmatrix}. \tag{5.3.18}$$

It is obvious that ρ_{aa} and ρ_{bb} are the probabilities of being in the upper and lower states, respectively. For the meaning of the off-diagonal elements we need to remember that the atomic polarization, see Eq. (5.2.25), of the two-level atom (at z) is

$$P(z,t) = C_aC_b^*\wp_{ba} + \text{c.c.} = \rho_{ab}(z,t)\wp_{ba} + \text{c.c.} \tag{5.3.19}$$

So we see that the off-diagonal elements determine the atomic polarization.

We could have found this form for ρ more directly from Eq. (5.3.5) by remembering that in spinor notation

$$|\psi\rangle = \begin{pmatrix} C_a \\ C_b \end{pmatrix}; \qquad \langle\psi| = (C_a^*\ C_b^*). \tag{5.3.20}$$

Then by matrix multiplication

$$\rho = \begin{pmatrix} C_a \\ C_b \end{pmatrix}(C_a^*\ C_b^*) = \begin{pmatrix} |C_a|^2 & C_aC_b^* \\ C_bC_a^* & |C_b|^2 \end{pmatrix}. \tag{5.3.21}$$

We can derive the equations of motion for the density matrix elements from Eq. (5.3.12) with the Hamiltonian given by Eqs. (5.2.4) and (5.2.5). The resulting equations are

$$\dot{\rho}_{aa} = -\gamma_a \rho_{aa} + \frac{i}{\hbar}[\wp_{ab} E \rho_{ba} - \text{c.c.}], \tag{5.3.22}$$

$$\dot{\rho}_{bb} = -\gamma_b \rho_{bb} - \frac{i}{\hbar}[\wp_{ab} E \rho_{ba} - \text{c.c.}], \tag{5.3.23}$$

$$\dot{\rho}_{ab} = -(i\omega + \gamma_{ab})\rho_{ab} - \frac{i}{\hbar}\wp_{ab} E(\rho_{aa} - \rho_{bb}), \tag{5.3.24}$$

where $\gamma_{ab} = (\gamma_a + \gamma_b)/2$ with γ_a and γ_b defined by Eq. (5.3.10) and $E(t)$ is given by Eq. (5.2.6). In the rotating-wave approximation, $\cos(\nu t)$ is replaced by $\exp(-i\nu t)/2$ in Eqs. (5.3.22)–(5.3.24).

5.3.3 *Inclusion of elastic collisions between atoms*

The physical interpretation of the elements of the density matrix allows us to include in these equations terms associated with certain processes. One such process is the elastic collision between atoms in a gas.

In particular, during an atom–atom collision the energy levels experience random Stark shifts without a change of state and the decay rate for ρ_{ab} is increased without much change in γ_a and γ_b. The change in the decay rate of ρ_{ab} may be computed in a simple way as follows.

We assume that the random Stark shifts are included in Eq. (5.3.24) by adding a random shift $\delta\omega(t)$ to the energy difference ω. Ignoring the atom–field interactions for simplicity, we can write the equation of motion for the density matrix element ρ_{ab} as

$$\dot{\rho}_{ab} = -[i\omega + i\delta\omega(t) + \gamma_{ab}]\rho_{ab}. \tag{5.3.25}$$

Integrating Eq. (5.3.25) formally, we have

$$\rho_{ab}(t) = \exp\left[-(i\omega + \gamma_{ab})t - i\int_0^t dt'\delta\omega(t')\right]\rho_{ab}(0). \tag{5.3.26}$$

We now perform an ensemble average of (5.3.26) over the random variations in $\delta\omega(t)$. This average affects only the $\delta\omega(t)$ factor, so that we find $\langle \exp[-i\int_0^t dt'\delta\omega(t')]\rangle$.

The function $\delta\omega$ is as often positive as negative. Hence the ensemble average $\langle\delta\omega(t)\rangle$ is zero. Furthermore, as the variations in $\delta\omega(t)$ are usually rapid compared to other changes which occur in times like $1/\gamma_{ab}$, we take

$$\langle\delta\omega(t)\delta\omega(t')\rangle = 2\gamma_{\text{ph}}\delta(t-t'), \tag{5.3.27}$$

where γ_{ph} is a constant. We also assume that $\delta\omega(t)$ is described by a Gaussian random process, so that the well-known moment theorem of Gaussian processes is valid. Under these conditions we obtain

$$\left\langle \exp\left[-i\int_0^t dt'\delta\omega(t')\right]\right\rangle = \exp(-\gamma_{\rm ph}t), \tag{5.3.28}$$

which gives for the average of (5.3.26)

$$\rho_{ab}(t) = \exp[-(i\omega + \gamma_{ab} + \gamma_{\rm ph})t]\rho_{ab}(0). \tag{5.3.29}$$

It follows, on differentiating this equation and including the interaction term, that we have the modified equation of motion for ρ_{ab}:

$$\dot{\rho}_{ab} = -(i\omega + \gamma)\rho_{ab} - \frac{i}{\hbar}\wp_{ab}E(z,t)(\rho_{aa} - \rho_{bb}), \tag{5.3.30}$$

where $\gamma = \gamma_{ab} + \gamma_{\rm ph}$ is the new decay rate. Equation (5.3.30) is an average equation with respect to collisions.

5.4 Maxwell–Schrödinger equations

The interaction of a single atom with the single-mode field, which was discussed in the previous sections, represents a simple, idealized system. In many problems of interest in quantum optics, one is interested in the interaction of the radiation field with a large number of atoms. The prime example of such a system is a single-mode laser where atoms pumped into the excited level interact with the electromagnetic field inside a cavity . Other examples include coherent pulse propagation and optical bistability.

In this section, we develop a mathematical framework to treat such problems based on a self-consistent set of equations for the matter and the field. This set of equations and its extensions enable us to deal with many semiclassical problems where the atoms are treated quantum mechanically and the field is treated classically.

In the present semiclassical atom–field interaction, the classical field induces electric dipole moments in the medium according to the laws of quantum mechanics. The density matrix is used to facilitate the statistical summations involved in obtaining the macroscopic polarization of the medium for the individual dipole moments. The semiclassical approach, though remarkably good for many problems of interest in the study of a given system, is however inadequate to provide information about the quantum statistical features of light. These aspects will be presented in later chapters where the radiation field will be treated quantum mechanically.

5.4.1 Population matrix and its equation of motion

We consider the interaction of an electromagnetic field with a medium which consists of two-level homogeneously broadened atoms. The individual atoms are described by the density operator (see Eqs. (5.3.14)–(5.3.17))

$$\rho(z,t,t_0) = \sum_{\alpha,\beta} \rho_{\alpha\beta}(z,t,t_0)|\alpha\rangle\langle\beta|, \tag{5.4.1}$$

where $\alpha, \beta = a, b$ and $\rho_{\alpha\beta}(z,t,t_0)$ are the density matrix elements for an individual atom at time t and position z, which starts interacting with the field at an initial time t_0. The initial time t_0 can be random. The single-atom density matrix elements $\rho_{\alpha\beta}(z,t,t_0)$ obey the equations of motion (5.3.22), (5.3.23), and (5.3.30). If the state of the atom at the time of injection is described by

$$\rho(z,t_0,t_0) = \sum_{\alpha,\beta} \rho_{\alpha\beta}^{(0)}|\alpha\rangle\langle\beta|, \tag{5.4.2}$$

then

$$\rho_{\alpha\beta}(z,t_0,t_0) = \rho_{\alpha\beta}^{(0)}. \tag{5.4.3}$$

The effect of all the atoms which are pumped at the rate $r_a(z,t_0)$ atoms per second per unit volume is obtained by summing over initial times. The resulting *population matrix* is defined as

$$\begin{aligned}\rho(z,t) &= \int_{-\infty}^{t} dt_0 r_a(z,t_0)\rho(z,t,t_0) \\ &= \sum_{\alpha,\beta}\int_{-\infty}^{t} dt_0 r_a(z,t_0)\rho_{\alpha\beta}(z,t,t_0)|\alpha\rangle\langle\beta|.\end{aligned} \tag{5.4.4}$$

Generally the excitation $r_a(z,t_0)$ varies slowly and can be taken to be a constant. The macroscopic polarization of the medium, $P(z,t)$, will be produced by an ensemble of atoms that arrive at z at time t, regardless of their time of excitation, i.e.,

$$\begin{aligned}P(z,t) &= \int_{-\infty}^{t} dt_0 r_a(z,t_0)\mathrm{Tr}[\hat{\wp}\rho(z,t,t_0)] \\ &= \sum_{\alpha,\beta}\int_{-\infty}^{t} dt_0 r_a(z,t_0)\rho_{\alpha\beta}(z,t,t_0)\wp_{\beta\alpha},\end{aligned} \tag{5.4.5}$$

where $\hat{\wp}$ is the dipole moment operator and, in the second line, we have substituted for $\rho(z,t,t_0)$ from Eq. (5.4.1). For a two-level atom, with $\wp_{ab} = \wp_{ba} = \wp$, we obtain

$$P(z,t) = \wp[\rho_{ab}(z,t) + \text{c.c.}]. \tag{5.4.6}$$

Thus the off-diagonal elements of the population matrix determine the macroscopic polarization.

The equations of motion for the elements of the population matrix $\rho(z,t)$ can be obtained by taking the time derivative of Eq. (5.4.4) and using Eqs. (5.3.22), (5.3.23), and (5.3.30). For example, if the atoms are incoherently excited to levels $|a\rangle$ and $|b\rangle$ at a constant rate r_a ($\rho_{ab}^{(0)} = \rho_{ba}^{(0)} = 0$), we obtain

$$\dot{\rho}_{aa} = \lambda_a - \gamma_a \rho_{aa} + \frac{i}{\hbar}(\wp E \rho_{ba} - \text{c.c.}), \tag{5.4.7}$$

$$\dot{\rho}_{bb} = \lambda_b - \gamma_b \rho_{bb} - \frac{i}{\hbar}(\wp E \rho_{ba} - \text{c.c.}), \tag{5.4.8}$$

$$\dot{\rho}_{ab} = -(i\omega + \gamma)\rho_{ab} - \frac{i}{\hbar}\wp E(\rho_{aa} - \rho_{bb}), \tag{5.4.9}$$

where $\lambda_a = r_a \rho_{aa}^{(0)}$ and $\lambda_b = r_a \rho_{bb}^{(0)}$. These equations for the two-level atomic medium are coupled to the field E. The condition of self-consistency requires that the equation of motion for the field E is driven by the atomic population matrix elements. In the following section, we derive such an equation for a single-mode running wave.

5.4.2 *Maxwell's equations for slowly varying field functions*

The electromagnetic field radiation is described by Maxwell's equations:

$$\nabla \cdot \mathbf{D} = 0, \qquad \nabla \times \mathbf{E} = -\frac{\partial \mathbf{B}}{\partial t}, \tag{5.4.10}$$

$$\nabla \cdot \mathbf{B} = 0, \qquad \nabla \times \mathbf{H} = \mathbf{J} + \frac{\partial \mathbf{D}}{\partial t}, \tag{5.4.11}$$

where

$$\mathbf{D} = \epsilon_0 \mathbf{E} + \mathbf{P}, \quad \mathbf{B} = \mu_0 \mathbf{H}, \quad \mathbf{J} = \sigma \mathbf{E}. \tag{5.4.12}$$

Here $\mathbf{P}$ is the macroscopic polarization of the medium. In order to avoid a complicated boundary-value problem, we assume the presence of a medium with conductivity σ. This conductivity is intended to take into account phenomenologically the linear losses due to any absorbing background medium, and also those losses due to diffraction and mirror transmission. Combining the curl equations, taking the appropriate time derivatives, and using Eq. (5.4.12), we get the wave equation

$$\nabla \times (\nabla \times \mathbf{E}) + \mu_0 \sigma \frac{\partial \mathbf{E}}{\partial t} + \mu_0 \epsilon_0 \frac{\partial^2 \mathbf{E}}{\partial t^2} = -\mu_0 \frac{\partial^2 \mathbf{P}}{\partial t^2}. \tag{5.4.13}$$

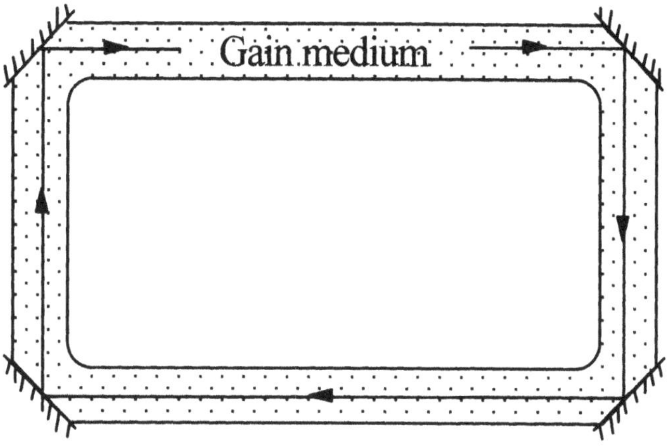

Fig. 5.4
Schematic diagram of a laser in a unidirectional ring configuration.

The polarization $\mathbf{P}(\mathbf{r},t)$ thus acts as a source term in the equation for the radiation field. We have in mind a situation in which the radiation field interacts with two-level atoms inside a unidirectional ring cavity as shown in Fig. 5.4. Usually both running waves exist inside the cavity. The unidirectional situation is achieved by the insertion of a device with high loss for one running wave. The variations in the field intensity transverse to the laser axis are typically slowly varying on the scale of the optical wavelength. Hence, we neglect the x- and y-dependence of $\mathbf{E}$, i.e.,

$$\mathbf{E}(\mathbf{r},t) = E(z,t)\hat{\mathbf{x}}. \tag{5.4.14}$$

Equation (5.4.13) then reduces to

$$-\frac{\partial^2 E}{\partial z^2} + \mu_0\sigma\frac{\partial E}{\partial t} + \frac{1}{c^2}\frac{\partial^2 E}{\partial t^2} = -\mu_0\frac{\partial^2 P}{\partial t^2}. \tag{5.4.15}$$

The field of frequency ν is represented as a running wave

$$E(z,t) = \frac{1}{2}\mathscr{E}(z,t)e^{-i[\nu t - kz + \phi(z,t)]} + \text{c.c.}, \tag{5.4.16}$$

where $\mathscr{E}(z,t)$ and $\phi(z,t)$ are slowly varying functions of position and time with $k = \nu/c$. For the problem of laser oscillation, $k = \nu_c/c$ where ν_c is the cavity frequency. In general, $\mathscr{E}(z,t)$ is a complex function; however, in the present and in the next section, we assume it to be real.

If the field is written as in Eq. (5.4.16) then the response of the medium, neglecting higher harmonics, is given by the polarization

$$P(z,t) = \frac{1}{2}\mathscr{P}(z,t)e^{-i[\nu t - kz + \phi(z,t)]} + \text{c.c.}, \tag{5.4.17}$$

where $\mathscr{P}(z,t)$ is a slowly varying function of position and time.

The slowly varying complex polarization $\mathscr{P}(z,t)$ is given in terms of the population matrix by identification of the positive frequency parts in Eqs. (5.4.6) and (5.4.17):

$$\mathscr{P}(z,t) = 2\wp\rho_{ab}e^{i[\nu t-kz+\phi(z,t)]}. \tag{5.4.18}$$

The expressions for $E(z,t)$ and $P(z,t)$ are substituted from Eqs. (5.4.16) and (5.4.17) in Eq. (5.4.15), and the following approximations are made

$$\frac{\partial\mathscr{E}}{\partial t} \ll \nu\mathscr{E}, \quad \frac{\partial\mathscr{E}}{\partial z} \ll k\mathscr{E}, \quad \frac{\partial\phi}{\partial t} \ll \nu, \quad \frac{\partial\phi}{\partial z} \ll k, \tag{5.4.19}$$

$$\frac{\partial\mathscr{P}}{\partial t} \ll \nu\mathscr{P}, \quad \frac{\partial\mathscr{P}}{\partial z} \ll k\mathscr{P}. \tag{5.4.20}$$

These slowly varying amplitude and phase approximations are justified when $\mathscr{E}$, ϕ, and $\mathscr{P}$ do not change appreciably in an optical frequency period. By noting that Eq. (5.4.15) can be rewritten as

$$\left(\frac{\partial}{\partial z} + \frac{1}{c}\frac{\partial}{\partial t}\right)\left(-\frac{\partial}{\partial z} + \frac{1}{c}\frac{\partial}{\partial t}\right)E = -\mu_0\sigma\frac{\partial E}{\partial t} - \mu_0\frac{\partial^2 P}{\partial t^2}, \tag{5.4.21}$$

and

$$\left(-\frac{\partial}{\partial z} + \frac{1}{c}\frac{\partial}{\partial t}\right)E \cong -2ikE, \tag{5.4.22}$$

we obtain

$$\frac{\partial\mathscr{E}}{\partial z} + \frac{1}{c}\frac{\partial\mathscr{E}}{\partial t} = -\kappa\mathscr{E} - \frac{1}{2\epsilon_0}k\,\mathrm{Im}\mathscr{P}, \tag{5.4.23}$$

$$\frac{\partial\phi}{\partial z} + \frac{1}{c}\frac{\partial\phi}{\partial t} = k - \frac{\nu}{c} - \frac{1}{2\epsilon_0}k\mathscr{E}^{-1}\mathrm{Re}\mathscr{P}, \tag{5.4.24}$$

where $\kappa = \sigma/2\epsilon_0 c$ is the linear loss coefficient.

Equations (5.4.7)–(5.4.9), (5.4.23), and (5.4.24) form a self-consistent set of equations. This set of equations is the starting point of the study of many systems involving the interaction of the radiation field with an ensemble of atoms. The generalization of this set of equations to a multi-level atomic system and a multi-mode field is straightforward.

As an important example of the applications of this set of equations, we present the semiclassical theory of the laser in the next section.

5.5 Semiclassical laser theory

In this section, we first outline the basic principle of laser operation and then present a theory of the laser as developed principally by Lamb and co-workers. The threshold condition for a laser and the evolution equation of the electromagnetic field is also derived.

5.5.1 Basic principle

In 1954, Gordon, Zeiger, and Townes showed that coherent electromagnetic radiation can be generated in the radio frequency range by the so-called maser (microwave amplification by stimulated emission of radiation). The first maser action was observed in ammonia.

The maser principle was extended by Schawlow and Townes, and also by Prokhorov, to the optical domain, thus obtaining a laser (light amplification by stimulated emission of radiation). A laser consists of a set of atoms interacting with an electromagnetic field inside a cavity. The cavity supports only a specific set of modes corresponding to a discrete sequence of eigenfrequencies. The active atoms, i.e., the ones that are pumped to the upper level of the laser transition, are in resonance with one of the eigenfrequencies of the cavity. A resonant electromagnetic field gives rise to stimulated emission, and the atoms transfer their excitation energy to the radiation field. The emitted radiation is still at resonance. If the upper level is sufficiently populated, this radiation gives rise to further transitions in other atoms. In this way all the excitation energy of the atoms is transferred to a single mode of the radiation field.

The first pulsed laser operation was demonstrated by Maiman in ruby. The first continuous wave (cw) laser, a He–Ne gas laser, was built by Javan. Since then, a large variety of systems have been demonstrated to exhibit lasing action; generating coherent light over a frequency domain ranging from infrared to ultraviolet. These include dye lasers, chemical lasers, and semiconductor lasers. A new class of lasers which uses electrons in a periodic magnetic field (called free-electron lasers) has also been developed.

From our discussion of the laser principle, it is clear that a laser theory should incorporate three basic elements, an active medium (two-level atoms with population inversion), pumping to the upper lasing level, and the radiation losses due to the cavity. A systematic semiclassical theory of the laser was developed by Lamb.

5.5.2 Lamb's semiclassical theory

We consider the semiclassical laser theory for the simple case of a linearly polarized electric field in a unidirectional ring cavity and two-level homogeneously broadened, active atoms.

The time dependence of the electric field $\mathscr{E}(z,t)$ can be separated from the spatial part by expanding the field in the normal modes of the cavity. In a ring cavity only certain discrete modes achieve appreciable magnitude, namely, those with the frequencies

$$\nu_m = \frac{m\pi c}{S} = k_m c, \tag{5.5.1}$$

where S is the circumference of the ring, m is a large integer (typically of the order 10^6), and k_m is the corresponding wave number. Here, we consider a single mode with unidirectional (running-wave) mode functions $U(z) = \exp(ikz)$ (Fig. 5.4).

The equations of motion for the field amplitude (5.4.23) and phase (5.4.24) for the present problem reduce to

$$\frac{\partial \mathscr{E}}{\partial t} = -\frac{1}{2}\mathscr{C}\mathscr{E} - \frac{1}{2}\left(\frac{\nu}{\epsilon_0}\right)\text{Im}\mathscr{P}, \tag{5.5.2a}$$

$$\frac{\partial \phi}{\partial t} = (\nu_c - \nu) - \frac{1}{2}\left(\frac{\nu}{\epsilon_0}\right)\mathscr{E}^{-1}\text{Re}\mathscr{P}, \tag{5.5.2b}$$

where ν_c is the cavity frequency and $\gamma = (\gamma_a + \gamma_b)/2$. In Eq. (5.5.2a), κ has been replaced by $\mathscr{C}/2c$ where $\mathscr{C} = \nu_c/Q$ (where Q is the quality factor of the cavity) to account for the field losses through the mirrors of the cavity. The driving polarization $\mathscr{P}$ (Eq. (5.4.18)) is determined by Eq. (5.4.9) which yields

$$\mathscr{P}(z,t) = \frac{-i\wp^2}{\hbar}\int_{-\infty}^{t} \exp[-\gamma(t-t') - i(\omega - \nu)(t-t')] \times \mathscr{E}(t')[\rho_{aa}(t') - \rho_{bb}(t')]dt'. \tag{5.5.3}$$

The integral (5.5.3) can be simply performed, provided the amplitude $\mathscr{E}(t')$ and the population difference $\rho_{aa} - \rho_{bb}$ do not change appreciably in the time $1/\gamma$, for then these terms can be factored outside the integral. This solution leads to rate equations for the atomic populations. These approximations are exact in steady state ($\dot{\mathscr{P}} = 0$). This gives

$$\text{Im}\mathscr{P}(t) = \frac{-\wp^2}{\hbar}\mathscr{E}(t)\gamma\frac{\rho_{aa}(t) - \rho_{bb}(t)}{\gamma^2 + (\omega - \nu)^2}, \tag{5.5.4a}$$

$$\text{Re}\mathscr{P}(t) = \frac{-\wp^2}{\hbar}\mathscr{E}(t)(\omega - \nu)\frac{\rho_{aa}(t) - \rho_{bb}(t)}{\gamma^2 + (\omega - \nu)^2}. \tag{5.5.4b}$$

On substituting Eqs. (5.5.4) into the equations of motion for ρ_{aa} and ρ_{bb} ((5.4.7) and (5.4.8)), we obtain the rate equations

$$\dot{\rho}_{aa} = \lambda_a - \gamma_a\rho_{aa} - R(\rho_{aa} - \rho_{bb}), \tag{5.5.5a}$$

$$\dot{\rho}_{bb} = \lambda_b - \gamma_b\rho_{bb} + R(\rho_{aa} - \rho_{bb}), \tag{5.5.5b}$$

where the rate constant is

$$R = \frac{1}{2}\left(\frac{\wp\mathscr{E}}{\hbar}\right)^2 \frac{\gamma}{\gamma^2 + (\omega - \nu)^2}. \tag{5.5.6}$$

It is evident that the rate constant R, which determines the rate at which the population difference varies in time, depends primarily on the rate at which the total field intensity varies. Hence, the rate-equation approximation consists of the assumption that the electric field envelope varies slowly in atomic lifetimes. We can determine the population difference in the steady state from Eqs. (5.5.5). This difference can be substituted in turn back into the equations for $\text{Im}\mathscr{P}$ and $\text{Re}\mathscr{P}$, thus determining the polarization components.

In the steady state ($\dot{\rho}_{aa} = \dot{\rho}_{bb} = 0$), Eqs. (5.5.5) yield

$$\rho_{aa} - \rho_{bb} = \frac{N_0}{1 + R/R_s}, \tag{5.5.7}$$

where $N_0 = \lambda_a \gamma_a^{-1} - \lambda_b \gamma_b^{-1}$ and $R_s = \gamma_a \gamma_b / 2\gamma$. The population difference is therefore given by N_0, which appears in the absence of the field, divided by a factor which increases as the intensity of the electric field increases.

Combining Eq. (5.5.7) with Eqs. (5.5.4) and (5.5.2) we obtain the amplitude and frequency determining equations

$$\dot{\mathscr{E}} = -\frac{\mathscr{C}}{2}\mathscr{E} + \frac{\mathscr{A}\mathscr{E}}{2\left[1 + \frac{\mathscr{B}}{\mathscr{A}}\left(\frac{\epsilon_0 V}{2\hbar\nu}\right)\mathscr{E}^2\right]}, \tag{5.5.8}$$

$$\nu + \dot{\phi} = \nu_c + \frac{(\omega - \nu)\mathscr{A}}{2\gamma\left[1 + \frac{\mathscr{B}}{\mathscr{A}}\left(\frac{\epsilon_0 V}{2\hbar\nu}\right)\mathscr{E}^2\right]}, \tag{5.5.9}$$

where V is the volume of the cavity and

$$\mathscr{A} = \left(\frac{\wp^2 \nu\gamma}{\epsilon_0 \hbar}\right)\frac{N_0}{\gamma^2 + (\omega - \nu)^2}, \tag{5.5.10a}$$

$$\mathscr{B} = \left(\frac{4\wp^2}{\hbar^2}\right)\left(\frac{\gamma^2}{\gamma_a\gamma_b}\right)\frac{\mathscr{A}}{\gamma^2 + (\omega - \nu)^2}\left(\frac{\hbar\nu}{2\epsilon_0 V}\right). \tag{5.5.10b}$$

Here $\mathscr{A}$ is the linear gain parameter and $\mathscr{B}$ is the saturation parameter.

We now define a dimensionless intensity

$$n = \frac{\epsilon_0 \mathscr{E}^2 V}{2\hbar\nu}, \tag{5.5.11}$$

which corresponds to the 'number of photons' in the laser. (Here, $\epsilon_0 \mathscr{E}^2 V/2$ is the total energy in the laser beam and $\hbar\nu$ is the energy associated with a single photon.) This statement will be sharpened when we study the quantum theory of the laser in Chapter 11. It follows from Eqs. (5.5.8) and (5.5.9) that

$$\dot{n} = -\mathscr{C}n + \frac{\mathscr{A}n}{1 + \frac{\mathscr{B}}{\mathscr{A}}n}, \tag{5.5.12}$$

$$\nu + \dot{\phi} = \nu_c + \frac{(\omega - \nu)\mathscr{A}}{2\gamma\left(1 + \frac{\mathscr{B}}{\mathscr{A}}n\right)}. \tag{5.5.13}$$

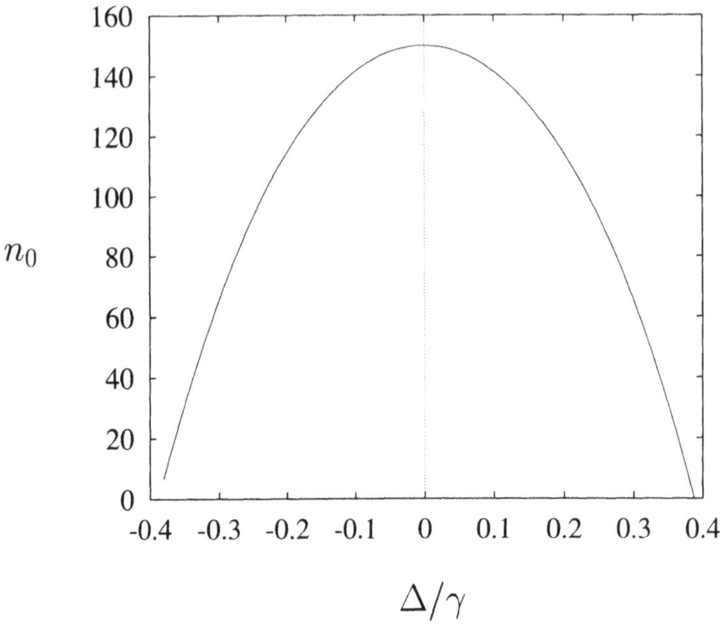

Fig. 5.5
A plot of the steady-state intensity n_0 versus the detuning δ/γ. Here $\mathscr{C} = 1\mu\text{sec}^{-1}$, $\wp^2 \nu N_0/\epsilon_0 \hbar \gamma = 1.15\mu\text{sec}^{-1}$, and $2\wp^4 \nu^2 N_0/\epsilon_0^2 \hbar^2 \gamma_a \gamma_b \gamma V = 10^{-3}\mu\text{sec}^{-1}$.

For small excitations ($\mathscr{B}n/\mathscr{A} \ll 1$), a perturbation theory is obtained by expanding the denominator in Eqs. (5.5.12) and (5.5.13), resulting in

$$\dot{n} = (\mathscr{A} - \mathscr{C})\, n - \mathscr{B} n^2, \tag{5.5.14}$$

$$\nu + \dot{\phi} = \nu_c + \left(\frac{\omega - \nu}{2\gamma}\right)(\mathscr{A} - \mathscr{B}n). \tag{5.5.15}$$

Equations (5.5.14) and (5.5.15) are the basic equations for the laser. As shown below, they yield the laser threshold condition, the steady-state and transient intensity of the laser, and the frequency pulling due to the presence of the gain medium.

It is easily seen from Eq. (5.5.14) that, in steady state ($\dot{n} = 0$), $n = 0$ unless $\mathscr{A} > \mathscr{C}$. When $\mathscr{A} > \mathscr{C}$, the steady-state intensity is given by

$$n_0 \equiv n = \frac{\mathscr{A} - \mathscr{C}}{\mathscr{B}}. \tag{5.5.16}$$

Thus, the laser threshold condition is $\mathscr{A} = \mathscr{C}$, i.e., when the gain is equal to the cavity losses.

In Fig. 5.5, the steady-state intensity is plotted against the detuning $\Delta = \omega - \nu$. According to Eq. (5.5.15), the oscillation frequency ν itself depends on the intensity. A good approximation, however, results from taking $\nu = \nu_c$ in the calculation of the various coefficients.

The frequency determining Eq. (5.5.15) predicts a pulling of the oscillation frequency from the passive cavity frequency towards line center. Specifically, in steady state ($\dot{\phi} = 0$)

$$\nu = \frac{\nu_c + \mathscr{S}\omega}{1 + \mathscr{S}}, \tag{5.5.17}$$

where the stabilization factor

$$\mathscr{S} = \frac{\mathscr{A} - \mathscr{B}n_0}{2\gamma} = \frac{\mathscr{C}}{2\gamma}. \tag{5.5.18}$$

Equation (5.5.17) can be interpreted as a center-of-mass equation in which the oscillation frequency ν assumes the average value of ν_c and ω with weights 1 and $\mathscr{S}$, respectively. In the typical case, $\mathscr{C} \ll 2\gamma$ and therefore $\nu \cong \nu_c$, but ν is pulled closer to the atomic frequency ω. This is called mode pulling.

5.6 A physical picture of stimulated emission and absorption

In order to better appreciate the physics behind stimulated emission and absorption, let us consider an atom at the point $z = 0$ interacting with the field $E(z,t) = \mathscr{E}(z,t)\cos(\nu t - kz)$. As before, the amplitudes C_a and C_b are determined by Eqs. (5.2.7) and (5.2.8), and the slowly varying amplitudes $c_a = C_a e^{i\omega_a t}$ and $c_b = C_b e^{i\omega_b t}$ are determined by Eqs. (5.2.12) and (5.2.13), respectively. For simplicity, we assume exact resonance $\Delta = \omega - \nu = 0$. Then the solution (5.2.21)–(5.2.22) becomes

$$c_a(t) = \left[c_a(0)\cos\left(\frac{\Omega_R t}{2}\right) + ic_b(0)\sin\left(\frac{\Omega_R t}{2}\right)\right], \tag{5.6.1a}$$

$$c_b(t) = \left[c_b(0)\cos\left(\frac{\Omega_R t}{2}\right) + ic_a(0)\sin\left(\frac{\Omega_R t}{2}\right)\right], \tag{5.6.1b}$$

where we have assumed a real dipole matrix element $\wp_{ab} = \wp_{ba} \equiv \wp$. Now, to the lowest order, we may trivially calculate $\rho_{ab} = c_a c_b^* e^{-i\omega t}$ for the cases of atom in the excited state and the ground state.

For the first case (stimulated emission), in which $c_a(0) = 1$ and $c_b(0) = 0$, we find to lowest order for an atom which passes through the laser cavity in a time τ

$$c_a(\tau) \cong 1, \tag{5.6.2a}$$

$$c_b(\tau) \cong i\frac{\Omega_R \tau}{2}, \tag{5.6.2b}$$

and the polarization is then (see Eq. (5.4.18))

$$\begin{aligned} \mathscr{P} &= 2\wp\rho_{ab}e^{i\nu\tau} \\ &\cong -i\wp\Omega_R\tau. \end{aligned} \tag{5.6.3}$$

For the case of absorption, initially $c_a(0) = 0$, $c_b(0) = 1$, to the lowest order one gets

$$c_a(\tau) \cong i\frac{\Omega_R \tau}{2}, \tag{5.6.4a}$$
$$c_b(\tau) \cong 1, \tag{5.6.4b}$$

and

$$\mathscr{P} \cong i\wp\Omega_R \tau. \tag{5.6.5}$$

Now, using Eq. (5.4.23), for the atom initially in the excited state we have

$$\frac{1}{c}\frac{\partial \mathscr{E}}{\partial t} = \frac{k}{2\epsilon_0}\frac{\wp^2}{\hbar}\mathscr{E}\tau, \tag{5.6.6}$$

where we have neglected the cavity loss. It follows from Eq. (5.6.6) that the change in the electric field during the time τ is

$$\Delta\mathscr{E} \cong \frac{ck}{2\epsilon_0}\frac{\wp^2}{\hbar}\mathscr{E}\tau^2, \tag{5.6.7}$$

i.e., the incident field experiences gain.

Likewise for the atom initially in the ground state, we have

$$\frac{1}{c}\frac{\partial \mathscr{E}}{\partial t} = -\frac{k}{2\epsilon_0}\frac{\wp^2}{\hbar}\mathscr{E}\tau, \tag{5.6.8}$$

and therefore

$$\Delta\mathscr{E} \cong -\frac{ck}{2\epsilon_0}\frac{\wp^2}{\hbar}\mathscr{E}\tau^2, \tag{5.6.9}$$

i.e., the incident field experiences loss. Thus the atom acts essentially as a tiny oscillating electronic current induced by the incident light field. Attenuation of an incident field is then the result of the radiation from this current interfering destructively with the incident light (see Fig. 5.6). This simple physical picture of stimulated emission and absorption can be expanded to explain more complicated phenomena, e.g., lasing without inversion, as we shall see in Section 7.3.1.

5.7 Time delay spectroscopy

In the previous section, we saw how simple and intuitively pleasing the concepts of stimulated emission and absorption are when treated within the framework of semiclassical radiation theory. As an example

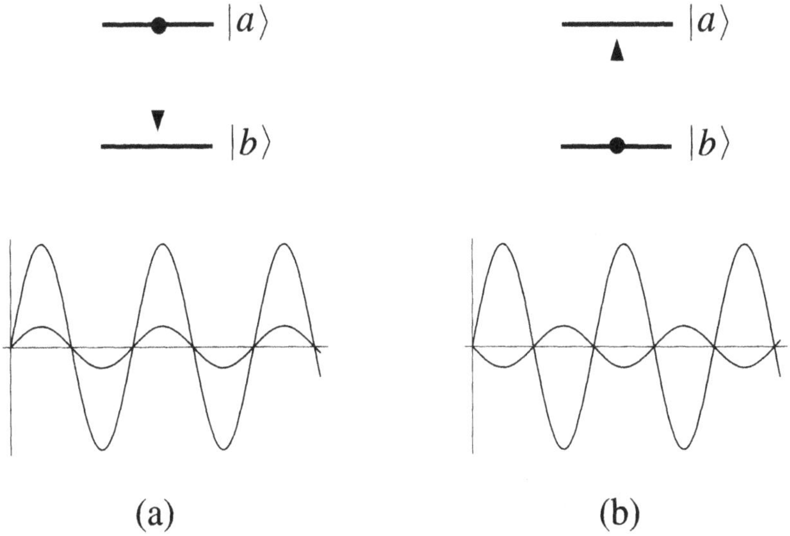

Fig. 5.6
(a) Emission: induced dipole radiation interferes constructively with incident radiation.
(b) Absorption: induced dipole radiation interferes destructively with incident radiation.

of unusual and counter-intuitive physics within the framework of semiclassical theory, we conclude this chapter with a discussion of time delay spectroscopy.

In conventional spectroscopy, the limit of resolution of the energy between two levels $|a\rangle$ and $|b\rangle$ is governed by the sum of the decay rates γ_a and γ_b out of these levels.

In this section we present a spectroscopic technique which provides resolution beyond the natural linewidth. These considerations are based on the fact that in the *transient regime*, the probability for induced transitions in a two-level system interacting with a monochromatic electromagnetic field is not governed by a Lorentzian of width $(\gamma_a + \gamma_b)/2 \equiv \gamma_{ab}$, but rather by $(\gamma_a - \gamma_b)/2 \equiv \delta_{ab}$. The Lorentzian width γ_{ab}, which usually appears in atomic physics, is regained only in the proper limits.

We proceed by considering the experimental situation in which an ensemble of two-level atoms is excited at time $t = t_0$ into the $|b\rangle$ state by some 'instantaneous' excitation mechanism, e.g., a picosecond optical pulse. The excited atoms are then driven by a monochromatic but tunable radiation field.

Consider the level scheme illustrated in Fig. 5.7. There we see an atom with two unstable levels $|a\rangle$ and $|b\rangle$ and a weak field driving the atom from the lower level $|b\rangle$ to the upper level $|a\rangle$. If one includes the lower levels ($|c\rangle$ and $|d\rangle$) to which $|a\rangle$ and $|b\rangle$ decay, this may be considered as a four-level atom. That is, we prepare the atom in level $|b\rangle$ at t_0, drive the atom to level $|a\rangle$, and count the number of atoms accumulating in level $|c\rangle$, starting a finite time t after the atom is

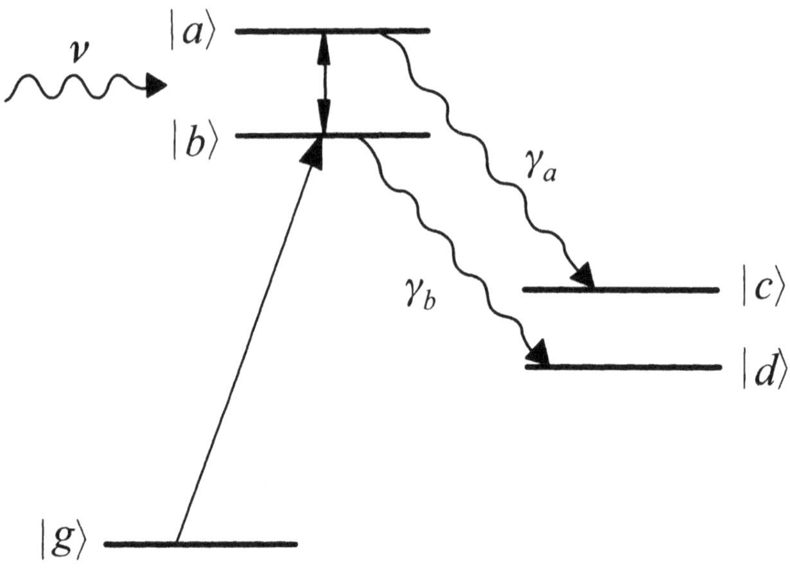

Fig. 5.7
Level diagram indicating excitation of atom from ground state to $|b\rangle$, subsequent interaction with resonant radiation promoting atom from $|b\rangle$ to $|a\rangle$ with atendant decays to states $|d\rangle$ and $|c\rangle$ at rates γ_b and γ_a, respectively.

prepared. The counting rate is measured as a function of the detuning between the laser and atomic frequencies.

We proceed by solving the density matrix equations of motion (5.3.22)–(5.3.24) for $\rho_{aa}(t)$ to lowest nonvanishing order. This yields

$$\rho_{aa}(t) = \frac{\Omega_R^2}{\Delta^2 + \delta_{ab}^2} \left[e^{-\gamma_a(t-t_0)} + e^{-\gamma_b(t-t_0)} - 2e^{-\gamma_{ab}(t-t_0)} \cos \Delta(t-t_0)\right], \tag{5.7.1}$$

where $\delta_{ab} = (\gamma_a - \gamma_b)/2$, Ω_R is the Rabi frequency of the driven transition and Δ is the detuning between the laser and ω_{ab}. The key point is that the Lorentzian factor in (5.7.1) goes as $\gamma_a - \gamma_b$ not $\gamma_a + \gamma_b$.

Now suppose we count the number of photons emitted when the excited atom makes the $|a\rangle \to |c\rangle$ transition. This will be equal to the total number of atoms accumulated in level $|c\rangle$ which is determined by the simple rate equation

$$\dot{\rho}_{cc}(t, t_0) = \gamma \rho_{aa}(t, t_0), \tag{5.7.2}$$

where the notation reminds us that the atoms are initially excited at time t_0. Then the total number of spontaneously emitted photons from time t_0 to a time long after the initial excitation to level $|b\rangle$ is given by

$$N(\Delta, t_0) = \eta \gamma_a \int_{t_0}^{\infty} \rho_{aa}(\Delta, t, t_0) dt, \tag{5.7.3}$$

where η is a constant determined by the efficiency of photon detection.

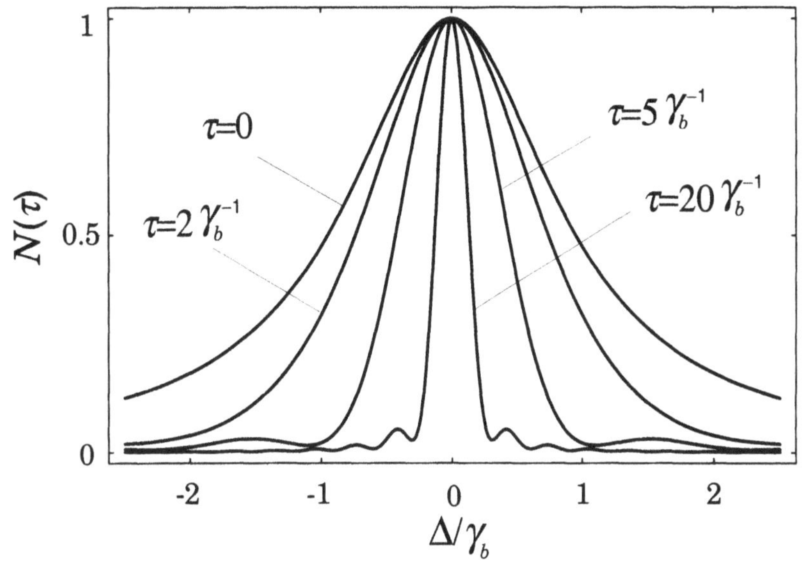

Fig. 5.8
Time delay spectroscopy signal $N(\tau)$ for different τ. The different curves have been normalized for simplicity. In fact the peak heights of the curves corresponding to larger τ are strongly reduced as indicated by Eq. (5.7.6). Nevertheless the line narrowing can be useful as discussed by Figger and Walther (1974).

Inserting (5.7.1) into (5.7.3) we find

$$N(\Delta, t_0) = \frac{\eta\gamma_a\Omega_R^2}{\Delta^2 + \gamma_{ab}^2}\frac{2\gamma_{ab}}{\gamma_a\gamma_b}. \tag{5.7.4}$$

That is, when we carry out the above procedure, collecting the $|a\rangle \rightarrow |c\rangle$ photons from t_0 onwards we regain the usual Lorentzian of width γ_{ab}. This is reassuring since in most experiments it is indeed γ_{ab} that governs the resolution of our experiments.

However, let us now wait for a time $t_0 + \tau$ before accepting any counts. That is let us measure

$$N(\Delta, t_0 + \tau) = \eta\gamma_a \int_{t_0+\tau}^{\infty} \rho_{aa}(\Delta, t, t_0)dt. \tag{5.7.5}$$

Inserting (5.7.1) into (5.7.5) we now find*

$$N(\Delta, t_o + \tau) = \frac{\eta\gamma_a\Omega_R^2}{\Delta^2 + \delta_{ab}^2}\left[\frac{\exp(-\gamma_a\tau)}{\gamma_a} + \frac{\exp(-\gamma_b\tau)}{\gamma_b} + \frac{2\exp(-\gamma_{ab}\tau)}{\Delta^2 + \gamma_{ab}^2}(\Delta\sin\Delta\tau - \gamma_{ab}\cos\Delta\tau)\right]. \tag{5.7.6}$$

The point is clear. When we delay observation we find a line narrowing as is seen by comparing Eqs. (5.7.4) and (5.7.6). Equation (5.7.6) is plotted for various values of time delay in Fig. 5.8.

We conclude by noting that, as pointed out explicitly by Figger and Walther, the line narrowing in time delay spectroscopy provides a

* See Meystre, Scully, and Walther [1980].

high spectral *resolution* in the sense that we can separate closely spaced lines. However, this higher resolution does not always lead to a higher experimental *accuracy* in the final result for the atomic transition frequencies ω_{ab}. The reason for this is the exponential damping of the signal with the time delay τ by means of the prefactors $\exp(-\gamma_a\tau)$ and $\exp(-\gamma_b\tau)$ in Eq. (5.7.6) which decrease the signal. We will return to the question of enhancing spectroscopic resolution in later chapters, e.g., in Section 21.7.

5.A Equivalence of the r · E and the p · A interaction Hamiltonians

In Section 5.1 we noted that in the radiation gauge (R-gauge) and in the dipole approximation $(\mathbf{A}(\mathbf{r},t), U(\mathbf{r},t)) \equiv (\mathbf{A}(t), 0)$, the gauge transformation

$$\chi(\mathbf{r},t) = -\frac{e}{\hbar}\mathbf{A}(t)\cdot\mathbf{r} \tag{5.A.1}$$

yields the gauge $(0, -\mathbf{E}(t)\cdot\mathbf{r})$. We observe that the gauge $(0, -\mathbf{E}(t)\cdot\mathbf{r})$ leads to the electric–dipole interaction $\mathscr{H}$ (Eq. (5.1.19)), and thus we call it the electric field gauge (E-gauge). The two Hamiltonians $\mathscr{H}$ (Eq. (5.1.19)) and $\mathscr{H}'$ (Eq. (5.1.21)) are therefore related via the gauge transformation (5.A.1). A gauge transformation requires a transformation of the potentials according to Eqs. (5.1.6) and (5.1.7) and of the wave functions according to Eq. (5.1.4). Nonidentical, *wrong* results are obtained for physically measurable quantities in different gauges if only one of these two transformations is carried out. We will discuss how we have to handle the wave functions in the two different gauges in order to obtain gauge-invariant physical predictions. Before this, however, let us briefly discuss some examples of physical quantities.

5.A.1 Form-invariant physical quantities

A form-invariant physical quantity is defined as a quantity whose corresponding operator $G_\chi = G(A_\chi, U_\chi)$ is form invariant under a unitary transformation $T(\mathbf{r},t) = \exp[i\chi(\mathbf{r},t)]$, i.e.,

$$G_{\chi'} = TG_\chi T^\dagger, \tag{5.A.2}$$

where the wave function in the gauge χ is transformed to the gauge χ' by the unitary transformation

$$\psi_{\chi'}(\mathbf{r},t) = T(\mathbf{r},t)\psi_\chi(\mathbf{r},t). \tag{5.A.3}$$

The difference between physical and nonphysical quantities lies in the gauge invariance of the eigenvalues. The eigenvalues of a physical quantity are identical in all gauges, whereas the eigenvalues of non-physical quantities depend on the chosen gauge. In order to show this, we denote the eigenvalues and eigenstates of the operator G_χ by g_n and $|\xi_{\chi,n}\rangle$, respectively:

$$G_\chi|\xi_{\chi,n}\rangle = g_n|\xi_{\chi,n}\rangle. \tag{5.A.4}$$

Only for physical quantities are the eigenvalues g_n gauge invariant, i.e.,

$$\begin{aligned} G_{\chi'}|\xi_{\chi',n}\rangle &= TG_\chi T^\dagger T|\xi_{\chi,n}\rangle \\ &= Tg_n|\xi_{\chi,n}\rangle \\ &= g_n|\xi_{\chi',n}\rangle. \end{aligned} \tag{5.A.5}$$

Hence, nonphysical quantities can only be used as calculational tools.

We next consider some examples of physical and nonphysical quantities. The starting point for these considerations is the fact that the operators $\mathbf{r}$ and $\mathbf{p}$ ($\mathbf{p} = -i\hbar\nabla$), associated with the position and the canonical momentum of the particle, are the same in all gauges, by which we mean that $\mathbf{p}$ is represented by $-i\hbar\nabla$ in all gauges. This ensures that, in any gauge, the commutation relation $[\mathbf{r}_j, \mathbf{p}_k] = i\hbar\delta_{jk}$ is satisfied. With this rule the operator for the mechanical momentum,

$$\pi_\chi = \mathbf{p} - e\mathbf{A}_\chi(\mathbf{r}, t), \tag{5.A.6}$$

is a physical, measurable quantity since

$$\begin{aligned} T\pi_\chi T^\dagger &= T[\mathbf{p} - e\mathbf{A}_\chi(\mathbf{r}, t)]T^\dagger \\ &= \mathbf{p} - e\mathbf{A}_\chi - \hbar\nabla\chi \\ &= \mathbf{p} - e\mathbf{A}_{\chi'} \\ &= \pi_{\chi'}. \end{aligned} \tag{5.A.7}$$

Similarly, the instantaneous energy operator of the system, consisting of the kinetic energy and the static potential (normally the atomic binding potential)

$$\mathscr{E}_\chi = \frac{1}{2m}[\mathbf{p} - e\mathbf{A}_\chi(\mathbf{r}, t)]^2 + V(r), \tag{5.A.8}$$

represents a physical quantity as well as any other operator which is only a function of other physical quantities like π_χ.

On the other hand, the canonical momentum $\mathbf{p}$ is not a physical quantity since

$$T\mathbf{p}T^\dagger = \mathbf{p} - \hbar\nabla\chi \neq \mathbf{p}. \tag{5.A.9}$$

In a similar way, the operator $\mathscr{H}_0 = p^2/2m$ (which does not depend on potentials) is not a physical quantity because

$$T\mathscr{H}_0T^\dagger = \mathscr{H}_0 - \frac{\hbar}{2m}[\mathbf{p}\cdot\nabla\chi + (\nabla\chi)\cdot\mathbf{p}] + \frac{\hbar^2}{2m}(\nabla\chi)^2 \neq \mathscr{H}_0. \tag{5.A.10}$$

In general, any operator which is a function of nonphysical quantities alone, like the canonical momentum $\mathbf{p}$ or the vector or the scalar potentials $\mathbf{A}_\chi$ or U_χ, represents a nonphysical quantity. The total Hamiltonian

$$\mathscr{H}_\chi = \frac{1}{2m}[\mathbf{p} - e\mathbf{A}_\chi(\mathbf{r}, t)]^2 + eU_\chi(\mathbf{r}, t) + V(r) \tag{5.A.11}$$

is also a nonphysical quantity, since it depends on the scalar potential U_χ.

We therefore conclude that the time evolution of a physical system is determined by Hamiltonians such as $\mathscr{H}_\chi$ or $\mathscr{H}_0$, which in general are not observable quantities. The physical quantities are, for example, the mechanical momentum and the instantaneous energy of the system.

5.A.2 Transition probabilities in a two-level atom

In this subsection we restrict the discussion to the large-wavelength dipole approximation (LWA) in which $\mathbf{A}$ may be considered to be independent of $\mathbf{r}$, i.e., $\mathbf{A}(\mathbf{r}, t) \equiv \mathbf{A}(t)$. Since the energy operator $\mathscr{E}_\chi$ (as given by Eq. (5.A.8)) is time dependent, its eigenstates $|\alpha_\chi(t)\rangle$, where $\alpha = a, b$, and its eigenvalues $E_\alpha = \hbar\omega_\alpha$ are also time dependent in general, namely

$$\mathscr{E}_\chi|\alpha_\chi(t)\rangle = E_\alpha|\alpha_\chi(t)\rangle. \tag{5.A.12}$$

However, in the LWA the eigenvalues of $\mathscr{E}_\chi$ are time independent. This can be seen with the help of the gauge transformation (5.A.1). In the LWA

$$\exp\left[-\frac{ie\mathbf{A}(t)\cdot\mathbf{r}}{\hbar}\right][\mathbf{p} - e\mathbf{A}(t)]^2 \exp\left[\frac{ie\mathbf{A}(t)\cdot\mathbf{r}}{\hbar}\right] = p^2, \tag{5.A.13}$$

so that

$$\exp\left[-\frac{ie\mathbf{A}(t)\cdot\mathbf{r}}{\hbar}\right]\mathscr{E}_\chi \exp\left[\frac{ie\mathbf{A}(t)\cdot\mathbf{r}}{\hbar}\right] = \mathscr{H}_0. \tag{5.A.14}$$

The eigenstate $|\alpha_\chi\rangle$ is then related to the eigenstate $|\alpha(t)\rangle$ of $\mathscr{H}_0$ by

$$|\alpha_\chi\rangle = \exp\left[\frac{ie\mathbf{A}_\chi(t)\cdot\mathbf{r}}{\hbar}\right]|\alpha(t)\rangle, \tag{5.A.15}$$

and the eigenvalues E_α of $\mathscr{E}_\chi$ coincide with the time-independent eigenvalues E_α of $\mathscr{H}_0$ since the eigenvalues of physical quantities are gauge independent.

In the E-gauge the unperturbed energy operator $\mathscr{E}_E$ is equal to the *unperturbed* Hamiltonian $\mathscr{H}_0$. Hence the eigenstates of $\mathscr{H}_0$ are also the eigenstates of $\mathscr{E}_E$. Therefore, only in the E-gauge is the wave function expanded in terms of energy eigenstates, and the coefficients $c_\alpha(t)$, where $\alpha = a, b$, in Eqs. (5.2.10) and (5.2.11) are interpreted as probability amplitudes for finding the system in an eigenstate of the observable energy. In any other gauge, $\mathscr{H}_0$ is a nonphysical quantity and its eigenstates are not the energy eigenstates of the system. The expansion coefficients $c_\alpha(t)$ in Eqs. (5.2.10) and (5.2.11) are then the probability amplitudes for finding the system in an eigenstate of $\mathscr{H}_0$. However, if $\mathscr{H}_0$ is a nonphysical quantity, this *probability* is gauge dependent and has to be distinguished from the measurable, gauge-invariant probability of finding the system in an energy eigenstate.

It is, therefore, useful to expand the wave function of the system in terms of eigenstates of the energy operator $\mathscr{E}_\chi$

$$|\psi_\chi(t)\rangle = d_a(t)e^{-i\omega_a t}|a_\chi\rangle + d_b(t)e^{-i\omega_b t}|b_\chi\rangle. \tag{5.A.16}$$

The expansion coefficients d_a and d_b then coincide with the probability amplitudes for transitions of the system to the eigenstates $|a_\chi\rangle$ and $|b_\chi\rangle$, respectively of the energy operator $\mathscr{E}_\chi$ with energies $\hbar\omega_a$ and $\hbar\omega_b$:

$$d_a(t) = \langle a_\chi|\psi_\chi(t)\rangle e^{i\omega_a t}, \tag{5.A.17}$$

$$d_b(t) = \langle b_\chi|\psi_\chi(t)\rangle e^{i\omega_b t}. \tag{5.A.18}$$

We will now show explicitly that these amplitudes are gauge invariant.

In the E-gauge, the probability amplitude $d_a(t)$ is given by

$$d_a^E(t) = \langle a|U_0(t)U_I^{(1)}(t)|b\rangle e^{i\omega_a t} \tag{5.A.19}$$

and, in the R-gauge, by

$$d_a^R(t) = \langle a|\exp\left[-\frac{ie}{\hbar}\mathbf{A}(t)\cdot\mathbf{r}\right] U_0(t)U_I^{(2)}(t)\exp\left[\frac{ie}{\hbar}\mathbf{A}(0)\cdot\mathbf{r}\right]|b\rangle e^{i\omega_a t}, \tag{5.A.20}$$

where $U_0(t) = \exp(-i\mathscr{H}_0 t/\hbar)$ and

$$U_I^{(i)}(t) = \mathscr{T}\exp\left[-\frac{i}{\hbar}\int_0^t d\tau U_0^\dagger(\tau)\mathscr{H}_i(\tau)U_0(\tau)\right]. \tag{5.A.21}$$

Here, we assume that the atom is initially in the ground state $|b\rangle$. Similar expressions exist for the amplitudes $d_b^E(t)$ and $d_b^R(t)$. In the first order of perturbation theory, the time-evolution operator $U_I^{(1)}$ becomes

$$U_I^{(1)} = 1 - \frac{i}{\hbar}\int_0^t d\tau U_0^\dagger(\tau)\mathscr{H}_1 U_0(\tau), \tag{5.A.22}$$

and the probability amplitude of the excited state in the E-gauge takes the form

$$\begin{aligned} d_a^E(t) &= -\frac{i}{\hbar}\langle a|\int_0^t d\tau U_0^\dagger(\tau)\mathscr{H}_1 U_0(\tau)|b\rangle \\ &= \frac{ie}{2\hbar}\mathscr{E}\cdot\langle a|\mathbf{r}|b\rangle\int_0^t d\tau e^{i(\omega-\nu)\tau} \\ &= \frac{e}{2\hbar}\mathscr{E}\cdot\mathbf{r}_{ab}\frac{e^{i\Delta t}-1}{\Delta}. \end{aligned} \tag{5.A.23}$$

This result is now compared to the corresponding result in the R-gauge. In first-order perturbation theory,

$$\begin{aligned} d_a^R(t) = \langle a|\left[1 - \frac{ie}{\hbar}\mathbf{A}(t)\cdot\mathbf{r}\right] U_0(t)\left[1 - \frac{i}{\hbar}\int_0^t d\tau U_0^\dagger(\tau)\mathscr{H}_2 U_0(\tau)\right] \\ \times\left[1 + \frac{ie}{\hbar}\mathbf{A}(0)\cdot\mathbf{r}\right]|b\rangle\exp(i\omega_a t). \end{aligned} \tag{5.A.24}$$

Using (5.1.26) and (5.2.32) and $A(t) = \frac{1}{2}\mathscr{A}e^{-i\nu t}$, to lowest order in $\mathscr{A}$, yields

$$d_a^R(t) = -\frac{ie}{\hbar}\frac{\mathscr{A}}{2}\left[-\mathbf{r}_{ab}e^{-i(\omega-\nu)t} + \mathbf{p}_{ab}\frac{e^{-i(\omega-\nu)t}-1}{i(\omega-\nu)} + \mathbf{r}_{ab}\right].$$

From (5.1.30) we have $\mathbf{p}_{ab} = +im\omega\mathbf{r}_{ab}$ and defining $\mathscr{E} = i\nu\mathscr{A}$ yields

$$d_a^R(t) = \frac{e}{2\hbar}\mathscr{E}\cdot\mathbf{r}_{ab}\frac{e^{i\Delta t}-1}{\Delta}. \tag{5.A.25}$$

Thus, the amplitudes $d_a^E(t)$ and $d_a^R(t)$ are seen to be identical. This resolves* the apparent contradiction pointed out at the end of Section 5.1.

* The present treatment is oversimplified in that the effects of atomic decay are not included. For the more general case, see Lamb, Schlicher and Scully [1987].

5.B Vector model of the density matrix

A physical picture of the density matrix is provided by reducing Eqs. (5.3.22), (5.3.23), and (5.3.30) into a form equivalent to the Bloch equations appearing in nuclear magnetic resonance. The present problem of a two-level atom interacting with an electromagnetic field is similar to that of a spin-1/2 magnetic dipole undergoing precession in a magnetic field. This formal similarity has led to the prediction, observation, and physical understanding of a number of phenomena associated with coherent pulse propagation in a system of two-level atoms.

We introduce the real quantities

$$R_1 = \rho_{ab}e^{i\nu t} + \text{c.c.}, \tag{5.B.1}$$

$$R_2 = i\rho_{ab}e^{i\nu t} + \text{c.c.}, \tag{5.B.2}$$

$$R_3 = \rho_{aa} - \rho_{bb}. \tag{5.B.3}$$

These quantities are components of the vector $\mathbf{R}$, given by

$$\mathbf{R} = R_1\hat{\mathbf{e}}_1 + R_2\hat{\mathbf{e}}_2 + R_3\hat{\mathbf{e}}_3. \tag{5.B.4}$$

where $\hat{\mathbf{e}}_1$, $\hat{\mathbf{e}}_2$, and $\hat{\mathbf{e}}_3$ form a set of mutually perpendicular unit vectors. Here, R_1 and R_2 represent the atom's dipole moment, and R_3 is the population difference between the levels $|a\rangle$ and $|b\rangle$.

It follows from Eqs. (5.3.22), (5.3.23), and (5.3.30) that, in the rotating-wave approximation, (with $\phi = 0$)

$$\dot{R}_1 = -\Delta R_2 - \frac{1}{T_2}R_1, \tag{5.B.5}$$

$$\dot{R}_2 = \Delta R_1 - \frac{1}{T_2}R_2 + \Omega_R R_3, \tag{5.B.6}$$

$$\dot{R}_3 = -\frac{1}{T_1}R_3 - \Omega_R R_2, \tag{5.B.7}$$

where we have assumed $\gamma_a = \gamma_b = 1/T_1$ and $\gamma = 1/T_2$. The quantities T_1 and T_2 are called the longitudinal and the transverse relaxation times, respectively, in analogy with the corresponding quantities in the Bloch equations. Equations (5.B.5)–(5.B.7) are referred to as the optical Bloch equations.

When $T_1 = T_2$, these equations can be written in the following compact form

$$\dot{\mathbf{R}} = -\frac{1}{T_1}\mathbf{R} + \mathbf{R} \times \mathbf{\Omega}, \tag{5.B.8}$$

where the effective field is given by

$$\mathbf{\Omega} = \Omega_R \hat{\mathbf{e}}_1 - \Delta \hat{\mathbf{e}}_3. \tag{5.B.9}$$

The time dependence of **R**, as given by Eq. (5.B.8), is well known from classical mechanics. The vector **R** precesses clockwise about the effective field $\mathbf{\Omega}$ with diminishing amplitude. The precessions for resonance and slightly off resonance are depicted in Fig. 5.9. Physically **R** pointing along $\hat{\mathbf{e}}_3$ ($R_3 = 1$, $R_1 = R_2 = 0$) represents a system in its upper level, $\rho_{aa} = 1$, $\rho_{bb} = 0$. Similarly, **R** pointing along $-\hat{\mathbf{e}}_3$ represents a system in its lower level.

5.C Quasimode laser physics based on the modes of the universe*

Most laser theories, e.g., that of Section 5.5, describe the electromagnetic field in terms of a discrete set of quasimodes of the laser cavity, each of which has a finite quality factor Q. In the present section, this theory is generalized for a laser with a cavity modeled by a semitransparent wall as one of the mirrors so that there are now many modes of the 'universe' corresponding to each quasimode. Here we show that the normal modes of the universe associated with a single 'mode' may, under proper conditions, lock together and the δ-function laser lineshape may be regained.

We consider the normal modes for a combined system of a laser cavity coupled to the outside world. We represent the 'universe' by a much larger cavity having perfectly reflecting walls. A simple one-dimensional model which carries the essential features of such a combined system is illustrated in Fig. 5.10. The mirrors at $z = L$ and $-L_0$ are completely reflective, while the one at $z = 0$ is semitransparent. Region 1 corresponds to a laser cavity and region 2 to the rest of the universe.

We represent a semitransparent mirror by a very thin plate with a very large dielectric constant. As an idealization of such a mirror we choose the dielectric constant around $z = 0$ to be

$$\epsilon(z) = \epsilon_0[1 + \eta\delta(z)], \tag{5.C.1}$$

where η is a parameter with the dimension of length which determines the transparency of this plate.

The normal mode functions of this system can be obtained by solving Maxwell's equations with the proper boundary conditions (see Problem 5.6). For those normal modes having frequency $\nu_k (= ck)$ close

* For further reading, see Lang, Scully and Lamb [1973].

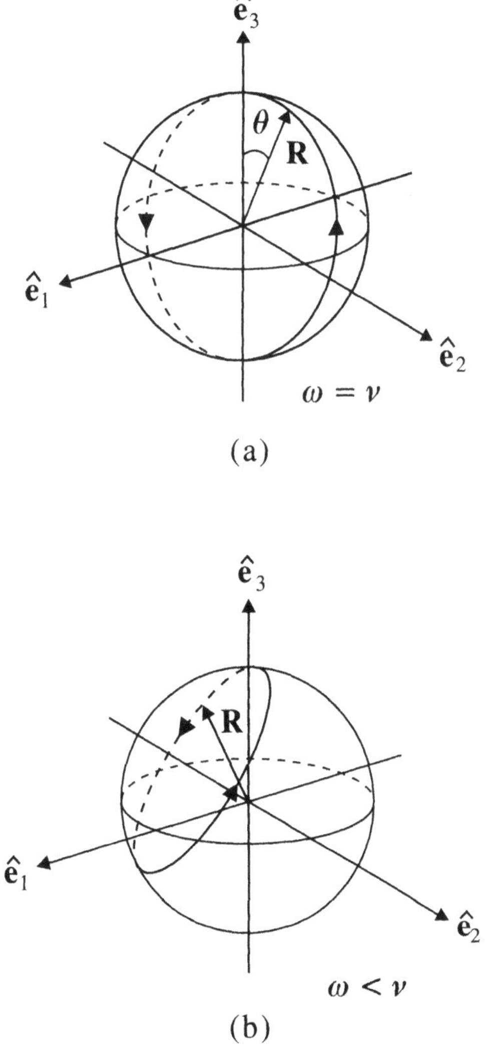

Fig. 5.9
Precession of Bloch vector **R** about the effective field **Ω** for (a) $\Delta = 0$ and (b) $\Delta \neq 0$.

to a 'resonant' frequency $\nu_0(= ck_0)$, the eigenfunctions of the entire cavity are

$$U_k(z) = \begin{cases} M_k \sin k(z - L) & (z > 0), \\ \xi_k \sin k(z + L) & (z < 0), \end{cases} \tag{5.C.2}$$

where ξ_k is a phase factor which alternates between 1 and -1 as k increases from one value to the next. The coefficients M_k in (5.C.2) are

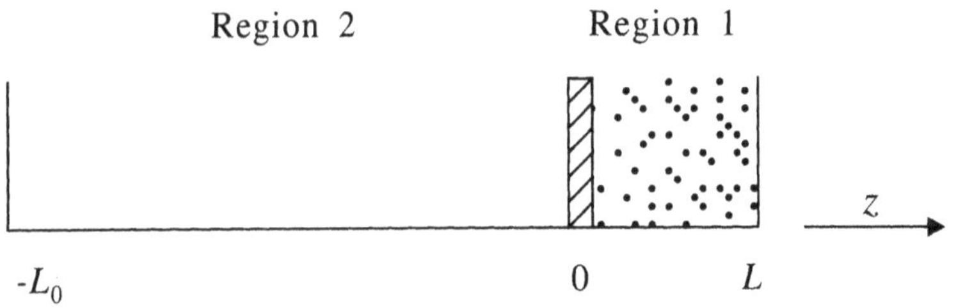

Fig. 5.10
Leaky cavity bounded by a perfect mirror at $z = L$ and a semitransparent mirror at $z = 0$. The auxiliary cavity which, along with the leaky cavity, constitutes the universe, is bounded by a perfect mirror at $z = -L_0$ ($L_0 \to \infty$) and the mirror at $z = 0$.

$$M_k = \frac{\mathscr{C}\Lambda}{2}\left[(\nu_k - \nu_0)^2 + \frac{\mathscr{C}^2}{4}\right]^{-1/2}, \tag{5.C.3}$$

where $\mathscr{C}$ is the bandwidth associated with the mirror transparency and is given by

$$\mathscr{C} = 2c/\eta^2 k_0^2 L = 2c/\Lambda^2 L, \tag{5.C.4}$$

with

$$\Lambda = \eta\nu_0/c = \eta k_0, \tag{5.C.5}$$

and the frequency ν_0 of the nth quasimode is given by

$$\nu_0 = ck_0 = (n\pi + 1/\Lambda)c/L. \tag{5.C.6}$$

An arbitrary undriven field in the entire cavity can be expressed as the positive frequency part of the field

$$E^{(+)}(z,t) = \sum_k \mathscr{E}_k(0)U_k(z)e^{-i\nu t} = \sum_k \mathscr{E}_k(t)U_k(z), \tag{5.C.7}$$

which is to be understood as a sum over modes of the large cavity, i.e., 'the universe'.

We now demonstrate that the semitransparency of the mirror leads to a damping of free oscillations in the laser cavity. Let us assume that, at $t = 0$, the laser cavity (region 1) contains a field of the form

$$E^{(+)}(z,0) = |\mathscr{E}_0|e^{-i\varphi}\sin k_0(z - L), \tag{5.C.8}$$

whereas no field exists outside the cavity, i.e., in region 2. The coefficients $\mathscr{E}_k(0)$ for this case are obtained by multiplying (5.C.8) by $U_k(z)$ defined in (5.C.2) and integrating over z. We find

$$\mathscr{E}_k(t) = (|\mathscr{E}_0|M_k L/L_0)e^{-i(\nu_k t+\varphi)}. \tag{5.C.9}$$

Therefore at later times, $t > 0$,

$$E(z,t) = (|\mathscr{E}_0|L/L_0)\sum_k M_k U_k(z)e^{-i(\nu_k t+\varphi)}. \tag{5.C.10}$$

The summation can be approximated by an integral if the frequency separation between the normal modes is small compared to $\mathscr{C}$. Carrying out the integration over k in (5.C.10), the explicit form of $E(z,t)$ in the maser cavity turns out to be

$$E^{(+)}(z,0) = |\mathscr{E}_0| \sin k_0(z-L)e^{-i(\nu_k t+\varphi)-\mathscr{C}t/2}. \tag{5.C.11}$$

Equation (5.C.11) indicates that the field localized in the maser cavity decays exponentially owing to leakage through the mirror at a rate $\mathscr{C}/2$.

Problems

5.1 Show that the Schrödinger equation (5.1.5) is invariant under local gauge transformations (5.1.4), (5.1.6), and (5.1.7).

5.2 The finite lifetime of the atomic levels can be described by adding phenomenological decay terms to the probability amplitude equations (5.2.12) and (5.2.13):

$$\dot{c}_a = -\frac{\gamma}{2}c_a + \frac{i\Omega_R}{2}e^{-i\phi}c_b,$$
$$\dot{c}_b = -\frac{\gamma}{2}c_b + \frac{i\Omega_R}{2}e^{i\phi}c_a,$$

where γ is the decay constant and $\omega = \nu$. For an atom initially in the state $|a\rangle$, show that the inversion at time t is

$$W(t) = e^{-\gamma t}\cos(\Omega_R t).$$

5.3 Find the solution of Eq. (5.B.8) (with $T_1 \to \infty$):

$$\dot{\mathbf{R}} = \mathbf{R} \times \mathbf{\Omega}$$

for $\mathbf{R}(0) = 0$. Give a physical interpretation of this solution.

5.4 Show that, in general,

$$\mathrm{Tr}(\rho^2) \leq 1,$$

where the equality is valid only for a pure state.

5.5 Consider a three-level atom interacting with a classical field of frequency ν. The transitions $|a\rangle \to |b\rangle$ and $|b\rangle \to |c\rangle$ are allowed whereas the transition $|a\rangle \to |c\rangle$ is forbidden. It is also assumed that $\omega_a - \omega_b = \omega_b - \omega_c = \nu$. Assuming the atom to be initially in level $|c\rangle$, find the probabilities for the atom to be in levels $|a\rangle$ and $|c\rangle$ after making the rotating-wave approximation.

5.6 The electromagnetic field in the entire cavity (region 1 and region 2 in Fig. 5.10) is governed by the Maxwell wave equation

$$\frac{\partial^2 E}{\partial z^2} - \mu_0 \epsilon_0 [1 + \eta \delta(z)] \frac{\partial^2 E}{\partial t^2} = 0,$$

where E can be written as

$$E = U_k(z) e^{-i\nu_k t}.$$

(a) Find $U_k(z)$ in the form (5.C.2) and prove that

$$\frac{M_k^2}{\xi_k^2} = \frac{\tan^2 kL + 1}{\tan^2 kL + (\Lambda \tan kL - 1)^2},$$

where Λ is given by Eq. (5.C.5). Derive Eq. (5.C.3).

(b) Show that

$$(\nu_k - \nu_{k'})^2 \int_{-L_0}^{L} dz\, U_k(z) U_{k'}(z) \epsilon(z) = 0,$$

where $\epsilon(z) = \epsilon_0[1 + \eta\delta(z)]$. (Hint: see R. Lang, M. O. Scully, and W. E. Lamb, Jr., *Phys. Rev. A* **7**, 1788 (1973).)

5.7 The $m = +1$ level of Fig. 5.3 is weakly coupled to the $\psi_b(r)$ level by the left-circular polarized light of Eq. (5.2.55) via counter rotating terms. Note that in such a case ($m = 0$ to $m = +1$) we normally rule out such coupling on the grounds that right-circularly polarized light is needed for the $m = 0$ to $m = +1$ transition.

(a) Show that if we define

$$\psi_{a'}(\mathbf{r}) = \psi_{n=2,l=1,m_l=+1}(\mathbf{r})$$
$$= \eta(x+iy)e^{-r/2a_0},$$

where η is the uninteresting constant $[\sqrt{64\pi a_0^3 a_0}]^{-1}$, then

$$\mathscr{V}_{a'b}(t) = -e\mathscr{E}\int d\mathbf{r}\psi_a^*(\mathbf{r})\mathbf{r}\cdot(\hat{x}\cos\nu t - \hat{y}\sin\nu t)$$
$$\psi_b(\mathbf{r})e^{i\omega_{a'b}t}$$
$$= -\wp\mathscr{E}e^{i(\omega_{a'b}+\nu)t}.$$

(b) Show, by specific example, that the counter terms associated with the $|b\rangle \to |a'\rangle$ transitions, which go like $[\omega_{a'b}+\nu]^{-1}$ can be much smaller than the usual counter terms $[\omega_{ab}+\nu]^{-1}$. Hint: consider a Rydberg atom in which $\omega_{ab} = \omega_{a'b} \cong 10^9$Hz. If we now apply a field of around 10^4 Gauss, we could arrange for the Zeeman shifted $\omega_{ab} \sim 10^3$Hz while $\omega_{a'b} \sim 10^{10}$Hz.

References and bibliography

Textbooks

B. E. A. Saleh and M. C. Teich, *Fundamentals of Photonics*, (Wiley-Interscience, New York 1991).

Atom–field interaction Hamiltonian

M. Göppert-Mayer, *Ann. Phys. (Leipzig)* **9**, 273 (1931).
E. A. Power and S. Zienau, *Philos. Tran. Roy. Soc. London* 251, 427 (1959).
C. Cohen-Tannoudji, J. Dupont-Roc, and G. Grynberg, *Atom-Photon Interactions*, (Wiley, New York 1992).

Interaction of a two-level atom with a single-mode field

H. A. Bethe, in *Quantentheorie*, Vol. 24/1 of *Handbuch der Physik*, 2nd ed. (Springer, Berlin 1933), p. 273.
W. E. Lamb and R. C. Retherford, *Phys. Rev.* **72**, 339 (1947).
R. P. Feynman, F. L. Vernon, and R. W. Hellworth, *J. Appl. Phys.* **28**, 49 (1957).
E. T. Jaynes and F. W. Cummings, *Proc. IEEE* **51**, 89 (1963).
L Allen and J. Eberly, *Optical Resonance and Two Level Atom*, (Wiley, New York 1975).
S. Stenholm, *Foundations of Laser Spectroscopy*, (Wiley, New York 1984).

Experimental realization of a two-level atom

J. A. Abate, *Opt. Commun.* **10**, 269 (1974).
R. E. Grove, F. Y. Wu, and S. Ezekiel, *Phys. Rev. A* **15**, 227 (1977).
D. Meschede, H. Walther, and G. Müller, *Phys. Rev. Lett.* **54**, 551 (1985).

Comparison of p · A and r · E Hamiltonians

W. E. Lamb, Jr., *Proc. Scuola Internazionale di Fisica Course XXXI*, Academic Press 1963.
H. Haken, *Z. Phys.* **181**, 96 (1964).
W. E. Lamb, Jr., *Phys. Rev.* **85**, 259 (1952).
K. H. Yang, *Ann. Phys. (NY)* **101**, 62 (1976).
D. H. Kobe and A. L. Smirl, *Am. J. Phys.* **46**, 624 (1978).
R. R. Schlicher, W. Becker, J. Bergou, and M. O. Scully, in *Quantum Electrodynamics and Quantum Optics*, ed. A. O. Barut (Plenum, New York 1984), p. 405.
W. Becker, *Opt. Commun.* **56**, 107 (1985).
W. E. Lamb, Jr., R. R. Schlicher, and M. O. Scully, *Phys. Rev. A* **36**, 2763 (1987).

Maser operation

J. P. Gordon, H. J. Zeiger, and C. H. Townes, *Phys. Rev.* **95**, 282 (1954); *ibid.* **99**, 1264 (1955).

Laser operation

T. H. Maiman, *Nature* **187**, 493 (1960).

A. Javan, W. R. Bennet, Jr., and D. R. Herriot, *Phys. Rev. Lett.* **6**, 106 (1961).

Semiclassical laser theory

A. M. Prokhorov, *Sov. Phys. JETP* **34**, 1658 (1958).

A. L. Schawlow and C. H. Townes, *Phys. Rev.* **112**, 1940 (1958).

H. Haken and H. Sauermann, *Z. Phys.* **173**, 261 (1963); *ibid.* **176**, 47 (1963).

Proc. Enrico Fermi School of Physics Course XXXI, p78 (1963)

W. E. Lamb, Jr., *Phys. Rev.* **134**, A1429 (1964).

M. Sargent III, W. E. Lamb, Jr., and R. L. Fork, *Phys. Rev.* **164**, 450 (1967).

S. Stenholm and W. E. Lamb, Jr., *Phys. Rev.* **181**, 618 (1969).

H. Haken, *Laser Theory*, (Springer, Berlin 1970).

W. Lamb and M. Scully, *OSA* Spring 1971, p. 66.

M. Sargent III, M. Scully, and W. E. Lamb, Jr., *Laser Physics* (Addison-Wesley, Reading, MA 1974).

A. E. Siegman, *Lasers* (Univ. Sci. Books, Mill Valley, CA 1986).

P. Milonni and J. Eberly, *Lasers* (Wiley, New York 1988).

Laser amplifier

F. T. Arecchi and R. Bonifacio, *IEEE J. Quan. Elect.* **QE-1**, 169 (1965).

F. A. Hopf and M. O. Scully, *Phys. Rev,* **179**, 399 (1969).

Laser cavity modes

A. G. Fox and T. Li, *Bell System Tech. J.* **40**, 453 (1961).

G. D. Boyd and J. P. Gordon, *Bell System Tech. J.* **40**, 489 (1961).

M. B. Spencer and W. E. Lamb, Jr., *Phys. Rev. A* **5**, 884 (1972).

R. Lang, M. O. Scully, and W. E. Lamb, Jr., *Phys. Rev. A* **7**, 1788 (1973).

R. Lang and M. O. Scully, *Opt. Commun.* **9**, 331 (1973).

K. Ujihara, *Phys. Rev. A* **12**, 148 (1975); *ibid.* **16**, 652 (1977).

Time delay spectroscopy

H. Figger and H. Walther, *Z. Phys.* **267**, 1 (1974).
P. Meystre, M. Scully, and H. Walther, *Opt. Commun.* **33**, 153 (1980).
W. Demtröder, *Laser Spectroscopy*, (Springer, Berlin 1995).

CHAPTER 6

Atom–field interaction – quantum theory

In the preceding chapters concerning the interaction of a radiation field with matter, we assumed the field to be classical. In many situations this assumption is valid. There are, however, many instances where a classical field fails to explain experimentally observed results and a quantized description of the field is required. This is, for example, true of spontaneous emission in an atomic system which was described phenomenologically in Chapter 5. For a rigorous treatment of the atomic level decay in free space, we need to consider the interaction of the atom with the vacuum modes of the universe. Even in the simplest system involving the interaction of a single-mode radiation field with a single two-level atom, the predictions for the dynamics of the atom are quite different in the semiclassical theory and the fully quantum theory. In the absence of the decay process, the semiclassical theory predicts Rabi oscillations for the atomic inversion whereas the quantum theory predicts certain *collapse* and *revival* phenomena due to the quantum aspects of the field. These interesting quantum field theoretical predictions have been experimentally verified.

In this chapter we discuss the interaction of the quantized radiation field with the two-level atomic system described by a Hamiltonian in the dipole and the rotating-wave approximations. For a single-mode field it reduces to a particularly simple form. This is a very interesting Hamiltonian in quantum optics for several reasons. First, it can be solved exactly for arbitrary coupling constants and exhibits some true quantum dynamical effects such as collapse followed by periodic revivals of the atomic inversion. Second, it provides the simplest illustration of spontaneous emission and thus explains the effects of various kinds of quantum statistics of the field in more complicated systems such as a micromaser and a laser, which we shall study in later chapters. Third, and perhaps most importantly,* it has become possible to realize it experimentally through the spectacular advances in the development of high-Q microwave cavities.

The spontaneous decay of an atomic level is treated by considering the interaction of the two-level atom with the modes of the universe in the vacuum state. We examine the state of the field that is generated in the process of emission of a quantum of energy equal to the energy difference between the atomic levels. Such a state may be regarded as a single-photon state.

6.1 Atom–field interaction Hamiltonian

The interaction of a radiation field $\mathbf{E}$ with a single-electron atom can be described by the following Hamiltonian in the dipole approximation:

$$\mathscr{H} = \mathscr{H}_A + \mathscr{H}_F - e\mathbf{r} \cdot \mathbf{E}. \tag{6.1.1}$$

Here $\mathscr{H}_A$ and $\mathscr{H}_F$ are the energies of the atom and the radiation field, respectively, in the absence of the interaction, and $\mathbf{r}$ is the position vector of the electron. In the dipole approximation, the field is assumed to be uniform over the whole atom.

The energy of the free field $\mathscr{H}_F$ is given in terms of the creation and destruction operators by

$$\mathscr{H}_F = \sum_{\mathbf{k}} \hbar\nu_k \left(a_{\mathbf{k}}^\dagger a_{\mathbf{k}} + \frac{1}{2} \right). \tag{6.1.2}$$

We can express $\mathscr{H}_A$ and $e\mathbf{r}$ in terms of the atom transition operators

$$\sigma_{ij} = |i\rangle\langle j|. \tag{6.1.3}$$

As before $\{|i\rangle\}$ represents a complete set of atomic energy eigenstates, i.e., $\sum_i |i\rangle\langle i| = 1$. It then follows from the eigenvalue equation $\mathscr{H}_A|i\rangle = E_i|i\rangle$ that

$$\mathscr{H}_A = \sum_i E_i|i\rangle\langle i| = \sum_i E_i\sigma_{ii}. \tag{6.1.4}$$

Also

$$e\mathbf{r} = \sum_{i,j} e|i\rangle\langle i|\mathbf{r}|j\rangle\langle j| = \sum_{i,j} \wp_{ij}\sigma_{ij}, \tag{6.1.5}$$

* Especially the micromaser of H. Walther and coworkers as discussed in Chapter 13. See also the Physics Today article by Haroche and Kleppner [1989] which presents the physics of cavity QED very nicely.

where $\wp_{ij} = e\langle i|\mathbf{r}|j\rangle$ is the electric–dipole transition matrix element. The electric field operator is evaluated in the dipole approximation at the position of the point atom. It follows from Eq. (1.1.27) that, for the atom at the origin, we have

$$\mathbf{E} = \sum_{\mathbf{k}} \hat{\epsilon}_{\mathbf{k}} \mathscr{E}_{\mathbf{k}} (a_{\mathbf{k}} + a_{\mathbf{k}}^{\dagger}), \tag{6.1.6}$$

where $\mathscr{E}_{\mathbf{k}} = (\hbar\nu_k/2\epsilon_0 V)^{1/2}$. Here, for simplicity, we have taken a linear polarization basis and the polarization unit vectors to be real.

It now follows, on substituting for $\mathscr{H}_F, \mathscr{H}_A, e\mathbf{r}$, and $\mathbf{E}$ from Eqs. (6.1.2), (6.1.4), (6.1.5), and (6.1.6) into Eq. (6.1.1), that

$$\mathscr{H} = \sum_{\mathbf{k}} \hbar\nu_k a_{\mathbf{k}}^{\dagger} a_{\mathbf{k}} + \sum_i E_i \sigma_{ii} + \hbar \sum_{i,j} \sum_{\mathbf{k}} g_{\mathbf{k}}^{ij} \sigma_{ij} (a_{\mathbf{k}} + a_{\mathbf{k}}^{\dagger}), \tag{6.1.7}$$

where

$$g_{\mathbf{k}}^{ij} = -\frac{\wp_{ij} \cdot \hat{\epsilon}_{\mathbf{k}} \mathscr{E}_{\mathbf{k}}}{\hbar}. \tag{6.1.8}$$

In Eq. (6.1.7), we have omitted the zero-point energy from the first term. For the sake of simplicity, we will assume $\wp_{ij}$ to be real throughout this chapter.

We now proceed to the case of a two-level atom. For $\wp_{ab} = \wp_{ba}$, we write

$$g_{\mathbf{k}} = g_{\mathbf{k}}^{ab} = g_{\mathbf{k}}^{ba}. \tag{6.1.9}$$

The following form of the Hamiltonian is obtained

$$\begin{aligned} \mathscr{H} = &\sum_{\mathbf{k}} \hbar\nu_k a_{\mathbf{k}}^{\dagger} a_{\mathbf{k}} + (E_a \sigma_{aa} + E_b \sigma_{bb}) \\ &+\hbar \sum_{\mathbf{k}} g_{\mathbf{k}} (\sigma_{ab} + \sigma_{ba})(a_{\mathbf{k}} + a_{\mathbf{k}}^{\dagger}). \end{aligned} \tag{6.1.10}$$

The second term in Eq. (6.1.10) can be rewritten as

$$E_a \sigma_{aa} + E_b \sigma_{bb} = \frac{1}{2}\hbar\omega(\sigma_{aa} - \sigma_{bb}) + \frac{1}{2}(E_a + E_b), \tag{6.1.11}$$

where we use $(E_a - E_b) = \hbar\omega$ and $\sigma_{aa} + \sigma_{bb} = 1$. The constant energy term $(E_a + E_b)/2$ can be ignored. If we use the notation

$$\sigma_z = \sigma_{aa} - \sigma_{bb} = |a\rangle\langle a| - |b\rangle\langle b|, \tag{6.1.12}$$

$$\sigma_+ = \sigma_{ab} = |a\rangle\langle b|, \tag{6.1.13}$$

$$\sigma_- = \sigma_{ba} = |b\rangle\langle a|, \tag{6.1.14}$$

the Hamiltonian (6.1.10) takes the form

$$\mathscr{H} = \sum_{\mathbf{k}} \hbar\nu_k a_{\mathbf{k}}^\dagger a_{\mathbf{k}} + \frac{1}{2}\hbar\omega\sigma_z + \hbar\sum_{\mathbf{k}} g_{\mathbf{k}}(\sigma_+ + \sigma_-)(a_{\mathbf{k}} + a_{\mathbf{k}}^\dagger). \quad (6.1.15)$$

It follows from the identity

$$[\sigma_{ij}, \sigma_{kl}] = \sigma_{il}\delta_{jk} - \sigma_{kj}\delta_{il}, \quad (6.1.16)$$

that σ_+, σ_-, and σ_z satisfy the spin-1/2 algebra of the Pauli matrices, i.e.,

$$[\sigma_-, \sigma_+] = -\sigma_z, \quad (6.1.17)$$

$$[\sigma_-, \sigma_z] = 2\sigma_-. \quad (6.1.18)$$

In the matrix notation, σ_-, σ_+, and σ_z are given by

$$\sigma_- = \begin{pmatrix} 0 & 0 \\ 1 & 0 \end{pmatrix}, \quad \sigma_+ = \begin{pmatrix} 0 & 1 \\ 0 & 0 \end{pmatrix}, \quad \sigma_z = \begin{pmatrix} 1 & 0 \\ 0 & -1 \end{pmatrix}. \quad (6.1.19)$$

The σ_- operator takes an atom in the upper state into the lower state whereas σ_+ takes an atom in the lower state into the upper state.

The interaction energy in Eq. (6.1.15) consists of four terms. The term $a_{\mathbf{k}}^\dagger\sigma_-$ describes the process in which the atom is taken from the upper state into the lower state and a photon of mode **k** is created. The term $a_{\mathbf{k}}\sigma_+$ describes the opposite process. The energy is conserved in both the processes. The term $a_{\mathbf{k}}\sigma_-$ describes the process in which the atom makes a transition from the upper to the lower level and a photon is annihilated, resulting in the loss of approximately $2\hbar\omega$ in energy. Similarly $a_{\mathbf{k}}^\dagger\sigma_+$ results in the gain of $2\hbar\omega$. Dropping the energy nonconserving terms corresponds to the rotating-wave approximation. The resulting simplified Hamiltonian is

$$\mathscr{H} = \sum_{\mathbf{k}} \hbar\nu_k a_{\mathbf{k}}^\dagger a_{\mathbf{k}} + \frac{1}{2}\hbar\omega\sigma_z + \hbar\sum_{\mathbf{k}} g_{\mathbf{k}}(\sigma_+ a_{\mathbf{k}} + a_{\mathbf{k}}^\dagger\sigma_-). \quad (6.1.20)$$

This form of the Hamiltonian describing the interaction of a single two-level atom with a multi-mode field is the starting point of many calculations in the field of quantum optics.

6.2 Interaction of a single two-level atom with a single-mode field

It follows from Eq. (6.1.20) that the interaction of a single-mode quantized field of frequency ν with a single two-level atom is described by the Hamiltonian

$$\mathscr{H} = \mathscr{H}_0 + \mathscr{H}_1, \tag{6.2.1}$$

where

$$\mathscr{H}_0 = \hbar \nu a^\dagger a + \frac{1}{2}\hbar\omega\sigma_z, \tag{6.2.2}$$

$$\mathscr{H}_1 = \hbar g(\sigma_+ a + a^\dagger \sigma_-). \tag{6.2.3}$$

Here we have removed the subscript from the coupling constant g. The Hamiltonian, given by Eqs. (6.2.1)–(6.2.3), describes the atom–field interaction in the dipole and rotating-wave approximations. As we show below, this important Hamiltonian of quantum optics provides us with an exactly solvable example of the field–matter interaction.

It is convenient to work in the interaction picture. The Hamiltonian, in the interaction picture, is given by

$$\mathscr{V} = e^{i\mathscr{H}_0 t/\hbar}\mathscr{H}_1 e^{-i\mathscr{H}_0 t/\hbar}. \tag{6.2.4}$$

Using

$$e^{\alpha A} B e^{-\alpha A} = B + \alpha[A, B] + \frac{\alpha^2}{2!}[A, [A, B]] + \ldots, \tag{6.2.5}$$

it can be readily seen that

$$e^{i\nu a^\dagger a t} a e^{-i\nu a^\dagger a t} = a e^{-i\nu t}, \tag{6.2.6}$$

$$e^{i\omega\sigma_z t/2}\sigma_+ e^{-i\omega\sigma_z t/2} = \sigma_+ e^{i\omega t}. \tag{6.2.7}$$

Combining Eqs. (6.2.1)–(6.2.3), (6.2.4), (6.2.6), and (6.2.7), we have

$$\mathscr{V} = \hbar g(\sigma_+ a e^{i\Delta t} + a^\dagger \sigma_- e^{-i\Delta t}), \tag{6.2.8}$$

where $\Delta = \omega - \nu$.

In this section, we present three different but equivalent methods to solve for the evolution of the atom–field system described by the Hamiltonian (6.2.1)–(6.2.3) based on the solutions of the probability amplitudes, the Heisenberg field and atomic operators, and the unitary time-evolution operator.

6.2.1 Probability amplitude method

We first proceed to solve the equation of motion for $|\psi\rangle$, i.e.,

$$i\hbar\frac{\partial|\psi\rangle}{\partial t} = \mathscr{V}|\psi\rangle. \tag{6.2.9}$$

At any time t, the state vector $|\psi(t)\rangle$ is a linear combination of the states $|a, n\rangle$ and $|b, n\rangle$. Here $|a, n\rangle$ is the state in which the atom is in the excited state $|a\rangle$ and the field has n photons. A similar description exists for the state $|b,n\rangle$. As we are using the interaction picture, we use the slowly varying probability amplitudes $c_{a,n}$ and $c_{b,n}$. The state vector is therefore

$$|\psi(t)\rangle = \sum_n [c_{a,n}(t)|a,n\rangle + c_{b,n}(t)|b,n\rangle]. \tag{6.2.10}$$

The interaction energy (6.2.8) can only cause transitions between the states $|a,n\rangle$ and $|b,n+1\rangle$. We therefore consider the evolution of the amplitudes $c_{a,n}$ and $c_{b,n+1}$. The equations of motion for the probability amplitudes $c_{a,n}$ and $c_{b,n+1}$ are obtained by first substituting for $|\psi(t)\rangle$ and $\mathscr{V}$ from Eqs. (6.2.10) and (6.2.8) in Eq. (6.2.9) and then projecting the resulting equations onto $\langle a,n|$ and $\langle b,n+1|$, respectively. We then obtain

$$\dot{c}_{a,n} = -ig\sqrt{n+1}\, e^{i\Delta t} c_{b,n+1}, \tag{6.2.11}$$

$$\dot{c}_{b,n+1} = -ig\sqrt{n+1}\, e^{-i\Delta t} c_{a,n}. \tag{6.2.12}$$

This coupled set of equations is very similar to that obtained in the semiclassical treatment (cf. Eqs. (5.2.12) and (5.2.13)). These equations can be solved exactly subject to certain initial conditions. A general solution for the probability amplitudes is

$$\begin{aligned} c_{a,n}(t) = \Bigg\{ & c_{a,n}(0)\left[\cos\left(\frac{\Omega_n t}{2}\right) - \frac{i\Delta}{\Omega_n}\sin\left(\frac{\Omega_n t}{2}\right)\right] \\ & - \frac{2ig\sqrt{n+1}}{\Omega_n} c_{b,n+1}(0)\sin\left(\frac{\Omega_n t}{2}\right)\Bigg\} e^{i\Delta t/2}, \end{aligned} \tag{6.2.13}$$

$$\begin{aligned} c_{b,n+1}(t) = \Bigg\{ & c_{b,n+1}(0)\left[\cos\left(\frac{\Omega_n t}{2}\right) + \frac{i\Delta}{\Omega_n}\sin\left(\frac{\Omega_n t}{2}\right)\right] \\ & - \frac{2ig\sqrt{n+1}}{\Omega_n} c_{a,n}(0)\sin\left(\frac{\Omega_n t}{2}\right)\Bigg\} e^{-i\Delta t/2}, \end{aligned} \tag{6.2.14}$$

where

$$\Omega_n^2 = \Delta^2 + 4g^2(n+1). \tag{6.2.15}$$

If initially the atom is in the excited state $|a\rangle$ then $c_{a,n}(0) = c_n(0)$ and $c_{b,n+1}(0) = 0$. Here $c_n(0)$ is the probability amplitude for the field alone. We then obtain

$$c_{a,n}(t) = c_n(0)\left[\cos\left(\frac{\Omega_n t}{2}\right) - \frac{i\Delta}{\Omega_n}\sin\left(\frac{\Omega_n t}{2}\right)\right] e^{i\Delta t/2}, \tag{6.2.16}$$

$$c_{b,n+1}(t) = -c_n(0)\frac{2ig\sqrt{n+1}}{\Omega_n}\sin\left(\frac{\Omega_n t}{2}\right) e^{-i\Delta t/2}. \tag{6.2.17}$$

Fig. 6.1
Behavior of $p(n)$, as given by Eq. (6.2.18), for an initially coherent state. The value of the various parameters are $\Delta = 0$, $\langle n \rangle = 25$, and $gt = 1$.

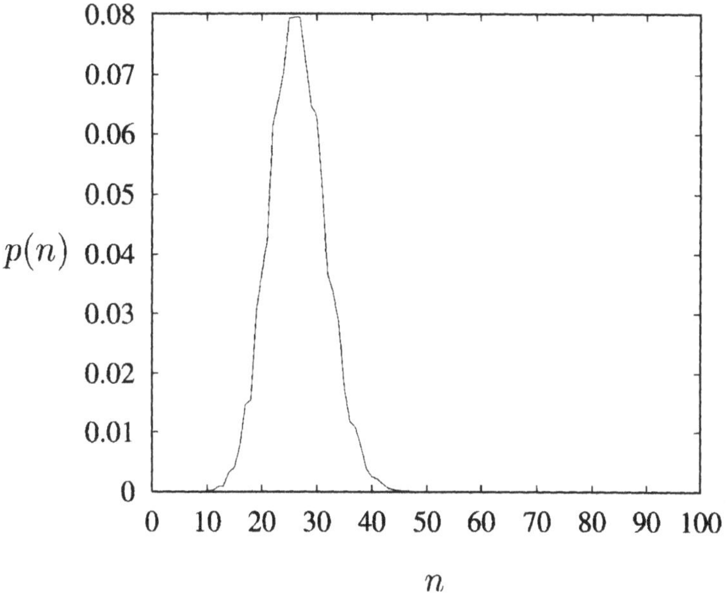

These equations give us a complete solution of the problem. All the physically relevant quantities relating to the quantized field and the atom can be obtained from them.

The expressions $|c_{a,n}(t)|^2$ and $|c_{b,n}(t)|^2$ represent the probabilities that, at time t, the field has n photons present and the atom is in levels $|a\rangle$ and $|b\rangle$, respectively. The probability $p(n)$ that there are n photons in the field at time t is therefore obtained by taking the trace over the atomic states, i.e.,

$$\begin{aligned} p(n) &= |c_{a,n}(t)|^2 + |c_{b,n}(t)|^2 \\ &= \rho_{nn}(0) \left[\cos^2\left(\frac{\Omega_n t}{2}\right) + \left(\frac{\Delta}{\Omega_n}\right)^2 \sin^2\left(\frac{\Omega_n t}{2}\right) \right] \\ &\quad + \rho_{n-1,n-1}(0) \left(\frac{4g^2 n}{\Omega_{n-1}^2}\right) \sin^2\left(\frac{\Omega_{n-1} t}{2}\right), \end{aligned} \tag{6.2.18}$$

where $\rho_{nn}(0) = |c_n(0)|^2$ is the probability that there are n photons present in the field at time $t = 0$. In Fig. 6.1, we plot $p(n)$ for an initial coherent state

$$\rho_{nn}(0) = \frac{\langle n \rangle^n e^{-\langle n \rangle}}{n!}. \tag{6.2.19}$$

Another important quantity is the inversion $W(t)$ which is related to the probability amplitudes $c_{a,n}(t)$ and $c_{b,n}(t)$ by the expression

$$W(t) = \sum_n \left[|c_{a,n}(t)|^2 - |c_{b,n}(t)|^2 \right]. \tag{6.2.20}$$

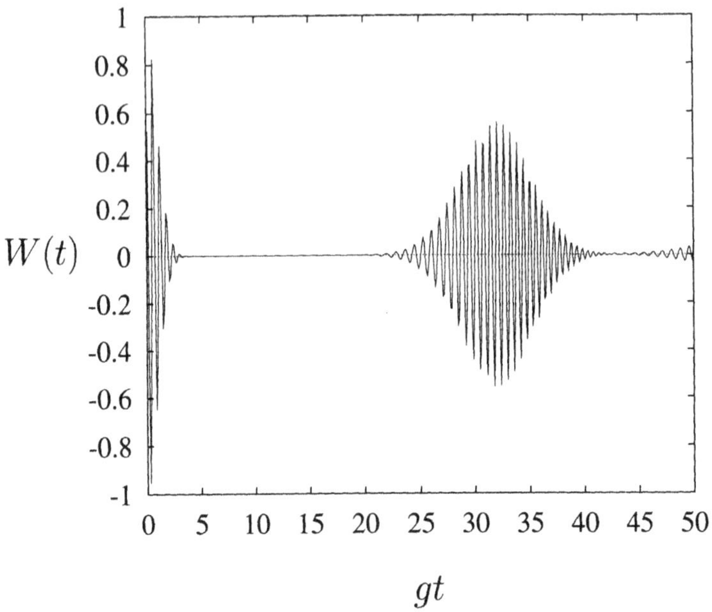

Fig. 6.2
Time evolution of the population inversion $W(t)$ for an initially coherent state with $\langle n \rangle = 25$ and $\Delta = 0$.

On substituting for $c_{a,n}(t)$ and $c_{b,n}(t)$ from Eqs. (6.2.16) and (6.2.17) and making some rearrangements, we obtain

$$W(t) = \sum_{n=0}^{\infty} \rho_{nn}(0) \left[\frac{\Delta^2}{\Omega_n^2} + \frac{4g^2(n+1)}{\Omega_n^2} \cos(\Omega_n t) \right]. \tag{6.2.21}$$

It is interesting to note that even for initial vacuum field ($\rho_{nn}(0) = \delta_{n0}$),

$$W(t) = \frac{1}{\Delta^2 + 4g^2} \left\{ \Delta^2 + 4g^2 \cos\left[\left(\Delta^2 + 4g^2\right)^{1/2} t \right] \right\}, \tag{6.2.22}$$

i.e., the Rabi oscillations take place. This result is drastically different from the predictions of the semiclassical theory of Chapter 5. In the semiclassical theory, the atom in the excited state cannot make a transition to the lower level in the absence of a driving field. In the fully quantum mechanical treatment, the transition from the upper level to the lower level in the *vacuum* becomes possible due to spontaneous emission. Equation (6.2.22) is the simplest example of spontaneous emission in which the spontaneously emitted photon contributes to the single mode of the field considered. A detailed analysis of spontaneous emission by an atom in free space due to the presence of infinitely many vacuum modes will be discussed in the next section.

In Fig. 6.2, $W(t)$ is plotted as a function of the normalized time $\tau = gt$ for an initial coherent state. The behavior of $W(t)$ is quite different from the corresponding curve (Fig. 5.2) in the semiclassical theory. In the present case the envelope of the sinusoidal Rabi oscillations 'collapses' to zero. However as time increases we encounter a 'revival' of the collapsed inversion. This behavior of collapse and revival of inversion is repeated with increasing time, with the amplitude of Rabi oscillations decreasing and the time duration in which revival takes place increasing and ultimately overlapping with the earlier revival.

The phenomena of collapse and revival can be physically understood from Eq. (6.2.21). Each term in the summation represents Rabi oscillations for a definite value of n. The photon distribution function $\rho_{nn}(0)$ determines the relative weight for each value of n. At the initial time, $t = 0$, the atom is prepared in a definite state and therefore all the terms in the summation are correlated. As time increases the Rabi oscillations associated with different excitations have different frequencies and therefore become uncorrelated leading to a collapse of inversion. As time is further increased, the correlation is restored and revival occurs. This behavior continues and an infinite sequence of revivals is obtained. The important thing is that revivals occur only because of the granular structure of the photon distribution. Revival is thus a pure quantum phenomenon. A continuous photon distribution (without zeros) would give a collapse, as would a classical random field, but no revivals.

Simple expressions for the times t_R, t_c, and t_r associated with the sinusoidal Rabi oscillations, the collapse of these oscillations and their revival, respectively, can be determined from Eq. (6.2.21) in the limit $\langle n\rangle \gg 1$. The time period t_R of the Rabi oscillations is given by the inverse of the Rabi frequency Ω_n at $n = \langle n\rangle$, i.e.,

$$t_R \sim \frac{1}{\Omega_{\langle n\rangle}} = \frac{1}{(\Delta^2 + 4g^2\langle n\rangle)^{1/2}}. \tag{6.2.23}$$

As mentioned earlier, these Rabi oscillations continue until a collapse time t_c, when the oscillations associated with different values of n become uncorrelated. Now, for the Poisson distribution (6.2.19) for the initial coherent field, the root-mean-square deviation in the photon number Δn is equal to $\sqrt{\langle n\rangle}$. An estimate of t_c can therefore be obtained from the condition

$$\left(\Omega_{\langle n\rangle+\sqrt{\langle n\rangle}} - \Omega_{\langle n\rangle-\sqrt{\langle n\rangle}}\right) t_c \sim 1. \tag{6.2.24}$$

Since $\langle n\rangle \gg \sqrt{\langle n\rangle}$ in the limit $\langle n\rangle \gg 1$, Eq. (6.2.24) yields

$$\begin{aligned} t_c &\sim \frac{1}{\Omega_{\langle n\rangle+\sqrt{\langle n\rangle}} - \Omega_{\langle n\rangle-\sqrt{\langle n\rangle}}} \\ &\simeq \frac{1}{\left[\Delta^2+4g^2\left(\langle n\rangle+\sqrt{\langle n\rangle}\right)\right]^{1/2} - \left[\Delta^2+4g^2\left(\langle n\rangle-\sqrt{\langle n\rangle}\right)\right]^{1/2}} \\ &\simeq \frac{1}{2g}\left(1+\frac{\Delta^2}{4g^2\langle n\rangle}\right)^{1/2}. \end{aligned} \tag{6.2.25}$$

Under the conditions of exact resonance, $\Delta = 0$, the collapse time t_c is equal to $1/2g$ and is independent of the mean number of photons $\langle n\rangle$. For nonzero detuning, t_c decreases with increasing $\langle n\rangle$. The interval between revivals, t_r, can be found from the condition

$$(\Omega_{\langle n\rangle} - \Omega_{\langle n\rangle-1})t_r = 2\pi m \quad (m = 1, 2, \ldots), \tag{6.2.26}$$

i.e., the revivals take place when the phases of oscillation of the neighboring terms in Eq. (6.2.21) differ by an integral multiple of 2π. Again, in the limit $\langle n\rangle \gg 1$, we obtain

$$\begin{aligned} t_r &= \frac{2\pi m}{\Omega_{\langle n\rangle} - \Omega_{\langle n\rangle-1}} \\ &\simeq \frac{2\pi m\sqrt{\langle n\rangle}}{g}\left(1+\frac{\Delta^2}{4g^2\langle n\rangle}\right)^{1/2}, \end{aligned} \tag{6.2.27}$$

where m is an integer. This shows that revivals take place at regular intervals.

6.2.2 Heisenberg operator method

So far we have considered the problem of the interaction of a single-mode quantized field with a single two-level atom in the interaction picture. In the following we give the solution of the same problem in the Heisenberg picture. In particular we solve the operator equations for the atomic and field operators $a(t)$ and $\sigma_\pm(t)$. These solutions may be particularly useful in the calculation of the multi-time correlation functions necessary in the study of the spectral properties of the field.

The Heisenberg equations for the operators a, σ_-, and σ_z are obtained from the atom–field Hamiltonian (6.2.1)

$$\dot{a} = \frac{1}{i\hbar}[a, \mathscr{H}] = -i\nu a - ig\sigma_-, \tag{6.2.28}$$

$$\dot{\sigma}_- = -i\omega\sigma_- + ig\sigma_z a, \tag{6.2.29}$$

$$\dot{\sigma}_z = 2ig(a^\dagger\sigma_- - \sigma_+ a). \tag{6.2.30}$$

In order to facilitate a solution of these coupled operator equations we define the following constants of motion:

$$N = a^\dagger a + \sigma_+\sigma_-, \tag{6.2.31}$$

$$C = \frac{1}{2}\Delta\sigma_z + g(\sigma_+ a + a^\dagger\sigma_-), \tag{6.2.32}$$

i.e., N and C commute with the Hamiltonian $[N, \mathscr{H}] = [C, \mathscr{H}] = 0$. Here N is an operator that represents the total excitation in the atom–field system, and C is an exchange constant.

We first derive an equation of motion for the atomic lowering operator σ_-. It follows from Eq. (6.2.29) that

$$\begin{aligned}\ddot{\sigma}_- &= -i\omega\dot{\sigma}_- + ig(\dot{\sigma}_z a + \sigma_z\dot{a}) \\ &= -i\omega\dot{\sigma}_- - 2g^2(a^\dagger\sigma_- a - \sigma_+ a^2) + \nu g\sigma_z a - g^2\sigma_-, \end{aligned} \tag{6.2.33}$$

where, in the second line, we substituted for $\dot{\sigma}_z$ and $\dot{a}$ from Eqs. (6.2.30) and (6.2.28), respectively. It is readily verified that

$$\begin{aligned} g^2(a^\dagger\sigma_- a - \sigma_+ a^2) = -i\left(\frac{\Delta}{2} + C\right)\dot{\sigma}_- \\ + \left(\nu C - \frac{1}{2}\Delta^2 + \frac{1}{2}\omega\Delta\right)\sigma_-, \end{aligned} \tag{6.2.34}$$

$$g\sigma_z a = -i\dot{\sigma}_- + \omega\sigma_-. \tag{6.2.35}$$

On substituting these expressions in Eq. (6.2.33), we obtain the desired equation for σ_-:

$$\ddot{\sigma}_- + 2i(\nu - C)\dot{\sigma}_- + (2\nu C - \nu^2 + g^2)\sigma_- = 0. \tag{6.2.36}$$

In a similar manner, we obtain

$$\ddot{a} + 2i(\nu - C)\dot{a} + (2\nu C - \nu^2 + g^2)a = 0. \tag{6.2.37}$$

These equations can be solved in a straightforward manner and the resulting expressions for $\sigma_-(t)$ and $a(t)$ are

$$\begin{aligned}\sigma_-(t) &= [\sigma_+(t)]^\dagger \\ &= e^{-i\nu t}e^{iCt}\left[\left(\cos\kappa t + iC\frac{\sin\kappa t}{\kappa}\right)\sigma_-(0) - ig\frac{\sin\kappa t}{\kappa}a(0)\right],\end{aligned} \tag{6.2.38}$$

$$a(t) = e^{-i\nu t}e^{iCt}\left[\left(\cos\kappa t - iC\frac{\sin\kappa t}{\kappa}\right)a(0) - ig\frac{\sin\kappa t}{\kappa}\sigma_-(0)\right], \tag{6.2.39}$$

where κ is a constant operator

$$\kappa = \left[\frac{\Delta^2}{4} + g^2(N+1) \right]^{1/2}, \tag{6.2.40}$$

which commutes with C, i.e., $[C, \kappa] = 0$. In deriving Eqs. (6.2.38) and (6.2.39), we used

$$C^2 = \frac{\Delta^2}{4} + g^2 N, \tag{6.2.41}$$

$$g\sigma_z a = 2C\sigma_- + \Delta\sigma_- - ga. \tag{6.2.42}$$

Equations (6.2.38) and (6.2.39) provide a complete solution of the problem involving interaction of a two-level atom with a single-mode field in the Heisenberg picture. All quantities of interest can be obtained from these solutions. For example, the expression for the inversion $W(t)$ (Eq. (6.2.21)) can be recovered from Eq. (6.2.38) via

$$\begin{aligned} W(t) &= \langle a, \alpha | \sigma_z(t) | a, \alpha \rangle, \\ &= 2\langle a, \alpha | \sigma_+(t)\sigma_-(t) | a, \alpha \rangle - 1. \end{aligned} \tag{6.2.43}$$

Here we have assumed that the atom is intially in the excited state $|a\rangle$ and the field is intially in the coherent state $|\alpha\rangle$.

As mentioned earlier, a particular advantage of working in the Heisenberg picture is that the evaluation of multi-time correlation functions is straightforward. As an example, we can use Eq. (6.2.38) to construct the dipole–dipole correlation function (Problem 6.5)

$$\begin{aligned} &\langle a, \alpha | \sigma_+(t)\sigma_-(t+\tau) | a, \alpha \rangle \\ &= e^{-i\nu\tau - |\alpha|^2} \sum_{n=0}^{\infty} \frac{|\alpha|^{2n}}{n!} \\ &\times \frac{1}{4\Omega_n^2} \left[\cos(\Omega_{n-1}\tau/2) - \frac{i\Delta}{2\Omega_{n-1}} \sin(\Omega_{n-1}\tau/2) \right] \\ &\times \{ (\Omega_n + \Delta)^2 e^{-i\Omega_n \tau/2} + (\Omega_n - \Delta)^2 e^{i\Omega_n \tau/2} \\ &+ 8g^2(n+1)\cos\left[\Omega_n(\tau + 2t)/2\right] \}, \end{aligned} \tag{6.2.44}$$

where Ω_n is given in Eq. (6.2.15).

6.2.3 *Unitary time-evolution operator method*

Another equivalent approach to deal with the problem of atom–field interaction is through the unitary time-evolution operator. In many problems where the evolution of the system is unitary, i.e., there is no dissipation, this approach may prove to be the simplest.

For the present problem of the interaction of a two-level atom with a single-mode quantized radiation field, the unitary time-evolution operation is given by

$$U(t) = \exp(-i\mathscr{V}t/\hbar), \tag{6.2.45}$$

where the interaction picture Hamiltonian $\mathscr{V}$, at exact resonance, is given by (Eq. (6.2.8) with $\Delta = 0$)

$$\mathscr{V} = \hbar g(\sigma_+ a + a^\dagger \sigma_-). \tag{6.2.46}$$

Here $\sigma_+ = |a\rangle\langle b|$ and $\sigma_- = |b\rangle\langle a|$. It follows, on using

$$(\sigma_+ a + a^\dagger \sigma_-)^{2\ell} = (aa^\dagger)^\ell |a\rangle\langle a| + (a^\dagger a)^\ell |b\rangle\langle b|, \tag{6.2.47}$$

$$(\sigma_+ a + a^\dagger \sigma_-)^{2\ell+1} = (aa^\dagger)^\ell a|a\rangle\langle b| + a^\dagger (aa^\dagger)^\ell |b\rangle\langle a|, \tag{6.2.48}$$

that

$$U(t) = \cos(gt\sqrt{a^\dagger a + 1})|a\rangle\langle a| + \cos(gt\sqrt{a^\dagger a})|b\rangle\langle b|$$
$$-i\frac{\sin(gt\sqrt{a^\dagger a + 1})}{\sqrt{a^\dagger a + 1}} a|a\rangle\langle b| - ia^\dagger \frac{\sin(gt\sqrt{a^\dagger a + 1})}{\sqrt{a^\dagger a + 1}}|b\rangle\langle a|. \tag{6.2.49}$$

The wave vector at time t in terms of the wave vector at time $t = 0$ is simply given by

$$|\psi(t)\rangle = U(t)|\psi(0)\rangle. \tag{6.2.50}$$

As an example of the equivalence of this method with earlier approaches we evaluate the probability amplitudes $c_{a,n}(t)$ and $c_{b,n+1}(t)$ for an atom initially in the excited state $|a\rangle$ and the field as a linear combination of number states, i.e.,

$$|\psi(0)\rangle = \sum_{n=0}^{\infty} c_n(0)|a, n\rangle. \tag{6.2.51}$$

On substituting for $U(t)$ and $|\psi(0)\rangle$ from Eqs. (6.2.49) and (6.2.51), respectively, in Eq. (6.2.50), we obtain

$$|\psi(t)\rangle = \sum_{n=0}^{\infty} c_n(0) \Big[\cos(gt\sqrt{n+1})|a, n\rangle$$
$$- i\sin(gt\sqrt{n+1})|b, n+1\rangle\Big]. \tag{6.2.52}$$

We thus have

$$c_{a,n}(t) = \langle a, n|\psi(t)\rangle = c_n(0)\cos(gt\sqrt{n+1}), \tag{6.2.53}$$

$$c_{b,n+1}(t) = -ic_n(0)\sin(gt\sqrt{n+1}), \tag{6.2.54}$$

in full agreement with Eqs. (6.2.16) and (6.2.17) for $\Delta = 0$.

6.3 Weisskopf–Wigner theory of spontaneous emission between two atomic levels

In the previous section, we showed that an atom in the upper level can make transitions back and forth to the lower state in time even in the absence of an applied field. However, it is seen experimentally that an atom in an excited state decays to the ground state with a characteristic lifetime but it does not make back and forth transitions. The atomic decay has been added into the atomic density matrix equations (see Problem 5.2) phenomenologically. In our model of spontaneous emission discussed in the previous section, the decay is not included because we have considered only one mode of the field. For a proper account of the atomic decay a continuum of modes, corresponding to a quantization cavity which is infinite in extent, needs to be included.

The interaction picture Hamiltonian, in the rotating-wave approximation, for this system is

$$\mathscr{V} = \hbar \sum_{\mathbf{k}} \left[g_{\mathbf{k}}^{*}(\mathbf{r}_0)\sigma_{+}a_{\mathbf{k}}e^{i(\omega-\nu_k)t} + \text{H.c.}\right], \tag{6.3.1}$$

where $g_{\mathbf{k}}(\mathbf{r}_0) = g_{\mathbf{k}}\exp(-i\mathbf{k}\cdot\mathbf{r}_0)$, i.e., we have included the spatial dependence explicitly. Here, $\mathbf{r}_0$ is the location of the atom. The interaction picture Hamiltonian is obtained following the same method as outlined in the beginning of Section 6.2. We assume that at time $t = 0$ the atom is in the excited state $|a\rangle$ and the field modes are in the vacuum state $|0\rangle$. The state vector is therefore

$$|\psi(t)\rangle = c_a(t)|a, 0\rangle + \sum_{\mathbf{k}} c_{b,\mathbf{k}}|b, 1_{\mathbf{k}}\rangle, \tag{6.3.2}$$

with

$$c_a(0) = 1, \quad c_{b,\mathbf{k}}(0) = 0. \tag{6.3.3}$$

We want to determine the state of the atom and the state of the radiation field at some later time when the atom begins to emit photons and we do so in the Weisskopf–Wigner approximation.

From the Schrödinger equation

$$|\dot{\psi}(t)\rangle = -\frac{i}{\hbar}\mathscr{V}|\psi(t)\rangle, \tag{6.3.4}$$

we get the equations of motion for the probability amplitudes c_a and $c_{b,\mathbf{k}}$:

$$\dot{c}_a(t) = -i\sum_{\mathbf{k}} g_{\mathbf{k}}^{*}(\mathbf{r}_0)e^{i(\omega-\nu_k)t}c_{b,\mathbf{k}}(t), \tag{6.3.5}$$

$$\dot{c}_{b,\mathbf{k}}(t) = -ig_{\mathbf{k}}(\mathbf{r}_0)e^{-i(\omega-\nu_k)t}c_a(t). \tag{6.3.6}$$

In order to get an equation that involves c_a only, we first integrate Eq. (6.3.6),

$$c_{b,\mathbf{k}}(t) = -ig_{\mathbf{k}}(\mathbf{r}_0)\int_0^t dt' e^{-i(\omega-\nu_k)t'}c_a(t'). \tag{6.3.7}$$

On substituting this expression of $c_{b,\mathbf{k}}(t)$ into Eq. (6.3.5), we obtain

$$\dot{c}_a(t) = -\sum_{\mathbf{k}} |g_{\mathbf{k}}(\mathbf{r}_0)|^2 \int_0^t dt' e^{i(\omega-\nu_k)(t-t')}c_a(t'). \tag{6.3.8}$$

This is still an exact equation. We have replaced two linear differential equations by one linear differential-integral equation. Next we make some approximations.

Assuming that the modes of the field are closely spaced in frequency, we can replace the summation over **k** by an integral:

$$\sum_{\mathbf{k}} \to 2\frac{V}{(2\pi)^3}\int_0^{2\pi} d\phi \int_0^{\pi} d\theta \sin\theta \int_0^{\infty} dk\, k^2, \tag{6.3.9}$$

where V is the quantization volume. It follows from Eq. (6.1.8) that

$$|g_{\mathbf{k}}(\mathbf{r}_0)|^2 = \frac{\nu_k}{2\hbar\epsilon_0 V}\wp_{ab}^2 \cos^2\theta, \tag{6.3.10}$$

where θ is the angle between the atomic dipole moment $\wp_{ab}$ and the electric field polarization vector $\hat{\epsilon}_{\mathbf{k}}$. Equation (6.3.8) therefore becomes

$$\dot{c}_a(t) = -\frac{4\wp_{ab}^2}{(2\pi)^2 6\hbar\epsilon_0 c^3}\int_0^{\infty} d\nu_k \nu_k^3 \int_0^t dt' e^{i(\omega-\nu_k)(t-t')}c_a(t'), \tag{6.3.11}$$

where integrations over θ and ϕ have been carried out and we have used $k = \nu_k/c$. In the emission spectrum, the intensity of light associated with the emitted radiation is going to be centered about the atomic transition frequency ω. The quantity ν_k^3 varies little around $\nu_k = \omega$ for which the time integral in Eq. (6.3.11) is not negligible. We can therefore replace ν_k^3 by ω^3 and the lower limit in the ν_k integration by $-\infty$. The integral

$$\int_{-\infty}^{\infty} d\nu_k e^{i(\omega-\nu_k)(t-t')} = 2\pi\delta(t-t'), \tag{6.3.12}$$

yields the following equation for $c_a(t)$, in the Weisskopf–Wigner approximation:

$$\dot{c}_a(t) = -\frac{\Gamma}{2}c_a(t), \tag{6.3.13}$$

where the decay constant

$$\Gamma = \frac{1}{4\pi\epsilon_0}\frac{4\omega^3\wp_{ab}^2}{3\hbar c^3}. \tag{6.3.14}$$

A solution of Eq. (6.3.13) gives

$$\rho_{aa} \equiv |c_a(t)|^2 = \exp(-\Gamma t), \tag{6.3.15}$$

i.e., an atom in the excited state $|a\rangle$ in vacuum decays exponentially in time with the lifetime $\tau = 1/\Gamma$.

During the process of spontaneous emission, the atom emits a quantum of energy equal to $E_a - E_b = \hbar\nu$. We now calculate the state of the field emitted during the spontaneous emission process.

We first calculate the coefficient $c_{b,\mathbf{k}}(t)$. Substituting the solution for $c_a(t)$ into Eq. (6.3.7) we find

$$\begin{aligned} c_{b,\mathbf{k}}(t) &= -ig_{\mathbf{k}}(\mathbf{r}_0)\int_0^t dt' e^{-i(\omega-\nu_k)t'-\Gamma t'/2} \\ &= g_{\mathbf{k}}(\mathbf{r}_0)\left[\frac{1-e^{-i(\omega-\nu_k)t-\Gamma t/2}}{(\nu_k-\omega)+i\Gamma/2}\right], \end{aligned} \tag{6.3.16}$$

so that

$$\begin{aligned} |\psi(t)\rangle &= e^{-\Gamma t/2}|a,0\rangle \\ &+ |b\rangle\sum_{\mathbf{k}} g_{\mathbf{k}} e^{-i\mathbf{k}\cdot\mathbf{r}_0}\left[\frac{1-e^{i(\omega-\nu_k)t-\Gamma t/2}}{(\nu_k-\omega)+i\Gamma/2}\right]|1_{\mathbf{k}}\rangle. \end{aligned} \tag{6.3.17}$$

Upon introducing the field state

$$|\gamma_0\rangle = \sum_{\mathbf{k}} g_{\mathbf{k}}\frac{e^{-i\mathbf{k}\cdot\mathbf{r}_0}}{(\nu_k-\omega)+i\Gamma/2}|1_{\mathbf{k}}\rangle, \tag{6.3.18}$$

for times long compared to the radiative decay $t \gg \Gamma^{-1}$ we have $|\psi\rangle \to |b\rangle|\gamma_0\rangle$. Here the index '0' in $|\gamma_0\rangle$ reminds us that this state corresponds to an atom located at position $\mathbf{r}_0$. This is a linear superposition of the single-photon states with different wave vectors associated with them.

The first-order correlation function $G^{(1)}(\mathbf{r},\mathbf{r};t,t)$ for large times is given by

$$\begin{aligned} G^{(1)}(\mathbf{r},\mathbf{r};t,t) &= \langle\psi|E^{(-)}(\mathbf{r},t)E^{(+)}(\mathbf{r},t)|\psi\rangle \\ &= \langle\gamma_0|E^{(-)}(\mathbf{r},t)E^{(+)}(\mathbf{r},t)|\gamma_0\rangle \\ &= \langle\gamma_0|E^{(-)}(\mathbf{r},t)|0\rangle\langle 0|E^{(+)}(\mathbf{r},t)|\gamma_0\rangle. \end{aligned} \tag{6.3.19}$$

Here a complete set of states is inserted and since only the vacuum state survives while the other states lead to zero, we keep the vacuum state only. We have also assumed, that the field is linearly polarized, say along the x-axis. As discussed in Section 4.2, $G^{(1)}(\mathbf{r},\mathbf{r};t,t)$ is proportional to the probability of registering a photon at time t by a photodetector located at the position $\mathbf{r}$. According to Eq. (6.3.19), $G^{(1)}(\mathbf{r},\mathbf{r};t,t) = |\langle 0|E^{(+)}(\mathbf{r},t)|\gamma_0\rangle|^2$. Thus the function

$$\Psi_\gamma(\mathbf{r},t) = \langle 0|E^{(+)}(\mathbf{r},t)|\gamma_0\rangle \tag{6.3.20}$$

can be interpreted as a kind of *wave function* for a photon. This is in analogy with the corresponding wave function for particles (see Section 1.5).

From the definitions of $E^{(+)}(\mathbf{r},t)$ and $|\gamma_0\rangle$ (Eqs. (1.1.30) and (6.3.18)), we find

$$\langle 0|E^{(+)}(\mathbf{r},t)|\gamma_0\rangle =$$
$$= \sqrt{\frac{\hbar}{2\epsilon_0 V}}\sum_{\mathbf{k},\mathbf{k}'}\langle 0|(\nu_{k'})^{1/2}a_{\mathbf{k}'}e^{-i\nu_{k'}t+i\mathbf{k}'\cdot\mathbf{r}}g_{\mathbf{k}}\frac{e^{-i\mathbf{k}\cdot\mathbf{r}_0}}{(\nu_k-\omega)+i\Gamma/2}|1_{\mathbf{k}}\rangle$$
$$= \sqrt{\frac{\hbar}{2\epsilon_0 V}}\sum_{\mathbf{k}}(\nu_k)^{1/2}g_{\mathbf{k}}e^{-i\nu_k t}e^{i\mathbf{k}\cdot(\mathbf{r}-\mathbf{r}_0)}\frac{1}{(\nu_k-\omega)+i\Gamma/2}. \tag{6.3.21}$$

We now evaluate this function by first converting the sum into an integral via Eq. (6.3.9). We however do not include the factor 2 from there as the field is assumed to be polarized along the x-axis. The ϕ- and θ-integrations can be carried out by choosing a coordinate system in which the vector $\mathbf{r}-\mathbf{r}_0$ points along the z-axis, the atomic dipole moment forms an angle η with the z-axis in the x-z plane, and the wave vector $\mathbf{k}$ has components

$$\mathbf{k} = k(\sin\theta\cos\phi\,\hat{x} + \sin\theta\sin\phi\,\hat{y} + \cos\theta\,\hat{z}). \tag{6.3.22}$$

The resulting expression for $\langle 0|E^{(+)}(\mathbf{r},t)|\gamma_0\rangle$ is*

$$\langle 0|E^{(+)}(\mathbf{r},t)|\gamma_0\rangle = \frac{ic\wp_{ab}\sin\eta}{8\pi^2\epsilon_0\Delta r} \times\int_0^\infty dk k^2\left(e^{ik\Delta r}-e^{-ik\Delta r}\right)\frac{e^{-i\nu_k t}}{(\nu_k-\omega)+i\Gamma/2}, \tag{6.3.23}$$

where $\Delta r = |\mathbf{r}-\mathbf{r}_0|$. In the above integral the term $\exp[-i(k\Delta r+\nu_k t)]$ represents an incoming wave and we will therefore neglect it. As in the Weisskopf–Wigner theory of spontaneous emission, we assume that the quantity ν_k^2 varies little around $\nu_k=\omega$ and therefore can be replaced by ω^2 and the lower limit of integration can be extended to $-\infty$. Making these approximations we are left with the integral

$$\int_{-\infty}^{\infty} d\nu_k \frac{e^{-i\nu_k t+i\nu_k\Delta r/c}}{(\nu_k-\omega)+i\Gamma/2}.$$

This integral is evaluated by using the contour method (see Fig. 6.3). For $t<\Delta r/c$, the contour lies in the upper half plane and if $t>\Delta r/c$, in the lower half-plane. On performing the integration, we find that

* Equation (6.3.23) can be derived in a more complete and rigorous way using the method in Appendix 10.A.

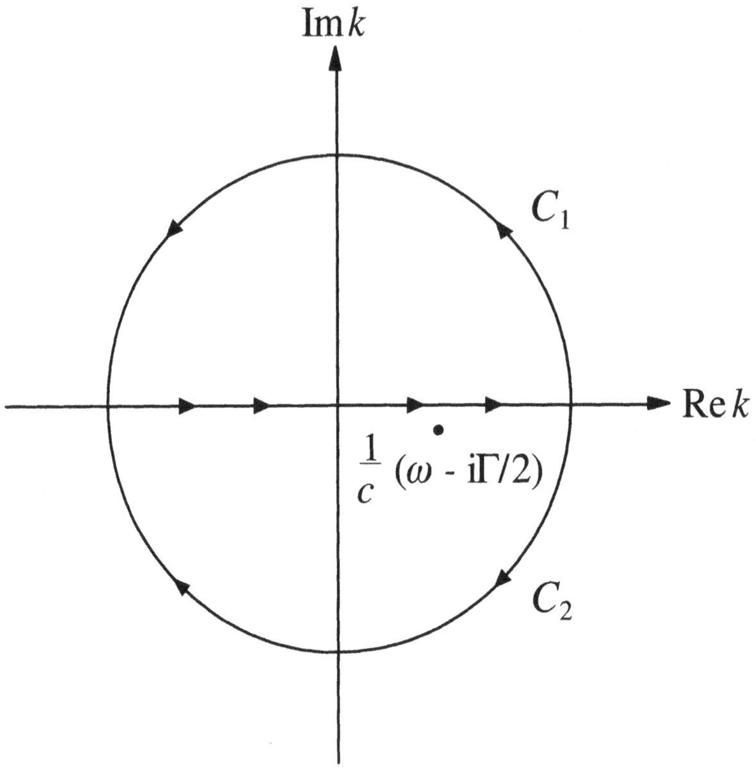

Fig. 6.3
Contours used for evaluating the integral in Eq. (6.3.23): C_1 if $t < \Delta r/c$ and C_2 if $t > \Delta r/c$.

$$\langle 0|E^{(+)}(\mathbf{r},t)|\gamma_0\rangle = \frac{\mathscr{E}_0}{\Delta r}\Theta\left(t - \frac{\Delta r}{c}\right) e^{-i\left(t-\frac{\Delta r}{c}\right)(\omega - i\Gamma/2)}, \tag{6.3.24}$$

where Θ is a unit step function and

$$\mathscr{E}_0 = -\frac{\omega^2 \wp_{ab}\sin\eta}{4\pi\epsilon_0 c^2 \cancel{\Delta r}}. \tag{6.3.25}$$

We then find that

$$G^{(1)}(\mathbf{r},\mathbf{r};t,t) = \frac{|\mathscr{E}_0|^2}{|\mathbf{r}-\mathbf{r}_0|^2}\Theta\left(t - \frac{|\mathbf{r}-\mathbf{r}_0|}{c}\right) e^{-\Gamma(t-|\mathbf{r}-\mathbf{r}_0|/c)}. \tag{6.3.26}$$

Here the step function is a manifestation of the fact that the signal cannot move faster than the speed of light.

6.4 Two-photon cascades

In this section we consider the spontaneous emission in a three-level atom in cascade configuration, as shown in Fig. 6.4. The atom in upper state $|a\rangle$ emits a $\mathbf{k}$ photon of frequency ν_k and decays to state

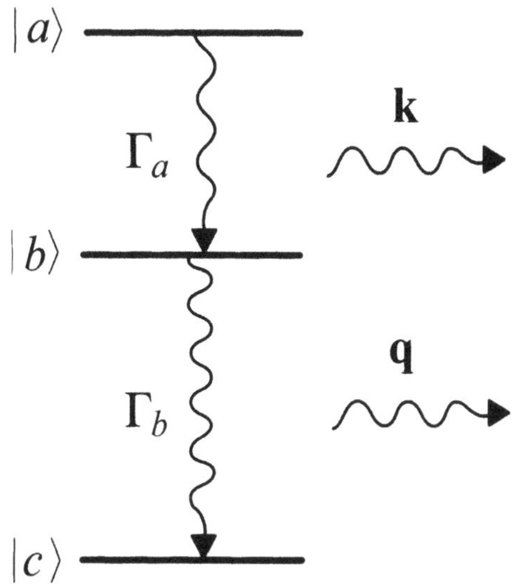

Fig. 6.4
Level scheme for atomic decay due to spontaneous emission in a three-level atom in cascade configuration.

$|b\rangle$ which decays to ground state $|c\rangle$, via emission of a $\mathbf{q}$ photon of frequency ν_q. The interaction picture Hamiltonian for the system is

$$\mathscr{V} = \hbar \sum_{\mathbf{k}} \left[g^*_{a,\mathbf{k}}(\mathbf{r}_0)\sigma^{(1)}_+ a_{\mathbf{k}} e^{i(\omega_{ab}-\nu_k)t} + \text{H.c.} \right]$$
$$+\hbar \sum_{\mathbf{q}} \left[g^*_{b,\mathbf{q}}(\mathbf{r}_0)\sigma^{(2)}_+ a_{\mathbf{q}} e^{i(\omega_{bc}-\nu_q)t} + \text{H.c.} \right], \qquad (6.4.1)$$

where $\sigma^{(1)}_+ = |a\rangle\langle b|$, $\sigma^{(2)}_+ = |b\rangle\langle c|$, and $g_{a,\mathbf{k}}(\mathbf{r}_0)$ and $g_{b,\mathbf{q}}(\mathbf{r}_0)$ are the appropriate coupling constants for $|a\rangle \to |b\rangle$ and $|b\rangle \to |c\rangle$ transitions, respectively.

The state of the atom–field system is now described by

$$|\psi(t)\rangle = c_a(t)|a,0\rangle + \sum_{\mathbf{k}} c_{b,\mathbf{k}}|b,1_{\mathbf{k}}\rangle + \sum_{\mathbf{k},\mathbf{q}} c_{c,\mathbf{k},\mathbf{q}}|c,1_{\mathbf{k}},1_{\mathbf{q}}\rangle. \qquad (6.4.2)$$

As in Section 6.3, the probability amplitudes c_a, $c_{b,\mathbf{k}}$, and $c_{c,\mathbf{k},\mathbf{q}}$ obey the equations of motion

$$\dot{c}_a = -i\sum_{\mathbf{k}} g^*_{a,\mathbf{k}}(\mathbf{r}_0)c_{b,\mathbf{k}}e^{i(\omega_{ab}-\nu_k)t}, \qquad (6.4.3)$$

$$\dot{c}_{b,\mathbf{k}} = -ig_{a,\mathbf{k}}(\mathbf{r}_0)c_a e^{-i(\omega_{ab}-\nu_k)t} - i\sum_{\mathbf{q}} g^*_{b,\mathbf{q}}(\mathbf{r}_0)c_{c,\mathbf{k},\mathbf{q}}e^{i(\omega_{bc}-\nu_q)t}, \qquad (6.4.4)$$

$$\dot{c}_{c,\mathbf{k},\mathbf{q}} = -ig_{b,\mathbf{q}}(\mathbf{r}_0)c_{b,\mathbf{k}}e^{-i(\omega_{bc}-\nu_q)t}. \qquad (6.4.5)$$

Following the lead of Section 6.3, we recognize that, in the Weisskopf–Wigner approximation,

$$-i\sum_{\mathbf{k}} g^*_{a,\mathbf{k}}(\mathbf{r}_0)c_{b,\mathbf{k}}e^{i(\omega_{ab}-\nu_k)t} = -\frac{\Gamma_a}{2}c_a, \qquad (6.4.6)$$

where Γ_a is the atomic decay rate from state $|a\rangle$ to state $|b\rangle$. Furthermore, it is clear that the second term in Eq. (6.4.4) represents decay from $|b\rangle$ to $|c\rangle$ and we may write

$$-i\sum_{\mathbf{q}} g^*_{b,\mathbf{q}}(\mathbf{r}_0)c_{c,\mathbf{k},\mathbf{q}}e^{i(\omega_{bc}-\nu_q)t} = -\frac{\Gamma_b}{2}c_{b,\mathbf{k}}, \tag{6.4.7}$$

where Γ_b is the decay rate from state $|b\rangle$ to state $|c\rangle$. Upon inserting Eqs. (6.4.6) and (6.4.7) into (6.4.3)–(6.4.5), we obtain the useful final form for the atom–field equations of motion

$$\dot{c}_a = -\frac{\Gamma_a}{2}c_a, \tag{6.4.8}$$

$$\dot{c}_{b,\mathbf{k}} = -ig_{a,\mathbf{k}}(\mathbf{r}_0)e^{-i(\omega_{ab}-\nu_k)t-\frac{\Gamma_a}{2}t} - \frac{\Gamma_b}{2}c_{b,\mathbf{k}}, \tag{6.4.9}$$

$$\dot{c}_{c,\mathbf{k},\mathbf{q}} = -ig_{b,\mathbf{q}}(\mathbf{r}_0)c_{b,\mathbf{k}}e^{-i(\omega_{bc}-\nu_q)t}, \tag{6.4.10}$$

where we have substituted $\exp(-\Gamma_a t/2)$ for $c_a(t)$ in the first term of Eq. (6.4.9).

We are most interested in the state of the field for times $t \gg \Gamma_a^{-1}$ and Γ_b^{-1}, i.e., we want to know $c_{c,\mathbf{k},\mathbf{q}}(\infty)$ as $c_a(\infty)$ and $c_{b,\mathbf{k}}(\infty)$ tend to zero.

It follows, on carrying out the simple integration implied by Eq. (6.4.9), that

$$\begin{aligned} c_{b,\mathbf{k}}(t) &= -ig_{a,\mathbf{k}}(\mathbf{r}_0)\int_0^t dt' e^{-i(\omega_{ab}-\nu_k)t'-\Gamma_a t'/2}e^{-\Gamma_b(t-t')/2} \\ &= -ig_{a,\mathbf{k}}(\mathbf{r}_0)\frac{e^{i(\nu_k-\omega_{ab})t-\Gamma_a t/2} - e^{-\Gamma_b t/2}}{i(\nu_k-\omega_{ab}) - \frac{1}{2}(\Gamma_a-\Gamma_b)}. \end{aligned} \tag{6.4.11}$$

This expression for $c_{b,\mathbf{k}}(t)$ can now be substituted into Eq. (6.4.10), and the resulting equation can be integrated to yield the following long time limit of $c_{c,\mathbf{k},\mathbf{q}}(t)$:

$$\begin{aligned} c_{c,\mathbf{k},\mathbf{q}}(\infty) &= g_{a,\mathbf{k}}g_{b,\mathbf{q}}e^{-i(\mathbf{k}+\mathbf{q})\cdot\mathbf{r}_0}\frac{1}{i(\nu_k-\omega_{ab})-\frac{1}{2}(\Gamma_a-\Gamma_b)} \\ &\quad\times\left[\frac{1}{i(\nu_k+\nu_q-\omega_{ac})-\frac{1}{2}\Gamma_a} - \frac{1}{i(\nu_q-\omega_{bc})-\frac{1}{2}\Gamma_b}\right] \\ &= \frac{-g_{a,\mathbf{k}}g_{b,\mathbf{q}}e^{-i(\mathbf{k}+\mathbf{q})\cdot\mathbf{r}_0}}{[i(\nu_k+\nu_q-\omega_{ac})-\frac{1}{2}\Gamma_a][i(\nu_q-\omega_{bc})-\frac{1}{2}\Gamma_b]}. \end{aligned} \tag{6.4.12}$$

As in the long time limit, both $c_a(t)$ and $c_{b,\mathbf{k}}(t)$ are zero, we insert $c_{c,\mathbf{k},\mathbf{q}}(\infty)$, as given by Eq. (6.4.12), into Eq. (6.4.2) and find that the state of the radiation field is given by

$$|\gamma,\phi\rangle = \sum_{\mathbf{k},\mathbf{q}} \frac{-g_{a,\mathbf{k}}g_{b,\mathbf{q}}e^{-i(\mathbf{k}+\mathbf{q})\cdot\mathbf{r}_0}}{[i(\nu_k+\nu_q-\omega_{ac})-\frac{1}{2}\Gamma_a][i(\nu_q-\omega_{bc})-\frac{1}{2}\Gamma_b]}|1_{\mathbf{k}},1_{\mathbf{q}}\rangle, \tag{6.4.13}$$

where $|\gamma,\phi\rangle$ represents the two-photon state.

We shall make detailed use of this result in Chapter 21 when we utilize two-photon correlation functions in order to gain insight into the foundations of quantum mechanics.

6.5 Excitation probabilities for single and double photoexcitation events

In Section 4.2 we presented heuristic arguments to show that the photodetection probability is governed by the normally ordered field correlation functions. Here we derive the excitation probability for single and double photoelectron events using the atom–field interaction formalism developed in this chapter.*

Consider the interaction of linearly polarized light, described by the field operators $E^{(+)}(\mathbf{r},t)$ and $E^{(-)}(\mathbf{r},t)$, with an atomic system consisting of a lower level $|b\rangle$ and a set of excited levels $|a_j\rangle$ (Fig. 6.5). We assume that the atom is initially in state $|b\rangle$ and the field is in state $|i\rangle$. The interaction picture Hamiltonian, in the rotating-wave approximation, is

$$\mathscr{V} = -\sum_j \wp_{a_jb}\sigma_{a_jb}E^{(+)}(\mathbf{r},t)\exp(i\omega_{a_j}t) + \text{H.c.} \tag{6.5.1}$$

The state of the atom–field system at time t is given by

$$\begin{aligned}|\psi(t)\rangle &= U_I(t)|\psi(0)\rangle \\ &\simeq \left[1-\frac{i}{\hbar}\int_0^t dt'\mathscr{V}(t')\right]|b\rangle\otimes|i\rangle.\end{aligned} \tag{6.5.2}$$

The probability of exciting the atom to level $|a_j\rangle$ is found by calculating the expectation value of the projection operator $|a_j\rangle\langle a_j|$, i.e.,

$$\begin{aligned}P_j(t) &= \langle\psi(t)|a_j\rangle\langle a_j|\psi(t)\rangle \\ &= \frac{\wp_{a_jb}^2}{\hbar^2}\int_0^t\int_0^t dt_1dt_2\exp[i\omega_{a_j}(t_1-t_2)]\langle i|E^{(-)}(\mathbf{r},t_2)E^{(+)}(\mathbf{r},t_1)|i\rangle,\end{aligned} \tag{6.5.3}$$

where we substitute for $|\psi(t)\rangle$ from Eq. (6.5.2). If we want only the probability of excitation, we should sum over all excited levels $|a_j\rangle$.

* For an excellent treatment see the Les Houches lectures of Glauber [1965].

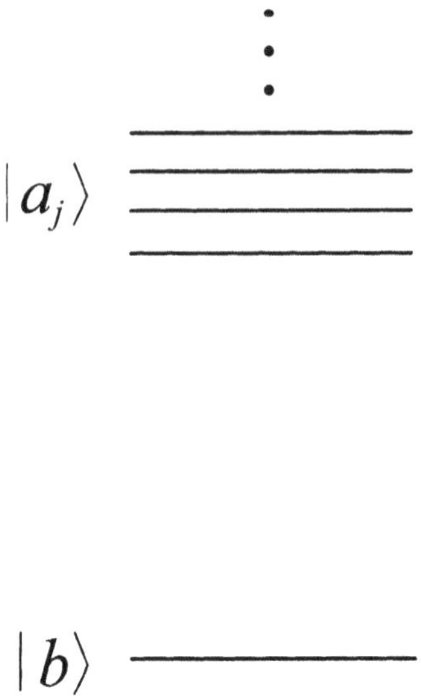

Fig. 6.5
Level scheme for photodetection. The atom makes a transition from state $|b\rangle$ to the manifold of excited states $|a_j\rangle$.

If $\wp^2_{a_jb}$ is largely independent of j, we can take $\wp^2_{a_jb} \simeq \wp^2$. Hence, for a broad-band detector the summation over j of the function $\exp[i\omega_{a_j}(t_1 - t_2)]$ introduces an effective $\delta(t_1 - t_2)$ function and we obtain

$$P(t) = \sum_j P_j(t)$$
$$= \kappa \int_0^t dt_1 \langle i|E^{(-)}(\mathbf{r}, t_1)E^{(+)}(\mathbf{r}, t_1)|i\rangle, \tag{6.5.4}$$

where κ is a constant. For mixed states, Eq. (6.5.4) becomes

$$P(t) = \kappa \int_0^t dt_1 \mathrm{Tr}\left[\rho E^{(-)}(\mathbf{r}, t_1)E^{(+)}(\mathbf{r}, t_1)\right]. \tag{6.5.5}$$

Next, we consider atoms at the points $\mathbf{r}_1$ and $\mathbf{r}_2$ and find the joint-count probability P_{12} of double photoexcitation, i.e., we want the expectation value of the photoexcitation operator

$$\left(\sum_j |a_j\rangle\langle a_j|\right)_\alpha$$

for both atoms, i.e., $\alpha = 1$ and 2. Similarly to Eq. (6.5.5), we obtain

$$P_{12} = \kappa' \int_0^t dt_1 \int_0^t dt_2$$
$$\times \mathrm{Tr}\left[\rho E^{(-)}(\mathbf{r}_1, t_1)E^{(-)}(\mathbf{r}_2, t_2)E^{(+)}(\mathbf{r}_2, t_2)E^{(+)}(\mathbf{r}_1, t_1)\right]. \tag{6.5.6}$$

Thus P_{12} is governed by the second-order correlation function of the field operators.

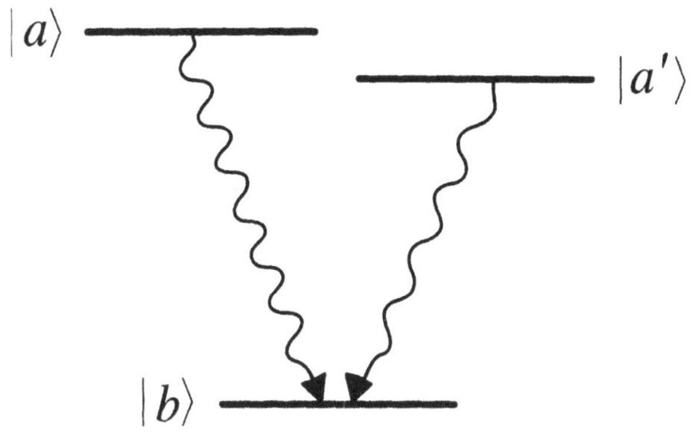

Fig. 6.6
Radiative decay of two closely lying levels $|a\rangle$ and $|a'\rangle$ to a common level $|b\rangle$.

Problems

6.1 A model sometimes considered to study the atom–field coupling in a lossless cavity is represented by the Hamiltonian

$$\mathscr{H} = \hbar \nu a^\dagger a + \hbar \omega \sigma_z + \hbar g \left[\sigma_+ a (a^\dagger a)^{1/2} + (a^\dagger a)^{1/2} a^\dagger \sigma_-\right],$$

in the usual notation. Note that the coupling is intensity dependent. Calculate the atomic inversion and discuss its evolution in terms of the various time scales, i.e., Rabi flopping time, the collapse time, and the revival time, for (a) an initial coherent state of the field and (b) an initial thermal state of the field. Note that the infinite series in the expression for inversion can be summed exactly in this case.

6.2 Calculate the population inversion for a two-level atom interacting with a single-mode quantized radiation field in the dipole and rotating-wave approximations for arbitrary time t when, at $t = 0$, the field is in a coherent state $|\alpha\rangle$, and the atomic state is $|\psi\rangle_{\text{atom}} = (|a\rangle + e^{-i\phi}|b\rangle)/\sqrt{2}$ ($|a\rangle$ and $|b\rangle$ are the upper and lower levels, respectively). Discuss the conditions under which the populations in the two levels remain 'trapped'.

6.3 Consider the atomic system shown in Fig. 6.6 with two closely spaced upper levels $|a\rangle$ and $|a'\rangle$ and a lower level $|b\rangle$. The selection rules and the energy spacing of levels $|a\rangle$ and $|a'\rangle$ is such that they interact with the same vacuum modes. The interaction of this system with a multi-mode vacuum field is described by the interaction picture Hamiltonian,

$$\mathscr{V} = \hbar \sum_{\mathbf{k}} \left[g_{\mathbf{k}}^{(ab)} a_{\mathbf{k}}^{\dagger} |b\rangle\langle a| e^{-i(\omega_{ab}-\nu_k)t} \right.$$
$$\left. + g_{\mathbf{k}}^{(a'b)} a_{\mathbf{k}}^{\dagger} |b\rangle\langle a'| e^{-i(\omega_{a'b}-\nu_k)t} \right]$$
$$+ \text{H.c.}$$

Here $a_{\mathbf{k}}^{\dagger}$ is the creation operator for the mode with wave vector $\mathbf{k}$, and $\omega_{ab} = \omega_a - \omega_b$, $\omega_{a'b} = \omega_{a'} - \omega_b$. Derive the amplitude equations of motion for the three levels and show that quantum interference effects arise due to the sharing of common vacuum modes by the upper two levels.

Hint: see Zhu, Narducci, and Scully, *Phys. Rev. A* **52**, 6 (1995).

6.4 If $C = \frac{1}{2}\Delta\sigma_z + g(\sigma_+ a + a^{\dagger}\sigma_-)$ and $N = a^{\dagger}a + \sigma_+\sigma_-$, show that

$$C^2 = \frac{\Delta^2}{4} + g^2 N.$$

6.5 Prove Eq. (6.2.44).

References and bibliography

Review articles and textbooks on the two-level systems interacting with single-mode quantized field

S. Stenholm, *Phys. Rep.* **6**, 1 (1973).

M. Sargent III, M. Scully, and W. E. Lamb, Jr., *Laser Physics*, (Addison-Wesley, Reading, MA, 1974).

L. Allen and J. H. Eberly, *Optical Resonance and Two-Level Atoms*, (John Wiley, New York, 1975).

P. L. Knight and P. W. Milonni, *Phys. Rep.* **66**, 21 (1980).

S. Stenholm, *Foundations of Laser Spectroscopy*, (Wiley, New York, 1984).

C. Cohen-Tannoudji, J. Dupont-Roc, and G. Grynberg, *Photons and Atoms, Introduction to Quantum Electrodynamics*, (Wiley, New York 1989).

S. Haroche and D. Kleppner, *Physics Today* **42**, Jan. 24 (1989).

C. Cohen-Tannoudji, J. Dupont-Roc, and G. Grynberg, *Atom-Photon Interactions*, (Wiley, New York 1992).

B. W. Shore and P. L. Knight, *J. Mod. Opt.* **40**, 1195 (1993).

P. Berman ed., *Cavity Quantum Electrodynamics* (Acad. Press, New York 1994).

D. F. Walls and G. J. Milburn, *Quantum Optics*, (Springer-Verlag, Berlin 1994).

General references

E. T. Jaynes and F. W. Cummings, *Proc. IEEE* **51**, 89 (1963).

R. J. Glauber, in *Quantum Optics and Electronics*, Les Houches, ed. C. DeWitt, A. Blandin, and C. Cohen-Tannoudji (Gordon and Breach, New York 1965).

M. O. Scully and W. E. Lamb, Jr., *Phys. Rev.* **159**, 208 (1967).

M. Tavis and F. W. Cummings, *Phys. Rev.* **188**, 692 (1969).

P. Meystre, A. Quattropani, and H. P. Baltes, *Phys. Lett.* **49A**, 85 (1974).

T. von Foerster, *J. Phys. A* **8**, 95 (1975).

J. R. Ackerhalt and K. Rzazewski, *Phys. Rev. A* **12**, 2549 (1975).

N. B. Narozhny, J. J. Sanchez-Mondragon, and J. H. Eberly, *Phys. Rev. A* **23**, 236 (1981).

P. L. Knight and P. M. Radmore, *Phys. Rev. A* **26**, 676 (1982).

K. Zaheer and M. S. Zubairy, *Phys. Rev. A* **37**, 1628 (1988).

J. Eiselt and H. Risken, *Opt. Commun.* **72**, 351 (1989); *Phys. Rev. A* **43**, 346 (1991).

T. Quang, P. L. Knight, and V. Bužek, *Phys. Rev. A* **44**, 6092 (1991).

Generalized models for quantized atom–field interactions

R. J. Cook and B. W. Shore, *Phys. Rev. A* **20**, 539 (1979).

S. Kumar and C. L. Mehta, *Phys. Rev. A* **21**, 1573 (1980).

B. Buck and C. V. Sukumar, *Phys. Lett. A* **81**, 132 (1981).

S. Singh, *Phys. Rev. A* **25**, 3206 (1982).

N. N. Bogolubov, Jr., F. Le Kien, and A. S. Shumovsky, *Phys. Lett. A* **101**, 201 (1984); *ibid.* **107**, 456 (1985).

Z. Deng, *Opt. Commun.* **54**, 222 (1985).

J. Seke, *J. Opt. Soc. Am. B* **2**, 968 (1985).
A. M. Abdel-Hafez, A. S. F. Obada, and M. M. A. Ahmad, *Phys. Rev. A* **35**, 1634 (1987).
P. Alsing and M. S. Zubairy, *J. Opt. Soc. Am. B* **4**, 177 (1987).
M. S. Iqbal, S. Mahmood, M. S. K. Razmi, and M. S. Zubairy, *J. Opt. Soc. Am. B* **5**, 1312 (1988).
R. R. Puri and R. K. Bullough, *J. Opt. Soc. Am. B* **5**, 2021 (1988).
V. Bužek and I. Jex, *J. Mod. Opt.* **36**, 1427 (1989).
M. P. Sharma, D. A. Cardimona, and A. Gavrielides, *J. Opt. Soc. Am. B* **6**, 1942 (1989).
V. Bužek, *Phys. Rev. A* **39**, 3196 (1989); *J. Mod. Opt.* **37**, 1033 (1990).
G. Adam, J. Seke, and O. Hittmair, *Phys. Rev. A* **42**, 5522 (1990).
A. Joshi and R. R. Puri, *Phys. Rev. A* **42**, 4336 (1990).
A. H. Toor and M. S. Zubairy, *Phys. Rev. A* **45**, 4951 (1992).

Collapse and revival phenomena

J. H. Eberly, N. B. Narozhny, and J. J. Sanchez-Mondragon, *Phys. Rev. Lett.* **44**, 1323 (1980).
G. Rempe and H. Walther, *Phys. Rev. Lett.* **58**, 353 (1987).
M. Fleischhauer and W. Schleich, *Phys. Rev. A* **47**, 4258 (1993).

Squeezing and photon antibunching in atom–field interactions

P. Meystre and M. S. Zubairy, *Phys. Lett. A* **89**, 390 (1982).
P. L. Knight, *Phys. Scripta* **33** (T12), 51 (1986).
N. N. Bogolubov, Jr., F. Le Kien, and A. S. Shumovsky, *Europhys. Lett.* **4**, 281 (1987).
P. K. Aravind and G. Hu, *Physica C* **150**, 427 (1988).
C. C. Gerry, *Phys. Rev. A* **37**, 2683 (1988).
J. R. Kuklinski and J. L. Madajczyk, *Phys. Rev. A* **37**, 3175 (1988).
M. S. Kim, F. A. M. De Oliveira, and P. L. Knight, *J. Mod. Optics* **37**, 659 (1990).
M. H. Mahran, *Phys. Rev. A* **42**, 4199 (1990).
M. A. Mir and M. S. K. Razmi, *Phys. Rev. A* **44**, 6071 (1991).

Coherent superposition in atom–field interactions

S. J. D. Phoenix and P. L. Knight, *Ann. Phys. (NY)* **186**, 381 (1988).
K. Zaheer and M. S. Zubairy, *Phys. Rev. A* **39**, 2000 (1989).
J. Gea-Banacloche, *Phys. Rev. Lett.* **65**, 3385 (1990).
B. Sherman and G. Kurizki, *Phys. Rev. A* **45**, 7674 (1992).

Weisskopf–Wigner approximation

V. Weisskopf and E. Wigner, *Z. Phys.* **63**, 54 (1930).
M. Sargent III, M. Scully, and W. E. Lamb, Jr., *Laser Physics*, (Addison-Wesley, Reading, MA 1974) p. 236.

W. Louisell, *Quantum Statistical Properties of Radiation*, (Wiley, New York 1974).
H. J. Kimble, A. Mezzacappa, and P. W. Milonni, *Phys. Rev. A* **31**, 3686 (1985).

Two-photon effects and cascade emission

H. Holt, *Phys. Rev. Lett.* **19**, 1275 (1967).
L. M. Narducci, M. O. Scully, G.-L. Oppo, P. Ru, and J. R. Tredicce, *Phys. Rev. A* **42**, 1630 (1990).
Y.-F. Zhu, D. J. Gauthier, and T. W. Mossberg, *Phys. Rev. Lett.* **66**, 2460 (1991).
H. Huang and J. H. Eberly, *J. Mod. Opt.* **40**, 915 (1993).

CHAPTER 7

Lasing without inversion and other effects of atomic coherence and interference

Quantum coherence and correlations in atomic and radiation physics have led to many interesting and unexpected consequences. For example, an atomic ensemble prepared in a coherent superposition of states yields the Hanle effect, quantum beats, photon echo, self-induced transparency, and coherent Raman beats.* In fact, in Section 1.4, we saw that the quantum beat effect provides one of the most compelling reasons for quantizing the radiation field.

A further interesting consequence of preparing an atomic system in a coherent superposition of states is that, under certain conditions, it is possible for atomic coherence to cancel absorption. Such atomic states are called *trapping states*.† The observation of nonabsorbing resonances via atomic coherence and interference impacts on the concepts of lasing without inversion (LWI),‡ enhancement of the index of refraction accompanied by vanishing absorption, and electromagnetically induced transparency.

In lasing without inversion, the essential idea is the absorption cancellation by atomic coherence and interference. This phenomenon is also the essence of electromagnetically induced transparency. Usually this is accomplished in three-level atomic systems in which there are two coherent routes for absorption that can destructively interfere, thus leading to the cancellation of absorption. A small population in the excited state can thus lead to net gain. A related phenomenon is

* The original treatments of self-induced transparency, McCall and Hahn [1969] and coherent transient phenomena such as coherent Raman beats, Brewer and Hahn [1973] are in the spirit of the present chapter and are recommended reading.

† For an excellent review of coherent population trapping, see Arimondo [1996].

‡ Although the LWI possibility was noted many years ago, Hänsch and Toschek [1970], Popov, Popov, and Ravtian [1970], and Arkhpkin and Heller [1983]; it has only recently been seriously pursued, see e.g. Kocharovskaya and Khanin [1988], Harris [1989],and Scully, Zhu, and Gavrielides [1989]. Proof of principle experiments now exist which clearly demonstrate the validity of the idea.

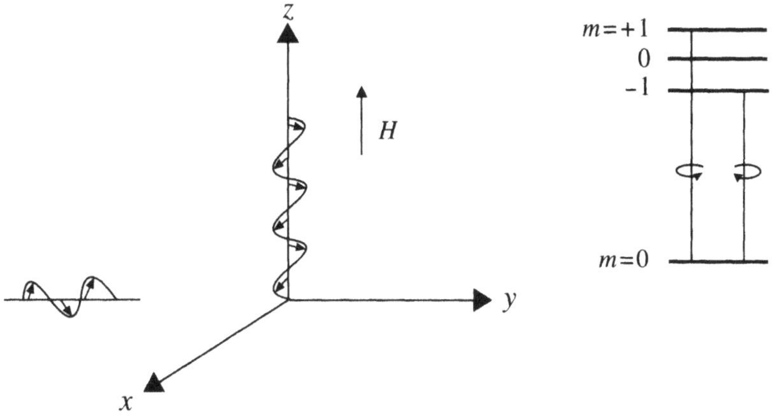

Fig. 7.1 Schematic illustration of Hanle's experiment and atomic level scheme.

that of resonantly enhanced refractive index without absorption in an ensemble of phase-coherent atoms (*phaseonium*). In a phaseonium gas with no population in the excited level, the absorption cancellation always coincides with vanishing refractivity. However, upon providing a small fraction of atoms in the excited state, absorption vanishes slightly off resonance, where the real part of the susceptibility has a substantial value. This gives rise to the possibility of high refractivity in a nonabsorbing medium.

In this chapter, we discuss these novel phenomena wherein the influence of the atomic coherence is clearly evident.

7.1 The Hanle effect

The experiment of Hanle provides one of the clearest and oldest demonstrations of a situation in which atomic coherence plays an important role. An ensemble of atoms, situated in a weak magnetic field, is illuminated with a pulse of $\hat{x}$-polarized light. The polarization of the light reradiated in the $\hat{z}$-direction is then detected. For a small magnetic field it is found that the reradiated light can be polarized in the $\hat{y}$-direction, as depicted in Fig. 7.1.

To understand how atoms excited by $\hat{x}$-polarized light reradiate $\hat{y}$-polarized light, we must calculate the dipole moment induced by the incident radiation. If we take an atom initially in the ground state

$$|\psi(0)\rangle = |0\rangle, \tag{7.1.1}$$

later in time the electric field

$$\mathbf{E}(\mathbf{r}, t) = \hat{x}\mathscr{E}_0 \cos(ky - \nu t), \tag{7.1.2}$$

induces transitions to the $m = \pm 1$ levels and the wave function becomes

$$|\psi(t)\rangle = c_+ \exp(i\omega_+ t)|+\rangle + c_- \exp(i\omega_- t)|-\rangle + c_0|0\rangle. \qquad (7.1.3)$$

The atomic frequencies $\omega_\pm$ are given by

$$\omega_\pm = v \pm \Delta, \qquad (7.1.4)$$

where Δ is the splitting of the levels due to the magnetic field. The atomic dipole is then

$$\begin{aligned}\langle \mathbf{P}(t)\rangle &= e\langle \psi(t)|(x\hat{x} + y\hat{y} + z\hat{z})|\psi(t)\rangle \\ &= p_+[\hat{x}\cos(v+\Delta)t + \hat{y}\sin(v+\Delta)t] \\ &\quad + p_-[\hat{x}\cos(v-\Delta)t - \hat{y}\sin(v-\Delta)t],\end{aligned} \qquad (7.1.5)$$

where $p_\pm$ is the polarization associated with $\rho_{\pm,0} = c_\pm c_0^*$ and

$$p_\pm = e\langle \pm|x|0\rangle(\rho_{\pm,0} + \text{c.c.}). \qquad (7.1.6)$$

Let us proceed by studying the reasonable case where $p_+ = p_- = p$. Then Eq. (7.1.5) becomes

$$\langle \mathbf{P}(t)\rangle = p\cos vt(\hat{x}\cos\Delta t + \hat{y}\sin\Delta t). \qquad (7.1.7)$$

This atomic dipole leads to a radiated field whose fluorescence and polarization is an interesting function of time. According to Eq. (7.1.7), if we now place a detector at a point $x = x_0$ on the x-axis, no scattered radiation will be detected if the applied magnetic field H, and therefore Δ, is equal to zero. For finite Δ, however, there will be a modulated field radiation along the $\hat{x}$-axis whose time dependence is given by

$$\mathscr{E}_{\text{scatt}} = \mathscr{E}_0 \cos vt \sin\Delta t. \qquad (7.1.8)$$

The reason for this modulated field is coherence induced between levels $|+\rangle$ and $|-\rangle$. Note that this is characterized by nonvanishing $\rho_\pm$. In general, we say that atomic coherence exists when the density matrix has off-diagonal elements.

7.2 Coherent trapping – dark states

An interesting phenomenon in which a coherent superposition of atomic states is responsible for a novel effect is coherent trapping. If an atom is prepared in a coherent superposition of states, it is possible to cancel absorption or emission under certain conditions. These atoms are then effectively transparent to the incident field even

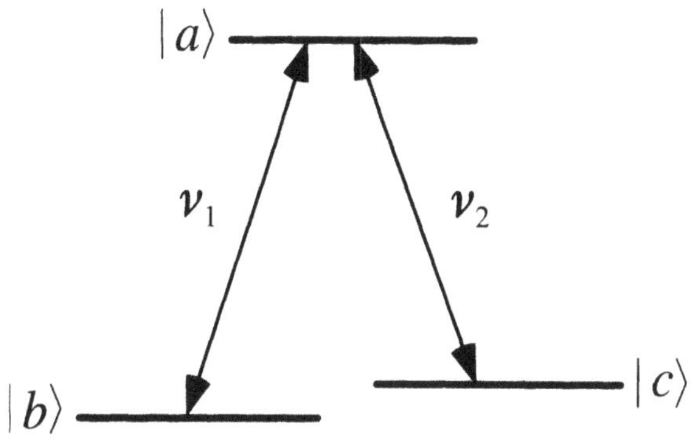

Fig. 7.2
Three-level atom in the Λ configuration interacting with two fields of frequencies ν_1 and ν_2.

in the presence of resonant transitions. In this section, we discuss the effect of coherent trapping in three-level atomic systems.

Coherent superpositions of atomic states in three-level atoms have many interesting applications. These include lasing without inversion and refractive index enhancement in a nonabsorbing medium as discussed later in this chapter. Further applications include quantum beats (see Section 1.4) and the correlated spontaneous emission laser (Chapter 14).

We consider coherent trapping in a three-level atom interacting with two fields of frequencies ν_1 and ν_2 as shown in Fig. 7.2. We assume the atom to be in the so-called Λ configuration in which two lower levels $|b\rangle$ and $|c\rangle$ are coupled to a single upper level $|a\rangle$. Other possible three-level schemes include V and cascade configurations.

The Hamiltonian for the system, in the rotating-wave approximation, is obtained by a suitable generalization of the Hamiltonian for a two-level atom interacting with a single-mode field (Eqs. (5.2.3)–(5.2.5)) to the present problem of a three-level atom interacting with a two-mode field:

$$\mathscr{H} = \mathscr{H}_0 + \mathscr{H}_1, \tag{7.2.1}$$

where

$$\mathscr{H}_0 = \hbar\omega_a|a\rangle\langle a| + \hbar\omega_b|b\rangle\langle b| + \hbar\omega_c|c\rangle\langle c|, \tag{7.2.2}$$

$$\mathscr{H}_1 = -\frac{\hbar}{2}(\Omega_{R1}e^{-i\phi_1}e^{-i\nu_1 t}|a\rangle\langle b| + \Omega_{R2}e^{-i\phi_2}e^{-i\nu_2 t}|a\rangle\langle c|) + \text{H.c.} \tag{7.2.3}$$

Here $\Omega_{R1}\exp(-i\phi_1)$ and $\Omega_{R2}\exp(-i\phi_2)$ are the complex Rabi frequencies associated with the coupling of the field modes of frequencies ν_1 and ν_2 to the atomic transitions $|a\rangle \to |b\rangle$ and $|a\rangle \to |c\rangle$, respectively. We have assumed that only $|a\rangle \to |b\rangle$ and $|a\rangle \to |c\rangle$ transitions are dipole allowed.

The atomic wave function can be written in the form

$$|\psi(t)\rangle = c_a(t)e^{-i\omega_a t}|a\rangle + c_b(t)e^{-i\omega_b t}|b\rangle + c_c(t)e^{-i\omega_c t}|c\rangle. \quad (7.2.4)$$

The equations of motion for the probability amplitudes $c_a(t)$, $c_b(t)$, and $c_c(t)$ can be derived from the Schrödinger equation $i\hbar|\dot{\psi}\rangle = \mathscr{H}|\psi\rangle$ to be

$$\dot{c}_a = \frac{i}{2}(\Omega_{R1}e^{-i\phi_1}c_b + \Omega_{R2}e^{-i\phi_2}c_c), \quad (7.2.5)$$

$$\dot{c}_b = \frac{i}{2}\Omega_{R1}e^{i\phi_1}c_a, \quad (7.2.6)$$

$$\dot{c}_c = \frac{i}{2}\Omega_{R2}e^{i\phi_2}c_a, \quad (7.2.7)$$

where we have assumed the fields to be resonant with the $|a\rangle \to |b\rangle$ and the $|a\rangle \to |c\rangle$ transitions respectively, i.e., $\omega_{ab} = \nu_1$ and $\omega_{ac} = \nu_2$.

We now assume the initial atomic state to be a superposition of the two lower levels $|b\rangle$ and $|c\rangle$

$$|\psi(0)\rangle = \cos(\theta/2)|b\rangle + \sin(\theta/2)e^{-i\psi}|c\rangle. \quad (7.2.8)$$

A solution of Eqs. (7.2.5)–(7.2.7) subject to the initial condition (7.2.8) is given by

$$c_a(t) = \frac{i\sin(\Omega t/2)}{\Omega}[\Omega_{R1}e^{-i\phi_1}\cos(\theta/2) + \Omega_{R2}e^{-i(\phi_2+\psi)}\sin(\theta/2)], \quad (7.2.9)$$

$$c_b(t) = \frac{1}{\Omega^2}\{[\Omega_{R2}^2 + \Omega_{R1}^2\cos(\Omega t/2)]\cos(\theta/2) - 2\Omega_{R1}\Omega_{R2}e^{i(\phi_1-\phi_2-\psi)}\sin^2(\Omega t/4)\sin(\theta/2)\}, \quad (7.2.10)$$

$$c_c(t) = \frac{1}{\Omega^2}\{-2\Omega_{R1}\Omega_{R2}e^{-i(\phi_1-\phi_2)}\sin^2(\Omega t/4)\cos(\theta/2) + [\Omega_{R1}^2 + \Omega_{R2}^2\cos(\Omega t/2)]e^{-i\psi}\sin(\theta/2)\} \quad (7.2.11)$$

with $\Omega = (\Omega_{R1}^2 + \Omega_{R2}^2)^{1/2}$. It is evident that coherent trapping occurs for

$$\Omega_{R1} = \Omega_{R2}, \qquad \theta = \pi/2, \qquad \phi_1 - \phi_2 - \psi = \pm\,\pi. \quad (7.2.12)$$

Under these conditions

$$c_a(t) = 0, \quad (7.2.13a)$$

$$c_b(t) = \frac{1}{\sqrt{2}}, \quad (7.2.13b)$$

$$c_c(t) = \frac{1}{\sqrt{2}}e^{-i\psi}, \quad (7.2.13c)$$

i.e., the population is *trapped* in the lower states and there is no absorption even in the presence of the field. In the present three-level atom, coherent trapping occurs due to the destructive quantum interference between the two transitions.

Finally we note that there is an interesting outgrowth of coherent population trapping (CPT) from adiabatically turning the field in Eq. (7.2.3) on and off. That is, if we consider the case in which we start with the atom in state $|b\rangle$ and $\Omega_{R1} = 0$ with Ω_{R2} finite and then proceed to turn Ω_{R2} off while slowly turning Ω_{R1} on, we will end up with the atom in the state $|c\rangle$. This is made clear by realizing that the atom is in the time-dependent trapping state

$$|\psi(t)\rangle = \frac{\Omega_{R2}(t)e^{-i\varphi_2}|b\rangle - \Omega_{R1}(t)e^{-i\varphi_1}|c\rangle}{\sqrt{\Omega_{R1}^2 + \Omega_{R2}^2}} \tag{7.2.14}$$

see Problem 7.2.

In this section, we discussed the case in which the atom is initially placed into the non-absorbing state. The interesting aspect of CPT is that it can occur even if the atom is not in a dark state at $t = 0$. In fact, the atom can be forced into this state by, e.g., continued action of EM fields and spontaneous emission (similar to the optical pumping mechanism) via adiabatic population transfer. An example of such a case is given in the following section.

7.3 Electromagnetically induced transparency

In the previous section, we discussed the phenomena of coherent population trapping via a three-level system in which the lower levels are prepared in a coherent superposition state. In this section, we discuss another related phenomenon, the electromagnetically induced transparency (EIT) of Harris and co-workers, in which quantum interference is introduced by driving the upper two levels of a three-level atomic system with a strong coherent field; see Fig. 7.3. Under appropriate conditions, the medium becomes effectively transparent (zero absorption) for a probe field.

We consider a closed three-level system as shown in Fig. 7.3. The levels $|a\rangle$ and $|b\rangle$ are coupled by a probe field of amplitude $\mathscr{E}$ and frequency ν, whose dispersion and absorption we are interested in. The upper level $|a\rangle$ is coupled to level $|c\rangle$ by a strong coherent field of frequency ν_μ, having complex Rabi frequency $\Omega_\mu \exp(-i\phi_\mu)$. The off-diagonal decay rates for ρ_{ab}, ρ_{ac}, and ρ_{cb} are denoted by γ_1, γ_2, and γ_3, respectively.

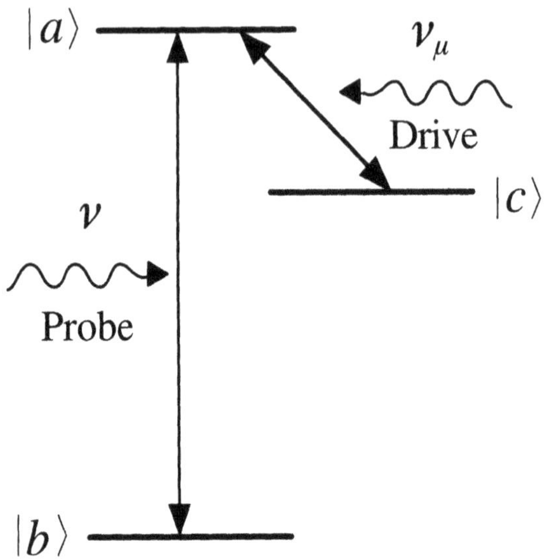

Fig. 7.3
Three-level atomic system for electromagnetically induced transparency.

The interaction Hamiltonian for the atom and the two fields is again given by Eqs. (7.2.1)–(7.2.3), but with the substitutions

$$\Omega_{R1}e^{-i\phi_1}e^{-i\nu_1 t} = \frac{\wp_{ab}\mathscr{E}}{\hbar}e^{-i\nu t}; \quad \Omega_{R2}e^{-i\phi_2}e^{-i\nu_2 t} = \Omega_\mu e^{-i\phi_\mu}e^{-i\nu_\mu t}. \tag{7.3.1}$$

The equations of motion for the density matrix elements ρ_{ab} and ρ_{cb} are given by

$$\dot{\rho}_{ab} = -(i\omega_{ab} + \gamma_1)\rho_{ab} - \frac{i}{2}\frac{\wp_{ab}\mathscr{E}}{\hbar}e^{-i\nu t}(\rho_{aa} - \rho_{bb}) + \frac{i}{2}\Omega_\mu e^{-i\phi_\mu}e^{-i\nu_\mu t}\rho_{cb}, \tag{7.3.2}$$

$$\dot{\rho}_{cb} = -(i\omega_{cb} + \gamma_3)\rho_{cb} - \frac{i}{2}\frac{\wp_{ab}\mathscr{E}}{\hbar}e^{-i\nu t}\rho_{ca} + \frac{i}{2}\Omega_\mu e^{i\phi_\mu}e^{i\nu_\mu t}\rho_{ab}, \tag{7.3.3}$$

$$\dot{\rho}_{ac} = -(i\omega_{ac} + \gamma_2)\rho_{ac} - \frac{i}{2}\Omega_\mu e^{-i\phi_\mu}e^{-i\nu_\mu t}(\rho_{aa} - \rho_{cc}) + \frac{i}{2}\frac{\wp_{ab}\mathscr{E}}{\hbar}e^{-i\nu t}\rho_{bc}. \tag{7.3.4}$$

As seen earlier, the dispersion and absorption are determined by $\rho_{ab}^{(1)}$, i.e., we only need to calculate the polarization to lowest order in $\mathscr{E}$. However, the coherent field coupling the levels $|a\rangle$ and $|c\rangle$ is large and we must treat this part of the problem exactly, keeping Ω_μ to all orders.

As the atoms are initially in the ground level $|b\rangle$,

$$\rho_{bb}^{(0)} = 1, \; \rho_{aa}^{(0)} = \rho_{cc}^{(0)} = \rho_{ca}^{(0)} = 0. \tag{7.3.5}$$

On substituting these values of the matrix elements into Eqs. (7.3.2) and (7.3.3), and making the substitutions

$$\rho_{ab} = \tilde{\rho}_{ab} e^{-i\nu t}, \tag{7.3.6}$$

$$\rho_{cb} = \tilde{\rho}_{cb} e^{-i(\nu+\omega_{ca})t}, \tag{7.3.7}$$

we obtain the following coupled set of equations:

$$\dot{\tilde{\rho}}_{ab} = -(\gamma_1 + i\Delta)\tilde{\rho}_{ab} + \frac{i}{2}\frac{\wp_{ab}\mathscr{E}}{\hbar} + \frac{i}{2}\Omega_\mu e^{-i\phi_\mu}\tilde{\rho}_{cb}, \tag{7.3.8}$$

$$\dot{\tilde{\rho}}_{cb} = -(\gamma_3 + i\Delta)\tilde{\rho}_{cb} + \frac{i}{2}\Omega_\mu e^{i\phi_\mu}\tilde{\rho}_{ab}, \tag{7.3.9}$$

where $\Delta = \omega_{ab} - \nu$ is the detuning of the probe field and we have assumed $\nu_\mu = \omega_{ac}$.

This set of equations can be solved, for example, by first writing in the matrix form,

$$\dot{R} = -MR + A, \tag{7.3.10}$$

with

$$R = \begin{bmatrix} \tilde{\rho}_{ab} \\ \tilde{\rho}_{cb} \end{bmatrix}, \quad M = \begin{bmatrix} \gamma_1 + i\Delta & -\frac{i}{2}\Omega_\mu e^{-i\phi_\mu} \\ -\frac{i}{2}\Omega_\mu e^{i\phi_\mu} & \gamma_3 + i\Delta \end{bmatrix}, \quad A = \begin{bmatrix} i\wp_{ab}\mathscr{E}/2\hbar \\ 0 \end{bmatrix}, \tag{7.3.11}$$

and then integrating

$$\begin{aligned} R(t) &= \int_{-\infty}^{t} e^{-M(t-t')} A dt' \\ &= M^{-1}A. \end{aligned} \tag{7.3.12}$$

This yields

$$\rho_{ab}(t) = \frac{i\wp_{ab}\mathscr{E}e^{-i\nu t}(\gamma_3 + i\Delta)}{2\hbar[(\gamma_1 + i\Delta)(\gamma_3 + i\Delta) + \Omega_\mu^2/4]}. \tag{7.3.13}$$

The relation $\mathscr{P} = \epsilon_0 \chi \mathscr{E}$, together with the definition (5.4.18) of the complex polarization, gives the following expression of the real and imaginary parts of the complex susceptibility $\chi = \chi' + i\chi''$:

$$\chi' = \frac{N_a|\wp_{ab}|^2\Delta}{\epsilon_0 \hbar Z}\left[\gamma_3(\gamma_1 + \gamma_3) + (\Delta^2 - \gamma_1\gamma_3 - \Omega_\mu^2/4)\right], \tag{7.3.14}$$

$$\chi'' = \frac{N_a|\wp_{ab}|^2}{\epsilon_0 \hbar Z}\left[\Delta^2(\gamma_1 + \gamma_3) - \gamma_3(\Delta^2 - \gamma_1\gamma_3 - \Omega_\mu^2/4)\right], \tag{7.3.15}$$

where N_a is the atom number density and

$$Z = (\Delta^2 - \gamma_1\gamma_3 - \Omega_\mu^2/4)^2 + \Delta^2(\gamma_1 + \gamma_3)^2. \tag{7.3.16}$$

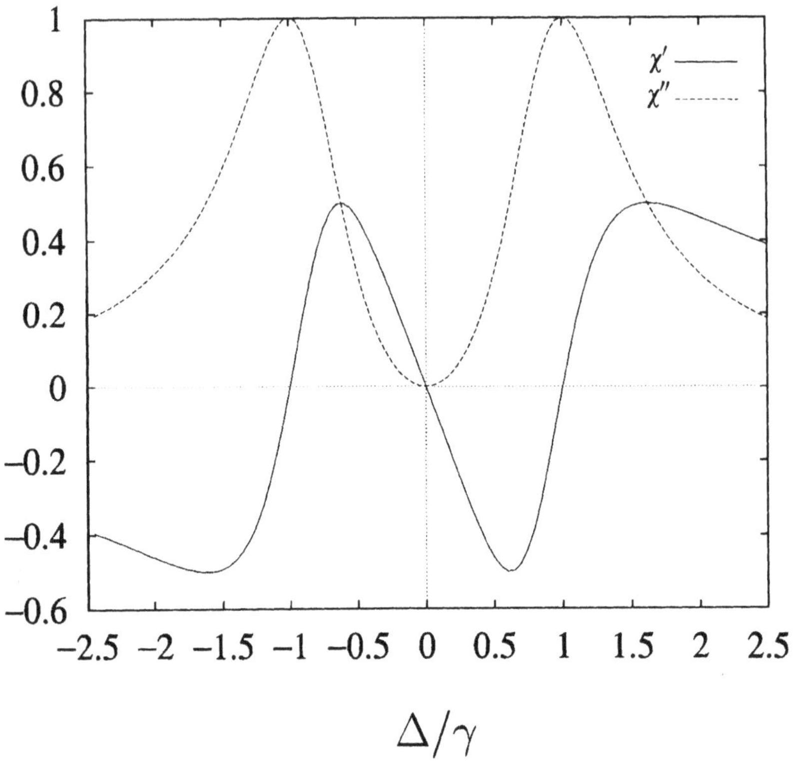

Fig. 7.4
Real (solid line) and imaginary (dashed line) parts of the linear susceptiblity (in arbitrary units) as a function of normalized detuning Δ/γ_1.

It is clear from Eqs. (5.4.23) and (5.4.24) that χ' and χ'' are related to dispersion and absorption, respectively. A more detailed discussion is given in Section 7.5 below.

In Fig. 7.4 the susceptibilities χ' and χ'' are plotted versus the detuning Δ in units of the atomic decay γ_1, for $\Omega_\mu = 2\gamma_1$ and $\gamma_1 \gg \gamma_3$ ($\gamma_3 = 10^{-4}\gamma_1$). It is seen that, at zero detuning, $\Delta = 0$, both χ' and χ'' are equal to zero, i.e., the absorption is almost zero where the index of refraction is unity. Thus the medium becomes transparent under the action of the strong coherent field. This is an example of electromagnetically induced transparency.

We note that on resonance, $\chi' = 0$ and χ'' is proportional to γ_3. Since the last quantity represents the relaxation rate of the dipole-forbidden transition, it can be made very small. The physical origin of EIT can be understood in terms of the dark states discussed in the last section. There is, however, a remarkable difference. In the previous section, the atom was assumed to be prepared initially in the dark state. In the example of EIT, however, the atom is pumped into the dark state by the combined action of the strong pump and weak probe. There exist two possible mechanisms for such pumping. The first is equivalent

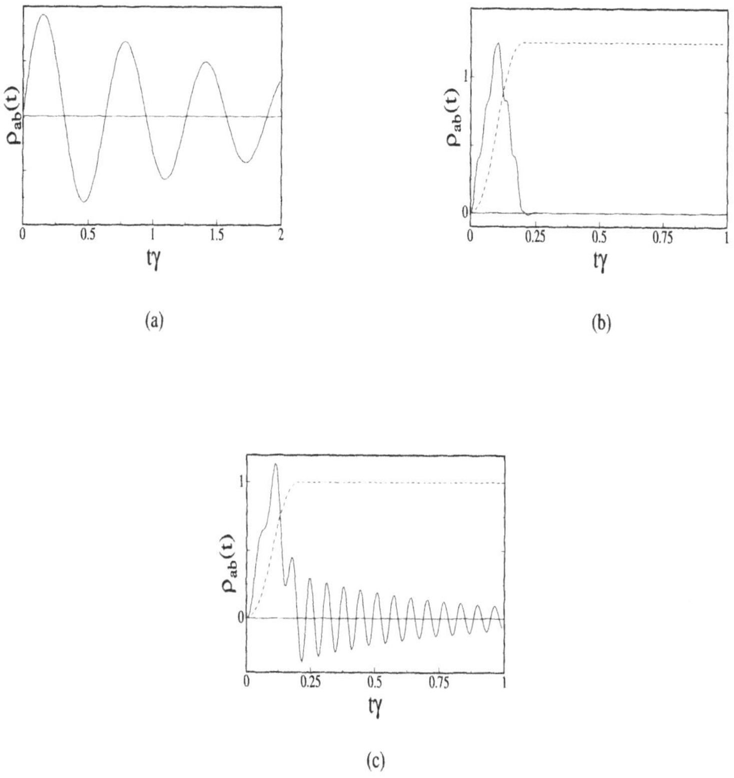

Fig. 7.5
Time scales for the establishment of EIT in a strongly driven Λ system with a weak probe. Plotted is the polarization of the probe transition (solid line) in arbitrary units as a function of time in units of atomic decay rate. All population is initially in the ground state of the probe transition, and the coupling laser is on for all times. (a) Sudden turn-on (step-function) of the probe leads to optical pumping effect on a time scale given by the atomic decay rate. (b) Slow turn-on of probe (dashed line) leads to adiabatic following of atom, and EIT is established when probe is on at $t = 0.2\gamma^{-1}$. The Rabi frequency of the drive is $\Omega = 200\gamma$. (c) For same conditions but weaker drive, $\Omega = 100\gamma$, adiabaticity is not perfect, and some population is left in absorbing state after probe is on. Optical pumping of this remaining population creates absorption after $t = 0.2\gamma^{-1}$.

to ordinary optical pumping into the trapped state. In this case, EIT is induced in an atom in a time of the order of a radiative lifetime, since this is the time for an excited atom to decay to the uncoupled ("trapped") state. This is indeed the case if we have a step function turn-on of the coupling field (between a and c) and the probe field. Then, from Fig. 7.5, we see that ρ_{ab}, which governs the absorption, decays in a few radiative lifetimes. But if we turn on the field slowly (as compared to the Rabi period of the coupling laser, Ω_{R2} of Eq. (7.2.14)) then the atomic state can be induced into the time-dependent trapping state (7.2.14) in a time of order Ω_{R2}^{-1}, as is shown in Fig. 7.5(b). Thus, we see that for large enough Ω_{R2}, the radiative decay time does not enter the problem.

However, we see in Fig. 7.5(c) that, even for $\Omega_{R2} \sim 100\gamma$, we are not completely in the EIT region. We have instead a situation which is "EIT-like" for short times $t \lesssim 2\gamma_{rad.}^{-1}$, but for longer times, radiative decay times dominate.*

* Figures courtesy of U. Rathe.

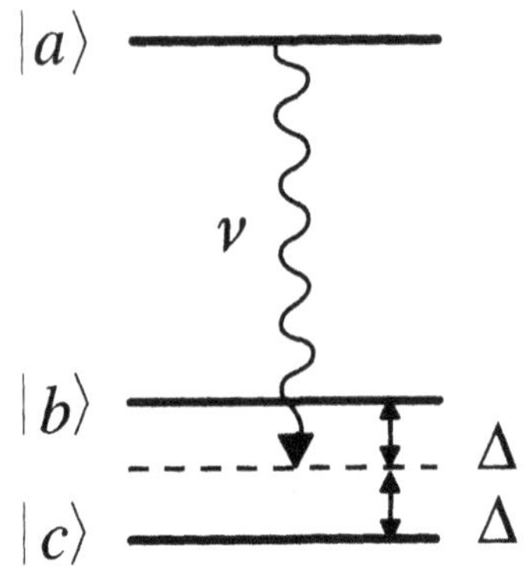

Fig. 7.6
Three-level atomic system interacting with a field of frequency ν.

7.4 Lasing without inversion

Having seen that a coherent superposition of a ground state doublet can cancel absorption, one is led to ask whether it might be possible to achieve lasing even when there are more atoms in the ground state doublet than in the excited $|a\rangle$ state. That is, it is generally the case (as seen in Section 5.5) that a laser requires population inversion in order to overcome the absorption from the lower level. But what if we can arrange things (i.e., 'phase' atoms) such that the absorption is canceled? Can we then lase without inversion? The answer is 'Yes'. In this section, we present the analysis of such problems in which lasing without inversion can be achieved using a coherently prepared three-level atomic system (see Fig. 7.6). To that end, we proceed by first presenting a very simple discussion of the physics behind lasing without inversion (LWI), and then proceed toward a more rigorous treatment of the problem.

7.4.1 The LWI concept

Let us consider again the Λ-system as shown in Fig. 7.2, where an upper level $|a\rangle$ is coupled to the lower levels $|b\rangle$ and $|c\rangle$ via two fields of frequencies ν_1 and ν_2. Only the atomic transitions $|a\rangle \to |b\rangle$ and $|a\rangle \to |c\rangle$ are dipole allowed and we assume, for simplicity, resonance conditions $\omega_{ab} = \nu_1$ and $\omega_{ac} = \nu_2$. The Hamiltonian for the system is given by Eqs. (7.2.1)–(7.2.3) and the equations of motion for the probability amplitudes are given by Eqs. (7.2.5)–(7.2.7).

Consider first the case in which the population is initially equally distributed with a fixed phase between the two lower states $|b\rangle$ and $|c\rangle$

$$c_a(0) = 0, \qquad c_b(0) = \frac{1}{\sqrt{2}}, \qquad c_c(0) = \frac{1}{\sqrt{2}}e^{-i\varphi}. \tag{7.4.1}$$

This is a particular case of the initial condition (7.2.8) with

$$\theta = \pi/2. \tag{7.4.2}$$

Hence, from the solution (7.2.9)–(7.2.11) it follows, to the lowest order, that

$$c_a(t) \cong i\frac{t}{2\sqrt{2}}\left[\Omega_{R1}e^{-i\phi_1} + \Omega_{R2}e^{-i(\phi_2+\psi)}\right]. \tag{7.4.3}$$

When

$$\Omega_{R1} = \Omega_{R2} \equiv \Omega_R, \tag{7.4.4}$$

Eq. (7.4.3) becomes

$$|c_a(t)|^2 = \frac{t^2\Omega_R^2}{4}\left[1 + \cos(\phi_1 - \phi_2 - \psi)\right]. \tag{7.4.5}$$

This means the absorption is canceled ($|c_a(t)|^2 = 0$) if

$$\phi_1 - \phi_2 - \psi = \pm\pi, \tag{7.4.6}$$

i.e., we recovered the conditions (7.2.12) for coherent trapping.

In the case of an initially excited atom

$$c_a(0) = 1, \qquad c_b(0) = c_c(0) = 0, \tag{7.4.7}$$

the solution of Eqs. (7.2.5)–(7.2.7) reads

$$c_a(t) = \cos\left(\frac{\Omega t}{2}\right), \tag{7.4.8}$$

$$c_b(t) = i\frac{\Omega_{R1}^*}{\Omega}\sin\left(\frac{\Omega t}{2}\right), \tag{7.4.9}$$

$$c_c(t) = i\frac{\Omega_{R2}^*}{\Omega}\sin\left(\frac{\Omega t}{2}\right). \tag{7.4.10}$$

For $\Omega t \ll 1$, one gets approximately

$$c_b(t) \cong i\frac{\Omega_{R1}^* t}{2}, \tag{7.4.11a}$$

$$c_c(t) \cong i\frac{\Omega_{R2}^* t}{2}. \tag{7.4.11b}$$

The emission probability is then

$$|c_b(t)|^2 + |c_c(t)|^2 = \frac{\Omega^2 t^2}{4}, \tag{7.4.12}$$

which is independent of the phases and is always positive. Thus if one can arrange the system such that the conditions (7.4.2), (7.4.4), and (7.4.6) for absorption cancelation are fulfilled, there will be net gain even in the absence of population inversion.

7.4.2 The laser physics approach to LWI: simple treatment

Consider a system of three-level atoms interacting with a laser field in a cavity. The simple model we will focus upon is that of Fig. 7.6. The atoms have one upper level $|a\rangle$ and two lower levels $|b\rangle$ and $|c\rangle$ with energies $\hbar\omega_a$, $\hbar\omega_b$, and $\hbar\omega_c$ and decay rates γ_a, γ_b, and γ_c, respectively. The transitions $|a\rangle \to |b\rangle$ and $|a\rangle \to |c\rangle$ are now induced by one classical light field of frequency ν. The transition $|b\rangle \to |c\rangle$ is dipole forbidden. The atoms are pumped at a rate r_a in a coherent superposition of states

$$\rho(t_i) = \rho_{aa}^{(0)}|a\rangle\langle a| + \rho_{bb}^{(0)}|b\rangle\langle b| + \rho_{cc}^{(0)}|c\rangle\langle c| + \rho_{bc}^{(0)}|b\rangle\langle c| + \rho_{cb}^{(0)}|c\rangle\langle b|. \tag{7.4.13}$$

Here $\rho_{\alpha\alpha}^{(0)}$ $(\alpha = a, b, c)$ are the level populations and $\rho_{\alpha\alpha'}^{(0)}$ $(\alpha \neq \alpha')$ are the atomic coherences.

Before presenting a detailed theory, we first give a simple argument to show how cancelation of absorption can lead to lasing without inversion in this scheme.

As the levels $|b\rangle$ and $|c\rangle$ are independent, the probability of emission is given by

$$\begin{aligned} P_{\text{emission}} &= P_b + P_c \\ &= (|\kappa_{a\to b}|^2\mathscr{E}^2 + |\kappa_{a\to c}|^2\mathscr{E}^2)\rho_{aa}^{(0)}, \end{aligned} \tag{7.4.14}$$

where $\kappa_{a\to b}$ and $\kappa_{a\to c}$ are constants which depend on the matrix element between the relevant levels and the coupling of the atom with the field. On the other hand, the absorption probability is given by

$$\begin{aligned} P_{\text{absorption}} &= \kappa|c_b + c_c|^2\mathscr{E}^2 \\ &= \kappa[\rho_{bb}^{(0)} + \rho_{cc}^{(0)} + \rho_{bc}^{(0)} + \rho_{cb}^{(0)}]\mathscr{E}^2. \end{aligned} \tag{7.4.15}$$

Therefore, the rate of growth of the laser field amplitude, under appropriate conditions, becomes

$$\dot{\mathscr{E}} = \frac{\mathscr{A}}{2}[\rho_{aa}^{(0)} - \rho_{bb}^{(0)} - \rho_{cc}^{(0)} - \rho_{bc}^{(0)} - \rho_{cb}^{(0)}]\mathscr{E}. \tag{7.4.16}$$

Here $\mathscr{A}$ is a constant. Thus, if the terms $\rho_{bc}^{(0)}$ and $\rho_{cb}^{(0)}$ cancel $\rho_{bb}^{(0)}$ and $\rho_{cc}^{(0)}$, we have

$$\dot{\mathscr{E}} = \frac{\mathscr{A}}{2}\rho_{aa}^{(0)}\mathscr{E}, \tag{7.4.17}$$

and we can have lasing even if only a small fraction of atoms is in the excited state $|a\rangle$, i.e., even if $\rho_{aa} < (\rho_{bb} + \rho_{cc})$.

Physically, the lack of absorption in the three-level system considered above is a manifestation of quantum coherence phenomena. When an atom makes a transition from the upper level to the two lower levels, the total transition probability is the sum of $|a\rangle \to |b\rangle$ and $|a\rangle \to |c\rangle$ probabilities. However, the transition probability from the two lower levels to the single upper level is obtained by squaring the sum of the two probability amplitudes. When there is coherence between the two lower levels, this can lead to interference terms yielding a null in the transition probability corresponding to photon absorption.

7.4.3 LWI analysis

In order to show the role of the atomic coherence in lasing without inversion in a more rigorous manner, we now present a semiclassical theory in which the field is treated classically and restrict ourselves to a linear analysis of the problem keeping terms up to the second order in the atom–field coupling constant. The equation of motion for the field amplitude for the present problem is (see Section 5.4)

$$\dot{\mathscr{E}}(t) = -\frac{\nu}{\epsilon_0}\mathrm{Im}\left\{e^{i\nu t}[\wp_{ca}\rho_{ac}(t) + \wp_{ba}\rho_{ab}(t)]\right\}. \tag{7.4.18}$$

The equations for the various elements of the population matrix can be obtained by generalizing the method developed in Section 5.5.

In the present situation, we have, however, a 3×3 population matrix

$$\rho(z,t) = \sum_{\alpha}\sum_{\beta}\int_{-\infty}^{t} dt_0 r_a(z,t_0)\rho_{z,\alpha\beta}(t,t_0)|\alpha\rangle\langle\beta|, \tag{7.4.19}$$

where the summations over α and β include the atomic levels $|a\rangle$, $|b\rangle$, and $|c\rangle$ and the atoms are pumped at a constant rate r_a in the coherent superposition of states (7.4.13). For the present problem of a single-mode field of frequency ν and complex amplitude $\mathscr{E}(t)$ interacting with a three-level atomic system, the atom–field intraction Hamiltonian is given by Eqs. (7.2.1)–(7.2.3) with the substitutions

$$\Omega_{R1}e^{-i\phi_1}e^{-i\nu_1 t} = \frac{\wp_{ab}\mathscr{E}(t)}{\hbar}e^{-i\nu t}; \quad \Omega_{R2}e^{-i\phi_2}e^{-i\nu_2 t} = \frac{\wp_{ac}\mathscr{E}(t)}{\hbar}e^{-i\nu t}. \tag{7.4.20}$$

The equations of motion for the elements of the population matrix are thus given by

$$\dot{\rho}_{ab} = -(i\omega_{ab} + \gamma_{ab})\rho_{ab} - \frac{i}{2}\frac{\wp_{ab}\mathscr{E}(t)}{\hbar}e^{-i\nu t}(\rho_{aa} - \rho_{bb}) + \frac{i}{2}\frac{\wp_{ac}\mathscr{E}(t)}{\hbar}e^{-i\nu t}\rho_{cb}, \tag{7.4.21}$$

$$\dot{\rho}_{ac} = -(i\omega_{ac} + \gamma_{ac})\rho_{ac} - \frac{i}{2}\frac{\wp_{ac}\mathscr{E}(t)}{\hbar}e^{-i\nu t}(\rho_{aa} - \rho_{cc}) + \frac{i}{2}\frac{\wp_{ab}\mathscr{E}(t)}{\hbar}e^{-i\nu t}\rho_{bc}, \tag{7.4.22}$$

$$\dot{\rho}_{aa} = r_a\rho_{aa}^{(0)} - \gamma_a\rho_{aa}, \tag{7.4.23}$$

$$\dot{\rho}_{bb} = r_a\rho_{bb}^{(0)} - \gamma_b\rho_{bb}, \tag{7.4.24}$$

$$\dot{\rho}_{cc} = r_a\rho_{cc}^{(0)} - \gamma_c\rho_{cc}, \tag{7.4.25}$$

$$\dot{\rho}_{bc} = r_a\rho_{bc}^{(0)} - (i\omega_{bc} + \gamma_{bc})\rho_{bc}. \tag{7.4.26}$$

Here we have not included the interaction terms in the equations for ρ_{aa}, ρ_{bb}, ρ_{cc}, and ρ_{bc} because we are interested only in the linear theory.

The zeroth-order solutions of these equations, namely,

$$\rho_{aa} = \int_{-\infty}^{t} dt_0 e^{-\gamma_a(t-t_0)} r_a\rho_{aa}^{(0)} = \frac{r_a}{\gamma_a}\rho_{aa}^{(0)}, \tag{7.4.27}$$

$$\rho_{bb} = \frac{r_a}{\gamma_b}\rho_{bb}^{(0)}, \tag{7.4.28}$$

$$\rho_{cc} = \frac{r_a}{\gamma_c}\rho_{cc}^{(0)}, \tag{7.4.29}$$

$$\rho_{bc} = \frac{r_a}{(\gamma_{bc} + i\omega_{bc})}\rho_{bc}^{(0)}, \tag{7.4.30}$$

can be substituted into Eqs. (7.4.21) and (7.4.22) for ρ_{ab} and ρ_{ac}. The resulting equations can then be integrated and we obtain

$$\begin{aligned}\rho_{ab}(t) &= -\frac{ir_a}{2\hbar}\int_{-\infty}^{t} dt_0 e^{-(i\omega_{ab}+\gamma_{ab})(t-t_0)}\mathscr{E}(t_0)e^{-i\nu t_0} \\ &\quad\times\left\{\wp_{ab}\left[\frac{\rho_{aa}^{(0)}}{\gamma_a} - \frac{\rho_{bb}^{(0)}}{\gamma_b}\right] - \wp_{ac}\frac{\rho_{cb}^{(0)}}{(\gamma_{bc} - i\omega_{bc})}\right\} \\ &= -\frac{ir_a}{2\hbar}\frac{\mathscr{E}(t)e^{-i\nu t}}{\gamma_{ab} + i(\omega_{ab} - \nu)} \\ &\quad\times\left\{\wp_{ab}\left[\frac{\rho_{aa}^{(0)}}{\gamma_a} - \frac{\rho_{bb}^{(0)}}{\gamma_b}\right] - \wp_{ac}\frac{\rho_{cb}^{(0)}}{(\gamma_{bc} - i\omega_{bc})}\right\},\end{aligned} \tag{7.4.31}$$

$$\begin{aligned}\rho_{ac}(t) &= -\frac{ir_a}{2\hbar}\frac{\mathscr{E}(t)e^{-i\nu t}}{\gamma_{ac} + i(\omega_{ac} - \nu)} \\ &\quad\times\left\{\wp_{ac}\left[\frac{\rho_{aa}^{(0)}}{\gamma_a} - \frac{\rho_{cc}^{(0)}}{\gamma_c}\right] - \wp_{ab}\frac{\rho_{bc}^{(0)}}{\gamma_{bc} + i\omega_{bc}}\right\}.\end{aligned} \tag{7.4.32}$$

In deriving these equations, we assume $\mathscr{E}(t)$ to be a slowly varying function of t during the atomic lifetime and therefore replace $\mathscr{E}(t_0)$ by $\mathscr{E}(t)$. On substituting back these expressions for ρ_{ab} and ρ_{ac} into Eq. (7.4.18) we obtain

$$\dot{\mathscr{E}}(t) = \frac{1}{2}(\mathscr{A}_{aa} - \mathscr{A}_{bb} - \mathscr{A}_{cc} + \mathscr{A}_{bc} + \mathscr{A}_{cb})\mathscr{E}(t), \tag{7.4.33}$$

where

$$\mathscr{A}_{aa} = \frac{\nu}{\epsilon_0 \hbar} r_a \left(\frac{\gamma_{ab}}{\gamma_{ab}^2 + \Delta^2} |\wp_{ab}|^2 + \frac{\gamma_{ac}}{\gamma_{ac}^2 + \Delta^2} |\wp_{ac}|^2 \right) \frac{\rho_{aa}^{(0)}}{\gamma_a}, \tag{7.4.34}$$

$$\mathscr{A}_{bb} = \frac{\nu}{\epsilon_0 \hbar} r_a \frac{\gamma_{ab}}{\gamma_{ab}^2 + \Delta^2} |\wp_{ab}|^2 \frac{\rho_{bb}^{(0)}}{\gamma_b}, \tag{7.4.35}$$

$$\mathscr{A}_{cc} = \frac{\nu}{\epsilon_0 \hbar} r_a \frac{\gamma_{ac}}{\gamma_{ac}^2 + \Delta^2} |\wp_{ac}|^2 \frac{\rho_{cc}^{(0)}}{\gamma_c}, \tag{7.4.36}$$

$$\mathscr{A}_{bc} = \frac{\nu}{\epsilon_0 \hbar} r_a \mathrm{Im} \left[\frac{(\Delta + i\gamma_{ac})(i\omega_{bc} - \gamma_{bc})}{(\gamma_{ac}^2 + \Delta^2)(\omega_{bc}^2 + \gamma_{bc}^2)} \wp_{ca} \wp_{ab} \rho_{bc}^{(0)} \right], \tag{7.4.37}$$

$$\mathscr{A}_{cb} = \frac{\nu}{\epsilon_0 \hbar} r_a \mathrm{Im} \left[\frac{(\Delta - i\gamma_{ab})(i\omega_{bc} + \gamma_{bc})}{(\gamma_{ab}^2 + \Delta^2)(\omega_{bc}^2 + \gamma_{bc}^2)} \wp_{ba} \wp_{ac} \rho_{cb}^{(0)} \right], \tag{7.4.38}$$

with $\Delta = \nu - \omega_{ab} = \omega_{ac} - \nu = \omega_{bc}/2$.

In Eq. (7.4.33), the term $\mathscr{A}_{aa}$ which is proportional to $\rho_{aa}^{(0)}$ is the gain term. It has two parts corresponding to the emission processes from level $|a\rangle$ to levels $|b\rangle$ and $|c\rangle$. The terms $\mathscr{A}_{bb}$ and $\mathscr{A}_{cc}$ which are proportional to $\rho_{bb}^{(0)}$ and $\rho_{cc}^{(0)}$, respectively, are the loss terms corresponding to absorption from levels $|b\rangle$ and $|c\rangle$ to the level $|a\rangle$. These are the usual terms for a semiclassical theory which will require population inversion for a net gain. However, due to atomic coherence, we now have phase-dependent terms $\mathscr{A}_{bc}$ and $\mathscr{A}_{cb}$ which are proportional to $\rho_{bc}^{(0)}$ and $\rho_{cb}^{(0)}$, respectively. It, therefore, appears possible that, for certain choices of parameters, the absorption terms $\mathscr{A}_{bb}$ and $\mathscr{A}_{cc}$ will cancel the coherence terms $\mathscr{A}_{bc}$ and $\mathscr{A}_{cb}$ leading to lasing without inversion. This happens, for example, in the two cases

$$\gamma_a = \gamma_b = \gamma_c = \gamma, \quad \wp_{ac} = \wp_{ab} = \wp, \quad \gamma \gg \omega_{bc}, \quad \rho_{bc}^{(0)} = |\rho_{cb}^{(0)}| e^{i\pi}, \tag{7.4.39}$$

and

$$\begin{aligned} &\gamma_a \ll \gamma_b, \quad \gamma_c = \gamma_b = \gamma, \quad \wp_{ac} = \wp_{ab} = \wp, \\ &\gamma = \omega_{ab}, \quad \rho_{bc}^{(0)} = |\rho_{cb}^{(0)}| e^{i3\pi/2}, \end{aligned} \tag{7.4.40}$$

with $2|\rho_{bc}| = \rho_{bb} + \rho_{cc}$. We then obtain

$$\dot{\mathscr{E}} = \frac{\mathscr{A}_{aa}}{2} \mathscr{E}, \tag{7.4.41}$$

with

$$\mathscr{A}_{aa} = \frac{2\nu}{\epsilon_0 \hbar} r_a \frac{|\wp|^2}{\gamma^2 + \Delta^2} \rho_{aa}^{(0)}, \tag{7.4.42}$$

and

$$\mathscr{A}_{aa} = \frac{4\nu}{\epsilon_0 \hbar} r_a \frac{|\wp|^2}{\gamma^2 + 4\Delta^2} \left(\frac{\gamma}{\gamma_a} \right) \rho_{aa}^{(0)}, \tag{7.4.43}$$

respectively. Thus, under conditions (7.4.39) or (7.4.40), any small amount of population in the upper level will lead to a gain.

7.5 Refractive index enhancement via quantum coherence

The index of refraction of an optical medium can reach values as high as 10 or 100 at frequencies near an atomic resonance. The price that must be paid for such high dispersion is usually an accompanying high absorption. However, atomic coherence and interference effects, which have led to phenomena such as the correlated emission laser and the lasing without inversion, result in the possibility of a transparent medium with an ultra-large index of refraction. In this section we discuss a scheme in which coherence and interference effects produce a high index of refraction, while at the same time the absorption can be very small or even vanishing.

The linear response of an atomic system to an electric field E is described by the complex polarization

$$P(z,t) = \epsilon_0 \int_0^\infty d\tau \tilde{\chi}(\tau) E(z, t-\tau), \tag{7.5.1}$$

which appears as the driving term in the wave equation for the electric field (Eq. (5.4.15) with $\sigma = 0$)

$$\frac{\partial^2 E}{\partial z^2} - \frac{1}{c^2}\frac{\partial^2 E}{\partial t^2} = \mu_0 \frac{\partial^2 P}{\partial t^2}. \tag{7.5.2}$$

In Eq. (7.5.1)

$$\tilde{\chi} = \tilde{\chi}' + i\tilde{\chi}'' \tag{7.5.3}$$

is the susceptibility with $\tilde{\chi}'$ and $\tilde{\chi}''$ being the real and imaginary parts, respectively.

For a plane wave of frequency ν,

$$E(z,t) = \frac{1}{2}\mathscr{E} e^{-i(\nu t - kz)} + \text{c.c.}, \tag{7.5.4}$$

we obtain, from Eq. (7.5.1),

$$P(z,t) = \frac{\epsilon_0}{2}\mathscr{E}\left[\chi(\nu) e^{-i(\nu t - kz)} + \chi(-\nu) e^{i(\nu t - kz)}\right], \tag{7.5.5}$$

where $\chi(\nu)$ is the Fourier transform of $\tilde{\chi}(t)$. A comparison with Eq. (5.4.17) (with $\phi = 0$) yields

$$\mathscr{P}(z,t) = \epsilon_0 \mathscr{E} \chi. \tag{7.5.6}$$

Thus, with $\chi = \chi' + i\chi''$,

$$\mathrm{Re}\mathscr{P} = \epsilon_0 \mathscr{E} \chi', \tag{7.5.7}$$

and

$$\mathrm{Im}\mathscr{P} = \epsilon_0 \mathscr{E} \chi''. \tag{7.5.8}$$

It follows, on substituting these expressions of $\mathrm{Re}\mathscr{P}$ and $\mathrm{Im}\mathscr{P}$ into Eqs. (5.4.23) and (5.4.24), that χ' and χ'' represent the dispersion and loss per unit wavelength, respectively.

Next, we relate the real and imaginary parts of the susceptibility to the refractive index and the absorption coefficient of the medium.

On substituting for E and P from Eqs. (7.5.4) and (7.5.5), respectively, into Eq. (7.5.2), we obtain the dispersion relation

$$k^2 - \frac{\nu^2}{c^2} n^2 = 0, \tag{7.5.9}$$

where

$$n^2(\nu) = 1 + \chi(\nu). \tag{7.5.10}$$

As usual, we set $k = \nu n/c$. If n' and n'' represent the real and imaginary parts of n, i.e.,

$$n = n' + in'', \tag{7.5.11}$$

then n' is the refractive index of the medium and n'' is the associated absorption coefficient. It is clear from the definition of n'' that the medium has absorption for $n'' > 0$ and gain for $n'' < 0$. It follows, on combining Eqs. (7.5.10) and (7.5.11), that

$$\begin{aligned} n' + in'' &= \left(1 + \chi' + i\chi''\right)^{1/2} \\ &= \left[(1+\chi')^2 + \chi''^2\right]^{1/4} \exp[i\,\mathrm{sgn}(\chi'')\theta/2], \end{aligned} \tag{7.5.12}$$

where $\theta = \tan^{-1}[|\chi''|/(1+\chi')]$. We then obtain

$$n' = \left\{ \frac{[(1+\chi')^2 + \chi''^2]^{1/2} + (1+\chi')}{2} \right\}^{1/2}, \tag{7.5.13}$$

$$n'' = \left\{ \frac{[(1+\chi')^2 + \chi''^2]^{1/2} - (1+\chi')}{2} \right\}^{1/2} \mathrm{sgn}(\chi''). \tag{7.5.14}$$

It may be noted that, for $\chi' > 0$ and $\chi' \gg |\chi''|$, we have

$$n' \simeq (1 + \chi')^{1/2}, \tag{7.5.15}$$

$$n'' \simeq 0, \tag{7.5.16}$$

i.e., a large refractive index with little absorption. We first show that these conditions are not satisfied in the usual two-level system before discussing the large index of refraction with vanishing absorption via quantum coherence.

For a two-level medium, the real and imaginary parts of the susceptibility can be determined by substituting for Re$\mathscr{P}$ and Im$\mathscr{P}$ from Eqs. (5.5.4) into Eqs. (7.5.7) and (7.5.8). We then obtain

$$\chi' = -\frac{\wp^2 r_a}{\epsilon_o \hbar \gamma_a} \frac{\Delta}{\gamma^2 + \Delta^2} \left[\rho_{aa}^{(0)} - \rho_{bb}^{(0)}\right], \tag{7.5.17a}$$

$$\chi'' = -\frac{\wp^2 r_a}{\epsilon_o \hbar \gamma_b} \frac{\gamma}{\gamma^2 + \Delta^2} \left[\rho_{aa}^{(0)} - \rho_{bb}^{(0)}\right], \tag{7.5.17b}$$

where $\Delta = \omega - \nu$ and we have used the linear approximation

$$\rho_{aa} = \frac{r_a}{\gamma_a} \rho_{aa}^{(0)}, \tag{7.5.18a}$$

$$\rho_{bb} = \frac{r_a}{\gamma_b} \rho_{bb}^{(0)}. \tag{7.5.18b}$$

Here $\gamma = (\gamma_a + \gamma_b)/2$ and we will take $\gamma_a = \gamma_b$. For a closed system of N_a atoms per unit volume, r_a/γ_a and r_a/γ_b in Eqs. (7.5.18a) and (7.5.18b) are replaced by N_a so that Eqs. (7.5.17a) and (7.5.17b) become

$$\chi' = -\frac{\wp^2 N_a}{\epsilon_0 \hbar} \frac{\Delta}{\gamma^2 + \Delta^2} \left[\rho_{aa}^{(0)} - \rho_{bb}^{(0)}\right], \tag{7.5.19a}$$

$$\chi'' = -\frac{\wp^2 N_a}{\epsilon_0 \hbar} \frac{\gamma}{\gamma^2 + \Delta^2} \left[\rho_{aa}^{(0)} - \rho_{bb}^{(0)}\right]. \tag{7.5.19b}$$

These equalities are plotted in Fig. 7.7. The conditions $\chi' \gg 1$ with $\chi' \gg |\chi''|$ are not satisfied for any value of detuning. For example, in the case of the $|b\rangle \to |a\rangle$ transition ($\rho_{aa}^{(0)} = 0, \rho_{bb}^{(0)} = 1$), $\chi' = \chi''$ when $\Delta = \gamma$ and we obtain

$$\chi' = \chi'' = \frac{3\pi c^3}{2\nu^3} N_a, \tag{7.5.20}$$

for $\gamma = \gamma_r$ where

$$\gamma_r = \frac{\wp^2 \nu^3}{3\pi \hbar \epsilon_0 c^3} \tag{7.5.21}$$

is the radiative decay rate between levels $|a\rangle$ and $|b\rangle$. Thus for a

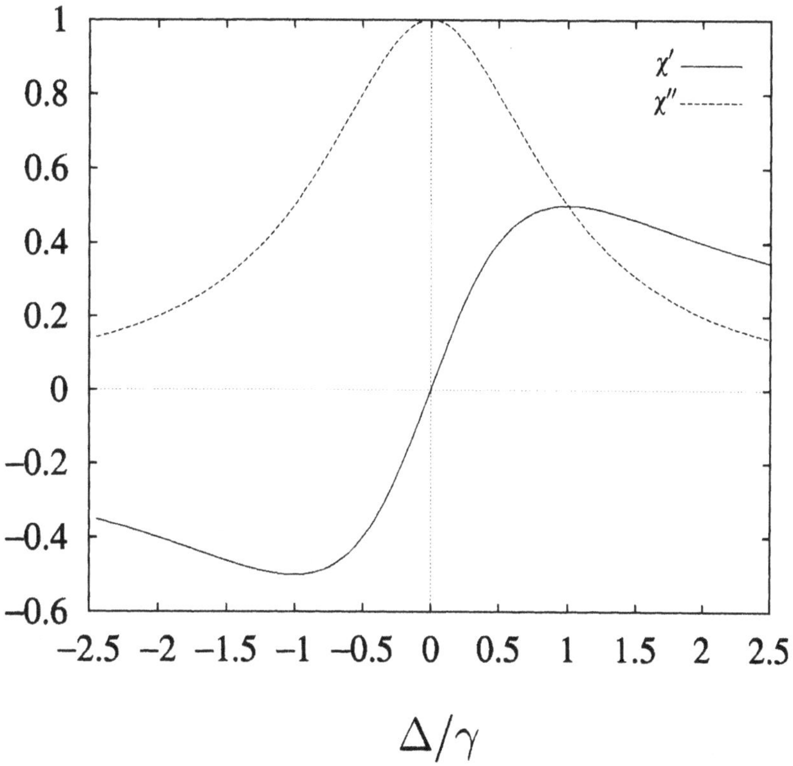

Fig. 7.7
Real (solid line) and imaginary (dashed line) parts of the linear susceptibility (in arbitrary units) as a function of normalized detuning Δ/γ of a gas of two-level atoms. Here $\rho_{aa}^{(0)} = 0, \rho_{bb}^{(0)} = 1$.

wavelength of 1 μm and a gas at one atmosphere such that $N_a \sim 10^{16}$ atoms/cm^3, we have $\chi' \sim 10^4$, i.e., large index of refraction. However, we also have $\chi'' \sim 10^4$ where χ'' is essentially the loss per unit wavelength. The light would therefore be totally absorbed in a small fraction of a wavelength. Thus a high refractive index is accompanied by large absorption.

The situation is completely different, however, for three-level schemes of the type considered in the previous section, in which atomic coherence is established and quantum interference effects occur.

We consider the case of a three-level structure with a pair of closely lying lower levels, in which coherence is established between the doublet states by some external means as discussed in Section 7.3. Assuming dipole allowed transitions between the single level $|a\rangle$ and the two closely spaced lower levels $|b\rangle$ and $|c\rangle$, the linear susceptibility of the system is

$$\chi = \frac{\mathscr{P}}{\epsilon_0 \mathscr{E}} = 2\left(\frac{\wp_{ba}\rho_{ab}}{\epsilon_0 \mathscr{E}} + \frac{\wp_{ca}\rho_{ac}}{\epsilon_0 \mathscr{E}}\right) e^{i\nu t}, \tag{7.5.22}$$

where the population matrix elements ρ_{ab} and ρ_{ac} are given by Eqs.

(7.4.31) and (7.4.32). It follows, on substituting these values of ρ_{ab} and ρ_{ac} into Eq. (7.5.22) that, after some rearrangement,

$$\chi' = -\frac{\wp^2 r_a}{\epsilon_0 \hbar}\left(\frac{1}{\gamma_{ac}^2+\Delta_{ac}^2}\left\{\Delta_{ac}\left[\frac{\rho_{aa}^{(0)}}{\gamma_a}-\frac{\rho_{cc}^{(0)}}{\gamma_c}\right]\right.\right.$$
$$\left. -\frac{|\rho_{cb}^{(0)}|}{\sqrt{\gamma_{cb}^2+\omega_{cb}^2}}(\Delta_{ac}\cos\phi+\gamma_{ac}\sin\phi)\right\}$$
$$+\frac{1}{\gamma_{ab}^2+\Delta_{ab}^2}\left\{\Delta_{ab}\left[\frac{\rho_{aa}^{(0)}}{\gamma_a}-\frac{\rho_{bb}^{(0)}}{\gamma_b}\right]\right.$$
$$\left.\left. -\frac{|\rho_{cb}^{(0)}|}{\sqrt{\gamma_{cb}^2+\omega_{cb}^2}}(\Delta_{ab}\cos\phi-\gamma_{ab}\sin\phi)\right\}\right), \tag{7.5.23a}$$

$$\chi'' = -\frac{\wp^2 r_a}{\epsilon_0 \hbar}\left(\frac{1}{\gamma_{ac}^2+\Delta_{ac}^2}\left\{\gamma_{ac}\left[\frac{\rho_{aa}^{(0)}}{\gamma_a}-\frac{\rho_{cc}^{(0)}}{\gamma_c}\right]\right.\right.$$
$$\left. -\frac{|\rho_{cb}^{(0)}|}{\sqrt{\gamma_{cb}^2+\omega_{cb}^2}}(\gamma_{ac}\cos\phi-\Delta_{ac}\sin\phi)\right\}$$
$$+\frac{1}{\gamma_{ab}^2+\Delta_{ab}^2}\left\{\gamma_{ab}\left[\frac{\rho_{aa}^{(0)}}{\gamma_a}-\frac{\rho_{bb}^{(0)}}{\gamma_b}\right]\right.$$
$$\left.\left. -\frac{|\rho_{cb}^{(0)}|}{\sqrt{\gamma_{cb}^2+\omega_{cb}^2}}(\gamma_{ab}\cos\phi+\Delta_{ab}\sin\phi)\right\}\right), \tag{7.5.23b}$$

where $\Delta_{ab} = \omega_{ab} - \nu$, $\Delta_{ac} = \omega_{ac} - \nu$, and the phase ϕ is defined by

$$\phi = \phi_{cb} + \tan^{-1}\left(\frac{\omega_{bc}}{\gamma_{bc}}\right), \tag{7.5.24}$$

with $\phi_{cb} = \arg[\rho_{cb}^{(0)}]$ and, for simplicity, we have taken $\wp_{ab} = \wp_{ac} = \wp_{ba} = \wp_{ca} \equiv \wp$. The phase ϕ therefore depends on the atomic coherence.

From Eqs. (7.5.23a) and (7.5.23b), we can see that it is possible to make the absorption χ'' vanish while maintaining a large χ' and hence a large refractive index. We define $\Delta = (\Delta_{ab} + \Delta_{ac})/2$, and adjust ω_{cb} by means of, e.g., a dc magnetic field in the Zeeman split levels, so that $\omega_{cb} = \gamma_{bc}$ and consider the reasonable case $\gamma_b = \gamma_c$. We prepare the levels $|b\rangle$ and $|c\rangle$ coherently so that $\phi = 5\pi/4$, and $\rho_{bb}^{(0)} = \rho_{cc}^{(0)} = |\rho_{bc}^{(0)}|$. The resulting polarization is plotted in Fig. 7.8. It can be seen that a high index of refraction can be obtained with zero absorption.

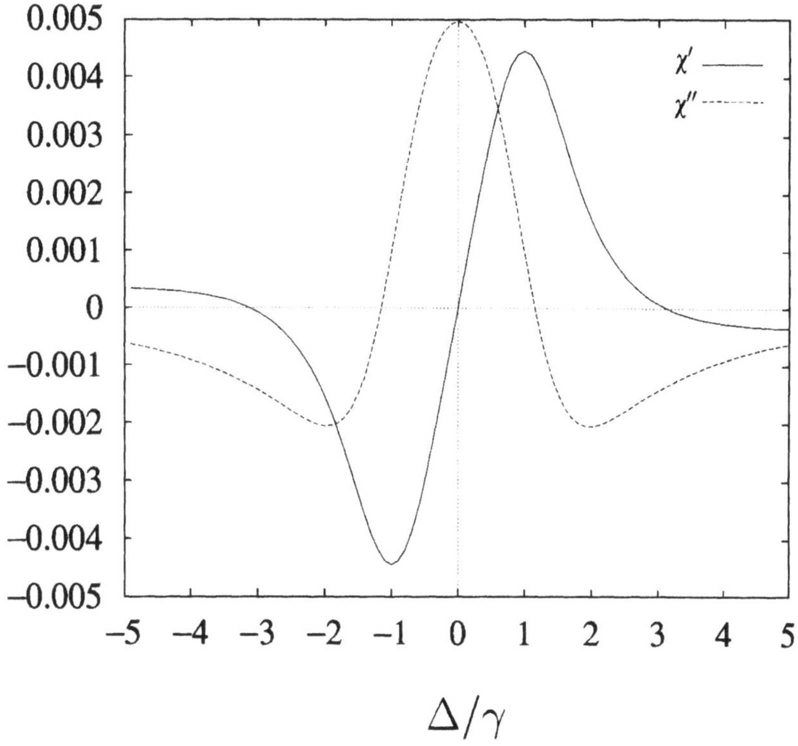

Fig. 7.8
Real (solid line) and imaginary (dashed line) parts of the linear susceptibility as a function of the normalized detuning Δ/γ for the case of injected coherence. Values of the parameters are $\gamma_a = 0.1$, $\gamma_b = \gamma_c = 2$, $\rho_{aa}^{(0)} = 0.01$, $\rho_{bb}^{(0)} = \rho_{cc}^{(0)} = 0.495$.

7.6 Coherent trapping, lasing without inversion, and electromagnetically induced transparency via an exact solution to a simple model

We have seen in the previous sections that population trapping, lasing without inversion, and electromagnetically induced transparency are consequences of quantum coherence and interference. In order to provide a unified treatment for these phenomena in a single system, we consider a system consisting of three-level atoms in the Λ configuration in which all levels decay at a rate γ (Fig. 7.9).

If we prepare our atoms in the initial state

$$|\psi(0)\rangle = c_a(0)|a\rangle + c_b(0)|b\rangle + c_c(0)|c\rangle, \tag{7.6.1}$$

then the atomic state at time t is given by (Problem 7.2)

$$|\psi(t)\rangle = A(t)|a\rangle + B(t)|b\rangle + C(t)|c\rangle, \tag{7.6.2a}$$

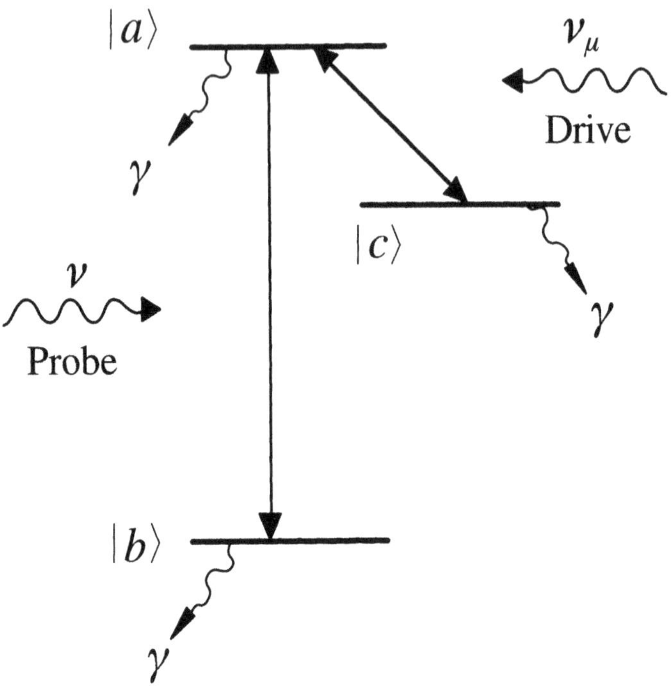

Fig. 7.9
Atomic configuration for a unified treatment of coherent trapping, lasing without inversion, and electromagnetically induced transparency.

where

$$A(t) = \left[c_a(0)\cos(\Omega t/2) - ic_b(0)\frac{\Omega_{R1}}{\Omega}\sin(\Omega t/2) - ic_c(0)\frac{\Omega_{R2}}{\Omega}\sin(\Omega t/2)\right]e^{-\gamma t/2}, \tag{7.6.2b}$$

$$B(t) = \left\{-ic_a(0)\frac{\Omega_{R1}}{\Omega}\sin(\Omega t/2) + c_b(0)\left[\frac{\Omega_{R2}^2}{\Omega^2} + \frac{\Omega_{R1}^2}{\Omega^2}\cos(\Omega t/2)\right] + c_c(0)\left[\frac{-\Omega_{R1}\Omega_{R2}}{\Omega^2} + \frac{\Omega_{R1}\Omega_{R2}}{\Omega^2}\cos(\Omega t/2)\right]\right\}e^{-\gamma t/2}, \tag{7.6.2c}$$

$$C(t) = \left\{-ic_a(0)\frac{\Omega_{R2}}{\Omega}\sin(\Omega t/2) + c_b(0)\left[-\frac{\Omega_{R1}\Omega_{R2}}{\Omega^2} + \frac{\Omega_{R1}\Omega_{R2}}{\Omega^2}\cos(\Omega t/2)\right] + c_c(0)\left[\frac{\Omega_{R1}^2}{\Omega^2} + \frac{\Omega_{R2}^2}{\Omega^2}\cos(\Omega t/2)\right]\right\}e^{-\gamma t/2}, \tag{7.6.2d}$$

Ω_{R1} and Ω_{R2} are the Rabi frequencies associated with the resonant fields driving the $|a\rangle \to |b\rangle$ and $|a\rangle \to |c\rangle$ transitions, respectively, and

$$\Omega = \sqrt{\Omega_{R1}^2 + \Omega_{R2}^2}. \tag{7.6.3}$$

This general solution for the three-level problem can be employed to obtain results relating to population trapping, lasing without inversion, and electromagnetically induced transparency under appropriate initial conditions.

(a) When atoms are prepared in a coherent superposition of lower states, i.e.,

$$|\psi(0)\rangle = \frac{\Omega_{R2}}{\Omega}|b\rangle - \frac{\Omega_{R1}}{\Omega}|c\rangle, \tag{7.6.4}$$

we obtain a trapped state for $\gamma = 0$. This can be verified by substituting $c_a(0) = 0$, $c_b(0) = \Omega_{R2}/\Omega$, and $c_c(0) = -\Omega_{R1}/\Omega$ into Eqs. (7.6.2).

(b) Next, we consider lasing without inversion. As we discussed earlier, the gain per unit length in a laser is proportional to the imaginary part of the polarization, i.e., $\text{Im}\mathscr{P}$. Now if we inject the atoms in the $|a\rangle$ state at the rate r_a, we find

$$\begin{aligned} \text{Im}\mathscr{P} &= \gamma r_a \int_{-\infty}^{t} dt_0 c_a(t,t_0)c_b^*(t,t_0) \\ &= \gamma r_a \int_{-\infty}^{t} dt_0 \frac{\Omega_{R1}}{2\Omega} \sin[\Omega(t-t_0)]e^{-\gamma(t-t_0)} \\ &= \gamma \frac{1}{2} r_a \frac{\Omega_{R1}}{\Omega^2+\gamma^2}. \end{aligned} \tag{7.6.5}$$

Here $c_a(t,t_0) = \cos[\Omega(t-t_0)/2]$ and $c_b(t,t_0) = -i\Omega_{R1}\sin[\Omega(t-t_0)/2]/\Omega$ are the probability amplitudes for the atom to be in states $|a\rangle$ and $|b\rangle$, respectively, subject to the initial condition that the atoms are injected in state $|a\rangle$ at time t_0. We can thus have lasing with inversion. If we also simultaneously inject atoms in the trapping state $(\Omega_{R2}|b\rangle - \Omega_{R1}|c\rangle)/\Omega$, we can have lasing without inversion since the trapped atoms will not affect the gain.

(c) Lastly we consider decay-free Λ electromagnetically induced transparency. Setting $\gamma = 0$ and taking the atoms to be injected in the $|b\rangle$ state, we have

$$\begin{aligned} |\psi(t)\rangle = &-i\frac{\Omega_{R1}}{\Omega}\sin(\Omega t/2)|a\rangle + \left[\frac{\Omega_{R2}^2}{\Omega^2} + \frac{\Omega_{R1}^2}{\Omega^2}\cos(\Omega t/2)\right]|b\rangle \\ &- \left[\frac{\Omega_{R1}\Omega_{R2}}{\Omega^2} - \frac{\Omega_{R1}\Omega_{R2}}{\Omega^2}\cos(\Omega t/2)\right]|c\rangle. \end{aligned} \tag{7.6.6}$$

Now the probability of absorbing a probe laser photon (i.e., exciting the atom to state $|a\rangle$) is given by

$$P_a = \frac{\Omega_{R1}^2}{\Omega^2} \sin^2(\Omega t/2), \tag{7.6.7}$$

which vanishes in the limit $\Omega_{R2} \gg \Omega_{R1}$. We then have electromagnetically induced transparency.

Problems

7.1 Consider a three-level atom in the Λ configuration as shown in Fig. 7.9. The atom–field Hamiltonian in the interaction picture and on resonance is

$$\mathscr{V} = -\frac{\hbar\Omega_{R1}}{2}|a\rangle\langle b| - \frac{\hbar\Omega_{R2}}{2}|a\rangle\langle c| + \text{H.c.},$$

where Ω_{R1} and Ω_{R2} are the Rabi frequencies associated with the field driving $|a\rangle \rightarrow |b\rangle$ and $|a\rangle \rightarrow |c\rangle$ transitions, respectively.

(1) Show that

$$|\psi_+\rangle = \frac{1}{\sqrt{2}}\left(|a\rangle + \frac{\Omega_{R1}}{\Omega}|b\rangle + \frac{\Omega_{R2}}{\Omega}|c\rangle\right),$$

$$|\psi_0\rangle = \frac{\Omega_{R2}}{\Omega}|b\rangle - \frac{\Omega_{R1}}{\Omega}|c\rangle,$$

$$|\psi_-\rangle = \frac{1}{\sqrt{2}}\left(|a\rangle - \frac{\Omega_{R1}}{\Omega}|b\rangle - \frac{\Omega_{R2}}{\Omega}|c\rangle\right),$$

with $\Omega = \sqrt{\Omega_{R1}^2 + \Omega_{R2}^2}$ are the eigenstates of the Hamiltonian. Find the corresponding eigenvalues.

(2) By including the equal decay constants γ for the three levels, show that the solution of the Schrödinger equation subject to the initial condition in Eq. (7.6.1) is given by Eqs. (7.6.2a-7.6.2d). (See M. O. Scully, *Quantum Optics* **6**, 203 (1994).)

7.2 Using adiabatic perturbation theory show that an atom in state (7.2.14) stays in that state. (Hint: see, J. R. Kuklinsky et al., *Phys. Rev. A* **40**, 6741 (1989).)

References and bibliography

Hanle effect, photon echo, self-induced transparency, and coherent Raman beats

W. Hanle, *Z. Phys.* **30**, 93 (1924).
G. Breit, *Rev. Mod. Phys.* **5**, 91 (1933).
I. Abella, N. Kurnit, and S. Hartmann, *Phys. Rev.* **141**, 391 (1966).
M. Scully, M. J. Stephen, and D. C. Burnham, *Phys. Rev.* **171**, 213 (1968).
S. McCall and E. Hahn, *Phys. Rev.* **183**, 457 (1969).
R. Shoemaker and R. Brewer, *Phys. Rev. Lett.* **28**, 1430 (1972).
R. Brewer and E. Hahn, *Phys. Rev.* **A8**, 464 (1973).
M. O. Scully, H. Fearn, and B. W. Atherton, in *New Frontiers in Quantum Electrodynamics and Quantum Optics*, ed. A. O. Barut (Plenum, New York 1989).

Coherent trapping

G. Alzetta, A. Gozzini, L. Moi, and G. Orriols, *Nuovo Cimento* **36B**, 5 (1976).
E. Arimondo and G. Orriols, *Nuovo Cimento Lett.* **17**, 333 (1976).
R. M. Whitley and C. R. Stroud, Jr., *Phys. Rev. A* **14**, 1498 (1976).
H. R. Gray, R. M. Whitley, and C. R. Stroud, Jr., *Opt. Lett.* **3**, 218 (1978).
G. Alzetta, L. Moi, and G. Orriols, *Nuovo Cimento* **52B**, 209 (1979).
J. D. Stettler, C. M. Bowden, N. M. Witriol, and J. H. Eberly, *Phys. Lett.* **73A**, 171 (1979).
P. M. Radmore and P. L. Knight, *J. Phys. B* **15**, 561 (1982); *Phys. Lett.* **102A**, 180 (1984).
B. J. Dalton and P. L. Knight, *Opt. Commun.* **42**, 411 (1982).
S. Swain, *J. Phys. B* **15**, 3405 (1982).
G. S. Agarwal and N. Nayak, *J. Phys. B* **19**, 3375 (1986).
K. Zaheer and M. S. Zubairy, *Phys. Rev. A* **39**, 2000 (1989).
E. Arimondo in *Progress in Optics* ed. E. Wolf, **35**, 257 (1996).

Lasing without inversion

A. Javan, *Phys. Rev.* **107**, 1579 (1956).
T. W. Hänsch and P. E. Toschek, *Z. Phys.* **236**, 213 (1970).
T. Popov, A Popov, and S. Ravtian, *JETP Lett.* **30**, 466 (1970).
V. Arkhipkin and Yu Heller, *Phys. Lett.* **48A**, 12 (1983).
O. Kocharovskaya and Ya. I. Khanin, *Pis'ma Zh. Eksp. Teor. Fiz.* **48**, 581 (1988) (*JETP Lett.* **48**, 630 (1988)).
S. E. Harris, *Phys. Rev. Lett.* **62**, 1033 (1989).
M. O. Scully, S.-Y. Zhu, and A. Gavrielides, *Phys. Rev. Lett.* **62**, 2813 (1989).
S. E. Harris and J. H. Macklin, *Phys. Rev. A* **40**, 4135 (1989).
A. Imamoğlu, *Phys. Rev. A* **40**, 2835 (1989).
A. Lyras, X. Tang, P. Lambropoulos, and J. Zhang, *Phys. Rev. A* **40**, 4131 (1989).

O. Kocharovskaya and P. Mandel, *Phys. Rev. A* **42**, 523 (1990).
E. E. Fill, M. O. Scully, and S.-Y. Zhu, *Opt. Commun.* **77**, 36 (1990).
O. Kocharovskaya, R.-D. Li, and P. Mandel, *Opt. Commun.* **77**, 215 (1990).
V. R. Blok and G. M. Krochik, *Phys. Rev. A* **41**, 1517 (1990).
G. S. Agarwal, S. Ravi, and J. Cooper, *Phys. Rev. A* **41**, 4721 (1990); *ibid.* **41**, 4727 (1990).
S. Basile and P. Lambropoulos, *Opt. Commun.* **78**, 163 (1990).
A. Imamoğlu, J. E. Field, and S. E. Harris, *Phys. Rev. Lett.* **66**, 1154 (1991).
G. S. Agarwal, *Phys. Rev. A* **44**, R28 (1991).
L. M. Narducci, H. M. Doss, P. Ru, M. O. Scully, and C. Keitel, *Opt. Commun.* **81**, 379 (1991).
J. A. Bergou and P. Bogár, *Phys. Rev. A* **43**, 4889 (1991).
O. Kocharovskaya, *Phys. Rep.* **219**, 175 (1992).
M. Scully, *Phys. Rep.* **219**, 191 (1992).
M. O. Scully, S.-Y. Zhu, and H. Fearn, *Z. Phys. D* **22**, 471 (1992).
M. Fleischhauer, C. H. Keitel, L. M. Narducci, M. O. Scully, S.-Y. Zhu, and M. S. Zubairy, *Opt. Commun.* **94**, 599 (1992).
M. O. Scully, *Quantum Optics* **6**, 203 (1994).
O. Kocharovskaya and P. Mandel, *Quantum Optics* **6**, 217 (1994).

Experimental demonstration of light amplification and laser oscillation without inversion

A. Nottelmann, C. Peters, and W. Lange, *Phys. Rev. Lett.* **70**, 1783 (1993).
E. S. Fry, X. Li, D. Nikonov, G. G. Padmabandu, M. O. Scully, A. V. Smith, F. K. Tittel, C. Wang, S. R. Wilkinson, and S.-Y. Zhu, *Phys. Rev. Lett.* **70**, 3235 (1993).
W. E. van der Veer, R. J. J. van Diest, A. Dönszelmann, and H. B. van Linden van den Heuvell, *Phys. Rev. Lett.* **70**, 3243 (1993).
A. S. Zibrov, M. D. Lukin, D. E. Nikonov, L. W. Hollberg, M. O. Scully, V. L. Velichansky, and H. G. Robinson, *Phys. Rev. Lett.* **75**, 1499 (1995).
G. G. Padmabandu, G. R. Welch, I. N. Shvbin, E. S. Fry, D. E. Nikonov, M. D. Lukin, and M. O. Scully, *Phys. Rev. Lett.* **76**, 2053 (1996).

Index enhancement via quantum coherence

M. O. Scully, *Phys. Rev. Lett.* **67**, 1855 (1991).
M. Fleischhauer, C. H. Keitel, M. O. Scully, and C. Su, *Opt. Commun.* **87**, 109 (1992).
M. Fleischhauer, C. H. Keitel, M. O. Scully, C. Su, B. T. Ulrich, and S.-Y. Zhu, *Phys. Rev. A* **46**, 1468 (1992).
M. O. Scully and S.-Y. Zhu, *Opt. Commun.* **87**, 134 (1992).
A. D. Wilson-Gordon and H. Friedmann, *Opt. Commun.* **94**, 238 (1992).
U. Rathe, M. Fleischhauer, S.-Y. Zhu, T. W. Hänsch, and M. O. Scully, *Phys. Rev. A* **47**, 4994 (1993).
S. Sultana and M. S. Zubairy, *Phys. Rev. A* **49**, 438 (1994).
O. Kocharovskaya, P. Mandel, and M. Scully, *Phys. Rev. Lett.* **74**, 2451 (1995).

Electromagnetically induced transparency (EIT)

S. E. Harris, J. E. Field, and A. Imamoğlu, *Phys. Rev. Lett.* **64**, 1107 (1990).
K. H. Hahn, D. A. King, and S. E. Harris, *Phys. Rev. Lett.* **65**, 2777 (1990).
K.-J. Boller, A. Imamoğlu, and S. E. Harris, *Phys. Rev. Lett.* **66**, 2593 (1991). This paper reports the first demonstration of EIT.
J. E. Field, K. H. Hahn, and S. E. Harris, *Phys. Rev. Lett.* **67**, 3062 (1991).
K. Hakuta, L. Marmet, and B. Stoicheff, *Phys. Rev. Lett.* **66**, 596, (1991).

Adiabatic population transfer in three-level systems

J. R. Kuklinski, U. Gaubatz, F. T. Hioe, and K. Bergmann, *Phys. Rev. A* **40**, 6741 (1989).
J. Oreg, F. T. Hioe, and J. H. Eberly, *Phys. Rev. A* **29**, 690 (1984).
S. R. Wilkinson, A. V. Smith, M. O. Scully, and E. S. Fry, *Phys. Rev. A* **53**, 126 (1996).

Fano interference phenomena

U. Fano, *Phys. Rev.* **124**, 1866 (1961).
U. Fano and A. Rau, *Atomic Collisions and Spectra*, (Academic Press, New York 1986).

CHAPTER 8

Quantum theory of damping – density operator and wave function approach

In many problems in quantum optics, damping plays an important role. These include, for example, the decay of an atom in an excited state to a lower state and the decay of the radiation field inside a cavity with partially transparent mirrors. In general, damping of a *system* is described by its interaction with a *reservoir* with a large number of degrees of freedom. We are interested, however, in the evolution of the variables associated with the system only. This requires us to obtain the equations of motion for the system of interest only after tracing over the reservoir variables. There are several different approaches to deal with this problem.

In this chapter, we present a theory of damping based on the density operator in which the reservoir variables are eliminated by using the *reduced* density operator for the system in the Schrödinger (or interaction) picture. We also present a 'quantum jump' approach to damping. In the next chapter, the damping of the system will be considered using the noise operator method in the Heisenberg picture.

An insight into the damping mechanism is obtained by considering the decay of an atom in an excited state inside a cavity. The atom may be considered as a single system coupled to the radiation field inside the cavity. Even in the absence of photons in the cavity, there are quantum fluctuations associated with the vacuum state. As discussed in Chapter 1, the field may be visualized as a large number of harmonic oscillators, one for each mode of the cavity. As the size of the cavity increases, the mode density increases, and, in free space, we get a continuum of modes. There is therefore a "cavity mode" which is resonant with the atomic transition.

We can also visualize the atom as an oscillator, with the excited atom corresponding to an oscillator in the excited state. The coupling of the atom to a large number of oscillators (associated with the large number of field modes) leads to decay. That is energy initially in the atom will distribute itself among damping oscillators, thus causing the decay of the atom to a lower energy state.

The dissipation is accompanied by fluctuations. We shall encounter this aspect of the damping mechanism, more formally put in the form of the so-called fluctuation–dissipation theorem, in the systems studied in this and the following chapters. We now start with a general reservoir theory before considering the atom and field damping by a reservoir of harmonic oscillator (bosonic) modes.

8.1 General reservoir theory

We consider in general a system denoted by S interacting with a reservoir denoted by R. The combined density operator is denoted by ρ_{SR}. The reduced density operator for the system ρ_S is obtained by taking a trace over the reservoir coordinates, i.e.,

$$\rho_S = \mathrm{Tr}_R(\rho_{SR}). \tag{8.1.1}$$

We assume that the system–reservoir interaction energy is given by $\mathscr{V}(t)$. The equation of motion for ρ_{SR} is then given by

$$i\hbar\dot{\rho}_{SR} = [\mathscr{V}(t), \rho_{SR}(t)]. \tag{8.1.2}$$

This equation can be formally integrated, and we obtain

$$\rho_{SR}(t) = \rho_{SR}(t_i) - \frac{i}{\hbar}\int_{t_i}^{t} [\mathscr{V}(t'), \rho_{SR}(t')]dt'. \tag{8.1.3}$$

Here t_i is an initial time when the interaction starts. On substituting $\rho_{SR}(t)$ back into Eq. (8.1.2), we find the equation of motion

$$\dot{\rho}_{SR} = -\frac{i}{\hbar}[\mathscr{V}(t), \rho_{SR}(t_i)] - \frac{1}{\hbar^2}\int_{t_i}^{t} [\mathscr{V}(t), [\mathscr{V}(t'), \rho_{SR}(t')]]dt'. \tag{8.1.4}$$

If the interaction energy $\mathscr{V}(t)$ is zero, the system and reservoir are independent and the density operator ρ_{SR} would factor as a direct product $\rho_{SR}(t) = \rho_S(t) \otimes \rho_R(t_i)$ where we assume the reservoir at equilibrium. Since $\mathscr{V}$ is small, we look for a solution of Eq. (8.1.4) of the form

$$\rho_{SR}(t) = \rho_S(t) \otimes \rho_R(t_i) + \rho_c(t), \tag{8.1.5}$$

where $\rho_c(t)$ is of higher order in $\mathscr{V}$. In order to satisfy (8.1.1), we require

$$\mathrm{Tr}_R[\rho_c(t)] = 0. \tag{8.1.6}$$

If we substitute for $\rho_{SR}(t)$ from Eq. (8.1.5) into the integrand of (8.1.4), and retain terms up to order $\mathcal{V}^2$, we have

$$\dot{\rho}_S = -\frac{i}{\hbar}\mathrm{Tr}_R[\mathcal{V}(t), \rho_S(t_i) \otimes \rho_R(t_i)] \\ -\frac{1}{\hbar^2}\mathrm{Tr}_R \int_{t_i}^{t} [\mathcal{V}(t), [\mathcal{V}(t'), \rho_S(t') \otimes \rho_R(t_i)]]\, dt'. \tag{8.1.7}$$

The reduced density operator $\rho_S(t)$, which determines the statistical properties of the system, depends on its past history from $t = t_i$ to t'. This can be seen in Eq. (8.1.7) as $\rho_S(t')$ occurs in the integrand. However, the reservoir is typically an extended open system having many degrees of freedom. Moreover, as is shown by specific example in the next section, the large number of reservoir degrees of freedom (modes, photons, etc.) leads to a delta function $\delta(t - t')$. Hence, the system density matrix $\rho_S(t')$ can be replaced by $\rho_S(t)$ and the process is said to be *Markovian*. This is a reasonable assumption since damping destroys memory of the past. Equation (8.1.7) now becomes

$$\dot{\rho}_S = -\frac{i}{\hbar}\mathrm{Tr}_R[\mathcal{V}(t), \rho_S(t_i) \otimes \rho_R(t_i)] \\ -\frac{1}{\hbar^2}\mathrm{Tr}_R \int_{t_i}^{t} [\mathcal{V}(t), [\mathcal{V}(t'), \rho_S(t) \otimes \rho_R(t_i)]]dt'. \tag{8.1.8}$$

This is a valid equation for a system represented by ρ_S interacting with a reservoir represented by ρ_R. In the next sections, we consider several examples of the system–reservoir interaction.

8.2 Atomic decay by thermal and squeezed vacuum reservoirs

The decay of an atom in an excited state may be understood from a simple model in which the atom is coupled to a reservoir of simple harmonic oscillators. In a very similar manner, the decay of the radiation field inside a cavity may be described by a model in which the mode of the field of interest is coupled to a whole set of reservoir modes. Such problems are of interest not only in maser and laser physics, but also in the quantum theory of passive interferometers such as those used in the detection of gravitational waves.

We first consider the radiative decay of a two-level atom damped by a reservoir of simple harmonic oscillators described by annihilation (and creation) operators $b_{\mathbf{k}}$ (and $b_{\mathbf{k}}^\dagger$) and density distributed frequencies $\nu_k = ck$. In the interaction picture and the rotating-wave approximation, the Hamiltonian is simply

$$\mathscr{V}(t) = \hbar \sum_{\mathbf{k}} g_{\mathbf{k}} \left[b_{\mathbf{k}}^\dagger \sigma_- e^{-i(\omega-\nu_k)t} + \sigma_+ b_{\mathbf{k}} e^{i(\omega-\nu_k)t} \right], \tag{8.2.1}$$

where $\sigma_- = |b\rangle\langle a|$ and $\sigma_+ = |a\rangle\langle b|$ in terms of the excited ($|a\rangle$) and ground ($|b\rangle$) states. The system now corresponds to the two-level atom ($\rho_S \equiv \rho_{\text{atom}}$). On inserting the interaction energy $\mathscr{V}$ (Eq. (8.2.1)) into the equation of motion (8.1.7) for $\rho_S \equiv \rho_{\text{atom}}$, we obtain

$$\begin{aligned}\dot{\rho}_{\text{atom}} = {} & -i \sum_{\mathbf{k}} g_{\mathbf{k}} \langle b_{\mathbf{k}}^\dagger \rangle [\sigma_-, \rho_{\text{atom}}(t_i)] e^{-i(\omega-\nu_k)t} \\ & - \int_{t_i}^{t} dt' \sum_{\mathbf{k},\mathbf{k}'} g_{\mathbf{k}} g_{\mathbf{k}'} \{ [\sigma_- \sigma_- \rho_{\text{atom}}(t') - 2\sigma_- \rho_{\text{atom}}(t') \sigma_- \\ & + \rho_{\text{atom}}(t') \sigma_- \sigma_-] \\ & \times e^{-i(\omega-\nu_k)t - i(\omega-\nu_{k'})t'} \langle b_{\mathbf{k}}^\dagger b_{\mathbf{k}'}^\dagger \rangle + [\sigma_- \sigma_+ \rho_{\text{atom}}(t') - \sigma_+ \rho_{\text{atom}}(t') \sigma_-] \\ & \times e^{-i(\omega-\nu_k)t + i(\omega-\nu_{k'})t'} \langle b_{\mathbf{k}}^\dagger b_{\mathbf{k}'} \rangle + [\sigma_+ \sigma_- \rho_{\text{atom}}(t') - \sigma_- \rho_{\text{atom}}(t') \sigma_+] \\ & \times e^{i(\omega-\nu_k)t - i(\omega-\nu_{k'})t'} \langle b_{\mathbf{k}} b_{\mathbf{k}'}^\dagger \rangle \} + \text{H.c.},\end{aligned} \tag{8.2.2}$$

where the expectation values refer to the initial state of the reservoir. At this point we choose a particular model for the state of the reservoir.

8.2.1 Thermal reservoir

As a first example, we assume that the reservoir variables are distributed in the uncorrelated thermal equilibrium mixture of states. The reservoir reduced density operator is the multi-mode extension of the thermal operator, namely,

$$\rho_R = \prod_{\mathbf{k}} \left[1 - \exp\left(-\frac{\hbar \nu_k}{k_B T} \right) \right] \exp\left(-\frac{\hbar \nu_k b_{\mathbf{k}}^\dagger b_{\mathbf{k}}}{k_B T} \right), \tag{8.2.3}$$

where k_B is the Boltzmann constant and T is the temperature. It can be shown easily that

$$\langle b_{\mathbf{k}} \rangle = \langle b_{\mathbf{k}}^\dagger \rangle = 0, \tag{8.2.4a}$$

$$\langle b_{\mathbf{k}}^\dagger b_{\mathbf{k}'} \rangle = \bar{n}_{\mathbf{k}} \delta_{\mathbf{k}\mathbf{k}'}, \tag{8.2.4b}$$

$$\langle b_{\mathbf{k}} b_{\mathbf{k}'}^\dagger \rangle = (\bar{n}_{\mathbf{k}} + 1) \delta_{\mathbf{k}\mathbf{k}'}, \tag{8.2.4c}$$

$$\langle b_{\mathbf{k}} b_{\mathbf{k}'} \rangle = \langle b_{\mathbf{k}}^\dagger b_{\mathbf{k}'}^\dagger \rangle = 0, \tag{8.2.4d}$$

where the thermal average boson number

$$\bar{n}_{\mathbf{k}} = \frac{1}{\exp\left(\frac{\hbar\nu_k}{k_B T}\right) - 1}. \tag{8.2.5}$$

On substituting for the various expectation values from Eqs. (8.2.4 into Eq. (8.2.2), we obtain

$$\begin{aligned}\dot{\rho}_{\text{atom}} &= -\int_{t_i}^{t} dt' \sum_{\mathbf{k}} g_{\mathbf{k}}^2 \{[\sigma_-\sigma_+\rho_{\text{atom}}(t') - \sigma_+\rho_{\text{atom}}(t')\sigma_-] \\ &\quad \bar{n}_{\mathbf{k}} e^{-i(\omega-\nu_k)(t-t')} \\ &\quad + [\sigma_+\sigma_-\rho_{\text{atom}}(t') - \sigma_-\rho_{\text{atom}}(t')\sigma_+](\bar{n}_{\mathbf{k}} + 1)e^{i(\omega-\nu_k)(t-t')}\} + \text{H.c.}\end{aligned} \tag{8.2.6}$$

We now carry out the same procedure as was used in the Weisskopf–Wigner theory of spontaneous emission.

The sum over **k** may be replaced by an integral through the standard prescription (Eq. (6.3.9))

$$\sum_{\mathbf{k}} \longrightarrow 2\frac{V}{(2\pi)^3} \int_0^{2\pi} d\phi \int_0^{\pi} d\theta \sin\theta \int_0^{\infty} dk\, k^2, \tag{8.2.7}$$

where V is the quantization volume. The integrations in Eq. (8.2.6) can be carried out in the Weisskopf–Wigner approximation as discussed in Section 6.3. In this way, we encounter integrals of the form (6.3.12). We thus find for the reduced density operator ρ_{atom}

$$\begin{aligned}\dot{\rho}_{\text{atom}}(t) &= -\bar{n}_{\text{th}}\frac{\Gamma}{2}[\sigma_-\sigma_+\rho_{\text{atom}}(t) - \sigma_+\rho_{\text{atom}}(t)\sigma_-] \\ &\quad -(\bar{n}_{\text{th}} + 1)\frac{\Gamma}{2}[\sigma_+\sigma_-\rho_{\text{atom}}(t) - \sigma_-\rho_{\text{atom}}(t)\sigma_+] + \text{H.c.},\end{aligned} \tag{8.2.8}$$

where $\bar{n}_{\text{th}} \equiv \bar{n}_{k_0}$ $(k_0 = \omega/c)$ and

$$\Gamma = \frac{1}{4\pi\epsilon_0}\frac{4\omega^3\wp_{ab}^2}{3\hbar c^3} \tag{8.2.9}$$

is the atomic decay rate which is identical to the decay constant (Eq. (6.3.14)) derived in the Weisskopf–Wigner theory of spontaneous emission. In deriving Eq. (8.2.8) we substituted the value of $g_{\mathbf{k}}$ from Eq. (6.1.8).

The equations of motion for the atomic density matrix elements can now be obtained from Eq. (8.2.8):

$$\dot{\rho}_{aa} = \langle a|\dot{\rho}_{\text{atom}}|a\rangle$$
$$= -(\bar{n}_{\text{th}} + 1)\Gamma\rho_{aa} + \bar{n}_{\text{th}}\Gamma\rho_{bb}, \tag{8.2.10a}$$
$$\dot{\rho}_{ab} = \dot{\rho}_{ba}^* = -\left(\bar{n}_{\text{th}} + \frac{1}{2}\right)\Gamma\rho_{ab}, \tag{8.2.10b}$$
$$\dot{\rho}_{bb} = -\bar{n}_{\text{th}}\Gamma\rho_{bb} + (\bar{n}_{\text{th}} + 1)\Gamma\rho_{aa}. \tag{8.2.10c}$$

It may be noted that $\dot{\rho}_{aa} + \dot{\rho}_{bb} = 0$. This is due to the fact that we are considering the decay from the upper level $|a\rangle$ to the lower level $|b\rangle$ only. The conservation of probability therefore implies $\rho_{aa} + \rho_{bb} = 1$. This situation is different from that discussed in Section 5.3, where atomic levels $|a\rangle$ and $|b\rangle$ decayed to some other levels via nonradiating transitions. For zero temperature ($\bar{n}_{\text{th}} = 0$), these equations simplify to

$$\dot{\rho}_{aa} = -\Gamma\rho_{aa}, \tag{8.2.11a}$$
$$\dot{\rho}_{ab} = -\frac{\Gamma}{2}\rho_{ab}, \tag{8.2.11b}$$
$$\dot{\rho}_{bb} = \Gamma\rho_{aa}. \tag{8.2.11c}$$

Equation (8.2.11a) is just the Weisskopf–Wigner result (6.3.15).

8.2.2 Squeezed vacuum reservoir

For our second example, we consider the situation where the atom is coupled to a squeezed vacuum field reservoir. The reservoir reduced density operator is given by

$$\rho_R = |\xi\rangle\langle\xi|$$
$$= \prod_{\mathbf{k}} S_{\mathbf{k}}(\xi)|0_{\mathbf{k}}\rangle\langle 0_{\mathbf{k}}|S_{\mathbf{k}}^\dagger(\xi), \tag{8.2.12}$$

where the squeeze operator (see Eq. (2.8.9) with $b_{\mathbf{k}} \equiv b(ck)$, etc.) is

$$S_{\mathbf{k}}(\xi) = \exp\left(\xi^* b_{\mathbf{k}_0+\mathbf{k}} b_{\mathbf{k}_0-\mathbf{k}} - \xi b_{\mathbf{k}_0+\mathbf{k}}^\dagger b_{\mathbf{k}_0-\mathbf{k}}^\dagger\right), \tag{8.2.13}$$

with $\xi = r\exp(i\theta)$, r being the squeeze parameter and θ being the reference phase for the squeezed field. A multi-mode squeezed field is not just a product of independently squeezed modes, rather there are correlations between modes symmetrically placed about the central, resonant frequency $\nu = ck_0$ of the squeezing device. Following the method used to derive Eqs. (2.7.6) and (2.7.7), we obtain

$$S^{\dagger}_{\mathbf{k}-\mathbf{k}_0} b_{\mathbf{k}} S_{\mathbf{k}-\mathbf{k}_0} = b_{\mathbf{k}} \cosh(r) - b^{\dagger}_{2\mathbf{k}_0-\mathbf{k}} e^{i\theta} \sinh(r), \tag{8.2.14a}$$

$$S^{\dagger}_{\mathbf{k}-\mathbf{k}_0} b^{\dagger}_{\mathbf{k}} S_{\mathbf{k}-\mathbf{k}_0} = b^{\dagger}_{\mathbf{k}} \cosh(r) - b_{2\mathbf{k}_0-\mathbf{k}} e^{-i\theta} \sinh(r). \tag{8.2.14b}$$

Similar expressions exist for $S^{\dagger}_{\mathbf{k}_0-\mathbf{k}} b_{\mathbf{k}} S_{\mathbf{k}_0-\mathbf{k}}$ and $S^{\dagger}_{\mathbf{k}_0-\mathbf{k}} b^{\dagger}_{\mathbf{k}} S_{\mathbf{k}_0-\mathbf{k}}$. The calculation of the expectation values, such as $\langle b^{\dagger}_{\mathbf{k}} b_{\mathbf{k}'} \rangle$, may therefore be simplified by writing

$$\langle b^{\dagger}_{\mathbf{k}} b_{\mathbf{k}'} \rangle = \prod_{\mathbf{q}} \langle 0_{\mathbf{q}} | S^{\dagger}_{\mathbf{q}} b^{\dagger}_{\mathbf{k}} S_{\mathbf{q}} S^{\dagger}_{\mathbf{q}} b_{\mathbf{k}'} S_{\mathbf{q}} | 0_{\mathbf{q}} \rangle. \tag{8.2.15}$$

It follows that

$$\langle b_{\mathbf{k}} \rangle = \langle b^{\dagger}_{\mathbf{k}} \rangle = 0, \tag{8.2.16a}$$

$$\langle b^{\dagger}_{\mathbf{k}} b_{\mathbf{k}'} \rangle = \sinh^2(r) \delta_{\mathbf{k}\mathbf{k}'}, \tag{8.2.16b}$$

$$\langle b_{\mathbf{k}} b^{\dagger}_{\mathbf{k}'} \rangle = \cosh^2(r) \delta_{\mathbf{k}\mathbf{k}'}, \tag{8.2.16c}$$

$$\langle b_{\mathbf{k}} b_{\mathbf{k}'} \rangle = -e^{i\theta} \sinh(r) \cosh(r) \delta_{\mathbf{k}',2\mathbf{k}_0-\mathbf{k}}, \tag{8.2.16d}$$

$$\langle b^{\dagger}_{\mathbf{k}} b^{\dagger}_{\mathbf{k}'} \rangle = -e^{-i\theta} \sinh(r) \cosh(r) \delta_{\mathbf{k}',2\mathbf{k}_0-\mathbf{k}}. \tag{8.2.16e}$$

On substituting Eqs. (8.2.16a-8.2.16e) into Eq. (8.2.2) and proceeding as in the derivation of Eq. (8.2.8), we obtain

$$\begin{aligned}
\dot{\rho}_{\text{atom}} = & -\frac{\Gamma}{2} \cosh^2(r)(\sigma_+\sigma_-\rho_{\text{atom}} - 2\sigma_-\rho_{\text{atom}}\sigma_+ + \rho_{\text{atom}}\sigma_+\sigma_-) \\
& -\frac{\Gamma}{2} \sinh^2(r)(\sigma_-\sigma_+\rho_{\text{atom}} - 2\sigma_+\rho_{\text{atom}}\sigma_- + \rho_{\text{atom}}\sigma_-\sigma_+) \\
& -\Gamma e^{-i\theta} \sinh(r) \cosh(r) \sigma_-\rho_{\text{atom}}\sigma_- \\
& -\Gamma e^{i\theta} \sinh(r) \cosh(r) \sigma_+\rho_{\text{atom}}\sigma_+.
\end{aligned} \tag{8.2.17}$$

In deriving Eq. (8.2.17) we used $\sigma_-\sigma_- = \sigma_+\sigma_+ = 0$.

From Eq. (8.2.17), equations of motion for the expectation value of the operators $\sigma_x = (\sigma_- + \sigma_+)/2$, $\sigma_y = (\sigma_- - \sigma_+)/2i$, and $\sigma_z = (2\sigma_+\sigma_- - 1)$ are

$$\langle \dot{\sigma}_x \rangle = -\frac{\Gamma}{2} e^{2r} \langle \sigma_x \rangle, \tag{8.2.18a}$$

$$\langle \dot{\sigma}_y \rangle = -\frac{\Gamma}{2} e^{-2r} \langle \sigma_y \rangle, \tag{8.2.18b}$$

$$\langle \dot{\sigma}_z \rangle = -\Gamma[2\sinh^2(r) + 1]\langle \sigma_z \rangle - \Gamma = -\Gamma_z \langle \sigma_z \rangle - \Gamma, \tag{8.2.18c}$$

where $\Gamma_z = \Gamma[2\sinh^2(r) + 1]$ and we have chosen the phase $\theta = 0$. It is therefore clear that a squeezed vacuum reservoir leads to a phase sensitive decay of the atom. The in-phase and in-quadrature components, $\langle \sigma_x \rangle$ and $\langle \sigma_y \rangle$, of the atomic dipole moment decay at different rates depending on its initial phase relative to the phase θ of the squeezed vacuum.

8.3 Field damping

We may apply the method developed in the last section to the decay of a mode of the electromagnetic field of frequency ν inside a cavity. Instead of Eq. (8.2.1), we now use an interaction Hamiltonian of the form

$$\mathscr{V} = \hbar \sum_{\mathbf{k}} g_{\mathbf{k}} [b_{\mathbf{k}}^{\dagger} a e^{-i(\nu - \nu_k)t} + a^{\dagger} b_{\mathbf{k}} e^{i(\nu - \nu_k)t}], \tag{8.3.1}$$

where a (and $a^{\dagger}$) are the destruction (and creation) operators of the mode of interest. The operators $b_{\mathbf{k}}$ and $b_{\mathbf{k}}^{\dagger}$ represent modes of the reservoir which damp the field. For transmission losses they actually represent the field outside the cavity.

The equation of motion for the reduced density operator for the field can now easily be obtained, since the calculation exactly parallels the one for the atomic system discussed in the last section. This is done by replacing σ_- and σ_+ by the field operators a and $a^{\dagger}$, respectively.

When the modes $b_{\mathbf{k}}$ are initially in the thermal equilibrium mixture of states (8.2.3), the result is

$$\begin{aligned} \dot{\rho} = &-\frac{\mathscr{C}}{2}\bar{n}_{\text{th}}(aa^{\dagger}\rho - 2a^{\dagger}\rho a + \rho a a^{\dagger}) \\ &-\frac{\mathscr{C}}{2}(\bar{n}_{\text{th}} + 1)(a^{\dagger}a\rho - 2a\rho a^{\dagger} + \rho a^{\dagger} a), \end{aligned} \tag{8.3.2}$$

where, as before, $\mathscr{C}$ is the decay constant and $\bar{n}_{\text{th}} = \bar{n}_{\mathbf{k}_0}$ is the mean number of quanta (at frequency ν) in the thermal reservoir. Here ρ denotes the reduced density operator for the field. In particular, at zero temperature ($\bar{n}_{\text{th}} = 0$),

$$\dot{\rho} = -\frac{\mathscr{C}}{2}(a^{\dagger}a\rho - 2a\rho a^{\dagger} + \rho a^{\dagger} a). \tag{8.3.3}$$

If all the losses are transmission losses, $\mathscr{C}$ may be related to the quality factor Q of the cavity by $\mathscr{C} = \nu/Q$.

When the modes $b_{\mathbf{k}}$ are initially in a squeezed vacuum (Eq. (8.2.12)), the resulting equation of motion for the reduced density matrix ρ is

$$\begin{aligned} \dot{\rho} = &-\frac{\mathscr{C}}{2}(N+1)(a^{\dagger}a\rho - 2a\rho a^{\dagger} + \rho a^{\dagger} a) \\ &-\frac{\mathscr{C}}{2}N(aa^{\dagger}\rho - 2a^{\dagger}\rho a + \rho a a^{\dagger}) \\ &+\frac{\mathscr{C}}{2}M(aa\rho - 2a\rho a + \rho a a) \\ &+\frac{\mathscr{C}}{2}M^{*}(a^{\dagger}a^{\dagger}\rho - 2a^{\dagger}\rho a^{\dagger} + \rho a^{\dagger}a^{\dagger}), \end{aligned} \tag{8.3.4}$$

where $N = \sinh^2(r)$ and $M = \cosh(r)\sinh(r)\exp(-i\theta)$. This equation describes, for instance, the evolution of the field in a cavity coupled through a partially transmitting mirror to an outside field which is in a squeezed vacuum state. The equation of motion for the thermal reservoir (Eq. (8.3.2)) can be recovered from Eq. (8.3.4) by the substitutions $N \to \bar{n}_{\text{th}}$, $M \to 0$. The parameters N and M are however related to each other via the equation $|M| = [N(N+1)]^{1/2}$ for a squeezed vacuum reservoir.

8.4 Fokker–Planck equation

A particularly interesting representation into which the density operator equation of motion can be transformed is the coherent state representation or P-representation discussed in Chapter 3. In this section, we derive an equation of motion for the P-representation corresponding to Eq. (8.3.2) for the density operator for a harmonic oscillator mode damped by a thermal bath full of harmonic oscillators. The resulting equation will have the form of a Fokker–Planck equation. The solution of this equation will reveal some interesting features about the temporal evolution of the field distribution.

We substitute the P-representation, see Eq. (3.1.16),

$$\rho = \int P(\alpha, \alpha^*, t)|\alpha\rangle\langle\alpha| d^2\alpha \tag{8.4.1}$$

into Eq. (8.3.2) and the resulting equation is

$$\begin{aligned}\int \dot{P}(\alpha, \alpha^*, t)|\alpha\rangle\langle\alpha| d^2\alpha = {} & -\frac{\mathscr{C}}{2}\bar{n}_{\text{th}} \int P(\alpha, \alpha^*, t)(aa^\dagger|\alpha\rangle\langle\alpha| \\ & -2a^\dagger|\alpha\rangle\langle\alpha|a + |\alpha\rangle\langle\alpha|aa^\dagger)d^2\alpha \\ & -\frac{\mathscr{C}}{2}(\bar{n}_{\text{th}}+1) \int P(\alpha, \alpha^*, t)(a^\dagger a|\alpha\rangle\langle\alpha| \\ & -2a|\alpha\rangle\langle\alpha|a^\dagger + |\alpha\rangle\langle\alpha|a^\dagger a)d^2\alpha. \end{aligned} \tag{8.4.2}$$

It follows from

$$a^\dagger|\alpha\rangle\langle\alpha| = \left(\frac{\partial}{\partial\alpha} + \alpha^*\right)|\alpha\rangle\langle\alpha|, \tag{8.4.3a}$$

$$a|\alpha\rangle\langle\alpha| = \alpha|\alpha\rangle\langle\alpha|, \tag{8.4.3b}$$

$$|\alpha\rangle\langle\alpha|a^\dagger = \alpha^*|\alpha\rangle\langle\alpha|, \tag{8.4.3c}$$

$$|\alpha\rangle\langle\alpha|a = \left(\frac{\partial}{\partial\alpha^*} + \alpha\right)|\alpha\rangle\langle\alpha|, \tag{8.4.3d}$$

that

$$
\begin{aligned}
&aa^\dagger|\alpha\rangle\langle\alpha| - 2a^\dagger|\alpha\rangle\langle\alpha|a + |\alpha\rangle\langle\alpha|aa^\dagger \\
&= \left[\left(\frac{\partial}{\partial\alpha} + \alpha^*\right)\alpha - 2\left(\frac{\partial}{\partial\alpha} + \alpha^*\right)\left(\frac{\partial}{\partial\alpha^*} + \alpha\right)\right. \\
&\quad \left. + \left(\frac{\partial}{\partial\alpha^*} + \alpha\right)\alpha^*\right]|\alpha\rangle\langle\alpha| \\
&= -\left(\alpha\frac{\partial}{\partial\alpha} + \alpha^*\frac{\partial}{\partial\alpha^*} + 2\frac{\partial^2}{\partial\alpha\partial\alpha^*}\right)|\alpha\rangle\langle\alpha|,
\end{aligned} \tag{8.4.4}
$$

and

$$
\begin{aligned}
&a^\dagger a|\alpha\rangle\langle\alpha| - 2a|\alpha\rangle\langle\alpha|a^\dagger + |\alpha\rangle\langle\alpha|a^\dagger a \\
&= \left[\alpha\left(\frac{\partial}{\partial\alpha} + \alpha^*\right) - 2|\alpha|^2 + \alpha^*\left(\frac{\partial}{\partial\alpha^*} + \alpha\right)\right]|\alpha\rangle\langle\alpha| \\
&= \left(\alpha\frac{\partial}{\partial\alpha} + \alpha^*\frac{\partial}{\partial\alpha^*}\right)|\alpha\rangle\langle\alpha|.
\end{aligned} \tag{8.4.5}
$$

We now substitute Eqs. (8.4.4) and (8.4.5) into Eq. (8.4.2) and integrate the result by parts. In doing so we encounter the integral

$$
\begin{aligned}
&\int P(\alpha,\alpha^*,t)\left(\alpha\frac{\partial}{\partial\alpha}|\alpha\rangle\langle\alpha|\right)d^2\alpha \\
&= \alpha P(\alpha,\alpha^*,t)|\alpha\rangle\langle\alpha|\Big|_{-\infty}^{\infty} - \int\left[\frac{\partial}{\partial\alpha}\alpha P(\alpha,\alpha^*,t)\right]|\alpha\rangle\langle\alpha|d^2\alpha.
\end{aligned} \tag{8.4.6}
$$

Since the distribution vanishes at the infinite limits, Eq. (8.4.6) becomes

$$
\int P(\alpha,\alpha^*,t)\left(\alpha\frac{\partial}{\partial\alpha}|\alpha\rangle\langle\alpha|\right)d^2\alpha = -\int\left[\frac{\partial}{\partial\alpha}\alpha P(\alpha,\alpha^*,t)\right]|\alpha\rangle\langle\alpha|d^2\alpha. \tag{8.4.7}
$$

Similarly

$$
\int P(\alpha,\alpha^*,t)\left(\frac{\partial^2}{\partial\alpha\partial\alpha^*}|\alpha\rangle\langle\alpha|\right)d^2\alpha = \int\left[\frac{\partial^2}{\partial\alpha\partial\alpha^*}P(\alpha,\alpha^*,t)\right]|\alpha\rangle\langle\alpha|d^2\alpha. \tag{8.4.8}
$$

Then we have from Eq. (8.4.2)

$$
\begin{aligned}
\int \dot{P}(\alpha,\alpha^*,t)|\alpha\rangle\langle\alpha|d^2\alpha = \frac{\mathscr{C}}{2}\int&\left[\left(\frac{\partial}{\partial\alpha}\alpha + \frac{\partial}{\partial\alpha^*}\alpha^* + 2\bar{n}_{\text{th}}\frac{\partial^2}{\partial\alpha\partial\alpha^*}\right)\right. \\
&\left.\times P(\alpha,\alpha^*,t)\right]|\alpha\rangle\langle\alpha|d^2\alpha.
\end{aligned} \tag{8.4.9}
$$

It follows on identifying the coefficients of $|\alpha\rangle\langle\alpha|$ in the integrands that the equation of motion for $P(\alpha,\alpha^*,t)$ is

$$\dot{P} = \frac{\mathscr{C}}{2}\left(\frac{\partial}{\partial\alpha}\alpha + \frac{\partial}{\partial\alpha^*}\alpha^*\right)P + \mathscr{C}\bar{n}_{\text{th}}\frac{\partial^2 P}{\partial\alpha\partial\alpha^*}. \tag{8.4.10}$$

This is the Fokker–Planck equation for the P-representation.

Next we find a solution of the Fokker–Planck equation. We assume that the field is initially in a coherent state $|\alpha_0\rangle$, i.e.,

$$P(\alpha, \alpha^*, 0) = \delta^{(2)}(\alpha - \alpha_0). \tag{8.4.11}$$

In the Gaussian representation of the δ-function,

$$P(\alpha, \alpha^*, 0) = \lim_{\epsilon\to 0}\frac{1}{\pi\epsilon}\exp\left(\frac{-|\alpha - \alpha_0|^2}{\epsilon}\right). \tag{8.4.12}$$

We therefore seek a solution of Eq. (8.4.10) in the form

$$P(\alpha, \alpha^*, t) = \exp[-a(t) + b(t)\alpha + c(t)\alpha^* - d(t)\alpha\alpha^*], \tag{8.4.13}$$

subject to the initial conditions

$$a(0) = \frac{|\alpha_0|^2}{\epsilon} + \ln(\pi\epsilon), \tag{8.4.14a}$$

$$b(0) = \frac{\alpha_0^*}{\epsilon}, \tag{8.4.14b}$$

$$c(0) = \frac{\alpha_0}{\epsilon}, \tag{8.4.14c}$$

$$d(0) = \frac{1}{\epsilon}. \tag{8.4.14d}$$

On substituting expression (8.4.13) for $P(\alpha, \alpha^*, t)$ into Eq. (8.4.10) and carrying out the necessary t and α differentiations, we obtain

$$\begin{aligned} -\dot{a} + \dot{b}\alpha + \dot{c}\alpha^* - \dot{d}|\alpha|^2 = \mathscr{C}\Bigg[1 + \bar{n}_{\text{th}}(bc - d) + \left(\frac{b}{2} - \bar{n}_{\text{th}}bd\right)\alpha \\ + \left(\frac{c}{2} - \bar{n}_{\text{th}}cd\right)\alpha^* - (d - \bar{n}_{\text{th}}d^2)|\alpha|^2\Bigg]. \end{aligned} \tag{8.4.15}$$

A comparison of the terms proportional to $|\alpha|^2$, α^*, α, and unity lead to the following set of differential equations:

$$\dot{d} = \mathscr{C}(d - \bar{n}_{\text{th}}d^2), \tag{8.4.16a}$$

$$\dot{c} = \mathscr{C}\left(\frac{c}{2} - \bar{n}_{\text{th}}cd\right), \tag{8.4.16b}$$

$$\dot{b} = \mathscr{C}\left(\frac{b}{2} - \bar{n}_{\text{th}}bd\right), \tag{8.4.16c}$$

$$\dot{a} = -\mathscr{C}[1 + \bar{n}_{\text{th}}(bc - d)]. \tag{8.4.16d}$$

The solution of these equations subject to the initial conditions (8.4.14a)–(8.4.14d) is given by

$$d(t) = \frac{1}{\bar{n}_{\rm th}(1 - e^{-\mathscr{C}t}) + \epsilon e^{-\mathscr{C}t}}, \tag{8.4.17a}$$

$$c(t) = \frac{\alpha_0 e^{-\mathscr{C}t/2}}{\bar{n}_{\rm th}(1 - e^{-\mathscr{C}t}) + \epsilon e^{-\mathscr{C}t}}, \tag{8.4.17b}$$

$$b(t) = \frac{\alpha_0^* e^{-\mathscr{C}t/2}}{\bar{n}_{\rm th}(1 - e^{-\mathscr{C}t}) + \epsilon e^{-\mathscr{C}t}}, \tag{8.4.17c}$$

$$a(t) = \frac{|\alpha_0|^2 e^{-\mathscr{C}t}}{\bar{n}_{\rm th}(1 - e^{-\mathscr{C}t}) + \epsilon e^{-\mathscr{C}t}} + \ln\left\{\pi\left[\bar{n}_{\rm th}\left(1 - e^{-\mathscr{C}t}\right) + \epsilon e^{-\mathscr{C}t}\right]\right\}. \tag{8.4.17d}$$

A substitution of these solutions into Eq. (8.4.13) results in the Gaussian form for $P(\alpha, \alpha^*, t)$:

$$P(\alpha, \alpha^*, t) = \frac{1}{\pi D(t)} \exp\left[-\frac{|\alpha - \alpha_0 U(t)|^2}{D(t)}\right], \tag{8.4.18}$$

where

$$D(t) = \bar{n}_{\rm th}(1 - e^{-\mathscr{C}t}) \tag{8.4.19}$$

is the dispersion of the Gaussian function about its mean value

$$\alpha_0 U(t) = \alpha_0 e^{-\mathscr{C}t/2 - i\nu t}. \tag{8.4.20}$$

In Eq. (8.4.20), we have included the factor $\exp(-i\nu t)$ by going back from the interaction picture to the Schrödinger picture.

The dispersion $D(t)$ increases from the initial value zero, while the center of the Gaussian distribution circles about on the exponential spiral given by Eq. (8.4.20). This is shown in Fig. 8.1 where the P-representation is plotted as a function of complex amplitude α. When the time t is much greater than the damping time, $\mathscr{C}^{-1}$, the field distribution comes to equilibrium with the heat bath oscillators. In the steady state, the dispersion has its limiting value $\bar{n}_{\rm th}$ and the Gaussian distribution is centered about the origin. Thus the field loses its initial excitation to the heat bath oscillators but acquires noise in the process of damping. This is a manifestation of the fluctuation–dissipation theorem, i.e., the dissipation via heat bath oscillators is accompanied by fluctuations. We will discuss it in the next chapter.

It is interesting to note that if we take the heat bath to be at zero temperature ($\bar{n}_{\rm th} = 0$), the dispersion $D(t)$ remains zero at all times and $P(\alpha, \alpha^*, t)$ always remains a δ-function, i.e.,

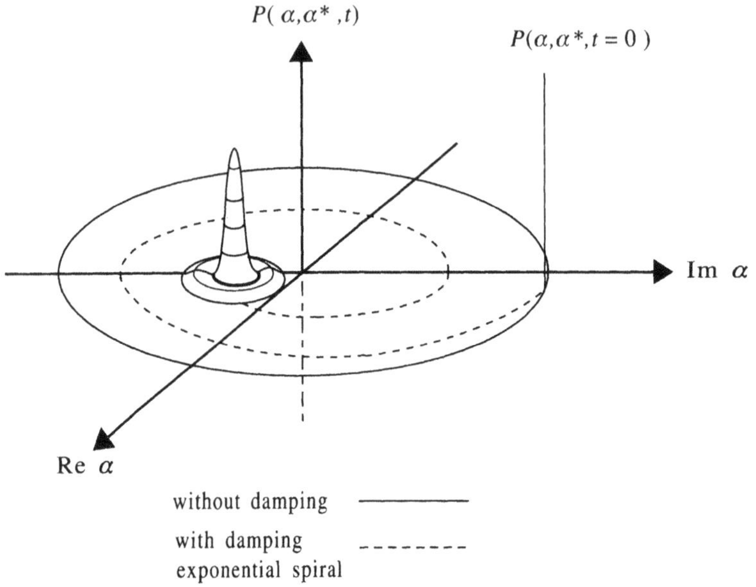

Fig. 8.1
The P-representation for the complex amplitude of a harmonic oscillator mode damped by a thermal bath. The harmonic oscillator mode starts at $t = 0$ in a pure coherent sate $|\alpha\rangle$ and the mean value of the amplitude moves on an exponential spiral decreasing steadily in modulus, while its dispersion increases.

$$P(\alpha, \alpha^*, t) = \delta^{(2)}[\alpha - \alpha_0 U(t)]. \tag{8.4.21}$$

The state of the field remains at all times in a pure coherent state. This form of dissipation is completely noise free.

8.5 The 'quantum jump' approach to damping

Historically, the notions of quantum jumps and instantaneous collapse of the wave function go back to the early days in which Einstein worried about outgoing spherical waves 'collapsing' when a photoelectron is detected; and the notion of Bohr concerning the emission of light when an atom 'jumped' between Bohr orbits.

However, with the coming of wave mechanics the whole question of quantum jumps took on a new perspective. Atomic transitions were 'induced' and one often encountered statements that 'there were no such a thing as quantum jumps'.

Recently, the work of Dehmelt and others clearly shows that sudden jumps are evident in many aspects of quantum optics, e.g., the spectacular work involving single ions in a Paul trap.

More recently a new 'quantum jump' approach to dissipation has developed, one can find names and concepts like: Monte Carlo simulation, quantum trajectories, collapse or reduction of the state vector,

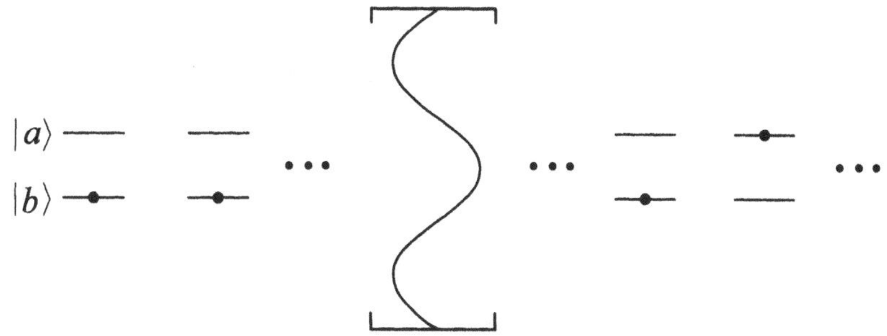

Fig. 8.2
Two-level atoms in their ground state $|b\rangle$ passing through a resonant cavity.

no count or 'null' measurement, and conditional density matrices. We will here give a short account of this interesting idea as it applies to damping or dissipation in quantum optics.

8.5.1 Conditional density matrices and the null measurement

In the previous sections of this chapter we have developed the theory of damping or dissipation in quantum mechanics from a density matrix perspective. The result is typically an expression of the form of (8.3.3) which describes the decay of a single mode of a resonant cavity at temperature $T = 0$. There we took the model of a large number of bath oscillators, e.g., phonons coupling energy out of the cavity mode. However the result, Eq. (8.3.3), is not specific to the model and we will here investigate the problem again using another model which will lead us naturally to a different point of view concerning dissipation processes.

Consider the model of Fig. 8.2 in which we are passing ground state atoms through a cavity which is resonant with the atoms, i.e., the Hamiltonian in the interaction picture is

$$\mathscr{V} = \hbar g(a^\dagger|b\rangle\langle a| + |a\rangle\langle b|a). \tag{8.5.1}$$

Consider the density matrix for the field at time $t+\tau$, $\rho(t+\tau)$, resulting from a ground state atom injected at time t, i.e.,

$$\begin{aligned}\rho(t+\tau) &= \mathrm{Tr}_{\mathrm{atom}}\left[e^{-i\mathscr{V}\tau/\hbar}\rho(t)\otimes|b\rangle\langle b|e^{i\mathscr{V}\tau/\hbar}\right]\\ &= \langle a|\rho_{\mathrm{atom-field}}(t+\tau)|a\rangle + \langle b|\rho_{\mathrm{atom-field}}(t+\tau)|b\rangle. \end{aligned}\tag{8.5.2}$$

It is natural to identify the two terms in (8.5.2) as 'conditional' density matrices, i.e.,

$$\begin{aligned}&\rho_a(t+\tau)\\ &= \text{conditional density matrix for field, atom excited}\\ &= \langle a|e^{-i\mathscr{V}\tau/\hbar}\rho(t)\otimes|b\rangle\langle b|e^{i\mathscr{V}\tau/\hbar}|a\rangle,\end{aligned}\tag{8.5.3a}$$

$$\begin{aligned}
&\rho_b(t+\tau)\\
&= \text{conditional density matrix for field, atom not excited}\\
&= \langle b|e^{-i\mathscr{V}\tau/\hbar}\rho(t)\otimes|b\rangle\langle b|e^{i\mathscr{V}\tau/\hbar}|b\rangle. \qquad (8.5.3b)
\end{aligned}$$

We may regard ρ_a and ρ_b as conditional density matrices corresponding to our having observed a count (excited atom) or no count (ground state atom) in our atomic beam. That is, the atomic beam serves two functions: it is a dissipation mechanism and it is also a kind of probe, or photodetector, of the field.

We proceed by noting that for small times τ, we may expand the $\exp(\pm i\mathscr{V}\tau/\hbar)$ factors and find

$$\rho_a(t+\tau) \cong g^2\tau^2 a\rho(t)a^\dagger, \qquad (8.5.4a)$$

$$\begin{aligned}
\rho_b(t+\tau) &\cong \rho(t) - \frac{1}{2}g^2\tau^2\left[a^\dagger a\rho(t) + \rho(t)a^\dagger a\right]\\
&\cong e^{-R\tau a^\dagger a}\rho(t)e^{-R\tau a^\dagger a}, \qquad (8.5.4b)
\end{aligned}$$

where $R = g^2\tau/2$.

Now we make the key step. We let the time $\tau \to 0$ and make the ansatz that Eq. (8.5.4a) is to be associated with a 'quantum jump' of photoabsorption at time t. Then if we consider a process in which n counts are observed at times $t_1, t_2, ..., t_n$ with no counts in between these times, we have the conditional density matrix

$$\begin{aligned}
\rho^{(n)} = &\left[e^{-S(t-t_n)}ae^{-S(t_n-t_{n-1})}...\,ae^{-S(t_2-t_1)}ae^{-St_1}\right.\\
&\left.\times\rho(0)e^{-St_1}a^\dagger e^{-S(t_2-t_1)}a^\dagger e^{-S(t_n-t_{n-1})}a^\dagger e^{-S(t-t_n)}\right]\\
&/\mathrm{Tr}, \qquad (8.5.5)
\end{aligned}$$

where $S = Ra^\dagger a$ and the trace factor in the denominator is the normalization factor. This may be simplified by taking account of the fact that, e.g.,

$$\begin{aligned}
e^{-S(t_2-t_1)}ae^{-St_1} &= e^{-Ra^\dagger a(t_2-t_1)}ae^{-Ra^\dagger at_1}\\
&= e^{-Ra^\dagger a(t_2-t_1)}e^{-Ra^\dagger at_1}e^{Ra^\dagger at_1}ae^{-Ra^\dagger at_1}\\
&= e^{-Ra^\dagger at_2}e^{-Rt_1}a, \qquad (8.5.6)
\end{aligned}$$

which may be used repeatedly to reduce Eq. (8.5.5) to the simple form

$$\rho^{(n)}(t) = \frac{e^{-Ra^\dagger at}a^n\rho(0)a^{\dagger n}e^{-Ra^\dagger at}}{\mathrm{Tr}[\rho(0)a^{\dagger n}e^{-2Ra^\dagger at}a^n]}, \qquad (8.5.7)$$

where the various factors of $\exp(-Rt_1)$ are canceled by the normalization. Equation (8.5.7) (and its generalizations) is the main result of this section. In particular, if we consider $\rho(0)$ to be a pure case density matrix $\rho(0) = |\psi(0)\rangle\langle\psi(0)|$, then Eq. (8.5.7) may be written as

$$\rho^{(n)}(t) = \frac{e^{-Ra^\dagger at}a^n|\psi(0)\rangle}{\sqrt{\langle\psi(0)|a^{\dagger n}e^{-2Ra^\dagger at}a^n|\psi(0)\rangle}} \frac{\langle\psi(0)|a^{\dagger n}e^{-Ra^\dagger at}}{\sqrt{\langle\psi(0)|a^{\dagger n}e^{-2Ra^\dagger at}a^n|\psi(0)\rangle}}. \tag{8.5.8}$$

Equation (8.5.8) provides a natural introduction to the wave function approach to dissipative processes.

8.5.2 The wave function Monte Carlo approach to damping

Motivated by the result of the previous section, i.e., Eq. (8.5.8), we present here a short account of damping via a wave function approach. In order to present the ideas we will continue to consider the simple problem of a damped single-mode field, but we will have a more general reservoir, such as that in Section 8.3, in mind. Thus, the decay rate R is no longer governed by the time τ but by the much shorter reservoir correlation times. From Eq. (8.5.8) we are led to write the 'conditional state vector'

$$|\psi^{(n)}(t+\delta t)\rangle = \frac{e^{-Ra^\dagger a\delta t}a^n|\psi(t)\rangle}{\sqrt{\langle\psi(t)|a^{\dagger n}e^{-2Ra^\dagger a\delta t}a^n|\psi(t)\rangle}}, \tag{8.5.9}$$

which represents the state of the field under the condition of n photons absorbed in time δt starting from $|\psi(t)\rangle$. In particular, the state involving only zero or one such event is of special interest. That is, the state at time $t + \delta t$ for $n = 0$ (a null measurement) is

$$\begin{aligned} |\psi^{(0)}(t+\delta t)\rangle &= \frac{e^{-Ra^\dagger a\delta t}|\psi(t)\rangle}{\sqrt{\langle\psi(t)|e^{-2Ra^\dagger a\delta t}|\psi(t)\rangle}} \\ &\cong \frac{(1 - Ra^\dagger a\delta t)}{\sqrt{1 - 2R\langle a^\dagger a\rangle\delta t}}|\psi(t)\rangle, \end{aligned} \tag{8.5.10a}$$

where $\langle a^\dagger a\rangle = \langle\psi(t)|a^\dagger a|\psi(t)\rangle$; and the state corresponding to $n = 1$ (quantum jump) is

$$\begin{aligned} |\psi^{(1)}(t+\delta t)\rangle &= \frac{e^{-Ra^\dagger a\delta t}a|\psi(t)\rangle}{\sqrt{\langle\psi(t)|a^\dagger e^{-2Ra^\dagger a\delta t}a|\psi(t)\rangle}} \\ &\cong \frac{a}{\sqrt{\langle a^\dagger a\rangle}}|\psi(t)\rangle. \end{aligned} \tag{8.5.10b}$$

For example, if the initial quantum state for the field mode is

$$|\psi(t)\rangle = c_0(t)|0\rangle + c_1(t)|1\rangle, \tag{8.5.11}$$

then (8.5.10a) and (8.5.10b) imply the conditional state vectors

$$|\psi^{(0)}(t+\delta t)\rangle = \frac{c_0|0\rangle + c_1(1-R\delta t)|1\rangle}{\sqrt{1-2c_1^*c_1R\delta t}}, \tag{8.5.12a}$$

and

$$|\psi^{(1)}(t+\delta t)\rangle = \frac{c_1}{\sqrt{|c_1|^2}}|0\rangle. \tag{8.5.12b}$$

However, we want to describe the evolution from $|\psi(t)\rangle$ as given by Eq. (8.5.11) to a general state at later times which must become $|0\rangle$ eventually. During the time δt the unnormalized 'no count' or ' null measurement' $|\tilde{\psi}\rangle$ is seen from Eq. (8.5.10a) to obey the equation of motion

$$\frac{|\tilde{\psi}^{(0)}(t+\delta t)\rangle - |\tilde{\psi}^{(0)}(t)\rangle}{\delta t} = -Ra^\dagger a|\tilde{\psi}^{(0)}(t)\rangle, \tag{8.5.13a}$$

that is

$$\frac{d}{dt}|\tilde{\psi}^{(0)}(t)\rangle = -\frac{i}{\hbar}(-i\hbar Ra^\dagger a)|\tilde{\psi}^{(0)}(t)\rangle. \tag{8.5.13b}$$

Thus we are motivated to describe the time evolution of the unnormalized state vector for the case of no absorption by a nonunitary Schrödinger equation

$$\frac{d}{dt}|\tilde{\psi}^{(0)}(t)\rangle = -\frac{i}{\hbar}\mathscr{V}_1|\tilde{\psi}^{(0)}(t)\rangle \tag{8.5.14}$$

governed by the non-Hermitian Hamiltonian

$$\mathscr{V}_1 = -i\hbar Ra^\dagger a. \tag{8.5.15}$$

The temporal development implied by Eq. (8.5.14) is, of course, interrupted by quantum jumps or collapses of the wave function at random times. When such a collapse occurs, the state is given by $|0\rangle$. This happens only once, from that time on the field is in the vacuum state. Continuing with our simple example, according to Eq. (8.5.14) the unnormalized state vector

$$|\tilde{\psi}(t)\rangle = \tilde{c}_0(t)|0\rangle + \tilde{c}_1(t)|1\rangle \tag{8.5.16}$$

obeys the simple equations of motion

$$\dot{\tilde{c}}_0(t) = 0, \tag{8.5.17a}$$

$$\dot{\tilde{c}}_1(t) = -R\tilde{c}_1(t), \tag{8.5.17b}$$

which imply

$$\tilde{c}_0(t) = \tilde{c}_0(0), \tag{8.5.18a}$$

$$\tilde{c}_1(t) = \tilde{c}_1(0)e^{-Rt}, \tag{8.5.18b}$$

and the corresponding normalized probability amplitudes

$$c_0(t) = \frac{c_0(0)}{\sqrt{|c_0(0)|^2 + |c_1(0)|^2 e^{-2Rt}}}, \tag{8.5.19a}$$

and

$$c_1(t) = \frac{c_1(0)e^{-Rt}}{\sqrt{|c_0(0)|^2 + |c_1(0)|^2 e^{-2Rt}}}. \tag{8.5.19b}$$

Thus we have the complete coherent evolution for the conditional state vector up to the point of collapse,

$$|\psi^{(0)}(t)\rangle = \frac{c_0(0)|0\rangle + c_1(0)e^{-Rt}|1\rangle}{\sqrt{|c_0(0)|^2 + |c_1(0)|^2 e^{-2Rt}}}. \tag{8.5.20}$$

Note that as $t \to \infty$ the state $|\psi^{(0)}(t)\rangle \to |0\rangle$. This is as it should be since the conditional state $|\psi^{(0)}(t)\rangle$ is that state which is conditioned on the premise that no photons are absorbed. Hence if after a long time we never see a 'count', then the conclusion is that we must have been in the vacuum state, $|0\rangle$, all along. To summarize: the field develops from $t = 0$ up to some time t according to Eq. (8.5.14), and between t and $t + \delta t$ a jump occurs, that is

$$|\psi(0)\rangle = c_0(0)|0\rangle + c_1(0)|1\rangle \tag{8.5.21a}$$

$$\Big\downarrow \text{ `no counts' from } 0 \to t$$

$$|\psi(t)\rangle = c_0(t)|0\rangle + c_1(t)|1\rangle \tag{8.5.21b}$$

$$\Big\downarrow \text{ collapse } t \to t + \delta t$$

$$\begin{aligned} |\psi(t+\delta t)\rangle &= \frac{a}{\sqrt{\langle\psi(t)|a^\dagger a|\psi(t)\rangle}}|\psi(t)\rangle \\ &= |0\rangle, \end{aligned} \tag{8.5.21c}$$

where $c_0(t)$ and $c_1(t)$ in (8.5.21b) are given by Eqs. (8.5.19a) and (8.5.19b) and Eq. (8.5.21c) follows from Eq. (8.5.10b). Now we recall that the probability of a collapse or jump at time t is governed by the density matrix conditional upon a single photon absorption, i.e., a 'count'. With that in mind, we write Eq. (8.3.3) for $R = \mathscr{C}/2$ as

$$\begin{aligned} \dot{\rho} &= -R\left(a^\dagger a\rho + \rho a^\dagger a\right) + 2Ra\rho a^\dagger \\ &= \underbrace{-\frac{i}{\hbar}\left(\mathscr{V}_1\rho - \rho\mathscr{V}_1^\dagger\right)} + \underbrace{2Ra\rho a^\dagger} \\ &= \quad \dot{\rho}(\text{no count}) \quad + \quad \dot{\rho}(\text{count}). \end{aligned} \tag{8.5.22}$$

Hence the probability for a collapse between t and $t + \delta t$ is given by

$$\begin{aligned}\mathrm{Tr}[\dot{\rho}(\mathrm{count})]\delta t &= 2R\delta t \mathrm{Tr}[\rho(t)a^\dagger a] \\ &= 2R\delta t \langle\psi(t)|a^\dagger a|\psi(t)\rangle \\ &= 2R\delta t \frac{\langle\tilde{\psi}(t)|a^\dagger a|\tilde{\psi}(t)\rangle}{\langle\tilde{\psi}(t)|\tilde{\psi}(t)\rangle}.\end{aligned} \tag{8.5.23}$$

Therefore, from Eqs. (8.5.16)–(8.5.19b) and (8.5.23), we have the jump probability for our present problem

$$P_{\mathrm{jump}}(t) = 2R\delta t \frac{|c_1(0)|^2 e^{-2Rt}}{|c_0(0)|^2 + |c_1(0)|^2 e^{-2Rt}}. \tag{8.5.24}$$

Finally we turn the above into a plot of the probability of finding a photon in the cavity after a time t given that $c_0(0) = 0$ and $c_1(0) = 1$. Then $P_{\mathrm{jump}}(t) = 2R\delta t$. This we do via a Monte Carlo procedure as follows. First, we start the field in state $|1\rangle$ with $c_1(0) = 1$ and we choose a number between 0 and 1 using a computer random number generator. If the number is smaller than $P_{\mathrm{jump}}(0)$, then a jump or collapse is taken to have occurred, and the photon number is set to zero. Most likely, however, the number will be larger than P_{jump} and we reevaluate $|\psi(t)\rangle$ from (8.5.20) and start again. We repeat this n times until a random number turns up which is smaller than $P_{\mathrm{jump}}(t)$ given by (8.5.24). At that point we make an entry in our table as follows:

$$\begin{array}{ll} t = 0 & |\psi(0)\rangle = c_0(0)|0\rangle + c_1(0)|1\rangle \\ & \quad\Big\downarrow \text{evolve} \\ t = \delta t & |\psi(\delta t)\rangle = c_0(\delta t)|0\rangle + c_1(\delta t)|1\rangle \\ & \quad\Big\downarrow \text{evolve} \\ t = 2\delta t & |\psi(2\delta t)\rangle = c_0(2\delta t)|0\rangle + c_1(2\delta t)|1\rangle \\ & \quad\vdots \\ & \quad\Big\downarrow \text{evolve} \\ t = n\delta t & |\psi(n\delta t)\rangle = c_0(n\delta t)|0\rangle + c_1(n\delta t)|1\rangle \\ & \quad\Big\downarrow \text{collapse} \\ t = (n+1)\delta t & |\psi[(n+1)\delta t]\rangle = |0\rangle. \end{array} \tag{8.5.25}$$

Needless to say, the preceding simple example was chosen for pedagogical purposes. Many more involved problems can be and have been solved by the quantum jump–Monte Carlo approach. These include spontaneous emission, resonance fluorescence, Doppler cooling, population trapping, and the dark line resonance, to name a few.

In conclusion we note that the approach of the present section is often referred to as the 'quantum trajectory method'. We also point to the interesting work of Willis Lamb in which the trajectories of Gaussian wave packets are calculated in order to treat the quantum theory of certain problems dealing with the measurement process. This work also uses a computer analysis to characterize the (random) outcomes of the experiment.

Problems

8.1 Derive Eqs. (8.2.14a) and (8.2.14b) and use these results to evaluate the correlation functions (8.2.16a)–(8.2.16e).

8.2 The equation of motion for the reduced density operator for a single-mode cavity field coupled to a vacuum reservoir through a partially transmitting mirror is

$$\dot{\rho} = -\frac{\mathscr{C}}{2}(a^\dagger a\rho - 2a\rho a^\dagger + \rho a^\dagger a).$$

Here $\mathscr{C}$ is the loss rate related to the Q-factor of the cavity by $\mathscr{C} = \nu/Q$. Derive the equations of motion for the relevant quantities, and then solve them to show that the variances $(\Delta X_1)_t^2$ and $(\Delta X_2)_t^2$ (with $X_1 = (a+a^\dagger)/2$ and $X_2 = (a-a^\dagger)/2i$) increase due to dissipation (fluctuation–dissipation theorem!). This situation can be viewed as a bosonic mode, uncorrelated to the cavity field, entering the cavity through the partially transmitting mirror, and hence adding the uncorrelated noise.

8.3 If the reservoir in the above problem is in a multi-mode squeezed vacuum state, the resulting equation of motion for the reduced density matrix is given by Eq. (8.3.4). As before, calculate the variances $(\Delta X_1)_t^2$ and $(\Delta X_2)_t^2$. Is it possible to suppress the added noise in this situation?

8.4 For a thermal reservoir

$$\dot{\rho} = -\frac{\mathscr{C}}{2}(\bar{n}_{\mathrm{th}} + 1)(a^\dagger a\rho - 2a\rho a^\dagger + \rho a^\dagger a) - \frac{\mathscr{C}}{2}\bar{n}_{\mathrm{th}}(aa^\dagger\rho - 2a^\dagger\rho a + \rho aa^\dagger),$$

where $\bar{n}_{\mathrm{th}}$ is the mean number of photons in the reservoir. Derive the corresponding equation for the Q-representation and solve it.

8.5 Derive Eqs. (8.2.18a)–(8.2.18c).

References and bibliography

Quantum theory of damping and dissipation

M. Lax, *Phys. Rev.* **145**, 110 (1966).

W. Weidlich, H. Risken, and H. Haken, *Z. Phys.* **201**, 369 (1967).

H. Haken, *Laser Theory*, (Springer, Berlin 1970).

W. Louisell, *Quantum Statistical Properties of Radiation*, (Wiley, New York 1974).

M. Sargent III, M. Scully, and W. E. Lamb, Jr., *Laser Physics* (Addison-Wesley, Reading, MA 1974).

Decay in squeezed vacuum

M. J. Collett and C. W. Gardiner, *Phys. Rev. A* **30**, 1386 (1984).

C. W. Gardiner, *Phys. Rev. Lett.* **56**, 1917 (1986).

J. Gea-Banacloche, M. O. Scully, and M. S. Zubairy, *Phys. Scripta* **T21**, 81 (1988).

P-representation for a harmonic oscillator mode damped by a thermal bath

R. J. Glauber, in *Quantum Optics and Electronics*, Les Houches, ed. C. DeWitt, A. Blanch, and C. Cohen-Tannoudji (Gordon and Breach, New York 1965), p. 331.

R. J. Glauber, in *New Techniques and Ideas in Quantum Measurement Theory, Annals of the New York Academy of Sciences*, Vol. 480, ed. D. M. Greenberger (1986), p. 336.

Quantum jumps

H. J. Dehmelt, *Bull. Am. Phys. Soc.* **20**, 60 (1975).

R. J. Cook and H. J. Kimble, *Phys. Rev. Lett.* **54**, 1023 (1985).

C. Cohen-Tanoudji and J. Dalibard, *Europhys. Lett.* **1**, 441 (1986).

W. Nagourney, J. Sandberg, and H. Dehmelt, *Phys. Rev. Lett.* **56**, 2797 (1986).

Th. Sauter, R. Blatt, W. Neuhauser, and P. E. Toschek, *Opt. Commun.* **60**, 287 (1986).

A. Schenzle, R. G. DeVoe, and R. G. Brewer, *Phys. Rev. A* **33**, 2127 (1986).

G. Nienhuis, *Phys. Rev. A* **35**, 4639 (1987).

P. Zoller, M. Marte, and D. F. Walls, *Phys. Rev. A* **35**, 198 (1987).

M. Porrati and S. Puttermann, *Phys. Rev. A* **36**, 925 (1987).

S. Reynaud, J. Dalibard, and C. Cohen-Tannoudji, *IEEE Journal of Quantum Electronics* **24**, 1395 (1988).

J. C. Bergquist, W. M. Itano, R. G. Hulet, and D. J. Wineland, *Phys. Scripta* **T22**, 79 (1988).

General theory of continuous measurements

M. D. Srinivas and E. B. Davies, *Opt. Acta* **28**, 981 (1981).
M. Ueda and M. K. Kitagawa, *Phys. Rev. Lett.* **68**, 3424 (1992).

Monte Carlo wave function approach

J. Dalibard, Y. Castin, and K. Mølmer, *Phys. Rev. Lett.* **68**, 580 (1992).
H. J. Carmichael, *An Open Systems Approach to Quantum Optics* (Springer-Verlag, Berlin 1993).

Quantum theory of measurement and trajectories

W. E. Lamb, Jr., *Physics Today*, **22**, 23 (1969); *Proc. 2nd Int. Symp. Found. of Quant. Mech.* (Tokyo 1986), p. 185.
W. E. Lamb, Jr., in *New Techniques and Ideas in Quantum Measurement Theory, Annals of the New York Academy of Sciences*, Vol. 480, ed. D. M. Greenberger (1986), p. 407.
W. E. Lamb, Jr., in *Quantum Measurement and Chaos*, ed. E. Pike and S. Sarkar (Plenum, New York 1987), p. 183.

CHAPTER 9

Quantum theory of damping – Heisenberg–Langevin approach

In the previous chapter, we developed the equation of motion for a system as it evolved under the influence of an unobserved (reservoir) system. We used the density matrix approach and worked in the interaction picture. In this chapter, we consider the same problem of the system–reservoir interaction using a quantum operator approach. We again eliminate the reservoir variables. The resulting equations for the system operators include, in addition to the damping terms, the *noise* operators which produce fluctuations. These equations have the form of classical Langevin equations, which describe, for example, the Brownian motion of a particle suspended in a liquid. The Heisenberg–Langevin approach discussed in this chapter is particularly suitable for the calculation of two-time correlation functions of the system operator as is, for example, required for the determination of the natural linewidth of a laser.

We first consider the damping of the harmonic oscillator by an interaction with a reservoir consisting of many other simple harmonic oscillators. This system describes, for example, the damping of a single-mode field inside a cavity with lossy mirrors. The reservoir, in this case, consists of a large number of phonon-like modes in the mirrors. We also consider the decay of the field due to its interaction with an atomic reservoir. An interesting application of the theory of the system–reservoir interaction is the evolution of an atom inside a damped cavity. It is shown that the spontaneous transition rate of the atom can be substantially enhanced if it is placed in a resonant cavity.

9.1 Simple treatment of damping via oscillator reservoir: Markovian white noise

We consider a system consisting of a single-mode field of frequency ν and annihilation operator $a(t)$ interacting with a reservoir. The reservoir may be taken as any large collection of systems with many degrees of freedom. We assume that the reservoir consists of many oscillators (e.g., phonons, other photon modes, etc) with closely spaced frequencies ν_k and annihilation (and creation) operators $b_{\mathbf{k}}$ (and $b_{\mathbf{k}}^\dagger$). This system therefore describes the damping of a harmonic oscillator by an interaction with a reservoir consisting of many other simple harmonic oscillators. The field–reservoir system evolves in time under the influence of the total Hamiltonian

$$\mathscr{H} = \mathscr{H}_0 + \mathscr{H}_1, \tag{9.1.1}$$

$$\mathscr{H}_0 = \hbar\nu a^\dagger a + \sum_{\mathbf{k}} \hbar\nu_k b_{\mathbf{k}}^\dagger b_{\mathbf{k}}, \tag{9.1.2}$$

$$\mathscr{H}_1 = \hbar \sum_{\mathbf{k}} g_{\mathbf{k}}(b_{\mathbf{k}}^\dagger a + a^\dagger b_{\mathbf{k}}). \tag{9.1.3}$$

As before, $\mathscr{H}_0$ consists of the energy of the free field and the reservoir modes, and $\mathscr{H}_1$ is the interaction energy. The field operators commute with the reservoir operators at a given time. We note that in Eq. (9.1.3) we have here made the usual rotating wave approximation.

The Heisenberg equations of motion for the operators are

$$\dot{a} = \frac{i}{\hbar}[\mathscr{H}, a] = -i\nu a(t) - i\sum_{\mathbf{k}} g_{\mathbf{k}} b_{\mathbf{k}}(t), \tag{9.1.4}$$

$$\dot{b}_{\mathbf{k}} = -i\nu_k b_{\mathbf{k}}(t) - i g_{\mathbf{k}} a(t). \tag{9.1.5}$$

We are interested in a closed equation for the harmonic oscillator operator $a(t)$. The equation for the reservoir operator $b_{\mathbf{k}}(t)$ can be formally integrated to yield

$$b_{\mathbf{k}}(t) = b_{\mathbf{k}}(0)e^{-i\nu_k t} - i g_{\mathbf{k}} \int_0^t dt' a(t') e^{-i\nu_k(t-t')}. \tag{9.1.6}$$

Here the first term represents the free evolution of the reservoir modes, whereas the second term arises from their interaction with the harmonic oscillator. The reservoir operators $b_{\mathbf{k}}(t)$ can be eliminated by substituting the formal solution of $b_{\mathbf{k}}(t)$ into Eq. (9.1.4). We find

$$\dot{a} = -i\nu a - \sum_{\mathbf{k}} g_{\mathbf{k}}^2 \int_0^t dt' a(t') e^{-i\nu_k(t-t')} + f_a(t), \tag{9.1.7}$$

$$f_a(t) = -i\sum_{\mathbf{k}} g_{\mathbf{k}} b_{\mathbf{k}}(0) e^{-i\nu_k t}. \tag{9.1.8}$$

In Eq. (9.1.7), $f_a(t)$ is a noise operator because it depends upon the reservoir operators $b_{\mathbf{k}}(0)$. The evolution of the expectation values involving the harmonic oscillator operator will therefore depend upon the fluctuations in the reservoir. The noise operator varies rapidly due to the presence of all the reservoir frequencies. The fast frequency dependence of $a(t)$ can be removed by transforming to the slowly varying annihilation operator

$$\tilde{a}(t) = a(t)e^{i\nu t}. \tag{9.1.9}$$

We see that

$$[\tilde{a}(t), \tilde{a}^\dagger(t)] = 1, \tag{9.1.10}$$

and Eq. (9.1.7) reduces to

$$\dot{\tilde{a}} = -\sum_{\mathbf{k}} g_{\mathbf{k}}^2 \int_0^t dt' \tilde{a}(t') e^{-i(\nu_k - \nu)(t-t')} + F_{\tilde{a}}(t), \tag{9.1.11}$$

$$F_{\tilde{a}}(t) = e^{i\nu t} f_a(t) = -i \sum_{\mathbf{k}} g_{\mathbf{k}} b_{\mathbf{k}}(0) e^{-i(\nu_k - \nu)t}. \tag{9.1.12}$$

The time integration in Eq. (9.1.11) is similar to that encountered in the Weisskopf–Wigner theory discussed in Section 6.3. As in the Weisskopf–Wigner approximation, the summation in Eq. (9.1.11) yields a $\delta(t - t')$ function and the integration can then be carried out. We obtain

$$\sum_{\mathbf{k}} g_{\mathbf{k}}^2 \int_0^t dt' \tilde{a}(t') e^{-i(\nu_k - \nu)(t-t')} \simeq \frac{1}{2}\mathscr{C}\tilde{a}(t), \tag{9.1.13}$$

where the damping constant

$$\mathscr{C} = 2\pi[g(\nu)]^2 D(\nu). \tag{9.1.14}$$

Here, $g(\nu) \equiv g_{\nu/c}$ is the coupling constant evaluated at $k = \nu/c$ and $D(\nu) = V\nu^2/\pi^2 c^3$ (with V being the quantization volume) is the density of states (see Eq. (1.1.26)). We can therefore replace Eq. (9.1.11) by the Langevin equation

$$\dot{\tilde{a}} = -\frac{1}{2}\mathscr{C}\tilde{a} + F_{\tilde{a}}(t), \tag{9.1.15}$$

where $F_{\tilde{a}}(t)$ is the noise operator which depends on the reservoir variables.

It is interesting to note that the presence of the noise operator in Eq. (9.1.15) is necessary to preserve the commutation relation (9.1.10) at all times. In the absence of the noise term ($F_{\tilde{a}}(t) = 0$), Eq. (9.1.15) can be solved and we get

$$\tilde{a}(t) = \tilde{a}(0)e^{-\mathscr{C}t/2}. \tag{9.1.16}$$

If the operator $\tilde{a}$ satisfies the commutation relation (9.1.10) at $t = 0$, then

$$[\tilde{a}(t), \tilde{a}^\dagger(t)] = e^{-\mathscr{C}t}, \tag{9.1.17}$$

representing a violation of the commutation relation. The noise operator with appropriate correlation properties helps to maintain the commutation relation (9.1.10) at all times. The presence of the noise term along with the damping term in Eq. (9.1.15) is a manifestation of the fluctuation–dissipation theorem of statistical mechanics, i.e., dissipation is always accompanied by fluctuations.

We suppose that the reservoir is in thermal equilibrium, so that

$$\langle b_{\mathbf{k}}(0)\rangle_R = \langle b_{\mathbf{k}}^\dagger(0)\rangle_R = 0, \tag{9.1.18}$$

$$\langle b_{\mathbf{k}}^\dagger(0) b_{\mathbf{k}'}(0)\rangle_R = \delta_{\mathbf{k}\mathbf{k}'}\bar{n}_{\mathbf{k}}, \tag{9.1.19}$$

$$\langle b_{\mathbf{k}}(0) b_{\mathbf{k}'}^\dagger(0)\rangle_R = (\bar{n}_{\mathbf{k}} + 1)\delta_{\mathbf{k}\mathbf{k}'}, \tag{9.1.20}$$

$$\langle b_{\mathbf{k}}(0) b_{\mathbf{k}'}(0)\rangle_R = \langle b_{\mathbf{k}}^\dagger(0) b_{\mathbf{k}'}^\dagger(0)\rangle_R = 0. \tag{9.1.21}$$

Using these relations with the noise operator value (9.1.12), we can evaluate various first- and second-order correlation functions involving $F_{\tilde{a}}(t)$ as follows:

(a) It follows trivially from Eq. (9.1.18) that the reservoir averages of $F_{\tilde{a}}(t)$ and its adjoint $F_{\tilde{a}}^\dagger(t)$ vanish, i.e.,

$$\langle F_{\tilde{a}}(t)\rangle_R = \langle F_{\tilde{a}}^\dagger(t)\rangle_R = 0. \tag{9.1.22}$$

(b) On using Eq. (9.1.19) we obtain

$$\begin{aligned}\langle F_{\tilde{a}}^\dagger(t) F_{\tilde{a}}(t')\rangle_R &= \sum_{\mathbf{k}}\sum_{\mathbf{k}'} g_{\mathbf{k}} g_{\mathbf{k}'} \langle b_{\mathbf{k}}^\dagger b_{\mathbf{k}'}\rangle_R \exp[i(\nu_k - \nu)t - i(\nu_{k'} - \nu)t'] \\ &= \sum_{\mathbf{k}} g_{\mathbf{k}}^2 \bar{n}_{\mathbf{k}} \exp[i(\nu_k - \nu)(t - t')] \\ &= \int_0^\infty D(\nu_k)[g(\nu_k)]^2 \bar{n}(\nu_k) e^{i(\nu_k - \nu)(t-t')} d\nu_k. \end{aligned} \tag{9.1.23}$$

In the last line, we have gone from a discrete representation to a continuous representation in the usual way. We can now pull out the slowly varying terms $D(\nu_k), g(\nu_k)$, and $\bar{n}(\nu_k)$ at $\nu_k = \nu$ and replace the integral by a δ-function. This gives

$$\langle F_{\tilde{a}}^\dagger(t) F_{\tilde{a}}(t')\rangle_R = \mathscr{C}\bar{n}_{\text{th}}\delta(t - t'), \tag{9.1.24}$$

where $\mathscr{C}$ is given by Eq. (9.1.14) and $\bar{n}_{\text{th}} = \bar{n}(\nu_k)$. In analogy with the classical Langevin theory, we define the diffusion coefficient $D_{\tilde{a}^\dagger \tilde{a}}$ for $\tilde{a}^\dagger \tilde{a}$ through the equation

$$\langle F_{\tilde{a}}^\dagger(t) F_{\tilde{a}}(t') \rangle_R = 2\langle D_{\tilde{a}^\dagger \tilde{a}} \rangle_R \delta(t - t'). \tag{9.1.25}$$

Hence, from Eq. (9.1.24), the diffusion coefficient is given by

$$2\langle D_{\tilde{a}^\dagger \tilde{a}} \rangle_R = \mathscr{C} \bar{n}_{\text{th}}. \tag{9.1.26}$$

In a similar manner, we can show that

$$\langle F_{\tilde{a}}(t) F_{\tilde{a}}^\dagger(t') \rangle_R = \mathscr{C}(\bar{n}_{\text{th}} + 1)\delta(t - t'), \tag{9.1.27}$$

$$\langle F_{\tilde{a}}(t) F_{\tilde{a}}(t') \rangle_R = \langle F_{\tilde{a}}^\dagger(t) F_{\tilde{a}}^\dagger(t') \rangle_R = 0, \tag{9.1.28}$$

so that

$$2\langle D_{\tilde{a}\tilde{a}^\dagger} \rangle_R = \mathscr{C}(\bar{n}_{\text{th}} + 1), \tag{9.1.29}$$

$$\langle D_{\tilde{a}\tilde{a}} \rangle_R = \langle D_{\tilde{a}^\dagger \tilde{a}^\dagger} \rangle_R = 0. \tag{9.1.30}$$

(c)We now determine $\langle F_{\tilde{a}}^\dagger(t) \tilde{a}(t) \rangle_R$. This quantity will be needed below in the derivation of the equation of motion for $\langle \tilde{a}^\dagger \tilde{a} \rangle_R$. It follows, on solving Eq. (9.1.15), that

$$\tilde{a}(t) = \tilde{a}(0) \exp\left(-\frac{\mathscr{C}}{2} t\right) + \int_0^t dt' \exp\left[-\frac{\mathscr{C}}{2}(t - t')\right] F_{\tilde{a}}(t'). \tag{9.1.31}$$

We then obtain

$$\langle F_{\tilde{a}}^\dagger(t) \tilde{a}(t) \rangle_R = \langle F_{\tilde{a}}^\dagger(t) \rangle_R \tilde{a}(0) \exp\left(-\frac{\mathscr{C}}{2} t\right) + \int_0^t dt' \exp\left[-\frac{\mathscr{C}}{2}(t - t')\right] \langle F_{\tilde{a}}^\dagger(t) F_{\tilde{a}}(t') \rangle_R. \tag{9.1.32}$$

Here, we assumed that $F_{\tilde{a}}(t)$ and $\tilde{a}(0)$ are statistically independent. From Eqs. (9.1.22) and (9.1.24), it follows that

$$\langle F_{\tilde{a}}^\dagger(t) \tilde{a}(t) \rangle_R = \frac{\mathscr{C}}{2} \bar{n}_{\text{th}} = \langle D_{\tilde{a}^\dagger \tilde{a}} \rangle_R. \tag{9.1.33}$$

Similarly, we can show that

$$\langle \tilde{a}^\dagger(t) F_{\tilde{a}}(t) \rangle_R = \frac{\mathscr{C}}{2} \bar{n}_{\text{th}}. \tag{9.1.34}$$

These correlation functions will be employed to derive equations of motion for the field correlation functions in Section 9.3. We next consider the damping of a single-mode field via an atomic reservoir and also extend and strengthen the present oscillator reservoir treatment. The main result of these consideration is a correlation function for the noise operator which is not a delta function, thus corresponding to 'colored' noise as opposed to the white noise presented in this section.

9.2 Extended treatment of damping via atom and oscillator reservoirs: non-Markovian colored noise

In this section we extend our approach to the problem of damping, this time involving finite (i.e., not delta function) correlation times. We first assume a field damping mechanism via two-level atoms in thermal distribution, passing through the cavity. The atoms are assumed to be long lived and monoenergetic so that they interact with the field inside the cavity for a fixed duration τ. We then return to the oscillator reservoir model extending the treatment of the oscillator reservoir problem beyond the Markovian limit.

9.2.1 An atomic reservoir approach*

We here consider the damping of a single-mode field by an ensemble of atoms. The Hamiltonian for the present problem is given by

$$\mathscr{H} = \mathscr{H}_0 + \mathscr{H}_1, \tag{9.2.1}$$

$$\mathscr{H}_0 = \hbar\nu a^\dagger a + \frac{1}{2}\hbar\nu \sum_i \sigma_z^i, \tag{9.2.2}$$

$$\mathscr{H}_1 = \hbar g \sum_i [f(t_i, t, \tau) a^\dagger \sigma_-^i + \text{H.c.}], \tag{9.2.3}$$

where σ_z^i and σ_-^i are the operators for the ith atom and $f(t_i, t, \tau)$ is a function which represents the injection of an atom at time t_i and its removal at a later time $t_i + \tau$. In this sense, $f(t_i, t, \tau)$ is a notch function which has the value

$$f(t_i, t, \tau) = \begin{cases} 1 & \text{for } t_i \le t < t_i + \tau, \\ 0 & \text{otherwise}. \end{cases} \tag{9.2.4}$$

For the sake of simplicity, we have assumed that the injected atoms are resonant with the field. Using this Hamiltonian, we write the equations for the field and atom operators in the interaction picture

$$\dot{a}(t) = -ig \sum_i f(t_i, t, \tau) \sigma_-^i(t), \tag{9.2.5}$$

$$\dot{\sigma}_-^i(t) = igf(t_i, t, \tau)\sigma_z^i a(t). \tag{9.2.6}$$

As before, we are interested in a closed equation for the operator $a(t)$. Integration of the atomic operator equation (9.2.6) yields

$$\sigma_-^i(t) = \sigma_-^i(t_i) + ig \int_{t_i}^{t} dt' f(t_i, t', \tau) \sigma_z^i(t') a(t'). \tag{9.2.7}$$

On substituting this expression for $\sigma_-^i(t)$ into the field operator

* The reader should consult Chapter 12 and Scully, Süssmann, and Benkert [1988] for further reading on the material of this section.

equation, we obtain

$$\dot{a} = g^2 \sum_i \int_{t_i}^{t} dt' f(t_i, t, \tau) f(t_i, t', \tau) \sigma_z^i(t') a(t')$$
$$-ig \sum_i f(t_i, t, \tau) \sigma_-^i(t_i). \tag{9.2.8}$$

If the field does not change appreciably during the transit time of the atoms, $a(t')$ in Eq. (9.2.8) can be replaced by $a(t)$. In a linear analysis, $\sigma_z^i(t')$ is also replaced by its value at the time of injection $\sigma_z^i(t_i)$. The resulting equation is

$$\dot{a} = -\frac{1}{2}\mathscr{C} a + F_a(t), \tag{9.2.9}$$

where

$$\mathscr{C} = -2g^2 \sum_i \int_{t_i}^{t} dt' f(t_i, t, \tau) f(t_i, t', \tau) \sigma_z^i(t_i), \tag{9.2.10}$$

$$F_a(t) = -ig \sum_i f(t_i, t, \tau) \sigma_-^i(t_i). \tag{9.2.11}$$

Here the decay constant $\mathscr{C}$ is positive as the inital inversion $\sigma_z^i(t_i)$ is negative in thermal equilibrium.

The noise operator $F_a(t)$ may be seen to have the moments

$$\langle F_a(t) \rangle = 0, \tag{9.2.12}$$

$$\langle F_a^\dagger(t) F_a(t') \rangle = g^2 \sum_{i,j} f(t_i, t, \tau) f(t_j, t', \tau) \langle \sigma_+^i(t_i) \sigma_-^j(t_j) \rangle$$
$$= g^2 [1 + \exp(\hbar\nu / k_B T)]^{-1} \sum_i f(t_i, t, \tau) f(t_i, t', \tau), \tag{9.2.13}$$

where we have used, with the atoms in a thermal equilibrium state at temperature T, (by solving Eqs. (8.2.10a) and (8.2.10c) in the steady state and using Eq. (8.2.5))

$$\langle \sigma_+^i(t_i) \sigma_-^j(t_j) \rangle = \delta_{ij} [1 + \exp(\hbar\nu / k_B T)]^{-1}. \tag{9.2.14}$$

After replacing the sum over i in Eq. (9.2.13) by an integral over the injection time,

$$\sum_i \rightarrow r_a \int_{-\infty}^{t} dt_i, \tag{9.2.15}$$

where r_a is the rate of injection of atoms into the cavity, we find

$$\langle F_a^\dagger(t)F_a(t')\rangle = r_a g^2\left[1+\exp\left(\frac{\hbar\nu}{k_B T}\right)\right]^{-1}\int_{-\infty}^{t} dt_i f(t_i,t,\tau)f(t_i,t',\tau). \tag{9.2.16}$$

The integration can be carried out, for example, by writing

$$f(t_i,t,\tau) = \Theta(t-t_i) - \Theta(t-\tau-t_i), \tag{9.2.17}$$

where Θ is the unit step function and using

$$\int_{-\infty}^{\infty} dt_i \Theta(t_1-t_i)\Theta(t_2-t_i) = \Theta(t_1-t_2)\int_{-\infty}^{t_2} dt_i + \Theta(t_2-t_1)\int_{-\infty}^{t_1} dt_i. \tag{9.2.18}$$

We then obtain

$$\begin{aligned}\int_{-\infty}^{t} dt_i f(t_i,t,\tau)f(t_i,t',\tau) = {} & [\Theta(t-t')-\Theta(t-t'-\tau)]\int_{-\infty}^{t'} dt_i \\ & +[\Theta(t-t')-\Theta(t-t'+\tau)]\int_{-\infty}^{t'-\tau} dt_i \\ & +[\Theta(t'-t)-\Theta(t'-t-\tau)]\int_{-\infty}^{t} dt_i \\ & +[\Theta(t'-t)-\Theta(t'-t+\tau)]\int_{-\infty}^{t-\tau} dt_i.\end{aligned} \tag{9.2.19}$$

A careful examination shows that the right hand-side of Eq. (9.2.19) is zero unless $\tau \geq |t-t'|$ in which case it is equal to $\tau - |t-t'|$. The correlation function (9.2.16) is therefore given by

$$\langle F_a^\dagger(t)F_a(t')\rangle = \begin{cases} \alpha_F(\tau-|t-t'|)/\tau^2 & \text{for } |t-t'|\leq\tau\,, \\ 0 & \text{otherwise}\,, \end{cases} \tag{9.2.20}$$

where $\alpha_F = r_a g^2\tau^2[1+\exp(\hbar\nu/k_B T)]^{-1}$. The correlation function is triangularly shaped as depicted in Fig. 9.1. This is one of the simplest examples of a 'colored' noise problem.

9.2.2 A generalized treatment of the oscillator reservoir problem*

We now present a treatment of the multi-oscillator heat bath problem. For an oscillator of momentum p and coordinate x coupled to a bath of

* This section follows the paper by Ford, Lewis, and O'Connell [1988].

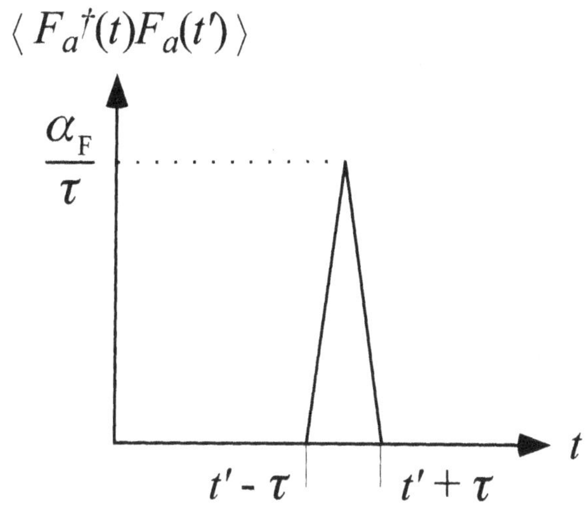

Fig. 9.1
Noise correlation function $\langle F_a^\dagger(t)F_a(t')\rangle$ as given in Eq. (9.2.20).

oscillators having momentum p_j and position q_j, the system–reservoir Hamiltonian can be written as

$$\mathscr{H} = \frac{p^2}{2m} + \frac{1}{2}mv^2x^2 + \sum_j \frac{p_j^2}{2m_j} + \frac{1}{2}m_j\omega_j^2(q_j - x)^2. \tag{9.2.21}$$

Note that in this form the Hamiltonian (9.2.21) does not make the rotating-wave approximation. Including the normal commutation rules $[x, p] = i\hbar$ and $[q_j, p_k] = i\hbar\delta_{jk}$, we find

$$\dot{x} = \frac{1}{i\hbar}[x, \mathscr{H}] = \frac{p}{m}, \tag{9.2.22a}$$

$$\dot{p} = \frac{1}{i\hbar}[p, \mathscr{H}] = -mv^2x + \sum_j m_j\omega_j^2(q_j - x), \tag{9.2.22b}$$

$$\dot{q}_j = \frac{1}{i\hbar}[q_j, \mathscr{H}] = \frac{p_j}{m_j}, \tag{9.2.22c}$$

$$\dot{p}_j = \frac{1}{i\hbar}[p_j, \mathscr{H}] = -m_j\omega_j^2(q_j - x). \tag{9.2.22d}$$

Differentiating Eqs. (9.2.22a) and (9.2.22c) and using Eqs. (9.2.22b) and (9.2.22d), we find

$$\ddot{x}(t) = -v^2x(t) + \sum_j \frac{m_j}{m}\omega_j^2[q_j(t) - x(t)], \tag{9.2.23a}$$

$$\ddot{q}_j(t) = -\omega_j^2[q_j(t) - x(t)]. \tag{9.2.23b}$$

As may be verified by direct substitution, the solution for $q_j(t)$ may be written in the form

$$q_j(t) - x(t) = q_j^0(t) - \int_{-\infty}^{t} dt' \cos\left[\omega_j(t-t')\right] \dot{x}(t'), \tag{9.2.24}$$

where $q_j^0(t)$ is the solution to the problem in the absence of coupling $x = 0$

$$q_j^0(t) = q_j \cos\omega_j t + p_j \frac{\sin\omega_j t}{m_j \omega_j}, \tag{9.2.25}$$

in which q_j and p_j are the usual time-independent position and momentum operators.

Substituting (9.2.24) into (9.2.23a) we find

$$m\ddot{x}(t) + \int_{-\infty}^{t} dt' \mu(t-t')\dot{x}(t') + m\nu^2 x(t) = F(t), \tag{9.2.26}$$

where the damping function is given by

$$\mu(t-t') = \sum_j m_j \omega_j^2 \cos\left[\omega_j(t-t')\right], \tag{9.2.27a}$$

and the noise operator takes the form

$$F(t) = \sum_j m_j \omega_j^2 q_j^0(t). \tag{9.2.27b}$$

As it stands, Eq. (9.2.26) is closely related to Eq. (9.1.11). However, the problem can be extended to include memory effects by writing the following general expression for a damped oscillator

$$m\ddot{x}(t) + \int_{-\infty}^{t} dt' \mu(t-t')\dot{x}(t') + m\nu^2 x = F(t), \tag{9.2.28}$$

where

$$\frac{1}{2}\langle F(t)F(t') + F(t')F(t)\rangle$$
$$= \frac{1}{\pi}\int_0^\infty d\omega \mathrm{Re}\left[\tilde{\mu}(\omega + i0^+)\right] \hbar\omega \coth\left(\frac{\hbar\omega}{2k_B T}\right) \cos\left[\omega(t-t')\right], \tag{9.2.29}$$

with $\tilde{\mu}$ being the Fourier transform of $\mu(t)$.

Now for the case of constant damping, which is the one of most interest to us, $\mathrm{Re}\left[\tilde{\mu}(\omega + i0^+)\right] = \Gamma$ and the correlation function takes the form

$$\frac{1}{2}\langle F(t)F(t') + F(t')F(t)\rangle$$
$$= \frac{\Gamma}{\pi}\int_0^\infty d\omega \hbar\omega \coth\left(\frac{\hbar\omega}{2k_B T}\right) \cos\left[\omega(t-t')\right]$$
$$= \Gamma k_B T \frac{d}{dt} \coth\left[\frac{\pi k_B T(t-t')}{\hbar}\right]. \tag{9.2.30}$$

We note that Eq. (9.2.30), while going to $\delta(t-t')$ in the limit, in general goes beyond the Markovian approximation, i.e., it implies colored noise.

9.3 Equations of motion for the field correlation functions

We can now derive the mean motion of $\tilde{a}(t)$ and of the number operator $\tilde{a}^\dagger\tilde{a}$. Since $\langle F_{\tilde{a}}(t)\rangle_R = 0$, it follows from Eq. (9.1.15), that

$$\frac{d}{dt}\langle\tilde{a}(t)\rangle_R = -\frac{1}{2}\mathscr{C}\langle\tilde{a}(t)\rangle_R. \tag{9.3.1}$$

Here, we see that the mean value of the system operator goes to zero in time. Note that Eq. (9.3.1) is only averaged over the reservoir coordinates It remains an operator in the field coordinates.

The mean time development of the field number operator is

$$\begin{aligned}\frac{d}{dt}\langle\tilde{a}^\dagger(t)\tilde{a}(t)\rangle_R &= \left\langle\frac{d\tilde{a}^\dagger(t)}{dt}\tilde{a}(t)\right\rangle_R + \left\langle\tilde{a}^\dagger(t)\frac{d\tilde{a}(t)}{dt}\right\rangle_R \\ &= -\mathscr{C}\langle\tilde{a}^\dagger(t)\tilde{a}(t)\rangle_R + \langle F_{\tilde{a}}^\dagger(t)\tilde{a}(t)\rangle_R + \langle\tilde{a}^\dagger(t)F_{\tilde{a}}(t)\rangle_R \\ &= -\mathscr{C}\langle\tilde{a}^\dagger(t)\tilde{a}(t)\rangle_R + \mathscr{C}\bar{n}_{\text{th}}.\end{aligned} \tag{9.3.2}$$

Thus, the steady-state value of the number operator $\langle\tilde{a}^\dagger(t)\tilde{a}(t)\rangle_R$ is $\bar{n}_{\text{th}}$ (times the field identity operator); this is nonzero in contrast to $\langle\tilde{a}^\dagger(t)\rangle_R$ and $\langle\tilde{a}(t)\rangle_R$, which decay to zero in time according to Eq. (9.3.1).

In a similar manner, it can be shown that

$$\frac{d}{dt}\langle\tilde{a}(t)\tilde{a}^\dagger(t)\rangle_R = -\mathscr{C}\langle\tilde{a}(t)\tilde{a}^\dagger(t)\rangle_R + \mathscr{C}(\bar{n}_{\text{th}} + 1). \tag{9.3.3}$$

On combining Eqs. (9.3.2) and (9.3.3), we see that the commutator $[\tilde{a}(t), \tilde{a}^\dagger(t)]$ retains its unity reservoir average in time instead of decaying to zero.

Using the same arguments as given for the derivation for the equations of motion for $\langle\tilde{a}(t)\rangle_R$ and $\langle\tilde{a}^\dagger(t)\tilde{a}(t)\rangle_R$, Eqs. (9.3.1) and (9.3.2), we can show that for arbitrary products of the creation and annihilation operators,

$$\frac{d}{dt}\langle(\tilde{a}^\dagger)^m\tilde{a}^n\rangle_R = -\frac{\mathscr{C}}{2}(m+n)\langle(\tilde{a}^\dagger)^m\tilde{a}^n\rangle_R + \mathscr{C}mn\bar{n}_{\text{th}}\langle(\tilde{a}^\dagger)^{m-1}\tilde{a}^{n-1}\rangle_R. \tag{9.3.4}$$

In terms of the operators a and $a^\dagger$ (Eq. (9.1.9)) this equation reads

$$\begin{aligned}\frac{d}{dt}\langle(a^\dagger)^m(a)^n\rangle_R &= \left[i\nu(m-n) - \frac{\mathscr{C}}{2}(m+n)\right]\langle(a^\dagger)^m a^n\rangle_R \\ &\quad + \mathscr{C}mn\bar{n}_{\text{th}}\langle(a^\dagger)^{m-1}a^{n-1}\rangle_R.\end{aligned} \tag{9.3.5}$$

This equation, in a general way, describes the effect of the reservoir.

As mentioned earlier, the present Heisenberg–Langevin approach to the quantum theory of damping is particularly suited for the calculation of multi-time correlation functions. This can be appreciated by considering the simple example of the damping of the field of frequency ν inside the cavity at the rate $\mathscr{C} = \nu/Q$. Here Q is the quality factor of the cavity.

The field operator $\tilde{a}(t) = a(t)\exp(i\nu t)$ obeys the equation

$$\dot{\tilde{a}} = -\frac{\nu}{2Q}\tilde{a} + F_{\tilde{a}}(t), \tag{9.3.6}$$

which can be solved to yield (with $\tau > 0$)

$$\tilde{a}(t_i+\tau) = \tilde{a}(t_i)\exp\left(-\frac{\nu}{2Q}\tau\right) + \int_{t_i}^{t_i+\tau} dt' \exp\left[-\frac{\nu}{2Q}(t_i+\tau-t')\right] F_{\tilde{a}}(t'). \tag{9.3.7}$$

It follows, on using $\langle\tilde{a}^\dagger(t_i)F_{\tilde{a}}(t')\rangle_R = \langle\tilde{a}^\dagger(t_i)\rangle_R\langle F_{\tilde{a}}(t')\rangle_R = 0$, that

$$\langle\tilde{a}^\dagger(t_i)\tilde{a}(t_i+\tau)\rangle_R = \langle\tilde{a}^\dagger(t_i)\tilde{a}(t_i)\rangle_R \exp\left(-\frac{\nu}{2Q}\tau\right), \tag{9.3.8}$$

i.e., the field correlation function decays exponentially with time. The field spectrum can be obtained by taking the Fourier transform of the correlation function

$$\langle a^\dagger(t_i)a(t_i+\tau)\rangle_R = \langle\tilde{a}^\dagger(t_i)\tilde{a}(t_i+\tau)\rangle_R e^{-i\nu\tau} \tag{9.3.9}$$

$$= \langle n\rangle \exp\left(-i\nu\tau - \frac{\nu}{2Q}\tau\right), \tag{9.3.10}$$

where $\langle n\rangle$ is the mean number of photons at the initial time t_i. We then obtain (see Eq. (4.3.14))

$$S(\omega) = \frac{1}{\pi}\mathrm{Re}\int_0^\infty \langle a^\dagger(t)a(t+\tau)\rangle_R e^{i\omega\tau}d\tau = \frac{\langle n\rangle}{\pi}\frac{\nu/2Q}{(\omega-\nu)^2+(\nu/2Q)^2}. \tag{9.3.11}$$

This is a Lorentzian distribution centered at $\omega = \nu$ with half-width $\nu/2Q$.

An approximate expression of the mode density of the empty cavity, $D_c(\omega)$, is obtained by dividing $S(\omega)$ by $\langle n\rangle$, i.e.,

$$D_c(\omega) = \frac{1}{\pi}\frac{\nu/2Q}{(\omega-\nu)^2+(\nu/2Q)^2}. \tag{9.3.12}$$

The density of states inside the cavity is therefore significantly different from its value in free space (see Eq. (1.1.26)).

9.4 Fluctuation–dissipation theorem and the Einstein relation

We now make a connection between the present quantum Langevin approach and the classical approach. In Section 9.1 we derived the second-order correlation function of the Langevin noise $F_{\tilde{a}}(t)$

$$\langle F_{\tilde{a}}^{\dagger}(t)F_{\tilde{a}}(t')\rangle_R = \mathscr{C}\bar{n}_{\text{th}}\delta(t-t'). \tag{9.4.1}$$

On integrating both sides, we obtain

$$\mathscr{C} = \frac{1}{\bar{n}_{\text{th}}}\int_{-\infty}^{\infty}\langle F_{\tilde{a}}^{\dagger}(t)F_{\tilde{a}}(t')\rangle_R dt'. \tag{9.4.2}$$

This states that the system damping $\mathscr{C}$ is determined from the fluctuating forces of the reservoir. Thus the fluctuations induced by the reservoir give rise to dissipation in the system. This is one formulation of the fluctuation–dissipation theorem.

Next we make use of Eqs. (9.1.15) and (9.1.26) to rewrite Eq. (9.3.2) as follows

$$\begin{aligned} 2\langle D_{\tilde{a}^{\dagger}\tilde{a}}\rangle_R = {} & \frac{d}{dt}\langle\tilde{a}^{\dagger}(t)\tilde{a}(t)\rangle_R - \left\langle\left[\frac{d\tilde{a}^{\dagger}}{dt} - F_{\tilde{a}}^{\dagger}(t)\right]\tilde{a}(t)\right\rangle_R \\ & - \left\langle\tilde{a}^{\dagger}(t)\left[\frac{d\tilde{a}}{dt} - F_{\tilde{a}}(t)\right]\right\rangle_R . \end{aligned} \tag{9.4.3}$$

This is the Einstein relation to determine the diffusion constant. We have derived this relation for the damped harmonic oscillator problem. It can, however, be shown that this relation is valid for many general system–reservoir problems. It can be similarly shown that

$$\begin{aligned} 2\langle D_{\tilde{a}\tilde{a}^{\dagger}}\rangle_R = {} & \frac{d}{dt}\langle\tilde{a}(t)\tilde{a}^{\dagger}(t)\rangle_R - \left\langle\tilde{a}(t)\left[\frac{d\tilde{a}^{\dagger}}{dt} - F_{\tilde{a}}^{\dagger}(t)\right]\right\rangle_R \\ & - \left\langle\left[\frac{d\tilde{a}}{dt} - F_{\tilde{a}}(t)\right]\tilde{a}^{\dagger}(t)\right\rangle_R . \end{aligned} \tag{9.4.4}$$

The Einstein relation relates the *drift* terms $[d\tilde{a}/dt - F_{\tilde{a}}(t)]$ and $[d\tilde{a}^{\dagger}/dt - F_{\tilde{a}}^{\dagger}(t)]$ to the diffusion coefficients. In many problems of interest, this relation provides an extremely simple way to calculate the diffusion constant.

The Einstein relation can be employed to determine the diffusion coefficients from the density matrix equations in a straightforward manner. In order to indicate the procedure, we consider the simple example of Eq. (8.3.2) which governs the damping of the field by an interaction with a thermal reservoir. It follows from this equation that

$$\left\langle \frac{da}{dt} \right\rangle = \mathrm{Tr}(a\dot{\rho}) = -\frac{\mathscr{C}}{2}\langle a \rangle, \tag{9.4.5}$$

$$\left\langle \frac{da^\dagger}{dt} \right\rangle = -\frac{\mathscr{C}}{2}\langle a^\dagger \rangle, \tag{9.4.6}$$

$$\frac{d}{dt}\langle a^\dagger a \rangle = -\mathscr{C}(\langle a^\dagger a \rangle - \bar{n}_{\mathrm{th}}), \tag{9.4.7}$$

where, in deriving these equations, we used the cyclic property of the trace (i.e., $\mathrm{Tr}(ABC) = \mathrm{Tr}(CAB)$, etc) and the commutation relation $[a, a^\dagger] = 1$. Now the quantities $[da/dt - F_a(t)]$ and $[da^\dagger/dt - F_a^\dagger(t)]$ can be obtained from Eqs. (9.4.5) and (9.4.6), respectively, by removing the expectation value sign on the right-hand side. We then obtain

$$\left[\frac{da}{dt} - F_a(t) \right] = -\frac{\mathscr{C}}{2} a, \tag{9.4.8}$$

$$\left[\frac{da^\dagger}{dt} - F_a^\dagger(t) \right] = -\frac{\mathscr{C}}{2} a^\dagger. \tag{9.4.9}$$

On substituting Eqs. (9.4.7)–(9.4.9) into Eq. (9.4.3), we get

$$2\langle D_{a^\dagger a} \rangle = \mathscr{C}\bar{n}_{\mathrm{th}}, \tag{9.4.10}$$

in agreement with Eq. (9.1.26).

9.5 Atom in a damped cavity

A very simple application of the mathematical framework developed in this chapter is the study of the evolution of a single two-level atom initially prepared in the upper level $|a\rangle$ of the transition resonant with the cavity mode. In particular, it is seen that the spontaneous emission rate of the atom inside a resonant cavity is substantially enhanced over its free-space value. The enhancement factor can be derived rigorously from a quantum mechanical analysis where the cavity damping is considered via interaction of the single-mode field with a reservoir consisting of a large number of simple harmonic oscillators. First, we present an heuristic argument to understand this interesting phenomenon.

We recall that, in Section 6.3, we considered the spontaneous emission of an atom in free space, so that the atom interacts with a continuum of modes of the electromagnetic field. The decay rate Γ, as given by Eq. (6.3.14) can be rewritten as

$$\Gamma = 2\pi\langle|g(\omega)|^2\rangle D(\omega), \tag{9.5.1}$$

where angle brackets represent an angular average, $g(\omega)$ is the vacuum Rabi frequency, and $D(\omega) = V\omega^2/\pi^2c^3$ is the density of states at the atomic transition frequency ω. The spontaneous decay rate is therefore proportional to the density of states. The mode structure of the vacuum field is dramatically altered in a cavity whose size is comparable to the wavelength. In a cavity of quality factor Q, the mode density $D_c(\omega)$ can be approximated by the Lorentzian (Eq. (9.3.12))

$$D_c(\omega) = \frac{1}{\pi}\frac{\nu/2Q}{(\omega-\nu)^2+(\nu/2Q)^2}. \tag{9.5.2}$$

The spontaneous decay rate of the atom inside the cavity is therefore obtained by replacing $D(\omega)$ by $D_c(\omega)$ in Eq. (9.5.1)

$$\Gamma_c = 2\pi\langle|g(\omega)|^2\rangle D_c(\omega). \tag{9.5.3}$$

For a cavity tuned near the atomic resonance frequency, we have $D_c(\omega) \simeq 2Q/\pi\omega$ and

$$\Gamma_c = \frac{2\pi}{3}\left(\frac{\omega\wp_{ab}^2}{2\hbar\epsilon_0 V}\right)\left(\frac{2Q}{\pi\omega}\right) = \Gamma Q\left(\frac{2\pi c^3}{V\omega^3}\right). \tag{9.5.4}$$

Thus, apart from the geometrical factor of order unity (for the lowest cavity mode $\omega = \pi c/L$, where L is the length of the side of the cavity, this factor is equal to $2/\pi^2$), the spontaneous decay rate inside the cavity is enhanced by a factor Q over its free-space value.

Another simple interpretation of the spontaneous emission enhancement can be given in terms of the image charges. We can simulate the effect of the cavity mirrors on the evolution of the atom by replacing them by the Q images of the atoms in these mirrors. As the cavity is resonant with the atomic transition, all the dipoles of these images are in phase with the atomic dipole. They therefore act as Q aligned antenna in phase. A given antenna in this array radiates Q times faster than an isolated antenna. The atomic energy is therefore dissipated Q times faster than in free space.

We now turn to a rigorous derivation of the atomic decay in a damped cavity. We consider a system of a two-level atom interacting with a single-mode electromagnetic field inside a cavity. The cavity is coupled to a thermal reservoir through the walls of the cavity. The atom–field reservoir Hamiltonian is therefore

$$\mathscr{H} = \mathscr{H}_F + \mathscr{H}_A + \mathscr{H}_{AF} + \mathscr{H}_R + \mathscr{H}_{FR}, \tag{9.5.5}$$

$$\mathscr{H}_F = \hbar\nu a^\dagger a, \tag{9.5.6}$$

$$\mathscr{H}_A = \frac{1}{2}\hbar\nu\sigma_z, \tag{9.5.7}$$

$$\mathscr{H}_{AF} = \hbar g(\sigma_+ a + a^\dagger\sigma_-), \tag{9.5.8}$$

$$\mathscr{H}_R = \sum_{\mathbf{k}} \hbar\nu_k b_{\mathbf{k}}^\dagger b_{\mathbf{k}}, \tag{9.5.9}$$

$$\mathscr{H}_{FR} = \hbar\sum_{\mathbf{k}} g_{\mathbf{k}}(b_{\mathbf{k}}^\dagger a + a^\dagger b_{\mathbf{k}}). \tag{9.5.10}$$

Here $\mathscr{H}_F$ and $\mathscr{H}_A$ are the free field and atom Hamiltonians, respectively, $\mathscr{H}_{AF}$ represents the interaction of the single-mode cavity field with the atom, $\mathscr{H}_R$ is the energy of the reservoir modes and $\mathscr{H}_{FR}$ represents the interaction of the field with the reservoir. For transmission losses, the reservoir modes correspond to the vacuum modes that enter the cavity through partially transmitting mirrors. We shall assume the reservoir modes to be in thermal equilibrium at temperature T.

The quantities of interest in the system are the energy of the field $\langle a^\dagger a\rangle$ and the atomic inversion $\langle\sigma_z\rangle$. The equation of motion for any operator of the form $(a^\dagger)^m a^n O_A$, (where O_A is an atomic operator, e.g., $\sigma_+, \sigma_-, \sigma_z$) is given by

$$\frac{d}{dt}[(a^\dagger)^m a^n O_A] = -\frac{i}{\hbar}[(a^\dagger)^m a^n O_A, \mathscr{H}_F + \mathscr{H}_A + \mathscr{H}_{AF}] + \left\langle \frac{d}{dt}[(a^\dagger)^m a^n] \right\rangle_R O_A, \tag{9.5.11}$$

where $\langle d[(a^\dagger)^m a^n]/dt\rangle_R$ is given by Eq. (9.3.5). Using this equation, we can derive the following equations of motion for $\langle a^\dagger a\rangle$ and $\langle\sigma_z\rangle$:

$$\frac{d\langle a^\dagger a\rangle}{dt} = ig\langle\sigma_+ a - a^\dagger\sigma_-\rangle - \mathscr{C}\langle a^\dagger a\rangle + \mathscr{C}\bar{n}_{\text{th}}, \tag{9.5.12}$$

$$\frac{d\langle\sigma_z\rangle}{dt} = -2ig\langle\sigma_+ a - a^\dagger\sigma_-\rangle. \tag{9.5.13}$$

The angle brackets denote the reservoir as well as the quantum mechanical average. These equations involve the average of the Hermitian operator $\langle\sigma_+ a - \sigma_- a^\dagger\rangle$ whose equation of motion in turn involves the quantity $\langle a^\dagger\sigma_z a\rangle$ and so on. In general, we get an infinite set of equations which may not be analytically solvable. However, the situation is considerably simpler if initially the atom is in the excited state $|a\rangle$, the field inside the cavity is in the vacuum state $|0\rangle$, and the cavity is at zero temperature ($\bar{n}_{th} = 0$). There can be at most one photon in the field and the state of the field inside the cavity at any time t will be a linear superposition of the vacuum state $|0\rangle$ and the one-photon state $|1\rangle$. The expectation value of the operators involving quadratic or higher powers in the field operators a and $a^\dagger$, e.g., $\langle (a^\dagger)^2 \sigma_z a^2 \rangle$, are therefore zero at all times. Under these conditions, we obtain the following closed set of equations

$$\frac{d\langle a^\dagger a\rangle}{dt} = gA_1 - \mathscr{C}\langle a^\dagger a\rangle, \tag{9.5.14}$$

$$\frac{d\langle \sigma_z\rangle}{dt} = -2gA_1, \tag{9.5.15}$$

$$\frac{dA_1}{dt} = g\langle \sigma_z\rangle + 2gA_2 + g - \frac{\mathscr{C}}{2}A_1, \tag{9.5.16}$$

$$\frac{dA_2}{dt} = -gA_1 - \mathscr{C}A_2, \tag{9.5.17}$$

where

$$A_1 = i\langle \sigma_+ a - a^\dagger \sigma_-\rangle, \tag{9.5.18}$$

$$A_2 = \langle a^\dagger \sigma_z a\rangle. \tag{9.5.19}$$

It may be noted that, in Eq. (9.5.17), we neglected the term proportional to $\langle \sigma_+ a^\dagger a^2 - (a^\dagger)^2 a \sigma_-\rangle$ in light of the above argument. The four equations (9.5.14)–(9.5.17) can be solved using, for example, the Laplace transform method. The resulting solutions for $\langle a^\dagger a\rangle_t$ and $\langle \sigma_z\rangle_t$, subject to the initial conditions $\langle a^\dagger a\rangle_0 = A_1(0) = A_2(0) = 0$ and $\langle \sigma_z\rangle_0 = 1$ are

$$\langle a^\dagger a\rangle_t = -\frac{8g^2 e^{-\mathscr{C}t/2}}{\mathscr{C}^2 - 16g^2}\left\{1 - \cosh\left[(\mathscr{C}^2 - 16g^2)^{1/2} t/2\right]\right\}, \tag{9.5.20}$$

$$\begin{aligned}\langle \sigma_z\rangle_t = -1 + \frac{4e^{-\mathscr{C}t/2}}{(\mathscr{C}^2 - 16g^2)}\Bigg\{ &- 4g^2 \\ &+ \left[\frac{\mathscr{C}^2}{4} - 2g^2 + \frac{\mathscr{C}}{4}(\mathscr{C}^2 - 16g^2)^{1/2}\right] \times e^{(\mathscr{C}^2 - 16g^2)^{1/2} t/2} \\ &+ \left[\frac{\mathscr{C}^2}{4} - 2g^2 - \frac{\mathscr{C}}{4}(\mathscr{C}^2 - 16g^2)^{1/2}\right] \times e^{-(\mathscr{C}^2 - 16g^2)^{1/2} t/2}\Bigg\}.\end{aligned} \tag{9.5.21}$$

In Fig. 9.2, the probability of the atom being in the upper level $P_a = (1 + \langle \sigma_z\rangle)/2$ is plotted for different values of $\mathscr{C}/4g$. Here we see a transition from damped Rabi oscillations to an overdamped situation.

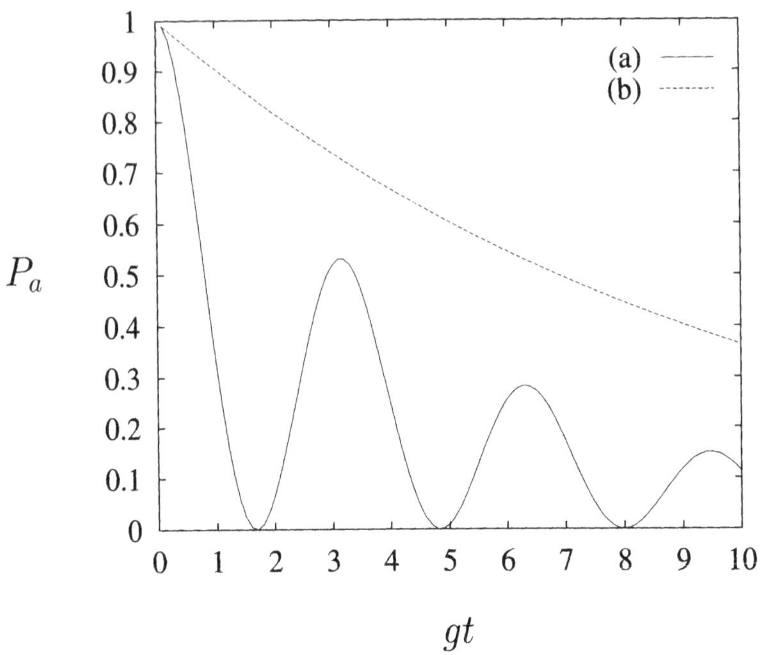

Fig. 9.2
A plot of P_a versus dimensionless time gt for (a) $\mathscr{C}/4g = 0.1$ and (b) $\mathscr{C}/4g = 10$.

This different behavior can be seen easily by considering two limiting cases of Eq. (9.5.21). When $\mathscr{C} \ll 4g$, the atomic inversion $\langle\sigma_z(t)\rangle$ and the probability P_a take the simple forms

$$\langle\sigma_z(t)\rangle = -1 + e^{-\mathscr{C}t/2}[1 + \cos(2gt)], \tag{9.5.22}$$

$$P_a(t) = \frac{e^{-\mathscr{C}t/2}}{2}[1 + \cos(2gt)]. \tag{9.5.23}$$

These damped Rabi oscillations are at the frequency $2g$. In the opposite limit $\mathscr{C} \gg 4g$, we obtain

$$\langle\sigma_z(t)\rangle = -1 + 2e^{-(4g^2t/\mathscr{C})}, \tag{9.5.24}$$

and

$$P_a(t) = e^{-(4g^2t/\mathscr{C})}, \tag{9.5.25}$$

i.e., the atom decays exponentially with a damping constant

$$\Gamma_c = \frac{4g^2}{\mathscr{C}} = \left(\frac{1}{4\pi\epsilon_0}\frac{4\nu^3\wp_{ab}^2}{3\hbar c^3}\right)\left(\frac{\nu}{\mathscr{C}}\right)\left(\frac{6\pi c^3}{V\nu^3}\right). \tag{9.5.26}$$

Apart from a trivial factor of 3, this expression is identical to Eq. (9.5.4), which was obtained using a heuristic argument based on the density of states. The factor of 3 disappears if, in Eq. (9.5.26), we replace g^2 by its average value over different orientations.

Problems

9.1 A single mode of frequency ν interacts with a thermal reservoir. The evolution of the field–reservoir system is described by the Langevin equation

$$\dot{\tilde{a}} = -\frac{1}{2}\mathscr{C}\tilde{a} + F_{\tilde{a}}(t),$$

where $\tilde{a}(t) = a(t)e^{i\nu t}$; a is the destruction operator for the field mode. Calculate the variance $(\Delta X_1)^2$ (with $X_1 = (\tilde{a} + \tilde{a}^\dagger)/2$) at a time t in terms of the variance at the initial time $t = 0$.

9.2 Find the correlation function $\langle F_a^\dagger(t)F_a(t')\rangle$ in Eq. (9.2.13) for

$$f(t_i, t, \tau) = \begin{cases} e^{-\Gamma(t-t_i)} & \text{for } t_i \le t < t_i + \tau\,, \\ 0 & \text{otherwise.} \end{cases}$$

9.3 Calculate the second-order correlation functions

$$\langle F_{\tilde{a}}^\dagger(t)F_{\tilde{a}}(t')\rangle_R\,, \qquad \langle F_{\tilde{a}}(t)F_{\tilde{a}}^\dagger(t')\rangle_R\,,$$
$$\langle F_{\tilde{a}}(t)F_{\tilde{a}}(t')\rangle_R, \text{ and } \langle F_{\tilde{a}}^\dagger(t)F_{\tilde{a}}^\dagger(t')\rangle_R$$

of the Langevin operator for a multi-mode squeezed vacuum reservoir.

9.4 Derive the equation of motion for arbitrary products of creation and destruction operators $\langle (a^\dagger)^m a^n\rangle$ for (a) a thermal reservoir and (b) a squeezed reservoir.

9.5 Consider the reservoir in a squeezed vacuum state. Use the equation of motion for the density matrix for the field mode and the Einstein relation to calculate the diffusion coefficient $D_{\tilde{a}\tilde{a}^\dagger}$. Verify your results from Langevin theory.

References and bibliography

Operator treatment of quantum theory of damping and related problems

M. Lax, *Phys. Rev.* **145**, 110 (1966).
G. W. Ford, J. T. Lewis, and R. F. O'Connell, *Phys. Rev. A* **37**, 4419 (1988).
M. O. Scully, G. Süssmann, and C. Benkert, *Phys. Rev. Lett.* **60**, 1014 (1988).

Atom in a damped cavity

E. M. Purcell, *Phys. Rev.* **69**, 681 (1946).
R. J. Cook and P. W. Milonni, *Phys. Rev.* **A35**, 5081 (1987).
S. Haroche, in *High Resolution Laser Spectroscopy*, ed. K. Shimoda (Springer-Verlag, Berlin 1976).
D. Kleppner, *Phys. Rev. Lett.* **47**, 233 (1981).
S. Sachdev, *Phys. Rev. A* **29**, 2627 (1984).
S. Haroche and J. M. Raimond, in *Advances in Atomic and Molecular Physics*, Vol. 20, ed. D. Bates and B. Bederson (Academic Press, New York 1985), p. 347.
J. A. Gallas, G. Leuchs, H. Walther, and H. Figger, in *Advances in Atomic and Molecular Physics*, Vol. 20, ed. D. Bates and B. Bederson (Academic Press, New York 1985), p. 413.

Experiments on enhanced or inhibited spontaneous emission in a cavity

P. Goy, J. M. Raimond, M. Gross, and S. Haroche, *Phys. Rev. Lett.* **50**, 1903 (1983).
G. Gabrielse, R. Van Dyck, Jr., P. Schwinberg, and H. Dehmelt, *Bull. Am. Phys. Soc.* **29**, 926 (1984).

CHAPTER 10

Resonance fluorescence

The phenomenon of resonance fluorescence provides an interesting manifestation of the quantum theory of light and is a "real world" application of the material of the Chapters 8 and 9. In this process, a two-level atom is typically driven by a resonant continuous-wave laser field and the spectral and quantum statistical properties of the fluorescent light emitted by the atom are measured. Experimentally this can be achieved by scattering a laser off a collimated atomic beam such that the directions of the laser beam, atomic beam, and detector axis are mutually perpendicular as shown in Fig. 10.1.

If the driving field is monochromatic, then at low excitation intensity the atom absorbs a photon at the excitation frequency and reemits it at the same frequency as a consequence of conservation of energy. The spectral width of the fluorescent light is therefore very narrow. The situation, however, is considerably more complicated when the excitation intensity increases and the Rabi frequency associated with the driving field becomes comparable to, or larger than, the atomic linewidth. At such intensity levels, the Rabi oscillations show up as a modulation of the quantum dipole moment and sidebands start emerging in the spectrum* of the emitted radiation. This so-called dynamic Stark splitting is an interesting feature of the atom–field interaction. In addition to that, the fluorescent light exhibits certain nonclassical properties including photon antibunching and squeezing.

In this chapter, we develop a theory of resonance fluorescence to explain these phenomena. We shall begin by relating the field operators required to determine the characteristics of the scattered light to the atomic dipole operators at an appropriately earlier time. The dipole

* The complete calculation of the spectrum was first given by Mollow [1969], and the beautiful dressed-state explanation of the physics was later given by C. Cohen-Tannoudji and co-workers, see e.g., Cohen-Tannoudji and Reynaud [1976].

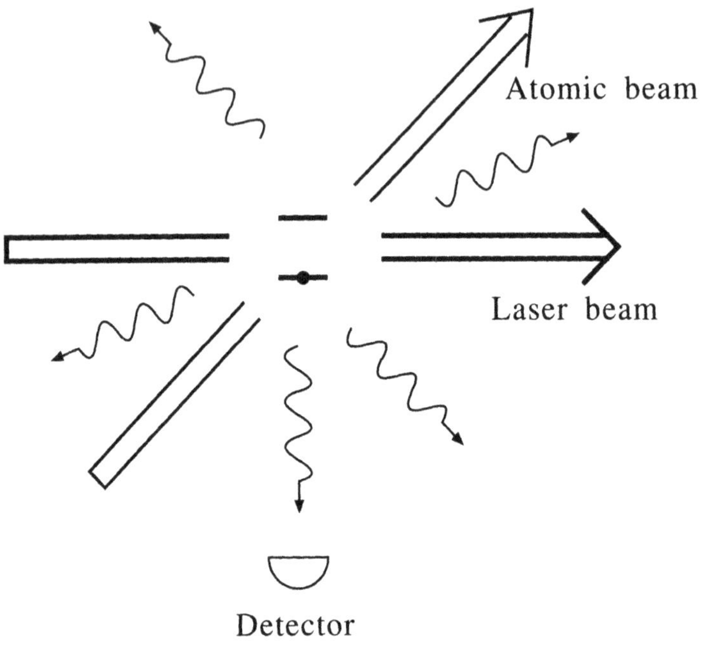

Fig. 10.1
Schematic setup of a resonance fluorescence experiment. The directions of the atomic beam, the laser beam, and the detector axis are mutually perpendicular.

correlation functions will then be evaluated in a semiclassical treatment in which the strong driving field is treated classically. The problem of resonance fluorescence from a driven V type system is also presented. In such a case, an interesting 'line narrowing' is observed. We here give a simplified treatment of the problem which contains the essential physics.

10.1 Electric field operator for spontaneous emission from a single atom

We begin by considering a two-level atom located at a point $\mathbf{r}_0$ which is driven by a strong continuous-wave laser field. This interaction can be treated semiclassically as in Chapter 5. The driven atom is excited to the higher energy state and then radiates spontaneously in all directions, see Chapter 6. The field operator at a point $\mathbf{r}$, associated with this fluorescent radiation field is related to the appropriate atomic operator at a retarded time in order to allow the field to propagate from the position $\mathbf{r}_0$ to $\mathbf{r}$. In Appendix 10.A, we determine this relationship, which will allow us to study the spectral properties of the fluorescent light by simply calculating the appropriate correlation functions involving atomic operators of the driven atom.

As is shown in Appendix 10.A, upon making essentially the Weisskopf–Wigner approximation, we find that the field operator $\mathbf{E}^{(+)}$ at the observation point $\mathbf{r}$ is given by

$$\mathbf{E}^{(+)}(\mathbf{r},t) = \frac{\omega^2 \wp \sin\eta}{4\pi\epsilon_0 c^2 |\mathbf{r}-\mathbf{r}_0|} \hat{x}\sigma_- \left(t - \frac{|\mathbf{r}-\mathbf{r}_0|}{c}\right) \tag{10.1.1}$$

with a similar expression for $\mathbf{E}^{(-)}(\mathbf{r},t)$. Equation (10.1.1) indicates that the positive frequency part of the field operator is proportional to the atomic lowering operator at a retarded time.

In Eq. (10.1.1), which is valid only in the far field, the dipole is assumed to be in the x-z plane and η is the angle the dipole makes with the z-axis, ω is the atomic transition frequency, and $\wp$ is the dipole matrix element between the two levels. It can be seen that, in the far-zone approximation, the scattered field is polarized in the x-direction.

10.2 An introduction to the resonance fluorescence spectrum

10.2.1 *Weak driving field limit*

Before embarking on the detailed calculations, we will employ simple arguments to understand the spectral properties of fluorescent light.

As depicted in Fig. 10.2, a field of spectral width D and central frequency ν is incident on an atom with spectral width Γ and central frequency ω. The field induces a dipole moment in the atom which governs the emitted or scattered light according to Eq. (10.1.1). Specifically, if we take the expectation value of (10.1.1), we find

$$\langle \mathbf{E}^{(+)}(\mathbf{r},t)\rangle = \frac{\omega^2 \wp \sin\eta}{4\pi\epsilon_0 c^2 |\mathbf{r}-\mathbf{r}_0|} \hat{x}\left\langle \sigma_-\left(t - \frac{|\mathbf{r}-\mathbf{r}_0|}{c}\right)\right\rangle. \tag{10.2.1}$$

The expectation value $\langle\sigma_-(t)\rangle$ may be calculated by noting that

$$\begin{aligned}\langle\sigma_-(t)\rangle &= \mathrm{Tr}\left[U^\dagger(t)\sigma_-(0)U(t)\rho(0)\right]\\ &= \mathrm{Tr}\left[\sigma_-(0)U(t)\rho(0)U^\dagger(t)\right]\\ &= \mathrm{Tr}\left[\sigma_-(0)\rho(t)\right],\end{aligned} \tag{10.2.2}$$

where $U(t)$ is the time-evolution operator for the atom driven by an intense classical field. Since the lowering operator is given by $\sigma_-(0) = |b\rangle\langle a|$, it follows from Eq. (10.2.2) that

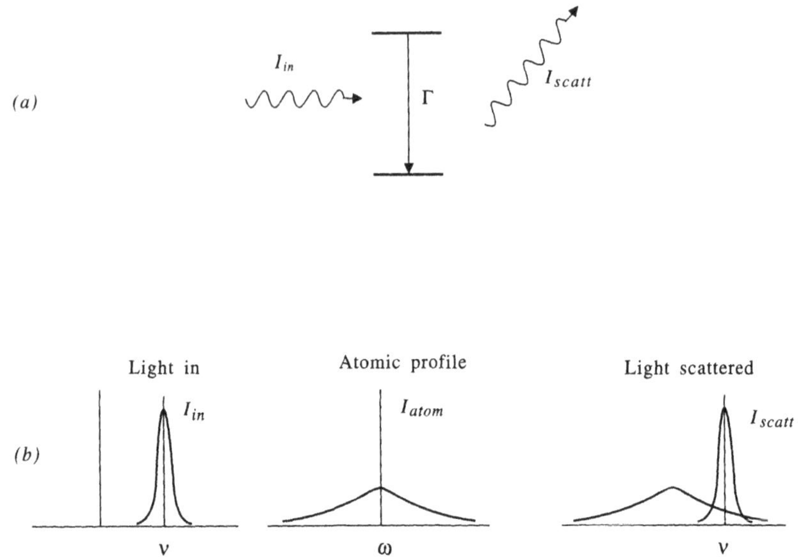

Fig. 10.2
(a) Incident light of frequency ν is scattered by an atom with a lifetime $1/\Gamma$.
(b) Incident light has a spectrum centered at frequency ν and badnwidth D. Scattered light has a bandwidth D, centered at ν for $\Gamma \gg D$.

$$\langle \sigma_-(t) \rangle = \text{Tr}\left[|b\rangle\langle a|\rho(t)\right] = \langle a|\rho(t)|b\rangle = \rho_{ab}(t). \tag{10.2.3}$$

Now from Eq. (5.3.24) we have for the element $\rho_{ab}(t)$ of the density matrix

$$\dot{\rho}_{ab}(t) = -(i\omega + \Gamma/2)\rho_{ab} - \frac{i\wp\mathscr{E}}{2\hbar}\left[\rho_{aa}(t) - \rho_{bb}(t)\right] e^{-i\nu t}. \tag{10.2.4}$$

Here, we used the rotating-wave approximation and took $\gamma_{ab} = \Gamma/2$. The solution of Eq. (10.2.4) reads

$$\begin{aligned}\rho_{ab}(t) &= \rho_{ab}(0)e^{-(i\omega+\Gamma/2)t} - i\frac{\Omega_R}{2}e^{-i\nu t} \\ &\quad \int_0^t dt' e^{-\left[i(\omega-\nu)+\Gamma/2\right](t-t')}\left[\rho_{aa}(t') - \rho_{bb}(t')\right] \\ &\cong -i\left(\frac{\Omega_R}{2}\right)\frac{e^{-i\nu t} - e^{-(i\omega+\Gamma/2)t}}{i(\omega-\nu)+\Gamma/2}\left[\rho_{aa}(0) - \rho_{bb}(0)\right],\end{aligned} \tag{10.2.5}$$

where Ω_R is the Rabi frequency of the driving field, $\Omega_R = \wp\mathscr{E}/\hbar$, and we have noted that $\rho_{ab}(0) = 0$, and that for weak fields $\rho_{aa}(t') - \rho_{bb}(t')$ may be replaced by $\rho_{aa}(0) - \rho_{bb}(0)$. Finally, in the long time limit such that $t \gg \Gamma^{-1}$, we obtain the result

$$\rho_{ab}(t) = \frac{-i(\Omega_R/2)e^{-i\nu t}}{i(\omega-\nu)+\Gamma/2}\left[\rho_{aa}(0) - \rho_{bb}(0)\right]. \tag{10.2.6}$$

Equation (10.2.6) indicates that the dipole oscillates at the driving frequency, not the atomic frequency. Thus from Eqs. (10.2.1) and

(10.2.6) we see that the field which is emitted into some new direction has the same spectrum as the incident field if its spectral width D is small compared to Γ (see Fig. 10.2). Alternatively, for a field whose spectral width is $1/D$ due to its finite duration, we understand the atomic response as that of a driven oscillator with the frequency of the driving field. The atomic oscillator scatters as long as it is driven, that is, for a time $1/D$. Thus the fluorescent light produced by a spectrally sharp driving field has a narrow spectral width.

10.2.2 The strong field limit: sidebands appear

The situation described above corresponds to a weak excitation intensity. When the Rabi frequency associated with the driving field Ω_R becomes comparable to or larger than the spectral width of the atom Γ, sidebands start emerging in the spectrum of the fluorescent light, leading to a three-peak spectrum. The emergence of the sidebands at frequencies $\nu + \Omega_R$ and $\nu - \Omega_R$ is due to the modulation of the dipole moment by the Rabi oscillations.

A physical understanding of this interesting behavior can be achieved by considering a dressed- atom picture of the atom–field interaction. The interaction Hamiltonian of a quantized field mode interacting resonantly with a two-level atom, in the rotating-wave approximation, is (see Eq. (6.2.8))

$$\begin{aligned}\mathscr{H} &= \mathscr{H}_0 + \mathscr{H}_1 \\ &= \frac{\hbar\omega}{2}\sigma_z + \hbar\nu a^\dagger a + \hbar g(\sigma_+ a + a^\dagger \sigma_-).\end{aligned} \tag{10.2.7}$$

We will consider the case in which $\omega = \nu$ and are therefore concerned only with the interaction picture Hamiltonian

$$\mathscr{V} = \hbar g(\sigma_+ a + a^\dagger \sigma_-). \tag{10.2.8}$$

As can be verified by direct substitution, the eigenstates of the Hamiltonian (10.2.8) are

$$|\pm, n\rangle = \frac{1}{\sqrt{2}}(|a, n\rangle \pm |b, n+1\rangle), \tag{10.2.9}$$

with eigenvalues $+\hbar\Omega_n/2$ and $-\hbar\Omega_n/2$, respectively, where the 'generalized' Rabi frequency is defined by $\Omega_n = 2g\sqrt{n+1}$. Thus, the previously degenerate states $|a, n\rangle$ and $|b, n+1\rangle$ are split into a doublet of dressed states separated by Ω_n as shown in Fig. 10.3. This is called *dynamic Stark splitting*. The dynamic Stark split doublets have almost equal spacing for $n \gg 1$. As indicated in Fig. 10.3, the single-photon spontaneous decay spectrum consists of a triplet of lines split by the

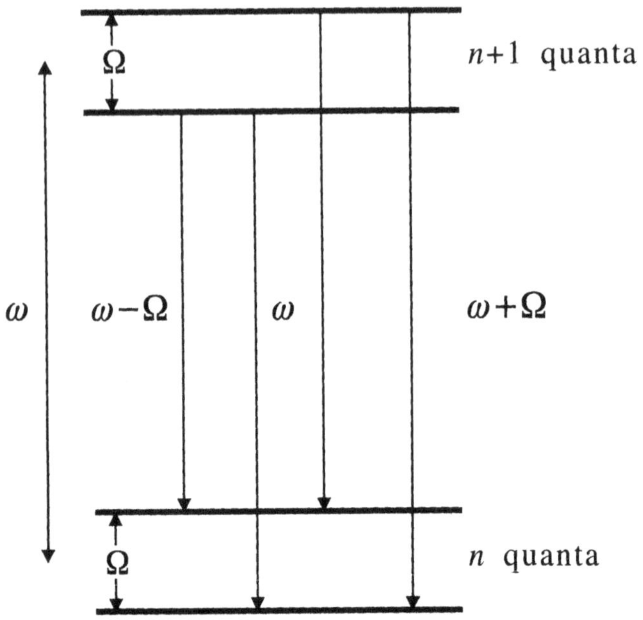

Fig. 10.3
Splitting of the atomic states by the dynamic Stark effect.

Rabi frequency Ω_n, with the central component being made of two equal contributions.

10.2.3 *The widths of the three peaks in the very strong field limit*

From Sections 10.2.1 and 10.2.2 we see that the gross spectral features, namely the delta function spectrum in the weak field limit and the appearance of three peaks in the strong field limit, are readily understood. What cannot be understood so easily is the widths of the peaks. As will be seen, the central peak has a width of $\Gamma/2$ and the 'sidebands' have widths of $3\Gamma/4$. Why are they not the same?

The usual derivation of the three peaks is quite involved (see Section 10.5 and Appendix 10.B). We proceed to give a simple derivation in the limit where the Rabi frequency is much larger than the atomic decay rate. In order to get a simple derivation of the linewidths, we first assume that the field photon distribution has a sharp peak around the mean photon number $\bar{n}$ so that we may write

$$\Omega_n \cong \Omega_{\bar{n}} \equiv \Omega_R. \tag{10.2.10}$$

In such a case, we may ignore the photon index in the eigenstates (10.2.9) to write

$$|\pm\rangle = \frac{1}{\sqrt{2}}(|a\rangle \pm |b\rangle). \tag{10.2.11}$$

These are eigenstates of the semiclassical interaction Hamiltonian (Eq. (5.2.54) with $\phi = \pi$)

$$\mathscr{V} = \frac{\hbar\Omega_R}{2}(|a\rangle\langle b| + |b\rangle\langle a|), \tag{10.2.12}$$

such that

$$\mathscr{V}|\pm\rangle = \pm\frac{\hbar\Omega_R}{2}|\pm\rangle. \tag{10.2.13}$$

With these definitions and assumptions at hand, we proceed to solve for the density matrix of the driven and damped atom. The atomic density matrix obeys the equation of motion (in the interaction picture)

$$\dot{\rho} = -\frac{i}{\hbar}[\mathscr{V}, \rho] + \mathscr{L}\rho, \tag{10.2.14}$$

where the damping term is found from Eq. (8.2.8) (in the limit $\bar{n}_{\text{th}} = 0$) to be

$$\mathscr{L}\rho = -\frac{\Gamma}{2}(\sigma_+\sigma_-\rho - 2\sigma_-\rho\sigma_+ + \rho\sigma_+\sigma_-). \tag{10.2.15}$$

To begin with, we seek the average $\langle\sigma_-(t)\rangle$ which in the dressed-state basis is given by (see Problem 10.2)

$$\langle\sigma_-(t)\rangle e^{i\omega t} = \frac{1}{2}[\rho_{++}(t) - \rho_{--}(t) - \rho_{+-}(t) + \rho_{-+}(t)]. \tag{10.2.16}$$

In view of the fact that $\rho_{--} = 1 - \rho_{++}$ and $\rho_{-+} = \rho_{+-}^*$ we need to find only $\rho_{++}(t)$ and $\rho_{+-}(t)$ to determine $\langle\sigma_-(t)\rangle$. Thus we write the equations of motion for ρ_{++} and ρ_{+-} using Eqs. (10.2.14) and (10.2.15) as (see Appendix 10.B)

$$\dot{\rho}_{++} = -\frac{\Gamma}{2}\rho_{++}(t) + \frac{\Gamma}{4}, \tag{10.2.17a}$$

$$\dot{\rho}_{+-} = -\left(i\Omega_R + \frac{3\Gamma}{4}\right)\rho_{+-} - \frac{\Gamma}{4}\rho_{-+} - \frac{\Gamma}{2}. \tag{10.2.17b}$$

Furthermore, in the secular approximation which holds for strong fields such that $\Omega_R \gg \Gamma$, we may neglect the last two terms in Eq. (10.2.17b) (because they will lead to rapidly oscillating terms) and write

$$\dot{\rho}_{+-} \cong -\left(i\Omega_R + \frac{3\Gamma}{4}\right)\rho_{+-}. \tag{10.2.18}$$

Solving (10.2.17a) and (10.2.18) we find

$$\rho_{++}(t) = \rho_{++}(0)e^{-\frac{\Gamma}{2}t} + \frac{1}{2}(1 - e^{-\frac{\Gamma}{2}t}), \tag{10.2.19a}$$

$$\rho_{+-}(t) = \rho_{+-}(0)e^{-i\Omega_R t - \frac{3\Gamma}{4}t}. \tag{10.2.19b}$$

Finally, we may insert Eqs. (10.2.19a,10.2.19b) into (10.2.16) to obtain

$$\langle\sigma_-(t)\rangle e^{i\omega t} = \frac{1}{4}\left\{[2\rho_{++}(0)-1]e^{-\frac{\Gamma}{2}t} - \left[\rho_{+-}(0)e^{-i\Omega_R t - \frac{3\Gamma}{4}t} - \text{c.c.}\right]\right\} \tag{10.2.20}$$

Equation (10.2.20) taken together with (10.2.1) suggests and leads to several interesting points.

First of all, there is a central component which goes as $\exp(-\Gamma t/2)$ and thus implies a width $\Gamma/2$ of the central peak together with the two sidebands at $\pm\Omega_R$ having width $3\Gamma/4$. This is pleasing in that we have a simple way to calculate the linewidths.

However, as can be seen from Eq. (10.2.19a), the steady-state value of $\rho_{++}(0)$ is 1/2 and therefore the first term in (10.2.20) vanishes. Likewise, in the limit $\Omega_R \gg \Gamma$, the steady-state value of $\rho_{+-}(0)$ tends to 0 and our dipole hence seems to have vanished. The problem is that we must think harder about what it means to calculate (and measure!) the spectrum of the scattered light. We turn to this problem in the next section.

10.3 Theory of a spectrum analyzer

In order to deal with the problem that was encountered in the last section, we reconsider and sharpen our treatment of the measurement of the spectrum of the scattered radiation. To that end, consider the model of a spectrum analyzer as illustrated in Fig. 10.4. There we see a detector atom having 'sharp' levels $|\alpha\rangle$ and $|\beta\rangle$ separated by an energy $\hbar\omega_\alpha$. When we open the shutter at time t_0, scattered light illuminates the atom until we close the shutter at time $t_0 + T$.

We require that the time T be much greater than the reciprocal of the spectral width $1/\Gamma$ of the scattered light. The detector atom is now excited to the upper level with a probability $P(\omega_\alpha)$. Finally we 'look' to see if the atom is excited, record the result, reset the detector atom to the ground state $|\beta\rangle$ and repeat the measurement many times.

Next we 'tune the detector atom' by changing ω_α (e.g., by varying an external magnetic field) and repeat the measurement sequence of the previous paragraph. Finally we plot the probability $P(\omega_\alpha)$ as a

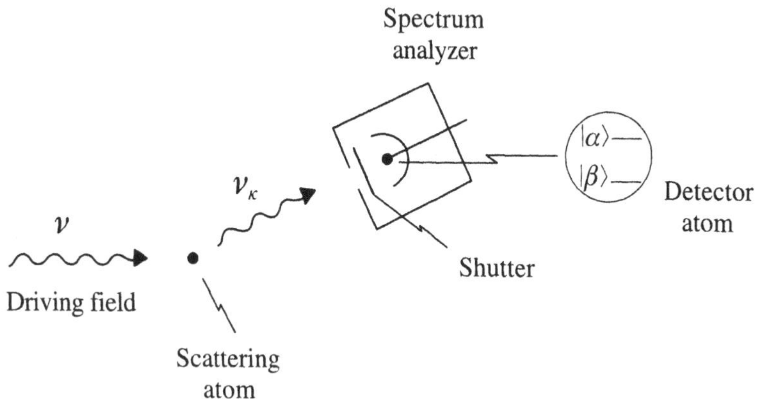

Fig. 10.4
Schematic setup for a 'gedanken spectrum analyzer'.

function of the frequency ω_α to obtain the spectrum of scattered light.

The above 'gedanken spectrum analyzer' provides us with an operational definition of the spectral profile. It is now a simple matter to calculate $P(\omega_\alpha)$, see Eq. (6.5.3), which yields in the present notation

$$P(\omega_\alpha, T) = \frac{\wp_{\alpha\beta}^2}{\hbar^2} \int_{t_0}^{T} dt_1 \int_{t_0}^{T} dt_2 \langle E^{(-)}(\mathbf{r}, t_1) E^{(+)}(\mathbf{r}, t_2) \rangle e^{-i\omega_\alpha (t_1 - t_2)}. \tag{10.3.1}$$

Here $\wp_{\alpha\beta}$ is the usual dipole matrix element, and the $\langle \cdots \rangle$ average over the creation and annihilation operators of the field stands for $\mathrm{Tr}\,[\rho_F \cdots]$, where ρ_F is the field density matrix. More to the point, the spectrum as defined by Eq. (10.3.1) can be written in terms of the correlation function $G^{(1)}(\mathbf{r}, \mathbf{r}; t_1, t_2)$ (Eq. (4.2.10)). In that notation, Eq. (10.3.1) may be written as

$$P(\omega_\alpha, T) = \frac{\wp_{\alpha\beta}^2}{\hbar^2} \int_{t_0}^{T} dt_1 \int_{t_0}^{T} dt_2 G^{(1)}(\mathbf{r}, \mathbf{r}; t_1, t_2) e^{-i\omega_\alpha (t_1 - t_2)}, \tag{10.3.2}$$

where $G^{(1)}(\mathbf{r}, \mathbf{r}; t_1, t_2) = \mathrm{Tr}\left[E^{(-)}(\mathbf{r}, t_1) E^{(+)}(\mathbf{r}, t_2) \rho_F\right]$. As discussed in Appendix 10.D, this probability is proportional to the power spectrum of the scattered light and is equal to (for stationary fields)

$$P(\omega_\alpha) = \kappa \int d\tau G^{(1)}(\mathbf{r}, \mathbf{r}; 0, \tau) e^{i\omega_\alpha \tau} + \text{c.c.}, \tag{10.3.3}$$

where κ is a constant depending on the efficiency of the detector, etc.

For the present resonance fluorescence problem we may use Eq. (10.1.1) to write (10.3.3) as

$$G^{(1)}(\mathbf{r},\mathbf{r};t_1,t_2)=\left(\frac{\omega^2\wp\sin\eta}{4\pi\epsilon_0c^2r}\right)^2\mathrm{Tr}\left[\sigma_+\left(t_1-\frac{r}{c}\right)\sigma_-\left(t_2-\frac{r}{c}\right)\rho\right], \tag{10.3.4}$$

where $r=|\mathbf{r}-\mathbf{r}_0|$, and ρ is the atomic density matrix.

But how do we calculate the two-time average (10.3.4)? In Section 10.2 we were always dealing with single-time expectation values

$$\langle\sigma_-(t)\rangle=\mathrm{Tr}\left[\sigma_-(0)\rho(t)\right]. \tag{10.3.5}$$

The question now is: can we use the knowledge of (10.3.5) to calculate two-time expectation values as in Eq. (10.3.4)? The next section shows how to accomplish this.

10.4 From single-time to two-time averages: the Onsager–Lax regression theorem

The two-time correlation functions which will be used in our study of the spectral density and photon antibunching properties of the fluorescent light are $\langle\sigma_+(t)\sigma_-(t+\tau)\rangle$ and $\langle\sigma_+(t)\sigma_+(t+\tau)\sigma_-(t+\tau)\sigma_-(t)\rangle$. In general, a solution of the density matrix is not sufficient to calculate the two-time correlation functions; we need to determine the transition probability distribution. However under certain conditions the quantum regression theorem allows us to calculate a two-time correlation function from a single-time correlation function.

The total density operator at a time τ with $\tau\geq 0$ is given in terms of the density operator at an earlier time $t=0$ by the expression

$$\rho(\tau)=U(\tau)\rho(0)U^\dagger(\tau), \tag{10.4.1}$$

where $U(\tau)$ is the unitary time-evolution operator for the total system. Here $\rho(t)$ is the density matrix for the total system of atom plus reservoir. It is assumed that the reservoir density matrix $\rho_R(0)$ is uncoupled from the atomic density matrix $\rho_{\mathrm{atom}}(0)$ at the initial (earlier) time $t=0$, i.e.,

$$\rho(0)=\rho_R(0)\otimes\rho_{\mathrm{atom}}(0). \tag{10.4.2}$$

The reduced density operator for the atom at time τ is thus obtained by taking the trace over reservoir variables, i.e.,

$$\rho_{\mathrm{atom}}(\tau)=\mathrm{Tr}_R[U(\tau)\rho_R(0)\otimes\rho_{\mathrm{atom}}(0)U^\dagger(\tau)]. \tag{10.4.3}$$

The expectation value of $\sigma_-(\tau)$ is then given by

$$\begin{aligned}\langle\sigma_-(\tau)\rangle &= \mathrm{Tr}_{\mathrm{atom}}\mathrm{Tr}_R[\sigma_-(0)\rho_R(\tau)\otimes\rho_{\mathrm{atom}}(\tau)] \\ &= \mathrm{Tr}_{\mathrm{atom}}\{\sigma_-(0)\mathrm{Tr}_R[U(\tau)\rho_R(0)\otimes\rho_{\mathrm{atom}}(0)U^\dagger(\tau)]\},\end{aligned} \tag{10.4.4}$$

while the two-time correlation function

$$\begin{aligned}&\langle\sigma_+(0)\sigma_-(\tau)\rangle \\ &= \mathrm{Tr}_{\mathrm{atom}}\mathrm{Tr}_R[\sigma_+(0)\sigma_-(\tau)\rho_R(0)\otimes\rho_{\mathrm{atom}}(0)] \\ &= \mathrm{Tr}_{\mathrm{atom}}\{\sigma_+(0)\mathrm{Tr}_R[U^\dagger(\tau)\sigma_-(0)U(\tau)\rho_R(0)\otimes\rho_{\mathrm{atom}}(0)]\} \\ &= \mathrm{Tr}_{\mathrm{atom}}\left(\sigma_-(0)\mathrm{Tr}_R\{U(\tau)\rho_R(0)\otimes[\rho_{\mathrm{atom}}(0)\sigma_+(0)]\,U^\dagger(\tau)\}\right).\end{aligned} \tag{10.4.5}$$

Comparing Eq. (10.4.5) with (10.4.4) we can simply calculate $\langle\sigma_+(0)\sigma_-(\tau)\rangle$ by employing $\rho_{\mathrm{atom}}(0)\sigma_+(0)$ instead of $\rho_{\mathrm{atom}}(0)$ in Eq. (10.4.4). It is clear that the two-time correlation function can be determined from a knowledge of single-time expectation values. The crucial assumption is Eq. (10.4.2) which is usually referred to as the *Markovian approximation.*

A more general statement of the *quantum regression theorem* is that if, for some operator $\hat{O}$,

$$\langle\hat{O}(t+\tau)\rangle = \sum_j a_j(\tau)\langle\hat{O}_j(t)\rangle, \tag{10.4.6}$$

then

$$\langle\hat{O}_i(t)\hat{O}(t+\tau)\hat{O}_k(t)\rangle = \sum_j a_j(\tau)\langle\hat{O}_i(t)\hat{O}_j(t)\hat{O}_k(t)\rangle. \tag{10.4.7}$$

With these premilinary steps in hand, we now go back to the problem of Section 10.2.2. Using Eqs. (10.B.28) and (10.B.29), we can write Eq. (10.2.20) in the bare-state representation as

$$\begin{aligned}\langle\sigma_-(\tau)\rangle e^{i\omega\tau} \;=\; &\frac{1}{2}\{\mathrm{Tr}[|b\rangle\langle a|\rho(0)] + \mathrm{Tr}[|a\rangle\langle b|\rho(0)]\}e^{-\frac{\Gamma}{2}\tau} \\ &-\frac{1}{4}\{2\mathrm{Tr}[|a\rangle\langle a|\rho(0)] - 1 - \mathrm{Tr}[|b\rangle\langle a|\rho(0)] + \mathrm{Tr}[|a\rangle\langle b|\rho(0)]\} \\ &e^{-\frac{3\Gamma}{4}\tau}e^{-i\Omega_R\tau} \\ &+\frac{1}{4}\{2\mathrm{Tr}[|a\rangle\langle a|\rho(0)] - 1 + \mathrm{Tr}[|b\rangle\langle a|\rho(0)] - \mathrm{Tr}[|a\rangle\langle b|\rho(0)]\} \\ &e^{-\frac{3\Gamma}{4}\tau}e^{i\Omega_R\tau},\end{aligned} \tag{10.4.8}$$

where we have also used the relation $\rho_{ij}(0) = \mathrm{Tr}(|j\rangle\langle i|\rho(0))$. If now we use $\rho(0)\sigma_+(0)$ instead of $\rho(0)$, as suggested above, we obtain

$$\begin{aligned}
&\langle\sigma_+(0)\sigma_-(\tau)\rangle e^{i\omega\tau} \\
&= \frac{1}{2}\mathrm{Tr}[|a\rangle\langle a|\rho(0)]e^{-\frac{\Gamma}{2}\tau} \\
&-\frac{1}{2}\{-\mathrm{Tr}[|a\rangle\langle b|\rho(0)] - \mathrm{Tr}[|a\rangle\langle a|\rho(0)]\}e^{-\frac{3\Gamma}{4}\tau}e^{-i\Omega_R\tau} \\
&+\frac{1}{2}\{-\mathrm{Tr}[|a\rangle\langle b|\rho(0)] + \mathrm{Tr}[|a\rangle\langle a|\rho(0)]\}e^{-\frac{3\Gamma}{4}\tau}e^{i\Omega_R\tau}. \qquad (10.4.9)
\end{aligned}$$

For the steady state, $\rho_{aa} = \rho_{bb} = 1/2$, $\rho_{ab} = \rho_{ba} = 0$, and we have the final form of the two-time correlation function as

$$\langle\sigma_+(0)\sigma_-(\tau)\rangle = \frac{1}{4}\left(e^{-\frac{\Gamma}{2}\tau} + \frac{1}{2}e^{-\frac{3\Gamma}{4}\tau}e^{-i\Omega_R\tau} + \frac{1}{2}e^{-\frac{3\Gamma}{4}\tau}e^{i\Omega_R\tau}\right)e^{-i\omega\tau}. \qquad (10.4.10)$$

This result demonstrates the basic physics of the three-peak resonance fluorescence spectrum including the 'widths' of the peaks. The present (strong field limit) result (10.4.10) is extended to a general field strength in Section 10.5.

10.5 The complete resonance fluorescence spectrum

In this section, we evaluate the complete power spectrum of the radiation scattered by a two-level atom driven by an incident field of arbitrary strength. The atom is assumed to be isolated and fixed in position. We look for the field emitted along the x-axis. The field operator in Eq. (10.1.1) can therefore be treated as scalar. The power spectrum $S(\mathbf{r}, \omega_0)$ of the fluorescent light at some suitably chosen point $\mathbf{r}$ in the far field is obtained by taking the Fourier transform of the normally ordered correlation function of the field $\langle E^{(-)}(\mathbf{r}, t)E^{(+)}(\mathbf{r}, t + \tau)\rangle$ with respect to τ (see Appendix 10.D)

$$S(\mathbf{r}, \omega_0) = \frac{1}{\pi}\mathrm{Re}\int_0^\infty d\tau\langle E^{(-)}(\mathbf{r}, t)E^{(+)}(\mathbf{r}, t + \tau)\rangle e^{i\omega_0\tau}. \qquad (10.5.1)$$

Here we have assumed that the field, in the steady state, is statistically stationary, i.e., the field correlation function is independent of the origin of time so the correlation function $\langle E^{(-)}(\mathbf{r}, t)E^{(+)}(\mathbf{r}, t + \tau)\rangle$ depends only on the time difference τ.

It follows from Eq. (10.1.1) that

$$\langle E^{(-)}(\mathbf{r}, t)E^{(+)}(\mathbf{r}, t + \tau)\rangle = I_0(\mathbf{r})\langle\sigma_+(t)\sigma_-(t + \tau)\rangle, \qquad (10.5.2)$$

where

$$I_0(\mathbf{r}) = \left(\frac{\omega^2 \wp \sin \eta}{4\pi\epsilon_0 c^2 |\mathbf{r} - \mathbf{r}_0|} \right)^2 . \tag{10.5.3}$$

The two-time correlation function $\langle \sigma_+(t)\sigma_-(t+\tau) \rangle$ can be calculated by using the quantum regression theorem if we know the appropriate single-time correlation functions. We are thus interested in the expectation values of the interaction picture dipole operators $\langle \sigma_+(t) \rangle = \rho_{ba}(t)\exp(i\omega t)$, $\langle \sigma_-(t) \rangle = \rho_{ab}(t)\exp(-i\omega t)$, and the inversion operator $\langle \sigma_z(t) \rangle = [\rho_{aa}(t) - \rho_{bb}(t)]$. All higher-order correlation functions can be determined from them. For example

$$\langle \sigma_\pm(t)\sigma_\pm(t) \rangle = 0, \tag{10.5.4a}$$

$$\langle \sigma_+(t)\sigma_-(t) \rangle = (\langle \sigma_z(t) \rangle + 1)/2. \tag{10.5.4b}$$

As is shown in Appendix 10.B (see also Appendix 10.C), an exact solution to the Liouville equation (10.2.14) yields

$$\begin{aligned} \langle \sigma_-(t+\tau) \rangle e^{i\omega(t+\tau)} &= \rho_{ab}(t+\tau) \\ &= a_1(\tau) + a_2(\tau)\langle \sigma_-(t) \rangle e^{i\omega t} + a_3(\tau)\langle \sigma_+(t) \rangle e^{-i\omega t} \\ &\quad + a_4(\tau)(\langle \sigma_z(t) \rangle + 1)/2, \end{aligned} \tag{10.5.5}$$

$$\begin{aligned} (\langle \sigma_z(t+\tau) \rangle + 1)/2 &= \rho_{aa}(t+\tau) \\ &= b_1(\tau) + b_2(\tau)\langle \sigma_-(t) \rangle e^{i\omega t} + b_3(\tau)\langle \sigma_+(t) \rangle e^{-i\omega t} \\ &\quad + b_4(\tau)(\langle \sigma_z(t) \rangle + 1)/2, \end{aligned} \tag{10.5.6}$$

where

$$a_1(\tau) = \frac{-i\Omega_R \Gamma}{\Gamma^2 + 2\Omega_R^2} \left\{ 1 - e^{-3\Gamma\tau/4} \left[\cos\mu\tau - \left(\frac{4\Omega_R^2 - \Gamma^2}{4\mu\Gamma} \right) \sin\mu\tau \right] \right\}, \tag{10.5.7}$$

$$a_2(\tau) = \frac{1}{2} e^{-\Gamma\tau/2} + \frac{e^{-3\Gamma\tau/4}}{8\mu} \left[\Gamma \sin(\mu\tau) + 4\mu \cos(\mu\tau) \right], \tag{10.5.8}$$

$$a_3(\tau) = \frac{1}{2} e^{-\Gamma\tau/2} - \frac{e^{-3\Gamma\tau/4}}{8\mu} \left[\Gamma \sin(\mu\tau) + 4\mu \cos(\mu\tau) \right], \tag{10.5.9}$$

$$a_4(\tau) = \frac{i\Omega_R}{\mu} e^{-3\Gamma\tau/4} \sin(\mu\tau), \tag{10.5.10}$$

$$b_1(\tau) = \frac{\Omega_R^2}{\Gamma^2 + 2\Omega_R^2} \left[1 - \left(\cos\mu\tau + \frac{3\Gamma}{4\mu} \sin\mu\tau \right) e^{-3\Gamma\tau/4} \right], \tag{10.5.11}$$

$$b_2(\tau) = \frac{i\Omega_R}{2\mu} e^{-3\Gamma\tau/4} \sin(\mu\tau), \tag{10.5.12}$$

$$b_3(\tau) = -\frac{i\Omega_R}{2\mu} e^{-3\Gamma\tau/4} \sin(\mu\tau), \tag{10.5.13}$$

$$b_4(\tau) = e^{-3\Gamma\tau/4} \left[\cos(\mu\tau) - \frac{\Gamma}{4\mu} \sin(\mu\tau) \right], \tag{10.5.14}$$

with

$$\mu = \left(\Omega_R^2 - \frac{\Gamma^2}{16}\right)^{1/2}. \tag{10.5.15}$$

We shall be interested in the steady-state properties of the scattered field. In the steady state ($t \to \infty$) the expectation values of various atomic operators are independent of the initial conditions. It therefore follows that

$$\lim_{t\to\infty}\langle\sigma_-(t)\rangle e^{i\omega t} = a_1(\infty) = \frac{-i\Omega_R\Gamma}{\Gamma^2 + 2\Omega_R^2}, \tag{10.5.16}$$

$$(\langle\sigma_z\rangle_{\text{ss}} + 1)/2 = b_1(\infty) = \frac{\Omega_R^2}{\Gamma^2 + 2\Omega_R^2}. \tag{10.5.17}$$

Now the evaluation of the two-time correlation function $\langle\sigma_+(t)\sigma_-(t+\tau)\rangle$ is, as we saw in Section 10.4, formally identical to the evaluation of the single-time expectation value $\langle\sigma_-(t+\tau)\rangle$ except that the non-Hermitian operator $\rho_{\text{atom}}(t)\sigma_+(t)$ must be used in place of the reduced density operator $\rho_{\text{atom}}(t)$. It therefore follows from Eq. (10.5.5) that

$$\begin{aligned}\langle\sigma_+(t)\sigma_-(t+\tau)\rangle e^{i\omega t} &= \{a_1(\tau)\langle\sigma_+(t)\rangle + a_2(\tau)\langle\sigma_+(t)\sigma_-(t)\rangle e^{i\omega t} \\ &\quad + a_3(\tau)\langle\sigma_+(t)\sigma_+(t)\rangle e^{-i\omega t} + a_4(\tau) \\ &\quad \times\langle\sigma_+(t)[\sigma_z(t) + 1]/2\rangle\} \\ &= \{a_1(\tau)\langle\sigma_+(t)\rangle + a_2(\tau)[\langle\sigma_z(t)\rangle + 1/2]\}.\end{aligned} \tag{10.5.18}$$

In the steady state, it follows from Eqs. (10.5.16) and (10.5.17) that

$$\langle\sigma_+(t)\sigma_-(t+\tau)\rangle_{\text{ss}} = [a_1(\tau)a_1^*(\infty) + a_2(\tau)b_1(\infty)]e^{-i\omega\tau}. \tag{10.5.19}$$

On substituting for various coefficients from Eqs. (10.5.7), (10.5.8), (10.5.16), and (10.5.17) in Eq. (10.5.19), we obtain the following explicit expression for the field two-time correlation function

$$\begin{aligned}&\langle E^{(-)}(\mathbf{r},t)E^{(+)}(\mathbf{r},t+\tau)\rangle \\ &= I_0(\mathbf{r})\langle\sigma_+(t)\sigma_-(t+\tau)\rangle_{\text{ss}} \\ &= I_0(\mathbf{r})e^{-i\omega\tau}\left(\frac{\Omega_R^2}{\Gamma^2 + 2\Omega_R^2}\right)\left\{\frac{\Gamma^2}{\Gamma^2 + 2\Omega_R^2} + \frac{e^{-\Gamma\tau/2}}{2}\right. \\ &\quad \left. + \frac{e^{-3\Gamma\tau/4}}{4}[e^{-i\mu\tau}(P + iQ) + e^{i\mu\tau}(P - iQ)]\right\}.\end{aligned} \tag{10.5.20}$$

where $I_0(\mathbf{r})$ is defined by Eq. (10.5.3) and the dimensionless constants P and Q are given by

$$P = \frac{2\Omega_R^2 - \Gamma^2}{2\Omega_R^2 + \Gamma^2}, \qquad Q = \frac{\Gamma}{4\mu}\frac{10\Omega_R^2 - \Gamma^2}{2\Omega_R^2 + \Gamma^2}. \tag{10.5.21}$$

A formula for the power spectrum can be derived by taking the Fourier transform of $\langle E^{(-)}(\mathbf{r},t)E^{(+)}(\mathbf{r},t+\tau)\rangle$ (see Eq. (10.5.1)). As we have already seen in Sections 10.2 and 10.4, the spectrum shows remarkably different behavior in the weak and strong field limits. We therefore treat them separately.

10.5.1 Weak field limit

Equation (10.5.20) simplifies considerably in the weak field limit, i.e., when

$$\Omega_R \ll \frac{\Gamma}{4}. \tag{10.5.22}$$

The Rabi frequency of the driving field is then much smaller than the rate of emission of spontaneously radiated photons. The atom therefore behaves as an overdamped quantum harmonic oscillator. It follows from the definition of μ, Eq. (10.5.15), that in this limit $\mu \cong i\Gamma/4$. The first term in the curly bracket in Eq. (10.5.20) is then of the order of unity, the second and third terms cancel each other, and the fourth term is zero. We obtain

$$\langle E^{(-)}(\mathbf{r},t)E^{(+)}(\mathbf{r},t+\tau)\rangle \cong I_0(\mathbf{r})\left(\frac{\Omega_R}{\Gamma}\right)^2 e^{-i\omega\tau}. \tag{10.5.23}$$

The power spectrum of the emitted field is therefore

$$S(\mathbf{r},\omega_0) = I_0(\mathbf{r})\left(\frac{\Omega_R}{\Gamma}\right)^2 \delta(\omega-\omega_0), \tag{10.5.24}$$

i.e., the spectrum is given by a δ-function. The result is not surprising because, for a monochromatic resonant driving field, the atom absorbs a photon at the excitation frequency, and energy conservation requires that the emitted photon has the same frequency, as predicted for elastic Rayleigh scattering. In this way, we regain the results of Section 10.2.1.

10.5.2 Strong field limit

The situation is more complex when the Rabi frequency of the driving field is comparable to or greater than the atomic decay rate. Under such circumstances the atom can coherently interact many times with the field before spontaneously radiating a photon.

It follows, on taking the Fourier transform of $\langle E^{(-)}(\mathbf{r},t)E^{(+)}(\mathbf{r},t+\tau)\rangle$ as given in Eq. (10.5.20), and using

$$\int_0^\infty d\tau e^{-i\omega\tau-\Gamma\tau/2+i\omega_0\tau} = \frac{1}{i(\omega-\omega_0)+\Gamma/2}, \tag{10.5.25}$$

$$\int_0^\infty d\tau e^{-i\omega\tau\mp i\mu\tau-3\Gamma\tau/4+i\omega_0\tau} = \frac{1}{i(\omega\pm\mu-\omega_0)+3\Gamma/4}, \tag{10.5.26}$$

that the power spectrum of fluorescent light is

$$\begin{aligned} S(\mathbf{r},\omega_0) = \frac{I_0(\mathbf{r})}{4\pi}\left(\frac{\Omega_R^2}{\Gamma^2+2\Omega_R^2}\right)\Bigg[&\frac{4\pi\Gamma^2}{\Gamma^2+2\Omega_R^2}\delta(\omega-\omega_0) \\ &+\frac{\Gamma}{(\omega-\omega_0)^2+(\Gamma/2)^2}+\frac{\alpha_+}{(\omega+\mu-\omega_0)^2+(3\Gamma/4)^2} \\ &+\frac{\alpha_-}{(\omega-\mu-\omega_0)^2+(3\Gamma/4)^2}\Bigg], \end{aligned} \tag{10.5.27}$$

where

$$\alpha_\pm = \frac{3\Gamma}{4}P \pm (\omega\pm\mu-\omega_0)Q. \tag{10.5.28}$$

In Fig. 10.5, $S(\mathbf{r},\omega_0)$ is plotted for various values of $4\Omega_R/\Gamma$. It is seen that, with the increasing driving field intensity, the single-peak spectrum $S(\mathbf{r},\omega_0)$ around $\omega_0=\omega$ is transformed into a three-peak spectrum, with peaks centered around $\omega_0=\omega$, $\omega\pm\Omega_R$. The relative heights of these peaks for $\Omega_R \gg \Gamma/4$ are 1 : 3 : 1. The elastic Rayleigh peak at $\omega_0=\omega$ disappears in this limit. This behavior is seen analytically from Eq. (10.5.26) in the limit $\Omega_R \gg \Gamma/4$. We then obtain

$$\begin{aligned} S(\mathbf{r},\omega_0) = \frac{I_0(\mathbf{r})}{8\pi}\Bigg[&\frac{3\Gamma/4}{(\omega-\Omega_R-\omega_0)^2+(3\Gamma/4)^2} \\ &+\frac{\Gamma}{(\omega-\omega_0)^2+(\Gamma/2)^2} \\ &+\frac{3\Gamma/4}{(\omega+\Omega_R-\omega_0)^2+(3\Gamma/4)^2}\Bigg], \end{aligned} \tag{10.5.29}$$

which is seen to be in agreement with Eq. (10.4.10). The width of the peaks centered at $\omega_0=\omega-\Omega_R$, ω, and $\omega+\Omega_R$ are $3\Gamma/4$, $\Gamma/2$, and $3\Gamma/4$, respectively. The integrated intensities in the peaks are however in the ratio 1 : 2 : 1.

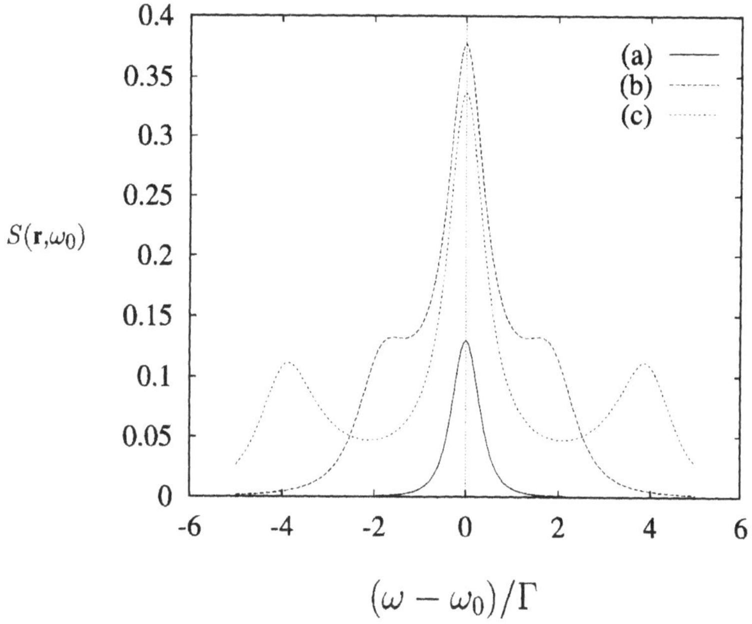

Fig. 10.5 Resonance fluorescence spectrum $S(\mathbf{r}, \omega_0)$ in arbitrary units, as given by Eq. (10.5.27), for (a) $4\Omega_R/\Gamma = 1$, (b) $4\Omega_R/\Gamma = 8$, and (c) $4\Omega_R/\Gamma = 16$. The elastic scattering term proportional to $\delta(\omega - \omega_0)$ has not been included.

10.6 Photon antibunching

We now show that the scattered field in resonance fluorescence exhibits the nonclassical effect associated with photon antibunching. To calculate the normalized second-order correlation function $g^{(2)}(\tau)$ (Eq. (4.2.21)), we first determine the two-time correlation function $\langle\sigma_+(t)\sigma_+(t+\tau)\sigma_-(t+\tau)\sigma_-(t)\rangle$ from Eq. (10.5.6). Using the quantum regression theorem, we obtain

$$\begin{aligned}
&\langle\sigma_+(t)\sigma_+(t+\tau)\sigma_-(t+\tau)\sigma_-(t)\rangle \\
&= \langle\sigma_+(t)\left(\sigma_z(t+\tau)+1/2\right)\sigma_-(t)\rangle \\
&= b_1(\tau)\langle\sigma_+(t)\sigma_-(t)\rangle + b_2(\tau)\langle\sigma_+(t)\sigma_-(t)\sigma_-(t)\rangle e^{i\omega t} \\
&\quad + b_3(\tau)\langle\sigma_+(t)\sigma_+(t)\sigma_-(t)\rangle e^{-i\omega t} \\
&\quad + b_4(\tau)\langle\sigma_+(t)(\sigma_z(t)+1)\sigma_-(t)\rangle/2 \\
&= b_1(\tau)(\langle\sigma_z(t)\rangle + 1)/2. \qquad (10.6.1)
\end{aligned}$$

In the steady state, we have from Eq. (10.5.6)

$$\langle\sigma_+(t)\sigma_+(t+\tau)\sigma_-(t+\tau)\sigma_-(t)\rangle_{ss} = b_1(\tau)b_1(\infty). \qquad (10.6.2)$$

On substituting for various coefficients from Eqs. (10.5.11) and (10.5.17) into Eq. (10.6.2), we obtain

$$\begin{aligned}
&\langle\sigma_+(t)\sigma_+(t+\tau)\sigma_-(t+\tau)\sigma_-(t)\rangle_{ss} \\
&= \left(\frac{\Omega_R^2}{\Gamma^2+2\Omega_R^2}\right)^2\left[1-\left(\cos\mu\tau+\frac{3\Gamma}{4\mu}\sin\mu\tau\right)e^{-3\Gamma\tau/4}\right]. \qquad (10.6.3)
\end{aligned}$$

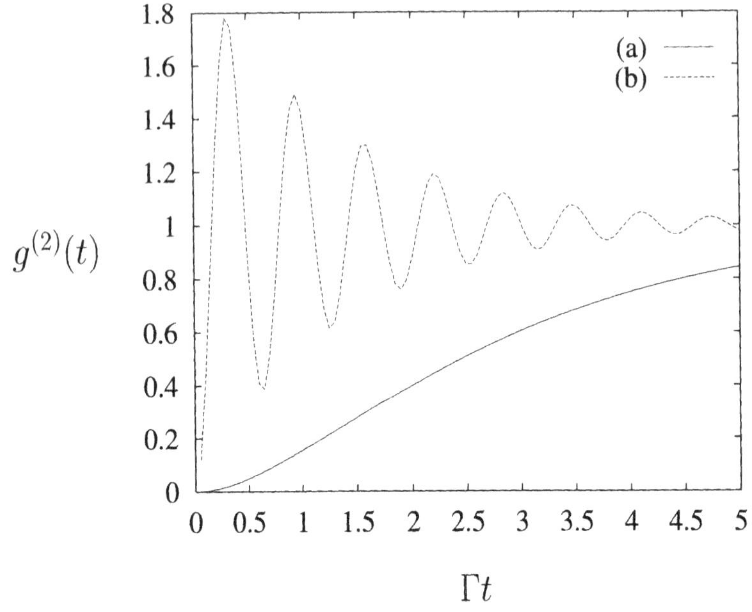

Fig. 10.6 Normalized second-order correlation function $g^{(2)}(\tau)$ versus the dimensionless delay time Γ_τ for (a) $4\Omega_R/\Gamma = 0.1$ and (b) $4\Omega_R/\Gamma = 10$.

Equation (10.6.3) together with Eqs. (10.1.1) and (4.2.21) gives

$$g^{(2)}(\tau) = 1 - \left(\cos\mu\tau + \frac{3\Gamma}{4\mu}\sin\mu\tau\right)e^{-3\Gamma\tau/4}. \tag{10.6.4}$$

In Fig. 10.6, $g^{(2)}(\tau)$ is plotted as a function of the time delay τ for different values of the normalized Rabi frequency of the driving field. It is seen that for $\tau = 0$ $g^{(2)}(\tau) = 0$, and as τ increases $g^{(2)}(\tau) > 0$. We thus have

$$g^{(2)}(\tau) > g^{(2)}(0), \tag{10.6.5}$$

which corresponds to the phenomenon of photon antibunching (see Eq. (4.4.71)).

For the weak driving field, $\Omega_R \ll \Gamma/4$, $g^{(2)}(\tau)$ increases monotonically from 0 to 1 as τ is increased. However, for the strong driving field, $\Omega_R \gg \Gamma/4$, $g^{(2)}(\tau)$ shows an oscillatory dependence on τ. The magnitude of these oscillations decreases as τ is increased and $g^{(2)}(\tau)$ approaches unity as $\tau \to \infty$.

It is easy to understand the physical reason for the photon antibunching in resonance fluorescence. Once a photon is emitted, the atom is found in the ground state and it takes the driving field some time to reexcite the atom to the upper level, from which the next photon can be emitted. On average, this delay is of the order of the Rabi period Ω_R^{-1} as can be seen from Eq. (10.6.4) in the limit $\Omega_R \gg \Gamma/4$. Hence, the spontaneously emitted photons show a tendency toward antibunching for small delay times τ.

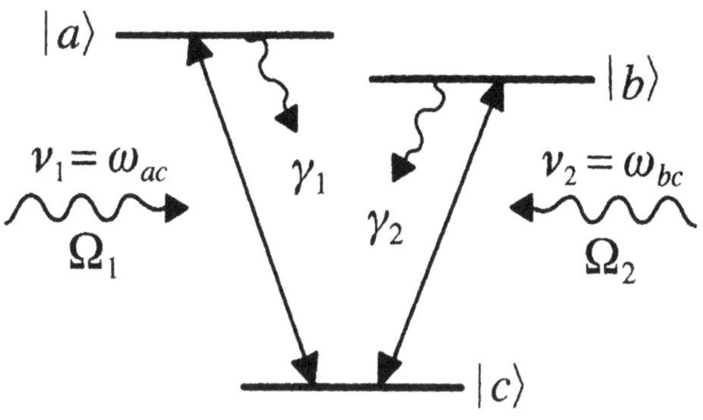

Fig. 10.7 Schematic representation of the V model; ν_1 and ν_2 are the carrier frequencies of the driving fields.

10.7 Resonance fluorescence from a driven V system

We conclude this chapter with a discussion of the fluorescence from a three-level atom* driven by two coherent fields, see Fig. 10.7. The major difference between this problem and the driven two-level atom problem is the addition of a second driving field to the traditional arrangement of resonance fluorescence. This not only causes qualitative changes in the shape of the emission and absorption spectra but also modifies the linewidth in a way that depends on the atomic parameters and the relative strength of the fields. In particular, the spectral components can acquire very different widths and peak heights relative to the case of the standard resonance fluorescence in which a two-level system is driven by a single near-resonant field, see Fig. 10.8.

The excited states in Fig. 10.7, $|a\rangle$ and $|b\rangle$, decay to level $|c\rangle$ at rates Γ_a and Γ_b, and we assume $\Gamma_a \gg \Gamma_b$. Furthermore, we assume the driving Rabi frequencies coupling $|a\rangle$ and $|c\rangle$ (Ω_{R1}) and $|b\rangle$ and $|c\rangle$ (Ω_{R2}) are such that $\Omega_{R2} \gg \Omega_{R1}$. In this case, a spectral narrowing is seen in Fig. 10.8. In the next few paragraphs we provide a heuristic treatment of the problem.

We approach this problem by expanding the atomic dipole operator in the most natural set of states, the dressed states of the atom–driving field system, and derive the relaxation rates of the dressed coherences which govern the linewidths of the resonance fluorescence spectrum in the strong driving field approximation. We derive an explicit expression for the damping rate that sets the width of one of the outer sidebands and identify the contributions made by the two excited atomic states to this dressed coherence decay. It turns out that the vacuum fluctuations at frequencies close to one of the two allowed atomic transitions may be dominated by those that affect the other transition because of the

* For a simple physical treatment see Keitel, Narducci, and Scully [1995]; the experimental paper of Zhu, Gauthier, and Mossberg [1991] is a classic.

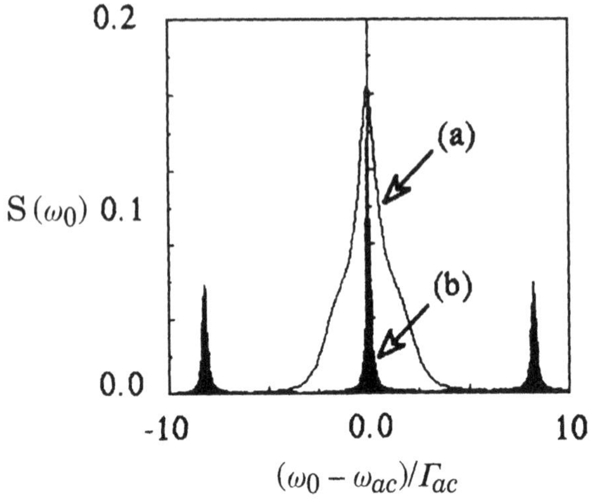

Fig. 10.8 Spectrum of spontaneous emission for the $|a\rangle \to |c\rangle$ transition, for resonant driving fields: (a) Standard spectrum with coherent field coupling $|a\rangle$ and $|c\rangle$, (b) same with strong $|b\rangle \to |c\rangle$ driving field. (From L. M. Narducci, M. O. Scully, G.-L. Oppo, P. Ru, and J. R. Tredicce, *Phys. Rev. A* **42**, 1630 (1990).

mutual coupling that the dressed states of the entire system impose between the bare excited atomic states.

The dressed states are eigenstates of the atom–driving field Hamiltonian

$$\mathscr{V} = \hbar g_1(|c\rangle\langle a|a_1^\dagger + |a\rangle\langle c|a_1) + \hbar g_2(|c\rangle\langle b|a_2^\dagger + |b\rangle\langle c|a_2) \quad (10.7.1)$$

where a_1 and a_2 are the annihilation operators for the field modes whose number states we denote by $|n\rangle$ and $|m\rangle$, respectively. Then the eigenstates are as given in Fig. 10.9 and the transition matrix elements between dressed states associated with the spontaneous emission of a photon are given in Fig. 10.10. As is depicted in Fig. 10.10, there are now five lines on both the $|a\rangle \to |c\rangle$ and the $|b\rangle \to |c\rangle$ transitions, though only three lines of each are depicted in Fig. 10.10.

The linewidths of the spectral components are assigned by the time evolution of the corresponding dressed coherences. As an explicit example, we consider the component of the dipole operator (see Eq. (10.2.16))

$$\begin{aligned}\langle\sigma_{ac}(t)\rangle &\equiv \mathrm{Tr}[|c\rangle\langle a|\rho(t)] \\ &= \frac{1}{2}(\ldots - \rho_{+-} + \ldots)e^{-i\omega_{ac}t},\end{aligned} \quad (10.7.2)$$

that oscillates with the frequency $\omega_{ac} + 2\Omega$, where

$$\Omega = \sqrt{\Omega_{R1}^2 + \Omega_{R2}^2}, \quad \Omega_{R1} = g_1\sqrt{\bar{n}}, \quad \Omega_{R2} = g_2\sqrt{\bar{m}}. \quad (10.7.3)$$

If the external driving fields are sufficiently strong, or more precisely if the effective Rabi frequency Ω is much larger than both spontaneous decay rates of the upper atomic states, we can apply the secular approximation and find that, see Appendix 10.E,

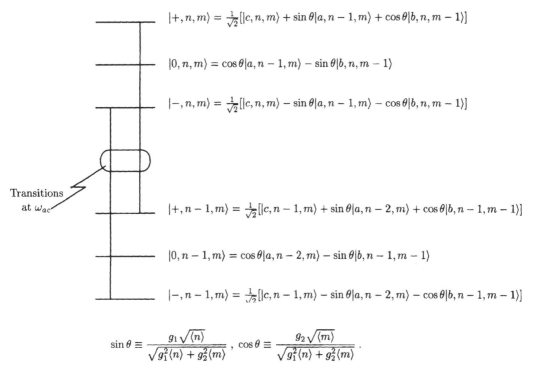

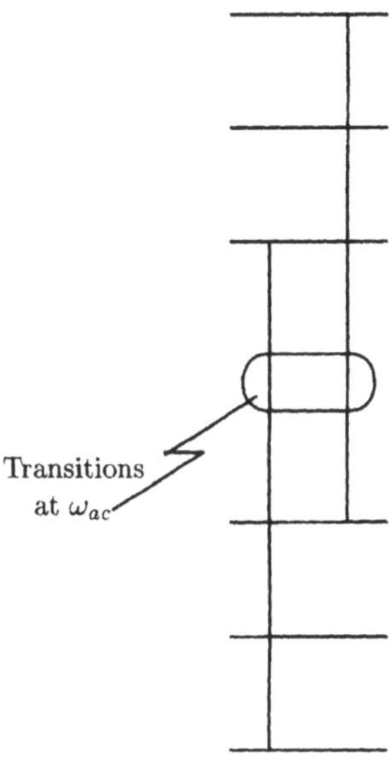

$$|+,n,m\rangle = \tfrac{1}{\sqrt{2}}[|c,n,m\rangle + \sin\theta|a,n-1,m\rangle + \cos\theta|b,n,m-1\rangle]$$

$$|0,n,m\rangle = \cos\theta|a,n-1,m\rangle - \sin\theta|b,n,m-1\rangle$$

$$|-,n,m\rangle = \tfrac{1}{\sqrt{2}}[|c,n,m\rangle - \sin\theta|a,n-1,m\rangle - \cos\theta|b,n,m-1\rangle]$$

$$|+,n-1,m\rangle = \tfrac{1}{\sqrt{2}}[|c,n-1,m\rangle + \sin\theta|a,n-2,m\rangle + \cos\theta|b,n-1,m-1\rangle]$$

$$|0,n-1,m\rangle = \cos\theta|a,n-2,m\rangle - \sin\theta|b,n-1,m-1\rangle$$

$$|-,n-1,m\rangle = \tfrac{1}{\sqrt{2}}[|c,n-1,m\rangle - \sin\theta|a,n-2,m\rangle - \cos\theta|b,n-1,m-1\rangle]$$

$$\sin\theta \equiv \frac{g_1\sqrt{\langle n\rangle}}{\sqrt{g_1^2\langle n\rangle + g_2^2\langle m\rangle}}, \quad \cos\theta \equiv \frac{g_2\sqrt{\langle m\rangle}}{\sqrt{g_1^2\langle n\rangle + g_2^2\langle m\rangle}}.$$

Fig. 10.9 Eigenstates corresponding to n photons in the first driving field and m photons in the second decaying to states with $n-1$ and m photons, respectively. Relevant portion of states associated with transitions at ω_{ac} are underscored with wavy lines.

$$\dot{\rho}_{+-} = -2i\Omega\rho_{+-} - \gamma_{+-}\rho_{+-}, \tag{10.7.4}$$

where

$$\gamma_{+-} = \frac{3}{4}\left(\Gamma_a\frac{\Omega_{R1}^2}{\Omega^2} + \Gamma_b\frac{\Omega_{R2}^2}{\Omega^2}\right). \tag{10.7.5}$$

It is clear that when $\Gamma_a \gg \Gamma_b$ but $\Omega_{R2} \gg \Omega_{R1}$, the width of the line at $\omega_{ac} + 2\Omega$ goes as Γ_b.

Thus we may say that the vacuum interaction associated with the $|b\rangle \to |c\rangle$ transition determines the width of the $|a\rangle \to |c\rangle$ peak at $\omega_{ac} + 2\Omega$. This effect has been experimentally observed, by Zhu, Gauthier and Mossberg [1991].

10.A Electric field operator in the far-zone approximation

We start by considering the interaction of the two-level atom with the radiation field, which is described by the following rotating-wave

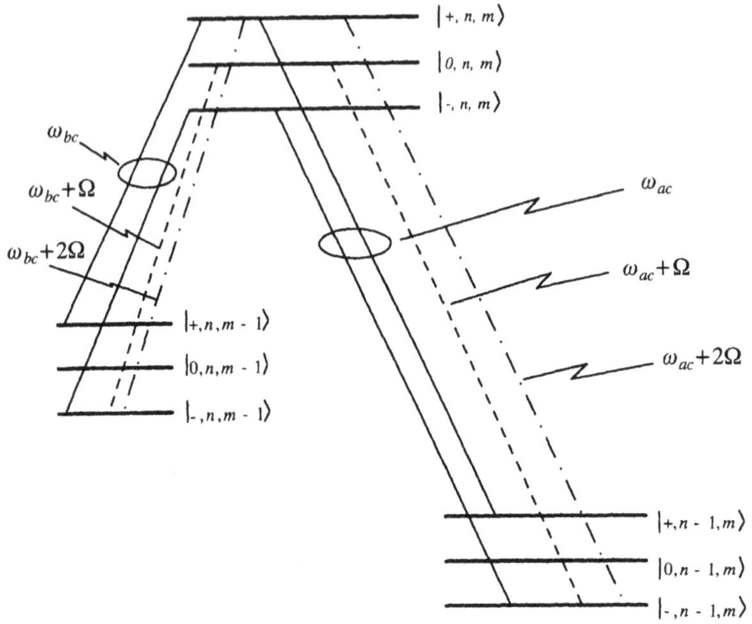

Fig. 10.10
Decay of states with n and m photons in driving fields to states having $n-1$, m and n, $m-1$ photons, respectively. Five peaks will be observed around both frequencies ω_{ac} and ω_{bc} of which only three each are shown.

$$\omega_{ac} \;:\; |\langle \pm, n, \,|\mathbf{r}|\pm, n-1, m\rangle|^2 = \frac{1}{4}\sin^2\theta\, |\langle a, n-1, m|\mathbf{r}|c, n-1, m\rangle|^2$$

$$\omega_{ac}+2\Omega \;:\; |\langle +, n, m|\mathbf{r}|-, n-1, m\rangle|^2 = \frac{1}{2}\sin^2\theta\, |\langle a, n-1, m|\mathbf{r}|c, n-1, m\rangle|^2$$

$$\omega_{ac}+\Omega \;:\; |\langle 0, n, m|\mathbf{r}|-, n-1, m\rangle|^2 = \frac{1}{2}\cos^2\theta\, |\langle a, n-1, m|\mathbf{r}|c, n-1, m\rangle|^2$$

approximation Hamiltonian (see Eq. (6.1.20))

$$\mathscr{H} = \frac{\hbar\omega}{2}\sigma_z + \sum_{\mathbf{k},\lambda} \hbar\nu_k a^\dagger_{\mathbf{k},\lambda} a_{\mathbf{k},\lambda} + \sum_{\mathbf{k},\lambda} \hbar g_{\mathbf{k},\lambda}\left(a_{\mathbf{k},\lambda}\sigma_+ e^{i\mathbf{k}\cdot\mathbf{r}_0} + a^\dagger_{\mathbf{k},\lambda}\sigma_- e^{-i\mathbf{k}\cdot\mathbf{r}_0}\right). \tag{10.A.1}$$

Here we have included the interaction between the atom and all the field modes characterized by wave vector $\mathbf{k}$ and polarization λ. We proceed by introducing the slowly varying operators $\tilde{a}_{\mathbf{k},\lambda}$ and $\tilde{\sigma}_-$ such that

$$a_{\mathbf{k},\lambda}(t) = \tilde{a}_{\mathbf{k},\lambda}(t) e^{-i\nu_{\mathbf{k}} t}, \tag{10.A.2a}$$

$$\sigma_-(t) = \tilde{\sigma}_-(t) e^{-i\omega t}. \tag{10.A.2b}$$

The Heisenberg equations of motion for these are

$$\dot{\tilde{a}}_{\mathbf{k},\lambda}(t) = -ig_{\mathbf{k},\lambda}\tilde{\sigma}_-(t) e^{-i(\omega-\nu_{\mathbf{k}})t - i\mathbf{k}\cdot\mathbf{r}_0}, \tag{10.A.3a}$$

$$\dot{\tilde{\sigma}}_-(t) = \sum_{\mathbf{k},\lambda} ig_{\mathbf{k},\lambda}\sigma_z(t)\tilde{a}_{\mathbf{k},\lambda}(t) e^{i(\omega-\nu_{\mathbf{k}})t + i\mathbf{k}\cdot\mathbf{r}_0}. \tag{10.A.3b}$$

These equations can be formally integrated to yield

$$\tilde{a}_{\mathbf{k},\lambda}(t) = \tilde{a}_{\mathbf{k},\lambda}(0) - ig_{\mathbf{k},\lambda}e^{-i(\omega-\nu_{\mathbf{k}})t-i\mathbf{k}\cdot\mathbf{r}_0}\int_0^t dt'\tilde{\sigma}_-(t')e^{i(\omega-\nu_{\mathbf{k}})(t-t')}, \tag{10.A.4a}$$

$$\tilde{\sigma}_-(t) = \tilde{\sigma}_-(0) + \sum_{\mathbf{k},\lambda} ig_{\mathbf{k},\lambda}e^{i(\omega-\nu_{\mathbf{k}})t+i\mathbf{k}\cdot\mathbf{r}_0} \times \int_0^t dt'\sigma_z(t')\tilde{a}_{\mathbf{k},\lambda}(t')e^{-i(\omega-\nu_{\mathbf{k}})(t-t')}. \tag{10.A.4b}$$

The first terms in these equations represent the free evolution of the field and atomic operators in the absence of interaction. In the following sections, we shall focus on the contribution to the field due to its interaction with the atom.

It follows from Eq. (1.1.30), that the positive frequency part of the electric field operator is

$$\mathbf{E}^{(+)}(\mathbf{r},t) = \sum_{\mathbf{k},\lambda} \mathscr{E}_{\mathbf{k}}\hat{\epsilon}_{\mathbf{k}}^{(\lambda)}a_{\mathbf{k},\lambda}(t)e^{i\mathbf{k}\cdot\mathbf{r}}, \tag{10.A.5}$$

where $\mathscr{E}_{\mathbf{k}} = (\hbar\nu_k/2\epsilon_0 V)^{1/2}$. On substituting for $a_{\mathbf{k},\lambda}(t)$ from Eqs. (10.A.2a) and (10.A.4a), we obtain

$$\mathbf{E}^{(+)}(\mathbf{r},t) = \left(\frac{i}{16\pi^3\epsilon_0}\right)e^{-i\omega t}\int d^3k\sum_\lambda \hat{\epsilon}_{\mathbf{k}}^{(\lambda)}[\hat{\epsilon}_{\mathbf{k}}^{(\lambda)}\cdot\wp]\nu_k e^{i\mathbf{k}\cdot(\mathbf{r}-\mathbf{r}_0)} \times \int_0^t dt'\tilde{\sigma}_-(t')e^{i(\omega-\nu_{\mathbf{k}})(t-t')}, \tag{10.A.6}$$

where we have recalled the definition of $g_{\mathbf{k}}$ from Eq. (6.1.8) and replaced the sum by an integral via

$$\sum_{\mathbf{k}} \to \frac{V}{(2\pi)^3}\int d^3k.$$

Recalling from Eq. (1.1.36), that

$$\sum_\lambda \hat{\epsilon}_{\mathbf{k}}^{(\lambda)}\hat{\epsilon}_{\mathbf{k}}^{(\lambda)} = 1 - \frac{\mathbf{kk}}{k^2},$$

we have the vector field operator in a useful form

$$\mathbf{E}^{(+)}(\mathbf{r},t) = \left(\frac{i}{16\pi^3\epsilon_0}\right)e^{-i\omega t}\int dk d\theta d\varphi k^2\sin\theta\left[\wp - \frac{\mathbf{k}(\mathbf{k}\cdot\wp)}{k^2}\right]\nu_k e^{i\mathbf{k}\cdot(\mathbf{r}-\mathbf{r}_0)} \times \int_0^t dt'\tilde{\sigma}_-(t')e^{i(\omega-\nu_{\mathbf{k}})(t-t')}. \tag{10.A.7}$$

Next we assume that the line joining the atom to the observation is

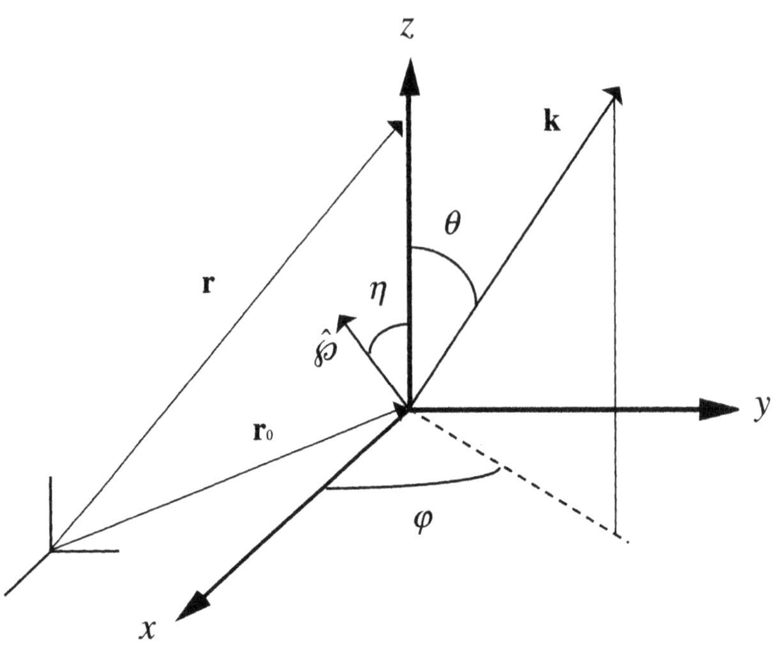

Fig. 10.11
The atomic dipole and the **k**-vector of the electric field in polar coordinates.

along the z-axis, which is parallel to $\mathbf{r}-\mathbf{r}_0$, and the electric dipole is in the x-z plane making an angle η with the z-axis, see Fig. 10.11. Then the vectors $\mathbf{k}$ and $\hat{\wp}$ are defined in polar coordinates by

$$\mathbf{k} = k(\hat{x}\sin\theta\cos\varphi + \hat{y}\sin\theta\sin\varphi + \hat{z}\cos\theta), \tag{10.A.8}$$

$$\hat{\wp} = \wp(\hat{x}\sin\eta + \hat{z}\cos\eta). \tag{10.A.9}$$

Consider first the angular integrations in Eq. (10.A.7). The φ-integration yields for the x-component

$$\begin{aligned}&\int_0^{2\pi} d\varphi \left[\hat{x}\cdot\hat{\wp} - \frac{(\hat{x}\cdot\mathbf{k})(\mathbf{k}\cdot\hat{\wp})}{k^2}\right]\\ &= \int_0^{2\pi} d\varphi\wp\,[\sin\eta - \sin\theta\cos\varphi(\sin\eta\sin\theta\cos\varphi + \cos\eta\cos\theta)]\\ &= 2\pi\wp\sin\eta(1 - \frac{1}{2}\sin^2\theta),\end{aligned} \tag{10.A.10}$$

for the y-component

$$\begin{aligned}&\int_0^{2\pi} d\varphi \left[\hat{y}\cdot\hat{\wp} - \frac{(\hat{y}\cdot\mathbf{k})(\mathbf{k}\cdot\hat{\wp})}{k^2}\right]\\ &= \int_0^{2\pi} d\varphi\wp\,[0 - \sin\theta\sin\varphi(\sin\eta\sin\theta\cos\varphi + \cos\eta\cos\theta)]\\ &= 0,\end{aligned} \tag{10.A.11}$$

and for the z-component

$$\int_0^{2\pi} d\varphi \left[\hat{z}\cdot\hat{\wp} - \frac{(\hat{z}\cdot\mathbf{k})(\mathbf{k}\cdot\hat{\wp})}{k^2}\right]$$
$$= \int_0^{2\pi} d\varphi\wp\,[\cos\eta - \cos\theta(\sin\eta\sin\theta\cos\varphi + \cos\eta\cos\theta)]$$
$$= 2\pi\wp\cos\eta(1-\cos^2\theta)$$
$$= 2\pi\wp\cos\eta\sin^2\theta. \tag{10.A.12}$$

Thus the y-component of the electric field vanishes. Next we consider the θ-integration in Eq. (10.A.7). By writing

$$e^{i\mathbf{k}\cdot(\mathbf{r}-\mathbf{r}_0)} = e^{ik|\mathbf{r}-\mathbf{r}_0|\cos\theta},$$

and using $\cos\theta$ as a new variable $\mu = \cos\theta$, we obtain for the x-component

$$2\pi\wp\sin\eta\int_0^{\pi} d\theta\sin\theta\left(1-\frac{1}{2}\sin^2\theta\right)e^{i\mathbf{k}\cdot(\mathbf{r}-\mathbf{r}_0)}$$
$$= 2\pi\wp\sin\eta\int_{-1}^{1} d\mu\left[1-\frac{1}{2}(1-\mu^2)\right]e^{ik|\mathbf{r}-\mathbf{r}_0|\mu}$$
$$= 2\pi\wp\sin\eta\left[\frac{e^{ik|\mathbf{r}-\mathbf{r}_0|}-e^{-ik|\mathbf{r}-\mathbf{r}_0|}}{ik|\mathbf{r}-\mathbf{r}_0|} + O\left(\frac{1}{|\mathbf{r}-\mathbf{r}_0|^2}\right)\right], \tag{10.A.13}$$

and for the z-component

$$2\pi\wp\cos\eta\int_0^{\pi} d\theta\sin^3\theta e^{i\mathbf{k}\cdot(\mathbf{r}-\mathbf{r}_0)}$$
$$= 2\pi\wp\cos\eta\int_{-1}^{1} d\mu(1-\mu^2)e^{ik|\mathbf{r}-\mathbf{r}_0|\mu}$$
$$\sim O\left(\frac{1}{|\mathbf{r}-\mathbf{r}_0|^2}\right), \tag{10.A.14}$$

where we have used the results (10.A.10) and (10.A.12). In the far-field region, the terms proportional to $O(1/|\mathbf{r}-\mathbf{r}_0|^2)$ can be neglected, and the z-component of the electric field also vanishes. On combining Eqs. (10.A.7) and (10.A.13), we obtain

$$\mathbf{E}^{(+)}(\mathbf{r},t) = \left(\frac{c\wp\sin\eta\hat{x}}{8\pi^2\epsilon_0|\mathbf{r}-\mathbf{r}_0|}\right)e^{-i\omega t}\int_0^{\infty} dk k^2\left(e^{ik|\mathbf{r}-\mathbf{r}_0|} - e^{-ik|\mathbf{r}-\mathbf{r}_0|}\right)$$
$$\times\int_0^t dt'\tilde{\sigma}_-(t')e^{i(\omega-\nu_\mathbf{k})(t-t')}. \tag{10.A.15}$$

We proceed as in Weisskopf–Wigner theory by replacing $k^2 \to (\omega/c)^2$, extend the limit of k-integration to $-\infty$ and upon performing the k-integration, obtain

$$\mathbf{E}^{(+)}(\mathbf{r},t) = \left(\frac{\omega^2 \wp \sin\eta\hat{x}}{4\pi\epsilon_0 c^2|\mathbf{r}-\mathbf{r}_0|}\right) \times \left[e^{-i\omega(t-|\mathbf{r}-\mathbf{r}_0|/c)} \int_0^t dt'\tilde{\sigma}_-(t')\delta\left(t-t'-\frac{|\mathbf{r}-\mathbf{r}_0|}{c}\right) - e^{-i\omega(t+|\mathbf{r}-\mathbf{r}_0|/c)} \int_0^t dt'\tilde{\sigma}_-(t')\delta\left(t-t'+\frac{|\mathbf{r}-\mathbf{r}_0|}{c}\right)\right]. \tag{10.A.16}$$

Ignoring the incoming wave contribution and going back to $\sigma_-(t)$, see Eq. (10.A.2b), we readily find Eq. (10.1.1).

10.B The equations of motion for and exact solution of the density matrix in a dressed-state basis

10.B.1 Deriving the equation of motion in the dressed-state basis

In Section 10.2, it has been shown that in the strong field limit and within the secular approximation the solution for the system in the dressed-state basis provides answers to the positions and widths of the spectral lines without involved calculations. Furthermore, as we will see below, an exact solution to the problem can also be obtained with less effort in the dressed-state basis than in the bare-state basis.

First we observe that, from Eq. (10.2.11),

$$|a\rangle = \frac{1}{\sqrt{2}}(|+\rangle + |-\rangle), \tag{10.B.1a}$$

$$|b\rangle = \frac{1}{\sqrt{2}}(|+\rangle - |-\rangle). \tag{10.B.1b}$$

Using Eqs. (10.B.1a,10.B.1b) we rewrite the interaction Hamiltonian $\mathscr{V}$ by expressing the lowering and raising operators σ_+ and σ_- in terms of $|\pm\rangle$ states

$$\mathscr{V} = \frac{\hbar\Omega_R}{2}\left(|+\rangle\langle+| - |-\rangle\langle-|\right), \tag{10.B.2}$$

$$\sigma_- = |b\rangle\langle a| = \frac{1}{2}\left(|+\rangle\langle+| + |+\rangle\langle-| - |-\rangle\langle+| - |-\rangle\langle-|\right), \tag{10.B.3}$$

$$\sigma_+\sigma_- = |a\rangle\langle a| = \frac{1}{2}(|+\rangle\langle+| + |+\rangle\langle-| + |-\rangle\langle+| + |-\rangle\langle-|). \tag{10.B.4}$$

Upon inserting Eqs. (10.B.2)–(10.B.4) into Eq. (10.2.14) and taking the appropriate matrix elements of the density operator, we obtain

$$\dot{\rho}_{++} = -\frac{\Gamma}{2}\rho_{++} + \frac{\Gamma}{4}, \tag{10.B.5a}$$

$$\dot{\rho}_{+-} = -\left(\frac{3\Gamma}{4} + i\Omega_R\right)\rho_{+-} - \frac{\Gamma}{4}\rho_{-+} - \frac{\Gamma}{2}, \tag{10.B.5b}$$

$$\dot{\rho}_{-+} = -\left(\frac{3\Gamma}{4} - i\Omega_R\right)\rho_{-+} - \frac{\Gamma}{4}\rho_{+-} - \frac{\Gamma}{2}. \tag{10.B.5c}$$

10.B.2 Solving the equations of motion

The advantage of the dressed-state basis is clearly seen in that Eq. (10.B.5a) is decoupled from Eqs. (10.B.5b) and (10.B.5c). The solution for $\rho_{++}(t)$ is already given in Eq. (10.2.19a). To solve for ρ_{+-} and ρ_{-+}, it is convenient to rewrite Eqs. (10.B.5b) and (10.B.5c) in the matrix form

$$\dot{R}(t) = -MR(t) + B, \tag{10.B.6}$$

where

$$R(t) = \begin{pmatrix} \rho_{+-}(t) \\ \rho_{-+}(t) \end{pmatrix}, \tag{10.B.7}$$

$$M = \begin{pmatrix} 3\Gamma/4 + i\Omega_R & \Gamma/4 \\ \Gamma/4 & 3\Gamma/4 - i\Omega_R \end{pmatrix}, \tag{10.B.8}$$

$$B = -\frac{\Gamma}{2}\begin{pmatrix} 1 \\ 1 \end{pmatrix}. \tag{10.B.9}$$

In order to solve Eq. (10.B.6), we seek the eigenstates and eigenvalues of M such that

$$M\mathbf{v}_i = \lambda_i \mathbf{v}_i \qquad (i = 1, 2), \tag{10.B.10}$$

and therefore the matrix made up of the $\mathbf{v}_i$ eigenvectors (column matrices) takes the form

$$V = (\,\mathbf{v}_1 \quad \mathbf{v}_2\,), \tag{10.B.11}$$

and

$$MV = (\,\lambda_1\mathbf{v}_1 \quad \lambda_2\mathbf{v}_2\,). \tag{10.B.12}$$

Likewise, we need the inverse of V, which we may find by conventional matrix methods, and write as a matrix of the row vectors

$$V^{-1} = \begin{pmatrix} \check{\mathbf{v}}_1 \\ \check{\mathbf{v}}_2 \end{pmatrix}. \tag{10.B.13}$$

It may help the student of modern quantum mechanics (who knows Dirac notation better than matrix mathematics!) to note that

$$\mathbf{v}_i \longleftrightarrow |i\rangle, \qquad \check{\mathbf{v}}_j \longleftrightarrow \langle j|. \tag{10.B.14}$$

Now by construction, (i.e., this is how we define V^{-1})

$$VV^{-1} = \mathbf{1}, \tag{10.B.15}$$

which in terms of our eigenvectors and reciprocal eigenvectors can be written as

$$VV^{-1} = (\mathbf{v}_1 \quad \mathbf{v}_2)\begin{pmatrix} \check{\mathbf{v}}_1 \\ \check{\mathbf{v}}_2 \end{pmatrix} = \mathbf{v}_1\check{\mathbf{v}}_1 + \mathbf{v}_2\check{\mathbf{v}}_2 = \mathbf{1}, \tag{10.B.16}$$

(again note that this is like $|1\rangle\langle 1| + |2\rangle\langle 2| = \mathbf{1}$).

Using (10.B.15) we may write (10.B.6) as

$$\dot{R}(t) = -VV^{-1}MVV^{-1}R(t) + B, \tag{10.B.17}$$

or

$$\frac{d}{dt}[V^{-1}R(t)] = -(V^{-1}MV)[V^{-1}R(t)] + V^{-1}B. \tag{10.B.18}$$

Since

$$V^{-1}MV = \begin{pmatrix} \lambda_1 & 0 \\ 0 & \lambda_2 \end{pmatrix} \equiv D, \tag{10.B.19}$$

we have

$$\frac{d}{dt}[V^{-1}R(t)] = -D[V^{-1}R(t)] + V^{-1}B, \tag{10.B.20}$$

which has the solution

$$V^{-1}R(t) = e^{-Dt}V^{-1}R(0) + \int_0^t dt'\, e^{-D(t-t')}V^{-1}B. \tag{10.B.21}$$

It follows readily from Eq. (10.B.21) that

$$R(t) = (Ve^{-Dt}V^{-1})R(0) + VD^{-1}(1 - e^{-Dt})V^{-1}B. \tag{10.B.22}$$

In view of the fact that D is diagonal, e^{-Dt} is simply equal to

$$e^{-Dt} = \begin{pmatrix} e^{-\lambda_1 t} & 0 \\ 0 & e^{-\lambda_2 t} \end{pmatrix}. \tag{10.B.23}$$

Now we can see that once the eigenvalues and eigenvectors are known, $\rho_{+-}(t)$ and $\rho_{-+}(t)$ can be derived from Eq. (10.B.22) by straightforward matrix multiplications.

For our problem, the eigenvalues of M are

$$\lambda_{1,2} = \frac{3\Gamma}{4} \pm i\mu, \qquad \mu = \sqrt{\Omega_R^2 - \left(\frac{\Gamma}{4}\right)^2}, \tag{10.B.24}$$

with corresponding eigenstates

$$\mathbf{v}_1 = \begin{pmatrix} \cos\theta \\ \sin\theta \end{pmatrix}, \qquad \mathbf{v}_2 = \begin{pmatrix} -\sin\theta \\ \cos\theta \end{pmatrix}, \tag{10.B.25}$$

where $\tan\theta = -(4i/\Gamma)(\Omega_R - \mu)$. With these, V, V^{-1}, and D can be constructed and after some algebra, we arrive at the following expression for $\rho_{+-}(t)$

$$\begin{aligned}
\rho_{+-} = &\frac{1}{2\mu}(2\mu\cos\mu t - 2i\Omega_R\sin\mu t)e^{-3\Gamma t/4}\rho_{+-}(0) \\
&-\frac{\Gamma}{4\mu}\sin\mu t e^{-3\Gamma t/4}\rho_{-+}(0) \\
&+\frac{\Gamma}{2\mu}\left\{\frac{2\mu(\Gamma/2 - i\Omega_R)}{\Gamma^2 + 2\Omega_R^2}\cos\mu t\right. \\
&\left.-\left[1 - \frac{(3\Gamma/2)(\Gamma/2 - i\Omega_R)}{\Gamma^2 + 2\Omega^2}\right]\sin\mu t\right\}e^{-3\Gamma t/4} \\
&-\frac{\Gamma(\Gamma/2 - i\Omega_R)}{\Gamma^2 + 2\Omega_R^2}.
\end{aligned} \tag{10.B.26}$$

On substituting for $\rho_{++}(t)$ from Eq. (10.2.19a) and $\rho_{+-}(t)$ from Eq. (10.B.26) into

$$\sigma_- = \rho_{ab} = \rho_{++} - i\,\mathrm{Im}(\rho_{+-}) - \frac{1}{2}, \tag{10.B.27}$$

and noting that

$$\rho_{++} = \frac{1}{2}(1 + \rho_{ab} + \rho_{ba}), \tag{10.B.28}$$

$$\rho_{+-} = \frac{1}{2}(2\rho_{aa} - 1 - \rho_{ab} + \rho_{ba}), \tag{10.B.29}$$

we obtain $\langle\sigma_-(t+\tau)\rangle$ and $\langle\sigma_z(t+\tau)\rangle$ in the form (10.5.5)–(10.5.14), which are needed for evaluating the power spectrum. For a comparison, we sketch the solution to the same problem in the 'conventional' bare-state basis in Appendix 10.C, using a slightly different technique.

10.C The equations of motion for and exact solution of the density matrix in the bare-state basis

It is instructive to solve the resonance fluorescence problem in the bare-state basis if only to show how much easier it is in the dressed-state basis. The equations of motion for the various density matrix elements can be obtained from Eqs. (10.2.14) and (10.2.15) by substituting for $\mathscr{V}$ from Eq. (10.2.12) and then taking the appropriate elements of the density operator. The resulting equations can be written in a compact matrix form as

$$\dot{R}(t) = -MR(t) + B, \tag{10.C.1}$$

where (with $\rho_{bb} = 1 - \rho_{aa}$)

$$R = \begin{pmatrix} \rho_{ab} \\ \rho_{aa} \\ \rho_{ba} \end{pmatrix}, \tag{10.C.2}$$

$$M = \begin{pmatrix} \Gamma/2 & -i\Omega_R & 0 \\ -i\Omega_R/2 & \Gamma & i\Omega_R/2 \\ 0 & i\Omega_R & \Gamma/2 \end{pmatrix}, \tag{10.C.3}$$

$$B = -\frac{i}{2}\begin{pmatrix} \Omega_R \\ 0 \\ -\Omega_R \end{pmatrix}. \tag{10.C.4}$$

The striking difference between Eqs. (10.B.5a-10.B.5c) (or (10.B.6)) and Eq. (10.C.1) above is that in the set of equations (10.B.5a -10.B.5c), (10.B.5a) is decoupled from the others, whereas in (10.C.1), all three equations for ρ_{ab}, ρ_{aa}, and ρ_{ba} are entangled. This means that in the bare-state basis, we have to deal with 3×3 matrices while in the dressed-state basis, we have only 2×2 matrices.

Equation (10.C.1) can be solved by first finding the eigenvalues of the matrix M, which read

$$\lambda_1 = \frac{\Gamma}{2}, \qquad \lambda_2 = \frac{3\Gamma}{4} + i\mu, \qquad \lambda_3 = \frac{3\Gamma}{4} - i\mu, \tag{10.C.5}$$

where as before $\mu = \sqrt{\Omega_R^2 - \Gamma^2/16}$. By denoting again the eigenvectors and reciprocal eigenvectors by $\mathbf{v}_l$ and $\check{\mathbf{v}}_l$ ($l = 1, 2, 3$), respectively, we have similarly to Eq. (10.B.16)

$$\sum_{l=1}^{3} \mathbf{v}_l \check{\mathbf{v}}_l = 1. \tag{10.C.6}$$

Using Eq. (10.C.6), it can be shown that

$$e^{-Mt}O = \sum_{l=1}^{3} (\check{\mathbf{v}}_l O)\mathbf{v}_l e^{-\lambda_l t}, \tag{10.C.7}$$

where O is a 3×1 column matrix. A solution can now be found as follows:

$$\begin{aligned} R(t+\tau) &= e^{-M\tau}R(t) + \int_t^{t+\tau} e^{-M(t+\tau-\tau')}Bd\tau' \\ &= e^{-M\tau}R(t) + (1 - e^{-M\tau})M^{-1}B \\ &= \sum_l \{\check{\mathbf{v}}_l[R(t) - M^{-1}B]\mathbf{v}_l e^{-\lambda_l \tau}\} + M^{-1}B. \end{aligned} \tag{10.C.8}$$

After finding the inverse matrix M^{-1} and substituting various quantities into equation (10.C.8), one can find Eqs. (10.5.5)–(10.5.14) for $\langle \sigma_-(t+\tau) \rangle$ and $\langle \sigma_z(t+\tau) \rangle$ (Problem 10.3).

10.D Power spectrum in the stationary regime

According to the Wiener–Khintchine theorem, the power spectrum $S(\omega_0)$ is given in terms of the two-time correlation function of the radiation field by

$$S(\omega_0) = \frac{1}{2\pi} \lim_{T\to\infty} \frac{1}{T} \int_0^T dt \int_0^T dt' \langle E^{(-)}(t)E^{(+)}(t') \rangle e^{-i\omega_0(t-t')}. \tag{10.D.1}$$

Under the stationary condition, the correlation function $\langle E^{(-)}(t)$ $E^{(+)}(t') \rangle$ depends only on the time difference $\tau = t - t'$ and Eq. (10.D.1) becomes

$$\begin{aligned} S(\omega_0) &= \frac{1}{2\pi} \lim_{T\to\infty} \frac{1}{T} \int_0^T dt \left(\int_0^t dt' + \int_t^T dt' \right) \\ &\quad \times \langle E^{(-)}(t)E^{(+)}(t') \rangle e^{-i\omega_0(t-t')} \\ &= \frac{1}{2\pi} \lim_{T\to\infty} \frac{1}{T} \int_0^T dt \left[\int_0^t d\tau \langle E^{(-)}(\tau)E^{(+)}(0) \rangle e^{-i\omega_0\tau} \right. \\ &\quad \left. + \int_0^{T-t} d\tau \langle E^{(-)}(0)E^{(+)}(\tau) \rangle e^{i\omega_0\tau} \right]. \end{aligned} \tag{10.D.2}$$

Provided that the field operators are correlated only over a short period of time, we can extend the upper limit of the τ-integrations to infinity with no significant change. Since we have

$$\langle E^{(-)}(\tau)E^{(+)}(0)\rangle = \langle E^{(-)}(0)E^{(+)}(\tau)\rangle^{*}, \tag{10.D.3}$$

it follows from Eq. (10.D.2) that

$$S(\omega_0) = \frac{1}{\pi}\text{Re}\int_0^{\infty} d\tau \langle E^{(-)}(0)E^{(+)}(\tau)\rangle e^{i\omega_0\tau}. \tag{10.D.4}$$

10.E Derivation of Eq. (10.7.5)

We note that (in an obvious notation) the damping operator in the dressed-state picture goes as

$$\begin{aligned}
&\left(\frac{\partial\rho_{+-}(t)}{\partial t}\right)_{\text{damping}} \\
&= \langle +|\left(\mathscr{L}_{ac}\rho + \mathscr{L}_{bc}\rho\right)|-\rangle \\
&= -\frac{\Gamma_a}{2}\langle +|\left(|a\rangle\langle a|\rho + \rho|a\rangle\langle a| - 2|c\rangle\langle a|\rho|a\rangle\langle c|\right)|-\rangle \\
&\quad -\frac{\Gamma_b}{2}\langle +|\left(|b\rangle\langle b|\rho + \rho|b\rangle\langle b| - 2|c\rangle\langle b|\rho|b\rangle\langle c|\right)|-\rangle,
\end{aligned} \tag{10.E.1}$$

and upon inserting a complete set of states

$$\begin{aligned}
&\left(\frac{\partial\rho_{+-}(t)}{\partial t}\right)_{\text{damping}} \\
&= -\frac{\Gamma_a}{2}\left(\sum_{\sigma}\langle +|a\rangle\langle a|\sigma\rangle\rho_{\sigma-} + \sum_{\sigma}\rho_{+\sigma}\langle\sigma|a\rangle\langle a|-\rangle\right. \\
&\quad \left. -2\sum_{\sigma\tau}\langle +|c\rangle\langle a|\sigma\rangle\rho_{\sigma\tau}\langle\tau|a\rangle\langle c|-\rangle\right) \\
&\quad -\frac{\Gamma_b}{2}\left(\sum_{\sigma}\langle +|b\rangle\langle b|\sigma\rangle\rho_{\sigma-} + \sum_{\sigma}\rho_{+\sigma}\langle\sigma|b\rangle\langle b|-\rangle\right. \\
&\quad \left. -2\sum_{\sigma\tau}\langle +|c\rangle\langle b|\sigma\rangle\rho_{\sigma\tau}\langle\tau|b\rangle\langle c|-\rangle\right).
\end{aligned} \tag{10.E.2}$$

In the strong field limit we keep only terms which go as ρ_{+-} and this yields

$$\left(\frac{\partial\rho_{+-}(t)}{\partial t}\right)_{\text{damping}}$$
$$= -\frac{\Gamma_a}{2}\left(\langle +|a\rangle\langle a|+\rangle + \langle -|a\rangle\langle a|-\rangle\right.$$
$$\left. - 2\langle +|c\rangle\langle a|+\rangle\langle -|a\rangle\langle c|-\rangle\right)\rho_{+-}$$
$$- -\frac{\Gamma_b}{2}\left(\langle +|b\rangle\langle b|+\rangle + \langle -|b\rangle\langle b|-\rangle\right.$$
$$\left. - 2\langle +|c\rangle\langle b|-\rangle\langle -|b\rangle\langle c|-\rangle\right)\rho_{+-}$$
$$= -\frac{3}{4}(\Gamma_a \sin^2\theta + \Gamma_b \cos^2\theta)\rho_{+-}$$
$$= -\frac{3}{4}\left(\frac{\Gamma_a\Omega_{R1}^2 + \Gamma_b\Omega_{R2}^2}{\Omega^2}\right)\rho_{+-} \equiv -\gamma_{+-}\rho_{+-}. \tag{10.E.3}$$

Thus $\rho_{+-}(t) \sim e^{-\gamma_{+-}t}\rho_{+-}(0)$, i.e., the dipole $P_{ac}(t) = \wp_{ac}\langle\sigma_{ac}(t)\rangle$ has a component which decays as $e^{-\gamma_{+-}t}$ and therefore (by the regression theorem) so does $\langle\sigma_{ac}^{\dagger}(0)\sigma_{ac}(t)\rangle$.

Problems

10.1 Show that the atom–field dressed states $|+,n\rangle$ and $|-,n\rangle$, as given in Eq. (10.2.9), are eigenstates of the Hamiltonian (10.2.8) with corresponding eigenvalues $E(\pm,n) = \pm\hbar\Omega_n/2$, where $\Omega_n = 2g\sqrt{n+1}$.

10.2 (a) Show that completeness in the form $|+\rangle\langle+| + |-\rangle\langle-| = \mathbf{1}$ is equivalent to $|a\rangle\langle a| + |b\rangle\langle b| = \mathbf{1}$.

(b) Using part (a) show that

$$\langle\sigma_-(t)\rangle e^{i\omega t}$$
$$= \mathrm{Tr}\,[\sigma_-(0)\rho(t)]$$
$$= \langle +|\sigma_-|+\rangle\langle +|\rho|+\rangle + \langle +|\sigma_-|-\rangle\langle -|\rho|+\rangle$$
$$+\langle -|\sigma_-|+\rangle\langle +|\rho|-\rangle + \langle -|\sigma_-|-\rangle\langle -|\rho|-\rangle.$$

Note that here in contrast with Eq. (10.2.2), there appears the factor $e^{i\omega t}$ since we are now working in the interaction picture.

(c) Show that

$$\langle +|\sigma_-|+\rangle = \langle +|b\rangle\langle a|+\rangle = 1/2,$$
$$\langle +|\sigma_-|-\rangle = 1/2,$$

and

$$\langle -|\sigma_-|+\rangle = \langle -|\sigma_-|-\rangle = -1/2.$$

(d) Collecting the above, prove Eq. (10.2.16).

10.3 Derive Eqs. (10.5.5)–(10.5.14) following the method presented in Appendix 10.C.

10.4 Consider the resonance fluorescence from a two-level atom which is damped by a multi-mode squeezed vacuum. In this case, the equation of motion for the atomic density operator is given by (see Eq. (8.2.17))

$$\begin{aligned}
&\dot{\rho} \\
&= -\frac{i}{\hbar}[\mathscr{V},\rho] - \frac{\Gamma}{2}\cosh^2(r)(\sigma_+\sigma_-\rho - 2\sigma_-\rho\sigma_+ + \rho\sigma_+\sigma_-) \\
&-\frac{\Gamma}{2}\sinh^2(r)(\sigma_-\sigma_+\rho - 2\sigma_+\rho\sigma_- + \rho\sigma_-\sigma_+) \\
&-\Gamma e^{-i\theta}\sinh(r)\cosh(r)\sigma_-\rho\sigma_- \\
&-\Gamma e^{i\theta}\sinh(r)\cosh(r)\sigma_+\rho\sigma_+,
\end{aligned}$$

where r is the squeeze parameter, θ is the reference phase for the squeezed field, and $\mathscr{V}$ is the interaction picture Hamiltonian (10.2.12). Derive an expression for the resonance fluorescence spectrum for a strong coherent driving field. Show that the width of the central peak in the three-peak spectrum can be made narrower than the corresponding width for the atomic decay in vacuum by an appropriate choice of θ. (Hint: see H. J. Carmichael, A. S. Lane, and D. F. Walls, *Phys. Rev. Lett.* **58**, 2539 (1987).)

References and bibliography

Review articles

P. L. Knight and P. W. Milonni, *Phys. Rep.* **66**, 23 (1980).

R. Loudon, *Rep. Prog. Phys.* **43**, 227 (1980).

B. R. Mollow, in *Progress in Optics*, Vol. 19, ed. E. Wolf (North-Holland, Amsterdam 1981).

Three-peak spectrum

A. I. Burshtein, *Sov. Phys. JETP* **21**, 567 (1965); *ibid.* **22**, 939 (1966).

M. Newstein, *Phys. Rev.* **167**, 89 (1968).

B. R. Mollow, *Phys. Rev.* **188**, 1969 (1969).

H. J. Kimble and L. Mandel, *Phys. Rev. A* **13**, 2123 (1976).

C. Cohen-Tannoudji and S. Reynaud, *J. Phys. B: Atom. Molec. Phys.* **10**, 345 (1976).

B. Renaud, R. M. Whitley, and C. R. Stroud, Jr., *J. Phys. B: Atom. Molec. Phys.* **19**, L9 (1976).

K. Wódkiewicz and J. H. Eberly, *Ann. Phys.* **101**, 574 (1976).

J. H. Eberly, C. V. Kunasz, and K. Wódkiewicz, *J. Phys. B* **13**, 217 (1980).

D. Cardimona, M. Raymer, and C. Stroud, *J. Phys. B.* **15**, 55 (1982).

M. Sargent III, M. S. Zubairy, and F. de Martini, *Opt. Lett.* **8**, 76 (1983).

Experimental observation of three-peak spectrum

F. Schuda, C. R. Stroud Jr., and M. Hercher, *J. Phys. B: Atom. Molec. Phys.* **198**, L7 (1974).

H. Walther, in *Laser Spectroscopy*, ed. S. Haroche, J. C. Pebay-Peyroula, T. W. Hänsch, and S. E. Harris (Springer-Verlag, Berlin 1975).

F. Y. Wu, R. E. Grove, and S. Ezekiel, *Phys. Rev. Lett.* **35**, 1426 (1975).

W. Hartig, W. Rasmussen, R. Schieder, and H. Walther, *Z. Physik A* **278**, 205 (1976).

R. E. Grove, F. Y. Wu, and S. Ezekiel, *Phys. Rev. A* **15**, 227 (1977).

Sub-natural line narrowing and fluorescence from a driven V system

L. M. Narducci, M. O. Scully, G.-L. Oppo, P. Ru, and J. R. Tredicce, *Phys. Rev. A* **42**, 1630 (1990).

Y.-F. Zhu, D. J. Gauthier, and T. W. Mossberg, *Phys. Rev. Lett.* **66**, 2460 (1991).

C. H. Keitel, L. M. Narducci, and M. O. Scully, *Appl. Phys. B* **60**, S153 (1995).

Antibunching and squeezing

H. J. Carmichael and D. F. Walls, *J. Phys. B* **9**, L43 (1976); *ibid.* **9**, 1199 (1976).

H. J. Kimble and L. Mandel, *Phys. Rev. A* **13**, 2133 (1976).
D. Cohen-Tannoudji, in *Frontiers in Laser Spectroscopy* Vol. 1, ed. R. Balian, S. Haroche, and S. Liberman (North-Holland, Amsterdam 1977).
D. F. Walls and P. Zoller, *Phys. Rev. Lett.* **47**, 709 (1981).
L. Mandel, *Phys. Rev. Lett.* **49**, 136 (1982).
R. Short and L. Mandel, *Phys. Rev. Lett.* **51**, 384 (1983).

Experimental observation of photon antibunching

H. J. Kimble, M. Dagenais, and L. Mandel, *Phys. Rev. Lett.* **39**, 691 (1977).
M. Dagenais and L. Mandel, *Phys. Rev. A* **18**, 2217 (1978).
J. D. Cresser, J. Häger, G. Leuchs, M. Rateike, and H. Walther, in *Dissipative Systems in Quantum Optics*, ed. R. Bonifacio, (Springer-Verlag, Berlin 1982).

Effect of pump fluctuations

H. J. Kimble and L. Mandel, *Phys. Rev. A* **15**, 689 (1977).
P. L. Knight, W. A. Molander, and C. R. Stroud, Jr., *Phys. Rev. A* **17**, 1547 (1979).
M. G. Raymer and J. Cooper, *Phys. Rev. A* **20**, 2238 (1979).

Quantum regression theorem

L. Onsager, *Phys. Rev.* **37**, 405 (1931).
M. Lax, *Phys. Rev.* **129**, 2342 (1963).

CHAPTER 11

Quantum theory of the laser – density operator approach

In Chapters 5 and 7 we presented a treatment of laser physics in which the light is described as a classical Maxwell field while the lasing medium is described as a collection of atoms whose dynamic evolution is governed by the Schrödinger equation. This semi-classical theory of laser behavior is sufficient to describe a rich variety of phenomena. However, there are many questions which require a fully quantized theory of the radiation. For example, the photon statistics and linewidth of the laser can be properly understood only via the full quantum theory of a laser.

The laser linewidth is an important quantity. For example, it determines the fundamental limit of operation of an active ring laser gyroscope. The first fully quantized derivation of the laser linewidth general enough to include even the semiconductor laser linewidth problem utilized a quantum noise operator approach,* and is presented in chapter 12.

The photon statistical distribution for the laser is of interest for several reasons. Historically, it was initially thought by some that the statistical photon distribution should be a Bose–Einstein distribution. A little reflection shows that this can not be, since the laser is operating far from thermodynamic equilibrium. However, a different paradigm recognizes many atoms oscillating in phase produce what is essentially a classical current, and this would generate a coherent state; the statistics of which is Possionian. But, for example, the photon statistics of a typical Helium–Neon laser is substantially different from a Possionian distribution. Of course, well above threshold, the steady-state laser photon statistical distribution is Poissonian. The first derivation of the

* The quantum noise operator treatment was presented by M. Lax at the 1965 Physics of Quantum Electronics Conference; see *Proceedings of the Int. Conf. on the Phys. of Quantum Electronics*, ed. P. Kelley, B. Lax, and P. Tannenwald (New York 1966).

photon statistics for the laser used a density matrix* formalism and is the subject of the present chapter.

There is a deep sense in which laser threshold behavior is analogous to a second-order phase transition.† This is properly understood within the context of a fully quantized theory of the laser as is discussed in this chapter. Finally we note that the micromaser is, in a real sense, the archetype device requiring a quantum theory of the laser for its proper understanding. This is presented in chapter 13.

11.1 Equation of motion for the density matrix

Having treated the semiclassical theory in Chapter 5, and having treated the damping of quantum systems in Chapters 8 and 9, we now proceed toward a 'photon' description of laser operation. The quantum theory of laser radiation is basically a problem in nonequilibrium quantum statistical mechanics. We seek a coarse-grained equation of motion for the laser radiation-density matrix as it evolves due to the addition (and subsequent removal) of many excited atoms. We will derive such an equation in two ways.

Method I

We consider a system of atoms inside a cavity, with the atomic level structure shown in Fig. 11.1. To describe laser oscillation, the theory must include pumping and damping mechanisms. To obtain laser pumping action, we assume that the atoms of the gain medium are pumped into their upper state $|a\rangle$ from a low lying level $|g\rangle$ at random times t_0 at a rate λ_a. The lasing levels $|a\rangle$ and $|b\rangle$ can decay to levels $|c\rangle$ and $|d\rangle$, respectively, at rates γ_a and γ_b, which in turn decay to level $|g\rangle$ at decay rates γ_c and γ_d, respectively, as shown in the Fig. 11.1. The decay rate from level $|a\rangle$ to level $|g\rangle$ is denoted by γ_g. For the sake of simplicity, we assume $\gamma_a = \gamma_b = \gamma$ and consider the laser to be tuned to atomic resonance. The details of the dissipation mechanism for the field inside the cavity are not very important for the theory of laser. In the semiclassical theory the damping was represented by Ohmic currents (Section 5.5), here we may assume that there are only

* The density matrix treatment as developed by Scully, Lamb, and Stephen was also presented at the 1965 Physics of Quantum Electronics Conf., *ibid* p. 75. Very clear treatments of the subject are to be found in Loudon [1973] and Pike and Sarkar [1995].

† Graham and Haken [1970] and DiGiorgio and Scully [1970]. For an excellent account of the subject, see Haken [1975].

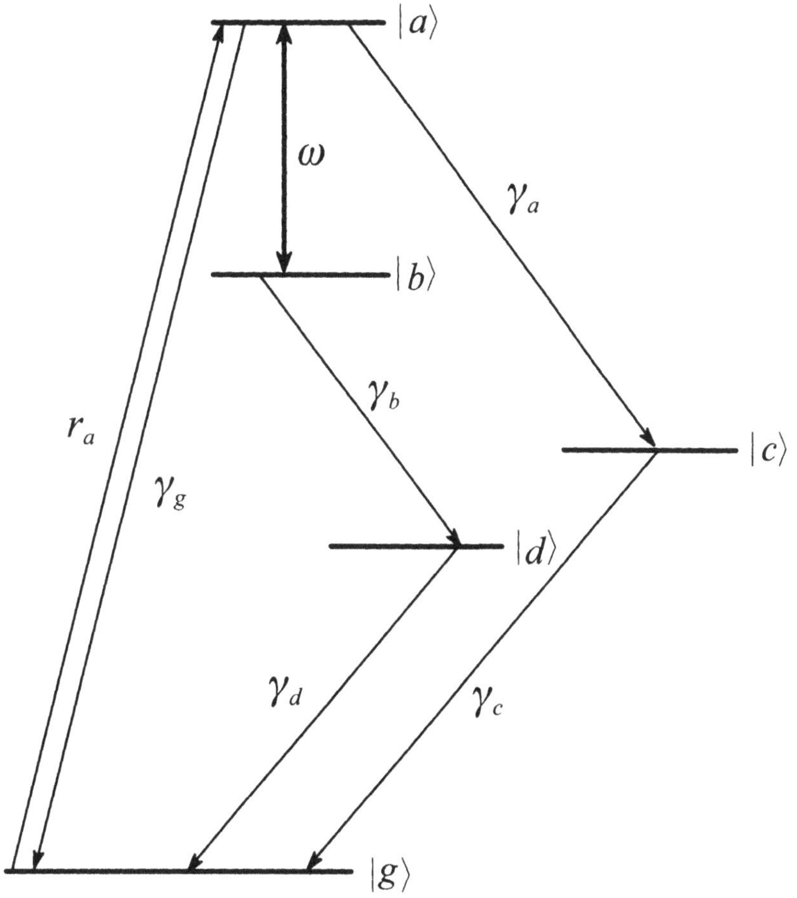

Fig. 11.1
Laser action takes place between the two excited energy levels $|a\rangle$ and $|b\rangle$ separated by a frequency ω. The level $|a\rangle$ is excited at a rate r_a, while levels $|a\rangle$ and $|b\rangle$ decay to levels $|c\rangle$ and $|d\rangle$, respectively, at rates γ_a and γ_b. Levels $|a\rangle$, $|c\rangle$, and $|d\rangle$ decay to level $|g\rangle$ at rates γ_g, γ_c, and γ_d, respectively.

transmission losses which can be accounted for by the interaction of the radiation field inside the cavity with a reservoir of vacuum modes representing the outside world (modes of the universe*) as seen through the partially transmitting mirrors.

In this approach we first seek the quantum mechanical analogue to Maxwell's equations for the field, and an expression corresponding to the atomic driving polarization. We will outline Method I emphasizing the relation with the semiclassical theory.

The Hamiltonian for the interaction of the active atoms, which are resonant with the single-mode laser field, in the interaction picture and the rotating-wave approximation (Eq. (6.2.8)) is

$$\mathscr{V} = \sum_i \hbar g(\sigma_+^i a + a^\dagger \sigma_-^i) = \sum_i \mathscr{V}^i,$$

* See R. Lang, M. O. Scully, and W. E. Lamb, Jr., *Phys. Rev. A* **7**, 1788 (1973) and R. Lang and M. O. Scully, *Opt. Commun.* **9**, 331 (1973).

where $\sigma_+^i = (|a\rangle\langle b|)^i$, $\sigma_-^i = (|b\rangle\langle a|)^i$ are the raising and lowering operators for the ith atom. Here we have included only the interaction of the lasing levels $|a\rangle$ and $|b\rangle$ with the field, the pumping from level $|g\rangle$ to $|a\rangle$ and the various level decays will enter phenomenologically as they did in the semiclassical laser theory.

The equation of motion for the laser field density matrix due to the interaction with the active lasing medium and the damping mechanism is

$$\dot{\rho}_{nn'} = -\frac{i}{\hbar}\mathrm{Tr}_{\mathrm{atoms}}\left[\mathscr{V},\rho\right]_{\mathrm{n,n'}} + (\mathscr{L}\rho)_{\mathrm{n,n'}}\,, \tag{11.1.1}$$

where from Eq. (8.3.3) we take

$$(\mathscr{L}\rho)_{n,n'} = -\frac{\mathscr{C}}{2}\,(n+n')\,\rho_{nn'} + \mathscr{C}\sqrt{(n+1)\,(n'+1)}\rho_{n+1,n'+1}. \tag{11.1.2}$$

The lasing term of (11.1.1) written in more detail is

$$\left(\frac{\delta\rho_{n,n'}}{\delta t}\right)_{\mathrm{gain}} = -\frac{i}{\hbar}\mathrm{Tr}_{\mathrm{atoms}}[\mathscr{V},\rho]_{\mathrm{n,n'}} = -\frac{i}{\hbar}\sum_{\{\alpha\}}[\mathscr{V},\rho]_{\{\alpha\},\mathrm{n};\{\alpha\},\mathrm{n'}} \tag{11.1.3a}$$

where $\{\alpha\} = \alpha_1, \alpha_2, ..., \alpha_N$ denotes the state of all N lasing atoms with $\alpha_i = a$ or b and this may be further simplified as

$$-\frac{i}{\hbar}\sum_{\alpha_i=a,b}\sum_i[\mathscr{V}^i,\rho^i]_{\alpha_i,n;\alpha_i,\,n'}\,, \tag{11.1.3b}$$

where the projection of the atom–field density matrix onto the space of the ith atom and the field is

$$\rho^i_{nn'}(t) = \sum_{\{\alpha'\}}\rho_{\{\alpha'\},n;\{\alpha'\},n'}(t) \tag{11.1.3c}$$

with $\{\alpha'\} = \alpha_1 \cdots \alpha_{i-1}, \alpha_{i+1} \cdots \alpha_N$. Using (11.1.3b,11.1.3c), we write the lasing driving contribution to $\dot{\rho}_{n,n'}$ in Eq. (11.1.3a) as

$$\begin{aligned}\left(\frac{\delta\rho_{n,n'}}{\delta t}\right)_{\mathrm{gain}} &= -\frac{i}{\hbar}\mathrm{Tr}_{\mathrm{atoms}}\left[\mathscr{V},\rho\right]_{\mathrm{n,n'}}\\ &= -\frac{i}{\hbar}\sum_i\Big[\mathscr{V}_{an,bn+1}\rho^i_{bn+1,an'} - \rho^i_{an,bn'+1}\mathscr{V}_{bn'+1,an'}\\ &\quad + \mathscr{V}_{bn,an-1}\rho^i_{an-1,bn'} - \rho^i_{bn,an'-1}\mathscr{V}_{an'-1,bn'}\Big],\end{aligned} \tag{11.1.3d}$$

where

$$\mathscr{V}_{an,bn+1} = \hbar g\sqrt{n+1}, \qquad \mathscr{V}_{bn'+1,an'} = \hbar g\sqrt{n'+1}.$$

At this point we note that terms like $\sum_i \rho^i_{an,bn+1}(t, t_i)$ (with t_i being the injection time of the ith atom) are the fully quantized analog of the population matrix (5.4.4).

Thus we are motivated to introduce the generalized population matrix with elements such as

$$\rho^{(t)}_{an,bn+1} = \sum_i \rho^i_{an,bn+1}(t, t_i).$$

Equation (11.1.3d) can be rewritten in terms of the elements of the population matrix as

$$\left(\frac{\delta \rho_{n,n'}}{\delta t}\right)_{\text{gain}} = -\frac{i}{\hbar}(\mathscr{V}_{an,bn+1}\rho_{bn+1,an} - \rho_{an,bn'+1}\mathscr{V}_{bn'-1,an'}$$
$$+\mathscr{V}_{bn,an-1}\rho_{an-1,bn'} - \rho_{bn,an'-1}\mathscr{V}_{an'-1,bn'}). \quad (11.1.4)$$

We proceed by assuming that the atom–field population matrix for the ith atom may be treated as if the effect of other atoms is felt only through their contributions to $\rho_{nn'}(t)$; just as in the semiclassical case the collective field $E(t)$ accounted for the influence of other atoms on the ith atom. In this spirit we write the equations of motion for the population matrix $\rho_{\alpha n,\beta n'} \equiv \langle \alpha, n|\rho_{\text{atom-field}}|\beta, n'\rangle$ as

$$\dot{\rho}_{an,an'} = \lambda_a \rho_{gn,gn'} - \gamma \rho_{an,an'} - \gamma_g \rho_{an,an'}$$
$$-\frac{i}{\hbar}(\mathscr{V}_{an,bn+1}\rho_{bn+1,an'} - \rho_{an,bn'+1}\mathscr{V}_{bn'+1,an'}), \quad (11.1.5a)$$

$$\dot{\rho}_{an,bn'+1} = -\gamma \rho_{an,bn'+1}$$
$$-\frac{i}{\hbar}(\mathscr{V}_{an,bn+1}\rho_{bn+1,bn'+1} - \rho_{an,an'}\mathscr{V}_{an',bn'+1}), \quad (11.1.5b)$$

$$\dot{\rho}_{bn+1,an'} = -\gamma \rho_{bn+1,an'}$$
$$-\frac{i}{\hbar}(\mathscr{V}_{bn+1,an}\rho_{an,an'} - \rho_{bn+1,bn'+1}\mathscr{V}_{bn'+1,an'}), \quad (11.1.5c)$$

$$\dot{\rho}_{bn+1,bn'+1} = -\gamma \rho_{bn+1,bn'+1}$$
$$-\frac{i}{\hbar}(\mathscr{V}_{bn+1,an}\rho_{an,bn'+1} - \rho_{bn+1,an'}\mathscr{V}_{an',bn'+1}), \quad (11.1.5d)$$

$$\dot{\rho}_{cn,cn'} = \gamma \rho_{an,an'} - \gamma_c \rho_{cn,cn'}, \quad (11.1.5e)$$

$$\dot{\rho}_{dn,dn'} = \gamma \rho_{bn,bn'} - \gamma_d \rho_{dn,dn'}, \quad (11.1.5f)$$

$$\dot{\rho}_{gn,gn'} = -\lambda_a \rho_{gn,gn'} + \gamma_g \rho_{an,an'} + \gamma_c \rho_{cn,cn'} + \gamma_d \rho_{dn,dn'}, \quad (11.1.5g)$$

where the term $\lambda_a \rho_{gn,gn'}$ in Eq. (11.1.5a) represents pumping only into the excited state $|a\rangle$ from level $|g\rangle$. It may be noted that Eq. (11.1.5g) is quite general and is valid for more complicated schemes, e.g., when atoms in levels $|a\rangle$ and $|b\rangle$ can decay to many other levels and the pumping to levels $|a\rangle$ and $|b\rangle$ takes place from any arbitrary level. As shown below, only Eqs. (11.1.5a)–(11.1.5d) and Eq. (11.1.4) are required to derive the equation of motion for the field density matrix. Thus an effective two-level picture is valid.

Next we determine the components of the atom–field population matrix elements $\rho_{bn+1,an'}$, $\rho_{an,bn'+1}$, $\rho_{an-1,bn'}$, and $\rho_{bn,an'-1}$. First, we assume that the decay rate from level $|a\rangle$ to level $|g\rangle$ is small, so that we can ignore the terms proportional to γ_g in Eqs. (11.1.5a) and (11.1.5g). It then follows from Eqs. (11.1.5e)–(11.1.5g) that, in steady state,

$$\rho_{gn,gn'} = \frac{\gamma}{\lambda_a}(\rho_{an,an'} + \rho_{bn,bn'}). \tag{11.1.6}$$

We next assume $\gamma_c, \gamma_d \gg \gamma$, so that $\rho_{cn,cn'} = \rho_{dn,dn'} \simeq 0$ in steady state. This condition ensures that as soon as the atoms have decayed from lasing levels $|a\rangle$ and $|b\rangle$ to levels $|c\rangle$ and $|d\rangle$, respectively, they decay quickly to level $|g\rangle$ from which they are excited to level $|a\rangle$ by the pumping process. Thus, from the fact that

$$\rho_{nn'} = \rho_{an,an'} + \rho_{bn,bn'} + \rho_{cn,cn'} + \rho_{dn,dn'} + \rho_{gn,gn'}\ , \tag{11.1.7a}$$

we then have

$$\rho_{an,an'} + \rho_{bn,bn'} \simeq \rho_{nn'} - \rho_{gn,gn'}\ , \tag{11.1.7b}$$

and Eq. (11.1.6) yields

$$\rho_{gn,gn'} \simeq \frac{\gamma}{\gamma + \lambda_a}\rho_{nn'}. \tag{11.1.8}$$

Equation (11.1.5a) can now be rewritten as

$$\dot{\rho}_{an,an'} \simeq r_a\rho_{nn'} - \gamma\rho_{an,an'} - \frac{i}{\hbar}(\mathscr{V}_{an,bn+1}\rho_{bn+1,an'} - \rho_{an,bn'+1}\mathscr{V}_{bn'+1,an'}), \tag{11.1.9}$$

where $r_a = \gamma\lambda_a/(\gamma+\lambda_a)$ is the effective pumping rate. Thus Eq. (11.1.9) together with Eqs. (11.1.5b)–(11.1.5d) form a closed set of equations for the matrix elements corresponding to the lasing levels $|a\rangle$ and $|b\rangle$. We solve the above equations of motion for $\rho_{an,bn+1}(t)$, etc, in Appendix 11.A.

On substituting the various components of the atom–field population matrix in Eq. (11.1.4), from Appendix 11.A, we find the equation of motion for $\rho_{nn'}$ due to the gain medium:

$$\left(\frac{\delta\rho_{nn'}}{\delta t}\right)_{\text{gain}} = -\left(\frac{\mathscr{N}'_{nn'}\mathscr{A}}{1+\mathscr{N}_{nn'}\mathscr{B}/\mathscr{A}}\right)\rho_{nn'} + \left(\frac{\sqrt{nn'}\mathscr{A}}{1+\mathscr{N}_{n-1,n'-1}\mathscr{B}/\mathscr{A}}\right)\rho_{n-1,n'-1}, \tag{11.1.10}$$

where we introduce the linear gain coefficient

$$\mathscr{A} = \frac{2r_a g^2}{\gamma^2}, \tag{11.1.11}$$

the self-saturation coefficient

$$\mathscr{B} = \frac{4g^2}{\gamma^2}\mathscr{A}, \tag{11.1.12}$$

and the dimensionless factors

$$\mathscr{N}'_{nn'} = \frac{1}{2}(n+1+n'+1) + \frac{(n-n')^2\mathscr{B}}{8\mathscr{A}}, \tag{11.1.13a}$$

$$\mathscr{N}_{nn'} = \frac{1}{2}(n+1+n'+1) + \frac{(n-n')^2\mathscr{B}}{16\mathscr{A}}. \tag{11.1.13b}$$

The gain and saturation coefficients $\mathscr{A}$ and $\mathscr{B}$ are seen to correspond to the coefficients in Eqs. (5.5.10) of the semiclassical laser theory in the case of upper level pumping and zero detuning. A connection between the present quantum treatment and the semiclassical results will be made at the end of this section.

It follows, on adding $\langle n|\mathscr{L}\rho|n'\rangle$ from (11.1.2), that

$$\begin{aligned}\dot{\rho}_{nn'} = &-\left(\frac{\mathscr{N}'_{nn'}\mathscr{A}}{1+\mathscr{N}_{nn'}\mathscr{B}/\mathscr{A}}\right)\rho_{nn'} \\ &+\left(\frac{\sqrt{nn'}\mathscr{A}}{1+\mathscr{N}_{n-1,n'-1}\mathscr{B}/\mathscr{A}}\right)\rho_{n-1,n'-1} \\ &-\frac{1}{2}\mathscr{C}(n+n')\rho_{nn'}+\mathscr{C}[(n+1)(n'+1)]^{1/2}\rho_{n+1,n'+1}.\end{aligned} \tag{11.1.14}$$

This equation constitutes our basic result. In particular, the diagonal elements $\rho_{nn} \equiv p(n)$, which represent the probability of n photons in the field, have the equation of motion

$$\begin{aligned}\dot{p}(n) = &-\left[\frac{(n+1)\mathscr{A}}{1+(n+1)\mathscr{B}/\mathscr{A})}\right]p(n) + \left(\frac{n\mathscr{A}}{1+n\mathscr{B}/\mathscr{A}}\right)p(n-1) \\ &-\mathscr{C}np(n) + \mathscr{C}(n+1)p(n+1).\end{aligned} \tag{11.1.15}$$

It is important to note that the diagonal elements are coupled only to diagonal elements and that, more generally, only off-diagonal elements with the same difference $(n-n')$ are coupled.

A simple physical meaning can be given to Eq. (11.1.15) for the photon distribution function in terms of a probability flow diagram (Fig. 11.2) by expanding the terms in the denominator of Eq. (11.1.15). There we see the 'flow' of probability in and out of the $|n\rangle$ state from and to the neighboring $|n+1\rangle$ and $|n-1\rangle$ states. For example, the $\mathscr{A}(n+1)p(n)$ term represents the flow of probability from the $|n\rangle$ state to the $|n+1\rangle$ state due to the emission of photons by lasing atoms initially in the upper states. Here $\mathscr{A}n$ is the rate of stimulated

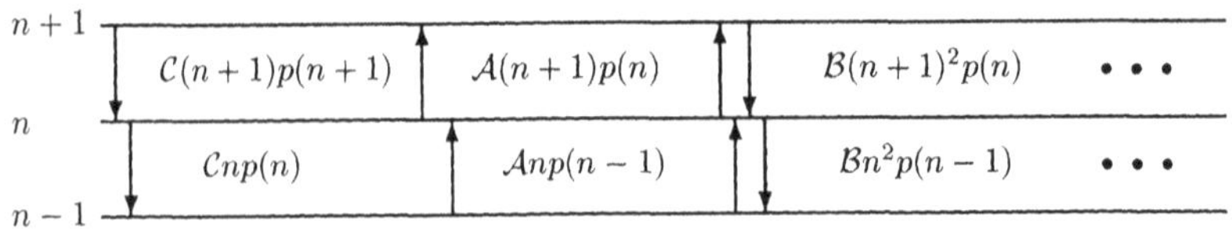

Fig. 11.2 Probability flow diagram for a laser.

emission, $\mathscr{A}$ is the spontaneous emission rate and these rates are multiplied by $p(n)$ to yield the total probability flow rate. Since the probability flows out of $p(n)$, this term is negative. The first term in the expansion of the square-bracketed term in (11.1.15), namely $\mathscr{B}(n+1)^2p(n) = \mathscr{A}(n+1)(\mathscr{B}/\mathscr{A})(n+1)p(n)$, corresponds to the process in which photons are emitted and then reabsorbed, the reabsorption rate being $(\mathscr{B}/\mathscr{A})(n+1)$. Similar explanations exist for the other terms including the loss terms.

Method II

As we have emphasized earlier, each atom in a gas laser contributes its energy to the field independently, except in so far as the other atoms have prepared an electromagnetic field with which it interacts. With this in mind we consider the change in the radiation field-density matrix due to the injection at time t_0 of a single pumping atom in the upper of the two atomic states $|a\rangle$ and $|b\rangle$. Working in the n representation, this change is given by

$$\delta\rho_{nn'} = \rho_{nn'}(t_0 + \tau) - \rho_{nn'}(t_0), \tag{11.1.16}$$

where τ is a time which is long compared with an atomic lifetime, but short compared to the time characterizing the growth or decay of the laser radiation. The macroscopic change in the density matrix $\Delta\rho_{nn'}$, due to N atoms acting on the field in a time Δt will then be

$$\Delta\rho_{nn'} = N\delta\rho_{nn'} = r_a\Delta t\delta\rho_{nn'}, \tag{11.1.17}$$

where r_a is the rate of atomic injection.

We now turn to the determination of $\delta\rho_{nn'}$ as it appears in Eq. (11.1.16). To obtain $\rho_{nn'}(t_0+\tau)$ we must follow the time development of the combined atom–laser field system to time $t_0+\tau$ and then form the trace of its density matrix over the atomic states

$$\rho_{nn'}(t_0 + \tau) = \sum_\alpha \rho_{\alpha,n;\alpha,n'}(t_0 + \tau) \qquad (\alpha = a, b). \tag{11.1.18}$$

A coarse-grained derivative for the laser radiation-density matrix, $\Delta\rho/\Delta t = r_a\delta\rho$, is obtained by combining Eqs. (11.1.16)–(11.1.18):

$$\left(\frac{d\rho_{nn'}}{dt}\right)_{\text{gain}} = r_a[\rho_{an,an'}(t_0+\tau) + \rho_{bn,bn'}(t_0+\tau) - \rho_{an,an'}(t_0)]. \tag{11.1.19}$$

In the adiabatic approximation, we replace t_0 in this equation by t. As the atoms are pumped to level $|a\rangle$ in the present treatment, $\rho_{bn,bn'}(t_0) = 0$.

The quantities $\rho_{an,an'}(t+\tau)$, $\rho_{bn,bn'}(t+\tau)$, and $\rho_{an,an'}(t)$ can be determined easily by solving the equations for the probability amplitudes c_{an} and c_{bn+1} (Eqs. (6.2.11) and (6.2.12)). For the resonant case, $\Delta = \omega - \nu = 0$, these solutions are

$$c_{an}(t+\tau) = c_{an}(t)\cos(g\tau\sqrt{n+1}), \tag{11.1.20}$$

$$c_{bn+1}(t+\tau) = -ic_{an}(t)\sin(g\tau\sqrt{n+1}), \tag{11.1.21}$$

yielding

$$\rho_{an,an'}(t+\tau) = \rho_{nn'}(t)\cos(g\tau\sqrt{n+1})\cos(g\tau\sqrt{n'+1}), \tag{11.1.22}$$

$$\rho_{bn,bn'}(t+\tau) = \rho_{n-1,n'-1}(t)\sin(g\tau\sqrt{n})\sin(g\tau\sqrt{n'}), \tag{11.1.23}$$

$$\rho_{an,an'}(t) = \rho_{nn'}(t), \tag{11.1.24}$$

since at time $\tau = 0$

$$\rho_{nn'}(t) = |c_{an}(t)|^2. \tag{11.1.25}$$

The resulting equation for the reduced density matrix of the field is then

$$\left(\frac{d\rho_{nn'}}{dt}\right)_{\text{gain}} = -r_a[1 - \cos(g\tau\sqrt{n+1})\cos(g\tau\sqrt{n'+1})]\rho_{nn'} + r_a\sin(g\tau\sqrt{n})\sin(g\tau\sqrt{n'})\rho_{n-1,n'-1}. \tag{11.1.26}$$

Notice that this equation has been obtained for a system in which two-level atoms are injected in their upper level $|a\rangle$ at random times at a rate r_a and they interact with the radiation field for a time τ before they are removed. This model, and consequently Eq. (11.1.26), is directly relevant to the study of the quantum statistics of a *micromaser* which we consider in Chapter 13.

It is possible to use Eq.(11.1.26) to obtain (11.1.10). To this end we consider an atom which is injected into the $|a\rangle$ state at time zero and decays from $|a\rangle$ and $|b\rangle$ to $|c\rangle$ and $|d\rangle$, as in Fig.1.1, with a rate $\gamma = \gamma_a = \gamma_b$. The probability that the atom will "live" in states $|a\rangle$ and $|b\rangle$ to a time τ is then

$$P(\tau)d\tau = \gamma e^{-\gamma\tau}d\tau. \tag{11.1.27}$$

We therefore take an average of Eq. (11.1.26) over this distribution to get the density coarse-grained matrix equation for the laser radiation

$$\left(\frac{d\rho_{nn'}}{dt}\right)_{\text{gain}}$$
$$= -r_a \int_0^\infty d\tau\gamma e^{-\gamma\tau}[1 - \cos(g\tau\sqrt{n+1})\cos(g\tau\sqrt{n'+1})]\rho_{nn'}$$
$$+ r_a \int_0^\infty d\tau\gamma e^{-\gamma\tau}\sin(g\tau\sqrt{n})\sin(g\tau\sqrt{n'})\rho_{n-1,n'-1}. \tag{11.1.28}$$

On carrying out the necessary integrations, we recover Eq. (11.1.10). Details of the calculation are to be found in Problem 11.7.

In order to make a connection between the present quantum theory of the laser and the semiclassical theory of Section 5.5, we derive an equation for the mean number of photons $\langle n\rangle$ from Eq. (11.1.15). For operation near threshold, $\mathscr{B}\langle n\rangle/\mathscr{A} \ll 1$, we obtain (after expanding the denominators in the first two terms on the right-hand side)

$$\frac{d\langle n\rangle}{dt} = \sum_{n=0}^{\infty} n\dot{p}(n)$$
$$= (\mathscr{A} - \mathscr{C})\langle n\rangle - \mathscr{B}\langle(n+1)^2\rangle + \mathscr{A}. \tag{11.1.29}$$

This equation reduces to Eq. (5.5.16) in the limits $\langle n\rangle \gg 1$ and $\langle n^2\rangle \cong \langle n\rangle^2$. Thus the semiclassical treatment is valid for large photon numbers and in situations where the decorrelation approximation

$$\langle n^2\rangle = \langle n\rangle^2 \tag{11.1.30}$$

is valid.

The term $\mathscr{A}$ in Eq. (11.1.29), which is absent in the semiclassical equation, gives the spontaneous emission into the laser field mode. We see that, for $n = 0$ initially, the semiclassical field remains zero for all times. In contrast, the quantum equation (11.1.29) can build up from zero because of the spontaneous emission term $\mathscr{A}$.

11.2 Laser photon statistics

As we saw in the last section, the probability $\rho_{nn}(t) \equiv p(n)$ of an n-photon laser field changes in time due to gain produced by stimulated emission and cavity losses. In this section, we consider the steady-state solutions of the equation of motion (11.1.15) for ρ_{nn} and obtain the photon number statistics for steady-state operation.

Before discussing the general solution of Eq. (11.1.15), we consider two limiting cases of laser operation.

11.2.1 Linear approximation ($\mathscr{B} = 0$)

The equation of motion for the photon distribution function $p(n)$ (cf. Eq. (11.1.15)) in the steady state ($\dot{p}(n) = 0$) reduces to

$$-\mathscr{A}(n+1)p(n) + \mathscr{A}np(n-1) - \mathscr{C}np(n) + \mathscr{C}(n+1)p(n+1) = 0. \tag{11.2.1}$$

It is clear that the detailed balance condition implies that the second-order difference equation (11.2.1) reduces to the equivalent system of two first-order difference equations

$$\mathscr{A}np(n-1) - \mathscr{C}np(n) = 0, \tag{11.2.2}$$

$$\mathscr{A}(n+1)p(n) - \mathscr{C}(n+1)p(n+1) = 0. \tag{11.2.3}$$

The solution of these equations is clearly

$$p(n) = p(0)\left(\frac{\mathscr{A}}{\mathscr{C}}\right)^n. \tag{11.2.4}$$

The constant $p(0)$ is determined from the normalization condition

$$\sum_{n=0}^{\infty} p(n) = 1. \tag{11.2.5}$$

For $\mathscr{A} < \mathscr{C}$, we obtain

$$p(0) = (1 - \mathscr{A}/\mathscr{C}), \tag{11.2.6}$$

so that

$$p(n) = \left(1 - \frac{\mathscr{A}}{\mathscr{C}}\right)\left(\frac{\mathscr{A}}{\mathscr{C}}\right)^n. \tag{11.2.7}$$

Since no solution for $p(n)$ exists for $\mathscr{A} \geq \mathscr{C}$ in the linear approximation, we interpret $\mathscr{A} = \mathscr{C}$ as the threshold condition. Hence, below threshold, the steady-state solution is essentially that of a black-body cavity

$$p(n) = \left[1 - \exp\left(\frac{-\hbar\nu}{k_B T}\right)\right] \exp\left(\frac{-n\hbar\nu}{k_B T}\right), \tag{11.2.8}$$

where the effective temperature T is defined by

$$\exp\left(\frac{-\hbar\nu}{k_B T}\right) = \frac{\mathscr{A}}{\mathscr{C}}. \tag{11.2.9}$$

11.2.2 Far above threshold ($\mathscr{A} \gg \mathscr{C}$)

In this situation, the saturation is so large that the quantity $\mathscr{B}\langle n\rangle/\mathscr{A} \gg 1$. We can then ignore the term unity in comparison with the term $\mathscr{B}n/\mathscr{A}$ in the denominators on the right-hand side of Eq. (11.1.15). The resulting steady-state equation of motion for $p(n)$ is

$$-\frac{\mathscr{A}^2}{\mathscr{B}}p(n)+\frac{\mathscr{A}^2}{\mathscr{B}}p(n-1)-\mathscr{C}np(n)+\mathscr{C}(n+1)p(n+1) = 0. \quad (11.2.10)$$

This equation is solved again by invoking the detailed balance condition which implies

$$\frac{\mathscr{A}^2}{\mathscr{B}}p(n-1) - \mathscr{C}np(n) = 0. \quad (11.2.11)$$

The normalized solution of these equations is

$$p(n) = e^{-\langle n\rangle}\frac{\langle n\rangle^n}{n!}, \quad (11.2.12)$$

with

$$\langle n\rangle = \frac{\mathscr{A}^2}{\mathscr{B}\mathscr{C}}. \quad (11.2.13)$$

Thus the photon statistics of the laser far above threshold are given by a Poisson distribution which is a characteristic of a coherent state.

11.2.3 Exact solution

After discussing the two limiting cases, we give the general steady-state solution of Eq. (11.1.15). It follows again from the detailed balance condition that

$$\frac{\mathscr{A}n}{1+\frac{\mathscr{B}}{\mathscr{A}}n}p(n-1) - \mathscr{C}np(n) = 0. \quad (11.2.14)$$

The solution of this equation is clearly

$$p(n) = p(0)\prod_{k=1}^{n}\frac{(\mathscr{A}/\mathscr{C})}{\left(1+\frac{\mathscr{B}}{\mathscr{A}}k\right)}, \quad (11.2.15)$$

where $p(0)$ is determined from the normalization condition $\sum_{n=0}^{\infty} p(n) = 1$. The quantity $p(n)$ is the product of n factors of the form $(\mathscr{A}/\mathscr{C})(1+\mathscr{B}k/\mathscr{A})^{-1}$. For

$$k < \left(\frac{\mathscr{A}}{\mathscr{C}}\right)\left(\frac{\mathscr{A}-\mathscr{C}}{\mathscr{B}}\right) = n_p, \quad (11.2.16)$$

these factors are each greater than unity, while for $k > n_p$, these factors are less than unity. Hence $p(n)$ increases for n up to n_p and goes monotonically to zero for $n > n_p$. Thus the distribution peaks at

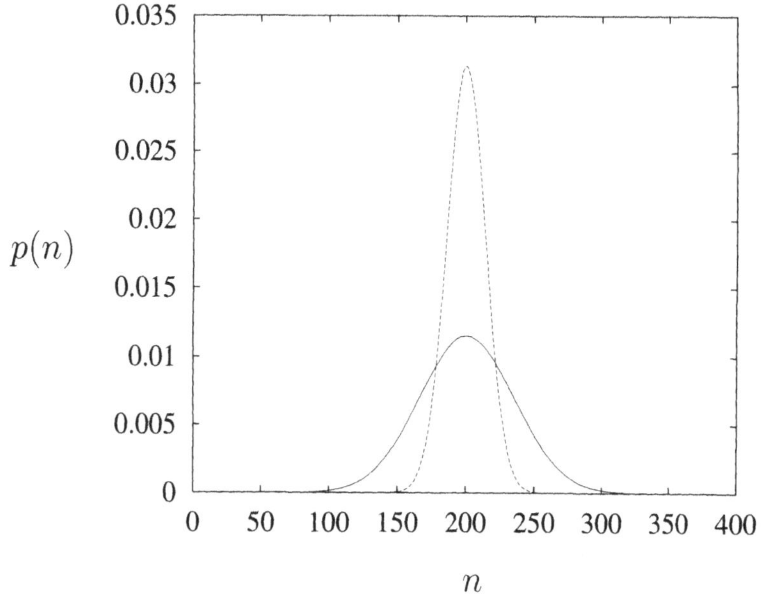

Fig. 11.3
Steady-state photon distribution function for coherent (....) and laser (—) radiation. The laser is taken to be 20 percent above threshold, with the parameter $\mathscr{B}$ chosen to give $\langle n \rangle = 200$.

$$n_p = \frac{\mathscr{A}}{\mathscr{C}}\left(\frac{\mathscr{A}-\mathscr{C}}{\mathscr{B}}\right). \tag{11.2.17}$$

We may write Eq. (11.2.15) in a more convenient form as

$$p(n) = p(0)\frac{\left(\frac{\mathscr{A}}{\mathscr{B}}\right)!\left(\frac{\mathscr{A}^2}{\mathscr{B}\mathscr{C}}\right)^n}{\left(n+\frac{\mathscr{A}}{\mathscr{B}}\right)!}. \tag{11.2.18}$$

The normalization constant $p(0)$ may be expressed in terms of confluent hypergeometric functions

$$\begin{aligned} p(0) &= \left[\sum_{n=0}^{\infty}\frac{\left(\frac{\mathscr{A}}{\mathscr{B}}\right)!\left(\frac{\mathscr{A}^2}{\mathscr{B}\mathscr{C}}\right)^n}{\left(n+\frac{\mathscr{A}}{\mathscr{B}}\right)!}\right]^{-1} \\ &= \left[F\left(1;\frac{\mathscr{A}}{\mathscr{B}}+1;\frac{\mathscr{A}^2}{\mathscr{B}\mathscr{C}}\right)\right]^{-1}. \end{aligned} \tag{11.2.19}$$

In Fig. 11.3, the photon distribution $p(n)$ (Eq. (11.2.18)) is shown for a laser operating 20 percent above threshold.

Calculating the average value of n, we find

$$\begin{aligned} \langle n \rangle &= p(0)\sum_{n=0}^{\infty} n\frac{\left(\frac{\mathscr{A}}{\mathscr{B}}\right)!\left(\frac{\mathscr{A}^2}{\mathscr{B}\mathscr{C}}\right)^n}{\left(n+\frac{\mathscr{A}}{\mathscr{B}}\right)!} \\ &= p(0)\left(\frac{\mathscr{A}}{\mathscr{B}}\right)!\sum_{n=1}^{\infty}\frac{\left(n+\frac{\mathscr{A}}{\mathscr{B}}-\frac{\mathscr{A}}{\mathscr{B}}\right)\left(\frac{\mathscr{A}^2}{\mathscr{B}\mathscr{C}}\right)^n}{\left(n+\frac{\mathscr{A}}{\mathscr{B}}\right)!} \end{aligned}$$

$$= p(0)\left(\frac{\mathscr{A}}{\mathscr{B}}\right)!\sum_{n=1}^{\infty}\frac{\left(\frac{\mathscr{A}^2}{\mathscr{B}\mathscr{C}}\right)^{n-1}}{\left(n+\frac{\mathscr{A}}{\mathscr{B}}-1\right)!}\left(\frac{\mathscr{A}^2}{\mathscr{B}\mathscr{C}}\right) - \frac{\mathscr{A}}{\mathscr{B}}[1-p(0)]$$

$$= \frac{\mathscr{A}^2}{\mathscr{B}\mathscr{C}}\sum_{n=0}^{\infty}p(n) - \frac{\mathscr{A}}{\mathscr{B}}[1-p(0)]$$

$$= \frac{\mathscr{A}}{\mathscr{C}}\left(\frac{\mathscr{A}-\mathscr{C}}{\mathscr{B}}\right) + \frac{\mathscr{A}}{\mathscr{B}}p(0). \tag{11.2.20}$$

For a laser appreciably above threshold, the last term in Eq. (11.2.20) is clearly insignificant because $p(0) \ll 1$. We then obtain

$$\langle n\rangle \simeq \frac{\mathscr{A}}{\mathscr{C}}\left(\frac{\mathscr{A}-\mathscr{C}}{\mathscr{B}}\right) = n_p. \tag{11.2.21}$$

A similar calculation for $\langle n^2\rangle$ yields

$$\langle n^2\rangle = \langle n\rangle^2 + \frac{\mathscr{A}^2}{\mathscr{B}\mathscr{C}}. \tag{11.2.22}$$

Far above threshold, $\langle n\rangle$ is given by Eq. (11.2.21). The normalized variance of the photon distribution is given by the so-called Mandel Q parameter for the field. For the laser it is given by

$$Q_f = \frac{\langle n^2\rangle - \langle n\rangle^2}{\langle n\rangle} - 1,$$

$$= \frac{\mathscr{C}}{\mathscr{A}-\mathscr{C}}. \tag{11.2.23}$$

Since $Q_f > 0$ above threshold ($\mathscr{A} > \mathscr{C}$), the field is super-Poissonian. However, very far above threshold ($\mathscr{A} \gg \mathscr{C}$), Q_f approaches zero which is a characteristic of the Poisson distribution and this agrees with our earlier analysis.

11.3 P-representation of the laser

In Chapter 3, we saw that the P-representation forms a correspondence between the classical and quantum coherence theory. It is, therefore, interesting to find the P-representation of the laser. All the normally ordered correlation functions of the field operators can be evaluated from it in a simple manner. There are, however, some other advantages of using P-representation. As we shall see in the next section, a calculation of the natural linewidth of the laser is highly facilitated by the equation of motion for the P-representation. This problem is relatively more complicated if the density matrix approach is followed. The P-representation also helps to make a correspondence between the density matrix approach discussed in Section 11.1 and the Langevin equation approach discussed in the next chapter.

As shown in Appendix 11.B the equation of motion for $P(\alpha, \alpha^*, t)$ is found to be

$$\frac{\partial P}{\partial t} = \frac{-\mathscr{A}}{2}\left(\frac{1}{r}\frac{\partial}{\partial r}r^2 - \frac{1}{2r}\frac{\partial}{\partial r}r\frac{\partial}{\partial r} - \frac{1}{2r^2}\frac{\partial^2}{\partial \theta^2} - \frac{\mathscr{B}}{4\mathscr{A}}\frac{\partial^2}{\partial \theta^2}\right)$$
$$\times \left\{1 - \frac{\mathscr{B}}{2\mathscr{A}}\left[\frac{1}{r}\frac{\partial}{\partial r}r^2 - 2(1+r^2)\right] - \frac{1}{16}\frac{\mathscr{B}^2}{\mathscr{A}^2}\frac{\partial^2}{\partial \theta^2}\right\}^{-1} P$$
$$+\frac{\mathscr{C}}{2}\left(\frac{1}{r}\frac{\partial}{\partial r}r^2\right)P. \tag{11.3.1}$$

Here, we have transformed from the independent complex variables α, α^* to polar coordinates r, θ using

$$\alpha = re^{i\theta}. \tag{11.3.2}$$

This equation contains the derivatives with respect to r and θ to all orders.

Near threshold, the steady-state solution of Eq. (11.3.1), as shown in Appendix 11.B, is

$$P(\alpha, \alpha^*) \simeq \frac{1}{\mathscr{N}'}\exp\left[\left(\frac{\mathscr{A}-\mathscr{C}}{\mathscr{A}}\right)|\alpha|^2 - \frac{\mathscr{B}}{2\mathscr{A}}|\alpha|^4\right], \tag{11.3.3}$$

where $\mathscr{N}'$ is a normalization constant.

11.4 Natural linewidth

So far we have discussed the photon statistics of the laser which are associated with the diagonal matrix elements ρ_{nn} of the reduced density matrix for the field. The quantum fluctuations which are responsible for the fluctuations in the number of photons in the field are also responsible for the phase fluctuations thus leading to a finite linewidth of the laser. In this section we shall derive expressions for the natural linewidth.

A detailed derivation of the natural linewidth would involve a calculation of the two-time correlation function $\langle E^{(-)}(t)E^{(+)}(t+\tau)\rangle$ of the field. According to Eq. (4.3.14), a Fourier transform of this two-time correlation function yields the field spectrum and hence the linewidth. This is a rather cumbersome process as it requires a time-dependent solution of Eq. (11.1.14) for the density matrix elements $\rho_{nn'}$ or of Eq. (11.3.1) for the corresponding P-representation. Here we follow a simpler approach in which an evaluation of the phase diffusion

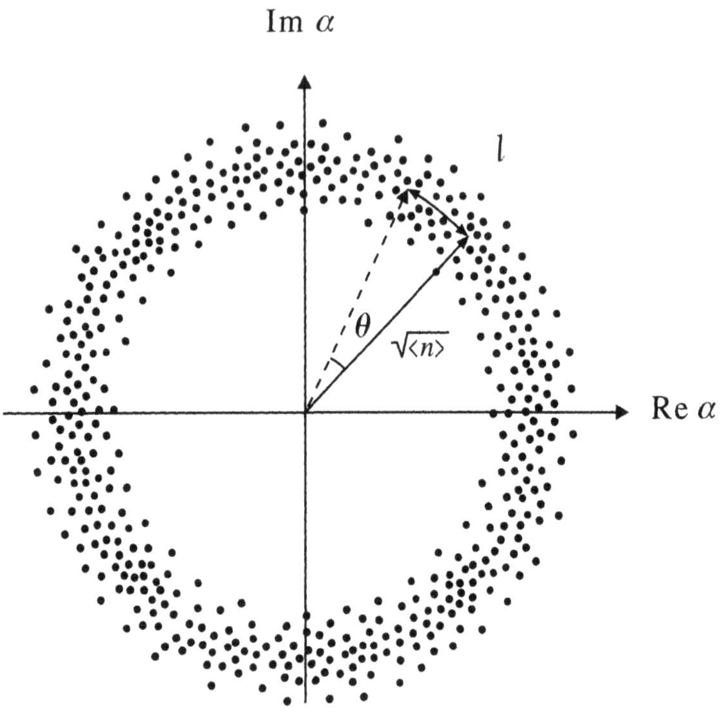

Fig. 11.4
Electromagnetic field as a complex vector. The phase diffuses due to spontaneous emission events.

coefficient in the Fokker–Planck equation for the P-representation gives the desired result. However, in order to motivate this approach, we first present the phase diffusion model for the laser linewidth.

11.4.1 Phase diffusion model

In order to study the phase fluctuations of the laser field, we describe the electromagnetic field as a complex vector (see Fig. 11.4) which experiences small changes in its phase due to spontaneous emission events. As these small changes are totally random, the phase gradually diffuses and becomes equally distributed over 2π. We assume that the small phase change due to a single spontaneous emission event takes place on a far shorter time scale than the overall evolution of the field.

We consider a situation in which the laser is operating sufficiently far above threshold that the amplitude fluctuations can be ignored. The field $E^{(+)}(t)$ can then be written as

$$E^{(+)}(t) = \sqrt{\langle n \rangle}\exp[i\theta(t) - i\nu_0 t], \tag{11.4.1}$$

where $\langle n \rangle$ is the mean number of photons of frequency ν_0 in the field in steady state. The spontaneous emission rate is equal to $\mathscr{A}$. Due to the random spontaneous emission events the electric field phasor will execute a random walk. Since the amplitude fluctuations are ignored, it will be a one-dimensional random walk along the angular direction. From the theory of random walk we find that the probability that a distance ℓ is travelled after $\mathscr{A}t$ steps is given by

$$P(\ell) = \frac{1}{(\pi \mathscr{A} t)^{1/2}} e^{-(\ell^2/\mathscr{A}t)}. \tag{11.4.2}$$

In terms of the angular displacement θ, we obtain

$$P(\theta) = \left(\frac{\langle n \rangle}{\pi \mathscr{A} t}\right)^{1/2} e^{-(\theta^2 \langle n \rangle / \mathscr{A} t)}, \tag{11.4.3}$$

where we use $\ell = \sqrt{\langle n \rangle}\theta$ (see Fig. 11.4). It is easily shown that $P(\theta)$ obeys the phase diffusion equation

$$\frac{\partial P}{\partial t} = D \frac{\partial^2 P}{\partial \theta^2}, \tag{11.4.4}$$

where

$$D = \frac{\mathscr{A}}{4\langle n \rangle}. \tag{11.4.5}$$

Thus we see that the simple model in which the spontaneous emission events give rise to a random walk for the tip of the electric field phasor leads to a phase diffusion equation for $P(\theta)$.

Next we look at the second-order correlation function $\langle E^{(-)}(t)E^{(+)}(t+\tau)\rangle$ for the field. We assume that the field is statistically stationary so that we need only to determine $\langle E^{(-)}(0)E^{(+)}(\tau)\rangle$. It is clear from Fig. 11.4 that

$$\langle E^{(-)}(t)E^{(+)}(t+\tau)\rangle \equiv \langle E^{(-)}(0)E^{(+)}(\tau)\rangle = \langle n \rangle e^{-i\nu_0 \tau} \langle e^{i\theta} \rangle. \tag{11.4.6}$$

It follows from Eq. (11.4.3) that

$$\langle \cos\theta \rangle = \int P(\theta) e^{i\theta} d\theta = e^{-D\tau}, \tag{11.4.7}$$

so that

$$\langle E^{(-)}(t)E^{(+)}(t+\tau)\rangle = \langle n \rangle e^{-i\nu_0 \tau - D\tau}. \tag{11.4.8}$$

The power spectrum is then obtained by taking the Fourier transform of the second-order correlation function (Eq. (4.3.14)), i.e.,

$$\begin{aligned} S(\nu) &= \frac{1}{\pi} \mathrm{Re} \int_0^\infty \langle E^{(-)}(t)E^{(+)}(t+\tau)\rangle e^{i\nu\tau} d\tau \\ &= \frac{\langle n \rangle}{\pi} \frac{D}{(\nu - \nu_0)^2 + D^2}. \end{aligned} \tag{11.4.9}$$

This is a Lorentzian distribution centered at $\nu = \nu_0$ with a linewidth

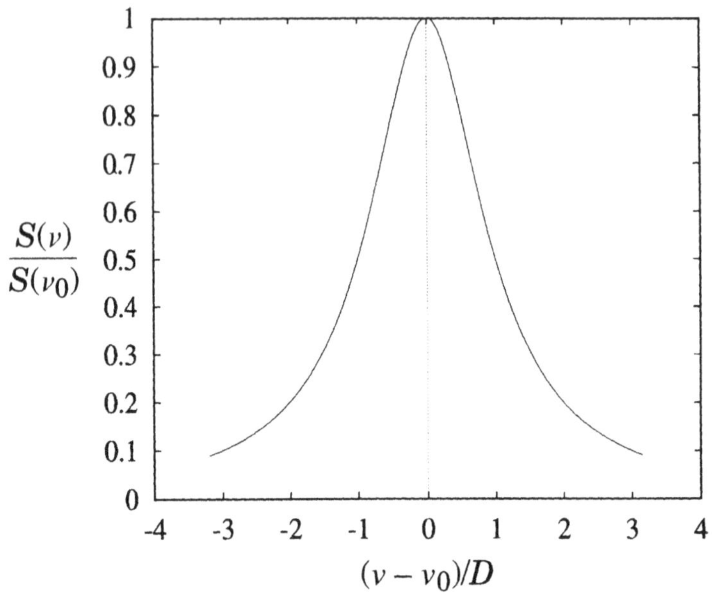

Fig. 11.5 Normalized frequency spectrum $S(\nu)/S(\nu_0)$ given by Eq. (11.4.9).

(full width at half maximum) of

$$2D = \frac{\mathscr{A}}{2\langle n \rangle}, \tag{11.4.10}$$

as depicted in Fig. 11.5.

This simple derivation of the laser linewidth based on a phase diffusion model indicates that the diffusion coefficient in the equation for the probability distribution gives the linewidth. In the next section we shall take advantage of this insight in the derivation of the laser linewidth based on the Fokker–Planck equation derived in the previous section.

The phase diffusion model discussed above motivates an extremely simple approach to many problems of interest. A field $E(t)$ with a finite linewidth $2D$ can be described by Eq. (11.4.1) where it is assumed that $\theta(t)$ is the random phase with Gaussian statistics which performs a Brownian motion described by the so-called Wiener–Levy stochastic process

$$\begin{aligned} \langle \theta(t) \rangle &= 0, \\ \langle \theta(t)\theta(t') \rangle &= D(t + t' - |t - t'|). \end{aligned} \tag{11.4.11}$$

The derivative of this diffusion process is a white noise

$$\langle \dot{\theta}(t)\dot{\theta}(t') \rangle = 2D\delta(t - t'). \tag{11.4.12}$$

It then follows from the moment theorem for a Gaussian distribution that

$$\begin{aligned}\langle E^{(-)}(t)E^{(+)}(t+\tau)\rangle &= \langle n\rangle e^{-i\nu_0\tau}\langle e^{-i[\theta(t)-\theta(t+\tau)]}\rangle \\ &= \langle n\rangle e^{-i\nu_0\tau}e^{-(1/2)\langle[\theta(t)-\theta(t+\tau)]^2\rangle} \\ &= \langle n\rangle e^{-i\nu_0\tau}e^{-D\tau},\end{aligned} \tag{11.4.13}$$

which is the same as Eq. (11.4.8). This model for a Lorentzian lineshape will be employed in Chapter 16 when we study the effect of the finite linewidth of the pump field in the parametric amplification process.

11.4.2 Fokker–Planck equation and laser linewidth

Following the classical approach to laser linewidth discussed above, we now turn back to the fully quantum mechanical equations of motion for the laser derived in previous sections. As we discussed in Chapter 4, the quantum–classical correspondence in quantum optics is made via P-representation. It is, therefore, reasonable to identify the diffusion coefficient with the laser linewidth if the equation of motion for the P-representation of the laser can be reduced to the form of (11.4.4).

As before we assume that, for steady-state operation sufficiently far above threshold, the changes in $P(\alpha,\alpha^*)$ along the radial (field amplitude) coordinate r are restricted by the steady-state operating condition. These changes can, therefore, be neglected compared to those along the θ coordinate and Eq. (11.3.1) becomes

$$\frac{\partial P}{\partial t} = D\frac{\partial^2 P}{\partial \theta^2}, \tag{11.4.14}$$

where

$$2D = \frac{\mathscr{A}(1+\mathscr{B}\langle n\rangle/2\mathscr{A})}{2\langle n\rangle\left(1+\frac{\mathscr{B}}{\mathscr{A}}\langle n\rangle\right)}. \tag{11.4.15}$$

In deriving this equation we replaced r^2 by $\langle n\rangle$ and assumed $\langle n\rangle \gg 1 \gg \mathscr{B}/\mathscr{A}$. It follows from Eq. (11.2.21) for $\langle n\rangle$ that

$$2D = \frac{\mathscr{A}+\mathscr{C}}{4\langle n\rangle}. \tag{11.4.16}$$

Near threshold when $\mathscr{A} \simeq \mathscr{C}$,

$$2D \simeq \frac{\mathscr{A}}{2\langle n\rangle}. \tag{11.4.17}$$

This expression for the diffusion coefficient, and hence the linewidth, is identical to that obtained using heuristic arguments based on a phase diffusion model.

It is interesting to note that Eq. (11.4.17) for the laser linewidth near threshold could be derived from the laser equation in the linear approximation ($\mathscr{B} = 0$) (see Eq. (11.4.15)). Furthermore, the loss term proportional to $\mathscr{C}$ does not directly enter the diffusion coefficient. Thus, for operation near threshold (but sufficiently far above threshold such that changes in P along the radial coordinate can be restricted by its steady-state value), the laser linewidth can simply be derived in the linear approximation without the loss term (as long as the cavity losses are described via interaction with a thermal reservoir at zero temperature). The mean number of photons $\langle n \rangle$ would, however, depend on the saturation and loss parameters. This simplicity will be exploited when we calculate the laser linewidth in more complicated laser systems such as those discussed in Chapter 14.

11.5 Off-diagonal elements and laser linewidth

As we discussed earlier, a rigorous derivation of laser linewidth requires a calculation of the two-time correlation function $\langle E^{(-)}(t)E^{(+)}(t+\tau)\rangle$ of the field. Due to the apparent difficulty in calculating this quantity we followed a simple approach based on a phase diffusion model in the previous section. Here we present an alternative approach to calculating the laser linewidth based on the evaluation of the off-diagonal elements of the reduced density matrix for the field.

We recall from Section 10.4 that, in the Markovian approximation (which is satisfied in the present problem of the laser), the quantum regression theorem is satisfied, i.e., the time dependence of the two-time correlation function is identical to the time dependence of the mean value of the field $\langle E^{(-)}(t)\rangle$. Thus the determination of $\langle E^{(-)}(t)\rangle$ would yield the laser linewidth. Now $\langle E^{(-)}(t)\rangle$ is related to the off-diagonal matrix element of the reduced density operator of the field by the expansion

$$\langle E^{(-)}(t)\rangle \sim \sum_{n=0}^{\infty} \sqrt{n+1}\,\rho_{n,n+1}(t). \qquad (11.5.1)$$

Hence the rate of decay of the off-diagonal matrix element $\rho_{n,n+1}(t)$ implies a simultaneous decay of the field, thus producing the laser linewidth. In this section we present a simple method to evaluate $\rho_{n,n+1}(t)$ and recover the results for the laser linewidth.

We begin with the equation of motion (11.1.14) for the density matrix $\rho_{n,n+k} \equiv \rho_n^{(k)}$

$$\dot{\rho}_n^{(k)} = -\left(\frac{\mathcal{N}'_{n,n+k}\mathcal{A}}{1+\mathcal{N}_{n,n+k}\mathcal{B}/\mathcal{A}}\right)\rho_n^{(k)} + \left[\frac{\sqrt{n(n+k)}\mathcal{A}}{1+\mathcal{N}_{n-1,n+k-1}\mathcal{B}/\mathcal{A}}\right]\rho_{n-1}^{(k)} - \frac{1}{2}\mathcal{C}(n+n+k)\rho_n^{(k)} + \mathcal{C}[(n+1)(n+k+1)]^{1/2}\rho_{n+1}^{(k)}$$

and rewrite it by adding and subtracting the appropriate terms:

$$\begin{aligned}\dot{\rho}_n^{(k)} &= -\left(\frac{\mathcal{N}'_{n,n+k}\mathcal{A}}{1+\mathcal{N}_{n,n+k}\mathcal{B}/\mathcal{A}}\right)\rho_n^{(k)} + \left[\frac{\sqrt{n(n+k)}\mathcal{A}}{1+\mathcal{N}_{n-1,n+k-1}\mathcal{B}/\mathcal{A}}\right]\rho_{n-1}^{(k)} \\ &\quad + \left[\frac{\sqrt{(n+1)(n+k+1)}\mathcal{A}}{1+\mathcal{N}_{n,n+k}\mathcal{B}/\mathcal{A}}\right]\rho_n^{(k)} - \left[\frac{\sqrt{(n+1)(n+k+1)}\mathcal{A}}{1+\mathcal{N}_{n,n+k}\mathcal{B}/\mathcal{A}}\right]\rho_n^{(k)} \\ &\quad - \frac{1}{2}\mathcal{C}(n+n+k)\rho_n^{(k)} + \mathcal{C}[(n+1)(n+k+1)]^{1/2}\rho_{n+1}^{(k)} \\ &\quad + \mathcal{C}[n(n+k)]^{1/2}\rho_n^{(k)} - \mathcal{C}[n(n+k)]^{1/2}\rho_n^{(k)} \\ &= -\frac{1}{2}\mu_n^{(k)}\rho_n^{(k)} + c_{n-1}^{(k)}\rho_{n-1}^{(k)} \\ &\quad - [c_n^{(k)} + d_n^{(k)}]\rho_n^{(k)} + d_{n+1}^{(k)}\rho_{n+1}^{(k)},\end{aligned} \tag{11.5.2}$$

where

$$\frac{1}{2}\mu_n^{(k)} = \frac{\left[\mathcal{N}'_{n,n+k} - \sqrt{(n+1)(n+k+1)}\right]\mathcal{A}}{1+\mathcal{N}_{n,n+k}\mathcal{B}/\mathcal{A}} + \mathcal{C}\left[n+\frac{k}{2}-\sqrt{n(n+k)}\right], \tag{11.5.3}$$

together with

$$c_n^{(k)} = \frac{\sqrt{(n+1)(n+k+1)}\mathcal{A}}{1+\mathcal{N}_{n,n+k}\mathcal{B}/\mathcal{A}}, \tag{11.5.4}$$

$$d_n^{(k)} = \mathcal{C}\sqrt{n(n+k)}. \tag{11.5.5}$$

Here the dimensionless factors $\mathcal{N}_{nn'}$ and $\mathcal{N}'_{nn'}$ are given by Eqs. (11.1.13a) and (11.1.13b). The equation of motion for $\rho_n^{(k)}$, Eq. (11.5.2), couples only the nearest neighbors of $\rho_n^{(k)}$, i.e., it couples $\rho_n^{(k)}$ to $\rho_{n+1}^{(k)}$ and $\rho_{n-1}^{(k)}$, but does not introduce a coupling in the k-index. We, therefore, face a three-term differential recurrence relation of complicated but time-independent coefficients. In the following we present an approximate but analytical approach to derive the lowest eigenvalue from a detailed balance condition.

The detailed balance condition $c_n^{(k)}\rho_n^{(k)} = d_{n+1}^{(k)}\rho_{n+1}^{(k)}$ (and hence $c_{n-1}^{(k)}\rho_{n-1}^{(k)} = d_n^{(k)}\rho_n^{(k)}$) suggests the ansatz

$$\begin{aligned}\rho_n^{(k)}(t) &= e^{-D_n^{(k)}(t)}\rho_n^{(k)}(0) \\ &= e^{-D_n^{(k)}(t)}\rho_0^{(0)}\prod_{j=1}^{n}\frac{c_{j-1}^{(k)}}{d_j^{(k)}} \\ &= e^{-D_n^{(k)}(t)}\rho_0^{(0)}\prod_{j=1}^{n}\frac{\mathscr{A}/\mathscr{C}}{1+\mathscr{N}_{j-1,j+k-1}\mathscr{B}/\mathscr{A}}.\end{aligned} \tag{11.5.6}$$

It thus follows from Eq. (11.5.2) that

$$\begin{aligned}\dot{D}_n^{(k)} &= \frac{1}{2}\mu_n^{(k)} + d_n^{(k)}\left\{1 - e^{-[D_{n-1}^{(k)} - D_n^{(k)}]}\right\} \\ &\quad + c_n^{(k)}\left\{1 - e^{-[D_{n+1}^{(k)} - D_n^{(k)}]}\right\}.\end{aligned} \tag{11.5.7}$$

Thus far the analysis is exact. We now expand the exponents and find to lowest order

$$D_n^{(k)}(t) \cong \frac{1}{2}\mu_n^{(k)}t. \tag{11.5.8}$$

Here we have assumed that

$$|D_n^{(k)} - D_{n-1}^{(k)}| \cong \frac{1}{2}|\mu_n^{(k)} - \mu_{n-1}^{(k)}|t \cong \frac{1}{2}\left|\frac{\partial \mu_n^{(k)}}{\partial t}\right| t \ll 1, \tag{11.5.9}$$

and analogously for $|D_{n+1}^{(k)} - D_n^{(k)}|$. This condition is satisfied when $\mu_n^{(k)}$ is a slowly varying function of n. Equations (11.5.6) and (11.5.8) then yield

$$\begin{aligned}\rho_n^{(1)}(t) &\cong \rho_n^{(1)}(0)\exp[-\frac{1}{2}\mu_n^{(1)}t] \\ &\cong \rho_n^{(1)}(0)\exp[-\frac{1}{2}\mu_{\langle n\rangle}^{(1)}t],\end{aligned} \tag{11.5.10}$$

where, in the last step, we have replaced n in $\mu_n^{(1)}$ by the average number of photons $\langle n\rangle$. The relation (11.5.10) reduces Eq. (11.5.1) to

$$\langle E^{(-)}(t)\rangle \sim e^{-Dt}\sum_{n=0}^{\infty}\sqrt{n+1}\rho_n^{(1)}(0). \tag{11.5.11}$$

With the help of Eq. (11.5.3), the linewidth $2D$ of the laser is, in the limit $\langle n\rangle \gg 1$,

$$2D = \mu_{\langle n\rangle}^{(1)} \cong \frac{1}{4}\left[\frac{\mathscr{A}+\mathscr{C}}{\langle n\rangle} - \frac{3\mathscr{B}/2 + \mathscr{B}^2/16\mathscr{A}}{\langle n\rangle(1+\langle n\rangle\mathscr{B}/\mathscr{A})}\right]. \tag{11.5.12}$$

The second term in Eq. (11.5.12) is however negligible as compared to the first term for all values of $\mathscr{B}\langle n\rangle/\mathscr{A}$. We thus obtain

$$2D \cong \frac{\mathscr{A}+\mathscr{C}}{4\langle n\rangle}, \tag{11.5.13}$$

in full agreement with Eq. (11.4.16).

11.6 Analogy between the laser threshold and a second-order phase transition

In this section, we consider the possibility of useful analogies between the laser theory and other problems in nonequilibrium statistical mechanics. In fact, an interesting comparison may be made between second-order phase transitions of ferromagnetic and superconducting materials and the laser near threshold. The purpose of the present section is to demonstrate that the laser threshold behavior is analogous to a second-order phase transition.

The basis of this similarity becomes evident when it is recalled that the usual treatment of laser behavior is a self-consistent field theory. In the laser analysis each atom develops a radiating dipole in an electromagnetic field due to (i.e., emitted by) all the other atoms. The radiation field produced by an ensemble of radiating atoms is then calculated in a self-consistent fashion. In this way, the physics of the laser problem is similar to that of a ferromagnet in which each spin sees a mean magnetic field due to all other spins and aligns itself accordingly, thus contributing to the average magnetic field. The formal similarity between the macroscopic equations describing the properties of the laser and those describing the ferromagnet is striking, and suggests the identification of the laser electric field as the variable corresponding to the ferromagnetic order parameter and the atomic population inversion as that corresponding to the temperature. Following this point of view, we show that the laser theory can be discussed using the language of second-order phase transitions.

The steady-state solution for the P-representation, $P(\alpha,\alpha^*)$, is given by Eq. (11.3.3). The expectation value $\langle E\rangle$ of the dimensionless electric field operator $E=(\alpha+\alpha^*)/2$ accordingly satisfies the equation of state

$$(\mathscr{A}-\mathscr{C})\langle E\rangle-\mathscr{B}\langle E^3\rangle=0. \tag{11.6.1}$$

If the laser is not too close to the threshold, we can replace $\langle E^3\rangle$ by $\langle E\rangle^3$ in the semiclassical limit, and obtain

$$(\mathscr{A}-\mathscr{C})\langle E\rangle-\mathscr{B}\langle E\rangle^3=0. \tag{11.6.2}$$

The threshold condition is given in Section 11.2 as $\mathscr{A} = \mathscr{C}$. Upon putting $\mathscr{A} = a\sigma$, $\mathscr{B} = b\sigma$, and $\mathscr{C} = a\sigma_t$, the steady-state solution of Eq. (11.6.2) is

$$\langle E \rangle = \begin{cases} 0 & \text{if } \sigma - \sigma_t \leq 0 \text{ (below threshold)}, \\ \left[\frac{a}{b}\left(\frac{\sigma-\sigma_t}{\sigma}\right)\right]^{1/2} & \text{if } \sigma - \sigma_t > 0 \text{ (above threshold)}. \end{cases} \tag{11.6.3}$$

We now consider the molecular field theory for a ferromagnet. The system contains N magnetic atoms per unit volume, and we assume that each atom has a magnetic moment μ. The average magnetization $\langle M \rangle$ will be a function of the absolute temperature T of the system and the external magnetic field H.

In the case of noninteracting spins, $\langle M \rangle$ is given by

$$\langle M \rangle = N\mu \tanh(\mu H/k_B T), \tag{11.6.4}$$

where k_B is the Boltzmann constant. The case of interacting spins is, of course, much more complicated and no exact solution is given for a three-dimensional system. The simplest approximate solution has been proposed by Weiss and is usually called the molecular field theory. In this model, the effect of interactions between the spins is taken into account simply by adding to the external field an internal magnetic field H, which can be computed self-consistently and is proportional to $\langle M \rangle$, through a constant λ. The equation of state of a ferromagnet can be derived from Eq. (11.6.4) by substituting for H the sum $H + \lambda M$ and performing a series expansion in powers of $\langle M \rangle$. We then have

$$\begin{aligned} \langle M \rangle &= N\mu \tanh[\mu(H + \lambda\langle M \rangle)/k_B T] \\ &\simeq N\mu \left[\frac{\mu}{k_B T}(H + \lambda\langle M \rangle) - \frac{1}{3}\left(\frac{\mu}{k_B T}\right)^3 (H + \lambda\langle M \rangle)^3\right], \end{aligned} \tag{11.6.5}$$

yielding

$$\begin{aligned} H &= a'(T - T_c)\langle M \rangle + \frac{\mu}{3Nk_B^2 T^2}(H + \lambda\langle M \rangle)^3 \\ &\simeq a'(T - T_c)\langle M \rangle + b'T\langle M \rangle^3, \end{aligned} \tag{11.6.6}$$

where

$$a' = \frac{k_B}{N\mu^2}, \qquad T_c = \frac{\lambda N \mu^2}{k_B}, \qquad b' = \frac{k_B}{3N^3\mu^4}. \tag{11.6.7}$$

Here we have substituted $H = a'(T - T_c)\langle M \rangle$ in the cubic term in the first line of Eq. (11.6.6).

From Eq. (11.6.6), we obtain the following relation between M and T for $H = 0$:

$$\langle M \rangle = \begin{cases} 0 & \text{if } T - T_c \geq 0, \\ \left[\frac{a'}{b'} \left(\frac{T_c - T}{T} \right) \right]^{1/2} & \text{if } T - T_c < 0. \end{cases} \tag{11.6.8}$$

Equations (11.6.3) for the laser and (11.6.8) for the ferromagnet are formally identical. The electric field envelope E corresponds to the static magnetization M which is the order parameter in the ferromagnetic transition. Apart from a change in sign, the population inversion σ corresponds to the temperature T.

In order to establish a completely satisfactory analogy, it is necessary to extend the laser analysis to include the injected signal. We recall that the ensemble equilibrium average of the magnetization can be zero only in the presence of an external magnetic field with a definite orientation. The analogous situation for the laser is realized by introducing an external classical field into the laser cavity. This leads to an additional polarization $\mathscr{S} \exp(-i\nu t)$ in the active medium. The signal strength $\mathscr{S}$ is a time-independent real quantity which designates the intensity of the external signal. It is also assumed that the external signal has the same direction of polarization as the laser mode.

The quantum theory of the laser can be extended to include an injected signal. The steady-state solution for the P-representation in the presence of an external field is found to be

$$P(\alpha, \alpha^*) \simeq \frac{1}{\mathscr{N}'} \exp\left[\left(\frac{\mathscr{A} - \mathscr{C}}{\mathscr{A}} \right) |\alpha|^2 - \frac{\mathscr{B}}{2\mathscr{A}} |\alpha|^4 + \frac{2\mathscr{S}}{A} (\alpha + \alpha^*) \right], \tag{11.6.9}$$

where $\mathscr{N}'$ is a normalization constant. For $\mathscr{S} = 0$, this equation reduces to Eq. (11.3.3). This expression for $P(\alpha, \alpha^*)$ may be rewritten in terms of an effective free energy G as

$$P(x, y) = \frac{1}{\mathscr{N}'} \exp[-G(x, y)/K\sigma], \tag{11.6.10}$$

where $x = (\alpha + \alpha^*)/2$ and $y = (\alpha - \alpha^*)/2i$ are the Cartesian coordinates,

$$G(x, y) = \frac{1}{4} a(\sigma - \sigma_t)(x^2 + y^2) + \frac{1}{8} b\sigma (x^2 + y^2)^2 - \mathscr{S} x + G_0, \tag{11.6.11}$$

and $K = a/4$ is one-fourth of the gain for one atom. Here we note that $\mathscr{A} - \mathscr{C} = a(\sigma - \sigma_t)$ and $\mathscr{B} = b\sigma$.

Let us now consider the corresponding expression for the probability density $P(M)$ for a ferromagnetic system of magnetization M near a phase transition. In thermal equilibrium, this density is given by

$$P(M) = \frac{1}{\mathscr{N}'} \exp[-F(M)/k_B T], \tag{11.6.12}$$

where

$$F(M) = \frac{1}{2}a'(T - T_c)M^2 + \frac{1}{4}b'TM^3 - HM + F_0, \qquad (11.6.13)$$

with F_0 being dependent on T and H. The similarity between the fluctuation probability density (11.6.12) for $P(M)$ and the expression (11.6.10) for $P(x, y)$ is apparent, and suggests the interpretation of $G(x, y)$ as a type of thermodynamic energy function. The fact that the probability distribution for the laser field (11.6.10) and the corresponding thermodynamical result (11.6.12) are in such close correspondence is a strong argument in support of the analogy between the laser threshold region and a second-order phase transition.

The close analogy between laser threshold behavior and a second-order phase transition clearly shows that cooperative phenomena can develop even far from thermal equilibrium. Other examples include soft modes in ferroelectrics, chemical reactions, and turbulance. Application of these ideas had led to new insights in fundamental and applied science from laser physics to computer science and even sociology. The notion of second order-disorder transitions far from thermal equilibrium has led to many interesting innovations. We conclude this chapter with a few examples.

In laser physics the study of symmetry breaking via an injected signal leads to the suggestion of a phase locked array of N low power lasers as a means of achieving high focal power (going as N^2) by overcoming Schawlow-Townes phase randomization. We also note that the order-disorder transition encounted in optical bistability is closely related laser threshold behavior in the presence of an injected signal. Furthermore the multiple bifurcation road to chaos was a natural extention of the laser phase-transition analogy. Finally we point to the recent beautiful research in hydrodynamics via the study of analogous behavior involving the transverse mode structure of a laser. The analogy between driven parametric systems and the dynamics of soft modes in a ferroelectric has led Landauer to new insights in computer science. Similarly the order-disorder transitions occuring in chemical reactions and even sociology have been fruitfully investigated from vantage of phase-transition physics far from thermodynamic equilibrium.

11.A Solution of the equations for the density matrix elements

A solution of the set of coupled equations (11.1.9) and (11.1.5b)–(11.1.5d) is facilitated by rewriting it in the following matrix form

$$\dot{R} = -MR + A, \tag{11.A.1}$$

where

$$R = \begin{bmatrix} \rho_{an,an'} \\ \rho_{an,bn'+1} \\ \rho_{bn+1,an'} \\ \rho_{bn+1,bn'+1} \end{bmatrix}, \tag{11.A.2}$$

$$M = \begin{bmatrix} \gamma & -ig\sqrt{n'+1} & ig\sqrt{n+1} & 0 \\ -ig\sqrt{n'+1} & \gamma & 0 & ig\sqrt{n+1} \\ ig\sqrt{n+1} & 0 & \gamma & -ig\sqrt{n'+1} \\ 0 & ig\sqrt{n+1} & -ig\sqrt{n'+1} & \gamma \end{bmatrix}, \tag{11.A.3}$$

$$A = r_a \rho_{nn'} \begin{bmatrix} 1 \\ 0 \\ 0 \\ 0 \end{bmatrix}. \tag{11.A.4}$$

Equation (11.A.1) can be formally integrated and we obtain

$$R(t) = \int_{-\infty}^{t} e^{-M(t-t_0)} A dt_0 = M^{-1} A. \tag{11.A.5}$$

Here we have made an adiabatic approximation by assuming that the field does not change appreciably during the lifetimes of the atomic levels. This allows us to treat A as independent of time in Eq. (11.A.5). Since we are interested in the matrix elements $\rho_{an,bn'+1}$ and $\rho_{bn+1,an'}$, we need only determine $(M^{-1})_{21}$ and $(M^{-1})_{31}$. The resulting expressions for the desired matrix elements are

$$\begin{aligned} \rho_{an,bn'+1} &= r_a \rho_{nn'} (M^{-1})_{21} \\ &= \frac{i r_a g \sqrt{n'+1}}{|M|} [g^2(n'-n) + \gamma^2] \rho_{nn'}, \end{aligned} \tag{11.A.6}$$

$$\begin{aligned} \rho_{bn+1,an'} &= r_a \rho_{nn'} (M^{-1})_{31} \\ &= \frac{-i r_a g \sqrt{n+1}}{|M|} [g^2(n-n') + \gamma^2] \rho_{nn'}, \end{aligned} \tag{11.A.7}$$

with

$$\begin{aligned} |M| &= \det M \\ &= \gamma^4 + 2g^2\gamma^2(n+1+n'+1) + g^4(n-n')^2. \end{aligned} \tag{11.A.8}$$

The matrix elements $\rho_{an-1,bn'}$ and $\rho_{bn,an'-1}$ can be determined from Eqs. (11.A.6) and (11.A.7), respectively, by replacing n, n' by $n-1$, $n'-1$.

11.B An exact solution for the P-representation of the laser

We consider the master equation (11.1.14) for the matrix elements. If we define the auxiliary quantity

$$\mu_{nn'} = \left(1 + \frac{\mathscr{B}}{\mathscr{A}}\mathscr{N}_{nn'}\right)^{-1} \rho_{nn'}, \tag{11.B.1}$$

Eq. (11.1.14) is equivalent to the following coupled equations

$$\begin{aligned}\frac{d\rho_{nn'}}{dt} = &-\mathscr{A}\left[\frac{1}{2}(n+1+n'+1) + \frac{\mathscr{B}}{8\mathscr{A}}(n-n')^2\right]\mu_{nn'} \\ &+\mathscr{A}\sqrt{nn'}\mu_{n-1,n'-1} \\ &+\mathscr{C}\sqrt{(n+1)(n'+1)}\rho_{n+1,n'+1} - \frac{1}{2}\mathscr{C}(n+n')\rho_{nn'},\end{aligned} \tag{11.B.2}$$

$$\left[1 + \frac{\mathscr{B}}{2\mathscr{A}}(n+1+n'+1) + \frac{\mathscr{B}^2}{16\mathscr{A}^2}(n-n')^2\right]\mu_{nn'} = \rho_{nn'}. \tag{11.B.3}$$

In terms of the P-representation, $\rho_{nn'}$ is given by

$$\rho_{nn'}(t) = \int d^2\alpha P(\alpha, \alpha^*, t) e^{-|\alpha|^2} \frac{\alpha^n (\alpha^*)^{n'}}{\sqrt{n!}\sqrt{n'!}}. \tag{11.B.4}$$

We define another auxiliary function $M(\alpha, \alpha^*)$ corresponding to $\mu_{nn'}$:

$$\mu_{nn'}(t) = \int d^2\alpha e^{-|\alpha|^2} \frac{\alpha^n (\alpha^*)^{n'}}{\sqrt{n!}\sqrt{n'!}} M(\alpha, \alpha^*). \tag{11.B.5}$$

Using these relationships, Eqs. (11.B.2) and (11.B.3) can be translated into an equivalent set of equations for P and M.

Various terms in Eqs. (11.B.2) and (11.B.3) can be written in terms of P and M as follows

$$\begin{aligned}(n+1+n'+1)\mu_{nn'} &= \int d^2\alpha \frac{1}{\sqrt{n!}\sqrt{n'!}} \\ &\times \left[\left(\frac{\partial}{\partial\alpha}\alpha + \frac{\partial}{\partial\alpha^*}\alpha^* + 2|\alpha|^2\right)\alpha^n(\alpha^*)^{n'}e^{-|\alpha|^2}\right] M(\alpha,\alpha^*) \\ &= \int d^2\alpha e^{-|\alpha|^2} \frac{\alpha^n(\alpha^*)^{n'}}{\sqrt{n!}\sqrt{n'!}} \\ &\times \left[\left(-\alpha\frac{\partial}{\partial\alpha} - \alpha^*\frac{\partial}{\partial\alpha^*} + 2|\alpha|^2\right) M(\alpha,\alpha^*)\right],\end{aligned} \tag{11.B.6}$$

$$
\begin{aligned}
&(n-n')^2 \mu_{nn'} \\
&= \int d^2\alpha \frac{1}{\sqrt{n!}\sqrt{n'!}} \left[\left(\alpha \frac{\partial}{\partial \alpha} - \alpha^* \frac{\partial}{\partial \alpha^*} \right)^2 \alpha^n (\alpha^*)^{n'} e^{-|\alpha|^2} \right] M(\alpha, \alpha^*) \\
&= \int d^2\alpha e^{-|\alpha|} \frac{\alpha^n (\alpha^*)^{n'}}{\sqrt{n!}\sqrt{n!}} \left[\left(\frac{\partial}{\partial \alpha}\alpha - \frac{\partial}{\partial \alpha^*}\alpha^* \right)^2 M(\alpha, \alpha^*) \right], \qquad (11.B.7)
\end{aligned}
$$

$$
\begin{aligned}
\sqrt{nn'}\mu_{n-1,n'-1} &= \int d^2\alpha \frac{1}{\sqrt{n!}\sqrt{n'!}} \\
&\times \left[\left(\frac{\partial^2}{\partial\alpha\partial\alpha^*} + \frac{\partial}{\partial\alpha}\alpha + \frac{\partial}{\partial\alpha^*}\alpha^* + |\alpha|^2 - 1 \right) \alpha^n (\alpha^*)^{n'} e^{-|\alpha|^2} \right] M(\alpha, \alpha^*) \\
&= \int d^2\alpha e^{-|\alpha|^2} \frac{\alpha^n (\alpha^*)^{n'}}{\sqrt{n!}\sqrt{n'!}} \\
&\times \left[\left(\frac{\partial^2}{\partial\alpha\partial\alpha^*} - \alpha\frac{\partial}{\partial\alpha} - \alpha^*\frac{\partial}{\partial\alpha^*} + |\alpha|^2 - 1 \right) M(\alpha, \alpha^*) \right], \qquad (11.B.8)
\end{aligned}
$$

$$
\sqrt{(n+1)(n'+1)}\rho_{n+1,n'+1} = \int d^2\alpha e^{-|\alpha|^2} \frac{\alpha^n (\alpha^*)^{n'}}{\sqrt{n!}\sqrt{n'!}} \left[|\alpha|^2 P(\alpha, \alpha^*) \right], \qquad (11.B.9)
$$

$$
\begin{aligned}
&(n+n')\rho_{n,n'} \\
&= \int d^2\alpha \frac{1}{\sqrt{n!}\sqrt{n'!}} \left[\left(\alpha\frac{\partial}{\partial\alpha} + \alpha^*\frac{\partial}{\partial\alpha^*} + 2|\alpha|^2 \right) \alpha^n (\alpha^*)^{n'} e^{-|\alpha|^2} \right] P(\alpha, \alpha^*) \\
&= \int d^2\alpha e^{|\alpha|^2} \frac{\alpha^n (\alpha^*)^{n'}}{\sqrt{n!}\sqrt{n'!}} \left[\left(-\frac{\partial}{\partial\alpha}\alpha - \frac{\partial}{\partial\alpha^*}\alpha^* + 2|\alpha|^2 \right) P(\alpha, \alpha^*) \right]. \qquad (11.B.10)
\end{aligned}
$$

On substituting for various terms in Eqs. (11.B.2) and (11.B.3) from Eqs. (11.B.6)–(11.B.8), we obtain after some rearrangement,

$$
\begin{aligned}
&\frac{\partial P(\alpha, \alpha^*)}{\partial t} \\
&= -\frac{\mathscr{A}}{2} \left[\frac{\partial}{\partial\alpha}\alpha + \frac{\partial}{\partial\alpha^*}\alpha^* - 2\frac{\partial^2}{\partial\alpha\partial\alpha^*} \right. \\
&\quad \left. + \frac{\mathscr{B}}{4\mathscr{A}} \left(\frac{\partial}{\partial\alpha}\alpha - \frac{\partial}{\partial\alpha^*}\alpha^* \right)^2 \right] M(\alpha, \alpha^*) \\
&\quad + \frac{\mathscr{C}}{2} \left(\frac{\partial}{\partial\alpha}\alpha + \frac{\partial}{\partial\alpha^*}\alpha^* \right) P(\alpha, \alpha^*), \qquad (11.B.11)
\end{aligned}
$$

$$\left\{1 - \frac{\mathscr{B}}{2\mathscr{A}}\left[\frac{\partial}{\partial\alpha}\alpha + \frac{\partial}{\partial\alpha^*}\alpha^* - 2(1+|\alpha|^2)\right] + \frac{1}{16}\frac{\mathscr{B}^2}{\mathscr{A}^2}\left(\frac{\partial}{\partial\alpha}\alpha - \frac{\partial}{\partial\alpha^*}\alpha^*\right)^2\right\} M(\alpha,\alpha^*) = P(\alpha,\alpha^*). \tag{11.B.12}$$

A convenient form of Eqs. (11.B.11) and (11.B.12) is obtained if we transform from the independent complex variables α, α^* to polar coordinates r, θ using

$$\alpha = re^{i\theta}. \tag{11.B.13}$$

It then follows that

$$\frac{\partial}{\partial\alpha}\alpha = \frac{1}{2r}\frac{\partial}{\partial r}r^2 - \frac{1}{2}i\frac{\partial}{\partial\theta}, \tag{11.B.14}$$

$$\frac{\partial^2}{\partial\alpha\partial\alpha^*} = \frac{1}{4r^2}\left(r\frac{\partial}{\partial r}r\frac{\partial}{\partial r} + \frac{\partial^2}{\partial\theta^2}\right). \tag{11.B.15}$$

The transformed Eqs. (11.B.11) and (11.B.12) are therefore

$$\frac{\partial P(r,\theta)}{\partial t} = -\frac{\mathscr{A}}{2}\left(\frac{1}{r}\frac{\partial}{\partial r}r^2 - \frac{1}{2r}\frac{\partial}{\partial r}r\frac{\partial}{\partial r} - \frac{1}{2r^2}\frac{\partial^2}{\partial\theta^2} - \frac{\mathscr{B}}{4\mathscr{A}}\frac{\partial^2}{\partial\theta^2}\right)M(r,\theta) + \frac{\mathscr{C}}{2}\left(\frac{1}{r}\frac{\partial}{\partial r}r^2\right)P(r,\theta), \tag{11.B.16}$$

$$\left\{1 - \frac{\mathscr{B}}{2\mathscr{A}}\left[\frac{1}{r}\frac{\partial}{\partial r}r^2 - 2(1+r^2)\right] - \frac{1}{16}\frac{\mathscr{B}^2}{\mathscr{A}^2}\frac{\partial^2}{\partial\theta^2}\right\} M(r,\theta) = P(r,\theta). \tag{11.B.17}$$

An equation for P can be obtained by substituting for M from Eq. (11.B.17) into Eq. (11.B.16):

$$\frac{\partial P}{\partial t} = \frac{-\mathscr{A}}{2}\left(\frac{1}{r}\frac{\partial}{\partial r}r^2 - \frac{1}{2r}\frac{\partial}{\partial r}r\frac{\partial}{\partial r} - \frac{1}{2r^2}\frac{\partial^2}{\partial\theta^2} - \frac{\mathscr{B}}{4\mathscr{A}}\frac{\partial^2}{\partial\theta^2}\right) \times \left\{1 - \frac{\mathscr{B}}{2\mathscr{A}}\left[\frac{1}{r}\frac{\partial}{\partial r}r^2 - 2(1+r^2)\right] - \frac{1}{16}\frac{\mathscr{B}^2}{\mathscr{A}^2}\frac{\partial^2}{\partial\theta^2}\right\}^{-1} P + \frac{\mathscr{C}}{2}\left(\frac{1}{r}\frac{\partial}{\partial r}r^2\right)P. \tag{11.B.18}$$

This equation contains the derivatives with respect to r and θ to all orders.

In the steady state ($\partial P/\partial t = 0$) the P-representation is a function of the modulus of α (i.e., $|\alpha| = r$) only, because all the off-diagonal elements of $\rho_{nn'}$ are zero. Equations (11.B.16) and (11.B.17) reduce to

$$\frac{-\mathscr{A}}{2}\left(\frac{1}{r}\frac{\partial}{\partial r}r^2 - \frac{1}{2r}\frac{\partial}{\partial r}r\frac{\partial}{\partial r}\right)M + \frac{\mathscr{C}}{2}\left(\frac{1}{r}\frac{\partial}{\partial r}r^2\right)P = 0, \quad (11.B.19)$$

$$\left\{1 - \frac{\mathscr{B}}{2\mathscr{A}}\left[\frac{1}{r}\frac{\partial}{\partial r}r^2 - 2(1+r^2)\right]\right\}M = P. \quad (11.B.20)$$

If we now introduce the intensity variable

$$I = r^2 \equiv |\alpha|^2, \quad (11.B.21)$$

Eqs. (11.B.19) and (11.B.20) simplify considerably and we obtain

$$-\mathscr{A}\left(\frac{\partial}{\partial I}I - \frac{\partial}{\partial I}I\frac{\partial}{\partial I}\right)M + \mathscr{C}\frac{\partial}{\partial I}(IP) = 0, \quad (11.B.22)$$

$$\left[1 - \frac{\mathscr{B}}{\mathscr{A}}\left(I\frac{\partial}{\partial I} - I\right)\right]M = P. \quad (11.B.23)$$

These equations can be solved for M by eliminating P; the result is

$$M(I) = \begin{cases} \text{const.}\ \left(1 - \frac{\mathscr{B}\mathscr{C}}{\mathscr{A}^2}I\right)^{\mathscr{A}/\mathscr{B}} e^I & \text{for } \frac{\mathscr{B}\mathscr{C}}{\mathscr{A}^2}I \le 1\,, \\ 0 & \text{otherwise}\,. \end{cases} \quad (11.B.24)$$

The P-representation is then found from Eq. (11.B.23):

$$P(I) = \begin{cases} \frac{1}{\mathscr{N}}\left(1 - \frac{\mathscr{B}\mathscr{C}}{\mathscr{A}^2}I\right)^{\mathscr{A}/\mathscr{B}-1} e^I & \text{for } \frac{\mathscr{B}\mathscr{C}}{\mathscr{A}^2} \le 1 \\ 0 & \text{otherwise}\,, \end{cases} \quad (11.B.25)$$

where $\mathscr{N}$ is an appropriate normalization constant. The presence of derivatives of all orders in Eq. (11.B.18) produces a steady state which is a nonanalytic function of $I = |\alpha|^2$.

The P-representation allows us to evaluate all the normal-ordered moments of the number operator $a^\dagger a$ from the first one:

$$\begin{aligned}\langle (a^\dagger)^m a^m\rangle &= \int_0^\infty dI\, I^m P(I) \\ &= \left[\frac{\mathscr{A}^2}{\mathscr{B}\mathscr{C}} - \frac{\mathscr{A}}{\mathscr{B}} - (m-1)\right]\langle (a^\dagger)^{m-1}a^{m-1}\rangle \\ &\quad + \frac{\mathscr{A}^2}{\mathscr{B}\mathscr{C}}(m-1)\langle (a^\dagger)^{m-2}a^{m-2}\rangle \qquad m > 1. \end{aligned} \quad (11.B.26)$$

Near threshold the average value of $I = |\alpha|^2$ is small and the inequality $(\mathscr{B}\mathscr{C}/\mathscr{A}^2)|\alpha|^2 \ll 1$ is satisfied. The factor in front of the exponential in Eq. (11.B.25) can be approximated as follows:

$$\left(1 - \frac{\mathscr{B}\mathscr{C}}{\mathscr{A}^2}|\alpha|^2\right)^{\mathscr{A}/\mathscr{B}-1} = \exp\left[\left(\frac{\mathscr{A}}{\mathscr{B}} - 1\right)\ln\left(1 - \frac{\mathscr{B}\mathscr{C}}{\mathscr{A}^2}|\alpha|^2\right)\right]$$
$$\simeq \exp\left(\frac{-\mathscr{C}}{\mathscr{A}}|\alpha|^2 - \frac{\mathscr{B}}{2\mathscr{A}}|\alpha|^4\right), \quad (11.B.27)$$

where we have approximated $\mathscr{A} \simeq \mathscr{C}$ in the $|\alpha|^4$ term in the exponent. The P-representation (Eq. (11.B.25)) is therefore

$$P(\alpha, \alpha^*) \simeq \frac{1}{\mathscr{N}'}\exp\left[\left(\frac{\mathscr{A} - \mathscr{C}}{\mathscr{A}}\right)|\alpha|^2 - \frac{\mathscr{B}}{2\mathscr{A}}|\alpha|^4\right], \quad (11.B.28)$$

where $\mathscr{N}'$ is a normalization constant. The restriction on P in Eq. (11.B.25) can now obviously be removed, for the distribution (11.B.28) never really sees the boundary and $|\alpha|$ can be taken to run from 0 to ∞.

Problems

11.1 Consider a single-mode two-photon laser in which the atoms in the excited state $|a\rangle$ make a transition to the lower level $|b\rangle$ by emitting two photons of frequency ν via a virtual level. The cavity mirror losses are assumed to be linear. Starting with the interaction picture effective Hamiltonian at exact resonance ($\omega_{ab} = 2\nu$)

$$\mathscr{V} = \hbar g[\sigma_+ a^2 + (a^\dagger)^2\sigma_-],$$

derive an equation of motion for the elements $\rho_{nn'}$ of the reduced density operator for the field. Show that

$$Q_f = \frac{1}{2}$$

for the laser operating high above threshold.

11.2 Consider a laser with a saturable absorber in which, in addition to a gain medium, there is an absorber medium inside the resonator cavity. The gain medium consists of two-level atoms with levels $|a\rangle$ and $|b\rangle$ in which atoms are pumped in the excited level $|a\rangle$ at a rate r_{a_1}. The absorber also consists of two-level atoms with levels $|c\rangle$ and $|d\rangle$ but atoms are pumped to the lower level $|d\rangle$ at a rate r_{a_2}. Assuming perfect resonance for both types of atoms ($\omega_{ab} = \omega_{cd} = \nu$), derive an equation of motion for the reduced density matrix of the field. Derive a steady-state solution for the photon distribution function $p(n)$, and then obtain an expression for the mean number of photons $\langle n\rangle$. (Hint: see R. Roy, *Phys. Rev. A* **20**, 2093 (1979).)

11.3 Show that, in the linear regime ($\mathscr{B}=0$), Eq. (11.1.14) for $\rho_{nn'}$ is equivalent to the following equation for the reduced density operator for the field

$$\dot{\rho}=-\frac{\mathscr{A}}{2}(aa^{\dagger}\rho-2a^{\dagger}\rho a+\rho aa^{\dagger})$$
$$-\frac{\mathscr{C}}{2}(a^{\dagger}a\rho-2a\rho a^{\dagger}+\rho a^{\dagger}a).$$

11.4 Derive Eq. (11.2.22).

11.5 Derive an equation of motion for $\rho_{nn'}$ for a single-mode laser when the field is not resonant with the atomic transitions, i.e., $\Delta=\omega-\nu\neq 0$.

11.6 Calculate the laser linewidth using the diffusion coefficient approach of Section 11.4 when the cavity losses are described via interaction with a thermal reservoir at temperature T.

11.7 Derive the equation of motion for diagonal elements of the density matrix by using the average over dwell times approach. To this end derive the relation (11.1.28) taking into account (11.1.27). Perform the integration of Eq. (11.1.28) assuming $n=n'$. Hint: starting from equation

$$\langle\frac{d\rho_{nn}}{dt}\rangle_{\tau}$$
$$=r_a\int_0^{\infty}d\tau\left[-(1-\cos^2(g\sqrt{n+1}\tau))\rho_{nn}\right.$$
$$\left.+\sin^2(g\sqrt{n}\tau)\rho_{n-1,n-1}\right]\gamma e^{-\gamma\tau},$$

show that

$$\gamma\int_0^{\infty}\sin^2(\Omega_n\tau)e^{-\gamma\tau}d\tau=\frac{2\Omega_n^2}{\gamma^2+4\Omega_n^2},$$

where $\Omega_n=g\sqrt{n}$. This would then give:

$$\langle\frac{d\rho_{nn}}{dt}\rangle_{\tau}=r_a\left(-\frac{2\Omega_n^2}{\gamma^2+4\Omega_n^2}\rho_{nn}+\frac{2\Omega_{n-1}^2}{\gamma^2+4\Omega_{n-1}^2}\rho_{n-1,n-1}\right).$$

References and bibliography

Laser theory – density operator approach

M. O. Scully, W. E. Lamb, Jr. and M. J. Stephen, *Proceedings of the International Conference on the Physics of Quantum Electronics, Puerto Rico*, 1965, ed. P. Kelley, B. Lax, and P. Tannanwald (New York, 1966), p. 75.
J. Fleck, Jr., *Phys. Rev.* **149**, 309 (1966).
M. O. Scully and W. E. Lamb, Jr., *Phys. Rev.* **159**, 208 (1967); *ibid.* **179**, 368 (1969).
E. R. Pine, *Riv. Nuovo Cimento*, Numero Speciale 1, 277 (1969).
M. Scully, *Proceedings of the International School of Physics "Enrico Fermi" Course XLII* ed. R. Glauber (Academic Press, New York 1969).
M. Sargent III, M. O. Scully, and W. E. Lamb, Jr., *Appl. Opt.* **9**, 2423 (1970).
M. O. Scully, D. M. Kim, and W. E. Lamb, Jr., *Phys. Rev. A* **2**, 2529 (1970); *ibid.* **2**, 2534 (1970).
Y. R. Wang and W. E. Lamb, Jr., *Phys. Rev. A* **8**, 866 (1973).
S. Stenholm, *Phys. Rep.* **6C**, 1 (1973).
R. Loudon, *The Quantum Theory of Light*, (Oxford University Press, New York 1973). 2nd ed.
L. A. Lugiato, *Physics* **81A**, 565 (1976).
E. R. Pike and S. Sarkar, *Quantum Theory of Radiation*, (Cambridge, London 1996).

Laser theory – Fokker–Planck approach

H. Risken, *Z. Phys.* **191**, 186 (1965); in *Progress in Optics*, Vol. 8, ed. E. Wolf (North-Holland, Amsterdam 1970).
R. D. Hemstead and M. Lax, *Phys. Rev.* **161**, 350 (1967).
H. Risken and H. D. Vollmer, *Z. Phys.* **201**, 323 (1967).
H. Haken, *Z. Phys.* **219**, 246 (1969).
S. Grossmann and P. H. Richter, *Z. Phys.* **249**, 43 (1971).
R. Graham and W. A. Smith, *Opt. Commun.* **7**, 289 (1973).
P. Mandel, *Physica (Utr.)* **77**, 174 (1974).
F. Casagrande and L. A. Lugiato, *Phys. Rev. A* **14**, 778 (1976).
F. T. Arecchi and A. M. Ricca, *Phys. Rev. A* **15**, 308 (1977).
M. M. Tehrani and L. Mandel, *Phys. Rev. A* **17**, 677 (1978); *ibid.* **17**, 694 (1978).

Laser–phase transition analogy and related works

V. DiGiorgio and M. O. Scully, *Phys. Rev. A* **2**, 1170 (1970).
R. Graham and H. Haken, *Z. Phys.* **237**, 31 (1970).
R. Landauer, *IBM Journal of Res. and Dev.* **14**, 152 (1970).
R. Landauer, *Ferroelectrics* **2**, 47 (1971).
P. Glandsdorff and I. Prigogine, *Thermodynamic Theory of Structure, Stability, and Fluctuations* (Wiley, New York 1971).

H. Haken, *Rev. Mod. Phys.* **47**, 67 (1975).
J. F. Scott, M. Sargent III, and C. Cantrell, *Opt. Commun.* **15**, 13 (1975).
W. W. Chow, M. O. Scully, and E. W. van Stryland, *Opt. Commun.* **15**, 6 (1975).
W. Weidlich, *Concepts and Models of a Quantitative Sociology: the Dynamics of Interacting Population*, (Springer, Berlin 1983).
A. Gatti and L. A. Lugiato, *Phys. Rev. A* **52**, 1675 (1995).

CHAPTER 12

Quantum theory of the laser – Heisenberg–Langevin approach

In this chapter, we present a theory of the laser based on the Heisenberg–Langevin* approach. This is a different, but completely equivalent approach to the density operator approach discussed in the previous chapter. In general, the density operator approach is better suited to study the photon statistics of the radiation field whereas the Heisenberg–Langevin approach has certain calculational advantages in the determination of phase diffusion coefficients, and consequently laser linewidth.

In Section 12.1, a simple approach to determine laser linewidth based on a linear theory is presented. This analysis is especially interesting and useful in that it includes atomic memory effects, something that is difficult to do within a density matrix theory. In Sections 12.2–12.4, we consider the complete nonlinear theory of the laser and rederive all the important quantities related to the quantum statistical properties of the radiation field.

12.1 A simple Langevin treatment of the laser linewidth including atomic memory effects†

The full nonlinear quantum theory of the laser discussed in the previous chapter yields most of the interesting quantum statistical properties of the radiation field. In many problems of interest, however, we do not need such an elaborate treatment. For example, as we saw in the previous chapter, the natural linewidth of the laser can

* The original treatment of Lax [1966,1968] is presented in a tutorial form in Louisell [1974]. The linewidth calculation of Haken, presented in his textbook [1970], is especially useful. See also Sargent, Scully, and Lamb [1974].

† This section is patterned after the treatment of Scully, Süssmann and Benkert [1988].

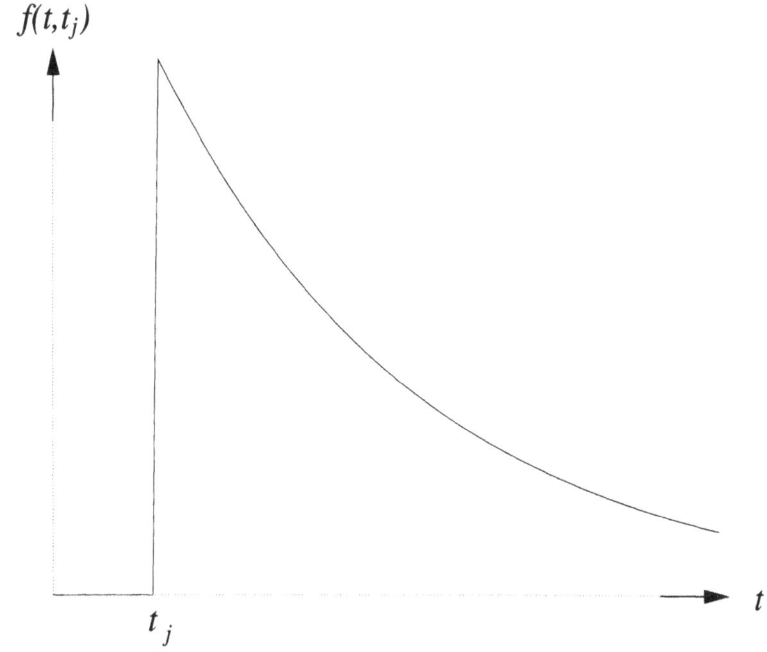

Fig. 12.1
Interaction function $f(t, t_j)$ which determines the interaction of the jth atom and the electromagnetic field.

be determined from a linearized theory of the laser. That is, the full nonlinear theory serves to determine the amplitude of the field but the phase fluctuations about this operating point are described by a linear theory. In this section we present a simple linear Langevin theory of the laser in which the atomic lifetime is included in an effective Hamiltonian. The advantage of this simple approach is that it can be easily extended to include interesting atomic memory effects as shown below and to calculate the natural linewidth in other quantum optical systems.

The effective Hamiltonian for a system of two-level atoms interacting with the single-mode field at exact resonance is given, in the interaction picture, by

$$\mathscr{H} = \hbar g \sum_j f(t, t_j)(a^\dagger \sigma_-^j + \sigma_+^j a). \tag{12.1.1}$$

Here we have introduced a function $f(t, t_j)$ that determines the interaction between the jth atom which is injected at time t_j in the upper level and the radiation field. To be more specific, we define $f(t, t_j)$ as

$$f(t, t_j) = \Theta(t - t_j) \exp[-\gamma(t - t_j)/\sqrt{2}], \tag{12.1.2}$$

where $\Theta(t)$ is the usual step function. The function $f(t, t_j)$ is depicted in Fig. 12.1. Here the exponential $\exp[-\gamma(t-t_j)/\sqrt{2}]$ models the decay of the lasing atoms. The atomic decay has thus been incorporated in the effective Hamiltonian.

The Heisenberg equations of motion for the radiation field and atomic dipole operators are

$$\dot{a} = -\frac{1}{2}\mathscr{C}a - ig\sum_j f(t, t_j)\sigma_-^j + F_\mathscr{C}(t), \tag{12.1.3}$$

$$\dot{\sigma}_-^j = igf(t, t_j)\sigma_z^j a. \tag{12.1.4}$$

In Eq. (12.1.3), we have included the effects of cavity losses through the cavity damping rate $\mathscr{C}$ and its corresponding Langevin noise operator $F_\mathscr{C}$. At zero temperature ($\bar{n}_{\text{th}} = 0$), all the normally ordered correlation functions of $F_\mathscr{C}$ are zero.

Formal integration of Eq. (12.1.4) then yields

$$\sigma_-^j = \sigma_-^j(t_j) + ig\int_{t_j}^t dt' f(t', t_j)\sigma_z^j(t')a(t'). \tag{12.1.5}$$

We note that the definition of the function $f(t', t_j)$ allows us to extend the lower integration index in Eq. (12.1.5) to $-\infty$. Substituting this equation into Eq. (12.1.3) for the radiation field results in

$$\dot{a} = -\frac{1}{2}\mathscr{C}a + g^2\int_{-\infty}^t dt' \sum_j f(t, t_j)f(t', t_j)\sigma_z^j(t')a(t')$$
$$+ F_\mathscr{C}(t) + F_a(t), \tag{12.1.6}$$

where the noise introduced into the field due to the coupling to the lasing atoms is given by

$$F_a(t) = -ig\sum_j f(t, t_j)\sigma_-^j(t_j). \tag{12.1.7}$$

Equation (12.1.6) for the field operator a can be simplified if we assume the linear gain to be mainly determined by the instantaneous value of the radiation field. Such an approximation holds if the electric field operator $a(t)$ changes slowly during the atomic lifetime γ^{-1}. This condition is met if $\gamma \gg \mathscr{C}$. We can then approximate $a(t')$ in the integral of Eq. (12.1.6) by $a(t)$ and take it out of the integration. We also restrict our analysis to linear order in the field operator. We can then approximate the population difference operator $\sigma_z^j(t')$ by its expectation value at the initial time, i.e., we take $\langle\sigma_z^j(t_j)\rangle = 1$. The resulting equation for $a(t)$ is

$$\dot{a} = \frac{1}{2}(\mathscr{A} - \mathscr{C})a + F_\mathscr{C}(t) + F_a(t), \tag{12.1.8}$$

in which the linear gain coefficient $\mathscr{A}$ is defined as

$$\mathscr{A} = 2g^2 \int_{-\infty}^{t} dt' \sum_j f(t, t_j) f(t', t_j). \tag{12.1.9}$$

If we assume the injection rate r_a of the atoms to be constant, we can substitute the sum over all atoms by an integration over injection times, i.e.,

$$\sum_j \to r_a \int_{-\infty}^{\infty} dt_j. \tag{12.1.10}$$

Using definition (12.1.2) of the interaction function $f(t, t_j)$ and evaluating the remaining integrals, the result for the linear gain coefficient is

$$\mathscr{A} = \frac{2g^2 r_a}{\gamma^2}, \tag{12.1.11}$$

which is identical to the linear gain coefficient obtained earlier (see Eq. (11.1.11)).

Next we turn to the calculation of the correlation function for the noise operators in Eq. (12.1.8). At zero temperature ($\bar{n}_{\text{th}} = 0$), the normally ordered correlation functions of $F_{\mathscr{C}}$ are zero (Eqs. (9.1.24), (9.1.27), and (9.1.28)). The noise operator $F_{\mathscr{C}}$ therefore does not contribute to the laser linewidth. This will be seen explicitly in the rigorous treatment of the later sections. We find from Eq. (12.1.7) that

$$\langle F_a^\dagger(t) F_a(t') \rangle = g^2 \sum_j \sum_k f(t, t_j) f(t', t_k) \langle \sigma_+^j(t_j) \sigma_-^k(t_k) \rangle. \tag{12.1.12}$$

All the atoms are initially prepared in the excited atom level $|a\rangle$ and are completely independent of each other. Thus

$$\langle \sigma_+^j(t_j) \sigma_-^k(t_k) \rangle = \delta_{jk}. \tag{12.1.13}$$

If we again substitute the remaining sum over all atoms by an integration over the injection times, we obtain

$$\langle F_a^\dagger(t) F_a(t') \rangle = \mathscr{A} \frac{\gamma}{2\sqrt{2}} \exp\left(-\frac{\gamma}{\sqrt{2}} |t - t'| \right). \tag{12.1.14}$$

It is interesting to note that the noise is not δ-correlated. The spontaneous emission events can no longer be taken as instantaneous *impulses*. Instead the spontaneous emission event is spread out over a characteristic time γ^{-1} during which the photon is emitted. Such *memory effects* due to the lasing atoms lead to colored noise in contrast to white noise.

In the limit $\gamma^{-1} \to 0$, we regain the white noise result

$$\langle F_a^\dagger(t) F_a(t') \rangle = \mathscr{A} \delta(t - t'). \tag{12.1.15}$$

This δ-correlation of the noise operators results from the assumption that the time evolution of the atom is much quicker than the evolution of the radiation field. In an analogous way, we find

$$\langle F_a(t)F_a(t')\rangle = \langle F_a^\dagger(t)F_a^\dagger(t')\rangle = 0. \tag{12.1.16}$$

The fact that both correlation functions in Eq. (12.1.16) are equal to zero is a direct consequence of the initial preparation of the atoms and our restriction to a linear analysis in the field.

In order to derive an expression for the phase diffusion of the electromagnetic field we next identify the Langevin operator equation (12.1.8) with a corresponding c-number equation. This can be done by substituting the operator a and $F_a(t)$ by the complex variables α and $\mathcal{F}_\alpha(t)$, respectively. The resulting equation is

$$\dot{\alpha} = \frac{1}{2}(\mathcal{A} - \mathcal{C})\alpha + \mathcal{F}_\alpha. \tag{12.1.17}$$

If we choose the classical noise function $\mathcal{F}_\alpha(t)$ to have the same correlation functions as those in Eqs. (12.1.15) and (12.1.16), all products of the complex variables α will correspond to normally ordered products of the operator a.

As we are interested in the phase fluctuations of the laser field, it is convenient to work in polar coordinates r and θ, defined via $\alpha = r\exp(i\theta)$. It then follows from Eq. (12.1.17) that

$$\dot{\theta} = \mathcal{F}_\theta(t), \tag{12.1.18}$$

where

$$\mathcal{F}_\theta(t) = \text{Im}\left(\frac{\mathcal{F}_\alpha}{\alpha}\right). \tag{12.1.19}$$

If we ignore the amplitude fluctuations, i.e., if we assume that $|\alpha|^2 = \langle n\rangle$, where $\langle n\rangle$ is the steady-state mean photon number in the laser field, then it follows from the correlation functions (12.1.15) and (12.1.16), that

$$\langle\dot{\theta}(t)\dot{\theta}(t')\rangle = \frac{\mathcal{A}}{2\langle n\rangle}\delta(t-t'). \tag{12.1.20}$$

A comparison with Eq. (11.4.12) yields

$$\langle 2D_{\theta\theta}\rangle = \frac{\mathcal{A}}{2\langle n\rangle}, \tag{12.1.21}$$

in agreement with the results of the previous chapter.*

* For further reading extending the results of the present section to include memory effects see Scully, Süssmann, and Benkert [1988].

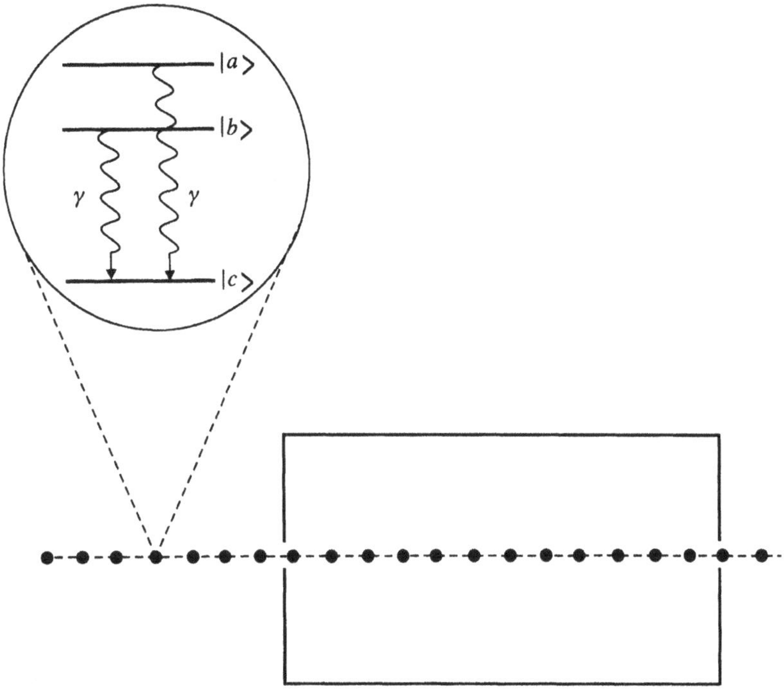

Fig. 12.2
Physical model of the laser: atoms (represented by dots) proceed through the laser cavity.

12.2 Quantum Langevin equations

In this section, we derive quantum Heisenberg–Langevin equations for the field and atomic variables with all the nonlinear saturation effects included. The photon statistics and the laser linewidth can be determined from this set of equations.

Our physical model of the laser is depicted in Fig. 12.2. A set of three-level atoms interacts with a single mode of the radiation field inside a laser cavity. The atoms are initially prepared in level $|a\rangle$ and the upper two atomic levels $|a\rangle$ and $|b\rangle$ constitute the lasing transition. The lowest atomic level $|c\rangle$, which is far from resonance, i.e., $\omega_{ac}, \omega_{bc} \gg \omega_{ab}$, is an inert ground state to which atoms decay with a decay rate γ. The Hamiltonian of such a system in the interaction picture and in the rotating-wave approximation, is given by

$$\mathscr{H} = \mathscr{H}_{AF} + \mathscr{H}_{R_1F} + \mathscr{H}_{R_2A} \tag{12.2.1}$$

where $\mathscr{H}_{AF}$ is the atom–field interaction Hamiltonian and $\mathscr{H}_{R_1F}$ and $\mathscr{H}_{R_2A}$ represent the interaction of the field and the atomic system, with reservoirs of harmonic oscillators of closely spaced frequencies ν_q and ν_k and annihilation operators $c_{\mathbf{q}}$ and $b_{\mathbf{k}}$, respectively:

$$\mathscr{H}_{AF} = \hbar g \sum_j \Theta(t - t_j)(a^\dagger \sigma_-^j + \sigma_+^j a), \tag{12.2.2}$$

$$\mathscr{H}_{R_1F} = \hbar \sum_{\mathbf{q}} g_{\mathbf{q}}^{(1)} c_{\mathbf{q}}^\dagger a \exp[-i(v - v_q)t] + \text{H.c.}, \tag{12.2.3}$$

$$\begin{aligned}\mathscr{H}_{R_2A} = \hbar \sum_j \sum_{\mathbf{k}} \{ & g_{\mathbf{k}}^{(2)} \sigma_{ac}^j b_{\mathbf{k}} \exp[i(\omega_{ac} - v_k)t] \\ & + g_{\mathbf{k}}^{(2)} \sigma_{bc}^j b_{\mathbf{k}} \exp[i(\omega_{bc} - v_k)t]\} + \text{H.c.},\end{aligned} \tag{12.2.4}$$

where $\Theta(t)$ is the unit step function, $\sigma_-^j = |b\rangle_{jj}\langle a|$ and $\sigma_+^j = |a\rangle_{jj}\langle b|$ are the lowering and raising operators, respectively, for the jth atom. We assume g, $g_{\mathbf{q}}^{(1)}$, and $g_{\mathbf{k}}^{(2)}$, the appropriate coupling constants, to be real, and

$$\sigma_{nm}^j = |n\rangle_{jj}\langle m| \qquad (n, m = a, b, c). \tag{12.2.5}$$

We have assumed, for simplicity, that the lasing transition $|a\rangle \to |b\rangle$ is resonant with the field, i.e., $\omega_{ab} = v$. Here we have included the effect of all the atoms in $\mathscr{H}_{AF}$ and the step function accounts for the fact that the jth atom starts the interaction with the field at its injection time t_j. This approach is somewhat different from the density operator method where the effect of a large number of atoms is accounted for by introducing the population matrix.

The Heisenberg equations of motion for the various field, atomic, and reservoir variables are given by

$$\dot{a} = -ig \sum_j \Theta(t - t_j)\sigma_-^j - i \sum_{\mathbf{q}} g_{\mathbf{q}}^{(1)} c_{\mathbf{q}} e^{i(v - v_q)t}, \tag{12.2.6}$$

$$\begin{aligned}\dot{\sigma}_-^j = {} & ig\Theta(t - t_j)(\sigma_{aa}^j - \sigma_{bb}^j)a \\ & + i \sum_{\mathbf{k}} g_{\mathbf{k}}^{(2)} b_{\mathbf{k}}^\dagger \sigma_{ca}^j e^{-i(\omega_{bc} - v_k)t} \\ & - i \sum_{\mathbf{k}} g_{\mathbf{k}}^{(2)} \sigma_{bc}^j b_{\mathbf{k}} e^{i(\omega_{ac} - v_k)t},\end{aligned} \tag{12.2.7}$$

$$\begin{aligned}\dot{\sigma}_{aa}^j = {} & ig\Theta(t - t_j)(a^\dagger \sigma_-^j - \sigma_+^j a) \\ & - i \sum_{\mathbf{k}} g_{\mathbf{k}}^{(2)} \sigma_{ac}^j b_{\mathbf{k}} e^{i(\omega_{ac} - v_k)t} \\ & + i \sum_{\mathbf{k}} g_{\mathbf{k}}^{(2)} b_{\mathbf{k}}^\dagger \sigma_{ca}^j e^{-i(\omega_{ac} - v_k)t},\end{aligned} \tag{12.2.8}$$

$$\begin{aligned}\dot{\sigma}_{bb}^j = {} & -ig\Theta(t - t_j)(a^\dagger \sigma_-^j - \sigma_+^j a) \\ & - i \sum_{\mathbf{k}} g_{\mathbf{k}}^{(2)} \sigma_{bc}^j b_{\mathbf{k}} e^{i(\omega_{bc} - v_k)t} \\ & + i \sum_{\mathbf{k}} g_{\mathbf{k}}^{(2)} b_{\mathbf{k}}^\dagger \sigma_{cb}^j e^{-i(\omega_{bc} - v_k)t},\end{aligned} \tag{12.2.9}$$

$$\dot{c}_{\mathbf{q}} = -ig_{\mathbf{q}}^{(1)} a e^{-i(\nu-\nu_q)t}, \tag{12.2.10}$$

$$\dot{b}_{\mathbf{k}} = -i \sum_j g_{\mathbf{k}}^{(2)} \left[\sigma_{cb}^j e^{-i(\omega_{bc}-\nu_k)t} + \sigma_{ca}^j e^{-i(\omega_{ac}-\nu_k)t} \right]. \tag{12.2.11}$$

On formally integrating Eqs. (12.2.10) and (12.2.11) and substituting for $c_{\mathbf{q}}$ and $b_{\mathbf{k}}$ in Eqs. (12.2.6)–(12.2.9) (see Section 9.1), we find the following quantum Langevin equations for the field and the atomic operators.

$$\dot{a} = -\frac{\mathscr{C}}{2} a - ig \sum_j \Theta(t - t_j)\sigma_-^j + F_{\mathscr{C}}(t), \tag{12.2.12}$$

$$\dot{\sigma}_-^j = -\gamma\sigma_-^j + ig\Theta(t - t_j)(\sigma_{aa}^j - \sigma_{bb}^j)a + F_{ba}^j(t), \tag{12.2.13}$$

$$\dot{\sigma}_{aa}^j = -\gamma\sigma_{aa}^j + ig\Theta(t - t_j)(a^\dagger\sigma_-^j - \sigma_+^j a) + F_{aa}^j(t), \tag{12.2.14}$$

$$\dot{\sigma}_{bb}^j = -\gamma\sigma_{bb}^j - ig\Theta(t - t_j)(a^\dagger\sigma_-^j - \sigma_+^j a) + F_{bb}^j(t), \tag{12.2.15}$$

where $\mathscr{C}$ is the usual cavity decay rate and γ is the atomic decay rate which, for simplicity, we have taken to be the same for both upper levels. We have made the usual Weisskopf–Wigner approximation (see Section 6.3) to obtain these decay constants.

The operators F on the right-hand side of Eqs. (12.2.12)–(12.2.15) are the Langevin noise operators which arise through the interaction with the heat bath and are given by

$$F_{\mathscr{C}}(t) = -i \sum_{\mathbf{q}} g_{\mathbf{q}}^{(1)} c_{\mathbf{q}}(0) e^{i(\nu-\nu_q)t}, \tag{12.2.16}$$

$$\begin{aligned} F_{ba}^j(t) = {} & i\sigma_{ca}^j(t) \sum_{\mathbf{k}} g_{\mathbf{k}}^{(2)} b_{\mathbf{k}}^\dagger(0) e^{-i(\omega_{bc}-\nu_k)t} \\ & -i\sigma_{bc}^j(t) \sum_{\mathbf{k}} g_{\mathbf{k}}^{(2)} b_{\mathbf{k}}(0) e^{i(\omega_{ac}-\nu_k)t}, \end{aligned} \tag{12.2.17}$$

$$F_{aa}^j(t) = -i\sigma_{ac}^j(t) \sum_{\mathbf{k}} g_{\mathbf{k}}^{(2)} b_{\mathbf{k}}(0) e^{i(\omega_{ac}-\nu_k)t} + \text{H.c.}, \tag{12.2.18}$$

$$F_{bb}^j(t) = -i\sigma_{bc}^j(t) \sum_{\mathbf{k}} g_{\mathbf{k}}^{(2)} b_{\mathbf{k}}(0) e^{i(\omega_{bc}-\nu_k)t} + \text{H.c.} \tag{12.2.19}$$

The normally ordered correlation functions of $F_{\mathscr{C}}$ are

$$\begin{aligned} \langle F_{\mathscr{C}}(t)\rangle &= 0, \\ \langle F_{\mathscr{C}}^\dagger(t)F_{\mathscr{C}}(t')\rangle &= \bar{n}_{\text{th}}\mathscr{C}\delta(t - t'), \\ \langle F_{\mathscr{C}}(t)F_{\mathscr{C}}(t')\rangle &= \langle F_{\mathscr{C}}^\dagger(t)F_{\mathscr{C}}^\dagger(t)\rangle = 0. \end{aligned} \tag{12.2.20}$$

For simplicity we assume that the heat reservoir is at zero temperature so that $\bar{n}_{\text{th}} = 0$. The nonvanishing moments of interest of the atomic noise operators at zero temperature are

$$\begin{aligned}
\langle F_{ba}^{j\dagger}(t)F_{ba}^{j}(t')\rangle &= \gamma\langle\sigma_{aa}^{j}(t)\rangle\delta(t-t'),\\
\langle F_{ll}^{j}(t)F_{ll}^{j}(t')\rangle &= \gamma\langle\sigma_{ll}^{j}(t)\rangle\delta(t-t') \qquad (l=a,b),\\
\langle F_{ba}^{j}(t)F_{aa}^{j}(t')\rangle &= \gamma\langle\sigma_{+}^{j}(t)\rangle\delta(t-t'),\\
\langle F_{ba}^{j\dagger}(t)F_{bb}^{j}(t')\rangle &= \gamma\langle\sigma_{-}^{j}(t)\rangle\delta(t-t').
\end{aligned} \tag{12.2.21}$$

As a next step we change from the operators for individual atoms to operators which describe the macroscopic atomic properties. This proves to be necessary for the approximation techniques employed later in this section. While the individual atomic operators are very sensitive to an adiabatic approximation, the averaged, macroscopic quantities can be treated by such a technique. Therefore we define the operators

$$M(t) = -i\sum_j \Theta(t-t_j)\sigma_{-}^{j}(t), \tag{12.2.22}$$

$$N_a(t) = \sum_j \Theta(t-t_j)\sigma_{aa}^{j}(t), \tag{12.2.23}$$

$$N_b(t) = \sum_j \Theta(t-t_j)\sigma_{bb}^{j}(t). \tag{12.2.24}$$

The operator M represents the macroscopic atomic polarization. The factor $(-i)$ in Eq. (12.2.22) has been chosen for mathematical convenience. Furthermore, N_a and N_b specify the number of atoms in the two excited atomic levels $|a\rangle$ and $|b\rangle$, respectively. With these definitions Eq. (12.2.12) for the electromagnetic field simplifies to

$$\dot{a} = -\frac{\mathscr{C}}{2}a + gM + F_{\mathscr{C}}(t). \tag{12.2.25}$$

The Langevin equations for the atomic operators can be found by differentiating Eqs. (12.2.22)–(12.2.24) and substituting from Eqs. (12.2.13)–(12.2.15), respectively. For example, for the operator N_a we obtain

$$\begin{aligned}
\dot{N}_a &= \sum_j [\delta(t-t_j)\sigma_{aa}^{j}(t) + \Theta(t-t_j)\dot{\sigma}_{aa}^{j}(t)]\\
&= \sum_j \delta(t-t_j)\sigma_{aa}^{j}(t_j) - \gamma N_a - g(a^\dagger M + M^\dagger a)\\
&\quad + \sum_j \Theta(t-t_j)F_{aa}^{j}(t).
\end{aligned} \tag{12.2.26}$$

The first term on the right-hand side of Eq. (12.2.26) corresponds to a pumping of the atoms into their upper excited state. To see this most clearly let us calculate the expectation value of this term,

$$\left\langle \sum_j \delta(t-t_j)\sigma_{aa}^j(t_j) \right\rangle = \left\langle \sum_j \delta(t-t_j)\langle\sigma_{aa}^j(t_j)\rangle \right\rangle_S$$
$$= \left\langle \sum_j \delta(t-t_j) \right\rangle_S . \tag{12.2.27}$$

Here we have made use of the fact that the atoms are initially prepared in their upper atomic levels so that $\langle\sigma_{aa}^j(t_j)\rangle = 1$. The index S on the brackets in Eq. (12.2.27) indicates that we still have to perform the statistical average over the injection times, i.e., the average over the pump statistics. If we assume a mean, time-independent atomic injection rate r_a, this average can be calculated as

$$\left\langle \sum_j \delta(t-t_j) \right\rangle_S = r_a \int_{-\infty}^{\infty} dt_j \delta(t-t_j)$$
$$= r_a. \tag{12.2.28}$$

Alternatively, Eq. (12.2.27) can be regarded as the definition for the averaged atomic injection rate r_a.

In order to separate the drift terms from the noise terms in Eq. (12.2.26) we add and subtract the expectation value of the first term and obtain

$$\dot{N}_a = r_a - \gamma N_a - g(a^\dagger M + M^\dagger a) + F_a(t), \tag{12.2.29}$$

with

$$F_a(t) = \sum_j \Theta(t-t_j)F_{aa}^j(t) + \sum_j \delta(t-t_j)\sigma_{aa}^j(t_j) - r_a. \tag{12.2.30}$$

The operator F_a is the total noise operator for the atomic quantity N_a. It incorporates the contributions from the reservoir-induced decay of the atoms and the influence of pump fluctuations. It is easy to verify that the expectation value of $F_a(t)$ is equal to zero at all times.

In a similar way we can derive the equations for the remaining atomic operators

$$\dot{N}_b = -\gamma N_b + g(a^\dagger M + M^\dagger a) + F_b(t), \tag{12.2.31}$$
$$\dot{M} = -\gamma M + g(N_a - N_b)a + F_M(t), \tag{12.2.32}$$

with

$$F_b(t) = \sum_j \Theta(t-t_j)F_{bb}^j(t) + \sum_j \delta(t-t_j)\sigma_{bb}^j(t_j), \tag{12.2.33}$$

$$F_M(t) = -i\sum_j \Theta(t-t_j)F_{ba}^j(t) - i\sum_j \delta(t-t_j)\sigma_-^j(t_j). \tag{12.2.34}$$

Note that there is no pumping term in the Eqs. (12.2.31) and (12.2.32) because we assume the atoms to be initially in the excited state $|a\rangle$.

The noise correlation functions can be calculated in the following way:

$$\begin{aligned}\langle F_a(t)F_a(t')\rangle = &\Big\langle \sum_{j,k} \Theta(t-t_j)\Theta(t-t_k)\langle F^j_{aa}(t)F^k_{aa}(t')\rangle \Big\rangle_S \\ &+\Big\langle \sum_{j,k} \delta(t-t_j)\delta(t'-t_k)\langle \sigma^j_{aa}(t_j)\sigma^k_{aa}(t_k)\rangle \Big\rangle_S \\ &-2r_a\Big\langle \sum_j \delta(t-t_j)\langle \sigma^j_{aa}(t_j)\rangle \Big\rangle_S + r_a^2. \end{aligned} \tag{12.2.35}$$

Here we have separated the statistical average over the injection times, denoted by the index S, from the quantum mechanical expectation value. For the evaluation of the different terms in Eq. (12.2.35) we note that the individual atoms are completely independent of each other. Therefore only the terms with $j = k$ contribute to the first term in Eq. (12.2.35). Furthermore, in the second term we can separate the expectation value $\langle \sigma^j_{aa}(t_j)\sigma^k_{aa}(t_k)\rangle$ for $j \neq k$ into the product $\langle \sigma^j_{aa}(t_j)\rangle\langle \sigma^k_{aa}(t_k)\rangle = 1$ ($\langle \sigma^j_{aa}(t_j)\rangle = 1$ for atoms initially injected in the excited state). We then obtain

$$\begin{aligned}\langle F_a(t)F_a(t')\rangle = &\,\gamma\Big\langle \sum_j \Theta(t-t_j)\sigma^j_{aa}(t)\Big\rangle_S \delta(t-t') \\ &+\Big\langle \sum_j \delta(t-t_j)\delta(t'-t_j)\langle \sigma^j_{aa}(t_j)\rangle \Big\rangle_S \\ &+\Big\langle \sum_{\substack{j,k\\ j\neq k}} \delta(t-t_j)\delta(t'-t_k)\Big\rangle_S \\ &-2r_a\Big\langle \sum_j \delta(t-t_j)\Big\rangle_S + r_a^2 \\ = &\,(\gamma\langle N_a\rangle + r_a)\delta(t-t'). \end{aligned} \tag{12.2.36}$$

In the last step we have used the definitions (12.2.23) and (12.2.28) for the operators N_a and r_a, respectively, and the relations

$$\Big\langle \sum_j \delta(t-t_j)\delta(t'-t_j)\Big\rangle_S = \Big\langle \sum_j \delta(t-t_j)\Big\rangle \delta(t-t') = r_a\delta(t-t'), \tag{12.2.37}$$

$$\Big\langle \sum_{\substack{j,k\\ j\neq k}} \delta(t-t_j)\delta(t'-t_k)\Big\rangle_S = r_a^2. \tag{12.2.38}$$

Analogously, the remaining nonvanishing noise correlation functions can be calculated to be

$$\langle F_b(t)F_b(t')\rangle = \gamma\langle N_b\rangle\delta(t-t'), \tag{12.2.39a}$$

$$\langle F_b(t)F_M(t')\rangle = \gamma\langle M\rangle\delta(t-t'), \tag{12.2.39b}$$

$$\langle F_M^\dagger(t)F_M(t')\rangle = (\gamma\langle N_a\rangle + r_a)\delta(t-t'), \tag{12.2.39c}$$

$$\langle F_a(t)F_M(t')\rangle = \gamma\langle M^\dagger\rangle\delta(t-t'). \tag{12.2.39d}$$

Before solving the set of equations (12.2.25), (12.2.29), (12.2.31), and (12.2.32) for the four macroscopic quantities we first convert the operator Langevin equations into corresponding c-number equations. This simplifies the analysis.

12.3 c-number Langevin equations

In order to convert the operator equations we have to define a certain ordering of the operators, to which the c-number equations correspond. This is necessary because the c numbers commute with each other while the operators do not. Therefore we obtain a unique relationship between operator and c-number Langevin equations only if we define the correspondence between a product of c-numbers and a product of operators. We here choose the normal ordering $a^\dagger, M^\dagger, N_a, N_b, M, a$, and can now derive four c-number Langevin equations for the variables α, $\mathscr{M}$, $\mathscr{N}_a$, and $\mathscr{N}_b$ such that the equations for their first and second moments are identical. Equations (12.2.25), (12.2.29), (12.2.31), and (12.2.32) are already in the chosen order so that we immediately obtain

$$\dot{\alpha} = -\frac{\mathscr{C}}{2}\alpha + g\mathscr{M} + \mathscr{F}_{\mathscr{C}}, \tag{12.3.1}$$

$$\dot{\mathscr{M}} = -\gamma\mathscr{M} + g(\mathscr{N}_a - \mathscr{N}_b)\alpha + \mathscr{F}_{\mathscr{M}}, \tag{12.3.2}$$

$$\dot{\mathscr{N}}_a = r_a - \gamma\mathscr{N}_a - g(\alpha^*\mathscr{M} + \mathscr{M}^*\alpha) + \mathscr{F}_a, \tag{12.3.3}$$

$$\dot{\mathscr{N}}_b = -\gamma\mathscr{N}_b + g(\alpha^*\mathscr{M} + \mathscr{M}^*\alpha) + \mathscr{F}_b. \tag{12.3.4}$$

The functions $\mathscr{F}$ in Eqs. (12.3.1)–(12.3.4) are again the typical Langevin noise forces with the expectation values

$$\langle\mathscr{F}_k(t)\rangle = 0, \tag{12.3.5}$$

$$\langle\mathscr{F}_k(t)\mathscr{F}_l(t')\rangle = \langle 2D_{kl}\rangle\delta(t-t'), \tag{12.3.6}$$

in which $\mathscr{F}_k$ and $\mathscr{F}_l$ can be any of the above noise forces. The diffusion coefficients D_{kl} are now determined by the requirement that the equations of motion for the second moments are also identical to the corresponding operator equations. It is easy to see that the diffusion coefficients for the noise force $\mathcal{F}_{\mathscr{C}}$ are the same as for the normally ordered noise operator $F_{\mathscr{C}}$, so that

$$D_{\mathscr{C}^*\mathscr{C}} = 0, \qquad D_{\mathscr{C}\mathscr{C}} = 0. \tag{12.3.7}$$

However, some of the atomic diffusion coefficients change in the transition from operator to c-number equation. As an example, let us calculate the diffusion coefficient $D_{\mathscr{M}\mathscr{M}}$. From the operator equation (12.2.32), we obtain

$$\begin{aligned}\frac{d}{dt}\langle M(t)M(t)\rangle = &-2\gamma\langle MM\rangle \\ &+ g[\langle (N_a - N_b)Ma\rangle + \langle M(N_a - N_b)a\rangle] \\ &+ \langle MF_M\rangle + \langle F_M M\rangle. \end{aligned}\tag{12.3.8}$$

We note that the second term in the square brackets is not in our chosen order because the operator M is to the left of N_a and N_b. Therefore we have to use the commutation relation $[M, N_a - N_b] = 2M$ to bring this term into chosen order. Also, the last two terms vanish so that we obtain

$$\frac{d}{dt}\langle M(t)M(t)\rangle = -2\gamma\langle MM\rangle + 2g\langle (N_a - N_b)Ma\rangle + 2g\langle Ma\rangle. \tag{12.3.9}$$

We now use Eq. (12.3.2) to obtain the corresponding c-number equation

$$\frac{d}{dt}\langle \mathscr{M}(t)\mathscr{M}(t)\rangle = -2\gamma\langle \mathscr{M}\mathscr{M}\rangle + 2g\langle (\mathscr{N}_a - \mathscr{N}_b)\mathscr{M}\alpha\rangle + \langle 2D_{\mathscr{M}\mathscr{M}}\rangle. \tag{12.3.10}$$

If we require the left-hand sides of Eqs. (12.3.9) and (12.3.10) to be equal we see that the diffusion coefficient $D_{\mathscr{M}\mathscr{M}}$ is given by

$$2D_{\mathscr{M}\mathscr{M}} = 2g\mathscr{M}\alpha. \tag{12.3.11}$$

The remaining nonvanishing c-number diffusion coefficients can be calculated in an analogous way and are given by

$$2D_{\mathscr{M}^*\mathscr{M}} = \gamma\mathscr{N}_a + r_a, \tag{12.3.12a}$$

$$2D_{a\mathscr{M}} = \gamma\mathscr{M}^* \tag{12.3.12b}$$

$$2D_{b\mathscr{M}} = \gamma\mathscr{M}, \tag{12.3.12c}$$

$$2D_{aa} = \gamma\mathscr{N}_a + r_a - g(\alpha^*\mathscr{M} + \mathscr{M}^*\alpha), \tag{12.3.12d}$$

$$2D_{bb} = \gamma\mathscr{N}_b - g(\alpha^*\mathscr{M} + \mathscr{M}^*\alpha), \tag{12.3.12e}$$

$$2D_{ab} = g(\alpha^*\mathscr{M} + \mathscr{M}^*\alpha). \tag{12.3.12f}$$

We are now in the position to solve the Langevin equations (12.3.1)–(12.3.4). Typically, the atomic decay rate γ is much larger than the photon decay rate $\mathscr{C}$, so that the evolution of the atomic variables happens on a much shorter time scale than the electromagnetic field. We can then adiabatically eliminate the atomic variables $\mathscr{M}$, $\mathscr{N}_a$, and $\mathscr{N}_b$ and derive an equation for the field α alone. Thus as a first step we set the time derivative of $\mathscr{M}$ in Eq. (12.3.2) equal to zero and obtain the adiabatic value for the atomic polarization,

$$\mathscr{M} = \frac{g}{\gamma}(\mathscr{N}_a - \mathscr{N}_b)\alpha + \frac{1}{\gamma}\mathscr{F}_{\mathscr{M}}. \tag{12.3.13}$$

Substituting this result into the equations for α, $\mathscr{N}_a$, and $\mathscr{N}_b$ yields

$$\dot{\alpha} = -\frac{\mathscr{C}}{2}\alpha + \frac{g^2}{\gamma}(\mathscr{N}_a - \mathscr{N}_b)\alpha + \mathscr{F}_{\mathscr{C}} + \frac{g}{\gamma}\mathscr{F}_{\mathscr{M}}, \tag{12.3.14}$$

$$\dot{\mathscr{N}}_a = r_a - \gamma\mathscr{N}_a - \frac{2g^2}{\gamma}(\mathscr{N}_a - \mathscr{N}_b)\alpha^*\alpha - \frac{g}{\gamma}(\mathscr{F}^*_{\mathscr{M}}\alpha + \alpha^*\mathscr{F}_{\mathscr{M}}) + \mathscr{F}_a, \tag{12.3.15}$$

$$\dot{\mathscr{N}}_b = -\gamma\mathscr{N}_b + \frac{2g^2}{\gamma}(\mathscr{N}_a - \mathscr{N}_b)\alpha^*\alpha + \frac{g}{\gamma}(\mathscr{F}^*_{\mathscr{M}}\alpha + \alpha^*\mathscr{F}_{\mathscr{M}}) + \mathscr{F}_b. \tag{12.3.16}$$

We next adiabatically eliminate the population variables $\mathscr{N}_a$ and $\mathscr{N}_b$ by setting their time derivative equal to zero. Equations (12.3.15) and (12.3.16) then reduce to a set of two coupled linear equations which can be easily solved. The result is

$$\mathscr{N}_a = \frac{1}{\gamma\left(1 + \frac{4g^2}{\gamma^2}I\right)} \left[r_a\left(1 + \frac{2g^2}{\gamma^2}I\right) + \left(1 + \frac{2g^2}{\gamma^2}I\right)\mathscr{G}_a + \frac{2g^2}{\gamma^2}I\mathscr{G}_b\right], \tag{12.3.17}$$

$$\mathscr{N}_b = \frac{1}{\gamma\left(1 + \frac{4g^2}{\gamma^2}I\right)} \left[r_a\frac{2g^2}{\gamma^2}I + \left(1 + \frac{2g^2}{\gamma^2}I\right)\mathscr{G}_b + \frac{2g^2}{\gamma^2}I\mathscr{G}_a\right], \tag{12.3.18}$$

in which I is the intensity $|\alpha|^2$ of the radiation field, and the noise functions $\mathscr{G}_a$ and $\mathscr{G}_b$ are defined by

$$\mathscr{G}_a = \mathscr{F}_a - \frac{g}{\gamma}(\mathscr{F}^*_{\mathscr{M}}\alpha + \alpha^*\mathscr{F}_{\mathscr{M}}), \tag{12.3.19}$$

$$\mathscr{G}_b = \mathscr{F}_b + \frac{g}{\gamma}(\mathscr{F}^*_{\mathscr{M}}\alpha + \alpha^*\mathscr{F}_{\mathscr{M}}). \tag{12.3.20}$$

We can now substitute the expressions (12.3.17) and (12.3.18) into Eq. (12.3.14) and obtain an equation of motion for the electromagnetic field alone,

$$\dot{\alpha} = -\frac{\mathscr{C}}{2}\alpha + \frac{\mathscr{A}\alpha}{2\left(1 + \frac{\mathscr{B}}{\mathscr{A}}I\right)} + \mathscr{F}_{\alpha}, \tag{12.3.21}$$

in which the noise force $\mathscr{F}_{\alpha}$ is given by

$$\mathscr{F}_{\alpha} = \mathscr{F}_{\mathscr{C}} + \frac{g}{\gamma}\mathscr{F}_{\mathscr{M}} + \frac{g^2\alpha}{\gamma^2\left(1 + \frac{\mathscr{B}}{\mathscr{A}}I\right)}(\mathscr{G}_a - \mathscr{G}_b). \tag{12.3.22}$$

The parameters $\mathscr{A}$ and $\mathscr{B}$ are the gain and saturation coefficients for the laser derived earlier (see Eqs. (11.1.11) and (11.1.12)). The noise force $\mathscr{F}_{\alpha}$ is characterized by the correlation functions

$$\langle \mathscr{F}_{\alpha}(t) \rangle = 0, \tag{12.3.23a}$$

$$\langle \mathscr{F}_{\alpha}^{*}(t)\mathscr{F}_{\alpha}(t') \rangle = \langle 2D_{\alpha^*\alpha} \rangle \delta(t - t'), \tag{12.3.23b}$$

$$\langle \mathscr{F}_{\alpha}(t)\mathscr{F}_{\alpha}(t') \rangle = \langle 2D_{\alpha\alpha} \rangle \delta(t - t'). \tag{12.3.23c}$$

The diffusion coefficients $D_{\alpha^*\alpha}$ and $D_{\alpha\alpha}$ determine the strength of the noise and can be calculated from the definition of $\mathscr{F}_{\alpha}$. A lengthy but straightforward calculation yields the results (see Problem 12.3)

$$2D_{\alpha^*\alpha} = \frac{\mathscr{A}}{\left(1 + \frac{\mathscr{B}}{\mathscr{A}}I\right)^2}\left[1 + \frac{\mathscr{B}}{4\mathscr{A}}I\left(3 + \frac{\mathscr{B}}{\mathscr{A}}I\right)\right], \tag{12.3.24}$$

$$2D_{\alpha\alpha} = (2D_{\alpha^*\alpha^*})^* = -\frac{\mathscr{B}\alpha^2}{4\left(1 + \frac{\mathscr{B}}{\mathscr{A}}I\right)^2}\left(3 + \frac{\mathscr{B}}{\mathscr{A}}I\right). \tag{12.3.25}$$

These results will now be used to discuss the steady-state operation and the quantum fluctuations in a laser.

12.4 Photon statistics and laser linewidth

We are interested in the properties of the intensity and the phase of the laser light. For this purpose we change into a polar coordinate system by defining

$$\alpha = \sqrt{I}e^{i\theta}. \tag{12.4.1}$$

On differentiating with respect to time and comparing the real and imaginary parts with Eq. (12.3.21), we obtain the following Langevin equations for the variables I and θ:

$$\dot{\theta} = F_\theta, \tag{12.4.2}$$

$$\dot{I} = -\mathscr{C}I + \frac{\mathscr{A}I}{1 + \frac{\mathscr{B}}{\mathscr{A}}I} + F_I, \tag{12.4.3}$$

where

$$F_\theta = \mathrm{Im}\left(\frac{\mathscr{F}_\alpha}{\alpha}\right), \tag{12.4.4}$$

$$F_I = 2I\,\mathrm{Re}\left(\frac{\mathscr{F}_\alpha}{\alpha}\right). \tag{12.4.5}$$

In Eq. (12.4.5), we have neglected the noise-induced drift terms which are much smaller than any of the remaining contributions. The diffusion coefficients for the noise operators F_θ and F_I are found to be

$$2D_{\theta\theta} = \frac{\mathscr{A}}{2I\left(1 + \frac{\mathscr{B}}{\mathscr{A}}I\right)}\left(1 + \frac{\mathscr{B}}{2\mathscr{A}}I\right), \tag{12.4.6}$$

$$2D_{II} = \frac{2\mathscr{A}I}{\left(1 + \frac{\mathscr{B}}{\mathscr{A}}I\right)^2}. \tag{12.4.7}$$

Before we start our discussion of the photon statistics and the natural linewidth, it is interesting to note that the diffusion coefficient $D_{\theta\theta}$ is identical to the diffusion coefficient as given in Eq. (11.4.15), if I is replaced by the steady-state expectation value $\langle n\rangle$. We shall establish a direct relationship between the density matrix and the corresponding P-representation approach discussed in the earlier sections with the present Langevin approach later in this section.

In order to establish a relationship between the quantity I and the photon statistics of the laser, we recall that we chose normal ordering in the process of going from the operator equations (12.2.25), (12.2.29), (12.2.31), and (12.2.32) to the c-number equations (12.3.1)–(12.3.4). The intensity I therefore corresponds to the normally ordered products of the operators of the field. The mean photon number $\langle n\rangle$ and the photon number variance are then given by

$$\langle n\rangle = \langle a^\dagger a\rangle = \langle I\rangle, \tag{12.4.8}$$

$$\begin{aligned}(\Delta n)^2 &= \langle a^\dagger a a^\dagger a\rangle - \langle a^\dagger a\rangle^2 \\ &= \langle a^\dagger a^\dagger a a\rangle + \langle a^\dagger a\rangle - \langle a^\dagger a\rangle^2 \\ &= \langle I^2\rangle + \langle I\rangle - \langle I\rangle^2 \\ &= (\Delta I)^2 + \langle I\rangle. \end{aligned} \tag{12.4.9}$$

It follows from Eq. (12.4.3) that, in steady state ($\langle\dot{I}\rangle = 0$),

$$-\mathscr{C}\langle I\rangle + \left\langle \frac{\mathscr{A}I}{1+\frac{\mathscr{B}}{\mathscr{A}}I} \right\rangle = 0, \tag{12.4.10}$$

where we use $\langle F_I(t)\rangle = 0$. If we make the assumption $\langle I^r\rangle = \langle I\rangle^r$ then we obtain

$$\langle n\rangle = I_0 = \frac{\mathscr{A}}{\mathscr{C}}\left(\frac{\mathscr{A}-\mathscr{C}}{\mathscr{B}}\right). \tag{12.4.11}$$

To determine the fluctuation in the quantity I, we first linearize Eq. (12.4.3) around its steady-state value. Defining $\Delta I = I - I_0$ and making use of expression (12.4.11) for the steady-state mean photon number, we find

$$\begin{aligned}\frac{d}{dt}(\Delta I) &= -\mathscr{C}(I_0+\Delta I) + \frac{\mathscr{A}(I_0+\Delta I)}{1+\frac{\mathscr{B}}{\mathscr{A}}(I_0+\Delta I)} + F_I \\ &\cong -\mathscr{C}(I_0+\Delta I) + \frac{\mathscr{A}(I_0+\Delta I)}{1+\frac{\mathscr{B}}{\mathscr{A}}I_0} - \frac{\mathscr{B}I_0\Delta I}{\left(1+\frac{\mathscr{B}}{\mathscr{A}}I_0\right)^2} + F_I \\ &= -\frac{\mathscr{C}}{\mathscr{A}}(\mathscr{A}-\mathscr{C})\Delta I + F_I. \end{aligned} \tag{12.4.12}$$

It then follows that

$$\frac{d}{dt}(\Delta I)^2 = -\frac{2\mathscr{C}}{\mathscr{A}}(\mathscr{A}-\mathscr{C})(\Delta I)^2 + 2\langle \Delta I(t)F_I(t)\rangle. \tag{12.4.13}$$

The correlation function $\langle \Delta I(t)F_I(t)\rangle$ can be determined using the methods developed in Section 9.1. It follows from the formal solution of Eq. (12.4.12) that

$$\begin{aligned}\langle \Delta I(t)F_I(t)\rangle &= \Delta I(0)\langle F_I(t)\rangle \exp\left[-\frac{\mathscr{C}}{\mathscr{A}}(\mathscr{A}-\mathscr{C})t\right] \\ &\quad + \int_0^t dt' \exp\left[-\frac{\mathscr{C}}{\mathscr{A}}(\mathscr{A}-\mathscr{C})(t-t')\right]\langle F_I(t')F_I(t)\rangle \\ &= \langle D_{II}\rangle, \end{aligned} \tag{12.4.14}$$

where we use $\langle F_I(t)\rangle = 0$ and $\langle F_I(t')F_I(t)\rangle = 2\langle D_{II}\rangle\delta(t-t')$. The steady-state solution of Eq. (12.4.13) is therefore

$$(\Delta I)^2 = \frac{\mathscr{A}}{\mathscr{C}(\mathscr{A}-\mathscr{C})}\langle D_{II}\rangle, \tag{12.4.15}$$

and, on substituting for $\langle(\Delta I)^2\rangle$ and $\langle I\rangle$ from Eqs. (12.4.15) and (12.4.11) in Eq. (12.4.9), we obtain

$$(\Delta n)^2 = \frac{\mathscr{A}}{\mathscr{A}-\mathscr{C}}\langle n\rangle. \tag{12.4.16}$$

The normalized variance of the photon distribution as given by the Mandel Q parameter for the field is given by

$$Q_f = \frac{(\Delta n)^2}{\langle n \rangle} - 1 = \frac{\mathscr{C}}{\mathscr{A} - \mathscr{C}}. \tag{12.4.17}$$

Equations (12.4.11) and (12.4.17) are identical to the corresponding results (11.2.21) and (11.2.23) obtained in Section 11.2 using the density matrix approach.

Next we look at the phase rate of diffusion which is calculated with the help of Eq. (12.4.2). On integrating Eq. (12.4.2), we get

$$\theta(t) = \int_0^t dt' F_\theta(t'), \tag{12.4.18}$$

so that

$$\begin{aligned} \frac{d}{dt}\langle \theta^2 \rangle &= \frac{d}{dt}\int_0^t dt' \int_0^t dt'' \langle F_\theta(t') F_\theta(t'') \rangle \\ &= \langle 2D_{\theta\theta} \rangle. \end{aligned} \tag{12.4.19}$$

On substituting for $\langle 2D_{\theta\theta} \rangle$ from Eq. (12.4.6) into Eq. (12.4.19) and using expression (12.4.11) for the steady-state mean photon number, we find, after time integration

$$\langle \theta^2 \rangle = \frac{1}{4\langle n \rangle} + \frac{\mathscr{A} + \mathscr{C}}{4\langle n \rangle} t. \tag{12.4.20}$$

The integration constant $1/4\langle n \rangle$ in Eq. (12.4.20) is due to the contribution of vacuum fluctuations, and will be discussed in detail in Section 14.5. The second term in Eq. (12.4.20) states that the phase diffuses linearly in time and the rate of diffusion $(\mathscr{A} + \mathscr{C})/4\langle n \rangle$ gives the natural linewidth of the laser.

A connection between the density matrix and the quantum Langevin approaches can be established via the equation for the probability distribution for the field. In Appendix 11.B we derived an equation of motion for the P-representation (Eqs. (11.B.11) and (11.B.12)) which is equivalent to the corresponding density matrix equation (11.1.14) for the laser. Now according to a theorem of stochastic processes, if the random complex variable α satisfies the Langevin equation (12.3.21) with the correlation functions of the form (12.3.23a)–(12.3.23c), then the probability distribution $P(\alpha, \alpha^*)$ satisfies the Fokker–Planck equation

$$\begin{aligned} \frac{\partial P}{\partial t} = \frac{1}{2}\frac{\partial}{\partial \alpha}\left[\left(\mathscr{C}\alpha + \frac{\mathscr{A}\alpha}{1 + \frac{\mathscr{B}}{\mathscr{A}}|\alpha|^2}\right)P\right] + \frac{\partial^2}{\partial \alpha^2}(D_{\alpha\alpha}P) \\ + \frac{1}{2}\frac{\partial^2}{\partial \alpha \partial \alpha^*}(D_{\alpha\alpha^*}P) + \text{c.c.} \end{aligned} \tag{12.4.21}$$

Since we chose the normal ordering to go from operator equations to c-number equations, we expect $P(\alpha, \alpha^*)$ to be associated with the evaluation of the normally ordered correlation functions. This corresponds to the P-representation of the field. It is a simple matter to show that if we ignore the third- and higher-order derivatives (which can be justified using scaling arguments), then Eqs. (11.B.11) and (11.B.12) are identical to Eq. (12.4.21). This therefore establishes the equivalence of the two methods.

Problems

12.1 (a) Derive the equations of motion for the c-number field and atom variables α, $\mathcal{M}$, $\mathcal{N}_a$, and $\mathcal{N}_b$ which correspond to antinormal ordering a, M, N_a, N_b, $M^\dagger$, $a^\dagger$ of the operators.

(b) Find all the nonzero diffusion coefficients associated with the Langevin noise forces in the equations of motion for α, $\mathcal{M}$, $\mathcal{N}_a$, and $\mathcal{N}_b$.

(c) By adiabatically eliminating the atomic variables, derive the equation of motion for the electromagnetic field α.

12.2 Derive Eqs. (12.3.12a)–(12.3.12f).

12.3 Derive Eqs. (12.3.24) and (12.3.25).

References and bibliography

Langevin theory of the laser

M. Lax, in *Physics of Quantum Electronics*, ed. P. L. Kelley, B. Lax, and P. E. Tannenwald (McGraw Hill, New York 1966), p. 735.

M. Lax, in *Statistical Physics, Phase Transition, and Superconductivity*, Vol. II, ed. M. Chrétien, E. P. Gross, and S. Dreser (Gordon and Breach, New York 1968).

W. Louisell, *Quantum Statistical Properties of Radiation*, (Wiley, New York 1974).

H. Haken, *Laser Theory*, (Springer, Berlin 1970).

H. Haken, *Light*, Vols. I and II, (North-Holland, Amsterdam 1981).

M. Fleischhauer, *Phys. Rev. A* **50**, 2773 (1994).

M. Sargent III, M. Scully, and W. E. Lamb, Jr., *Laser Physics*, (Addison-Welsley, Reading, MA, 1974).

Atomic memory effects in a laser

F. Haake, *Z. Phys.* **227**, 179 (1969).

M. O. Scully, G. Süssmann, and C. Benkert, *Phys. Rev. Lett.* **60**, 1014 (1988).

M. O. Scully, M. S. Zubairy, and K. Wódkiewicz, *Opt. Commun.* **65**, 440 (1988).

C. Benkert, M. O. Scully, and G. Süssmann, *Phys. Rev. A* **41**, 6119 (1990).

Effects of mode locking and Colored noise

W. W. Chow, M. O. Scully, and E. Van Stryland, *Opt. Commun.* **15**, 6 (1975).

J. D. Cresser, W. H. Louisell, P. Meystre, W. Schleich, and M. O. Scully, *Phys. Rev. A* **25**, 2214 (1982).

R. Graham, M. Höhnenback, and A. Schenzle, *Phys. Rev. Lett.* **48**, 1396 (1982).

R. F. Fox, G. E. James, and R. Roy, *Phys. Rev. Lett.* **52**, 1778 (1984).

Role of pump statistics

M. Golubev and I. V. Sokolov, *Zh. Eksp. Teor. Fiz.* **87**, 408 (1984) (*Sov. Phys. JETP* **60**, 234 (1984)).

Y. Yamamoto, S. Machida, and O. Nilsson, *Phys. Rev. A* **34**, 4025 (1986).

S. Machida, Y. Yamamoto, and K. Itaya, *Phys. Rev. Lett.* **58**, 1000 (1987).

M. A. Marte, H. Ritsch, and D. F. Walls, *Phys. Rev. Lett.* **61**, 1093 (1988).

J. Bergou, L. Davidovich, M. Orszag, C. Benkert, M. Hillery, and M. O. Scully, *Phys. Rev. A* **40**, 5073 (1989) (density operator approach).

F. Haake, S. M. Tan, and D. F. Walls, *Phys. Rev. A* **40**, 7121 (1989).

C. Benkert, M. O. Scully, J. Bergou, L. Davidovich, M. Hillery, and M. Orszag, *Phys. Rev. A* **41**, 2756 (1990) (quantum Langevin approach).
H.-J. Briegel, B.-G. Englert, C. Ginzel, and A. Schenzle, *Phys. Rev. A* **49**, 5019 (1994).
M. T. Fontenelle and L. Davidovich, *Phys. Rev. A* **51**, 2560 (1995).

CHAPTER 13

Theory of the micromaser

The development of a single-atom maser or a micromaser* allows a detailed study of the atom–field interaction. The situation realized is very close to the ideal case of a single two-level atom interacting with a single-mode quantized field as treated in Section 6.2. In a micromaser a stream of two-level atoms is injected into a superconducting cavity with a high quality factor. The injection rate can be such that only one atom is present inside the resonator at any time. Due to the high quality factor of the cavity, the radiation decay time is much larger than the characteristic time of the atom–field interaction, which is given by the inverse of the single-photon Rabi frequency. Therefore, a field is built up inside the cavity when the mean time between the atoms injected into the cavity is shorter than the cavity decay time. A micromaser, therefore, allows sustained oscillations with less than one atom on the average in the cavity.

The realization of a single-atom maser or a micromaser has been made possible due to the enormous progress in the construction of superconducting cavities together with the laser preparation of highly excited atoms called Rydberg atoms. The quality factor of the superconducting cavities is high enough for periodic energy exchanges between atom and cavity field to be observed. The interesting properties of the Rydberg atoms make them ideal for micromasers. In Rydberg atoms the probability of induced transitions between adjacent states becomes very large and scales as n^4, where n denotes the principle quantum number. Consequently, a few photons are enough to saturate the transition between adjacent levels. In addition, the lifetime for spontaneous transition is very large.

* The first micromaser was realized by Meschede, Walther, and Müller [1985]. For a review of key earlier work leading up to the micromaser, see Haroche and Raimond [1985].

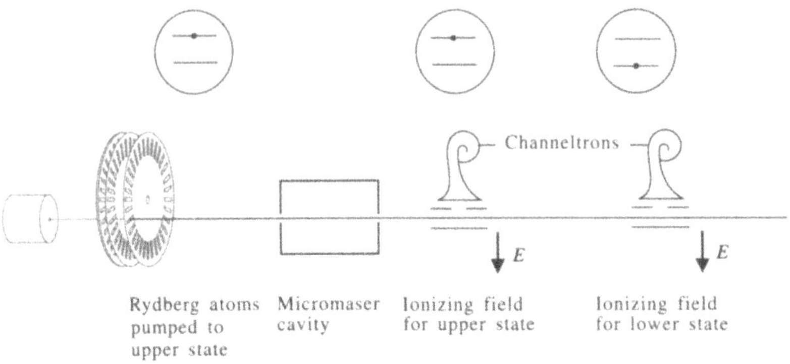

Fig. 13.1 Experimental setup of a micromaser. (From G. Rempe, F. Schmidt-Kaler, and H. Walther, *Phys. Rev. Lett.* **64**, 2783 (1990).)

A sketch of the beautiful micromaser experiment at the Max-Planck Institut für Quantenoptik in Garching is shown in Fig. 13.1. A highly collimated beam of rubidium atoms passes through a Fizeau velocity selector. Before entering the cavity, the atoms are excited into the upper maser level $63p_{3/2}$. The superconducting niobium cavity is cooled down to a temperature of 0.5 K by means of a ^{3}He cryostat. At such a low temperature the number of thermal photons is reduced to about 0.15 at a frequency of 21.5 GHz. The quality factor of the cavity is 3×10^{10} corresponding to a photon lifetime of about 0.2 seconds. The maser transitions from the $63p_{3/2}$ level to the $61d_{3/2}$ level at 21.5 GHz are studied. The average transit time of Rydberg atoms through the cavity is 50 μs and the atomic flux is as small as 1750 atoms/s. After passing through the cavity, the Rydberg atoms in the upper and lower levels are detected in two separate field ionization detectors. The field strength is adjusted so as to ensure that in the first detector the atoms in the upper level are ionized, but not those in the lower level.

The photon statistics of a micromaser exhibit many interesting effects including sub-Poissonian statistics. Even a number state can be generated using a cavity with a high enough quality factor. If there are no thermal photons in the cavity, a condition which is achieved by cooling the cavity to an extremely low temperature, interesting features such as trapping states occur.

In this chapter, we study these interesting features of the photon statistics as well as the linewidth of the micromaser. The application of micromasers to the quantum measurement problem will be described in Chapters 19 and 20.

13.1 Equation of motion for the field density matrix

We consider a single-mode resonator into which excited two-level atoms are injected at a rate low enough that at most one atom at a time is present inside the resonator cavity. The cavity damping is considered so weak as to be negligible during the interaction time. The lifetimes of all the levels are also assumed to be much larger than the interaction time of the atom with the field in a maser cavity. The spontaneous decay processes, to other levels or other modes, can therefore be neglected, which means that the joint evolution of the single-mode field and the atom is unitary.

The contribution of all the atoms which enter the cavity at random times at a rate r_a and interact for a fixed time τ with the radiation field inside the cavity before leaving the cavity have to be added to obtain the coarse-grained equation of motion for the reduced density matrix for the field, as is done in Method II, Section 11.1. The resulting equation is (Eq. (11.1.26))

$$\left(\frac{d\rho_{nn'}}{dt}\right)_{\text{gain}} = -r_a[1-\cos(g\tau\sqrt{n+1})\cos(g\tau\sqrt{n'+1})]\rho_{nn'} \\ +r_a\sin(g\tau\sqrt{n})\sin(g\tau\sqrt{n'})]\rho_{n-1,n'-1}, \qquad (13.1.1)$$

to which we add the contribution due to the cavity losses (Eq. (8.3.2))

$$\mathscr{L}\rho = -\frac{\mathscr{C}}{2}(\bar{n}_{\text{th}}+1)(a^\dagger a\rho - 2a\rho a^\dagger + \rho a^\dagger a) \\ -\frac{\mathscr{C}}{2}\bar{n}_{\text{th}}(aa^\dagger\rho - 2a^\dagger\rho a + \rho aa^\dagger). \qquad (13.1.2)$$

This leads to the following equation of motion of the density matrix in the photon number representation (with $\rho_{nn'} = \langle n|\rho|n'\rangle$; $n, n' = 0, 1, \ldots$):

$$\dot{\rho}_{nn'} = a_{n,n'}\rho_{nn'} + b_{n-1,n'-1}\rho_{n-1,n'-1} + c_{n+1,n'+1}\rho_{n+1,n'+1}, \qquad (13.1.3)$$

where

$$a_{n,n'} = -r_a\left[1-\cos(g\tau\sqrt{n+1})\cos(g\tau\sqrt{n'+1})\right] \\ -\frac{\mathscr{C}}{2}\left[2\bar{n}_{\text{th}}(n+n'+1)+(n+n')\right], \qquad (13.1.4)$$

$$b_{n,n'} = r_a\sin(g\tau\sqrt{n+1})\sin(g\tau\sqrt{n'+1}) \\ +\mathscr{C}\bar{n}_{\text{th}}[(n+1)(n'+1)]^{1/2}, \qquad (13.1.5)$$

$$c_{n,n'} = \mathscr{C}(\bar{n}_{\text{th}}+1)\sqrt{nn'}. \qquad (13.1.6)$$

This master equation forms the basis of most studies on the quantum statistical properties of radiation in a micromaser. The diagonal elements $p(n) = \rho_{nn}$, which represent the probability of n photons in the field, satisfy the equation of motion

$$\dot{p}(n) = a_{n,n}p(n) + b_{n-1,n-1}p(n-1) + c_{n+1,n+1}p(n+1), \qquad (13.1.7)$$

with

$$a_{n,n} = -r_a \sin^2(g\tau\sqrt{n+1}) - \mathscr{C}[\bar{n}_{\text{th}}(2n+1)+n], \tag{13.1.8}$$

$$b_{n,n} = r_a \sin^2(g\tau\sqrt{n+1}) + \mathscr{C}\bar{n}_{\text{th}}(n+1), \tag{13.1.9}$$

$$c_{n,n} = \mathscr{C}(\bar{n}_{\text{th}}+1)n. \tag{13.1.10}$$

It may be noted that

$$a_{n,n} + b_{n,n} + c_{n,n} = 0. \tag{13.1.11}$$

13.2 Steady-state photon statistics

The steady-state photon number distribution can be obtained from Eq. (13.1.7) by taking $\dot{p}(n) = 0$ just as in Chapter 11. The resulting equation

$$\begin{aligned} &-\{r_a \sin^2(g\tau\sqrt{n+1}) + \mathscr{C}[\bar{n}_{\text{th}}(2n+1)+n]\}p(n) \\ &+ [r_a \sin^2(g\tau\sqrt{n}) + \mathscr{C}\bar{n}_{\text{th}}n]p(n-1) \\ &+ [\mathscr{C}(\bar{n}_{\text{th}}+1)(n+1)]p(n+1) = 0 \end{aligned} \tag{13.2.1}$$

leads to the following equivalent recursion relations:

$$\begin{aligned} &[r_a \sin^2(g\tau\sqrt{n}) + \mathscr{C}\bar{n}_{\text{th}}n]p(n-1) \\ &= \mathscr{C}(\bar{n}_{\text{th}}+1)np(n), \end{aligned} \tag{13.2.2}$$

$$\begin{aligned} &[r_a \sin^2(g\tau\sqrt{n+1}) + \mathscr{C}\bar{n}_{\text{th}}(n+1)]p(n) \\ &= \mathscr{C}(\bar{n}_{\text{th}}+1)(n+1)p(n+1), \end{aligned} \tag{13.2.3}$$

and we obtain

$$p(n) = p(0)\prod_{\ell=1}^{n} \frac{\bar{n}_{\text{th}}\mathscr{C} + r_a \sin^2(g\tau\sqrt{\ell})/\ell}{\mathscr{C}(\bar{n}_{\text{th}}+1)}, \tag{13.2.4}$$

where $p(0)$ is determined from the normalization relation

$$\sum_{n=0}^{\infty} p(n) = 1. \tag{13.2.5}$$

Equation (13.2.4) is the central result of this section as all the features of the photon statistics for the micromaser can be extracted from it.

Before discussing the behavior of the steady-state photon distribution function $p(n)$, we find a threshold condition for a micromaser. A linear analysis is adequate for this purpose. In this approximation,

$$\sin^2(g\tau\sqrt{n}) \simeq (g\tau)^2 n, \tag{13.2.6}$$

and Eq. (13.1.7) simplifies to

$$\dot{p}(n) = -[r_a(g\tau)^2(n+1) + \mathscr{C}n]p(n) + r_a(g\tau)^2 np(n-1) + \mathscr{C}(n+1)p(n+1). \tag{13.2.7}$$

Here we have assumed $\bar{n}_{\text{th}} = 0$. The growth of the field can now be looked at from the equation for the mean number of photons:

$$\langle \dot{n} \rangle = \sum_{n=0}^{\infty} n\dot{p}(n) = [r_a(g\tau)^2 - \mathscr{C}]\langle n \rangle + r_a(g\tau)^2. \tag{13.2.8}$$

The last term arises due to spontaneous emission. It is clear that the field builds up when

$$r_a(g\tau)^2 > \mathscr{C}. \tag{13.2.9}$$

The condition $r_a(g\tau)^2 = \mathscr{C}$, therefore, describes the maser threshold. Below threshold ($r_a(g\tau)^2 < \mathscr{C}$) the field just dies down. This suggests the use of the normalized interaction time

$$\tau_{\text{int}} = g\tau\sqrt{\frac{r_a}{\mathscr{C}}}, \tag{13.2.10}$$

which is equal to unity at threshold.

In Fig. 13.2, the normalized mean photon number

$$\langle N \rangle = \frac{1}{N_{\text{ex}}}\langle n \rangle = \frac{1}{N_{\text{ex}}}\sum_{n=0}^{\infty} np(n) \tag{13.2.11}$$

has been plotted as a function of τ_{int} for $N_{\text{ex}} \equiv r_a/\mathscr{C} = 20$ and 2000, and for $\bar{n}_{\text{th}} = 0.1$. Here the parameter $N_{\text{ex}} = r_a/\mathscr{C}$ represents the average number of atoms that pass through the cavity during the lifetime of the field. The mean number of photons remains virtually zero for small τ_{int}, but at threshold $\tau_{\text{int}} = 1$, $\langle n \rangle$ becomes finite and increases rapidly to almost unity ($\langle n \rangle = N_{\text{ex}}$) with increasing τ_{int}. It then decreases to reach a minimum at about $\tau_{\text{int}} \simeq 2\pi$, where the field abruptly jumps to a higher intensity. This oscillatory behavior continues, but becomes less pronounced for increasing τ_{int}.

In Fig. 13.3, the Mandel Q parameter

$$Q_f = \frac{\langle n^2 \rangle - \langle n \rangle^2}{\langle n \rangle} - 1 \tag{13.2.12}$$

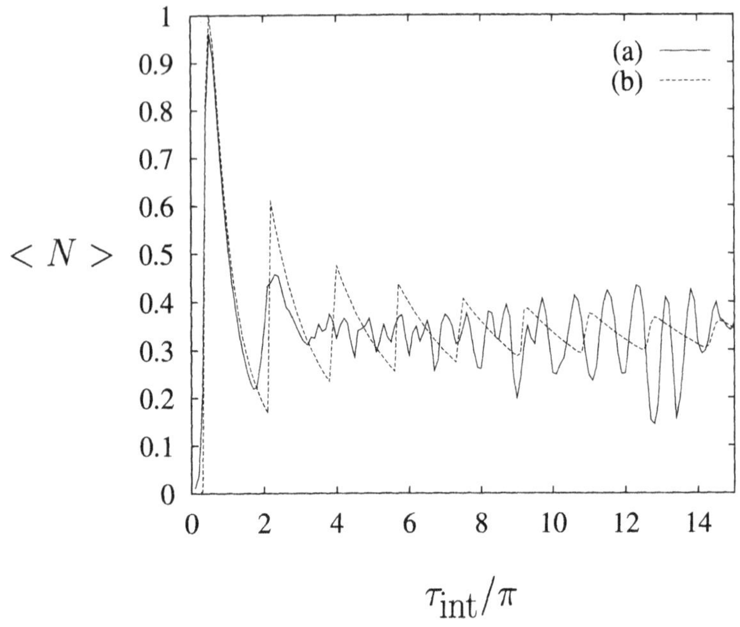

Fig. 13.2 Normalized steady-state mean photon number $\langle N \rangle$ as a function of the normalized interaction time τ_{int} for $\bar{n}_{\text{th}} = 0.1$ and (a) $N_{\text{ex}} = 20$ and (b) $N_{\text{ex}} = 2000$.

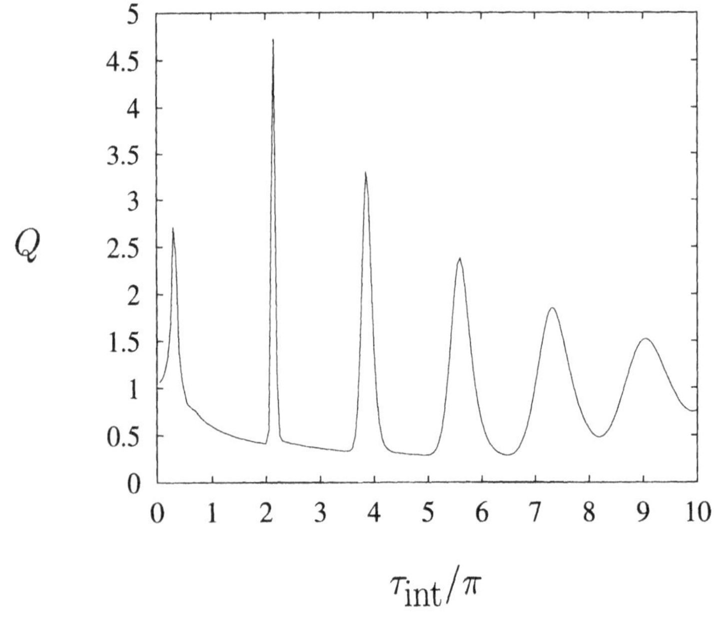

Fig. 13.3 Mandel Q parameter as a function of the normalized interaction time τ_{int} for $N_{\text{ex}} = 200$ and $\bar{n}_{\text{th}} = 0.1$.

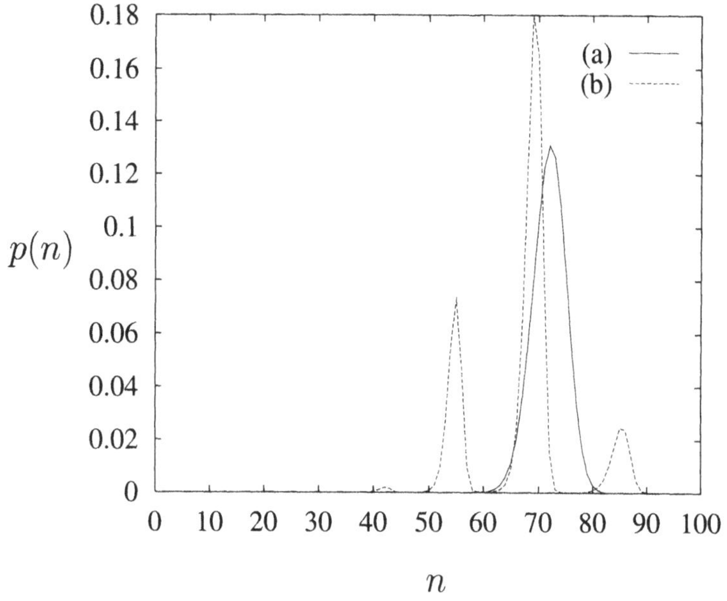

Fig. 13.4 Steady-state photon statistics for $N_{ex} = 200$, $\bar{n}_{th} = 0.1$, and (a) $\tau_{int} = 3\pi$ and (b) $\tau_{int} = 15\pi$. The photon distribution function is three peaked for $\tau_{int} = 15\pi$.

is plotted as a function of τ_{int}. Just above the threshold the photon distribution is first strongly super-Poissonian ($Q_f \gg 0$) but then the Q parameter drops below the Poisson level $Q_f = 0$. Super-Poissonian peaks are obtained at approximately 2π-intervals followed by sub-Poissonian statistics with increasing τ_{int}. The generation of sub-Poissonian statistics in steady state is an interesting feature of micromasers. In the next section we shall show how a pure number state can be generated in a micromaser.

So far we have examined the behavior of the first two moments of the steady-state photon distribution function. In Fig. 13.4 we plot $p(n)$ as a function of n for different values of interaction times. For $\tau_{int} = 3\pi$ the sub-Poissonian behavior is evident. However, for $\tau_{int} = 15\pi$ the distribution has no longer a single peak. The multi-peak distribution is another special feature of a micromaser.

13.3 Preparation of number state in a high-Q micromaser

In this section, we discuss two possible ways of creating a number state in a high-Q micromaser, namely, via a state reduction scheme in which we infer the photon statistics by *looking* at the Rydberg atoms as they exit the cavity and via a trapping state scheme in which the interaction time is chosen such that the emission probability at a particular photon number becomes small and the field is truncated at that photon number.

To experimentally achieve a state of the field where the number of photons is exactly known, two conditions have to be fulfilled. The first condition concerns the temperature. Thermal photons have to be suppressed because they induce decay and influence the statistics so that a superposition of number states is obtained. We can eliminate thermal photons by cooling the cavity to a low enough temperature. The mean number of thermal photons for a frequency of about 20 GHz is 3×10^{-5} at $T = 0.1$ K. As a second condition, we must not lose photons stored in the cavity for the duration of the experiment, i.e., we need a cavity in which losses can be neglected for this time. The photon lifetime is determined by a decay rate $\mathscr{C} = v/Q$. The quality factor Q of the cavity can reach values of up to 10^{11}, and with a microwave frequency of about 20 GHz this results in photon lifetimes of several seconds.

13.3.1 State reduction

We consider an experimental setup as shown in Fig. 13.1. Atoms in their excited state are injected into an empty cavity, i.e., the field is initially in the vacuum state $|0\rangle$. After they leave the cavity, they are probed by a static electric field which ionizes all atoms in the upper level. All the atoms that are not ionized have emitted a photon in the cavity. When these atoms are counted (via electron detection as in Fig. 13.1), the total number of photons in the maser field can be inferred. It may be noted that state reduction and the connected ideas of measurement theory are essential to this logic. By the determination of the state of the outgoing atoms, the photon number in the field is exactly known, i.e., the state of the field is reduced to a pure number state. Since there is initally no radiation in the cavity, the field is always in a number state when an atom enters the cavity. By the interaction of the atom with the field, which is in a state $|n\rangle$, the field state will be changed to a superposition of states $|n\rangle$ and $|n+1\rangle$. Due to the measurement of the atomic state afterwards, this superposition is reduced to one of the states $|n\rangle$ or $|n+1\rangle$, depending on the result of the measurement.

Now, with zero cavity losses, we will maintain a number state since no radiation will be lost from the cavity in the present experiment. However, we will have only *a priori* probabilities as to which number state we actually generate in the present state reduction scheme. These *a priori* probabilities should not be confused with photon statistical distributions. For example, if one considers a coherent state of the radiation field, then every laser or every system being considered would be in an indefinite superposition of number states. Whereas in the present case, every system is in a specific number state; however, we do not know prior to the experiment which state that will be. Therefore, we have to perform the experiment repeatedly with a constant total number of atoms, thus generating a large number of different number states. The distribution of the photon numbers will be given by the *a priori* probability distribution, which we are going to calculate. It should be emphasized that the number of atoms leaving the cavity in the lower state is equal to the number of photons in the cavity only for a lossless cavity.

We now turn to the calculation of the probability $P_n(m)$ of having n photons in the field after m atoms have passed the cavity. To obtain this probability, we derive a recursion relation. The time-evolution operator $U(\tau)$ for the interaction of one two-level atom with the field is given by Eq. (6.2.49), i.e.,

$$\begin{aligned} U(\tau) = & \cos(g\tau\sqrt{a^\dagger a+1})|a\rangle\langle a| + \cos(g\tau\sqrt{a^\dagger a})|b\rangle\langle b| \\ & -i\frac{\sin(g\tau\sqrt{a^\dagger a+1})}{\sqrt{a^\dagger a+1}}a|a\rangle\langle b| - ia^\dagger\frac{\sin(g\tau\sqrt{a^\dagger a+1})}{\sqrt{a^\dagger a+1}}|b\rangle\langle a|. \end{aligned} \tag{13.3.1}$$

We assume that initially the atom is in the upper level $|a\rangle$ and the field is in the number state $|n\rangle$. The combined atom–field density operator is therefore $|a, n\rangle\langle a, n|$. After the interaction time τ we have

$$\begin{aligned} \rho(\tau) &= U(\tau)|a,n\rangle\langle a,n|U^\dagger(\tau) \\ &= \cos^2(g\tau\sqrt{n+1})|a,n\rangle\langle a,n| \\ &\quad + \sin^2(g\tau\sqrt{n+1})|b,n+1\rangle\langle b,n+1| \\ &\quad -i\sin(g\tau\sqrt{n+1})\cos(g\tau\sqrt{n+1}) \\ &\quad \times[|b,n+1\rangle\langle a,n| - |a,n\rangle\langle b,n+1|]. \end{aligned} \tag{13.3.2}$$

The state of the radiation field is now determined via state reduction. That is, if we determine that the atom is in the upper state $|a\rangle$, then the density matrix (13.3.2) is reduced to the state

$$\rho(\tau) = \cos^2(g\tau\sqrt{n+1})|a,n\rangle\langle a,n|, \tag{13.3.3}$$

and if the atom is found to be in the state $|b\rangle$, the system density matrix is given by

$$\rho(\tau) = \sin^2(g\tau\sqrt{n+1})|b,n+1\rangle\langle b,n+1|. \tag{13.3.4}$$

From this we find that the probability for the field to remain in the state $|n\rangle$ is

$$c(n) \equiv \cos^2(g\tau\sqrt{n+1})$$

and the probability for a transition to the state $|n+1\rangle$ is

$$s(n) \equiv \sin^2(g\tau\sqrt{n+1}).$$

When $m-1$ atoms have passed, the field is in a state $|n\rangle$ with the probability $P_n(m-1)$ and in the state $|n-1\rangle$ with a probability $P_{n-1}(m-1)$. The probability for the field to be in the state $|n\rangle$ after m atoms have been in the cavity is then simply

$$P_n(m) = c(n)P_n(m-1) + s(n-1)P_{n-1}(m-1). \tag{13.3.5}$$

We assume that the field is initially in the vacuum state $|0\rangle$, i.e., $P_0(0) = 1$. Then we have for one atom $P_0(1) = c(0)$ and $P_1(1) = s(0)$; for two atoms

$$\begin{aligned} P_0(2) &= c(0)P_0(1) = [c(0)]^2, \\ P_1(2) &= c(1)P_1(1) + s(0)P_0(1) = s(0)[c(0) + c(1)], \\ P_2(2) &= s(1)P_1(1) = s(0)s(1); \end{aligned} \tag{13.3.6}$$

and so on.

The probability distribution $P_n(m)$ can be evaluated numerically for different values of $g\tau$ as a function of the number of passing atoms. In Fig. 13.5, we show results for up to 1000 atoms. Obviously, the probability $P_n(m)$ is very strongly dependent on the value of $g\tau$. This parameter can be varied experimentally by changing the velocity of the atomic beam. When $g\tau < 1$, then a peak in the photon distribution develops and moves towards higher photon numbers as the number of passing atoms grows. In theory, $s(n_0)$ could become exactly 0, so that the probability distribution will be a δ-function in the steady state, a case discussed below. In any experiment, however, the velocity distribution of the atomic beam is never that sharply defined. Therefore, the height of the peak in the probability distribution diminishes as more atoms are injected, and a new peak develops in front of the next barrier at about $(2\pi/g\tau)^2$. Thus the realization of a number state is coupled to the detection via the outgoing atoms.

Atomic velocity itself is not a complicating factor in the present scheme of n-state preparation. It leads to a change in the probability of emitting a photon, but for the experiment the only important fact is whether a photon has been emitted or not. The basic notion of

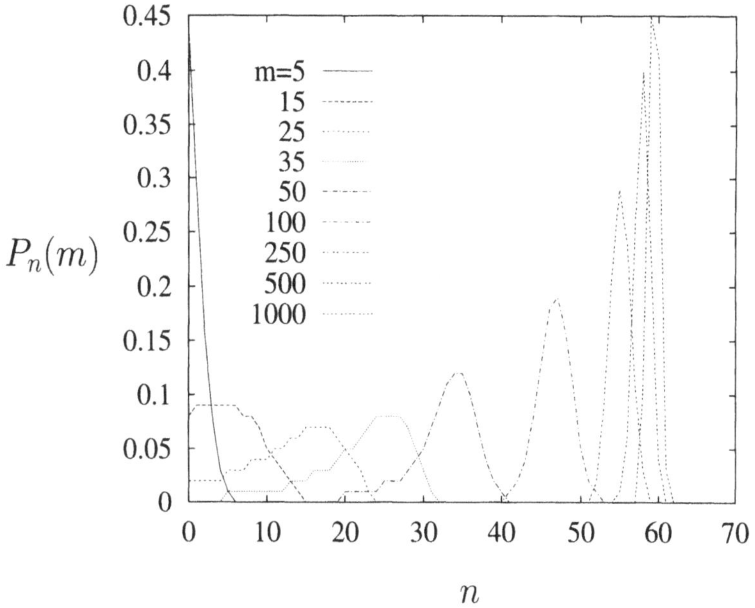

Fig. 13.5 Probability of obtaining n photons in the cavity after m atoms have passed for $g\tau = 0.4$. (From J. Krause, M. O. Scully, and H. Walther, *Phys. Rev. A* **36**, 4547 (1987).)

atomic observation leading to field information is in force regardless of complicating influences such as atomic motion. If the experiment is performed repeatedly, however, the statistics of the obtained photon numbers in the number states will be influenced.

13.3.2 Trapping states

In general a steady state is reached in a micromaser due to the presence of cavity losses as shown in Section 13.2. In such a situation, the field statistics essentially depend on the duration of interaction τ. Generally, a steady state does not exist for a lossless cavity because atoms in excited states are constantly being pumped inside the cavity. However, a steady state can be obtained in a lossless cavity by choosing the interaction time for which the emission probability at a particular photon number becomes small, so that the field is truncated at that photon number. Such a state is known as a trapping state. In this section we discuss a scheme based on the manipulation of interaction times to generate a pure number state.

We consider a setup in which atoms in their excited state are injected into an empty cavity. If the interaction time of the atoms is chosen such that, for n_q photons inside the cavity,

$$g\tau\sqrt{n_q + 1} = q\pi \qquad (q = 1, 2, 3, \ldots), \tag{13.3.7}$$

an excited atom will undergo q Rabi oscillations and will leave the cavity in an excited state. The maser is said to be in an upward trapping state since no probability flows from state $|n_q\rangle$ to state $|n_q+1\rangle$ with the injection of inverted atoms, the maser field evolves to a number state $|n_q\rangle$.

On the other hand, if

$$g\tau\sqrt{n_q} = q\pi \qquad (q = 1, 2, 3, \ldots), \tag{13.3.8}$$

an atom entering the cavity in its ground state will undergo complete q Rabi oscillations to leave the cavity in the same state. The maser is said to be in a downward trapping state. Clearly, the number state immediately following a downward trapping state is an upward trapping state. In both cases, the atoms entering the cavity see the field as a $2q\pi$ pulse, and a *steady state* is achieved even in the absence of losses, so that

$$\rho(t_{i+1}) = \rho(t_i), \qquad t_{i+1} = t_i + r^{-1}. \tag{13.3.9}$$

Equation (13.3.9) represents the conventional steady-state criterion for a micromaser despite the fact that the field changes during the interaction time τ.

We now show how a number state results from an exact solution of the master equation (13.1.1) for the micromaser.

In matrix notation, Eq. (13.1.1) (without the loss terms) for the photon distribution function can be rewritten as

$$\dot{R} = -MR, \tag{13.3.10}$$

where

$$R(t) = \begin{bmatrix} p(0,t) \\ p(1,t) \\ \vdots \\ p(\ell,t) \\ \vdots \end{bmatrix}, \tag{13.3.11}$$

$$M = \begin{bmatrix} a_0 & & & & & & \\ -a_0 & a_1 & & 0 & & & \\ & -a_1 & \bullet & & & & \\ & & \bullet & \bullet & & & \\ & 0 & & \bullet & a_\ell & & \\ & & & & -a_\ell & \bullet & \\ & & & & & & \bullet \end{bmatrix} \tag{13.3.12}$$

with $p(\ell,t) = \langle \ell|\rho_F(t)|\ell\rangle$ and $a_\ell = -r_a \sin^2(g\tau\sqrt{\ell+1})$.

We define the eigenvalue λ_ℓ of the matrix M, corresponding to the right eigenstate

$$A_\ell = \begin{bmatrix} \alpha_0^\ell \\ \alpha_1^\ell \\ \vdots \end{bmatrix} \tag{13.3.13}$$

and the left eigenstate

$$B_\ell = [\beta_0^\ell \beta_1^\ell \ldots] \tag{13.3.14}$$

by the following relations

$$MA_\ell = \lambda_\ell A_\ell, \tag{13.3.15}$$

$$B_\ell M = \lambda_\ell B_\ell. \tag{13.3.16}$$

If we multiply Eq. (13.3.15) by $B_{\ell'}$ from the left and Eq. (13.3.16) for ℓ' by A_ℓ from the right and subtract the two equations, we obtain an orthogonality relation:

$$B_{\ell'} A_\ell = \delta_{\ell\ell'}, \tag{13.3.17}$$

provided $\lambda_\ell \neq \lambda_{\ell'}$ or λ_ℓ, $\lambda_{\ell'}$ nondegenerate.

In view of Eq. (13.3.15), the solution of Eq. (13.3.10) is given by

$$R(t) = \sum_{\ell=0}^{\infty} C_\ell A_\ell e^{-\lambda_\ell t}. \tag{13.3.18}$$

In cases where the eigenvalues are nondegenerate, the coefficient C_ℓ can be determined using the orthogonality condition (13.3.17), and we obtain

$$C_\ell = B_\ell R(0). \tag{13.3.19}$$

It follows from Eqs. (13.3.18) and (13.3.19) that

$$p(n, t) = \sum_{\ell=0}^{\infty} \sum_{q=0}^{\infty} \beta_q^\ell \alpha_n^\ell e^{-\lambda_\ell t} p(q, 0). \tag{13.3.20}$$

A determination of the eigenvalues λ_ℓ and the elements of the right and the left eigenstates A_ℓ and B_ℓ of the matrix will completely determine the photon distribution function.

The eigenvalues λ_ℓ of M, which satisfy the equation

$$\det(M - \lambda_\ell I) = 0 \tag{13.3.21}$$

(I being the unit matrix), are given by

$$\lambda_\ell = a_\ell = -r_a \sin^2(g\tau\sqrt{\ell + 1}). \tag{13.3.22}$$

For $0 \leq \ell \leq n_0$ such that $g\tau\sqrt{n_0+1} = \pi$, we have $\lambda_\ell \neq \lambda_{\ell'}$ for $\ell \neq \ell'$. A substitution from Eq. (13.3.22) into Eqs. (13.3.15) and (13.3.16) leads to the following recursion relations for the matrix elements α_n^ℓ and β_q^ℓ:

$$-a_{n-1}\alpha_{n-1}^\ell + a_n\alpha_n^\ell = a_\ell\alpha_n^\ell, \tag{13.3.23}$$

$$a_{q-1}\beta_{q-1}^\ell - a_{q-1}\beta_q^\ell = a_\ell\beta_{q-1}^\ell. \tag{13.3.24}$$

By iterating the recursion relations (13.3.23) and (13.3.24), we obtain

$$\alpha_n^\ell = \begin{cases} \prod_{r=\ell+1}^{n} a_{r-1}/(a_r - a_\ell) & n > \ell, \\ 1 & n = \ell, \\ 0 & n < \ell, \end{cases} \tag{13.3.25}$$

(13.3.26)

$$\beta_q^\ell = \begin{cases} \prod_{r=q}^{\ell-1} a_r/(a_r - a_\ell) & q < \ell, \\ 1 & q = \ell, \\ 0 & q > \ell. \end{cases} \tag{13.3.27}$$

Equation (13.3.20) combined with the expressions for λ_ℓ, α_n^ℓ, and β_q^ℓ completely determines the time evolution of the density matrix.

We consider the initial state of the field to be vacuum. Under this condition we have $p(q,0) = \delta_{q0}$. The interaction time of the atom is chosen such that $g\tau\sqrt{n_0+1} = \pi$. We then get $a_{n_0} = 0$. Under these conditions, the expression of the photon distribution function simplifies considerably and is given by

$$p(n,t) = \begin{cases} e^{a_0 t} & n = 0, \\ \sum_{\ell=0}^{n} \left[\prod_{r=0}^{n-1} a_r e^{-a_\ell t} \Big/ \prod_{\substack{r=0 \\ r\neq\ell}}^{n} (a_\ell - a_r)\right] & n \geq 1. \end{cases} \tag{13.3.28}$$

In the steady state ($t \to \infty$), $p(n,\infty)$ is zero for $n \neq n_0$. The only term that gives a nonvanishing contribution in the summation is $\ell = n_0$. The steady-state photon distribution $p_{ss}(n)$ is, therefore,

$$p_{ss}(n) = \delta_{nn_0}, \tag{13.3.29}$$

which is the photon distribution function for a number state.

13.4 Linewidth of a micromaser

We now turn to the calculation of the micromaser spectrum. The approach followed in this section will be based on the evaluation of the off-diagonal elements of field density matrix elements, as was done in Section 11.5 for the laser case.

The equation of motion for the density matrix $\rho_{n,n+k} = \rho_n^{(k)}$ (Eq. (13.1.3)) can be rewritten in the form of Eq. (11.5.2), but with

$$\frac{1}{2}\mu_n^{(k)} = 2r_a \sin^2\left[\frac{g\tau}{2}(\sqrt{n+1+k} - \sqrt{n+1})\right]$$
$$+\mathscr{C}(\bar{n}_{\text{th}}+1)\left[n+\frac{k}{2} - \sqrt{n(n+k)}\right]$$
$$+\mathscr{C}\bar{n}_{\text{th}}\left[n+1+\frac{k}{2} - \sqrt{(n+1)(n+k+1)}\right], \qquad (13.4.1)$$

$$c_n^{(k)} = r_a \sin(g\tau\sqrt{n+1})\sin(g\tau\sqrt{n+k}) + \mathscr{C}\bar{n}_{\text{th}}\sqrt{n(n+k)}, \qquad (13.4.2)$$

$$d_n^{(k)} = \mathscr{C}(\bar{n}_{\text{th}}+1)\sqrt{n(n+k)}. \qquad (13.4.3)$$

Following the same procedure as in Section 11.5, we obtain the following expression for the micromaser linewidth

$$2D \cong \mu_{\langle n\rangle}^{(1)} = 4r_a \sin^2\left(\frac{g\tau}{4\langle n\rangle}\right) + \frac{\mathscr{C}(2\bar{n}_{\text{th}}+1)}{4\langle n\rangle}, \qquad (13.4.4)$$

where we expand the square roots in Eq. (13.4.1) in the limit $\langle n\rangle \gg 1$.

In Fig. 13.6, we depict the detailed behavior of this approximate phase diffusion constant D as a function of the normalized interaction time $\tau_{\text{int}} = g\tau\sqrt{r_a/\mathscr{C}}$ for $r_a/\mathscr{C} = 50$ atoms and $\bar{n}_{\text{th}} = 10^{-4}$ thermal photons. The sharp resonances in the monotonic increase of D are reminiscent of the trapping states. We note that the phase diffusion is especially large when the maser is locked to a trapping state, that is, when $\langle n\rangle$ is caught in one of the sharp minima. Equation (13.4.4) reveals this behavior in the limit of short interaction times or large photon numbers, i.e., when $g\tau/4\langle n\rangle^{1/2} \ll 1$. We expand the sine function and arrive at the Schawlow–Townes linewidth, Eq. (11.5.13),

$$2D = \frac{\mathscr{A} + \mathscr{C}(2\bar{n}_{\text{th}}+1)}{4\langle n\rangle}, \qquad (13.4.5)$$

where

$$\mathscr{A} = r_a g^2\tau^2. \qquad (13.4.6)$$

The complicated pattern of the micromaser linewidth results from the complicated dependence of $\langle n\rangle$ on the pump parameter which enters in the denominator. The sine function in Eq. (13.4.4) suggests in the limit of large τ_{int} an oscillatory behavior of the linewidth. This is confirmed by the exact numerical results shown in the inset of Fig. 13.6.

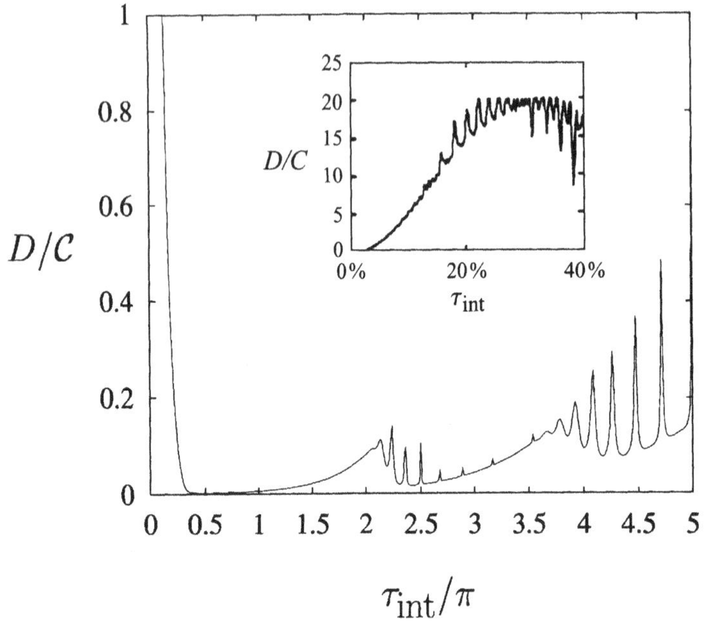

Fig. 13.6
The relative linewidth $D/\mathscr{C}$ based on Eq. (13.4.4) as a function of the interaction time $\tau_{\rm int}$ for $r_a/\mathscr{C} = 50$ and $\bar{n}_{\rm th} = 10^{-4}$. The inset shows the exact linewidth $D/\mathscr{C}$ based on the numerical solution of the density matrix Eq. (13.1.3) for large interaction times $\tau_{\rm int}$ for $r_a/\mathscr{C} = 20$ and $\bar{n}_{\rm th} = 1$.

Problems

13.1 Show that, in the semiclassical approximation

$$\langle n^r \rangle = \langle n \rangle^r \qquad \langle n \rangle \gg 1,$$

the normalized mean number of photons, $\langle N \rangle = \langle n \rangle / N_{\rm ex}$, in a micromaser is governed by

$$\sin^2(\tau_{\rm int}\sqrt{\langle N \rangle}) = \langle N \rangle.$$

Hence show that, for increasing values of $\tau_{\rm int}$, there can be more than one steady-state value of $\langle N \rangle$.

13.2 Show that the probability $P_n(m)$ that there are n photons in the micromaser cavity after m excited two-level atoms have passed through an initially empty cavity is given by the general expression

$$P_n(m) = \prod_{i=0}^{n-1} s(i) \sum_{\substack{i_j=0 \\ (i_{m-1}\le\cdots\le i_n)}}^{n} \prod_{j=n}^{m-1} c(i_j),$$

where we define

$$\sum_{\substack{i_j=0 \\ (i_{m-1}\le\cdots\le i_n)}}^{n} \prod_{j=n}^{m-1} c(i_j)$$

$$\equiv \begin{cases} \sum_{i_n=0}^{n} \sum_{i_{n+1}=0}^{i_n} \cdots \sum_{i_{m-1}=0}^{i_{m-2}} \prod_{j=n}^{m-1} c(i_j) & \text{for } m > n,\ n \ge 0, \\ 1 & \text{for } m = n,\ n \ge 0, \\ 0 & \text{for } m < n \text{ or } n < 0, \end{cases}$$

with $\prod_{i=0}^{-1} = 1$. (Hint: See J. Krause, M. O. Scully, T. Walther, and H. Walther, *Phys. Rev. A* **39**, 1915 (1989).)

References and bibliography

Experiment

D. Meschede, H. Walther, and G. Müller, *Phys. Rev. Lett.* **54**, 551 (1985).

S. Haroche and J. M. Raimond, in *Advances in Atomic and Molecular Physics*, Vol. 20, ed. D. R. Bates and B. Bederson (Academic, New York 1985), p. 350.

G. A. C. Gallas, G. Leuchs, H. Walther, and H. Figger, in *Advances in Atomic and Molecular Physics*, Vol. 20, ed. D. R. Bates and B. Bederson (Academic, New York 1985), p. 414.

G. Rempe, H. Walther, and N. Klein, *Phys. Rev. Lett.* **58**, 353 (1987).

Theory

P. Filipowicz, J. Javanainen, and P. Meystre, *Phys. Rev. A* **34**, 3077 (1986).

L. A. Lugiato, M. O. Scully, and H. Walther, *Phys. Rev. A* **36**, 740 (1987).

Photon statistics via statistics of atomic detection

W. Vogel, D.-G. Welsch, and L. Leine, *J. Opt. Soc. Am. B.* **4**, 1633 (1987).

G. Rempe and H. Walther, *Phys. Rev. A* **42**, 1650 (1990).

H. Paul and Th. Richter, *Opt. Commun.* **85**, 508 (1991).

H.-J. Briegel, B.-G. Englert, N. Sterpi, and H. Walther, *Phys. Rev. A* **49**, 2962 (1994).

C. Wagner, A. Schenzle, and H. Walther, *Opt. Commun.* **107**, 318 (1994).

P. J. Bardroff, E. Mayr, and W. Schleich, *Phys. Rev. A* **51**, 4369 (1995).

P. J. Bardroff, E. Mayr, W. P. Schleich, P. Domokos, M. Brune, J.M. Raimond, and S. Haroche, *Phys. Rev. A* **53** 2736 (1996).

Two-photon micromaser

M. Brune, J. M. Raimond, and S. Haroche, *Phys. Rev. A* **35**, 154 (1987).

L. Davidovich, J. M. Raimond, M. Brune, and S. Haroche, *Phys. Rev. A* **36**, 3771 (1987).

I. Ashraf, J. Gea-Banacloche, and M. S. Zubairy, *Phys. Rev. A* **42**, 6704 (1990).

I. Ashraf and M. S. Zubairy, *Opt. Commun.* **77**, 85 (1990).

Generation of number states and trapping states

P. Filipowicz, J. Javanainen, and P. Meystre, *J. Opt. Soc. Am. B* **3**, 906 (1986).

J. Krause, M. O. Scully, and H. Walther, *Phys. Rev. A* **36**, 4547 (1987).

J. Krause, M. O. Scully, T. Walther, and H. Walther, *Phys. Rev. A* **39**, 1915 (1989).

J. J. Slosser, P. Meystre, and S. L. Braunstein, *Phys. Rev. Lett.* **63**, 934 (1989).

S. Qamar, K. Zaheer, and M. S. Zubairy, *Opt. Commun.* **78**, 341 (1990).

Micromaser spectrum

S. Qamar and M. S. Zubairy, *Phys. Rev. A* **44**, 7804 (1991).

M. O. Scully, H. Walther, G. S. Agarwal, T. Quang, and W. Schleich, *Phys. Rev. A* **44**, 5992 (1991).

N. Lu, *Phys. Rev. Lett.* **70**, 912 (1993); *Phys. Rev. A* **47**, 1347 (1993).

T. Quang, G. S. Agarwal, J. Bergou, M. O. Scully, H. Walther, K. Vogel, and W. Schleich, *Phys. Rev. A* **48**, 803 (1993).

Microlaser

K. An, J. J. Childs, R. R. Desari, and M. S. Feld, *Phys. Rev. Lett.* **73**, 3375 (1994).

K. An and M. S. Feld, *Phys. Rev. A* **52**, 1691 (1995).

CHAPTER 14

Correlated emission laser: concept, theory, and analysis

As discussed in the last three chapters, the fundamental source of noise in a laser is spontaneous emission. A simple pictorial model for the origin of the laser linewidth envisions it as being due to the random phase diffusion process arising from the addition of spontaneously emitted photons with random phases to the laser field. In this chapter we show that the quantum noise leading to the laser linewidth can be suppressed below the standard, i.e., Schawlow–Townes limit by preparing the atomic systems in a coherent superposition of states as in the Hanle effect and quantum beat experiments discussed in Chapter 7. In such coherently prepared atoms the spontaneous emission is said to be correlated. Lasers operating via such a phase coherent atomic ensemble are known as *correlated spontaneous emission lasers* (CEL).* An interesting aspect of the CEL is that it is possible to eliminate the spontaneous emission quantum noise in the relative linewidths by correlating the two spontaneous emission noise events.

A number of schemes exist in which quantum noise quenching below the standard limit can be achieved. In two-mode schemes a correlation between the spontaneous emisson events in two different modes of the radiation field is established via atomic coherence so that the relative phase between them does not diffuse or fluctuate. In a Hanle laser and a quantum beat laser this is achieved by pumping the atoms coherently such that every spontaneously emitting atom contributes equally to the two modes of the radiation, leading to a reduction and even vanishing of the noise in the phase difference. In a two-photon CEL, a cascade transition involving three-level atoms is coupled to only one mode of the radiation field. A well-defined

* The simplest CEL treatment is via the quantum Langevin approach, Scully [1985]. We have presented a density matrix Fokker–Planck analysis, Scully and Zubairy [1987], since it is more readily extended to include, for example, nonlinear effects Krause and Scully [1987].

coherence between the upper and lower levels leads to a correlation between the light emitted by an $|a\rangle \to |b\rangle$ and a subsequent $|b\rangle \to |c\rangle$ transition, see Fig. 14.8.

In this chapter, we present the microscopic theories of these correlated spontaneous emission laser schemes and derive the conditions under which the CEL quantum noise quenching takes place. CEL operation leads to a vanishing diffusion constant in the relative phase for the Hanle and quantum beat lasers and also to phase squeezing in the two-photon CEL.

For the sake of simplicity, we restrict our analysis to the linear theory. In our discussion on the natural linewidth of a single-mode laser (see Section 11.4) we have already seen that a linear analysis is sufficient for the linewidth calculations in above-threshold regions. This brings out the physics most directly, and the fully nonlinear analysis of Krause and Scully [1987] verifies this approach.

14.1 Correlated spontaneous emission laser concept

There is a good deal of interest in high precision laser interferometric measurements in many areas of modern science. For example, the heart of today's efforts to see gravitational radiation is the Michelson interferometer, while the laser gyro using a Sagnac ring interferometer often operates at the standard quantum "limit". As discussed in Section 4.1, a Sagnac ring interferometer is used to measure rotation rates. Here, in order to set the stage for the correlated spontaneous emission laser, we discuss the quantum limits of passive and active laser gyros.

We recall from Section 4.1.2, that in an optical ring of radius b rotating at a rate Ω_r, the phase difference $\Delta\theta$ between the counter-propagating laser beams after one round trip is given by (Eq. (4.1.14))

$$\Delta\theta = \frac{4\pi b^2 \Omega_r}{c\bar{\lambda}}, \qquad (14.1.1)$$

where $\bar{\lambda} = \lambda/2\pi$ is the reduced wavelength.

However, since we are operating in a high quality optical cavity characterized by a cavity decay rate $\mathscr{C}$, the light will make N circuits around the ring where $N \cong c\mathscr{C}^{-1}/2\pi b$, and the actual phase shift between the co- and counter-propagating waves is given by

$$\Delta\theta_{\text{sig}} = N\Delta\theta = S\mathscr{C}^{-1}\Omega_r \qquad \text{(passive)}, \qquad (14.1.2)$$

where $S = 2b/\bar{\lambda}$.

We recall that the quantum noise (for unit quantum efficiency), on the other hand, is such that $\Delta\theta_{\text{err}} \sim 1/\sqrt{\bar{n}}$ where $\bar{n}$, the measured photon (photoelectron) number, is given by $Pt_m/\hbar\nu$ where P is the laser power and t_m is the measurement time. So that

$$\Delta\theta_{\text{err}} \cong \sqrt{\frac{\hbar\nu}{Pt_m}} \qquad \text{(passive)}, \tag{14.1.3}$$

and by equating (14.1.2) to (14.1.3) we find the usual quantum limit for the laser gyro

$$\Omega_{\min} \cong S^{-1}\mathscr{C}\sqrt{\frac{\hbar\nu}{Pt_m}} \qquad \text{(passive)}. \tag{14.1.4}$$

Now comes an interesting point. An active laser gyro operates on the change of frequency associated with co- and counter-propagating light. If the round-trip transit time of the clockwise (CW) and counter-clockwise (CCW) propagating beams are denoted by t^+ and t^-, respectively, then the frequencies ν_+ and ν_- associated with the CW and CCW beams are given by the resonance conditions $ct^{\pm} = m\pi c/\nu_{\pm}$, where m is an even integer. It follows, on substituting for t^+ and t^- from Eqs. (4.1.9), that to lowest order in $b\Omega_r/c$ we have

$$\nu_{\pm} = \nu(1 \mp b\Omega_r/c), \tag{14.1.5}$$

where $\nu = m\pi c/2\pi b$. Then the frequency difference $\Delta\nu = 2\nu b\Omega_r/c$ yields the active signal phase

$$\Delta\theta_{\text{sig}} = \Delta\nu t_m = St_m\Omega_r \qquad \text{(active)}. \tag{14.1.6}$$

Comparing (14.1.2) and (14.1.6) we see that the phase signal in an active gyro is many orders of magnitude larger than that of a passive one since, for example, we could have $t_m \sim 1$ sec but $\mathscr{C}^{-1} \sim 10^{-4}$ to 10^{-6} sec. It is for this reason that commercial laser gyros are commonly active devices.

But what is the signal-to-noise ratio for an active gyro? One might be tempted to argue (and many people have fallen into this type of trap) that we should take the shot-noise error (14.1.3) together with the signal (14.1.6) to find

$$\Omega_{\min} \cong S^{-1}\frac{1}{t_m}\sqrt{\frac{\hbar\nu}{Pt_m}} \qquad \text{(active gyro shot noise limit)}. \tag{14.1.7}$$

The good news is that expression (14.1.7) is much superior to the passive gyro limit (14.1.4), the bad news is that it is wrong. As we recall from the discussion of the spontaneous emission induced phase diffusion, the error in the phase θ built up in a time t_m is given by $\Delta\theta_{\rm err} = \sqrt{D\, t_m}$ where D is the phase diffusion rate (Eq. (11.4.11)). Furthermore, since $D \simeq \mathscr{C}/\langle n\rangle$ and $\langle n\rangle = P\mathscr{C}^{-1}/\hbar\nu$ we have

$$\Delta\theta_{\rm err} \cong \mathscr{C}t_m\sqrt{\frac{\hbar\nu}{Pt_m}} \qquad \text{(active spontaneous emission noise).} \tag{14.1.8}$$

Thus, we see that due to the $\mathscr{C}t_m$ factor the phase noise (14.1.8) is much larger than (14.1.3). Finally we note that when we use (14.1.8) and (14.1.6) together (as we should) we regain the standard limit (14.1.4). That is, the limit is the same for passive and active devices. Yet, we are naturally led to ask: is there any way we can make an active device but avoid spontaneous emission noise? This question is the starting point for study of the correlated (spontaneous) emission laser (CEL) to which we now turn.

14.2 Hanle effect correlated emission laser via density matrix analysis

In the last section we recalled that in active laser interferometer experiments, the limiting source of quantum noise is often spontaneous emission fluctuations in the relative phase angle. We will now show that diffusion of the relative phase angle between two such laser modes may be eliminated by preparing a laser medium consisting of 'three-level' atoms, and arranging that the two transitions $|a\rangle \leftrightarrow |c\rangle$ and $|b\rangle \leftrightarrow |c\rangle$ drive a doubly resonant cavity; see Fig. 14.1. In this way the optical paths may be differently affected by an external influence of interest (e.g., a gravity wave or a Sagnac frequency shift).

The atomic transitions driving the two optical paths are strongly correlated when the upper levels $|a\rangle$ and $|b\rangle$ are prepared in a coherent superposition as in Hanle effect or quantum beat experiments. In the Hanle effect example, the levels $|a\rangle$ and $|b\rangle$ can be taken to be the 'linear polarization' states formed from a single 'elliptical polarization' state as shown in Fig. 14.1(a). In the quantum beat case the coherent mixing is produced by a strong external microwave signal as in Fig. 14.2. The fields emitted by the atoms of Fig. 14.1 will differ in polarization while fields produced by the atoms of Fig. 14.2 will differ in frequency.

In both cases discussed above the heterodyne beat note between the spontaneously emitted fields 1 and 2 shows that they are strongly

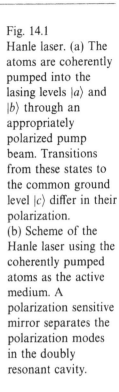

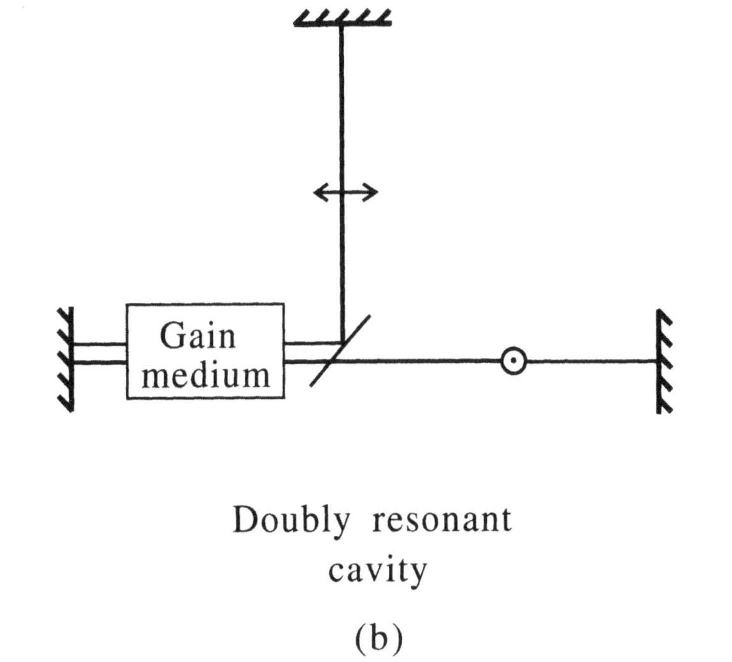

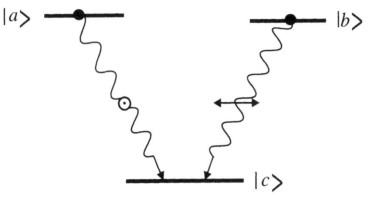

Fig. 14.1
Hanle laser. (a) The atoms are coherently pumped into the lasing levels $|a\rangle$ and $|b\rangle$ through an appropriately polarized pump beam. Transitions from these states to the common ground level $|c\rangle$ differ in their polarization.
(b) Scheme of the Hanle laser using the coherently pumped atoms as the active medium. A polarization sensitive mirror separates the polarization modes in the doubly resonant cavity.

Fig. 14.2
Quantum beat laser. (a) The atoms are prepared in a coherent superposition of upper levels $|a\rangle$ and $|b\rangle$ by an external field with an effective Rabi frequency Ω_R. The two laser transitions at frequencies ν_1 and ν_2 share a common lower level $|c\rangle$. (b) Scheme of the quantum beat laser with doubly resonant cavity.

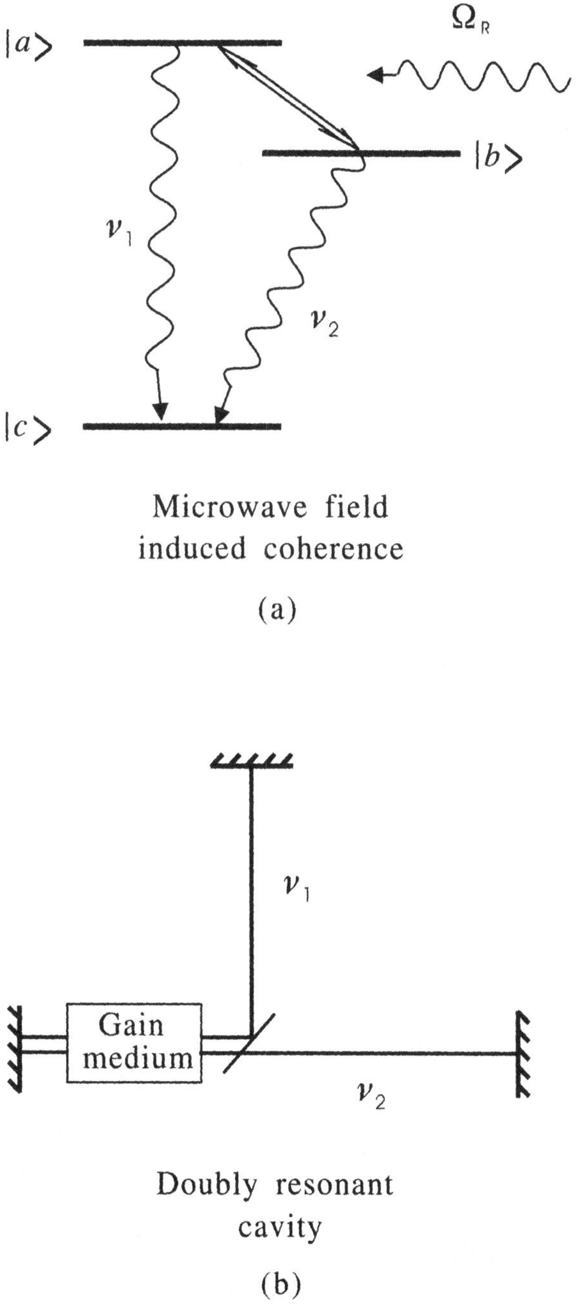

correlated. To see this, consider the atoms of Fig. 14.2 interacting with a quantized field. The state vector is given by

$$|\Psi\rangle = \alpha e^{-i\phi_a}|a,0_1,0_2\rangle + \beta e^{-i\phi_b}|b,0_1,0_2\rangle + \gamma_1|c,1_1,0_2\rangle + \gamma_2|c,0_1,1_2\rangle, \tag{14.2.1}$$

where $|1_i\rangle$ is the state $a_i^\dagger|0_i\rangle$, $i = 1,2$, and $a_i^\dagger$ (and a_i) is the creation (and annihilation) operators for photons having frequency ν_i. Now the expectation value for the electric field operator $E_1^{(+)}$,

$$E_1^{(+)}(\mathbf{r},t) = \mathscr{E}_1 a_1 e^{i(\mathbf{k}_1\cdot\mathbf{r}-\nu_1 t)}, \tag{14.2.2}$$

calculated using Eq. (14.2.1) is easily seen to vanish since the states $|a\rangle$, $|b\rangle$, and $|c\rangle$ are orthogonal. Similar arguments show that $\langle E_2^{(+)}\rangle$ likewise vanishes. However, the cross term does not vanish:

$$\langle\Psi|E_1^{(-)}E_2^{(+)}|\Psi\rangle = \mathscr{E}_1^*\mathscr{E}_2\gamma_1^*\gamma_2\langle c|c\rangle \exp[-i(\mathbf{k}_1-\mathbf{k}_2)\cdot\mathbf{r} + i(\nu_1-\nu_2)t]. \tag{14.2.3}$$

That is, the spontaneously emitted photons at ν_1 and ν_2 are correlated.

Motivated by the preceding arguments we are led to investigate diffusion in the relative phase angle of the doubly resonant cavities in Fig.'s 14.1b and 14.2b.

In the Hanle laser, we have an active medium of three-level atoms such that the upper levels $|a\rangle$ and $|b\rangle$ are prepared in a coherent superposition and decay to the state $|c\rangle$ via emission of radiation of different polarization states (Fig. 14.1). In a doubly resonant cavity, the two waves interact with the active medium with the help of a polarization sensitive mirror.

The Hamiltonian for the atom–field system, in the rotating-wave approximation, is given by

$$\mathscr{H} = \mathscr{H}_0 + \mathscr{H}_1, \tag{14.2.4}$$

where

$$\mathscr{H}_0 = \sum_{i=a,b,c} \hbar\omega_i|i\rangle\langle i| + \hbar\nu_1 a_1^\dagger a_1 + \hbar\nu_2 a_2^\dagger a_2, \tag{14.2.5}$$

$$\mathscr{H}_1 = \hbar g_1 a_1|a\rangle\langle c| + \hbar g_2 a_2|b\rangle\langle c| + \text{H.c.} \tag{14.2.6}$$

Here g_1 and g_2 are coupling constants for the transitions $|a\rangle \to |c\rangle$ and $|b\rangle \to |c\rangle$, respectively. If the upper levels $|a\rangle$ and $|b\rangle$ are the Zeeman sublevels of a degenerate upper state then, in the absence of an applied magnetic field, $\omega_a \simeq \omega_b$, and thus the fields emitted by the atoms in the $|a\rangle \to |c\rangle$ and $|b\rangle \to |c\rangle$ transitions will differ in polarization, but their frequencies will be essentially the same ($\nu_1 = \nu_2 = \nu$). If we now transform the Hamiltonian (14.2.4) into an interaction picture with respect to $\mathscr{H}_0$, we obtain

$$\mathscr{V} = \hbar g_1 a_1 e^{i\Delta t}|a\rangle\langle c| + \hbar g_2 a_2 e^{i\Delta t}|b\rangle\langle c| + \text{H.c.}, \qquad (14.2.7)$$

where $\Delta = \omega_a - \omega_c - \nu = \omega_b - \omega_c - \nu$ is the detuning which we have taken to be identical for the two transitions.

We assume that the atoms have initially been prepared in a coherent superposition of the two upper states. The atomic density operator is therefore given by

$$\begin{aligned}\rho_{\text{atom}}(0) &= (e^{i\phi}c_a|a\rangle + c_b|b\rangle)(e^{-i\phi}c_a^*\langle a| + c_b^*\langle b|)\\ &= \rho_{aa}^{(0)}|a\rangle\langle a| + \rho_{ab}^{(0)}|a\rangle\langle b| + \rho_{ba}^{(0)}|b\rangle\langle a| + \rho_{bb}^{(0)}|b\rangle\langle b|. \qquad (14.2.8)\end{aligned}$$

Here ϕ is some fixed relative phase between the upper states and $\rho_{ab}^{(0)} = |\rho_{ab}^{(0)}|\exp(i\phi)$ is the initial coherence.

An equation for the reduced density operator of the field ρ is obtained by taking a trace over atoms,* which leads to

$$\begin{aligned}\dot{\rho} &= -\frac{i}{\hbar}\text{Tr}_{\text{atom}}[\mathscr{V}, \rho_{\text{atom-field}}]\\ &= -\frac{i}{\hbar}([\mathscr{V}_{ac}, \rho_{ca}] + [\mathscr{V}_{bc}, \rho_{cb}] + [\mathscr{V}_{ca}, \rho_{ac}] + [\mathscr{V}_{cb}, \rho_{bc}]), \qquad (14.2.9)\end{aligned}$$

where $\rho_{\text{atom-field}}$ denotes the full atom–field density operator and $\mathscr{V}_{ac} = \hbar g_1 a_1 \exp(i\Delta t)$, $\mathscr{V}_{bc} = \hbar g_2 a_2 \exp(i\Delta t)$. The atomic matrix elements ρ_{ac} and ρ_{bc} can be evaluated to the first order in the coupling constants g_i by solving the equations for the corresponding theory.

The equations of motion for ρ_{ac} and ρ_{bc} are

$$\dot{\rho}_{ac} = -\gamma\rho_{ac} - \frac{i}{\hbar}(\mathscr{V}_{ac}\rho_{cc} - \rho_{aa}\mathscr{V}_{ac} - \rho_{ab}\mathscr{V}_{bc}), \qquad (14.2.10)$$

$$\dot{\rho}_{bc} = -\gamma\rho_{bc} - \frac{i}{\hbar}(\mathscr{V}_{bc}\rho_{cc} - \rho_{bb}\mathscr{V}_{bc} - \rho_{ba}\mathscr{V}_{ac}). \qquad (14.2.11)$$

Here γ is the atomic decay constant which, for simplicity, we have taken to be the same for all levels. The zeroth-order equations of motion for $\rho_{cc}, \rho_{aa}, \rho_{bb}$, and ρ_{ab} are

$$\dot{\rho}_{cc} = -\gamma\rho_{cc}, \qquad (14.2.12)$$

$$\dot{\rho}_{aa} = r_a\rho_{aa}^{(0)}\rho - \gamma\rho_{aa}, \qquad (14.2.13)$$

$$\dot{\rho}_{bb} = r_a\rho_{bb}^{(0)}\rho - \gamma\rho_{bb}, \qquad (14.2.14)$$

$$\dot{\rho}_{ab} = r_a\rho_{ab}^{(0)}\rho - \gamma\rho_{ab}, \qquad (14.2.15)$$

* From this point on, ρ without indices refers to the reduced density matrix $\rho_{\text{field}} \equiv \text{Tr}_{\text{atom}}\rho_{\text{atom-field}}$. With indices, such as $\rho_{ac} \equiv \langle a|\rho_{\text{atom-field}}|c\rangle$, it refers to the element of the full density matrix between atomic states.

where r_a is the rate of injection of atoms in the coherent superposition (14.2.8). A steady-state solution of these equations yields

$$\rho_{cc} = 0, \tag{14.2.16}$$

$$\rho_{aa} = \frac{r_a}{\gamma}\rho_{aa}^{(0)}\rho, \tag{14.2.17}$$

$$\rho_{bb} = \frac{r_a}{\gamma}\rho_{bb}^{(0)}\rho, \tag{14.2.18}$$

$$\rho_{ab} = \frac{r_a}{\gamma}\rho_{ab}^{(0)}\rho. \tag{14.2.19}$$

It follows, on substituting these expressions for the zeroth-order matrix elements in Eqs. (14.2.10) and (14.2.11) and then integrating, that

$$\begin{aligned}\rho_{ac}(t) &= \frac{ir_a}{\gamma}\int_{-\infty}^{t} dt_0 e^{-\gamma(t-t_0)}\left[\rho_{aa}^{(0)}\rho a_1 g_1 e^{i\Delta t_0}\right.\\ &\quad \left.+\rho_{ab}^{(0)}\rho a_2 g_2 e^{i\Delta t_0}\right]\\ &= \frac{ir_a}{\gamma(\gamma+i\Delta)}\left[\rho_{aa}^{(0)}g_1\rho a_1 e^{i\Delta t} + \rho_{ab}^{(0)}g_2\rho a_2 e^{i\Delta t}\right],\end{aligned} \tag{14.2.20}$$

$$\rho_{bc}(t) = \frac{ir_a}{\gamma(\gamma+i\Delta)}\left[\rho_{bb}^{(0)}g_2\rho a_2 e^{i\Delta t} + \rho_{ba}^{(0)}g_1\rho a_1 e^{i\Delta t}\right]. \tag{14.2.21}$$

The other matrix elements in Eq. (14.2.9) can be determined using $\rho_{ca} = \rho_{ac}^{\dagger}$ and $\rho_{cb} = \rho_{bc}^{\dagger}$.

On substituting for ρ_{ac} and ρ_{bc} from Eqs. (14.2.20) and (14.2.21) into Eq. (14.2.9), the following master equation for the field-density operator ρ is obtained

$$\begin{aligned}\dot{\rho} = &-\frac{1}{2}\alpha_{11}(\rho a_1 a_1^{\dagger} - a_1^{\dagger}\rho a_1) - \frac{1}{2}\alpha_{22}(\rho a_2 a_2^{\dagger} - a_2^{\dagger}\rho a_2)\\ &-\frac{1}{2}\alpha_{12}(\rho a_2 a_1^{\dagger} - a_1^{\dagger}\rho a_2)e^{i\phi} - \frac{1}{2}\alpha_{21}(\rho a_1 a_2^{\dagger} - a_2^{\dagger}\rho a_1)e^{-i\phi}\\ &+\text{H.c.}\end{aligned} \tag{14.2.22}$$

with

$$\alpha_{11} = \frac{2r_a g_1^2}{\gamma(\gamma+i\Delta)}\rho_{aa}^{(0)}, \tag{14.2.23}$$

$$\alpha_{12} = \frac{2r_a g_1 g_2}{\gamma(\gamma+i\Delta)}|\rho_{ab}^{(0)}|, \tag{14.2.24}$$

$$\alpha_{21} = \frac{2r_a g_2 g_1}{\gamma(\gamma+i\Delta)}|\rho_{ba}^{(0)}|, \tag{14.2.25}$$

$$\alpha_{22} = \frac{2r_a g_2^2}{\gamma(\gamma+i\Delta)}\rho_{bb}^{(0)}. \tag{14.2.26}$$

In this equation, the terms proportional to α_{11} and α_{22} are the gain terms for the two modes which would yield the usual linewidth expressions discussed in Section 11.4. The terms proportional to α_{12} and α_{21} are, however, phase sensitive as they arise due to the coherent preparation of the upper levels $|a\rangle$ and $|b\rangle$. In Eq. (14.2.22), we have not included the cavity decay terms as they do not contribute to the phase diffusion coefficients in the following calculations.

In order to study the phase noise, we convert Eq. (14.2.22) into an equivalent Fokker–Planck equation for the P-representation $P(\alpha_1,\alpha_1^*,\alpha_2,\alpha_2^*)$ by the following substitutions

$$\begin{aligned} a_i\rho &\leftrightarrow \alpha_i P, \\ a_i^\dagger\rho &\leftrightarrow \left(\alpha_i^* - \frac{\partial}{\partial\alpha_i}\right)P, \quad \rho a_i^\dagger \leftrightarrow \alpha_i^* P, \\ \rho a_i &\leftrightarrow \left(\alpha_i - \frac{\partial}{\partial\alpha_i^*}\right)P, \end{aligned} \tag{14.2.27}$$

with $i = 1, 2$. The resulting Fokker–Planck equation then reads

$$\frac{\partial P}{\partial t} = \sum_{j,k=1,2} \frac{\alpha_{jk}}{2}\left[-\frac{\partial}{\partial\alpha_j}(\alpha_k P) + \frac{\partial^2 P}{\partial\alpha_j\partial\alpha_k^*}\right] e^{i\phi(k-j)} + \text{c.c.} \tag{14.2.28}$$

Next we define the polar coordinates r_j, θ_j ($j = 1, 2$) via the relation $\alpha_j = r_j \exp(i\theta_j)$. We also define the difference and mean angle variables $\theta = \theta_1 - \theta_2$ and $\mu = (\theta_1 + \theta_2)/2$. We then have

$$\frac{\partial}{\partial\alpha_1} = \frac{e^{-i\theta_1}}{2}\left(\frac{\partial}{\partial r_1} + \frac{1}{ir_1}\frac{\partial}{\partial\theta_1}\right), \tag{14.2.29}$$

$$\frac{\partial}{\partial\alpha_2} = \frac{e^{-i\theta_2}}{2}\left(\frac{\partial}{\partial r_2} + \frac{1}{ir_2}\frac{\partial}{\partial\theta_2}\right), \tag{14.2.30}$$

$$\frac{\partial}{\partial\theta_1} = \frac{1}{2}\frac{\partial}{\partial\mu} + \frac{\partial}{\partial\theta}, \tag{14.2.31}$$

$$\frac{\partial}{\partial\theta_2} = \frac{1}{2}\frac{\partial}{\partial\mu} - \frac{\partial}{\partial\theta}. \tag{14.2.32}$$

Above threshold the amplitude fluctuations are small and can be neglected. This amounts to the assumption that P is independent of r_1 and r_2 and it only depends on θ and μ. The variables r_1 and r_2 can be replaced by their mean values $\sqrt{\langle n_1\rangle}$ and $\sqrt{\langle n_2\rangle}$ in the steady state. The exact expressions for $\langle n_1\rangle$ and $\langle n_2\rangle$ can be determined by a nonlinear analysis. Under these conditions Eq. (14.2.28) reduces to

$$\begin{aligned} \frac{\partial P}{\partial t} = &-\frac{\partial}{\partial\theta}(d_\theta P) - \frac{\partial}{\partial\mu}(d_\mu P) + \frac{\partial^2}{\partial\theta^2}(D_{\theta\theta}P) \\ &+\frac{\partial^2}{\partial\theta\partial\mu}(D_{\theta\mu}P) + \frac{\partial^2}{\partial\mu^2}(D_{\mu\mu}P), \end{aligned} \tag{14.2.33}$$

where

$$d_\theta = -\frac{i}{4}\left[\alpha_{11} - \alpha_{22} + \alpha_{12}\left(\frac{\langle n_2\rangle}{\langle n_1\rangle}\right)^{1/2} e^{-i\psi} - \alpha_{21}\left(\frac{\langle n_1\rangle}{\langle n_2\rangle}\right)^{1/2} e^{i\psi}\right] + \text{c.c.}, \tag{14.2.34}$$

$$d_\mu = -\frac{i}{8}\left[\alpha_{11} + \alpha_{22} + \alpha_{12}\left(\frac{\langle n_2\rangle}{\langle n_1\rangle}\right)^{1/2} e^{-i\psi} + \alpha_{21}\left(\frac{\langle n_1\rangle}{\langle n_2\rangle}\right)^{1/2} e^{i\psi}\right] + \text{c.c.}, \tag{14.2.35}$$

$$D_{\theta\theta} = \frac{1}{8}\left(\frac{\alpha_{11}}{\langle n_1\rangle} + \frac{\alpha_{22}}{\langle n_2\rangle} - \frac{\alpha_{12}}{\sqrt{\langle n_1\rangle\langle n_2\rangle}}e^{-i\psi} - \frac{\alpha_{21}}{\sqrt{\langle n_1\rangle\langle n_2\rangle}}e^{i\psi}\right) + \text{c.c.}, \tag{14.2.36}$$

$$D_{\theta\mu} = \frac{1}{8}\left(\frac{\alpha_{11}}{\langle n_1\rangle} - \frac{\alpha_{22}}{\langle n_2\rangle}\right) + \text{c.c.}, \tag{14.2.37}$$

$$D_{\mu\mu} = \frac{1}{32}\left(\frac{\alpha_{11}}{\langle n_1\rangle} + \frac{\alpha_{22}}{\langle n_2\rangle} + \frac{\alpha_{12}}{\sqrt{\langle n_1\rangle\langle n_2\rangle}}e^{-i\psi} + \frac{\alpha_{21}}{\sqrt{\langle n_1\rangle\langle n_2\rangle}}e^{i\psi}\right) + \text{c.c.}, \tag{14.2.38}$$

with $\psi = \theta - \phi$.

The physical meaning of the terms in Eq. (14.2.33) is the following. The coefficients d_θ and d_μ are the drift coefficients with respect to the variables $\theta = \theta_1 - \theta_2$ and $\mu = (\theta_1 + \theta_2)/2$ and $D_{\theta\theta}$ and $D_{\mu\mu}$ are the corresponding diffusion coefficients. The key feature of the diffusion and the drift coefficients is that they are explicitly phase dependent and these phase dependences arise due to the injected coherence. A much simpler set of coefficients is obtained when

$$\rho_{aa}^{(0)} = \rho_{bb}^{(0)} = |\rho_{ab}^{(0)}| = |\rho_{ba}^{(0)}| = \frac{1}{2}, \tag{14.2.39}$$

$g_1 = g_2 = g$ and $\langle n_1\rangle = \langle n_2\rangle = \langle n\rangle$. Under these conditions

$$\alpha_{11} = \alpha_{12} = \alpha_{21} = \alpha_{22} = \frac{r_a g^2}{\gamma(\gamma + i\Delta)}, \tag{14.2.40}$$

and

$$d_\theta = -\frac{\mathscr{A}}{2}\sin\psi, \tag{14.2.41}$$

$$d_\mu = -\frac{\mathscr{A}\Delta}{4\gamma}(1 + \cos\psi), \tag{14.2.42}$$

$$2D_{\theta\theta} = \frac{\mathscr{A}}{2\langle n\rangle}(1 - \cos\psi), \tag{14.2.43}$$

$$2D_{\theta\mu} = 0, \tag{14.2.44}$$

$$2D_{\mu\mu} = \frac{\mathscr{A}}{8\langle n\rangle}(1 + \cos\psi), \tag{14.2.45}$$

where $\mathscr{A} = 2r_a g^2/(\gamma^2 + \Delta^2)$ is the linear gain coefficient.

From the Fokker–Planck equation (14.2.33), we derive the following equation of motion for the relative phase

$$\frac{d}{dt}\langle\theta\rangle = \langle d_\theta\rangle. \tag{14.2.46}$$

Phase locking ($\langle\theta\rangle$ = constant) therefore takes place for those values of θ for which the drift coefficient vanishes. This happens when $\psi = 0$, i.e., $\theta = \phi$. We also see that the diffusion coefficient for the relative phase angle (Eq. (14.2.43)), which is proportional to $(1 - \cos\psi)$, vanishes when the angle ψ itself vanishes. A correlated spontaneous emission laser (CEL) operation is therefore obtained in a Hanle laser.

It is interesting to note that the conditions, under which the diffusion in relative phase $\theta = \theta_1 - \theta_2$ vanishes, do not lead to a vanishing of $D_{\mu\mu}$ where $\mu = (\theta_1 + \theta_2)/2$.

Physically we can understand the quenching of the spontaneous emission fluctuations in the relative phase θ by referring to Fig. 14.3. Here we consider the 'random walk' of the tips of the electric field phases of the two modes in the complex α-plane. If we ignore the amplitude fluctuations, the phase fluctuations in the field associated with the spontaneous emission allow the tips of the fields to diffuse out around a circle in the complex plane. When $D_{\theta\theta} = 0$, the spontaneous emission in the two modes becomes highly correlated so that the relative phase angle θ is 'locked' to a particular value. The average phase variable has, however, nonvanishing diffusion.

14.3 Quantum beat laser via pictorial treatment

Like the Hanle laser, a quantum beat laser consists of three-level atoms in the V configuration which are pumped in the upper level $|a\rangle$ inside a doubly resonant cavity (Fig. 14.2). A coherence is introduced between the upper levels $|a\rangle$ and $|b\rangle$ by an external field which is characterized by the Rabi frequency $\Omega_R \exp(-i\phi)$ where Ω_R and ϕ are the real amplitude and phase. The transitions $|a\rangle \to |c\rangle$ and $|b\rangle \to |c\rangle$ are assumed dipole-allowed. The $|a\rangle \to |b\rangle$ transition is therefore dipole-forbidden. The external field leading to a coherence between these levels could be a strong magnetic field for a magnetic dipole-allowed transition. We shall treat the $|a\rangle \to |b\rangle$ transition semiclassically and to all orders in the Rabi frequency. The $|a\rangle \to |c\rangle$ and $|b\rangle \to |c\rangle$ transitions will be treated fully quantum mechanically, but only to the second order in the corresponding coupling constants. The analysis is given in Appendix 14.A, where it is again found that the spontaneous

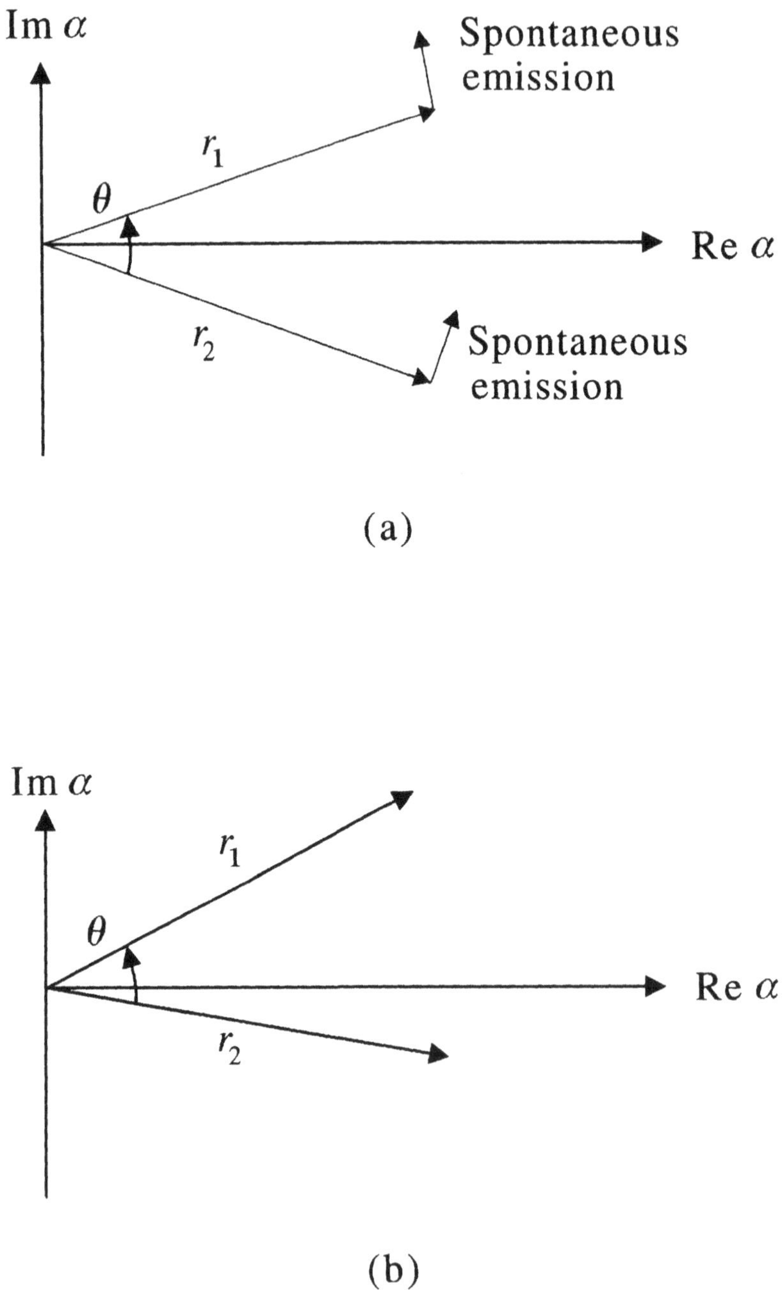

Fig. 14.3 Geometrical representation of the CEL effect. The spontaneous emission events in the two modes are highly correlated and the relative phase remains the same (a) before and (b) after the spontaneous emission event.

emission quantum noise in the relative phase angle can be made to vanish. We will here present a simple pictorial* treatment of the problem yielding this result in an intuitively appealing fashion.

In the quantum beat CEL configuration two electric fields $\mathscr{E}_1 = r_1 \exp(i\theta_1)$ and $\mathscr{E}_2 = r_2 \exp(i\theta_2)$ with slowly varying amplitudes r_j and phases θ_j $(j = 1, 2)$ are locked to a constant relative phase angle θ_0. This phase locking is described in its simplest form by the following Adler equation for the phase difference (see Problem 14.3)

$$\dot{\theta} = a - b \sin\theta, \tag{14.3.1}$$

where $a = \nu_1 - \nu_2$ denotes the difference between the eigenfrequencies ν_j of the two cavities and b is proportional to the gain coefficient.

We choose a rotating coordinate system in which the fields $\mathscr{E}_1$ and $\mathscr{E}_2$ are slowly varying having a relative phase angle $\theta_0 = \sin^{-1}(a/b)$ given by Eq. (14.3.1) such that $\mathscr{E}_1 = \sqrt{\langle n \rangle} \exp(-i\theta_0/2)$ and $\mathscr{E}_2 = \sqrt{\langle n \rangle} \exp(i\theta_0/2)$. Here we have assumed a symmetric configuration such that $r_1 = r_2 = \sqrt{\langle n \rangle}$ and $\langle n \rangle$ denotes the mean number of photons in steady state.

Due to spontaneous emission, the electric fields $\mathscr{E}_j$ $(j=1,2)$ fluctuate,

$$\delta\dot{\mathscr{E}}_j = F_j(t) \tag{14.3.2}$$

with Gaussian noise sources F_j such that

$$\langle F_j \rangle = \langle F_j(t) F_k(t') \rangle = 0 \qquad (j,k = 1,2), \tag{14.3.3}$$

$$\langle F_j^*(t) F_k(t') \rangle = 2D_{jk}\delta(t - t') \qquad (j,k = 1,2). \tag{14.3.4}$$

Note that the spontaneous emission events from two coherently excited states are strongly correlated as demonstrated in the Hanle effect laser. As a result the cross-correlation diffusion coefficient D_{12} can be made nonvanishing.

We now consider the effect of fluctuating forces F_1 and F_2 on the relative phase difference θ shown in Fig. 14.4. The phase shift $\delta\theta_1$ is caused by a spontaneous emission event $\delta\mathscr{E}_1$, and is given for $|\delta\mathscr{E}_1| \ll \sqrt{\langle n \rangle}$ by

$$\begin{aligned}
\delta\theta_1 &\cong \frac{|\delta\mathscr{E}_1|}{\sqrt{\langle n \rangle}} \sin\left(\delta\phi_1 + \frac{\theta}{2}\right) \\
&= \frac{1}{\sqrt{\langle n \rangle}} [|\delta\mathscr{E}_1| \sin(\delta\phi_1) \cos(\theta/2) + |\delta\mathscr{E}_1| \cos(\delta\phi_1) \sin(\theta/2)] \\
&= \frac{1}{\sqrt{\langle n \rangle}} [\mathrm{Im}(\delta\mathscr{E}_1) \cos(\theta/2) + \mathrm{Re}(\delta\mathscr{E}_1) \sin(\theta/2)].
\end{aligned} \tag{14.3.5}$$

* For further reading see Schleich and Scully [1988].

Similarly we arrive at

$$\begin{aligned}\delta\theta_2 &\cong \frac{|\delta\mathscr{E}_2|}{\sqrt{\langle n\rangle}} \sin\left(\delta\phi_2 - \frac{\theta}{2}\right) \\ &= \frac{1}{\sqrt{\langle n\rangle}}[|\delta\mathscr{E}_2|\sin(\delta\phi_2)\cos(\theta/2) - |\delta\mathscr{E}_2|\cos(\delta\phi_2)\sin(\theta/2)] \\ &= \frac{1}{\sqrt{\langle n\rangle}}[\mathrm{Im}(\delta\mathscr{E}_2)\cos(\theta/2) - \mathrm{Re}(\delta\mathscr{E}_2)\sin(\theta/2)]. \qquad (14.3.6)\end{aligned}$$

Hence the total fluctuation $\delta\theta = \delta\theta_1 - \delta\theta_2$ in the phase difference θ is

$$\delta\theta = \frac{1}{\sqrt{\langle n\rangle}}[\cos(\theta/2)\mathrm{Im}(\delta\mathscr{E}_1 - \delta\mathscr{E}_2) + \sin(\theta/2)\mathrm{Re}(\delta\mathscr{E}_1 + \delta\mathscr{E}_2)] \qquad (14.3.7)$$

and in view of Eq. (14.3.2),

$$\delta\theta = \frac{1}{\sqrt{\langle n\rangle}}[\cos(\theta/2)\mathrm{Im}(F_1\delta t - F_2\delta t) + \sin(\theta/2)\mathrm{Re}(F_1\delta t + F_2\delta t)]. \qquad (14.3.8)$$

Defining $\delta\theta/\delta t \equiv (\partial\theta/\partial t)|_{\text{fluct}}$ we thus find

$$\left.\frac{\partial\theta}{\partial t}\right|_{\text{fluct}} = \frac{1}{\sqrt{\langle n\rangle}}[\cos(\theta/2)\mathrm{Im}(F_1 - F_2) + \sin(\theta/2)\mathrm{Re}(F_1 + F_2)]. \qquad (14.3.9)$$

Adding this to the deterministic equation (14.3.1) we arrive at the geometrically motivated equation of motion

$$\dot{\theta} = a - b\sin\theta + \cos(\theta/2)F_- + \sin(\theta/2)F_+. \qquad (14.3.10)$$

In the last step we have introduced the Gaussian Langevin forces $F_- = \mathrm{Im}(F_1 - F_2)/\sqrt{\langle n\rangle}$ and $F_+ = \mathrm{Re}(F_1 + F_2)/\sqrt{\langle n\rangle}$ which according to Eqs. (14.3.3) and (14.3.4) have the properties

$$\langle F_-\rangle = \langle F_+\rangle = 0, \qquad (14.3.11a)$$

and

$$\langle F_-(t)F_-(t')\rangle = \frac{1}{\langle n\rangle}[D_{11} + D_{22} - 2\mathrm{Re}(D_{12})]\delta(t - t'), \qquad (14.3.11b)$$

$$\langle F_+(t)F_+(t')\rangle = \frac{1}{\langle n\rangle}[D_{11} + D_{22} + 2\mathrm{Re}(D_{12})]\delta(t - t'), \qquad (14.3.11c)$$

$$\langle F_+(t)F_-(t')\rangle = -\frac{2}{\langle n\rangle}\mathrm{Im}(D_{12})\delta(t - t'). \qquad (14.3.11d)$$

Here we have used the fact that $D_{12} = D_{21}^*$ (which follows from Eq. (14.3.4)). According to Eqs. (14.3.11b) and (14.3.11c) a correlation of

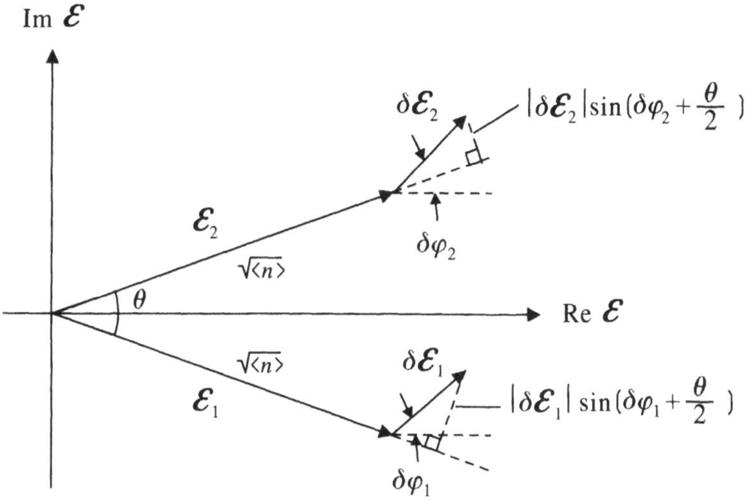

Fig. 14.4
Phase diagram of spontaneous emission in a phase-locked laser. The electric fields $\mathscr{E}_1$ and $\mathscr{E}_2$ are locked to a phase angle θ. A fluctuation $\delta\mathscr{E}_j$ $(j = 1, 2)$ causes a phase change $\delta\theta_j$ given by Eqs. (14.3.5) and (14.3.6).

the noise sources F_1 and F_2, that is, $D_{12} \neq 0$, leads to a reduction of noise strength of F_- and to a corresponding increase in F_+. Thus the Langevin forces F_- and F_+ are correlated as expressed by Eq. (14.3.11d). Depending on the amount of correlation between F_1 and F_2 either the cosine or the sine contribution in the equation of motion for θ (Eq. (14.3.10)) gains more weight.

In the absence of correlation, that is, $D_{12} = 0$, the two noise forces F_- and F_+ have equal weight and Eq. (14.3.7) can be simplified by introducing the Gaussian noise $F = \cos(\theta/2)F_- + \sin(\theta/2)F_+$ which according to Eqs. (14.3.11a)–(14.3.11d) has the properties

$$\langle F\rangle = 0, \tag{14.3.12}$$

$$\langle F(t)F(t')\rangle = 2\mathscr{D}\delta(t-t'), \tag{14.3.13}$$

where $\mathscr{D} = D/\langle n\rangle$. Here and in the remainder of the section we have set $D_{11} = D_{22} \equiv D$. The equation of motion for θ (Eq. (14.3.10)) thus reduces to an equation for the so-called phase-locked laser (PLL)

$$\dot{\theta} = a - b\sin\theta + F(t). \tag{14.3.14}$$

We now turn to the case of maximum correlation, that is, $D_{11} + D_{22} = 2\mathrm{Re}\, D_{12}$, and thus $\langle F_-(t)F_-(t')\rangle = 0$. For the sake of simplicity we assume $\mathrm{Im}\, D_{12} = 0$ and therefore $\langle F_-F_+\rangle = 0$. Since $\langle F_-\rangle = 0$ and F_- is Gaussian all higher correlation functions are zero as well; therefore, $F_- = 0$. As a result, Eq. (14.3.10) simplifies to the equation of motion for the phase difference in the correlated spontaneous emission laser (CEL)

$$\dot{\theta} = a - b\sin\theta + \sin(\theta/2)F_{+}(t), \tag{14.3.15}$$

where according to Eq. (14.3.11c)

$$\langle F_{+}(t)F_{+}(t')\rangle = 2(2\mathscr{D})\delta(t - t'). \tag{14.3.16}$$

Comparing Eq. (14.3.16) with Eq. (14.3.13) we note that due to the noise correlation, the noise strength is twice that associated with F. Moreover, we emphasize that the two equations of motion for θ, Eqs. (14.3.14) and (14.3.15), are distinctly different. In Eq. (14.3.15) the noise F_{+} is multiplied by a nonlinear function of the stochastic variable θ. In particular we note that $\langle(\delta\theta)^2\rangle = 4(D/\langle n\rangle)\sin^2(\theta/2)\delta t = 2(D/\langle n\rangle)(1 - \cos\theta)\delta t$ which is the desired result.

14.4 Holographic laser

In previous sections, we discussed Hanle effect and quantum beat lasers in which three-level atoms sustain the two laser modes which correspond to transitions from two coherently prepared upper levels to a common ground level. Here we discuss a *holographic laser*, in which the active medium consists of two-level atoms in a ring cavity. The CEL operation can be achieved in the two oppositely directed running waves via a spatial modulation of the active medium. The noise due to spontaneous emission is then suppressed just as in the *three-level lasers* discussed earlier.

The motivation for the present CEL device derives from the realm of coherent Fourier optics and holography. In particular, we recall that in the process of preparing a hologram, one radiates a film with two beams of light as indicated in Fig. 14.5(a). These two beams of light (the reference beam and the incident beam) interfere to produce a holographic grading or modulation in the film. We then read out the information stored in this film by probing with the original light beam which is now scattered from the striated layers of developed film to produce our new signal (Fig. 14.5(b)). In this way we note that the read beam scatters from the striated medium to produce the new signal of interest.

In a similar way we anticipate that a striated gain medium will produce a strong coupling between the two counter-propagating modes of the ring laser. This correlation will be such that the two modes are strongly correlated and this correlation is anticipated to carry over

Fig. 14.5
(a) To create a hologram, an object beam and a reference (write) beam interfere. The interference pattern is recorded on the file. (b) After development the reference (now read) beam is scattered from the atomic layers in the hologram. From the scattered light a virtual image of the object is obtained.

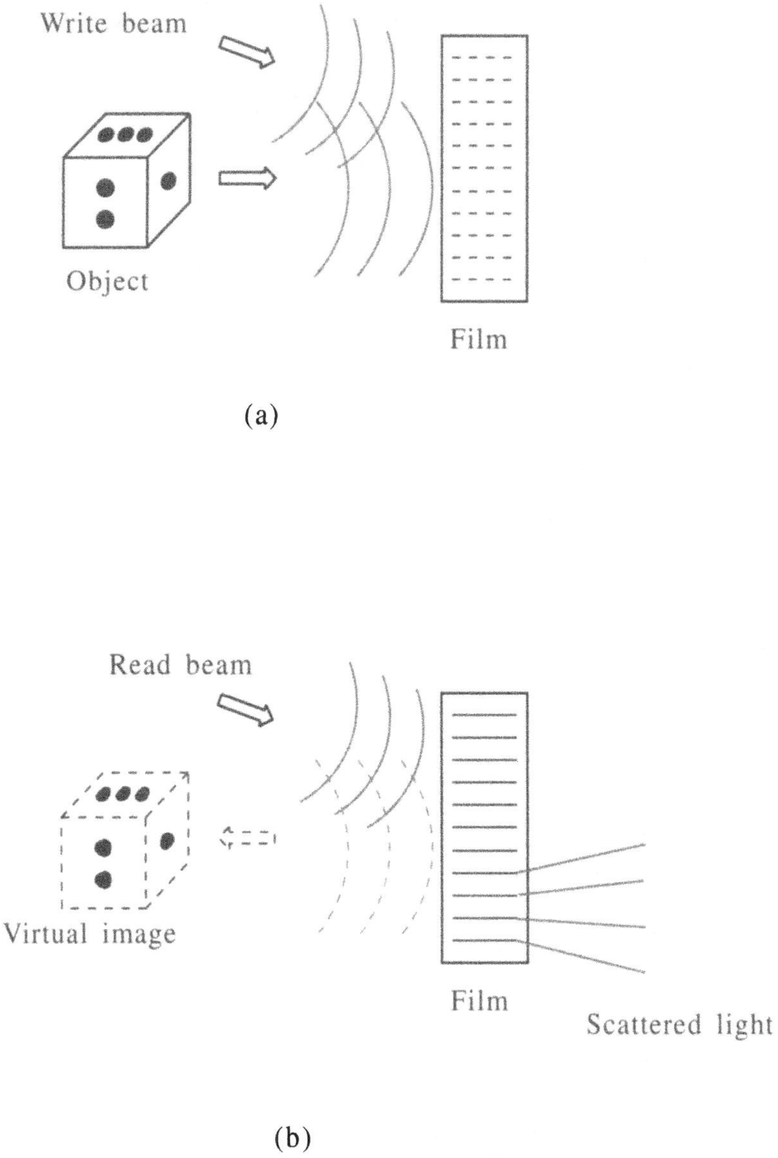

into the quantum character of the fields as well. This type of ring laser is therefore called a holographic laser (HL).

In the HL the active medium in the ring cavity consists of thin layers* with a constant spacing (see Fig. 14.6). The coherent coupling of the two counter-propagating modes in the ring cavity occurs by backscattering. When light of one mode is backscattered from a layer of the gain medium, constructive interference is achieved when the phases of the reflected part of the beam and the counter-propagating beam match. The two fields in the ring cavity can be described by their wave vectors k_1, k_2, their frequencies ν_1, ν_2, and their amplitude $\mathscr{E}_0$ which we assume to be equal for both beams,

$$E_1 = \mathscr{E}_0 \exp[i(k_1 z - \nu_1 t)], \quad E_2 = \mathscr{E}_0 \exp[-i(k_2 z + \nu_2 t)]. \quad (14.4.1)$$

At the reflection at time t_0 the phases of both beams have to be equal, up to an integer multiple of 2π,

$$k_1 z_0 - \nu_1 t_0 - 2\pi j = -k_2 z_0 - \nu_2 t_0 \quad (j = 0, \pm 1, \pm 2 \ldots), \quad (14.4.2)$$

where z_0 is the coordinate of the reflecting layer. From this we get

$$(k_1 + k_2) z_0 - (\nu_1 - \nu_2) t_0 = 2\pi j. \quad (14.4.3)$$

Since the frequencies lock such that $\nu_1 = \nu_2$ the condition for z_0 is

$$z_0 = 2\pi j/(k_1 + k_2) = \pi j/k, \quad (14.4.4)$$

since $k_1 \simeq k_2$.

We conclude from this heuristic derivation that the layers of the gain medium have to be located at $z = (\pi/k)j$ in order to get maximum coupling between the beams. In fact, the same result is obtained from our detailed analysis.

As in the preceding linear theories of Hanle effect and quantum beat lasers, we describe both modes by annihilation and creation operators $a_1, a_1^\dagger$ and $a_2, a_2^\dagger$. Both modes interact with the same two-level atoms in the gain medium. Here we allow the 'bare-cavity' eigenfrequencies ν_{c1} and ν_{c2} to be different from the operating frequencies ν_1 and ν_2 of the laser fields, since as we see from (4.1.19) this is necessary for a laser gyroscope. The interaction between the beams with the laser medium is described by the Hamiltonian

$$\begin{aligned} \mathscr{H} &= \hbar\nu_{c1} a_1^\dagger a_1 + \hbar\nu_{c2} a_2^\dagger a_2 + \hbar\omega_a |a\rangle\langle a| + \hbar\omega_b |b\rangle\langle b| \\ &+ \hbar g \sum_{j=1,2} [U_j(z)|a\rangle\langle b| a_j + U_j^*(z) a_j^\dagger |b\rangle\langle a|], \end{aligned} \quad (14.4.5)$$

where $|a\rangle$ and $|b\rangle$ are the upper and lower atomic states with energy

* Scully [1987], Raja, Brueck, Scully, and Lee [1991].

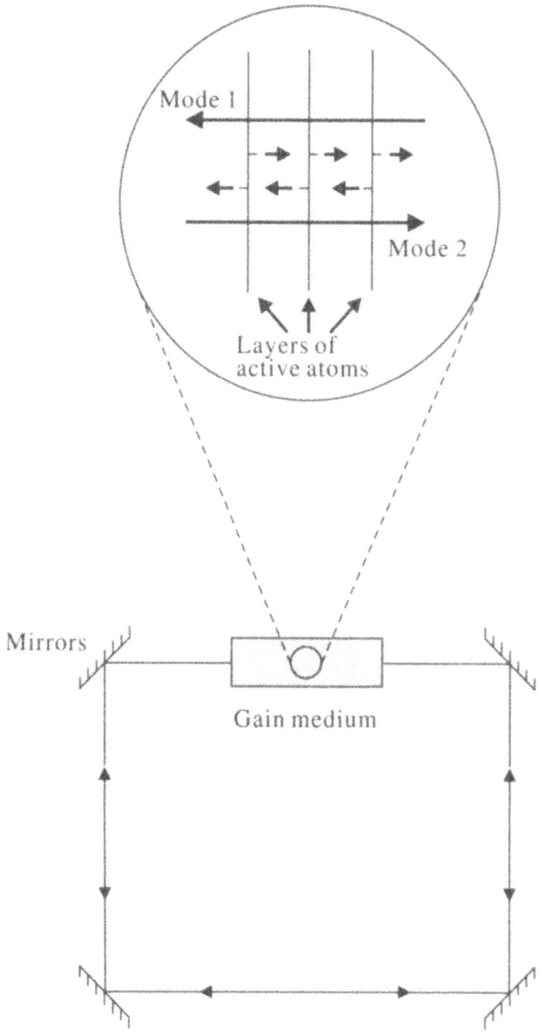

Fig. 14.6
Holographic laser. A stratified medium is inside a ring cavity. The two travelling modes are partially reflected in the various atomic layers. The reflected light can interfere constructively with the counter-propagating mode, quenching the phase noise. (From J. Krause and M. O. Scully, *Phys. Rev. A* **36**, 1771 (1987).)

separation $\omega = \omega_a - \omega_b$ and $U_j(z)$ $(j = 1, 2)$ are normal mode functions. In the interaction picture, this reduces to

$$\mathscr{V} = \hbar(\nu_{c1} - \nu_1)a_1^\dagger a_1 + \hbar(\nu_{c2} - \nu_2)a_2^\dagger a_2 + \hbar g[|a\rangle\langle b|A(z,t) + A^\dagger(z,t)|b\rangle\langle a|], \tag{14.4.6}$$

where the combined operator $A(z, t)$ is given by

$$A(z,t) = a_1 \exp(i\Delta_1 t)U_1(z) + a_2 \exp(i\Delta_2 t)U_2(z), \tag{14.4.7}$$

with $\Delta_1 = \omega - \nu_1$ and $\Delta_2 = \omega - \nu_2$. The coordinate z is defined parallel to the beams so that the gain medium extends from $z = -\ell/2$ to $z = \ell/2$, and we assume that the extension of the medium perpendicular to the beam direction is independent of z.

Following the method outlined in Section 14.2 for the Hanle effect laser, we can derive the following equation of motion for the reduced density operator for the field:

$$\dot{\rho} = -i(\nu_{c1} - \nu_1)[a_1^\dagger a_1, \rho] - i(\nu_{c2} - \nu_2)[a_2^\dagger a_2, \rho]$$
$$-\frac{\mathscr{A}}{2}\int_{-\ell/2}^{\ell/2} dz n(z)[A(z)A^\dagger(z)\rho - 2A^\dagger(z)\rho A(z) + \rho A(z)A^\dagger(z)], \tag{14.4.8}$$

where $\mathscr{A} = 2r_a g^2/\gamma^2$ is the usual gain parameter, and the summation over all atoms inside the cavity has been carried out by a spatial integration over the density $n(z)$ of the gain atoms. For the sake of simplicity, we assume $\nu_1 = \nu_2 = \nu c_1 = \nu c_2 = \omega$ in the following.

For travelling waves the normal mode functions $U_1(z)$ and $U_2(z)$ are given by

$$U_1(z) = e^{ikz}, \tag{14.4.9a}$$
$$U_2(z) = e^{-ikz}. \tag{14.4.9b}$$

Then Eq. (14.4.8) reduces to the form (14.2.22) with $\phi = 0$. The coefficients α_{ij} for the holographic laser are given by

$$\alpha_{jj} = \mathscr{A}\int_{-\ell/2}^{\ell/2} n(z)|U_j(z)|^2 dz$$
$$= \mathscr{A}\int_{-\ell/2}^{\ell/2} n(z)dz \qquad (j = 1, 2), \tag{14.4.10}$$

$$\alpha_{12} = \alpha_{21}^*$$
$$= \mathscr{A}\int_{-\ell/2}^{\ell/2} n(z)U_1(z)U_2^*(z)dz$$
$$= \mathscr{A}\int_{-\ell/2}^{\ell/2} n(z)e^{2ikz}dz. \tag{14.4.11}$$

As shown for the Hanle effect laser, the diffusion coefficient $D_{\theta\theta}$ for the relative phase $\theta_1 - \theta_2$ is given by (Eq. (14.2.36))

$$2D_{\theta\theta} = \frac{1}{2\langle n\rangle}(\alpha_{11} + \alpha_{22} - \alpha_{12}e^{-i\theta} - \alpha_{21}e^{i\theta}), \tag{14.4.12}$$

where we assume $\langle n_1\rangle = \langle n_2\rangle = \langle n\rangle$.

If the gain medium is not spatially modulated, i.e., $n(z) = n_0$, α_{12} and α_{21} are of the order α_{11}/k which is very small at optical frequencies as compared to α_{11} and can be neglected in Eq. (14.4.12). In this case we no longer obtain noise quenching in the relative angle, and from Eq. (14.4.12) we have

$$2D_{\theta\theta} = \frac{\alpha_{11}}{\langle n \rangle}. \tag{14.4.13}$$

From Eqs. (14.4.10) and (14.4.11) we conclude that a modulation for which the gain coefficients are equal must have the density of atoms mainly at points where $e^{2ikz} \simeq 1$, at $z = \pi j/k$ where j is an integer. In principle we would like to have δ-functions there but in reality there will be peaks in the density with a width Δ. When we assume that these peaks are of Gaussian* type, the density function $n(z)$ between $z = -\ell/2$ and $\ell/2$ looks as follows:

$$n(z) = \sum_{j=-j_0}^{j_0} \frac{n_0}{2j_0+1} \left(\frac{\ln 2}{\pi}\right)^{1/2} \frac{1}{\Delta} \exp\left[-\frac{\ln 2}{\Delta^2}\left(z - \frac{\pi}{k}j\right)^2\right], \tag{14.4.14}$$

with $j_0 = \ell k/2\pi$ and $n_0 = $ constant.

With this equation we obtain for the integrals of Eqs. (14.4.10) and (14.4.11),

$$\int_{-\ell/2}^{\ell/2} n(z)dz = n_0, \tag{14.4.15a}$$

$$\int_{-\ell/2}^{\ell/2} n(z)e^{2ikz}dz = n_0 \exp\left(-\frac{\Delta^2 k^2}{\ln 2}\right). \tag{14.4.15b}$$

We see that all integrals are approximately equal if the width Δ is much smaller than $1/k$. This result (Eqs. (14.4.15)) is still valid when only every nth peak in Eq. (14.4.14) is nonzero (n is a positive integer). From Eqs. (14.4.10) and (14.4.11) we see immediately that all coefficients for gain are equal ($\alpha_{11} = \alpha_{22} = \alpha_{12} = \alpha_{21}$), which is required to obtain a vanishing diffusion coefficient $D(\theta)$.

14.5 Quantum phase and amplitude fluctuations

As seen in Chapter 2, squeezed quantum fluctuations of the radiation field are associated with the decomposition of the electric field amplitude into its '$\cos\omega t$' and '$\sin\omega t$' phases. This suggests that the electric field annihilation operator be written as: $a = a_1 + ia_2$, where a_1 and a_2 are the Hermitian amplitudes of the two quadrature phases. The quantum mechanical properties of these amplitudes imply the uncertainty relation:

$$\Delta a_1 \Delta a_2 \geq \frac{1}{4}. \tag{14.5.1}$$

* See Krause and Scully [1987].

Squeezed states of light are those for which

$$(\Delta a_1)^2 < \frac{1}{4} \text{ or } (\Delta a_2)^2 < \frac{1}{4}. \tag{14.5.2}$$

These are the standard definitions of squeezing.

In laser physics amplitude and phase fluctuations are fundamental quantities of interest. As shown in Chapter 11, far above threshold, the amplitude fluctuations are quite small, and occur around a constant value which, in a semiclassical approximation, is given by $\sqrt{\langle n\rangle}$ where $\langle n\rangle$ is the steady-state number of emitted photons. In the following we will relate the laser phase and amplitude fluctuations to the Hermitian operators a_1 and a_2. We present the arguments using first a simple semiclassical picture of the laser radiation.

From Fig. 14.7 it is clear that the fluctuation $\delta a_{\parallel}$ is associated with pure amplitude fluctuations of a. Simple trigonometry leads to the following relations:

$$\delta a_{\parallel} = |\delta a| \cos(\theta - \theta_0) = \frac{\delta a e^{-i\theta_0} + \delta a^* e^{i\theta_0}}{2}, \tag{14.5.3}$$

$$\delta a_{\perp} = |\delta a| \sin(\theta - \theta_0) = \frac{\delta a e^{-i\theta_0} - \delta a^* e^{i\theta_0}}{2i}, \tag{14.5.4}$$

where $\delta a = |\delta a| e^{i\theta}$. In these formulas θ_0 is the instantaneous phase of the semiclassical electric field amplitude and θ is the phase of the fluctuating displacement δa. The terms $\delta a_{\parallel}$ and $\delta a_{\perp}$ are related to the following θ_0-dependent amplitudes:

$$a_1(\theta_0) = \frac{a e^{-i\theta_0} + a^* e^{i\theta_0}}{2}, \tag{14.5.5}$$

$$a_2(\theta_0) = \frac{a e^{-i\theta_0} - a^* e^{i\theta_0}}{2i}. \tag{14.5.6}$$

We recognize in these amplitudes the standard a_1 and a_2 quadratures but rotated by an angle θ_0 towards the direction fixed by the electric amplitude a. For fluctuations leading to a small change of the phase $\delta\theta$ and amplitude δr, we have

$$\delta a_{\parallel} = \delta a_1(\theta_0) = \delta r, \tag{14.5.7}$$

$$\delta a_{\perp} = \delta a_2(\theta_0) = a \tan \delta\theta \sim \sqrt{\langle n\rangle}\delta\theta. \tag{14.5.8}$$

Here in the last step we have replaced a by the semiclassical expression $\sqrt{\langle n\rangle}$ since in the rotated frame its phase is zero, and have approximated the tangent by the arc. These relations allow us to approximately identify amplitude and phase fluctuations with the fluctuations of the phase-dependent quantities $a_1(\theta_0)$ and $a_2(\theta_0)$.

Fig. 14.7
Amplitude $\delta a_{\parallel}$ and phase fluctuation $\delta\theta$ of the complex amplitude a.

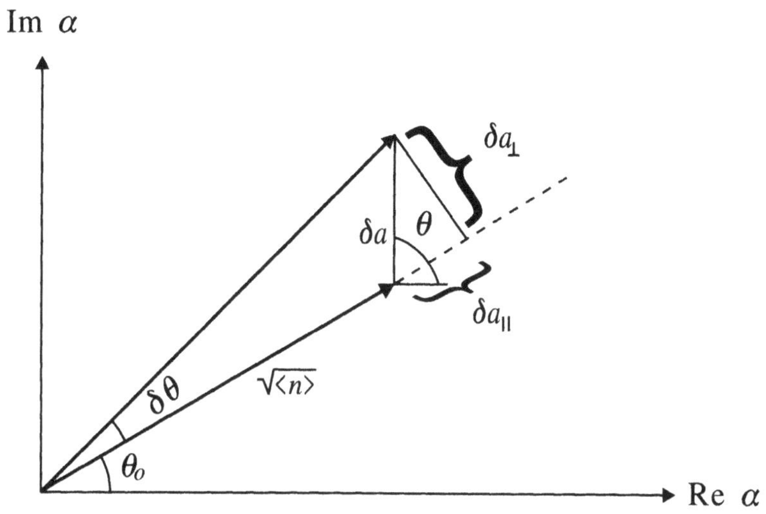

Up to this point our arguments have been semiclassical. We now extend our conclusions to the quantum case. To this end we replace, in Eqs. (14.5.5) and (14.5.6), the classical amplitudes a and a^* by boson annihilation and creation operators a and $a^\dagger$ and associate the phase θ_0 with a rotation of Hermitian observables. Such a unitary transformation leaves the commutation relation intact, i.e., $[a_1(\theta_0), a_2(\theta_0)] = i/2$ and the uncertainty relation (14.5.1) is seen to be invariant under arbitrary rotations by the angle θ_0. This means that by generalizing the semiclassical amplitude and phase fluctuations we obtain the following quantum mechanical expressions:

$$\langle(\delta a_{\parallel})^2\rangle = \langle\Delta a_1^2(\theta_0)\rangle, \tag{14.5.9}$$

$$\langle(\delta\theta)^2\rangle = \frac{1}{\langle n\rangle}\langle\Delta a_2^2(\theta_0)\rangle, \tag{14.5.10}$$

where now the right-hand sides of Eqs. (14.5.9) and (14.5.10) are the quantum mechanical variances of the $a_1(\theta_0)$ and the $a_2(\theta_0)$ operators. From these definitions and the relation (14.5.1) we obtain the following phase–amplitude uncertainty relation:

$$\langle(\delta\theta)^2\rangle\langle(\delta a_{\parallel})^2\rangle \geq \frac{1}{16\langle n\rangle}. \tag{14.5.11}$$

Note that this definition is free from problems associated with attempts to construct a quantum phase operator. It has a clear physical interpretation and relates in a simple way the phase and the amplitude fluctuations to the well behaved quantum mechanical observables a_1 and a_2.

We can rewrite the quantum mechanical variances $\Delta a_i^2(\theta_0)(i = 1, 2)$ in a form that contains only the normally ordered operators (normally ordered variances) and the commutator contribution

$$\Delta a_i^2(\theta_0) =: \Delta a_i^2(\theta_0) : + \frac{1}{4}. \tag{14.5.12}$$

This formulation of quantum mechanical fluctuations will be seen to be extremely useful in the discussion of CEL operation when squeezing is also present. We recall that squeezed states of light are those for which $: \Delta a_1^2(\theta_0) :$ or $: \Delta a_2^2(\theta_0) :$ is less than zero.

Using (14.5.12) we may rewrite Eqs. (14.5.9) and (14.5.10) in the following form:

$$\langle (\delta\theta)^2 \rangle = \frac{1}{4\langle n \rangle} + \langle : (\delta\theta)^2 : \rangle, \tag{14.5.13}$$

$$\langle (\delta a_{\parallel})^2 \rangle = \frac{1}{4} + \langle : (\delta a_{\parallel})^2 : \rangle, \tag{14.5.14}$$

where the symbolic expressions $\langle : (\delta\theta)^2 : \rangle$ and $\langle : (\delta a_{\parallel})^2 : \rangle$ denote the normally ordered variances of $a_1(\theta_0)$ and $a_2(\theta_0)$ as in (14.5.9) and (14.5.10). For squeezed states these 'normally ordered' phase and amplitude fluctuations become negative.

The appearance of normally ordered operators in our equations is useful in that whenever a normally ordered quantum expectation value is involved we naturally use the P-representation.

14.6 Two-photon correlated emission laser

So far we have considered correlated spontaneous emission schemes in which a correlation is established between spontaneous emission events in two different modes of the radiation field so that the relative phase between them does not diffuse. In this section we discuss a different type of correlated spontaneous emission laser in which a cascade transition of three-level atoms (Fig. 14.8) is coupled to only one mode of the radiation field. In this scheme, it is not only possible to quench the spontaneous emission phase diffusion noise below the Schawlow–Townes limit but also to obtain phase-noise squeezing in a laser.

14.6.1 Theory

The scheme involves injection of three-level atoms in a cascade configuration in a laser cavity at a rate r_a with populations $\rho_{ii}^{(0)}$ $(i = a, c)$

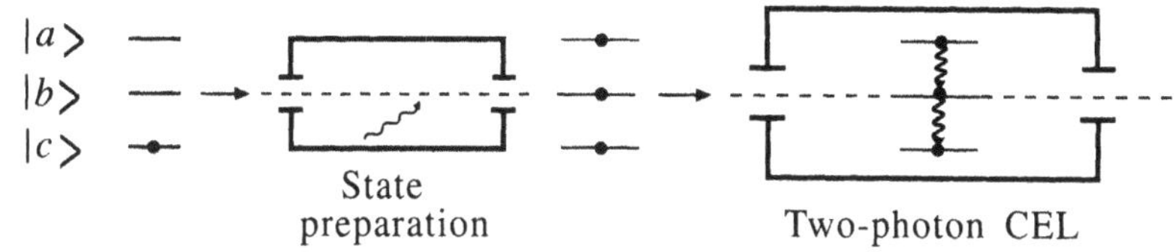

Fig. 14.8
Scheme of the two-photon correlated emission laser. State preparation (first cavity) is separated from the laser operation (second cavity). Atoms in the first cavity are prepared in a coherent superposition by, e.g., passing through a foil or near a knife edge or optical pumping, and are injected into the second cavity where laser operation takes place. The two transitions $|a\rangle \to |b\rangle$ and $|b\rangle \to |c\rangle$ are coupled to the same mode of the radiation field.

and initial coherence $\rho_{ac}^{(0)} = |\rho_{ac}^{(0)}| \exp(i\phi)$. It is assumed that the intermediate level $|b\rangle$ is detuned with respect to the laser frequency ν by an amount Δ, i.e., $\Delta = \omega_{bc} - \nu$ and $\omega_{ac} = 2\nu$. The interaction picture Hamiltonian is given by

$$\mathscr{V} = \hbar(\nu_c - \nu)a^\dagger a + \hbar g(ae^{-i\Delta t}|a\rangle\langle b| + ae^{i\Delta t}|b\rangle\langle c| + \text{H.c.}) \quad (14.6.1)$$

where ν_c is the bare cavity eigenfrequency.

Following the method used in the derivation of the master equation for the reduced density matrix for the field ρ in a Hanle effect laser, we obtain the following equation for the present scheme.

$$\begin{aligned}\dot{\rho} = \Big\{ &-\frac{\mathscr{A}}{2}\mathscr{L}\big[\rho_{aa}^{(0)}(\rho a a^\dagger - a^\dagger \rho a) + \rho_{cc}^{(0)}(\rho a^\dagger a - a\rho a^\dagger) \\ &+ \rho_{ac}^{(0)}(\rho a^\dagger a^\dagger - a^\dagger \rho a^\dagger) + \rho_{ca}^{(0)}(\rho a a - a\rho a)\big] + \text{H.c.}\Big\} \\ &-i(\nu_c - \nu)[a^\dagger a, \rho],\end{aligned} \quad (14.6.2)$$

where $\mathscr{L} = \gamma/(\gamma - i\Delta)$ and $\mathscr{A} = 2r_a g^2/\gamma^2$.

In order to study the phase noise, we convert Eq. (14.6.2) into an equivalent Fokker–Planck equation for the P-representation $P(\alpha, \alpha^*)$. As in the Hanle effect laser we assume that the amplitude fluctuations can be neglected. The corresponding Fokker–Planck equation in the phase variable θ may be obtained with the substitution $\alpha \simeq \sqrt{\langle n\rangle}\exp(i\theta)$, $\langle n\rangle$ being the mean photon number in the field. The resulting equation is (Problem 14.4)

$$\frac{\partial P}{\partial t} = -\frac{\partial}{\partial \theta}(d_\theta P) + \frac{\partial^2}{\partial \theta^2}(D_{\theta\theta}P), \quad (14.6.3)$$

where

$$\begin{aligned} d_\theta = \text{Im}\Big\{ &\frac{-\mathscr{A}}{2}\mathscr{L}^*[\rho_{aa}^{(0)} - \rho_{cc}^{(0)}] - i(\nu_c - \nu) \\ &+ \frac{\mathscr{A}}{2}(\mathscr{L} - \mathscr{L}^*)\rho_{ca}^{(0)}e^{2i\theta}\Big\},\end{aligned} \quad (14.6.4)$$

$$D_{\theta\theta} = \frac{1}{\langle n\rangle}\frac{\mathscr{A}}{8}\{(\mathscr{L} + \mathscr{L}^*)\rho_{aa}^{(0)} + 2\text{Re}[\rho_{ca}^{(0)}\mathscr{L}e^{2i\theta}]\}. \quad (14.6.5)$$

A phase-locking condition can be obtained from the following equation for the phase

$$\frac{d}{dt}\langle\theta\rangle = \langle d_\theta\rangle$$
$$= -\frac{\mathscr{A}\Delta}{2\gamma}|\mathscr{L}|^2[\rho_{aa}^{(0)} - \rho_{cc}^{(0)} - 2|\rho_{ca}^{(0)}|\langle\cos(\phi - 2\theta)\rangle] + \nu - \nu_c. \tag{14.6.6}$$

From Eq. (14.6.6) we get both a frequency pulling equation

$$\nu = \nu_c + \frac{\mathscr{A}\Delta}{2\gamma}|\mathscr{L}|^2[\rho_{aa}^{(0)} - \rho_{cc}^{(0)}], \tag{14.6.7}$$

and a phase-locking equation

$$\frac{d}{dt}\langle\theta\rangle = \frac{\mathscr{A}\Delta}{\gamma}|\mathscr{L}|^2|\rho_{ca}^{(0)}|\langle\cos(\phi - 2\theta)\rangle. \tag{14.6.8}$$

Phase locking occurs for

$$\theta = \theta_0 = \frac{\phi}{2} + \frac{\pi}{4}\mathrm{sgn}(\Delta). \tag{14.6.9}$$

This choice of θ ensures that $d\langle d_\theta\rangle/dt < 0$ for $\theta = \theta_0$.

Under the phase-locking condition (14.6.9) the diffusion coefficient becomes

$$D_{\theta\theta}|_{\theta=\theta_0} \equiv D(\theta_0) = \frac{\mathscr{A}}{4\langle n\rangle}|\mathscr{L}|^2\left[\rho_{aa}^{(0)} - \frac{|\rho_{ca}^{(0)}\Delta|}{\gamma}\right]. \tag{14.6.10}$$

It is clear that the contribution from the atomic coherence is affected by the detuning differently from that of the population. A CEL action is obtained when $\rho_{aa}^{(0)} = |\rho_{ca}^{(0)}\Delta|/\gamma$. For $\rho_{aa}^{(0)} < |\rho_{ca}^{(0)}\Delta|/\gamma$, $D(\theta_0) < 0$ and phase squeezing may be obtained as shown below.

As discussed before, the total phase uncertainty in the steady state is

$$\langle(\delta\theta)^2\rangle = \frac{1}{4\langle n\rangle} + \langle:(\delta\theta)^2:\rangle, \tag{14.6.11}$$

where the first term represents the shot noise due to vacuum fluctuations and the second term is due to spontaneous emission noise. In order to connect the phase-diffusion coefficient to noise in the phase variable of a laser, we obtain the equation of motion for $\langle:(\delta\theta)^2:\rangle$ from the Fokker–Planck equation as

$$\frac{d}{dt}\langle:(\delta\theta)^2:\rangle = 2\langle: d_\theta\delta\theta :\rangle + 2\langle D_{\theta\theta}\rangle, \tag{14.6.12}$$

where $\delta\theta = \theta - \langle\theta\rangle$. In the steady state, the phase locks to the mean value $\theta = \theta_0$ for which $d_{\theta_0} = 0$. This value is stable if $\partial d_{\theta_0}/\partial\theta < 0$. Expanding d_θ around θ_0 up to first order in Eq. (14.6.12), i.e.,

$$d_\theta \simeq d_{\theta_0} + \frac{\partial d_{\theta_0}}{\partial \theta}\delta\theta = \frac{\partial d_{\theta_0}}{\partial \theta_0}\delta\theta, \tag{14.6.13}$$

we obtain

$$\langle : (\delta\theta)^2 : \rangle = \langle D_{\theta\theta} \rangle \left| \frac{\partial d_{\theta_0}}{\partial \theta} \right|^{-1}. \tag{14.6.14}$$

It follows, on substituting this value in Eq. (14.6.11), that the steady-state value of the phase variance is

$$\langle (\delta\theta)^2 \rangle = \frac{1}{4\langle n \rangle} + \langle D_{\theta\theta} \rangle \left| \frac{\partial d_{\theta_0}}{\partial \theta} \right|^{-1}. \tag{14.6.15}$$

Under stable phase locking, the sign of $\langle D_{\theta\theta} \rangle$ therefore decides whether the additional noise due to spontaneous emission adds to or subtracts from the vacuum noise. For $\langle D_{\theta\theta} \rangle < 0$ the phase fluctuations are *squeezed* below the vacuum level. On substituting for $D(\theta_0)$ and $\langle d_\theta \rangle$ from Eqs. (14.6.10) and (14.6.4) respectively, into Eq. (14.6.15), we obtain the total phase noise in steady state

$$\langle (\delta\theta)^2 \rangle = \frac{1}{8\langle n \rangle} \left[1 + \frac{\rho_{aa}^{(0)}\gamma}{|\rho_{ac}^{(0)}\Delta|} \right]. \tag{14.6.16}$$

Thus, for $\rho_{aa}^{(0)}\gamma < |\rho_{ac}^{(0)}|\Delta$, phase squeezing is obtained in the laser.

In order to show that squeezing is compatible with net gain in the laser, we note that the linear gain G of the two-photon CEL, defined by

$$\frac{d\langle n \rangle}{dt} = \langle Gn \rangle + \mathscr{A}|\mathscr{L}|^2 \rho_{aa}^{(0)}, \tag{14.6.17}$$

can be determined from Eq. (14.6.2) and we obtain

$$G = \mathscr{A}|\mathscr{L}|^2 \left\{ [\rho_{aa}^{(0)} - \rho_{cc}^{(0)}] - \frac{2\Delta}{\gamma}|\rho_{ac}^{(0)}| \sin(\phi - 2\theta) \right\}. \tag{14.6.18}$$

The gain is composed of two types of terms: $\mathscr{A}|\mathscr{L}|^2[\rho_{aa}^{(0)} - \rho_{cc}^{(0)}]$ is the usual laser gain and the rest is an extra correlated emission gain which has a phase dependence. Under the phase-locking condition, G becomes

$$G = \mathscr{A}|\mathscr{L}|^2 \left\{ [\rho_{aa}^{(0)} - \rho_{cc}^{(0)}] + \frac{2|\rho_{ac}^{(0)}\Delta|}{\gamma} \right\}. \tag{14.6.19}$$

This expression can be optimized with respect to $|\Delta|/\gamma$ so that the maximum gain is obtained for

$$\frac{|\Delta|}{\gamma} = \frac{\rho_{cc}^{(0)}}{|\rho_{ac}^{(0)}|}. \tag{14.6.20}$$

In deriving this condition, we recall that $|\mathscr{L}|^2$ depends on $|\Delta|/\gamma$ and we choose $|\rho_{ac}^{(0)}|^2 = \rho_{aa}^{(0)}\rho_{cc}^{(0)}$ with $\rho_{aa}^{(0)} + \rho_{cc}^{(0)} = 1$. When this condition is satisfied

$$G = \mathscr{A}\rho_{aa}^{(0)}, \tag{14.6.21}$$

and

$$\langle(\delta\theta)^2\rangle = \frac{1}{8\langle n\rangle}\left[1 + \frac{\rho_{aa}^{(0)}}{1-\rho_{aa}^{(0)}}\right]. \tag{14.6.22}$$

These expressions for gain and phase noise are remarkably simple. The gain depends only on the population of the upper level and it does not depend on the initial population density of the lower level $|c\rangle$. It is, therefore, possible to have net stimulated emission gain and, hence, lasing even in the noninversion regime. This result has its physical origin in the quantum interference which is brought about by the initial preparation of the atomic system in a coherent superposition of levels. This interference eliminates the absorption of radiation by the atoms while still allowing emission. A look at the expression for the phase noise (14.6.22) indicates that phase squeezing takes place for $0 < \rho_{aa}^{(0)} < 0.5$ with a maximum 50 percent squeezing taking place for $\rho_{aa}^{(0)} \simeq 0$. The squeezing is, therefore compatible with a net gain, leading to a bright source of squeezed light.

14.6.2 Heuristic account of a two-photon CEL

As described above, the two-photon CEL consists of coherently prepared three-level atoms which are coupled to one mode of the cavity field. For example, the atoms may be injected into the cavity after being prepared in a coherent superposition of atomic levels (Fig. 14.8). This atomic coherence leads to 'correlated emission' on the successive transitions $|a\rangle \to |b\rangle$ and $|b\rangle \to |c\rangle$. When summing the contributions of the $|a\rangle \to |b\rangle$ and $|b\rangle \to |c\rangle$ transitions, we find that the phase of the total radiated electric field is independent of the phase of level $|b\rangle$ and the (spontaneously) emitted field is solely determined by the coherence between upper and lower levels. In a spontaneous emission event, the atom undergoes a spontaneous transition from, e.g., the upper level $|a\rangle$ to $|b\rangle$. Because of the randomness of such a transition, the acquired phase of the level $|b\rangle$ is not determined by the atomic coherence but is totally arbitrary. However, in the subsequent transition of the atom from level $|b\rangle$ to level $|c\rangle$, the atom 'remembers' the arbitrary phase of level $|b\rangle$ and, in a CEL, total phase coherence in the two-photon

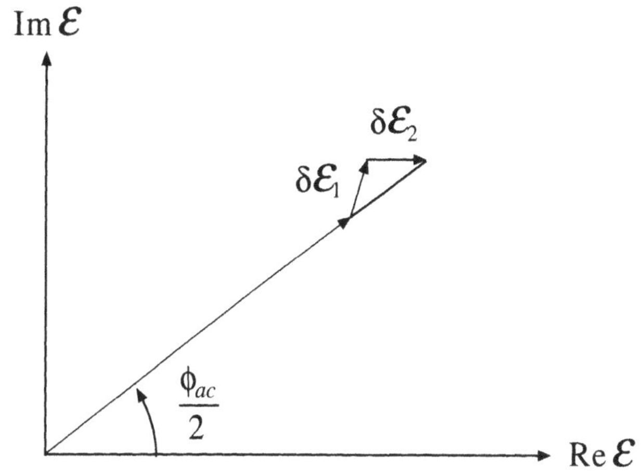

Fig. 14.9
In the two-photon CEL, if we represent the electric field contributions of two correlated spontaneous emission events by $\delta\mathscr{E}_1$ and $\delta\mathscr{E}_2$ in the complex $\mathscr{E}$-plane, then the phase of the final electric field remains unchanged. Thus no phase noise is added.

transition is preserved. In other words, the noise which is created by a spontaneous transition from level $|a\rangle$ to level $|b\rangle$ is compensated by a subsequent transition from level $|b\rangle$ to $|c\rangle$ in such a way that the combined electric field always has the same phase (see Fig. 14.9). This phase is completely determined by the atomic coherence between the upper and lower levels. Therefore the 'noise' created by spontaneous emission events in the CEL is quenched, and we understand the origin of phase-noise reduction in a two-photon correlated emission laser.

In an attempt to make these considerations more concrete we next explain, using simple arguments, how atomic coherence can lead to a significant reduction of phase noise in the two-photon correlated emission laser and derive equations which support our pictorial, physically intuitive arguments.

We start with a three-level atom as in Fig. 14.8, in which the atomic transitions from level $|a\rangle$ to $|b\rangle$ and from $|b\rangle$ to $|c\rangle$ are resonantly coupled to one mode of the radiation field. We assume that we have created a coherence between the three levels prior to the interaction with the radiation field such that the atomic state vector can be written as

$$|\psi(t)\rangle = c_a e^{i\phi_a} e^{-i\omega_a t}|a\rangle + c_b e^{i\phi_b} e^{-i\omega_b t}|b\rangle + c_c e^{i\phi_c} e^{-i\omega_c t}|c\rangle \quad (14.6.23)$$

with c_a, c_b, and c_c being real numbers. The polarization for the transition between levels $|a\rangle$ and $|b\rangle$ is then given by (see Eq. (5.4.6))

$$\begin{aligned} P_{ab}(t) &= \wp|\rho_{ab}|e^{-i(\omega t - \phi_{ab})} + \text{c.c.} \\ &= 2\wp|\rho_{ab}|\cos(\omega t - \phi_{ab}) \end{aligned} \quad (14.6.24)$$

with

$$\wp = er_{ba}. \tag{14.6.25}$$

In the last step of Eq. (14.6.24) we have assumed the matrix element r_{ba} to be real. Furthermore, we have used the notations $\phi_{ab} = \phi_a - \phi_b$ and $\omega = \omega_a - \omega_b$. Analogously we find for the transition between the levels $|b\rangle$ and $|c\rangle$

$$P_{bc}(t) = 2\wp|\rho_{bc}|\cos(\omega t - \phi_{bc}). \tag{14.6.26}$$

For simplicity we analyze the resonant case in which the energy differences between the lasing levels $|a\rangle \to |b\rangle$ and $|b\rangle \to |c\rangle$ are equal to the energy of a photon in the radiation field. Additionally, we have set $r_{ba} = r_{cb}$. The total polarization per atom, is then given by the sum of $P_{ab}(t)$ and $P_{bc}(t)$:

$$\begin{aligned} P(t) &= 2\wp[|\rho_{ab}|\cos(\omega t - \phi_{ab}) + |\rho_{bc}|\cos(\omega t - \phi_{bc})] \\ &= 2\wp\ell\cos(\omega t - \phi_0) \end{aligned} \tag{14.6.27}$$

with

$$\ell = \sqrt{|\rho_{ab}|^2 + |\rho_{bc}|^2 + 2|\rho_{ab}||\rho_{bc}|\cos(\phi_{ab} - \phi_{bc})}. \tag{14.6.28}$$

The phase ϕ_0 of the polarization is given by

$$\cos\phi_0 = (|\rho_{ab}|\cos\phi_{ab} + |\rho_{bc}|\cos\phi_{bc})/\ell, \tag{14.6.29}$$

$$\sin\phi_0 = (|\rho_{ab}|\sin\phi_{ab} + |\rho_{bc}|\sin\phi_{bc})/\ell. \tag{14.6.30}$$

In the special case that $|\rho_{ab}|$ is equal to $|\rho_{bc}|$, Eqs. (14.6.28) and (14.6.30) can be further simplified:

$$\ell = 2|\rho_{ab}|\cos[(\phi_{ab} - \phi_{bc})/2]. \tag{14.6.31}$$

Using the trigonometric identities $\cos\alpha + \cos\beta = 2\cos[(\alpha+\beta)/2]\cos[(\alpha-\beta)/2]$ and $\sin\alpha + \sin\beta = 2\sin[(\alpha+\beta)/2]\cos[(\alpha-\beta)/2]$, we find

$$\begin{aligned} \phi_0 &= \frac{1}{2}(\phi_{ab} + \phi_{bc}) \\ &= \frac{1}{2}\phi_{ac}. \end{aligned} \tag{14.6.32}$$

We thus see that the phase of the atomic polarization is completely independent of the phase ϕ_b. This result will become important when we discuss the influence of spontaneous emission events on the system. Since the emitted electromagnetic field is proportional to the atomic polarization, we can conclude that in a two-photon CEL the electric field vector is governed by the phase ϕ_0 given by Eq. (14.6.30). This is depicted in Fig. 14.9.

We now turn to the analysis of spontaneous emission noise. Although a purely semiclassical theory does not account for spontaneous emission, we will model such an event as follows. When an excited atom makes a spontaneous transition from the upper level $|a\rangle$ to the middle level $|b\rangle$ it develops a polarization

$$P_{ab}(t) = \delta P \cos(\omega t - \xi_{ab}). \tag{14.6.33}$$

Here δP is the atomic polarization per photon and ξ_{ab} is given by $\xi_{ab} = \phi_a - \xi_b$. Note that in contrast to an induced transition the phase ξ_b is not given by a specific atomic phase ϕ_b but is totally arbitrary. This models the randomness of spontaneous emission events. A consecutive spontaneous emission event by the atom from level $|b\rangle$ to $|c\rangle$ has the analogous atomic polarization

$$P_{bc}(t) = \delta P \cos(\omega t - \xi_{bc}) \tag{14.6.34}$$

with $\xi_{bc} = \xi_b - \phi_c$. The total polarization and hence the total contribution to the electromagnetic field due to spontaneous emission is then obtained by adding (14.6.33) and (14.6.34) and is found to be proportional to

$$P(t) \propto \delta P \cos(\omega t - \phi_{ac}/2). \tag{14.6.35}$$

We see that the random phase ξ_b has canceled and that the combined polarization always has the phase ϕ_{ac}, which is fixed by the atomic coherence. This can be visualized as in Fig. 14.9. Every noise contribution due to a spontaneous emission event from level $|a\rangle \to |b\rangle$ is compensated by a consecutive spontaneous emission event from $|b\rangle \to |c\rangle$ in such a way, that the net phase for the (correlated) two-photon event is ϕ_{ac} for all atoms. Therefore, all two-photon noise contributions will have the same phase, i.e., are noise 'free', when the phase of the two-photon noise is the same as that of the radiation field. This is the case of complete noise quenching in the two-photon CEL.

14.A Spontaneous emission noise in the quantum beat laser

The Hamiltonian for the atom–field system in the quantum beat laser is the same as given for Hanle effect laser in Eqs. (14.2.4), (14.2.5), and (14.2.6). There is, however, an additional term in $\mathscr{H}_1$ arising due to the interaction of the coherent field with the atomic system given by

$$-\frac{\hbar\Omega_R}{2}(e^{-i\phi-i\nu_3 t}|a\rangle\langle b| + e^{i\phi+i\nu_3 t}|b\rangle\langle a|) \tag{14.A.1}$$

where Ω_R is the Rabi frequency and ϕ is the phase of the dipole matrix element. The major difference between the present scheme and that of the Hanle effect laser is that atoms, pumped incoherently in the excited state $|a\rangle$, are driven into a coherent superposition of upper levels by an external field whereas, in the Hanle effect laser, atoms are already in a coherent superposition of states before their interaction with the laser field starts. The Hamiltonian, in the interaction picture, is given by

$$\mathscr{V} = \hbar g(a_1 e^{i\Delta t}|a\rangle\langle c| + a_2 e^{i\Delta t}|b\rangle\langle c|) - \frac{\hbar\Omega_R}{2}e^{-i\phi}|a\rangle\langle b| + \text{H.c.} \tag{14.A.2}$$

where $\Delta = \omega_a - \omega_c - \nu_1 = \omega_b - \omega_c - \nu_2$ is the atomic detuning with respect to the field and, for simplicity, we have assumed $\nu_1 - \nu_2 = \nu_3 = \omega_{ab}$. An equation for the reduced density operator of the field can be obtained, as before, by taking a trace over atoms, i.e.,

$$\dot{\rho} = -\frac{i}{\hbar}([\mathscr{V}_{ac}, \rho_{ca}] + [\mathscr{V}_{bc}, \rho_{cb}] + [\mathscr{V}_{ca}, \rho_{ac}] + [\mathscr{V}_{cb}, \rho_{bc}]). \tag{14.A.3}$$

The matrix elements of the density operator ρ_{ac} and ρ_{bc} can be obtained by solving the following matrix equation

$$\dot{R} = -MR - igAe^{i\Delta t}, \tag{14.A.4}$$

where

$$R = \begin{pmatrix} \rho_{ac} \\ \rho_{bc} \end{pmatrix} \tag{14.A.5}$$

$$M = \begin{pmatrix} \gamma & -\frac{i\Omega_R}{2}e^{-i\phi} \\ -\frac{i\Omega_R}{2}e^{i\phi} & \gamma \end{pmatrix}, \tag{14.A.6}$$

$$A = \begin{pmatrix} a_1\rho_{cc} - \rho_{aa}a_1 - \rho_{ab}a_2 \\ a_2\rho_{cc} - \rho_{bb}a_2 - \rho_{ba}a_1 \end{pmatrix}. \tag{14.A.7}$$

It is clear that here, unlike the Hanle effect laser, the elements ρ_{ac} and ρ_{bc} are coupled even in the zeroth order in the coupling constant g by the external field $\Omega_R \exp(i\phi)$. A solution of Eq. (14.A.4) which is linear in the coupling constant g is given by

$$\begin{aligned} R(t) &= -ig\int_{-\infty}^{t} dt_0 e^{-M(t-t_0)}Ae^{i\Delta t_0} \\ &= -ig(M + i\Delta I)^{-1}Ae^{i\Delta t}, \end{aligned} \tag{14.A.8}$$

where I is the unit matrix. Here

$$(M + i\Delta I)^{-1} = \frac{1}{D}\begin{pmatrix} \gamma + i\Delta & \frac{i\Omega_R}{2}e^{-i\phi} \\ \frac{i\Omega_R}{2}e^{i\phi} & \gamma + i\Delta \end{pmatrix}, \tag{14.A.9}$$

where $D = (\gamma+i\Delta)^2+\Omega_R^2/4$. It then follows, from Eqs. (14.A.5)–(14.A.7), that

$$\rho_{ac}(t) = -\frac{ig}{D}\Big[(\gamma + i\Delta)(a_1\rho_{cc} - \rho_{aa}a_1 - \rho_{ab}a_2) + \frac{i\Omega_R}{2}e^{-i\phi}(a_2\rho_{cc} - \rho_{bb}a_2 - \rho_{ba}a_1)\Big]e^{i\Delta t}, \tag{14.A.10}$$

$$\rho_{bc}(t) = -\frac{ig}{D}\Big[\frac{i\Omega_R}{2}e^{i\phi}(a_1\rho_{cc} - \rho_{aa}a_1 - \rho_{ab}a_2) + (\gamma + i\Delta)(a_2\rho_{cc} - \rho_{bb}a_2 - \rho_{ba}a_1)\Big]e^{i\Delta t}. \tag{14.A.11}$$

In the next step, we determine ρ_{aa}, ρ_{bb}, ρ_{cc}, and ρ_{ab} in the zeroth order in g. Since atoms are initially pumped in the state $|a\rangle$ at a rate r_a, we have $\rho_{cc} = 0$ and ρ_{aa}, ρ_{bb}, and ρ_{ab} are determined by treating the $|a\rangle \to |b\rangle$ transition semiclassically to all orders in Ω_R. The equations of motion for the elements of the density operator are

$$\dot{\widetilde{R}} = -\widetilde{M}\widetilde{R} + \widetilde{B}, \tag{14.A.12}$$

where

$$\widetilde{R} = \begin{pmatrix} \rho_{aa} \\ \rho_{ab} \\ \rho_{ba} \\ \rho_{bb} \end{pmatrix}, \qquad \widetilde{B} = r_a\rho\begin{pmatrix} 1 \\ 0 \\ 0 \\ 0 \end{pmatrix}, \tag{14.A.13}$$

$$\widetilde{M} = \begin{pmatrix} \gamma & \frac{i\Omega_R}{2}e^{i\phi} & -\frac{i\Omega_R}{2}e^{-i\phi} & 0 \\ \frac{i\Omega_R}{2}e^{-i\phi} & \gamma & 0 & -\frac{i\Omega_R}{2}e^{-i\phi} \\ -\frac{i\Omega_R}{2}e^{i\phi} & 0 & \gamma & \frac{i\Omega_R}{2}e^{i\phi} \\ 0 & -\frac{i\Omega_R}{2}e^{i\phi} & \frac{i\Omega_R}{2}e^{-i\phi} & \gamma \end{pmatrix}. \tag{14.A.14}$$

A solution of Eq. (14.A.12) is

$$\widetilde{R}(t) = \int_{-\infty}^{t} dt_0 e^{-\widetilde{M}(t-t_0)}\widetilde{B} = \widetilde{M}^{-1}\widetilde{B}. \tag{14.A.15}$$

We need only to determine $\widetilde{M}_{11}^{-1}, \widetilde{M}_{21}^{-1}, \widetilde{M}_{31}^{-1}$, and $\widetilde{M}_{41}^{-1}$ in order to determine $\widetilde{R}$. On evaluating these elements and then substituting in Eq. (14.A.15) we obtain

$$\rho_{aa} = r_a\widetilde{M}_{11}^{-1}\rho = \frac{r_a}{|\widetilde{M}|}\gamma(\gamma^2 + \Omega_R^2/2)\rho, \tag{14.A.16}$$

$$\rho_{ab} = r_a\widetilde{M}_{21}^{-1}\rho = -\frac{ir_a}{2|\widetilde{M}|}\gamma^2\Omega_R e^{-i\phi}\rho, \tag{14.A.17}$$

$$\rho_{bb} = r_a\widetilde{M}_{41}^{-1}\rho = \frac{r_a}{2|\widetilde{M}|}\gamma\Omega_R^2\rho, \tag{14.A.18}$$

where $|\widetilde{M}| = \gamma^2(\gamma^2 + \Omega_R^2)$.

We can now substitute for ρ_{aa}, ρ_{bb}, and ρ_{ab} from Eqs. (14.A.16)–(14.A.18) in addition to $\rho_{cc} = 0$ in Eqs. (14.A.10) and (14.A.11). The resulting expressions for ρ_{ac} and ρ_{bc} and their complex conjugates can, in turn, be substituted in Eq. (14.A.3). We then obtain the following equation of motion for the reduced density matrix for the field:

$$\begin{aligned}\dot{\rho} = &-\frac{1}{2}\alpha_{11}(\rho a_1 a_1^\dagger - a_1^\dagger \rho a_1) - \frac{1}{2}\alpha_{22}(\rho a_2 a_2^\dagger - a_2^\dagger \rho a_2)\\ &-\frac{1}{2}\alpha_{12}(\rho a_2 a_1^\dagger - a_1^\dagger \rho a_2)e^{i\phi} - \frac{1}{2}\alpha_{21}(\rho a_1 a_2^\dagger - a_2^\dagger \rho a_1)e^{-i\phi}\\ &+\text{H.c.},\end{aligned} \tag{14.A.19}$$

where

$$\begin{aligned}\alpha_{11} = &\frac{g^2 r_a}{2\gamma(\gamma^2+\Omega_R^2)}\left\{\frac{(2\gamma^2+\Omega_R^2+i\Omega_R\gamma)[\gamma - i(\Delta-\Omega_R/2)]}{[\gamma^2+(\Delta-\Omega_R/2)^2]}\right.\\ &\left.+\frac{(2\gamma^2+\Omega_R^2-i\Omega_R\gamma)[\gamma-i(\Delta+\Omega_R/2)]}{[\gamma^2+(\Delta+\Omega_R/2)^2]}\right\},\end{aligned} \tag{14.A.20}$$

$$\begin{aligned}\alpha_{12} = &\frac{g^2 r_a \Omega_R}{2\gamma(\gamma^2+\Omega_R^2)}\left\{\frac{[\gamma - i(\Delta-\Omega_R/2)]}{[\gamma^2+(\Delta-\Omega_R/2)^2]}(\Omega - i\gamma)\right.\\ &\left.-\frac{[\gamma-i(\Delta+\Omega_R/2)]}{[\gamma^2+(\Delta+\Omega_R/2)^2]}(\Omega + i\gamma)\right\},\end{aligned} \tag{14.A.21}$$

$$\begin{aligned}\alpha_{21} = &\frac{g^2 r_a}{2\gamma(\gamma^2+\Omega_R^2)}\left\{\frac{(2\gamma^2+\Omega_R^2+i\Omega_R\gamma)[\gamma - i(\Delta-\Omega_R/2)]}{[\gamma^2+(\Delta-\Omega_R/2)^2]}\right.\\ &\left.-\frac{(2\gamma^2+\Omega_R^2-i\Omega_R\gamma)[\gamma-i(\Delta+\Omega_R/2)]}{[\gamma^2+(\Delta+\Omega_R/2)^2]}\right\},\end{aligned} \tag{14.A.22}$$

$$\begin{aligned}\alpha_{22} = &\frac{g^2 r_a \Omega_R}{2\gamma(\gamma^2+\Omega_R^2)}\left\{\frac{[\gamma - i(\Delta-\Omega_R/2)]}{[\gamma^2+(\Delta-\Omega_R/2)^2]}(\Omega - i\gamma)\right.\\ &\left.+\frac{[\gamma-i(\Delta+\Omega_R/2)]}{[\gamma^2+(\Delta+\Omega_R/2)^2]}(\Omega + i\gamma)\right\}.\end{aligned} \tag{14.A.23}$$

We note that the form of Eq. (14.A.19) for the quantum beat laser is identical to Eq. (14.2.22) for the Hanle effect laser. The coefficients α_{ij} $(i = 1, 2)$ for the quantum beat laser are, however, more complicated. As in the Hanle effect laser, we can transform Eq. (14.A.19) into an equivalent Fokker–Planck equation in the P-representation. When the amplitude fluctuations are ignored, an equation of the form Eq. (14.2.33) in terms of the relative and mean phases $\theta = \theta_1 - \theta_2$ and $\mu = (\theta_1 + \theta_2)/2$ is obtained.

A CEL action is obtained in a quantum beat laser when the following conditions are satisfied:

$$\Delta = \frac{\Omega_R}{2}, \qquad \Omega_R \gg \gamma, \tag{14.A.24}$$

i.e., the field detuning from the corresponding atomic lines is equal to half the Rabi frequency of the driving field that coherently mixes the levels $|a\rangle$ and $|b\rangle$, and they are much larger than the atomic decay constant. Under these conditions

$$\mathrm{Re}\alpha_{11} = \mathrm{Re}\alpha_{22} \simeq \frac{g^2 r_a}{2\gamma^2}, \tag{14.A.25}$$

$$\alpha_{12} + \alpha_{21}^* \simeq \frac{g^2 r_a}{\gamma^2}. \tag{14.A.26}$$

We then obtain from Eq. (14.2.36) (for $\langle n_1\rangle = \langle n_2\rangle = \langle n\rangle$)

$$D_{\theta\theta} \simeq \frac{\mathscr{A}}{8\langle n\rangle}(1 - \cos\psi), \tag{14.A.27}$$

where $\mathscr{A} = 2g^2 r_a/\gamma^2$ is the usual gain coefficient and $\psi = \theta - \phi$. When $\psi = 0$, the diffusion coefficient vanishes.

Problems

14.1 Consider a two-photon CEL in which atoms are injected into the cavity with initial populations $\rho_{aa}^{(0)}$, $\rho_{bb}^{(0)}$, and $\rho_{cc}^{(0)}$ and initial coherences

$$\rho_{ab}^{(0)} = \rho_{ba}^{(0)*} = |\rho_{ab}^{(0)}|\exp(i\phi_{ab}),$$
$$\rho_{bc}^{(0)} = \rho_{cb}^{(0)*} = |\rho_{bc}^{(0)}|\exp(i\phi_{bc}),$$
$$\rho_{ac}^{(0)} = \rho_{ca}^{(0)*} = |\rho_{ac}^{(0)}|\exp(i\phi_{ac}),$$

where $|a\rangle$, $|b\rangle$, and $|c\rangle$ refer to the top, middle, and bottom levels, respectively of a three-level atomic system in a cascade configuration. Assume the detunings $\omega_{ab} - \nu$ and $\omega_{bc} - \nu$ to be zero. Find the phase-locking condition and the steady-state phase noise $\langle(\delta\theta)^2\rangle$ at the locked phase. Determine the conditions under which CEL action and phase squeezing are obtained. (Hint: see M. O. Scully et al., *Phys. Rev. Lett.* **60**, 1832 (1988).)

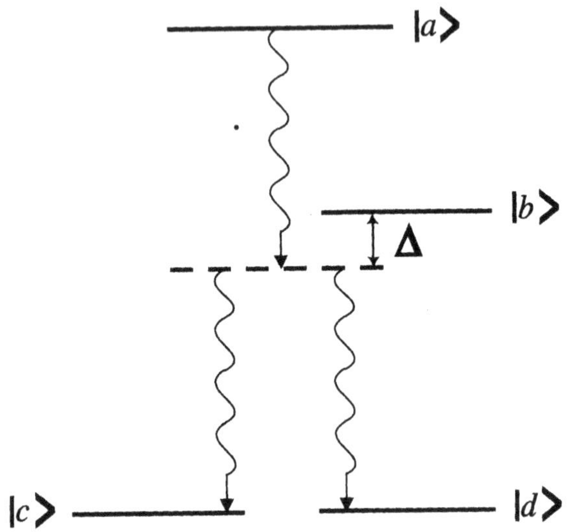

Fig. 14.10
Atomic level scheme relevant to Problem 14.2

14.2 Consider a double two-photon CEL in which four-level atoms are injected into the cavity with initial populations $\rho_{aa}^{(0)}$, $\rho_{cc}^{(0)}$, and $\rho_{dd}^{(0)}$ and initial coherences $\rho_{ac}^{(0)}$, $\rho_{ad}^{(0)}$, and $\rho_{cd}^{(0)}$ (with $\rho_{ij}^{(0)} = \rho_{ji}^{(0)*}$ and $\rho_{ij}^{(0)} = |\rho_{ij}^{(0)}|\exp(i\phi_{ij})$) (see Fig. 14.10). Levels $|c\rangle$ and $|d\rangle$ are considered to be almost degenerate. Two-photon resonance is assumed, but the level $|b\rangle$ is detuned by Δ with respect to the one-photon resonance. Derive an equation of motion for the reduced density operator of the field. Find the phase-locking condition and the steady-state phase noise $\langle(\delta\theta)^2\rangle$. Show that, under optimum conditions, phase noise squeezing is compatible with population inversion. (Hint: see J. Bergou et al., *Phys. Rev. A* **42**, 5544 (1990)).

14.3 Consider the following general Fokker–Planck equation for the P-representation:

$$\frac{\partial}{\partial t}P(\alpha,\alpha^*,t) = -\frac{\partial}{\partial\alpha}(d_\alpha P) - \frac{\partial}{\partial\alpha^*}(d_\alpha^* P) + 2\frac{\partial^2}{\partial\alpha\partial\alpha^*}(D_{\alpha^*\alpha}P) + \frac{\partial^2}{\partial\alpha^2}(D_{\alpha\alpha}P) + \frac{\partial^2}{\partial\alpha^{*2}}(D_{\alpha^*\alpha^*}P),$$

with

$$d_\alpha = R\alpha + G^*\alpha^*,$$
$$D_{\alpha\alpha} = (D_{\alpha^*\alpha^*})^* = |D|e^{i\phi_\alpha},$$

where R and G are completely arbitrary complex parameters. In polar coordinates, this equation reduces to

$$\frac{\partial P}{\partial t} = -\frac{1}{r}\frac{\partial}{\partial r}(rd_r P) - \frac{\partial}{\partial \theta}(d_\theta P) + \frac{1}{r}\frac{\partial}{\partial r}(rD_{rr}\frac{\partial P}{\partial r})$$
$$+\frac{\partial^2}{\partial \theta^2}(D_{\theta\theta}P) + \frac{2}{r}\frac{\partial^2}{\partial r \partial \theta}(rD_{r\theta}P).$$

Find d_r, d_θ, D_{rr}, $D_{\theta\theta}$, and $D_{r\theta}$ in terms of R, G, $|D|$, and ϕ_α. Find the condition for phase locking and then a condition for complete quenching of the phase noise. Ignore amplitude fluctuations, i.e., replace r by $\sqrt{\langle n \rangle}$.

14.4 Derive the diffusion coefficients $D_{\alpha\alpha}$, $D_{\alpha^*\alpha}$, and $D_{\alpha^*\alpha^*}$ corresponding to the field density operator equation (14.6.2) using Einstein's relations as well as a Fokker–Planck equation for $P(\alpha, \alpha^*)$. Use these results to derive $D_{\theta\theta}$.

14.5 (a) Show that, in a quantum beat laser described by the field density operator equation (14.A.19) with $\phi = 0$,

$$\frac{d}{dt}\langle a_1 \rangle = \frac{1}{2}(\alpha_{11}\langle a_1 \rangle + \alpha_{12}\langle a_2 \rangle),$$
$$\frac{d}{dt}\langle a_2 \rangle = \frac{1}{2}(\alpha_{22}\langle a_2 \rangle + \alpha_{21}\langle a_1 \rangle).$$

(b) Use these equations to derive an equation for the relative phase $\theta = \theta_1 - \theta_2$ of the form (14.3.1) by writing

$$\langle a_1 \rangle = \sqrt{\langle n \rangle} e^{i\theta_1 - i\nu_1 t},$$
$$\langle a_2 \rangle = \sqrt{\langle n \rangle} e^{i\theta_2 - i\nu_2 t}.$$

Also assume all α_{ij} $(i, j = 1, 2)$ to be equal.

References and bibliography

Quantum beat laser

M. O. Scully, *Phys. Rev. Lett.* **55**, 2802 (1985).
M. O. Scully and M. S. Zubairy, *Phys. Rev. A* **35**, 752 (1987).
J. Bergou, M. Orszag, and M. O. Scully, *Phys. Rev. A* **38**, 754 (1988).
K. Zaheer and M. S. Zubairy, *Phys. Rev. A* **38**, 5227 (1988).
S. Swain, *J. Mod. Opt.* **35**, 1 (1988).
N. A. Ansari and M. S. Zubairy, *Phys. Rev. A* **40**, 5690 (1989).
U. W. Rathe and M. O. Scully, *Phys. Rev. A* **52**, 3193 (1995).

Hanle laser

J. Bergou, M. Orszag, and M. O. Scully, *Phys. Rev. A* **38**, 768 (1988).
B. J. Dalton in *New Frontiers in Quantum Electrodynamics and Quantum Optics*, ed. A. O. Barut (Plenum Press, New York 1990), p. 167.

Holographic laser

J. Krause and M. O. Scully, *Phys. Rev. A* **36**, 1771 (1987).
M. O. Scully, *Phys. Rev. A* **35**, 452 (1987).
S. R. J. Brueck, M. Y. A. Raja, M. Orsinki, C. F. Schaus, M. Mahbodzadeh, and K. J. Dahlhaus, *SPIE* **1043**, 111 (1988).
M. Y. A. Raja, S. R. J. Brueck, M. O. Scully, and C. Lee, *Phys. Rev. A* **44**, 4599 (1991).

Two-photon CEL

M. O. Scully, K. Wódkiewicz, M. S. Zubairy, J. Bergou, N. Lu, and J. Meyer ter Vehn, *Phys. Rev. Lett.* **60**, 1832 (1988).
N. Lu and S. Y. Zhu, *Phys. Rev. A* **40**, 5735 (1989); *ibid.* **41**, 2865 (1990).
N. A. Ansari, J. Gea-Banacloche, and M. S. Zubairy, *Phys. Rev. A* **41**, 5179 (1990).
J. Bergou, C. Benkert, L. Davidovich, M. O. Scully, S.-Y. Zhu, and M. S. Zubairy, *Phys. Rev. A* **42**, 5544 (1990).

Geometrical picture for CEL

W. Schleich and M. O. Scully, *Phys. Rev. A* **37**, 1261 (1988).
W. Schleich, M. O. Scully, and H.-G. von Garssen, *Phys. Rev. A* **37**, 3010 (1988).

Phase squeezing

J. Bergou, M. Orszag, M. O. Scully, and K. Wódkiewicz, *Phys. Rev. A* **39**, 5136 (1989).

CEL experimental papers

M. Ohtsu and K.-Y. Liou, *Appl. Phys. Lett.* **52**, 10 (1988).

M. P. Winters, J. L. Hall, and P. E. Toschek, *Phys. Rev. Lett.* **65**, 3116 (1990).

I. Steiner, L. Schanz, and P. E. Toschek, in *Proceedings of the Twelfth International Conference on Laser Spectroscopy*, Island of Capri, Italy (June 11–16, 1995).

CHAPTER 15

Phase sensitivity in quantum optical systems: applications

In the previous chapter, we saw that phase sensitivity through atomic coherence can lead to the quenching of spontaneous emission noise. Earlier, in Chapter 8, the role of squeezed vacuum in introducing phase-sensitive damping was discussed. In this chapter, we discuss some potential applications of these concepts.

As a first example, we show that the sensitivity of a ring laser interferometer, in which a gain medium is modulated (the holographic laser), is potentially improved beyond the usual quantum limit. We then discuss the quantum theory of linear amplifiers and show that *noise-free amplification* can be obtained if the amplifier consists of three-level atoms prepared in a coherent superposition of atomic states, as in a two-photon CEL. Finally, we show a reduction in the natural linewidth of a laser if the field inside the laser cavity is coupled to a reservoir of harmonic oscillators in a squeezed vacuum state. These examples illustrate the significant role that phase sensitivity can play in the suppression of noise.

15.1 The CEL gyro

Having developed the CEL ring laser in Section 14.4 we now return to the question posed at the beginning of Chapter 14, namely, to what extent can we make an active gyro free from spontaneous emission noise? As discussed in Appendix 15.A, the relative phase angle between the two modes of the CEL ring gyro of Fig. 15.1(a)

$$\psi = (\nu_1 - \nu_2)t - \theta_1 + \theta_2, \tag{15.1.1}$$

obeys the locking equation (Eq. (15.A.16) with $\mathscr{C}_c = 0$)

$$\dot{\psi} = S\Omega_r - \mathscr{A} \sin \psi + \mathscr{F}_\psi, \tag{15.1.2}$$

where S is the scale factor of Section 14.1, namely, $S = 4A/\bar{\lambda}p$ where A and p are the gyro enclosed area and perimeter, $\bar{\lambda} = \lambda/2\pi$ is the reduced wavelength, and Ω_r is the rotation rate. In Eq. (15.1.2), $\mathscr{A}$ is the linear gain which is the same for both modes, and, as is further discussed in Appendix 15.A, the noise source $\mathscr{F}_\psi(t)$ is defined by

$$\langle \mathscr{F}_\psi^*(t) \mathscr{F}_\psi(t') \rangle = 2D(\psi)\delta(t - t'), \tag{15.1.3}$$

where

$$D(\psi) = \frac{\mathscr{A}}{4\langle n \rangle}(1 - \cos \psi), \tag{15.1.4}$$

and $\langle n \rangle$ is the average photon number which is taken to be the same for both modes.

We now write ψ as

$$\psi = \psi_0 + \delta\psi(t), \tag{15.1.5}$$

where ψ_0 is the average value and $\delta\psi$ is the noise induced fluctuation about ψ_0. Now since $S\Omega_r \ll \mathscr{A}$ we have $\psi_0 \ll 1$ so that $\sin \psi_0 \cong \psi_0$ to a very good approximation and Eq. (15.1.2) yields

$$\psi_0 = \frac{S\Omega_r}{\mathscr{A}} \qquad \text{when } t_m \simeq \mathscr{A}^{-1}. \tag{15.1.6}$$

Furthermore Eq. (15.1.2) implies that the fluctuations $\delta\psi$ about ψ_0 are given by

$$\delta\psi(t) = \int_0^t dt' \exp[-\mathscr{A}(t - t')] \mathscr{F}_\psi(t'), \tag{15.1.7}$$

and from Eqs. (15.1.7), (15.1.3), and (15.1.4) we have

$$\begin{aligned} \langle \delta\psi^2 \rangle &= \frac{\mathscr{A}/\langle n \rangle}{4\mathscr{A}} \left(1 - e^{-2\mathscr{A}t}\right)(1 - \cos \psi_0) \\ &\cong 0, \end{aligned} \tag{15.1.8}$$

since $\psi_0 \ll 1$. Hence, the limiting source of noise is now shot noise. The phase error is then $\delta\psi \sim 1/\langle n \rangle^{1/2}$, where $\langle n \rangle$ is the average photon number detected in time t_m, that is, $\langle n \rangle = P_d t_m/\hbar\nu$, where P_d is the laser power at the detector. For $\mathscr{A} \simeq \mathscr{C}, P_d$ is equal to the total emitted power P. Equating the shot-noise error $\delta\psi$ to the signal ψ_0 as given by Eq. (15.1.6) and solving for the minimum detectable rotation rate, we find

$$\Omega_{\min} \cong S^{-1} \mathscr{A} \sqrt{\frac{\hbar\nu}{P t_m}}. \tag{15.1.9}$$

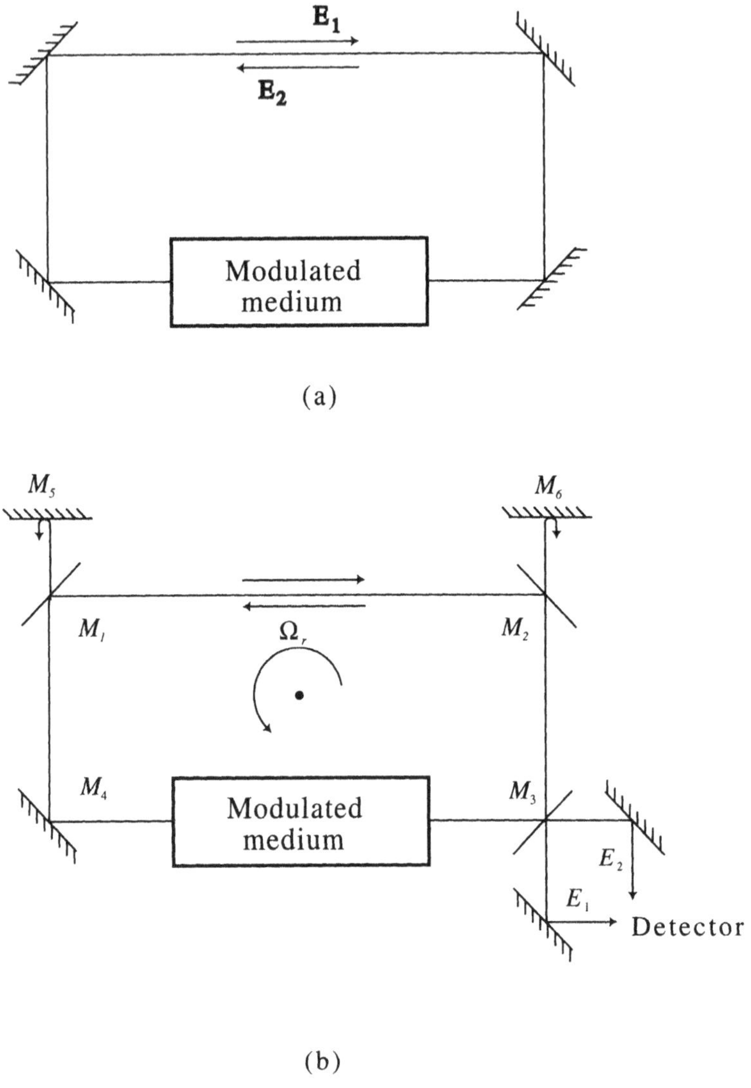

Fig. 15.1
(a) CEL ring laser in which correlation is produced by a striated or modulated gain medium.
(b) CEL laser gyroscope in which light is extracted by mirrors M_1 and M_2 and reinjected by mirrors M_5 and M_6. This leads to an enhanced gyroscope sensitivity. (From M. O. Scully, *Phys. Rev. A* **35**, 452 (1987).)

Evidently the CEL gyro limit (15.1.9) is the usual quantum limit given by Eqs. (14.1.4) or (14.1.8) since $\mathscr{A} \approx \mathscr{C}$. However, we note that the measurement time t_m as it appears in (15.1.9) is of order $\mathscr{A}^{-1}$, while in the usual quantum limit t_m is not constrained. Thus it might look as if the CEL gyro has an inferior sensitivity. But this is not the case because we could arrange for the gyro to attain its value ψ_0 in a time $\cong \mathscr{A}^{-1}$, record the value of ψ_0, unlock the gyro by, e.g., dithering or other means and then 'turn off' the dither and allow the gyro to again return to its locked value ψ_0. This we could do N times in a time $\tau \gg \mathscr{A}^{-1}$ where $N \approx \tau/\mathscr{A}^{-1} = \mathscr{A}\tau$. So that if we rewrite (15.1.9) for $t_m \cong \mathscr{A}^{-1}$

$$\Omega_{\min} \cong S^{-1}\mathscr{C}\sqrt{\frac{\hbar\nu}{P\mathscr{A}^{-1}}}, \tag{15.1.10}$$

we can improve on this by a factor of $1/\sqrt{N} \cong 1/\sqrt{\mathscr{A}\tau}$ to obtain

$$\begin{aligned}\Omega_{\min} &\cong S^{-1}\mathscr{C}\sqrt{\frac{\hbar\nu}{P\mathscr{A}^{-1}}}\frac{1}{\sqrt{\mathscr{A}\tau}} \\ &= S^{-1}\mathscr{C}\sqrt{\frac{\hbar\nu}{P\tau}}.\end{aligned} \tag{15.1.11}$$

Hence, we see that the 'many measurement' CEL gyro limit (15.1.11) is not restricted to short times and the result is the same as the usual quantum limit.

However, we note that we can improve on the CEL sensitivity (15.1.9) if we extend the problem to include controlled backscatter as in Fig. 15.1(b). As shown in Appendix 15.A, the equation for ψ now becomes

$$\dot{\psi} = S\Omega_r - (\mathscr{A} - \mathscr{C}_c)\sin\psi + \mathscr{F}_\psi(t) \tag{15.1.12}$$

where

$$\mathscr{C}_c = \frac{c}{p}T\sqrt{R_c}. \tag{15.1.13}$$

In the above, p is the perimeter of the ring, T is the transmittance of mirrors M_1 and M_2 in Fig. 15.1(b), and R_c is the reflectivity of mirrors M_5 and M_6. We now find a larger signal

$$\psi_0 = \frac{S\Omega_r}{\mathscr{A} - \mathscr{C}_c}, \tag{15.1.14}$$

and therefore

$$\Omega_{\min} \simeq \left(\frac{\mathscr{A} - \mathscr{C}_c}{\mathscr{C}}\right) S^{-1}\mathscr{C}\sqrt{\frac{\hbar\nu}{P_d t_m}}. \tag{15.1.15}$$

Finally, we note that the power P_d is not the total emitted power P but only $(\mathscr{C}_d/\mathscr{C})P$ (with $\mathscr{C}_d \cong \mathscr{A} - \mathscr{C}_c$) since the detector port M_3 extracts only a fraction $\mathscr{C}_d/\mathscr{C}$ of the total emitted power. Furthermore, the prefactor $(\mathscr{A} - \mathscr{C}_c)/\mathscr{C}$ in Eq. (15.1.15) is just $\mathscr{C}_d/\mathscr{C}$, which taken together with the fact that $P_d = (\mathscr{C}_d/\mathscr{C})P$, leads to

$$\Omega_{\min} \cong \epsilon S^{-1}\mathscr{C}\sqrt{\frac{\hbar\nu}{P t_m}}, \tag{15.1.16}$$

where $\epsilon = 1/\sqrt{\mathscr{C}t_m}$ and where we have noted that $t_m^{-1} \sim (\mathscr{A}-\mathscr{C}_c) = \mathscr{C}_d$. Thus, for our idealized model, gyroscope sensitivity is improved by the factor ϵ. We emphasize that the present calculation ignores mirror losses due to absorption $\mathscr{C}_a$. When such losses are included in the simplest models, the factor ϵ is governed by $\mathscr{C}_a/(\mathscr{C}_a+\mathscr{C})$. In conclusion, we see that: (1) the CEL gyroscope has, in principle, a sensitivity superseding the conventional quantum limit,* and (2) this device is not hampered by the usual 'dead-band' locking problem associated with low rotation rates.

15.2 Linear amplification process: general description

A linear amplifier is an amplifier whose output is linearly related to its input. This broad definition requiring only the linearity of operation enables one to give a unified account of the quantum limits for such devices without going into the details of their internal working.

Amplifiers are important in the measurement process as they bridge the gap between the quantum and classical worlds. Apart from this, there are numerous other applications. In communication systems, for example, it may be advantageous to replace a repeater unit in an optical fiber link by a direct amplification unit or an amplifier unit could be used in a predetection capacity at the receiver.

A related question of interest is whether photon cloning by an amplifier is possible. Such precise duplication of photons would present opportunities for overcoming the limitations imposed by the uncertainty principle in interference and photon correlation experiments. Since amplification is achieved through an emissive process in optical amplifiers, it is invariably influenced by spontaneous emission. This imposes fundamental quantum limits on the signal-to-noise ratio at the output.

The following operator transformation equation relating the annihilation operators a_{in} and a_{out} in the input and the output, respectively, describes a linear amplification process:

$$a_{\text{out}} = \sqrt{G}a_{\text{in}} + \sqrt{G-1}F^{\dagger}, \qquad (15.2.1)$$

where G is the amplification factor and the second term is required for consistency so that the commutation relation at the output is

* A similar type of reinjection scheme applied to a CEL gravity wave detection also improves sensitivity, see Scully and Gea-Banacloche [1986].

satisfied. Here $F^\dagger$ is an operator obeying a boson commutation relation, $[F, F^\dagger] = 1$, which commutes with a_{in}. We now define the two Hermitian quadratures X_θ and $X_{\theta+\pi/2}$ for the bosonic signals

$$X_\theta = \frac{1}{2}(ae^{-i\theta} + a^\dagger e^{i\theta}). \tag{15.2.2}$$

The variances in the two quadratures are

$$(\Delta X_\theta^{\text{out}})^2 = G(\Delta X_\theta^{\text{in}})^2 + (G-1)(\Delta F_\theta)^2, \tag{15.2.3}$$

$$(\Delta X_{\theta+\pi/2}^{\text{out}})^2 = G(\Delta X_{\theta+\pi/2}^{\text{in}})^2 + (G-1)(\Delta F_{\theta+\pi/2})^2, \tag{15.2.4}$$

where

$$F_\theta = \frac{1}{2}(Fe^{-i\theta} + F^\dagger e^{i\theta}). \tag{15.2.5}$$

The first term in the preceding equations represents the amplified input noise and the second term represents the *additive noise* due to amplification. Such a gain-dependent noise is a manifestation of the fluctuation–dissipation theorem discussed in Section 9.4.

In this connection, two questions arise. First, can the output of a linear amplifier still display nonclassical features if these are imposed at the input? Second, can an amplifier generate a squeezed output? Answers to these questions divide the study of amplifiers into two general classes, namely, the phase-insensitive and the phase-sensitive amplifiers. In a phase-insensitive amplifier, the two quadratures are treated equally, i.e., the quadratures are amplified by the same factor and equal (phase-insensitive) noise is added to the two quadratures. In such an amplifier, there is no preferred phase and a phase shift at the input results in an equal phase shift at the output. In contrast, a phase-sensitive amplifier treats the two quadratures differently in the form of unequal added noise. For example, the added noise in one quadrature can be reduced at the expense of a larger added noise in the conjugate quadrature.

In order to amplify a signal, one needs a reservoir of energy that can be supplied to the signal. However, a reservoir at high temperature would simply feed thermal noise into the signal. The model involves the interaction of a boson mode with a bath of *inverted* harmonic oscillators. Such a reservoir has no ground state but the states of the inverted harmonic oscillators have an upper bound. This amplifier still adds thermal noise to the system since the state of the reservoir resembles a canonical distribution with negative temperature. A practical implementation of this model consists of a group of inverted two-level atoms. In this case, the linearity of operation means that only the one-photon processes are taking place. This situation could correspond to a laser with the end mirrors removed, being perturbed by a weak external field. In the next section, we will discuss the characteristics of this system and show explicitly that it adds a phase-insensitive noise to the signal.

A phase-sensitive amplifier can be constructed by *rigging** or by modifying the reservoir by preparing it in a squeezed vacuum centered around the frequency of the mode that is to be amplified. The boson modes of the bath obey correlation functions (8.2.16). The master equation for the amplifier whose reservoir consists of 'squeezed white noise' is of the form

$$\begin{aligned}\dot{\rho} = &-\frac{\mathscr{A}}{2}N(a^\dagger a\rho - 2a\rho a^\dagger + \rho a^\dagger a)\\ &-\frac{\mathscr{A}}{2}(N+1)(aa^\dagger\rho - 2a^\dagger\rho a + \rho aa^\dagger)\\ &+\frac{\mathscr{A}}{2}M^*(aa\rho - 2a\rho a + \rho aa)\\ &+\frac{\mathscr{A}}{2}M(a^\dagger a^\dagger\rho - 2a^\dagger\rho a^\dagger + \rho a^\dagger a^\dagger),\end{aligned} \tag{15.2.6}$$

where $\mathscr{A}$ is an amplification constant. The terms proportional to M and M^* are responsible for the phase sensitivity of the amplifier. An example of a physical implemenation of a phase-sensitive amplifier is provided by a two-photon CEL. This system is discussed in Section 15.4.

15.3 Phase-insensitive amplification in a two-level system†

The interaction Hamiltonian for any given atom in the interaction picture, at exact resonance and in the rotating-wave approximation, is

$$\mathscr{V} = \hbar g(a^\dagger\sigma_- + \sigma_+ a). \tag{15.3.1}$$

The atoms are being injected into the cavity at a rate r_a in an incoherent superposition of upper and lower levels $|a\rangle$ and $|b\rangle$, i.e., at an initial time t_i,

$$\rho_{\text{atom}}(t_i) = \rho_{aa}^{(0)}|a\rangle\langle a| + \rho_{bb}^{(0)}|b\rangle\langle b|, \tag{15.3.2}$$

where $\rho_{aa}^{(0)}$ and $\rho_{bb}^{(0)}$ are the excitation probabilities to levels $|a\rangle$ and $|b\rangle$ respectively. Following the methods developed in Chapter 11, we

* Stenholm [1986].
† See e.g., Caves [1982].

obtain the usual equation of motion for the reduced density operator of the field, from the quantum theory of the laser:

$$\dot{\rho} = -\frac{\mathscr{A}}{2}[\rho_{aa}^{(0)}(aa^\dagger\rho - 2a^\dagger\rho a + \rho aa^\dagger) + \rho_{bb}^{(0)}(a^\dagger a\rho - 2a\rho a^\dagger + \rho a^\dagger a)], \tag{15.3.3}$$

where $\mathscr{A} = 2r_a g^2/\gamma^2$ is the gain coefficient with r_a being the rate of injection of atoms in the state (15.3.2) and γ being the atomic decay rate. In the above equation, the terms proportional to $\rho_{aa}^{(0)}$ and $\rho_{bb}^{(0)}$ correspond to gain (due to emission) and loss (due to absorption), respectively. As will be seen in the following, net amplification is achieved when $\rho_{aa}^{(0)} > \rho_{bb}^{(0)}$.

Equation (15.3.3) can be solved for arbitrary times. For the present purpose however, we can use it to obtain equations of motion for various moments of the field operators a and $a^\dagger$,

$$\frac{d}{dt}\langle a\rangle = \frac{\mathscr{A}}{2}[\rho_{aa}^{(0)} - \rho_{bb}^{(0)}]\langle a\rangle, \tag{15.3.4}$$

$$\frac{d}{dt}\langle a^\dagger a\rangle = \mathscr{A}[\rho_{aa}^{(0)} - \rho_{bb}^{(0)}]\langle a^\dagger a\rangle + \mathscr{A}\rho_{aa}^{(0)}, \tag{15.3.5}$$

$$\frac{d}{dt}\langle a^2\rangle = \mathscr{A}[\rho_{aa}^{(0)} - \rho_{bb}^{(0)}]\langle a^2\rangle. \tag{15.3.6}$$

The set of linear equations (15.3.4)–(15.3.6) can be solved exactly to obtain

$$\langle a\rangle_t = \sqrt{G}\langle a\rangle_0, \tag{15.3.7}$$

$$\langle a^\dagger a\rangle_t = G\langle a^\dagger a\rangle_0 + (G-1)\frac{\rho_{aa}^{(0)}}{\rho_{aa}^{(0)} - \rho_{bb}^{(0)}}, \tag{15.3.8}$$

$$\langle a^2\rangle_t = G\langle a^2\rangle_0, \tag{15.3.9}$$

where

$$G = \exp\{\mathscr{A}[\rho_{aa}^{(0)} - \rho_{bb}^{(0)}]t\}, \tag{15.3.10}$$

is the gain factor. Equations (15.3.7)–(15.3.10) can be viewed as input–output equations for the amplifier; the expectation value at time t being the output from the amplifier in terms of the input (expectation value at time $t = 0$). It is clear that amplification takes place ($G > 1$) when $\rho_{aa}^{(0)} > \rho_{bb}^{(0)}$ and the system acts as an attenuator ($G < 1$) if $\rho_{aa}^{(0)} < \rho_{bb}^{(0)}$.

We now define the two quadratures X_θ and $X_{\theta+\pi/2}$, similar to the definitions (15.2.3) and (15.2.4), through the relation

$$X_\theta = \frac{1}{2}(a^\dagger e^{i\theta} + ae^{-i\theta}). \tag{15.3.11}$$

Here the multiplication by the exponential factors merely means an arbitrary rotation in the complex X-plane and it leaves the commutation relations unaltered. It follows from Eqs. (15.3.7)–(15.3.9) that

$$\langle X_\theta \rangle_t = \sqrt{G}\langle X_\theta \rangle_0, \tag{15.3.12}$$

$$(\Delta X_\theta^2)_t = G(\Delta X_\theta^2)_0 + (G-1)\mathcal{N}, \tag{15.3.13}$$

where

$$\mathcal{N} = \frac{1}{4[\rho_{aa}^{(0)} - \rho_{bb}^{(0)}]}. \tag{15.3.14}$$

We see that the two quadrature components X_θ and $X_{\theta+\pi/2}$ experience equal gains ($G_1 = G_2 = G$) and that the added noise in the two quadratures is equal and independent of the phase angle θ, i.e., $\Delta\mathcal{F}_1^2 = \Delta\mathcal{F}_2^2 = (G-1)\mathcal{N}$. The amplifier under consideration is therefore a phase-preserving, phase-insensitive amplifier. From Eq. (15.3.13), we also see that the maximum gain preserving any squeezing at the output is

$$G_{\max} = \frac{\frac{1}{4} + \mathcal{N}}{(\Delta X_\theta^2)_0 + \mathcal{N}}, \tag{15.3.15}$$

which for a highly squeezed input, $(\Delta X_\theta^2)_0 = 0$, and for $\rho_{aa}^{(0)} = 1$ give $G_{\max} = 2$. This is usually referred to as the *cloning limit*. Any squeezing imposed at the input therefore, disappears during the process of amplification due to the phase-insensitive *added noise*.

15.4 Phase-sensitive amplification via the two-photon CEL: noise-free amplification*

In this section, we discuss a two-photon linear amplifier in which phase sensitivity is introduced by preparing the gain medium, i.e., the atoms, in a coherent superposition of states.

The system consists of three-level atoms in cascade configurations with levels $|a\rangle$, $|b\rangle$, and $|c\rangle$. The boson mode of frequency ν is assumed to be in resonance with the two atomic transitions. The interaction Hamiltonian in the interaction representation and in the rotating-wave approximation is

$$\mathscr{V} = \hbar g[a^\dagger(|b\rangle\langle a| + |c\rangle\langle b|) + (|a\rangle\langle b| + |b\rangle\langle c|)a]. \tag{15.4.1}$$

* This section follows Scully and Zubairy [1988].

We consider a situation in which the atoms are injected at a rate r_a, in a coherent superposition of the states $|a\rangle$ and $|c\rangle$. The atomic density operator at the initial time is

$$\rho_{\text{atom}}(t_i) = \rho_{aa}^{(0)}|a\rangle\langle a| + \rho_{ac}^{(0)}|a\rangle\langle c| + \rho_{ca}^{(0)}|c\rangle\langle a| + \rho_{cc}^{(0)}|c\rangle\langle c|. \quad (15.4.2)$$

Here $\rho_{ac}^{(0)} = \rho_{ca}^{(0)*}$ is the initial coherence between the levels $|a\rangle$ and $|c\rangle$ and $\rho_{aa}^{(0)}$ and $\rho_{cc}^{(0)}$ are the corresponding populations for the two levels.

Following the same procedure as discussed in Section 11.1, the equation of motion for the reduced density operator is obtained as

$$\begin{aligned}\dot{\rho} = &-\frac{\mathscr{A}}{2}\rho_{aa}^{(0)}(aa^\dagger\rho - 2a^\dagger\rho a + \rho aa^\dagger)\\ &-\frac{\mathscr{A}}{2}\rho_{ca}^{(0)}(aa\rho - 2a\rho a + \rho aa)\\ &-\frac{\mathscr{A}}{2}\rho_{ac}^{(0)}(a^\dagger a^\dagger\rho - 2a^\dagger\rho a^\dagger + \rho a^\dagger a^\dagger)\\ &-\frac{\mathscr{A}}{2}\rho_{cc}^{(0)}(a^\dagger a\rho - 2a\rho a^\dagger + \rho a^\dagger a).\end{aligned} \quad (15.4.3)$$

Similarly to Eq. (15.3.3), the terms proportional to $\rho_{aa}^{(0)}$ and $\rho_{cc}^{(0)}$ correspond to the usual gain and absorption. But now, the anomalous terms proportional to $\rho_{ac}^{(0)}$ and $\rho_{ca}^{(0)}$ are also present and are responsible for the phase-sensitive operation of the amplifier.

It follows from Eq. (15.4.3) that

$$(d/dt)\langle a\rangle_t = \frac{1}{2}\mathscr{A}[\rho_{aa}^{(0)} - \rho_{cc}^{(0)}]\langle a\rangle_t, \quad (15.4.4)$$

$$(d/dt)\langle a^\dagger a\rangle_t = \mathscr{A}[\rho_{aa}^{(0)} - \rho_{cc}^{(0)}]\langle a^\dagger a\rangle_t + \mathscr{A}\rho_{aa}^{(0)}, \quad (15.4.5)$$

$$(d/dt)\langle a^2\rangle_t = \mathscr{A}[\rho_{aa}^{(0)} - \rho_{cc}^{(0)}]\langle a^2\rangle_t - \mathscr{A}\rho_{ac}^{(0)}. \quad (15.4.6)$$

These equations can be solved exactly and we obtain

$$\langle a\rangle_t = \sqrt{G}\langle a\rangle_0, \quad (15.4.7)$$

$$\langle a^\dagger a\rangle_t = G\langle a^\dagger a\rangle_0 + \frac{\rho_{aa}^{(0)}}{\rho_{aa}^{(0)} - \rho_{cc}^{(0)}}(G-1), \quad (15.4.8)$$

$$\langle a^2\rangle_t = G\langle a^2\rangle_0 - \frac{\rho_{ac}^{(0)}}{\rho_{aa}^{(0)} - \rho_{cc}^{(0)}}(G-1), \quad (15.4.9)$$

with $G = \exp\{\mathscr{A}[\rho_{aa}^{(0)} - \rho_{cc}^{(0)}]t\}$ as the gain factor.

The evolution of the quadrature components defined in Eq. (15.3.11) can now be obtained. It follows from Eq. (15.4.7) and its complex conjugate that

$$\langle X_\theta\rangle_t = \sqrt{G}\langle X_\theta\rangle_0. \quad (15.4.10)$$

The noise in the quadrature X_θ is

$$(\Delta X_\theta^2)_t = G(\Delta X_\theta^2)_0 + (G-1)\mathcal{N}_\theta, \tag{15.4.11}$$

where

$$\mathcal{N}_\theta = \frac{\rho_{aa}^{(0)} + \rho_{cc}^{(0)} - 2|\rho_{ac}^{(0)}|\cos(2\theta - \phi)}{4[\rho_{aa}^{(0)} - \rho_{cc}^{(0)}]}, \tag{15.4.12}$$

with

$$\rho_{ac}^{(0)} = |\rho_{ac}^{(0)}|\exp(i\phi). \tag{15.4.13}$$

Equation (15.4.11) is similar in form, to Eq. (15.3.13), but now the added noise, i.e., the second term, depends upon the phase angle θ. Defining a parameter ϵ such that

$$\rho_{aa}^{(0)} = \frac{(1+\epsilon)}{2}; \quad \rho_{cc}^{(0)} = \frac{(1-\epsilon)}{2}; \quad \rho_{ac}^{(0)} = \frac{(1-\epsilon^2)^{1/2}}{2}, \tag{15.4.14}$$

$$\mathcal{N}_\theta = \frac{1-(1-\epsilon^2)^{1/2}\cos(2\theta-\phi)}{4\epsilon}. \tag{15.4.15}$$

It follows that the signal $\langle X_\theta \rangle$ can be amplified ($G \gg 1$) without introducing added noise ($\mathcal{N}_\theta \ll 1$) under the following limits:

$$(2\theta - \phi) = 0; \quad \epsilon \to 0; \quad \mathcal{A}t \to \infty; \quad \mathcal{A}t\epsilon = \text{finite}. \tag{15.4.16}$$

Under these conditions, $\mathcal{N}_{\theta+\pi/2} \to \infty$, i.e., the added noise in one quadrature is reduced at the expense of increased added noise in the second quadrature such that $(\mathcal{N}_\theta + \mathcal{N}_{\theta+\pi/2}) = 1/2\epsilon > 1/2$.

15.5 Laser with an injected squeezed vacuum

In Chapter 11, we saw that, in the absence of all sources of noise (such as thermal and mechanical), the laser linewidth is limited by spontaneous emission. A simple pictorial model in this regard (see Section 11.4) envisions it as being due to a random phase diffusion process arising due to the addition of spontaneously emitted photons with random phases to the laser field. In an ordinary laser, vacuum is entering the cavity through the out-coupling mirror and some of the spontaneously emitted photons result from amplifying the fluctuations of this vacuum. If the vacuum is replaced by a *squeezed vacuum*, the phases of the spontaneously emitted photons will be biased resulting in reduced phase diffusion and, consequently a reduced laser linewidth.

Here we consider a ring laser with one running-wave mode (Fig. 15.2) operating above threshold. An external field is coupled to the intracavity field through an end mirror. The state of the external field is assumed to be squeezed vacuum centered around the operating frequency of the laser $\nu = ck_0$, and is given by (see Eqs. (8.2.12) and (8.2.13))

Fig. 15.2
Ring laser arrangement to couple squeezed vacuum (dashed line) to the intracavity field.

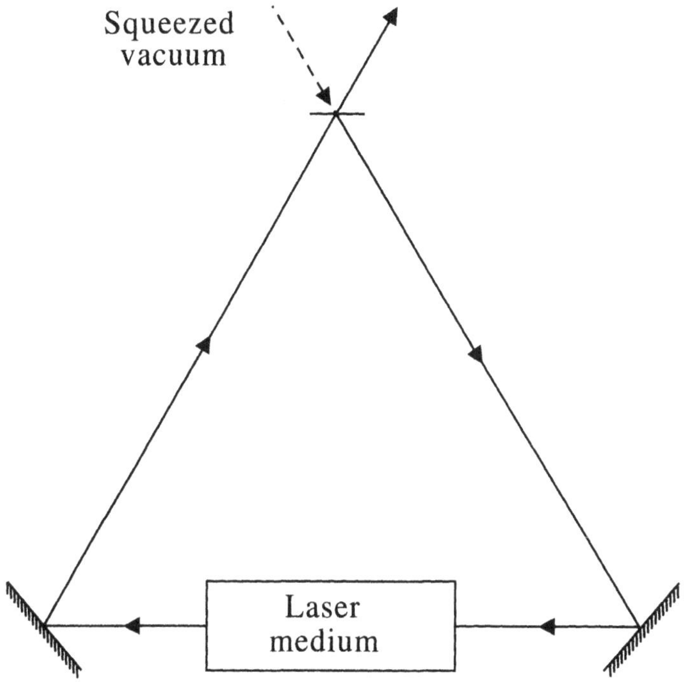

$$|\psi\rangle = \prod_{\mathbf{k}} \exp\left(\xi^* b_{\mathbf{k}_0+\mathbf{k}} b_{\mathbf{k}_0-\mathbf{k}} - \xi b^\dagger_{\mathbf{k}_0+\mathbf{k}} b^\dagger_{\mathbf{k}_0-\mathbf{k}}\right) |0_{\mathbf{k}}\rangle, \tag{15.5.1}$$

where $\xi = r\exp(i\phi)$. The radiation loss through the transmitting mirror is therefore obtained by coupling the radiation inside the cavity to a reservoir of harmonic oscillator modes in squeezed vacuum states. This is done in Eq. (8.3.4).

The effect of a squeezed vacuum is thus incorporated in the master equation of the laser by replacing the cavity loss term (8.3.3) by (8.3.4). In the linear approximation, we obtain*

$$\begin{aligned}\dot{\rho} = &-\frac{\mathscr{A}}{2}(aa^\dagger\rho - 2a^\dagger\rho a + \rho aa^\dagger)\\ &-\frac{\mathscr{C}}{2}(N+1)(a^\dagger a\rho - 2a\rho a^\dagger + \rho a^\dagger a)\\ &-\frac{\mathscr{C}}{2}N(aa^\dagger\rho - 2a^\dagger\rho a + \rho aa^\dagger)\\ &+\frac{\mathscr{C}}{2}M(aa\rho - 2a\rho a + \rho aa)\\ &+\frac{\mathscr{C}}{2}M^*(a^\dagger a^\dagger\rho - 2a^\dagger\rho a^\dagger + \rho a^\dagger a^\dagger),\end{aligned} \tag{15.5.2}$$

* This interesting result was first obtained by Gea-Banacloche [1987]. It is important to note, however, that this only leads to a 50% reduction of the ordinary laser linewidth. However, when a squeezed vacuum is injected into a CEL, the system can yield for example, a bright source of squeezed light; see, e.g., Bergou, Lu, and Scully [1989].

where $N = \sinh^2(r)$ and $M = \cosh(r)\sinh(r)\exp(-i\phi)$. In Eq. (15.5.2) $\mathscr{A}$ is the usual linear gain coefficient.

The associated diffusion coefficient $D_{\theta\theta}$ can be determined using either the Einstein relation (Section 9.4) or by converting the density operator equation into an equivalent Fokker–Planck equation for the P-representation using Eq. (14.2.27). The resulting expression for the phase diffusion coefficient is

$$2D_{\theta\theta} = \frac{1}{2\langle n\rangle}\{\mathscr{A}+\mathscr{C}[\sinh^2 r-\sinh r\cosh r\cos(\phi-2\theta)]\}. \tag{15.5.3}$$

If we choose the phase of the squeezed field $\phi = 2\theta$, we obtain

$$\begin{aligned} 2D_{\theta\theta} &= \frac{1}{2\langle n\rangle}\left(\mathscr{A} - \frac{\mathscr{C}}{2} + \frac{\mathscr{C}}{2}e^{-2r}\right) \\ &= \frac{\mathscr{A}}{4\langle n\rangle}\left(1 + e^{-2r}\right), \end{aligned} \tag{15.5.4}$$

where, in the last line, we replace $\mathscr{C}$ by $\mathscr{A}$ which is a good approximation for a laser operating near threshold. It follows, on comparing Eq. (15.5.4) with Eq. (11.4.17), that a reduction of up to fifty percent of the laser's phase diffusion rate (and the corresponding laser linewidth) is possible when the laser is coupled out to a squeezed vacuum as opposed to an ordinary vacuum.

15.A Analysis of the CEL gyro with reinjection*

We recall from Section 14.4 that the laser radiation density matrix for the holographic CEL is given in terms of the annihilation (and creation) operators a_j (and $a_j^\dagger$), $j = 1, 2$, by (Eq. (14.4.8))

$$\begin{aligned} &\dot{\rho}(a_1, a_1^\dagger, a_2, a_2^\dagger) \\ &= -i(\nu_{c1} - \nu_1)[a_1^\dagger a_1, \rho] - i(\nu_{c2} - \nu_2)[a_2^\dagger a_2, \rho] \\ &\quad -\frac{\mathscr{A}}{2}\int_{-\ell/2}^{\ell/2} dz\, n(z)[A(z)A^\dagger(z)\rho - 2A^\dagger(z)\rho A(z) + \rho A(z)A^\dagger(z)] \\ &\quad +\mathscr{L}_1\rho + \mathscr{L}_2\rho, \end{aligned} \tag{15.A.1}$$

where the above laser models have 'bare-cavity' eigenfrequencies ν_{c1} and ν_{c2}, operating frequencies ν_1 and ν_2, $\mathscr{A}$ represents the gain, and $n(z)$ is the density of lasing atoms. The field operator $A(z)$, written in terms of the resonator normal-mode function $U_j(z)$ is

$$A(z) = a_1 \exp(i\Delta_1 t)U_1(z) + a_2 \exp(i\Delta_2 t)U_2(z), \tag{15.A.2}$$

* See Scully [1987].

where $\Delta_1 = \omega_{ab} - \nu_1$ and $\Delta_2 = \omega_{ab} - \nu_2$. Finally, we have included the cavity losses due to transmission as described by the Liouville operator

$$\mathscr{L}_j\rho = -\frac{1}{2}\mathscr{C}_j(a_j^\dagger a_j\rho - 2a_j\rho a_j^\dagger + \rho a_j^\dagger a_j). \tag{15.A.3}$$

It is convenient to summarize the information contained in Eq. (15.A.1) using the quantum Langevin equation, that is,

$$\dot{a}_1 = -i(\nu_{c1} - \nu_1)a_1 + \frac{1}{2}\alpha_{11}a_1 + \frac{1}{2}\alpha_{12}a_2e^{i\Phi} - \frac{1}{2}\mathscr{C}_1a_1 + F_1, \tag{15.A.4}$$

$$\dot{a}_2 = -i(\nu_{c2} - \nu_2)a_2 + \frac{1}{2}\alpha_{22}a_2 + \frac{1}{2}\alpha_{21}a_1e^{-i\Phi} - \frac{1}{2}\mathscr{C}_2a_2 + F_2. \tag{15.A.5}$$

The Langevin noise operators appearing in Eqs. (15.A.4) and (15.A.5) are defined by

$$\langle F_i^\dagger(t)F_j(t')\rangle = 2D_{ij}\delta(t - t'), \tag{15.A.6}$$

where the matrix of the diffusion coefficients D_{ij} is

$$[D_{ij}] = \frac{1}{4}\begin{pmatrix} \alpha_{11} + \alpha_{11}^* & (\alpha_{12}^* + \alpha_{21})e^{-i\Phi} \\ (\alpha_{21}^* + \alpha_{12})e^{i\Phi} & \alpha_{22} + \alpha_{22}^* \end{pmatrix}. \tag{15.A.7}$$

The phase angle Φ is given by $(\nu_1 - \nu_2)t$. For the present discussion we will consider the gain coefficients to be equal, $\alpha_{11} = \alpha_{22} = \alpha$.

The cross-coupling coefficients α_{12} and α_{21} depend upon the spatial distribution of the gain medium. For example, when the active medium is spread uniformly over the region $-\ell/2 \leq x \leq \ell/2$ we find $\alpha_{12} = \alpha_{21} = 0$. However, if we consider a modulated gain medium such that the active atomic medium is distributed according to Eq. (14.4.14) as in a holographic laser (see Fig. 15.1(a)), we find that $\alpha_{12} = \alpha_{21} = \alpha$. That is, we have a strong correlation between modes 1 and 2. Note in particular that the cross-coupling diffusion coefficients D_{12} and D_{21} are nonvanishing in the case of the modulated gain medium but are zero when the lasing medium is uniformly pumped. As was shown in Section 14.4, finite D_{12} can lead to a quenching of spontaneous emission fluctuations in the relative phase angle. We now consider the application of such a holographic CEL to the laser gyroscope.

We note that the essential ingredients in conventional gyroscope operation are gain, loss, and mode coupling due to backscattering. Extending the CEL dynamics as given by Eqs. (15.A.4) and (15.A.5) to include the effects of reinjection, as in Fig. 15.1(b), and rewriting the equations of motion in terms of amplitude and phase variables defined by $a_i = r_i \exp(i\theta_i)$, we have our working CEL gyroscope equations

$$2\dot{r}_1 = \alpha_{11}r_1 + \alpha_{12}r_2 \cos\psi - \mathscr{C}_1 r_1 + \mathscr{C}_{12} r_2 \cos(\psi + \phi), \quad (15.A.8)$$

$$2\dot{r}_2 = \alpha_{22}r_2 + \alpha_{21}r_1 \cos\psi - \mathscr{C}_2 r_2 + \mathscr{C}_{21} r_1 \cos(\psi + \phi), \quad (15.A.9)$$

$$\dot{\psi} = \nu_{c1} - \nu_{c2} - \frac{1}{2}\left(\alpha_{12}\frac{r_2}{r_1} + \alpha_{21}\frac{r_1}{r_2}\right) \sin\psi$$
$$-\frac{1}{2}\left(\mathscr{C}_{12}\frac{r_2}{r_1} + \mathscr{C}_{21}\frac{r_1}{r_2}\right) \sin(\psi + \phi) + \mathscr{F}_\psi(t), \quad (15.A.10)$$

where the relative phase angle ψ is defined as

$$\psi = (\nu_1 - \nu_2)t - \theta_1 + \theta_2, \quad (15.A.11)$$

and the loss rate and backscatter cross-coupling rate (see Fig. 15.1(b)) are given by

$$\mathscr{C}_1 = \mathscr{C}_2 = \mathscr{C} = \frac{c}{p}(1 - R), \quad (15.A.12)$$

$$\mathscr{C}_{12} = \mathscr{C}_{21} = \mathscr{C}_c = \frac{c}{p}\, T\sqrt{R_c}. \quad (15.A.13)$$

Here, p is the perimeter of the ring, R and T are the reflectivity and the transmittivity of mirrors M_1 and M_2 (Fig. 15.1(b)), respectively, and R_c is the reflectivity of mirrors M_5 and M_6. The extra phase ϕ accumulated in the backscattering depends on the position of the external mirrors M_5 and M_6. Finally, the noise source in Eq. (15.A.10) is obtained from Eqs. (15.1.4) and (15.1.5) and is defined by

$$\langle \mathscr{F}_\psi^*(t)\mathscr{F}_\psi(t')\rangle = 2D(\psi)\delta(t - t'), \quad (15.A.14)$$

where, in the physically interesting case of $r_1 = r_2 \cong \sqrt{\langle n\rangle}$, the phase diffusion rate is given by

$$D(\psi) = \frac{\mathscr{A}}{4\langle n\rangle}(1 - \cos\psi). \quad (15.A.15)$$

In this case (choosing $\phi = \pi$), our basic working equation (15.A.10) becomes

$$\dot{\psi} = S\Omega_r - (\mathscr{A} - \mathscr{C}_c)\sin\psi + \mathscr{F}_\psi(t), \quad (15.A.16)$$

where we recall that $\nu_{c1} - \nu_{c2} = S\Omega_r$ with Ω_r being the rotation rate of the gyroscope and S being the gyroscope scale factor of Section 4.1.

Problems

15.1 Consider a two-photon linear amplifier in which a coherence between the upper level $|a\rangle$ and the lower level $|c\rangle$ is introduced by a driving classical field of Rabi frequency Ω. The Hamiltonian for the atom–field system, in the interaction picture, is

$$\mathscr{V} = \hbar g[a^\dagger(|b\rangle\langle a| + |c\rangle\langle b|) + (|a\rangle\langle b| + |b\rangle\langle c|)a] - i\frac{\hbar\Omega}{2}(|a\rangle\langle c| - |c\rangle\langle a|),$$

where we have set the phase of the driving field $\phi = -\pi/2$. We assume equal decay rates γ for all three levels.

(a) Derive an equation of motion for the reduced density matrix of the field ρ.

(b) If we define the quadratures as $X_1 = (a + a^\dagger)/2$ and $X_2 = (a - a^\dagger)/2i$, then show that

$$(\Delta X_1)_t^2 = G_1(\Delta X_1)_0^2 + \mathscr{N}_1(G_1 - 1),$$
$$(\Delta X_2)_t^2 = G_2(\Delta X_2)_0^2 + \mathscr{N}_2(G_2 - 1).$$

Determine $G_1, G_2, \mathscr{N}_1$, and $\mathscr{N}_2$.

(c) Show that when $\gamma \gg \Omega$, we have a phase-insensitive amplifier. (Hint: see N. A. Ansari, J. Gea-Banacloche, and M. S. Zubairy, *Phys. Rev. A* **41**, 5179 (1990).)

15.2 Consider a phase-insensitive amplifier corresponding to the model discussed in Section 15.3. If the input field is thermal then show that the output field is also thermal.

15.3 Calculate the normally ordered photon number fluctuations

$$(:\Delta n:)_t^2 = \langle a^\dagger a^\dagger aa\rangle_t - (\langle a^\dagger a\rangle_t)^2$$

in the phase-sensitive amplifier discussed in Section 15.4. Show that the additive noise depends not only on the initial atomic parameters but also on the state of the input field. Can the additive noise become negative? If so, derive the conditions on the input field and the parameters of the amplifiers.

References and bibliography

CEL gyro and gravity wave detector

M. O. Scully and J. Gea-Banacloche, *Phys. Rev. A* **34**, 4043 (1986).
M. O. Scully, *Phys. Rev. A* **35**, 452 (1987).

Quantum theory of linear amplification

C. M. Caves, *Phys. Rev. D* **26**, 1817 (1982).
K. Zaheer and M. S. Zubairy, in *New Frontiers in Quantum Electrodynamics and Quantum Optics*, ed. A. O. Barut (Plenum, New York, 1990), p. 203.

Models for phase-insensitive amplifiers

K. Shimoda, H. Takahasi and C. H. Townes, *J. Phys. Soc. Japan* **12**, 687 (1957).
H. A. Haus and J. A. Mullen, *Phys. Rev.* **128**, 2407 (1962).
M. O. Scully and W. E. Lamb, Jr., *Phys. Rev.* **159**, 208 (1967).
S. Carusotto, *Phys. Rev. A* **11**, 1629 (1975).
N. B. Abraham and S. R. Smith, *Phys. Rev. A* **15**, 421 (1977).
E. B. Rockower, N. B. Abraham, and S. R. Smith, *Phys. Rev. A* **17**, 1100 (1978).
S. Friberg and L. Mandel, *Opt. Commun.* **46**, 141 (1983).
R. Loudon and T. J. Shepherd, *Opt. Acta* **31**, 1243 (1984).
C. K. Hong, S. Friberg, and L. Mandel, *J. Opt. Soc. Am. B* **2**, 494 (1985).
R. J. Glauber, in *Frontiers in Quantum Optics*, ed. E. R. Pike and S. Sarkar (Hilger, London 1986).
S. Stenholm, *Opt. Commun.* **58**, 177 (1986).

Models for phase-sensitive amplifiers

M. A. Dupertuis, S. M. Barnett, and S. Stenholm, *J. Opt. Soc. Am. B* **4**, 1102 (1987); *ibid.* **4**, 1124 (1987).
M. A. Dupertuis and S. Stenholm, *J. Opt. Soc. Am. B* **4**, 1094 (1987); *Phys. Rev. A* **37**, 1226 (1988).
G. J. Milburn, M. L. Steyn-Ross, and D. F. Walls, *Phys. Rev. A* **35**, 4443 (1987).
M. O. Scully and M. S. Zubairy, *Opt. Commun.* **66**, 303 (1988).
K. Zaheer and M. S. Zubairy, *Opt. Commun.* **69**, 37 (1988).
N. A. Ansari, J. Gea-Banacloche, and M. S. Zubairy, *Phys. Rev. A* **41** 5179 (1990).
M. S. Kim and V. Buzek, *Phys. Rev. A* **47**, 610 (1993).

Laser with injected squeezed vacuum

N. Imoto, H. A. Haus, and Y. Yamamoto, *Phys. Rev. A* **32**, 2287 (1985).

S. Stenholm, *Phys. Scripta* **T12**, 56 (1986)
J. Gea-Banacloche, *Phys. Rev. Lett.* **59**, 543 (1987).
M. A. M. Marte and D. F. Walls, *Phys. Rev. A* **37**, 1235 (1988).
M. A. M. Marte, H. Ritsch, and D. F. Walls, *Phys. Rev. A* **38**, 3577 (1988).
J. Bergou, N. Lu, and M. Scully *Opt. Commun.* **73**, 57 (1989).
C. Ginzel, R. Schack, and A. Schenzle, *J. Opt. Soc. Am. B* **8**, 1704 (1991).

CHAPTER 16

Squeezing via nonlinear optical processes

When light beams interact inside a nonlinear medium, new harmonics can be generated. Such is the case in the optical parametric and four-wave mixing processes. In a parametric amplifier, a pump beam generates signal and idler beams by interacting with a $\chi^{(2)}$ nonlinearity, whereas, in a four-wave mixing process, two beams interact with a signal beam in a $\chi^{(3)}$ nonlinearity and give rise to a conjugate beam. These processes have long been considered as important sources of the squeezed state of the radiation field, results that have been verified experimentally.

In this chapter, we present the quantum statistical properties of radiation in these nonlinear optical processes with special reference to squeezing.*

16.1 Degenerate parametric amplification

A parametric amplifier or a parametric down-converter essentially consists of two modes, usually called the signal and idler modes at frequencies ν_s and ν_i, respectively, coupled through a nonlinearity in, e.g., a nonlinear crystal having a $\chi^{(2)}$ coefficient by a pump mode at frequency ν_p such that

$$\nu_p = \nu_s + \nu_i. \tag{16.1.1}$$

The pump is usually assumed to be in a large amplitude coherent state and hence to produce a classically modulated interaction between the signal and the idler modes. If the signal and the idler frequencies are equal, the amplifier is said to operate in a degenerate mode. The fully

* An excellent treatment of the subject matter is to be found in the textbook of Walls and Milburn [1994].

Fig. 16.1
Parametric amplifier geometry.

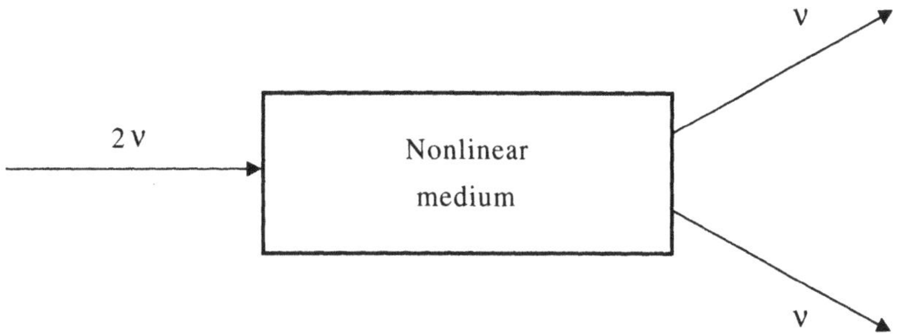

quantum mechanical interaction picture Hamiltonian for the nondegenerate parametric amplifier in the rotating-wave approximation is

$$\mathscr{V} = \hbar\kappa(a_s^\dagger a_i^\dagger b + a_s a_i b^\dagger). \tag{16.1.2}$$

Here b and $a_s(a_i)$ are the annihilation operators for the pump and signal (idler) modes, respectively, and κ is a coupling constant which depends upon the second-order susceptibility tensor that mediates the interaction.

Due to the underlying two-photon nature of the interaction, the parametric process has long been envisaged as a source of squeezed light, and this has experimentally been found to be the case. In nondegenerate operation, the field produced is a two-mode squeezed state, whereas a degenerate parametric amplifier (DPA) in which $\nu_p = 2\nu$ and $\nu_s = \nu_i = \nu$ (see Fig. 16.1) produces a single-mode squeezed state.

In this section, we consider degenerate parametric amplification and discuss the characteristics of and limitations on the generation of squeezed states.

The Hamiltonian for degenerate parametric amplification, in the interaction picture, is

$$\mathscr{V} = \hbar\kappa(a^{\dagger 2} b + a^2 b^\dagger). \tag{16.1.3}$$

In the parametric approximation, the pump field is treated classically and pump depletion is neglected. The Hamiltonian in Eq. (16.1.3) becomes

$$\mathscr{V} = \hbar\kappa\beta_p(a^{\dagger 2} e^{-i\phi} + a^2 e^{i\phi}), \tag{16.1.4}$$

where β_p and ϕ are the real amplitude and phase of the coherent pump field. This approximation is valid in the limits

$$\kappa t \to 0, \quad \beta_p \to \infty, \quad \kappa\beta_p t = \text{constant}. \tag{16.1.5}$$

The Heisenberg equations of motion for the signal mode are

$$\dot{a} = -i\Omega_p a^\dagger e^{-i\phi}, \tag{16.1.6}$$

$$\dot{a}^\dagger = i\Omega_p a e^{i\phi}. \tag{16.1.7}$$

Here $\Omega_p = 2\kappa\beta_p$ is the effective Rabi frequency. Solution of Eqs. (16.1.6) and (16.1.7) gives the following equations for operator expectation values:

$$a(t) = a_0 \cosh(\Omega_p t) - ia_0^\dagger \sinh(\Omega_p t)e^{-i\phi}, \tag{16.1.8}$$

$$a^\dagger(t) = a_0^\dagger \cosh(\Omega_p t) + ia_0 \sinh(\Omega_p t)e^{i\phi}, \tag{16.1.9}$$

where $a_0 = a(0)$. Note that for $\phi = \pi/2$, the above equations are the same as the transformation equations (2.7.6) and (2.7.7). For the signal initially in a vacuum state, the variances in the two quadratures $X_1 = (a + a^\dagger)/2$ and $X_2 = (a - a^\dagger)/2i$ are therefore given by (see Eqs. (2.7.15) and (2.7.16))

$$(\Delta X_1)_t^2 = \frac{1}{4}e^{-2u}, \tag{16.1.10}$$

$$(\Delta X_2)_t^2 = \frac{1}{4}e^{2u}, \tag{16.1.11}$$

where $u = \Omega_p t$ is the effective squeeze parameter. Equations (16.1.10) and (16.1.11) show that the output from a DPA can be squeezed to 100 percent and is in an ideal squeezed state. This makes it a particularly important source of squeezed radiation.

In the above analysis, we have assumed a perfectly coherent, monochromatic pump with a stabilized intensity. This is an ideal situation and in practice the quantum (as well as classical) noise in the laser pump leads to fluctuations in amplitude and phase. In Appendix 16.A we consider the effect of pump phase fluctuations associated with a finite linewidth on the squeezing properties of the signal mode. For the pump linewidth D, the variances in the two quadratures are given by Eqs. (16.A.19) and (16.A.20). In Fig. 16.2, we have plotted $(\Delta X_1)^2$ versus $\Omega_p t$ for various values of D/Ω_p. The fluctuations in the amplitude X_1 increase due to the phase fluctuations in the pump field and $(\Delta X_1)^2$ exhibits a minimum which decreases with increasing D/Ω_p. Equations (16.A.19) and (16.A.20) simplify considerably in the limit $D \ll t^{-1} \ll \Omega_p$. We then obtain

$$(\Delta X_1)_t^2 = \frac{1}{4}e^{-2u} + \frac{1}{4}e^{2u}\left(\frac{1}{2}Dt\right), \tag{16.1.12}$$

$$(\Delta X_2)_t^2 = \frac{1}{4}e^{2u}(1 - 2Dt). \tag{16.1.13}$$

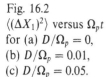

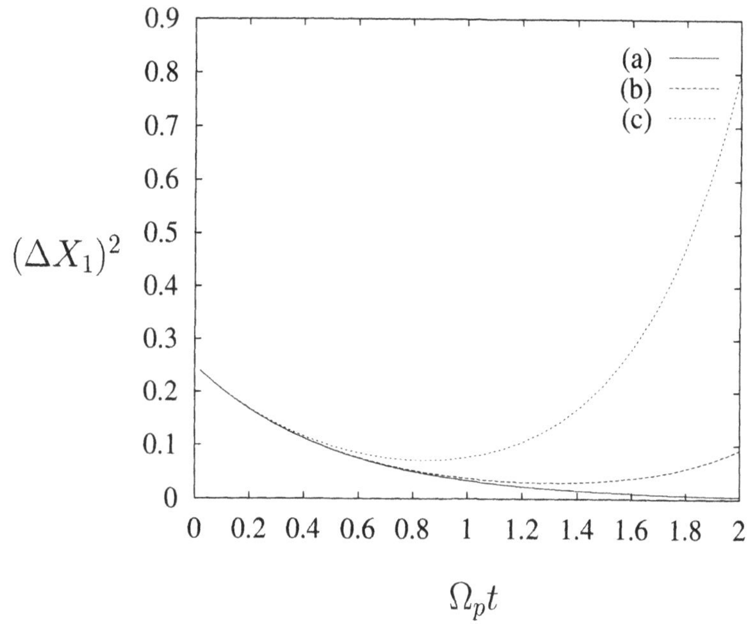

Fig. 16.2 $\langle(\Delta X_1)^2\rangle$ versus $\Omega_p t$ for (a) $D/\Omega_p = 0$, (b) $D/\Omega_p = 0.01$, (c) $D/\Omega_p = 0.05$.

It is clear from Eq. (16.1.12) that if the pump phase is off by ϕ, then the large uncertainty $\exp(2u)/4$ in the amplified quadrature $(\Delta X_2)_t^2$ is mixed into the uncertainty of the squeezed quadrature $(\Delta X_1)_t^2$ with the phase angle ϕ; Dt is, roughly speaking, the amount by which ϕ random walks in the time t.

16.2 Squeezing in an optical parametric oscillator

So far we have considered the open-ended model of the amplifier. When the nonlinear medium is placed within an optical cavity, oscillations build up inside and we have an optical parametric oscillator (OPO). This is the preferred method to generate squeezing, since the interaction is typically very weak and confining the light in a cavity helps to obtain a sizable effect by increasing considerably the interaction time. Discussion of the signal field should now include the losses from one or two end mirrors. For simplicity, we consider in the following one of the end mirrors to be partially transmitting.

Using the methods developed in Chapter 9, we can derive the following Heisenberg–Langevin equation for the field operators a and $a^\dagger$:

$$\dot{a} = -\Omega_p a^\dagger - \frac{\mathscr{C}}{2} a + F(t), \tag{16.2.1}$$

$$\dot{a}^\dagger = -\Omega_p a - \frac{\mathscr{C}}{2}a^\dagger + F^\dagger(t). \tag{16.2.2}$$

Here $\mathscr{C}$ represents the cavity decay and $F(t)$ is the associated noise operator (cf. Eq. (9.1.15)) and we have chosen the phase of the pump field $\phi = \pi/2$. The expectation value of the noise operator is zero, i.e.,

$$\langle F(t)\rangle = 0. \tag{16.2.3}$$

Also

$$\langle F(t)F(t')\rangle = \langle F^\dagger(t)F(t')\rangle = \langle F^\dagger(t)F^\dagger(t')\rangle = 0, \tag{16.2.4}$$

and

$$\langle F(t)F^\dagger(t')\rangle = \mathscr{C}\delta(t-t'). \tag{16.2.5}$$

In order to study the squeezing properties of the signal mode in the steady state we evaluate $\langle a\rangle$, $\langle a^2\rangle$, and $\langle a^\dagger a\rangle$. It follows from Eqs. (16.2.1) and (16.2.2) that the expectation values $\langle a\rangle$ and $\langle a^\dagger\rangle$ satisfy the following set of coupled equations:

$$\frac{d}{dt}\langle a\rangle = -\Omega_p\langle a^\dagger\rangle - \frac{\mathscr{C}}{2}\langle a\rangle, \tag{16.2.6}$$

$$\frac{d}{dt}\langle a^\dagger\rangle = -\Omega_p\langle a\rangle - \frac{\mathscr{C}}{2}\langle a^\dagger\rangle, \tag{16.2.7}$$

where we have used $\langle F(t)\rangle = \langle F^\dagger(t)\rangle = 0$. A solution of these equations is given by

$$\langle a\rangle_t = [\langle a_0\rangle\cosh(\Omega_p t) - \langle a^\dagger\rangle_0\sinh(\Omega_p t)]e^{-\mathscr{C}t/2}, \tag{16.2.8}$$

$$\langle a^\dagger\rangle_t = [\langle a_0^\dagger\rangle\cosh(\Omega_p t) - \langle a\rangle_0\sinh(\Omega_p t)]e^{-\mathscr{C}t/2}. \tag{16.2.9}$$

It is clear from the above equations that, for an OPO operating below threshold ($\mathscr{C}/2 > \Omega_p$), in the steady state we have

$$\langle a\rangle_{ss} = \langle a^\dagger\rangle_{ss} = 0. \tag{16.2.10}$$

Next we look at the bilinear quantities $\langle a^2\rangle$, $\langle a^{\dagger 2}\rangle$, and $\langle a^\dagger a\rangle$. It follows from Eqs. (16.2.3) and (16.2.4) that the quantities

$$A_1 = \langle a^2\rangle, \tag{16.2.11}$$

$$A_2 = \langle(aa^\dagger + a^\dagger a)\rangle, \tag{16.2.12}$$

$$A_3 = \langle a^{\dagger 2}\rangle, \tag{16.2.13}$$

satisfy the following set of equations

$$\dot{A}_1 = -\Omega_p A_2 - \mathscr{C}A_1 + \langle(aF + Fa)\rangle, \tag{16.2.14}$$

$$\dot{A}_2 = -2\Omega_p A_3 - 2\Omega_p A_1 - \mathscr{C}A_2 + \langle(aF^\dagger + F^\dagger a + a^\dagger F + Fa^\dagger)\rangle, \tag{16.2.15}$$

$$\dot{A}_3 = -\Omega_p A_2 - \mathscr{C}A_3 + \langle(a^\dagger F^\dagger + F^\dagger a^\dagger)\rangle. \tag{16.2.16}$$

In order to determine the quantities involving the noise operators F and $F^\dagger$, we first rewrite Eqs. (16.2.1) and (16.2.2) in the matrix form

$$\dot{\mathscr{A}} = -\mathscr{M}\mathscr{A} + \mathscr{F}, \tag{16.2.17}$$

$$\mathscr{A} = \begin{bmatrix} a \\ a^\dagger \end{bmatrix}, \tag{16.2.18}$$

$$\mathscr{M} = \begin{bmatrix} \frac{\mathscr{C}}{2} & \Omega_p \\ \Omega_p & \frac{\mathscr{C}}{2} \end{bmatrix}, \tag{16.2.19}$$

$$\mathscr{F} = \begin{bmatrix} F \\ F^\dagger \end{bmatrix}. \tag{16.2.20}$$

A formal solution of Eq. (16.2.17) is given by

$$\mathscr{A}(t) = e^{-\mathscr{M}t}\mathscr{A}(0) + \int_0^t e^{-\mathscr{M}(t-t')}\mathscr{F}(t')dt'. \tag{16.2.21}$$

On multiplying Eq. (16.2.21) by $\mathscr{F}^\dagger(t)$ from the left and using Eqs. (16.2.3)–(16.2.5) along with the assumption that the field operators at initial time $t = 0$ are statistically independent of the fluctuations, i.e., $\langle a(0)F(t)\rangle = 0$ etc., we obtain

$$\begin{aligned} \langle \mathscr{F}^\dagger(t)\mathscr{A}(t)\rangle &= \begin{pmatrix} \langle F^\dagger a\rangle & \langle Fa\rangle \\ \langle F^\dagger a^\dagger\rangle & \langle Fa^\dagger\rangle \end{pmatrix} \\ &= \frac{\mathscr{C}}{2}\begin{pmatrix} 0 & 0 \\ 0 & 1 \end{pmatrix}. \end{aligned} \tag{16.2.22}$$

In a similar manner

$$\begin{aligned} \langle \mathscr{A}^\dagger(t)\mathscr{F}(t)\rangle &= \begin{pmatrix} \langle a^\dagger F\rangle & \langle a^\dagger F^\dagger\rangle \\ \langle aF\rangle & \langle aF^\dagger\rangle \end{pmatrix} \\ &= \frac{\mathscr{C}}{2}\begin{pmatrix} 0 & 0 \\ 0 & 1 \end{pmatrix}. \end{aligned} \tag{16.2.23}$$

This means that all the correlation functions involving the noise operators in Eqs. (16.2.14)–(16.2.16) are zero except $\langle Fa^\dagger\rangle = \langle aF^\dagger\rangle = \mathscr{C}/2$. Equations (16.2.14)–(16.2.16) can therefore be simplified and we obtain

$$\dot{A}_1 = -\Omega_p A_2 - \mathscr{C}A_1, \tag{16.2.24}$$

$$\dot{A}_2 = -2\Omega_p A_3 + -2\Omega_p A_1 - \mathscr{C}A_2 + \mathscr{C}, \tag{16.2.25}$$

$$\dot{A}_3 = -\Omega_p A_2 - \mathscr{C}A_3. \tag{16.2.26}$$

These are three linear differential equations with three unknowns which can be solved exactly. We are interested here only in the steady state. Simple algebra leads to

$$A_1 = \langle a^2 \rangle_{ss} = \frac{-\mathscr{C}\Omega_p}{4\left[\left(\frac{\mathscr{C}}{2}\right)^2 - \Omega_p^2\right]}, \tag{16.2.27}$$

$$A_2 = \langle (aa^\dagger + a^\dagger a) \rangle_{ss} = \frac{\mathscr{C}^2}{4\left[\left(\frac{\mathscr{C}}{2}\right)^2 - \Omega_p^2\right]}, \tag{16.2.28}$$

$$A_3 = \langle a^{\dagger 2} \rangle_{ss} = \frac{-\mathscr{C}\Omega_p}{4\left[\left(\frac{\mathscr{C}}{2}\right)^2 - \Omega_p^2\right]}. \tag{16.2.29}$$

To see the squeezing, the field is expressed in terms of the Hermitian operators

$$X_1 = \frac{1}{2}(ae^{-i\theta/2} + a^\dagger e^{i\theta/2}), \tag{16.2.30}$$

$$X_2 = \frac{1}{2i}(ae^{-i\theta/2} - a^\dagger e^{i\theta/2}). \tag{16.2.31}$$

The variances of these operators in the steady state are

$$\begin{aligned}(\Delta X_1^2)_{ss} &= \frac{1}{4}\langle (aa^\dagger + a^\dagger a + a^2 e^{-i\theta} + a^{\dagger 2} e^{i\theta}) \rangle \\ &\quad - \frac{1}{4}\langle (ae^{-i\theta/2} + a^\dagger e^{i\theta/2}) \rangle^2 \\ &= \frac{1}{8}\frac{\mathscr{C}}{\left(\frac{\mathscr{C}}{2} + \Omega_p\right)},\end{aligned} \tag{16.2.32}$$

$$(\Delta X_2^2)_{ss} = \frac{1}{8}\frac{\mathscr{C}}{\left(\frac{\mathscr{C}}{2} - \Omega_p\right)}, \tag{16.2.33}$$

where we have taken $\theta = 0$. It is clear that the best squeezing in an OPO is achieved on the oscillation threshold ($\Omega_p = \mathscr{C}/2$) giving

$$(\Delta X_1)^2_{ss} = \frac{1}{8}. \tag{16.2.34}$$

This however represents only 50 percent squeezing below the vacuum level. The reason for this moderate squeezing can be attributed to the vacuum fluctuations that enter the cavity through the partially transmitting mirror. Alternatively, one may think of the OPO as producing pairs of correlated (signal + idler) photons and the cavity mirror as letting some single photons escape from each pair, so that some of the quantum correlation (and with it the squeezing) is lost.

A theoretical limit of 50 percent squeezing is not only unattractive, it does not reflect the true experimental situation, which is concerned with the field outside the cavity, whose degree of squeezing may actually be quite different from the inside. We address this problem in the next section.

16.3 Squeezing in the output of a cavity field

A theoretical limit of 50 percent on intracavity squeezing in an optical parametric oscillator (OPO) is somewhat disappointing. In this section we show that the field emitted from the OPO is however almost perfectly squeezed.* To begin with, we show that the relationship between the cavity field and the field outside is not trivial especially when one is interested in the antinormally ordered expectation values involved in squeezing. A partially reflecting out-coupling mirror in a cavity, not only lets the cavity field out, it allows the field from outside (vacuum) to leak into the cavity. The cavity field and the input field eventually become correlated over time. This correlation makes it possible for the residual fluctuations in the spectrum of the transmitted cavity field to cancel out with the corresponding fluctuations in the reflected input field and lead to almost perfect multi-mode, or *spectral* squeezing at an appropriate frequency.

The relationship between the fields outside and inside the cavity may be understood as follows. Consider first a field of amplitude $E_{\text{in}}(t)$ leaking into an empty cavity, of length L, through a single semitransparent mirror of reflectivity R and transmittivity T. The buildup of the cavity field amplitude E_{cav} is, at resonance, described by the following equation,

$$E_{\text{cav}}(t) = \sqrt{R} E_{\text{cav}}(t - 2L/c) + \sqrt{T} E_{\text{in}}(t), \tag{16.3.1}$$

where $2L/c$ is the cavity round-trip time and it has been assumed that there are no other losses than those due to the semitransparent mirror. Here, we have used the fact that there is a π phase shift when the reflection takes place from low to high index of refraction, and no phase shift for the opposite sequence. If the amplitudes in Eq. (16.3.1) are slowly varying $(E_{\text{cav}}(t - 2L/c) \simeq E_{\text{cav}}(t) - (2L/c)dE_{\text{cav}}/dt)$, and the reflectivity of the mirror is high $(R \simeq 1)$, this equation may be approximated by

$$\frac{dE_{\text{cav}}}{dt} = -\frac{\mathscr{C}}{2} E_{\text{cav}} + \frac{c\sqrt{T}}{2L} E_{\text{in}}, \tag{16.3.2}$$

where $\mathscr{C} \simeq cT/2L$. Equation (16.3.2) is the basic working equation for the so-called 'inside–outside' problem. A rigorous calculation of this important result in terms of the 'modes of the universe' method of Appendix 5.C is given in Appendix 16.B.

The resemblance between this equation and (16.2.1) suggests immediately that, in the quantized-field case, the Langevin force operator

* This important point was first made by Yurke [1984].

$F(t)$ may be identified with the vacuum field outside as it leaks into the cavity. Indeed, writing for the positive-frequency part of the input field $E_{in}(t)$

$$E_{in}^{(+)}(t) = \left(\frac{\hbar\nu_0}{4\pi\epsilon_0 cA}\right)^{1/2} \int b_{in}(\nu)e^{-i(\nu-\nu_0)t}d\nu, \tag{16.3.3}$$

where the $b_{in}(\nu)$ are operators for a multi-mode field in the vacuum state, it is easy to verify that $E_{in}^{(+)}(t)$ does indeed have the properties (16.2.3)–(16.2.5) of the Langevin operator F. In Eq. (16.3.3), the normalization adopted is appropriate for a one-dimensional travelling field, with a transverse quantization area A; the creation and annihilation operators obey the commutation relations $[b(\nu'), b^{\dagger}(\nu)] = \delta(\nu-\nu')$. The assumption has been made that all the integrals over frequency need only extend over an interval large compared to the cavity bandwidth, but small compared to the actual central frequency ν_0; hence $\sqrt{\nu_0}$ has been factored out of the integral.

When Eq. (16.3.3) is substituted in Eq. (16.3.2) and the positive-frequency part of the cavity field is written as $E_{cav}^{(+)} = (\hbar\nu_0/4\epsilon_0 AL)^{1/2}a$, we obtain

$$\frac{da}{dt} = -\frac{\mathscr{C}}{2}a + \sqrt{\frac{\mathscr{C}}{2\pi}} \int b_{in}(\nu)e^{-i(\nu-\nu_0)t}d\nu, \tag{16.3.4}$$

where we have used the approximation $\mathscr{C} \simeq cT/2L$ (implying large mirror reflectivity). This agrees with (16.2.1)(with $\Omega_p = 0$, i.e., for an empty cavity) if

$$F(t) = \sqrt{\frac{\mathscr{C}}{2\pi}} \int b_{in}(\nu)e^{-i(\nu-\nu_0)t}d\nu, \tag{16.3.5}$$

which, as mentioned above, yields the right correlation properties for $F(t)$ if the input field, on which the operators $b_{in}(\nu)$ act, is assumed to be in the vacuum state.

To understand the far-reaching consequences of this identification of the Langevin operator $F(t)$ with the incoming vacuum field, we observe that the output field is a combination of the reflected input field and the transmitted cavity field, i.e.,

$$E_{out}(t) = -\sqrt{R}E_{in}(t) + \sqrt{T}E_{cav}(t). \tag{16.3.6}$$

The possibility of observing greater squeezing outside the cavity than inside arises because the two fields to be added on the right-hand side of Eq. (16.3.6) are in fact correlated by virtue of Eq. (16.3.1). Equations (16.2.22) and (16.2.23) display the correlations between the Langevin operator F, proportional to E_{in}, and the intracavity field mode operator $a(t)$ (proportional to $E_{cav}(t)$) in the case when the cavity contains a parametric amplifier medium. The squeezing in the output field results from a partial cancelation of the fluctuations in the spectral components of E_{in} with the (already partly squeezed) fluctuations in $a(t)$.

Writing the equivalent of Eq. (16.3.6) for the positive-frequency parts of the field, and taking the Fourier transform, yields

$$b_{\text{out}}(\nu) = -\sqrt{R}b_{\text{in}}(\nu) + \sqrt{2\pi\mathscr{C}}\bar{a}(\nu), \tag{16.3.7}$$

where $a(t)$ has been written as

$$a(t) = \int \bar{a}(\nu)e^{-i(\nu-\nu_0)t}d\nu. \tag{16.3.8}$$

The explicit form of the Fourier components $\bar{a}(\nu)$ of the intracavity field is easily obtained from the exact solution (16.2.21). First, assume operation in steady state, so that the first term, proportional to the initial field inside the cavity, is gone. Then Eq. (16.2.21), with the identification (16.3.5), explicitly shows that the field inside the cavity grows over time from the input vacuum field. We obtain

$$\begin{aligned} a(t) &= \frac{1}{2}\int_0^t \left[e^{-\left(\frac{\mathscr{C}}{2}-\Omega_p\right)(t-t')} + e^{-\left(\frac{\mathscr{C}}{2}+\Omega_p\right)(t-t')}\right] F(t')dt' \\ &\quad -\frac{1}{2}\int_0^t \left[e^{-\left(\frac{\mathscr{C}}{2}-\Omega_p\right)(t-t')} - e^{-\left(\frac{\mathscr{C}}{2}+\Omega_p\right)(t-t')}\right] F^\dagger(t')dt'. \end{aligned} \tag{16.3.9}$$

Substituting Eq. (16.3.5) and carrying out the integration over time,

$$\begin{aligned} a(t) &= \frac{1}{2}\sqrt{\frac{\mathscr{C}}{2\pi}}\int d\nu \\ &\times\left\{\left[\frac{e^{-i(\nu-\nu_0)t} - e^{-\left(\frac{\mathscr{C}}{2}-\Omega_p\right)t}}{\frac{\mathscr{C}}{2}-\Omega_p - i(\nu-\nu_0)} + \frac{e^{-i(\nu-\nu_0)t} - e^{-\left(\frac{\mathscr{C}}{2}+\Omega_p\right)t}}{\frac{\mathscr{C}}{2}+\Omega_p - i(\nu-\nu_0)}\right] b_{\text{in}}(\nu)\right. \\ &\quad \left. - \left[\frac{e^{i(\nu-\nu_0)t} - e^{-\left(\frac{\mathscr{C}}{2}-\Omega_p\right)t}}{\frac{\mathscr{C}}{2}-\Omega_p + i(\nu-\nu_0)} - \frac{e^{i(\nu-\nu_0)t} - e^{-\left(\frac{\mathscr{C}}{2}+\Omega_p\right)t}}{\frac{\mathscr{C}}{2}+\Omega_p + i(\nu-\nu_0)}\right] b_{\text{in}}^\dagger(\nu)\right\}. \end{aligned} \tag{16.3.10}$$

In the steady-state limit, i.e., the long time limit, the decaying exponentials in (16.3.10) may be neglected. We can also change the variable of integration in the second integral, from ν to the reflected frequency $2\nu_0 - \nu$, with the result

$$\begin{aligned} a(t) &= \sqrt{\frac{\mathscr{C}}{2\pi}} \\ &\int\left\{\frac{\left[\frac{\mathscr{C}}{2} - i(\nu-\nu_0)\right] b_{\text{in}}(\nu) - \Omega_p b_{\text{in}}^\dagger(2\nu_0-\nu)}{\left[\frac{\mathscr{C}}{2} - i(\nu-\nu_0)\right]^2 - \Omega_p^2}\right\} e^{-i(\nu-\nu_0)t}d\nu. \end{aligned} \tag{16.3.11}$$

From Eqs. (16.3.11) and (16.3.8) we can find $\bar{a}(\nu)$ and substitute it in Eq. (16.3.7). The result is

$$b_{\text{out}}(\nu) = -\sqrt{R}\, b_{\text{in}}(\nu) + \mathscr{C}\frac{\left[\frac{\mathscr{C}}{2} - i(\nu - \nu_0)\right] b_{\text{in}}(\nu) - \Omega_p b_{\text{in}}^\dagger(2\nu_0 - \nu)}{\left[\frac{\mathscr{C}}{2} - i(\nu - \nu_0)\right]^2 - \Omega_p^2}. \quad (16.3.12)$$

In what follows, consistently with the earlier approximations, we assume that the reflectivity is so large that R in the first term may be set equal to 1. Equation (16.3.12) exhibits the possibility of the cancelations of the fluctuations in the spectral components of the intracavity field (last term) with those of the reflected input field (first term). The resultant squeezing may be much greater than the intracavity one.

Taking the Hermitian conjugate of (16.3.12), evaluated at the reflected frequency $2\nu_0 - \nu$, and adding it to (16.3.12) we obtain

$$b_{\text{out}}(\nu) + b_{\text{out}}^\dagger(2\nu_0 - \nu) = \left[\frac{\mathscr{C}}{\frac{\mathscr{C}}{2} - i(\nu - \nu_0) + \Omega_p} - 1\right] [b_{\text{in}}(\nu) + b_{\text{in}}^\dagger(2\nu_0 - \nu)], \quad (16.3.13)$$

from which the degree of squeezing at any particular frequency of the output field follows immediately from that of the input. Note that this equation is quite general, in that it holds even if the input field is not in the vacuum state.

Next, we define the quadrature variables

$$X_{1\text{out}}(\nu_0) = \frac{1}{2}[b_{\text{out}}(\nu_0) + b_{\text{out}}^\dagger(\nu_0)],$$

$$X_{1\text{in}}(\nu_0) = \frac{1}{2}[b_{\text{in}}(\nu_0) + b_{\text{in}}^\dagger(\nu_0)], \quad (16.3.14)$$

at the central frequency $\nu = \nu_0$. It follows from Eq. (16.3.13), that the ratio of the variances in the output and the input quadratures $X_{1\text{out}}$ and $X_{1\text{in}}$ is given by

$$\frac{[\Delta X_{1\text{out}}(\nu_0)]^2}{[\Delta X_{1\text{in}}(\nu_0)]^2} = \left(\frac{\mathscr{C}/2 - \Omega_p}{\mathscr{C}/2 + \Omega_p}\right)^2. \quad (16.3.15)$$

For the OPO near threshold, $\Omega_p \simeq \mathscr{C}/2$, and Eq. (16.3.15) shows that essentially perfect squeezing is obtained in this particular quadrature. (Note that this holds regardless of the input.) The noise reduction for other frequencies, i.e., the spectrum of squeezing, is derived from Eq. (16.3.13):

$$\frac{S_{\text{out}}}{S_{\text{in}}} = 1 - \frac{2\mathscr{C}\Omega_p}{\left(\frac{\mathscr{C}}{2} + \Omega_p\right)^2 + (\nu - \nu_0)^2}. \quad (16.3.16)$$

This is an inverted Lorentzian whose width is of the order of the cavity linewidth (since, below threshold, $\Omega_p \leq \mathscr{C}/2$).

16.4 Four-wave mixing

Four-wave mixing is a nonlinear process in which two planar counter-propagating pump waves interact in a nonlinear medium with a probe field entering at an arbitrary angle to the pump waves and yield a fourth (output) wave. The nature of the nonlinear medium in four-wave mixing enters through the nonlinear susceptibility $\chi^{(3)}$. The two pump waves and the probe wave couple through $\chi^{(3)}$ to produce the fourth wave, which is proportional to the spatial complex conjugate of the probe wave.

A significant interest in the four-wave mixing process stems from the possibility of generating phase conjugate waves with applications in adaptive optics. The phenomenon of phase conjugation involves a generation of the wave which contains the complex conjugate of only the spatial part of the incident wave, leaving the temporal part unchanged, i.e., a probe wave $E(\mathbf{r},t) = \mathrm{Re}\{\mathscr{E}(\mathbf{r})\exp[i(\mathbf{k}\cdot\mathbf{r} - \nu t)]\}$ is converted into a phase conjugate wave $E_{pc}(\mathbf{r},t) = \mathrm{Re}\{\mathscr{E}^*(\mathbf{r})\exp[-i(\mathbf{k}\cdot\mathbf{r} + \nu t)]\}$. Equivalently, the spatial part of $E(\mathbf{r},t)$ remains unchanged and the sign of t is reversed; phase conjugation is thus equivalent to *time reversal*. This property of phase conjugate *mirrors* can be used to restore severely aberrated waves to their original state on passing through the distorting medium twice.

Four-wave mixing is also an important source of squeezed light. The first ever generation of a squeezed state was in this system.

In this section, we present a theory of four-wave mixing in a $\chi^{(3)}$ medium and discuss the generation of a squeezed state.

16.4.1 Amplification and oscillation in four-wave mixing

Consider the geometry shown in Fig. 16.3. Two intense pump waves E_2 and $E_{2'}$ travelling in opposite directions interact with two weak fields E_1 and E_3, all of the same frequency ν, inside a nonlinear medium characterized by a third-order nonlinear susceptibility $\chi^{(3)}$. The fields E_1 and E_3 also travel in opposite directions, but different from those of E_2 and $E_{2'}$. The fields E_j ($j = 1, 2, 2', 3$) are assumed to be linearly polarized and are given by

$$E_j(\mathbf{r},t) = \frac{1}{2}\mathscr{E}_j(\mathbf{r})e^{i(\mathbf{k}_j\cdot\mathbf{r}-\nu t)} + \text{c.c.}, \tag{16.4.1}$$

where $\mathscr{E}_j(\mathbf{r})$ are slowly varying quantities which are in general complex. The wave directions imply that

$$\mathbf{k}_1 + \mathbf{k}_3 = 0, \qquad \mathbf{k}_2 + \mathbf{k}_{2'} = 0. \tag{16.4.2}$$

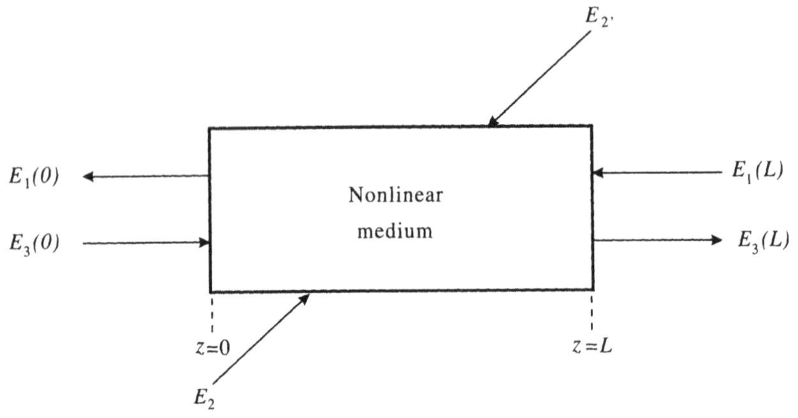

Fig. 16.3
Four-wave mixing geometry.

We now show that when, in the presence of the pump fields E_2 and $E_{2'}$, a signal field E_1 is incident on the nonlinear crystal, a new field E_3 is created that is the *phase conjugate* of E_1.

We start with the wave equation (see Eq. (5.4.15))

$$\nabla^2 E - \frac{1}{c^2}\frac{\partial^2 E}{\partial t^2} = \mu_0 \frac{\partial^2 P}{\partial t^2}, \tag{16.4.3}$$

where

$$E = E_1 + E_2 + E_{2'} + E_3, \tag{16.4.4}$$

is the total field and P is the nonlinear polarization that couples the waves. For a $\chi^{(3)}$ medium the nonlinear polarization within the medium is given by

$$P = \chi^{(3)} E^3. \tag{16.4.5}$$

In view of Eq. (16.4.4), it is evident that the polarization contains a large number of terms with different spatial dependences. We, however, retain only those terms with spatial dependence $\exp(i\mathbf{k}_j \cdot \mathbf{r})$ ($j = 1, 2, 2', 3$) because they can act as the phase-matched source terms in Eq. (16.4.3) for the four interacting fields (16.4.1).

We now substitute for $E(\mathbf{r}, t)$ and $P(\mathbf{r}, t)$ from Eqs. (16.4.1), (16.4.4), and (16.4.5) in Eq. (16.4.3). Also, we let z be the spatial coordinate measured in the direction of propagation of the E_1 field. We also use the slowly varying approximation

$$\left|\frac{d^2\mathscr{E}_i}{dz^2}\right| \ll \left|k_i \frac{d\mathscr{E}_i}{dz}\right|. \tag{16.4.6}$$

Under these conditions, the signal and conjugate fields obey the following coupled-amplitude equations:

$$\frac{d\mathscr{E}_1}{dz} = \left(\frac{i\nu}{\epsilon_0 c}\right)\tilde{P}_1, \tag{16.4.7}$$

$$\frac{d\mathscr{E}_3}{dz} = -\left(\frac{i\nu}{\epsilon_0 c}\right)\tilde{P}_3, \tag{16.4.8}$$

where

$$\tilde{P}_1 = \frac{3\chi^{(3)}}{8}(\mathscr{E}_1^2\mathscr{E}_1^* + 2\mathscr{E}_1\mathscr{E}_3\mathscr{E}_3^* + 2\mathscr{E}_1\mathscr{E}_2\mathscr{E}_2^* + 2\mathscr{E}_1\mathscr{E}_{2'}\mathscr{E}_{2'}^* + 2\mathscr{E}_2\mathscr{E}_{2'}\mathscr{E}_3^*), \tag{16.4.9}$$

$$\tilde{P}_3 = \frac{3\chi^{(3)}}{8}(\mathscr{E}_3^2\mathscr{E}_3^* + 2\mathscr{E}_3\mathscr{E}_1\mathscr{E}_1^* + 2\mathscr{E}_3\mathscr{E}_2\mathscr{E}_2^* + 2\mathscr{E}_3\mathscr{E}_{2'}\mathscr{E}_{2'}^* + 2\mathscr{E}_2\mathscr{E}_{2'}\mathscr{E}_1^*). \tag{16.4.10}$$

Since the signal and conjugate fields $\mathscr{E}_1$ and $\mathscr{E}_3$, respectively, are assumed much weaker than the pump fields $\mathscr{E}_2$ and $\mathscr{E}_{2'}$, we can neglect the first two terms in Eqs. (16.4.9) and (16.4.10). We also assume that the pump fields are not depleted. We can then write Eqs. (16.4.7) and (16.4.8) as

$$\frac{d\mathscr{E}_1}{dz} = i\kappa_1\mathscr{E}_1 + i\kappa\mathscr{E}_3^*, \tag{16.4.11}$$

$$\frac{d\mathscr{E}_3}{dz} = -i\kappa_1\mathscr{E}_3 - i\kappa\mathscr{E}_1^*, \tag{16.4.12}$$

where

$$\kappa_1 = \frac{3\nu\chi^{(3)}}{4\epsilon_0 c}(|\mathscr{E}_2|^2 + |\mathscr{E}_{2'}|^2), \tag{16.4.13}$$

$$\kappa = \frac{3\nu\chi^{(3)}}{4\epsilon_0 c}\mathscr{E}_2\mathscr{E}_{2'}. \tag{16.4.14}$$

This is a coupled set of linear equations which can be solved exactly subject to the boundary conditions.

We can however reduce this set of equations to a simpler form by a change of variables

$$\tilde{\mathscr{E}}_1 = \mathscr{E}_1 e^{-i\kappa_1 z}, \tag{16.4.15}$$

$$\tilde{\mathscr{E}}_3 = \mathscr{E}_3 e^{i\kappa_1 z}. \tag{16.4.16}$$

Equations (16.4.11) and (16.4.12) then reduce to

$$\frac{d\tilde{\mathscr{E}}_1}{dz} = i\kappa\tilde{\mathscr{E}}_3^*, \tag{16.4.17}$$

$$\frac{d\tilde{\mathscr{E}}_3}{dz} = -i\kappa\tilde{\mathscr{E}}_1^*. \tag{16.4.18}$$

The generated field $\tilde{\mathscr{E}}_3$ is driven only by the complex conjugate of the signal field amplitude, thus leading to phase conjugation.

If the fields are specified at the ends of the nonlinear crystal, $z = 0$ and $z = L$, then the solution of the set of equations (16.4.17) and (16.4.18) is

$$\tilde{\mathscr{E}}_1^*(z) = -\frac{i|\kappa|\sin(|\kappa|z)}{\kappa\cos(|\kappa|L)}\tilde{\mathscr{E}}_3(L) + \frac{\cos[|\kappa|(z-L)]}{\cos(|\kappa|L)}\tilde{\mathscr{E}}_1^*(0), \quad (16.4.19)$$

$$\tilde{\mathscr{E}}_3(z) = \frac{\cos(|\kappa|z)}{\cos(|\kappa|L)}\tilde{\mathscr{E}}_3(L) - \frac{i\kappa\sin[|\kappa|(z-L)]}{|\kappa|\cos(|\kappa|L)}\tilde{\mathscr{E}}_1^*(0). \quad (16.4.20)$$

A particularly interesting case is that of a single input $\tilde{\mathscr{E}}_1(0)$ at $z = 0$ with $\tilde{\mathscr{E}}_3(L) = 0$. The output field amplitudes are then given by

$$\tilde{\mathscr{E}}_1^*(L) = \frac{\tilde{\mathscr{E}}_1^*(0)}{\cos(|\kappa|L)}, \quad (16.4.21)$$

$$\tilde{\mathscr{E}}_3(0) = \frac{i\kappa}{|\kappa|}\tan(|\kappa|L)\tilde{\mathscr{E}}_1^*(0). \quad (16.4.22)$$

It is interesting to note that, for $\pi/4 < |\kappa|L < 3\pi/4$,

$$|\tilde{\mathscr{E}}_3(0)| > |\tilde{\mathscr{E}}_1(0)|, \quad (16.4.23)$$

i.e., the reflected wave amplitude is larger than that of the input. Thus the device acts as a reflection amplifier. It also follows from Eq. (16.4.21) that

$$|\tilde{\mathscr{E}}_1(L)| > |\tilde{\mathscr{E}}_1(0)|, \quad (16.4.24)$$

i.e., the transmitted signal wave is always more intense than the incident wave. The amplifications of the reflected phase conjugate wave and the transmitted signal wave take place because the medium is actively pumped by externally applied waves, which can supply energy. In the *photon picture*, we can describe the process of four-wave mixing as a process in which one photon is annihilated from each of the two pump waves and one photon is added to each of the signal and phase conjugate field.

When $|\kappa|L = \pi/2$

$$\frac{|\tilde{\mathscr{E}}_3(0)|}{|\tilde{\mathscr{E}}_1(0)|} \to \infty, \qquad \frac{|\tilde{\mathscr{E}}_1(L)|}{|\tilde{\mathscr{E}}_1(0)|} \to \infty, \quad (16.4.25)$$

which corresponds to oscillation. This however requires a careful consideration as the assumption relating to the no-pump-depletion ceases to apply as $|\kappa|L = \pi/2$.

16.4.2 Squeezing in four-wave mixing

We can extend the present classical analysis to a phenomenological quantum treatment. This can be done if we replace the field variables $\tilde{\mathscr{E}}_1$ and $\tilde{\mathscr{E}}_3$ in Eqs. (16.4.17) and (16.4.18) by the operators a_1 and a_3. We then obtain

$$\frac{da_1}{dz} = i\kappa a_3^\dagger, \tag{16.4.26}$$

$$\frac{da_3}{dz} = -i\kappa a_1^\dagger. \tag{16.4.27}$$

For simplicity, we assume $\kappa = |\kappa|$. The solution of these operator equations can be obtained from the classical solutions (16.4.19) and (16.4.20):

$$a_1(L) = i\tan(\kappa L)a_3^\dagger(L) + \sec(\kappa L)a_1(0), \tag{16.4.28}$$

$$a_3(0) = \sec(\kappa L)a_3(L) + i\tan(\kappa L)a_1^\dagger(0). \tag{16.4.29}$$

These solutions are analogous to the corresponding solutions for the field operators in the parametric amplification process discussed in Section 16.1. We are therefore tempted to study the squeezing properties of the radiation field in the four-wave mixing process.

We define the quadrature components of the signal and the conjugate fields:

$$a_{j1} = \frac{1}{2}(a_j + a_j^\dagger), \tag{16.4.30}$$

$$a_{j2} = \frac{1}{2i}(a_j - a_j^\dagger), \tag{16.4.31}$$

where $j = 1, 3$. We assume the input fields $a_1(0)$ and $a_3(L)$ to be in the coherent state. For the fluctuations in the quadrature components for the output fields, we get

$$\begin{aligned}\Delta a_{1\ell}^2(L) &= \Delta a_{3\ell}^2(0)\\ &= \frac{1}{4}[1 + 2\tan^2(\kappa L)],\end{aligned} \tag{16.4.32}$$

where $\ell = 1, 2$. Equation (16.4.32) shows that the fluctuations in the output field are increased in the process of four-wave mixing. This is not surprising because the output fields are amplified and we have already learned that, in general, amplification adds noise.

If, instead, we consider the modes d_1 and d_3 which are formed by a linear combination of the input modes:

$$d = \frac{1}{\sqrt{2}}(a_1 + a_3)e^{i\theta}, \tag{16.4.33}$$

then the canonically conjugate Hermitian amplitude operators are given by

$$d_1 = \frac{1}{2}(d + d^\dagger), \tag{16.4.34}$$

$$d_2 = \frac{1}{2i}(d - d^\dagger). \tag{16.4.35}$$

The squared variance of the operator d_1 in terms of the correlation functions of the field operators $a_1(L)$ and $a_3(0)$ is

$$\begin{aligned}(\Delta d_1)^2 &= \langle d_1^2\rangle - \langle d_1\rangle^2 \\ &= \frac{1}{8}\Big\{1 + [\langle a_1^2(L)\rangle + \langle a_3^2(0)\rangle + \langle a_1(L)a_3(0)\rangle + \langle a_3(0)a_1(L)\rangle]e^{2i\theta} \\ &\quad + \langle a_1^\dagger(L)a_1(L)\rangle + \langle a_3^\dagger(0)a_3(0)\rangle + \langle a_1^\dagger(L)a_3(0)\rangle \\ &\quad + \langle a_3^\dagger(0)a_1(L)\rangle\Big\} - \frac{1}{8}\Big\{[\langle a_1(L)\rangle + \langle a_3(0)\rangle]^2 e^{2i\theta} \\ &\quad + [\langle a_1(L)\rangle + \langle a_3(0)\rangle][\langle a_1^\dagger(L)\rangle + \langle a_3^\dagger(0)\rangle]\Big\} + \text{c.c.}\end{aligned} \tag{16.4.36}$$

A similar expression can be written for $(\Delta d_2)^2$. It follows, on substituting for $a_1(L)$ and $a_3(0)$ from Eqs. (16.4.28) and (16.4.29), respectively, in Eq. (16.4.36) that, when $\theta = \pi/4$ and the input fields are in coherent states,

$$(\Delta d_1)^2 = \frac{1}{4}[\sec(\kappa L) - \tan(\kappa L)]^2. \tag{16.4.37}$$

Similarly

$$(\Delta d_2)^2 = \frac{1}{4}[\sec(\kappa L) + \tan(\kappa L)]^2. \tag{16.4.38}$$

As κL grows, the fluctuations in d_1 are reduced below 1/4, and eventually vanish as $\kappa L \to \pi/2$. The amplitude d_1 therefore is squeezed. From Eqs. (16.4.37) and (16.4.38), we obtain

$$\Delta d_1 \Delta d_2 = \frac{1}{4}, \tag{16.4.39}$$

i.e., the squeezed state is an ideal squeezed state.

16.A Effect of pump phase fluctuations on squeezing in degenerate parametric amplification*

In Section 16.1, we pointed out that, in degenerate parametric amplification, the assumption of a perfectly coherent monochromatic pump with a stabilized intensity is an idealized situation. In practice, the laser pump has a finite bandwidth which arises due to the phase

* This section follows Wódkiewicz and Zubairy [1983].

fluctuations of the field. In this appendix we consider the effect of phase fluctuations of the laser pump in a parametric amplifier on the quadrature variances, especially its effect on the squeezing of the signal field.

The phase diffusion can be represented by

$$\phi(t) = \phi_0 + \delta\phi(t), \tag{16.A.1}$$

where ϕ_0 is a fixed phase which in the present case we take to be equal to $\pi/2$ and $\delta\phi(t)$ is the random phase with zero mean, i.e., $\langle\delta\phi(t)\rangle = 0$. We assume that $\delta\phi(0) = 0$. As discussed in Section 11.4, a Lorentzian lineshape of the pump field with phase-induced bandwidth D is obtained if the derivative of the random phase $\delta\phi(t)$ is white noise, i.e.,

$$\langle\delta\dot{\phi}(t)\delta\dot{\phi}(t')\rangle = 2D\delta(t-t'). \tag{16.A.2}$$

We first determine the expectation value of the operator as $aa^\dagger + a^\dagger a$. From the Heisenberg equations of motion (16.1.6) and (16.1.7) we obtain the following stochastic multiplicative equation:

$$\dot{\Phi} = [M_0 + i\delta\dot{\phi}(t)M_1]\Phi, \tag{16.A.3}$$

with

$$\Phi = \begin{pmatrix} a^\dagger a + aa^\dagger \\ a^2 e^{i\delta\phi} \\ (a^\dagger)^2 e^{-i\delta\phi} \end{pmatrix}, \tag{16.A.4}$$

and

$$M_0 = \begin{pmatrix} 0 & -2\Omega_p & -2\Omega_p \\ -\Omega_p & 0 & 0 \\ -\Omega_p & 0 & 0 \end{pmatrix}, \tag{16.A.5}$$

$$M_1 = \begin{pmatrix} 0 & 0 & 0 \\ 0 & 1 & 0 \\ 0 & 0 & -1 \end{pmatrix}. \tag{16.A.6}$$

Note that the operator $a^\dagger a + aa^\dagger$ is not coupled directly to a^2 and $(a^\dagger)^2$. In order to calculate the stochastic expectation value of $a^\dagger a + aa^\dagger$ we need to evaluate two auxiliary quantities $\langle a^2 e^{i\delta\phi}\rangle$ and $\langle(a^{\dagger 2})e^{-i\delta\phi}\rangle$. For the fluctuating random phase $\delta\phi(t)$ given by the Wiener–Levy stochastic process it can be shown that the following exact equation is satisfied:

$$\langle\dot{\Phi}\rangle = (M_0 - DM_1^2)\langle\Phi\rangle \tag{16.A.7}$$

for arbitrary form of the time-independent matrices M_0 and M_1. This matrix equation specified for Φ, M_0, and M_1 given by Eqs.

(16.A.4)–(16.A.6) can be solved exactly using, for example, the Laplace-transform techniques. For the vacuum initial state of the signal mode, we obtain

$$\langle a^\dagger a + a a^\dagger \rangle_t = \int_C \frac{dz}{2\pi i} \frac{e^{zt}(z+D)}{z + zD - 4\Omega_p^2}. \tag{16.A.8}$$

Computing the roots of the algebraic equation in the denominator in Eq. (16.A.8) and choosing properly the contour of integration C we find the explicit time evolution as follows:

$$\langle a^\dagger a + a a^\dagger \rangle_t = \left[\frac{1}{2\beta}\sinh(\mu t) + \cosh(\mu t)\right] e^{-Dt/2}, \tag{16.A.9}$$

where

$$\mu = \left(\frac{1}{4}D^2 + 4\Omega_p^2\right)^{1/2}. \tag{16.A.10}$$

Finally, the last operator required for squeezing amplitudes is the stochastic average of $\langle a^2 \rangle$. Again from the Heisenberg equations of motion with fluctuating phase we generate a multiplicative stochastic equation of the form given by Eq. (16.A.3) with

$$\Phi = \begin{pmatrix} a^2 \\ e^{-i\delta\phi}(a^\dagger a + a a^\dagger) \\ e^{-2i\delta\phi}(a^\dagger)^2 \end{pmatrix}, \tag{16.A.11}$$

and different forms of M_0 and M_1,

$$M_0 = \begin{pmatrix} 0 & -\Omega_p & 0 \\ -2\Omega_p & 0 & -2\Omega_p \\ 0 & -\Omega_p & 0 \end{pmatrix}, \tag{16.A.12}$$

$$M_1 = \begin{pmatrix} 0 & 0 & 0 \\ 0 & -1 & 0 \\ 0 & 0 & -2 \end{pmatrix}. \tag{16.A.13}$$

As in the previous case the stochastic expectation value of $\langle \Phi \rangle$ satisfies an exact differential equation (16.A.7) with matrices M_0 and M_1 given now by expressions (16.A.12) and (16.A.13). With our specific initial condition, the Laplace-transform solution has the following exact form:

$$\langle a^2 \rangle_t = -\int_C \frac{dz}{2\pi i} \frac{e^{zt}\Omega_p(z+4D)}{[z^3 + 5Dz^2 + (4D^2 - 4\Omega_p^2)z - 8\Omega_p^2 D]}. \tag{16.A.14}$$

The exact time dependence, accordingly, has the form

$$\langle a^2 \rangle_t = -\sum_{\substack{i,j,k \\ i \neq j \neq k}} \frac{e^{\lambda_i t}\Omega_p(\lambda_i + 4D)}{(\lambda_i - \lambda_j)(\lambda_i - \lambda_k)}, \tag{16.A.15}$$

where λ_i are the roots of the following cubic equation:

$$\lambda^3 + 5D\lambda^2 + 4(D^2 - \Omega_p^2)\lambda - 8\Omega_p^2 D = 0. \tag{16.A.16}$$

These roots can be obtained exactly using the Cardano formula. We can, however, get a reasonable understanding of the physics by solving this cubic equation in the realistic limit of small phase fluctuations (as compared with the driving Rabi frequency, i.e., $D \ll \Omega_p$). In this limit,

$$\begin{aligned} \lambda_1 &\simeq -2D, \\ \lambda_2 &\simeq 2\Omega_p - \frac{3}{2}D, \\ \lambda_3 &\simeq -2\Omega_p - \frac{3}{2}D, \end{aligned} \tag{16.A.17}$$

and accordingly

$$\begin{aligned} \langle a^2 \rangle_t \simeq\ & e^{-2Dt}\frac{D\Omega_p}{(2\Omega_p^2 - D^2/8)} \\ & - e^{(2\Omega_p - 3D/2)t}\frac{(2\Omega_p + 5D/2)}{4(2\Omega_p + D/2)} \\ & + e^{-(2\Omega_p + 3D/2)t}\frac{(2\Omega_p - 5D/2)}{4(2\Omega_p - D/2)}. \end{aligned} \tag{16.A.18}$$

From Eqs. (16.A.9) and (16.A.18) we obtain the following formulas for the variance of the Hermitian amplitudes with laser phase fluctuations (with approximated roots λ_i):

$$\begin{aligned} (\Delta X_1)_t^2 \simeq \frac{1}{4}\Bigg[& e^{-2Dt}\frac{2D\Omega_p}{(2\Omega_p^2 - D^2/8)} - e^{(2\Omega_p - 3D/2)t}\frac{(2\Omega_p + 5D/2)}{2(2\Omega_p + D/2)} \\ & + e^{-(2\Omega_p + 3D/2)t}\frac{(2\Omega_p - 5D/2)}{2(2\Omega_p - D/2)} + \frac{e^{-Dt/2}D\sinh(2\Omega_p t)}{(D^2 + 16\Omega_p^2)^{1/2}} \\ & + e^{-Dt/2}\cosh(2\Omega_p t)\Bigg], \end{aligned} \tag{16.A.19}$$

$$\begin{aligned} (\Delta X_2)_t^2 \simeq -\frac{1}{4}\Bigg[& e^{-2Dt}\frac{2D\Omega_p}{(2\Omega_p^2 - D^2/8)} - e^{(2\Omega_p - 3D/2)t}\frac{(2\Omega_p + 5D/2)}{2(2\Omega_p + D/2)} \\ & + e^{-(2\Omega_p + 3D/2)t}\frac{(2\Omega_p - 5D/2)}{2(2\Omega_p - D/2)} \\ & - \frac{e^{-Dt/2}D\sinh(2\Omega_p t)}{(D^2 + 16\Omega_p^2)^{1/2}} + e^{-Dt/2}\cosh(2\Omega_p t)\Bigg]. \end{aligned} \tag{16.A.20}$$

These are the required expressions for the variance of the Hermitian amplitudes in the presence of the pump fluctuations of the pump field.

16.B Quantized field treatment of input–output formalism leading to Eq. (16.3.4)*

Consider a one-sided cavity that is bounded by a perfectly reflecting mirror at $z = L$ and by a partially transmitting mirror at $z = 0$. Let the latter mirror be characterized by reflection and transmission coefficients $-\sqrt{R}$ and $\sqrt{T}$ for fields normally incident on it from the left and by $\sqrt{R}$ and $\sqrt{T}$ for fields normally incident on it from the right. Because $T \neq 0$, normal modes entirely confined to the interior of the cavity cannot, strictly speaking, be defined. A way to retain the usefulness of the concept of normal modes for the present problem is to introduce an auxiliary perfect mirror at $z = -L_0 < 0$ and to take $L_0 \to \infty$ in the end, as in Appendix 5.C (see Fig. 5.8). The positive frequency part of the quantized field can now be defined as (see Eq. (1.1.15))

$$E^{(+)}(z,t) = \sum_k \mathscr{E}_k a_k(t) U_k(z), \tag{16.B.1}$$

where $\mathscr{E}_k$ is given by

$$\mathscr{E}_k = \left(\frac{\hbar\nu_k}{\epsilon_0 A L_0}\right)^{1/2}, \tag{16.B.2}$$

and the mode functions $U_k(z)$ are (Eq. (5.C.2))

$$U_k(z) = \begin{cases} \xi_k \sin k(z + L_0) & (z < 0), \\ M_k \sin k(z - L) & (z > 0). \end{cases} \tag{16.B.3}$$

We note that Eq. (16.B.1) defines the quantized field so that it is an easy matter to identify the incoming, outgoing, and intracavity field operators. The rightward and leftward travelling parts of (16.B.1) in the region $z < 0$, say at $z = 0^-$, are the input field and the output field, respectively, while the field in the region $z > 0$ (say its rightward travelling part at $z = 0^+$), is the intracavity field. The positive-frequency parts of these fields are

$$E_{\text{in}}^{(+)}(t) = \frac{1}{2i}\sum_k \left(\frac{\hbar\nu_k}{\epsilon_0 A L_0}\right)^{1/2} \xi_k a_k(t) e^{ikL_0}, \tag{16.B.4}$$

$$E_{\text{out}}^{(+)}(t) = -\frac{1}{2i}\sum_k \left(\frac{\hbar\nu_k}{\epsilon_0 A L_0}\right)^{1/2} \xi_k a_k(t) e^{-ikL_0}, \tag{16.B.5}$$

* The material of this section follows Gea-Banacloche *et al.* [1990].

and

$$E_{\text{cav}}^{(+)}(t) = \frac{1}{2i} \sum_k \left(\frac{\hbar \nu_k}{\epsilon_0 A L_0} \right)^{1/2} M_k a_k(t) e^{-ikL}, \tag{16.B.6}$$

$$E_{\text{out}}^{(+)}(t) = -\sqrt{R}. \tag{16.B.7}$$

It is customary, as in Appendix 5.C, to define an intracavity annihilation operator a at a quasimode frequency ν_0 such that over the full quasimode a behaves as a single-mode annihilation operator in the sense that $[a, a^\dagger] = 1$. If we restrict the quasicontinuous sum over k in (16.B.6) to a single quasimode,

$$\nu_0 - c\pi/2L < \nu_k < \nu_0 + c\pi/2L, \tag{16.B.8}$$

then we may write that restricted sum, denoted by a prime superscript, as

$$E_{\text{cav}}^{(+)'}(t) = \beta a(t) e^{-ik_0 L} \quad (k_0 = \nu_0/c), \tag{16.B.9}$$

in which

$$|\beta|^2 = \left[E_{\text{cav}}^{(+)'}(t), E_{\text{cav}}^{(-)'}(t)\right] = \sum_k{}' \frac{\hbar \nu_k}{4\epsilon_0 A L_0} |M_k|^2. \tag{16.B.10}$$

In the limit $L_0 \to \infty$, this becomes (with $\delta\nu_k = \nu_k - \nu_0$)

$$|\beta|^2 \cong \frac{\hbar \nu_0}{4\epsilon_0 A L_0} \frac{L_0}{c\pi} \int d(\delta\nu_k) \frac{c\mathscr{C}/(2L)}{(\delta\nu_k)^2 + \mathscr{C}^2/4} = \frac{\hbar \nu_0}{4\epsilon_0 A L}, \tag{16.B.11}$$

where we substitute for M_k from Eq. (5.C.3). In other words, we may write $E_{\text{cav}}^{(+)'}(t)$ (up to an arbitrary phase factor) as

$$E_{\text{cav}}^{(+)'}(t) = \left(\frac{\hbar \nu_0}{4\epsilon_0 A L} \right)^{1/2} a(t). \tag{16.B.12}$$

From (16.B.6), this enables us to write the expression for $a(t)$ directly in terms of the annihilation operators a_k of the modes of the whole space,

$$a(t) \simeq i \left(\frac{L}{L_0} \right)^{1/2} \sum_k{}' M_k a_k(t) e^{-ikL}. \tag{16.B.13}$$

Now we may derive the operator equation for the decay of the intracavity field when the cavity is empty. In that case we can write

$$a_k(t) = a_k(0) e^{-i\nu_k t}, \tag{16.B.14}$$

and thus

$$a(t - 2L/c) = i \left(\frac{L}{L_0} \right)^{1/2} \sum_k{}' M_k a_k e^{ikL}, \tag{16.B.15}$$

where we use $2\nu_0 L/c = 2n\pi$. Then using

$$M_k e^{-ikL} = \sqrt{R}e^{ikL}M_k + \sqrt{T}\xi_k e^{ikL_0}, \tag{16.B.16}$$

we obtain

$$\begin{aligned} a(t-2L/c) &= \frac{i}{\sqrt{R}}\left(\frac{L}{L_0}\right)^{1/2}\sum_k{}' M_k a_k e^{-ikL} \\ &\quad -i\sqrt{\frac{T}{R}}\left(\frac{L}{L_0}\right)^{1/2}\sum_k{}' \xi_k a_k e^{ikL_0} \\ &= \frac{1}{\sqrt{R}}a(t) + 2\sqrt{\frac{T}{R}}\left(\frac{\epsilon_0 AL}{\hbar\nu_0}\right)^{1/2} E_{\text{in}}^{(+)}(t). \end{aligned} \tag{16.B.17}$$

Finally the approximation

$$a(t-2L/c) \simeq a(t) - \frac{2L}{c}\frac{da}{dt} \tag{16.B.18}$$

enables us to write (for $R \simeq 1$)

$$\frac{da}{dt} = -\frac{\mathscr{C}}{2}a + \sqrt{T}\frac{c}{L}\left(\frac{\epsilon_0 AL}{\hbar\nu_0}\right)^{1/2} E_{\text{in}}^{(+)}(t), \tag{16.B.19}$$

which reduces to Eq. (16.3.4) by using Eq. (16.B.12).

Problems

16.1 Consider a degenerate parametric amplification process in the parametric approximation (Eq. (16.1.4)). Calculate the variances in the two quadratures for an arbitrary choice of the phase ϕ of the pump field. Assume the signal field to be initially in the vacuum state. Show that the quadratures are squeezed only for a limited choice of phase around $\phi = \pi/2$. What is the physical significance of this result?

16.2 Consider a non-degenerate parametric amplification process (Eq. (16.1.2)) in the parametric approximation ($b \to \beta_p \exp(-i\phi)$). Show that the Cauchy–Schwarz inequality

$$\langle a_s^\dagger a_s^\dagger a_s a_s\rangle_t \langle a_i^\dagger a_i^\dagger a_i a_i\rangle_t \geq (\langle a_s^\dagger a_i^\dagger a_i a_s\rangle_t)^2$$

is violated in this process. Assume the signal and idler fields to be initially in the vacuum state.

16.3 Consider a degenerate parametric oscillator such that one end mirror of the optical cavity is partially transmitting, through which, instead of an ordinary vacuum, a multi-mode squeezed vacuum centered around the signal frequency ν_s couples to the field inside the cavity. Write down the equations of motion for $\langle a\rangle$, $\langle a^2\rangle$, and $\langle a^\dagger a\rangle$. Show that the steady-state intracavity squeezing at the pump threshold can be larger than 50 percent.

16.4 In the density operator approach, the equation of motion for the density operator ρ for the signal field for the degenerate parametric oscillator is given by

$$\dot{\rho} = -\frac{i}{\hbar}[\mathscr{V}, \rho] - \frac{\mathscr{C}}{2}(a^\dagger a\rho - 2a\rho a^\dagger + \rho a^\dagger a),$$

where $\mathscr{V}$ is the interaction Hamiltonian (16.1.4).

(a) Derive the equations of motion for $\langle a\rangle$, $\langle a^2\rangle$, and $\langle a^\dagger a\rangle$ when the phase of the pump field is chosen to be $\pi/2$. Compare these equations with the corresponding equations derived in Section 16.2 using a Heisenberg–Langevin approach.

(b) By solving these equations, show that

$$(\Delta X_1)_t^2 = G_1(\Delta X_1)_0^2 + N_1(1 - G_1),$$
$$(\Delta X_2)_t^2 = G_2(\Delta X_2)_0^2 + N_2(1 - G_2).$$

Find G_1, G_2, N_1, and N_2.

16.5 Consider a degenerate parametric amplification process in the parametric approximation (Eq. (16.1.4)). It is assumed that the pump field has no phase fluctuations $\phi = \phi_0$, but it has amplitude fluctuations $\beta_p = \beta_0 + \beta_1(t)$. The random amplitude $\beta_1(t)$, with zero mean, $\langle\beta_1(t)\rangle = 0$, is described by the so-called Ornstein–Uhlenbeck process,

$$\langle\beta_1(t)\beta_1(t')\rangle = \Gamma e^{-\Gamma(t-t')}.$$

Find $\langle a\rangle_t$, $\langle a^2\rangle_t$, and $\langle a^\dagger a\rangle_t$ and show that squeezing decreases with increasing Γ.

References and bibliography

Textbooks

N. Bloemberger, *Nonlinear Optics*, (Benjamin, New York 1965).
Y. R. Shen, *The Principles of Nonlinear Optics*, (Wiley-Interscience, New York 1984).
D. F. Walls and G. J. Milburn, *Quantum Optics*, (Springer, New York 1994).

General references on parametric amplification

W. H. Louisell, A. Yariv, and A. E. Siegmann, *Phys. Rev.* **124**, 1646 (1961).
J. P. Gordon, W. H. Louisell, and L. P. Walker, *Phys. Rev.* **129**, 481 (1963).
J. P. Gordon, L. R. Walker, and W. H. Louisell, *Phys. Rev.* **130**, 806 (1963).
B. R. Mollow and R. J. Glauber, *Phys. Rev.* **160**, 1076 (1967); *ibid.* **160**, 1097 (1967).

Sub-Poissonian statistics and squeezed light in parametric processes

M. T. Raiford, *Phys. Rev. A* **2**, 1541 (1970); *ibid.* **9**, 2060 (1974).
D. Stoler, *Phys. Rev. Lett.* **33**, 1397 (1974).
H. P. Yuen, *Phys. Rev. A* **13**, 2226 (1976).
L. Mista, V. Perinova, J. Perina, and Z. Brounerova, *Act. Phys. Pol. A* **51**, 739 (1977).
G. J. Milburn and D. F. Walls, *Opt. Commun.* **39**, 401 (1981).
E. J. Jakeman and J. G. Walker, *Opt. Commun.* **55**, 219 (1985); *Opt. Acta* **32**, 1303 (1985).
Z. Rauf and M. S. Zubairy, *Phys. Rev. A* **36**, 1481 (1987).
G. Bjork and Y. Yamamoto, *Phys. Rev. A* **37**, 125 (1988); *ibid.* **37**, 1991 (1988).
S.-Y. Zhu, *Phys. Lett. A* **151**, 529 (1990).
J. Anwar and M. S. Zubairy, *Phys. Rev. A* **45**, 1804 (1992).

Squeezing in the output of a cavity field

B. Yurke, *Phys. Rev. A* **29**, 408 (1984).
M. J. Collett and C. W. Gardiner, *Phys. Rev. A* **30**, 1386 (1984).
C. W. Gardiner and M. J. Collett, *Phys. Rev. A* **31**, 3761 (1985).
H. J. Carmichael, *J. Opt. Soc. Am. B* **4**, 1588 (1987).
M. J. Collett, R. Loudon, and C. W. Gardiner, *J. Mod. Opt.* **34**, 881 (1987).
J. Gea-Banacloche, N. Lu, L. M. Pedrotti, S. Prasad, M. O. Scully, and K. Wódkiewicz, *Phys. Rev. A* **41**, 369 (1990); *ibid.* **41**, 381 (1990).

Role of fluctuations in the pump

M. Hillery and M. S. Zubairy, *Phys. Rev. A* **26**, 451 (1982).
K. Wódkiewicz and M. S. Zubairy, *Phys. Rev. A* **27**, 2003 (1983).
G. Scharf and D. F. Walls, *Opt. Commun.* **50**, 245 (1984).
C. M. Caves and D. D. Crouch, *J. Opt. Soc. Am. B* **4**, 1535 (1987).
D. D. Crouch and S. L. Braunstein, *Phys. Rev. A* **36**, 4696 (1988).
J. Gea-Banacloche and M. S. Zubairy, *Phys. Rev. A* **42**, 1742 (1990).

Semiclassical theory of four-wave mixing

A. Yariv and D. M. Pepper, *Opt. Lett.* **1**, 16 (1977).
R. L. Abrams and R. C. Lind, *Opt. Lett.* **2**, 94 (1978).
A. Yariv, *Opt. Commun.* **25**, 23 (1978).
T. Fu and M. Sargent III, *Opt. Lett.* **5**, 433 (1980).
R. W. Boyd, M. G. Raymer, P. Narum, and D. J. Harter, *Phys. Rev. A* **24**, 411 (1981).
R. Saxena and G. S. Agarwal, *Phys. Rev. A* **31**, 877 (1985).

Four-wave mixing in atomic medium

M. D. Reid and D. F. Walls, *Phys. Rev. A* **30**, 343 (1984).
M. Sargent III, D. A. Holm, and M. S. Zubairy, *Phys. Rev. A* **31**, 3112 (1985).
S. Stenholm, D. A. Holm, and M. Sargent III, *Phys. Rev. A* **31**, 3124 (1985).
D. A. Holm, M. Sargent III, and L. M. Hoffer, *Phys. Rev. A* **32**, 963 (1985).
G. S. Agarwal, *Phys. Rev. A* **34**, 4055 (1986).

Squeezing and nonclassical effects in four-wave mixing

H. P. Yuen and J. H. Shapiro, *Opt. Lett.* **4**, 334 (1979).
P. Kumar and J. H. Shapiro, *Phys. Rev. A* **30**, 1508 (1984).
M. D. Reid and F. D. Walls, *Opt. Commun.* **50**, 406 (1984); *Phys. Rev. A* **31**, 1622 (1985); *ibid.* **32**, 392 (1985).
D. A. Holm, M. Sargent III, and B. A. Capron, *Opt. Lett.* **7**, 443 (1986).
N. A. Ansari and M. S. Zubairy, *Phys. Rev. A* **38**, 2380 (1988).

Experimental observation of squeezing

R. E. Slusher, L. W. Hollberg, B. Yurke, J. C. Mertz, and J. F. Valley, *Phys. Rev. Lett.* **55**, 2409 (1985).
R. M. Shelby, M. D. Levenson, S. Parlmutter, R. Devoe, and D. F. Walls, *Phys. Rev. Lett.* **57**, 691 (1986).

R. E. Slusher, B. Yurke, P. Grangier, A. LaPorta, D. F. Walls, and M. D. Reid, *J. Opt. Soc. Am. B* **4**, 1453 (1987).
L. A. Wu, M. Xiao, and H. J. Kimble, *J. Opt. Soc. Am. B* **4**, 1465 (1987).
L. A. Orozco, M. G. Raizen, M. Xiao, R. J. Brecha, and H. J. Kimble, *J. Opt. Soc. Am. B* **4**, 1490 (1987).

CHAPTER 17

Atom optics

Matter–wave interferometry dates from the inception of quantum mechanics, i.e., the early electron diffraction experiments. More recent neutron interferometry experiments have yielded new insights into many fundamental aspects of quantum mechanics. Presently, atom interferometry has been demonstrated and holds promise as a new field of optics – matter–wave optics. This field is particularly interesting since the potential sensitivity of matter–wave interferometers far exceeds that of their light-wave or 'photon' antecedents.

In this chapter we consider the physics of light-induced forces on the center-of-mass motion of atoms and their application to atom optics (Fig. 17.1). The most obvious being the recoil associated with the emission and absorption of light. This 'radiation pressure' is the basis for laser induced cooling.*

Another very important mechanical effect is the gradient force due to, e.g., transverse variation in the laser beam. These, essentially semiclassical, forces are useful in guiding and trapping neutral atoms.

After considering the basic forces which allow us to cool, guide, and trap atoms, we turn to the optics of atomic center-of-mass de Broglie waves, i.e., atom optics. In keeping with the spirit of the present text, we will focus on the quantum limits to matter–wave interferometry. An analysis of a matter–wave gyro in an obvious extension of the laser gyro and the similarity and relative merits of the two will be compared and contrasted.

Finally we derive the "recoil limit" to laser cooling; and show that it is possible to supersede this limit via atomic coherence effects.

* For further reading, see the proceedings of the CXIII Enrico Fermi School, edited by Arimondo, Phillips, and Strumia (1992).

17.1 Mechanical effects of light

As a consequence of the conservation of energy and momentum, atoms can experience light-induced forces during their interaction with a radiation field. In this section, we discuss the application of these forces in causing the deflection, cooling, and diffraction of the atomic beams. We also discuss the gradient force due to transverse variation of the laser field.

17.1.1 Atomic deflection

When an atom absorbs or emits a photon of frequency ν from a light beam, a transfer of recoil momentum $\Delta p = \hbar k = \hbar \nu / c$ takes place between the atom and the field. If absorption is followed by stimulated emission, no net momentum is transferred to the atom as the momentum transferred in the process of absorption is canceled by an equal but opposite transfer of momentum in the process of stimulated emission. If, however, absorption is followed by spontaneous emission, there is a net momentum transfer to the atom as the spontaneous emission in arbitrary directions gives no average contribution to the momentum. If this process (absorption followed by spontaneous emission) takes place a large number of times, a substantial transfer of momentum can occur, from the light beam to the atom, leading to atomic deflection. In the following, we derive an expression for the deflection or recoil force on the atoms.

As discussed above, an atom experiences a momentum recoil of $\Delta p = \hbar k$ upon each radiative event. Hence the absorptive force of the atom F_a is given by

$$F_a = r\hbar k, \tag{17.1.1}$$

where r is the rate of radiation decay or the net fluorescence rate. For a two-level atom at rest, with a transition frequency ω, the rate r is proportional to the upper level occupancy ρ_{aa} of the atom, i.e.,

$$r = \Gamma \rho_{aa}, \tag{17.1.2}$$

where Γ is the spontaneous emission rate from the excited state $|a\rangle$ to the ground state $|b\rangle$.

The interaction of a two-level atom with a radiation field of frequency ν is described by the following set of equations for the density matrix elements:

$$\dot{\rho}_{ab} = -\left(i\Delta + \frac{\Gamma}{2}\right)\rho_{ab} + i\Omega_R\rho_{aa} - \frac{i}{2}\Omega_R, \tag{17.1.3}$$

$$\dot{\rho}_{aa} = -\Gamma\rho_{aa} + \frac{i\Omega_R}{2}(\rho_{ab} - \rho_{ba}), \tag{17.1.4}$$

$$\dot{\rho}_{ba} = \left(i\Delta - \frac{\Gamma}{2}\right)\rho_{ba} - i\Omega_R\rho_{aa} + \frac{i}{2}\Omega_R. \tag{17.1.5}$$

These equations are obtained by generalizing Eqs. (10.C.1)–(10.C.4) to include the detuning $\Delta = \omega - \nu$. Here Ω_R is the Rabi frequency associated with the light beam. A steady-state solution of Eqs. (17.1.3)–(17.1.5) yields

$$\rho_{aa} = \frac{\Omega_R^2}{4\Delta^2 + \Gamma^2 + 2\Omega_R^2}. \tag{17.1.6}$$

The absorptive force is thus given by

$$F_r = \hbar k\Gamma\frac{\Omega_R^2}{4\Delta^2 + \Gamma^2 + 2\Omega_R^2}, \tag{17.1.7}$$

and is in the same direction as the light beam.

17.1.2 Laser cooling*

So far we have considered the force of a light beam on an atom at rest. If the atom is moving with a velocity v along the light beam, it sees a Doppler shifted frequency, $\nu \pm kv$, of the light beam. Here the $+$ (or $-$) sign corresponds to a situation when the atom is moving in the opposite (or same) direction to the light beam. The expression for the absorptive force F_a then becomes

$$F_a = \hbar k\Gamma\frac{\Omega_R^2}{4(\Delta \mp kv)^2 + \Gamma^2 + 2\Omega_R^2}. \tag{17.1.8}$$

In the limit of no saturation ($\Omega_R = 0$ in the denominator) and a small velocity, we can expand the denominator. The resulting expression for F_a is

$$F_a = F_o \pm \beta m v, \tag{17.1.9}$$

* The laser cooling concept was first proposed by Hänsch and Schawlow [1975] for free atoms and by Wineland and Dehmelt [1975] for trapped ions.

where

$$F_o = \hbar k \Gamma \frac{\Omega_R^2}{4\Delta^2 + \Gamma^2}, \tag{17.1.10}$$

$$\beta = 8\hbar k^2 \Gamma \frac{\Omega_R^2 \Delta}{m(4\Delta^2 + \Gamma^2)^2}. \tag{17.1.11}$$

The first term in Eq. (17.1.9) is a constant deflecting force, whereas the second term, proportional to atomic velocity, acts like a friction term.

If the atom is located in a standing wave, it sees two oppositely moving light waves, one in the same direction as the velocity of the atom and other in the opposite direction. We assume that the forces due to the two beams can be superimposed. Hence the total force on the atom in a standing wave is

$$\begin{aligned} F_{\text{standing wave}} &= (F_o - \beta m v) - (F_o + \beta m v) \\ &= -2\beta m v, \end{aligned} \tag{17.1.12}$$

i.e., the deflection forces F_o cancel and the friction forces from the two beams remain. The friction force is responsible for the slowing down of the atom leading to *laser cooling*.

Physically, we can understand the process of laser cooling as follows. If $\Delta > 0$, i.e., $\omega > \nu$, the field moving in the opposite direction to the atom will be Doppler up-shifted, thus compensating the detuning. The atom will therefore be decelerated. By this mechanism the atoms can be slowed down to the pace of extremely sluggish atomic molasses.

17.1.3 Atomic diffraction

When a beam of atoms interacts with the periodic structure of a standing wave, a diffractive scattering takes place. This sends the atomic beam in many directions as shown in Fig. 17.1(a). This phenomenon is analogous to the scattering of a light wave from an optical grating. Under suitable conditions, the atomic beam can be diffracted into two directions only, resulting in an atomic beam-splitter. Such a beam-splitter can be used in atomic interferometers.

Here we give a simple derivation of this effect, which neglects the internal two-level structure of the atom.

The dipole interaction of an atom interacting resonantly with a standing-wave field in the z-direction is given by

$$\mathscr{H}_1 = \wp \mathscr{E}_0 \sin kz, \tag{17.1.13}$$

Fig. 17.1
(a) Interaction of the atomic beam with a standing wave can result in atomic diffraction. (b) Field gradient force can make atoms rebound like a light beam reflected from a mirror. (c) Atomic beam may be focussed by the gradient force of the electromagnetic field.

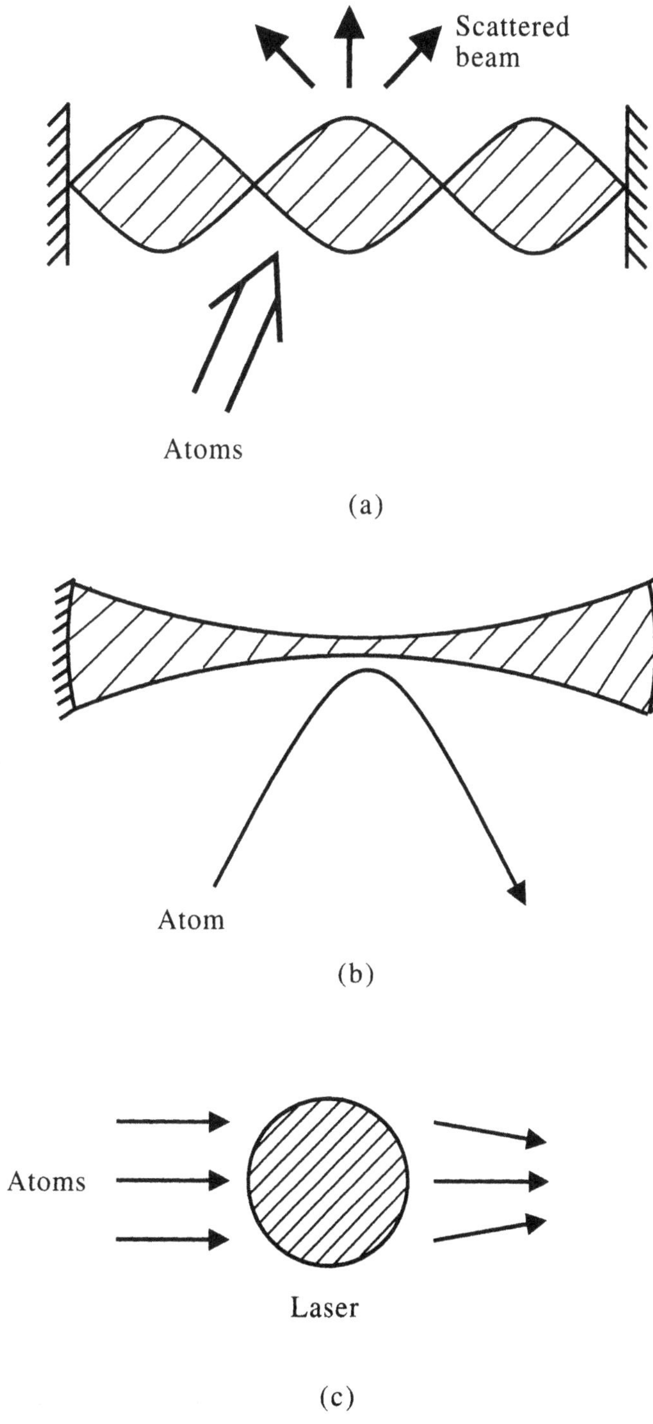

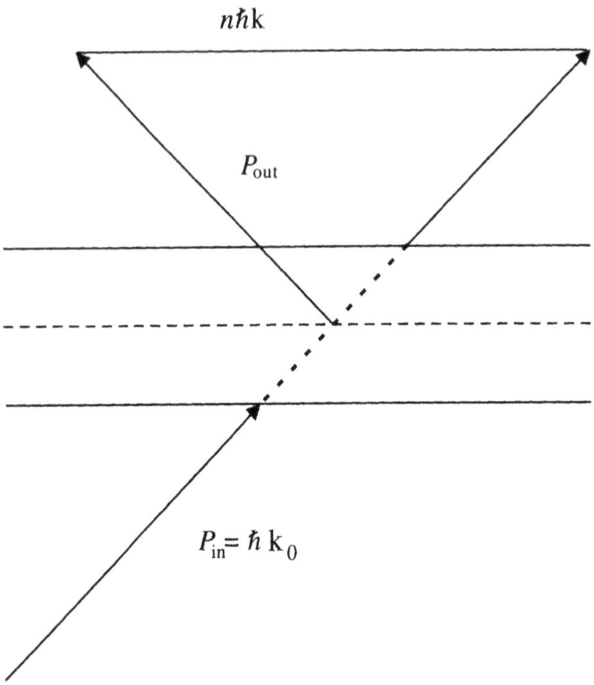

Fig. 17.2
An atomic beam of wave vector $\mathbf{k}_0$ can acquire a momentum $n\hbar\mathbf{k}$ during passage through a standing wave resulting in atomic diffraction.

where $\wp$ is the atomic dipole moment. If the atom is initially in a momentum eigenstate with momentum $\hbar k_0$, i.e., $\psi(z,0) = \exp(ik_0 z)$, we find the time-dependent wave function

$$\begin{aligned}\psi(z,t) &= \exp\left(-\frac{i}{\hbar}\mathscr{H}_1 t\right)\psi(z,0)\\ &= \exp(-i\Omega_R t \sin kz)\exp(ik_0 z)\\ &= \sum_{n=-\infty}^{\infty} J_n(\Omega_R t)\exp[i(k_0+nk)z],\end{aligned} \tag{17.1.14}$$

where $\Omega_R = \wp\mathscr{E}_0/\hbar$ is the Rabi frequency and J_n is the nth order Bessel function. The momentum representation of $\psi(z,t)$, i.e., $\tilde{\psi}(p,t)$, is obtained by taking the Fourier transformation of Eq. (17.1.14):

$$\tilde{\psi}(p,t) = \hbar\sum_{n=-\infty}^{\infty} J_n(\Omega_R t)\delta[p - \hbar(k_0+nk)]. \tag{17.1.15}$$

This represents the diffractive scattering into many momentum components spaced by the photon momenta $\hbar k$. This diffractive scattering is due to the fact that the atom, during its passage through the standing wave, can exchange an integral number of photons of momenta $\hbar k$ in the $+$ or $-z$-direction (see Fig. 17.2).

17.1.4 Semiclassical gradient force

We now consider the force of an atom which enters the cavity in a field region having a strong transverse variation. This is, for example, the situation in a highly focussed Gaussian laser beam. We calculate the magnitude of this gradient force.

The interaction Hamiltonian for an atom interacting with the field can be written in the dipole approximation as

$$\mathscr{H}_1 = -\hat{\wp} E(\mathbf{r}, t), \tag{17.1.16}$$

where, for a light beam propagating in the z-direction,

$$E(\mathbf{r}, t) = \frac{1}{2}\mathscr{E}_0(x, y)e^{-i(\nu t - kz)} + \text{c.c.} \tag{17.1.17}$$

In the semiclassical approximation, we can replace the dipole moment operator $\hat{\wp}$ in Eq. (17.1.16) by its expectation value

$$\langle \hat{\wp} \rangle = \wp_{ab}\rho_{ab}e^{i(\nu t - kz)} + \text{c.c.} \tag{17.1.18}$$

In the rotating-wave approximation, the interaction energy is thus equal to

$$W_1 = \frac{-\hbar\Omega_R}{2}(\rho_{ba} + \rho_{ab}), \tag{17.1.19}$$

where $\Omega_R = \wp_{ab}\mathscr{E}_0/\hbar$ is the Rabi frequency, which we assume to be real.

A steady-state solution of Eqs. (17.1.3)–(17.1.5) yields

$$\rho_{ab} = \frac{-2\Omega_R(\Delta + i\Gamma/2)}{4\Delta^2 + \Gamma^2 + 2\Omega_R^2}, \tag{17.1.20}$$

which gives

$$W_1 = \frac{2\hbar\Delta\Omega_R^2(x, y)}{4\Delta^2 + \Gamma^2 + 2\Omega_R^2}. \tag{17.1.21}$$

Now the atom entering the field region will experience a force (for $\Omega_R \ll \Gamma$)

$$\begin{aligned} \mathbf{F} &= -\nabla_\perp W_1 \\ &= -\frac{2\hbar\Delta}{4\Delta^2 + \Gamma^2}\nabla_\perp\Omega_R^2(x, y), \end{aligned} \tag{17.1.22}$$

where $\nabla_\perp$ is the transverse gradient. For a plane wave, Ω_R is independent of transverse coordinates, and this force vanishes. However for a focussed beam of width a

$$|\nabla_\perp\Omega_R^2| \simeq \frac{\Omega_R^2}{a}, \tag{17.1.23}$$

the force (17.1.22) then becomes

$$F = \frac{2\hbar\Delta\Omega_R^2}{a(4\Delta^2 + \Gamma^2)}. \tag{17.1.24}$$

The semiclassical gradient force is thus proportional to the detuning and its direction depends on the sign of Δ. For positive detuning the force is in the direction of the field gradient. This force may thus make the atoms rebound like a light beam reflected from a mirror (Fig. 17.1(b)). Alternatively, an atomic beam may experience focussing and the situation corresponds to that of a cylindrical lens (Fig. 17.1(c)).

17.2 Atomic interferometry

In this section, we develop the theory of atomic interferometers cast in an operator formalism. This formalism will be employed to study the quantum limit on the overall sensitivity of the device in the next section. We also consider the use of an atom interferometer as a rotation detector or a gyroscope.

17.2.1 Atomic Mach–Zehnder interferometer

As depicted in Fig. 17.3, we consider a scheme whereby a stream of N atoms are sent through a Mach–Zehnder interferometer during a measurement time t_m. The atoms are split at beam-splitter 1, follow paths α or β, are reflected off mirrors, and are then recombined at beam-splitter 2. The recombined atoms are detected at upper detector a or lower detector b where interference fringes are recorded.

We assume that, upon reflection from a beam-splitter surface, the particles undergo an unimportant phase shift that we take to be $\pi/2$, but that, in reality, depends upon the structure of the beam-splitter. Upon passage *through* a beam-splitter, however, the atom undergoes a phase shift of φ_i ($i = 1, 2$ for the first and second beam-splitter, respectively). The cumulative effect in the interferometer of these various processes on the atomic wave function ψ is depicted in Fig. 17.3(b), and leads to a wave function ψ_a corresponding to the upper detector, namely

$$\psi_a = \frac{\psi}{2} e^{i\theta_a} \left[1 - e^{-ik(l_\alpha - l_\beta)}\right], \tag{17.2.1}$$

and

$$\psi_b = \frac{\psi}{2} e^{i\theta_b} \left[1 + e^{-ik(l_\alpha - l_\beta)}\right], \tag{17.2.2}$$

where $\theta_a \equiv \pi/2 + kl_\alpha + \varphi_2$, and $\theta_b \equiv kl_\alpha + \varphi_1 + \varphi_2$, and where, without

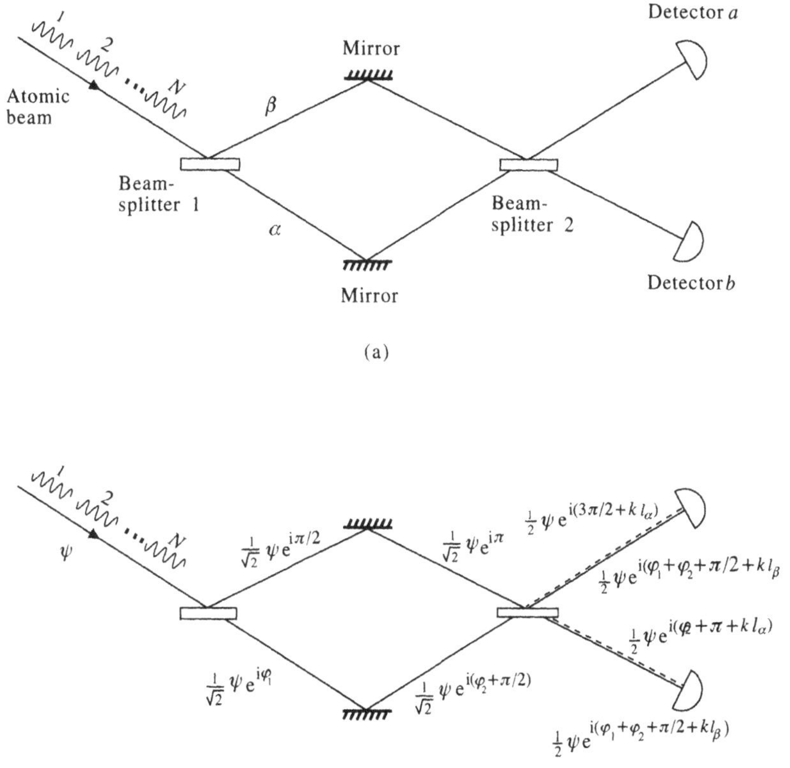

Fig. 17.3
(a) Schematic of the atomic Mach–Zehnder interferometer.
(b) Phase changes by the beam-splitters and mirrors account for accumulated phase shifts in the upper or lower branches. (From M. O. Scully and J. P. Dowling, *Phys. Rev. A* **48**, 3186 (1993).)

loss of generality, we let $\varphi_1 = \varphi_2 = \pi$. Here, k is the atomic wave number and l_α and l_β are the path lengths through the upper and the lower branches, respectively. We imagine now that the beam is recombined by the second beam-splitter and the detectors a and b shown in Fig. 17.3(a) count the number of atoms as they arrive in the recombined upper beam or lower beam, respectively. If we label N atoms with the index $i = 1, \ldots, N$, as those sent through the interferometer during the measurement time t_m, then the appropriate state vector $|\varphi\rangle_i$ for the ith atom in the interferometer, after recombination, is given by

$$|\varphi\rangle_i = \frac{e^{i\theta_a}}{2}\left(1 - e^{-i\varphi_{\alpha\beta}}\right)|1_a, 0_b\rangle_i + \frac{e^{i\theta_b}}{2}\left(1 + e^{-i\varphi_{\alpha\beta}}\right)|0_a, 1_b\rangle_i, \tag{17.2.3}$$

where here $\varphi_{\alpha\beta} \equiv k(l_\alpha - l_\beta)$. We see that this state is an appropriate superposition of the number states $|1_a, 0_b\rangle$ and $|0_a, 1_b\rangle$ corresponding to an atom incident on the upper or lower detector, respectively. The state vector $|\Phi\rangle_N$ for the N-atom state is then constructed via a direct product of the individual atomic states, namely

$$|\Phi\rangle_N \equiv \prod_{i=1}^{N} |\varphi\rangle_i. \tag{17.2.4}$$

Let $c^\dagger_{\sigma,i}$ and $c_{\sigma,i}$, where $\sigma = a, b$, be the creation and annihilation operators, respectively, for the number states $|n_a, n_b\rangle_i$, where, corresponding to number operators $n_{\sigma,i} \equiv c^\dagger_{\sigma,i} c_{\sigma,i}$, the eigenvalues n_a and n_b are 0 or 1. Then the number operator N_σ for the number of upper or lower atoms is determined by

$$N_\sigma = \sum_{i=1}^{N} n_{\sigma,i} \qquad (\sigma = a, b), \tag{17.2.5}$$

and the operator c obeys the commutation relations

$$\left[c_{\sigma,i} c^\dagger_{\sigma,j} \pm c^\dagger_{\sigma,j} c_{\sigma,i}\right] = \delta_{ij}, \tag{17.2.6}$$

where the plus or minus sign indicates Bose or Fermi statistics, respectively. The statistical nature of the atoms will be important in circumstances where the density of particles in the interferometer is so large that there is more than one atom at a time within a single coherence length, or if the atoms are injected in a correlated manner into the input port. The expectation values $\langle N_\sigma\rangle_N$ of these number operators, Eq. (17.2.5), are given by

$${}_N\langle\Phi|N_a|\Phi\rangle_N = \sum_{i=1}^{N} \left|\frac{1 - e^{-i\varphi_{\alpha\beta}}}{2}\right|^2 {}_i\langle 1_a, 0_b|n_{a,i}|1_a, 0_b\rangle_i, \tag{17.2.7}$$

$${}_N\langle\Phi|N_b|\Phi\rangle_N = \sum_{i=1}^{N} \left|\frac{1 + e^{-i\varphi_{\alpha\beta}}}{2}\right|^2 {}_i\langle 0_a, 1_b|n_{b,i}|0_a, 1_b\rangle_i. \tag{17.2.8}$$

This yields the expressions for the mean number of atoms in the α and β branches as

$$\langle N_a\rangle_N = N \sin^2 \varphi_{\alpha\beta}/2, \qquad \langle N_b\rangle_N = N \cos^2 \varphi_{\alpha\beta}/2. \tag{17.2.9}$$

These expectations constitute the signal.

17.2.2 Atomic gyroscope

We now consider an idealized atomic interferometer used as a rotation sensor or gyroscope. The atom interferometer consists of semicircular arms as depicted in Fig. 17.4. If the loop rotates with an angular

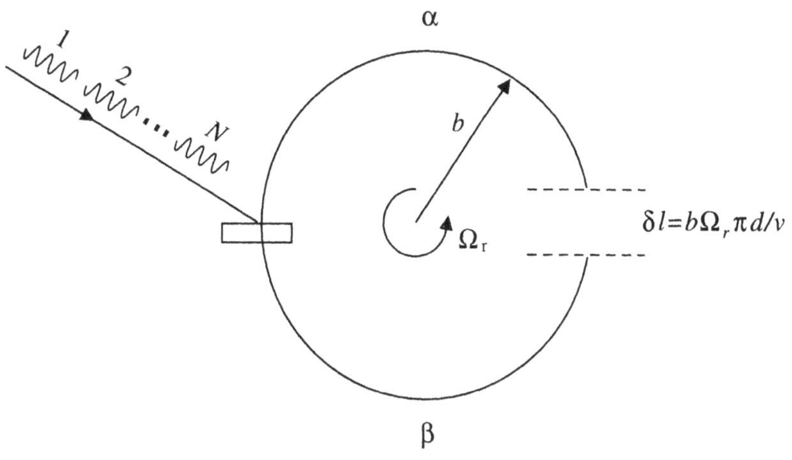

Fig. 17.4
Schematic of an atomic interferometer with semicircular arms to be used as a rotation sensor or gyroscope.

frequency Ω_r about an axis through its center and normal to the loop plane, the path difference between counter-propagating and co-propagating beams can be seen to be (see Eq. (4.1.12))

$$\Delta L = \frac{2\pi b^2 \Omega_r}{v}, \tag{17.2.10}$$

where b is the radius of the beam path and v is the atomic velocity. This path difference translates into a Sagnac phase difference of

$$\Delta\theta = \frac{2\pi b^2 \Omega_r}{v\bar{\lambda}} = \frac{2A\Omega_r}{v\bar{\lambda}}, \tag{17.2.11}$$

where A is the area enclosed by the arms and

$$\bar{\lambda} = \frac{\hbar}{mv} \tag{17.2.12}$$

is the atomic de Broglie wavelength. The phase difference is then given by

$$\Delta\theta = \frac{2Am\Omega_r}{\hbar}. \tag{17.2.13}$$

This expression holds for both atom and light interferometers, if, in the photon case, we define an effective photon mass m_γ implicitly by

$$m_\gamma c^2 = \hbar\nu. \tag{17.2.14}$$

Now, since the 'mass' of a photon is governed by optical energies of a few electron volts whereas atomic masses are of the order 10^3 Mev, we see that the matter–wave gyroscopes potentially have a signal that is enhanced by many orders of magnitude, compared to the laser gyroscopes. However, in order to determine the minimum rotation rate, we have to consider the noise in an atom interferometer, to which we turn in the next section.

17.3 Quantum noise in an atomic interferometer

We compute the quantum noise fluctuations using the second quantized formalism developed earlier. Recalling the definitions for the number operator N_σ, Eq. (17.2.5), and the state vector $|\Phi\rangle_N$, Eq. (17.2.4), and using the commutation relations, Eq. (17.2.6), we may write,

$$\begin{aligned}
\langle \Delta N_\sigma^2 \rangle &= {}_N\langle \Phi | N_\sigma^2 | \Phi \rangle_N - \left({}_N\langle \Phi | N_\sigma | \Phi \rangle_N \right)^2 \\
&= {}_N\langle \Phi | \sum_{i=1}^{N} n_{\sigma,i} \sum_{j=1}^{N} n_{\sigma,j} | \Phi \rangle_N - \left({}_N\langle \Phi | \sum_{i=1}^{N} n_{\sigma,i} | \Phi \rangle_N \right)^2 \\
&= \sum_{i=1}^{N} {}_i\langle \varphi | n_{\sigma,i} | \varphi \rangle_i \sum_{\substack{j=1 \\ j \neq i}}^{N} {}_j\langle \varphi | n_{\sigma,j} | \varphi \rangle_j \\
&\quad + \sum_{i=1}^{N} {}_i\langle \varphi | c_{\sigma,i}^\dagger (1 \pm n_{\sigma,i}) c_{\sigma,i} | \varphi \rangle_i - \left(\sum_{i=1}^{N} {}_i\langle \varphi | n_{\sigma,i} | \varphi \rangle_i \right)^2 \\
&= \frac{N}{4} \sin^2 \varphi_{\alpha\beta} \pm \sum_{i=1}^{N} {}_i\langle \varphi | c_{\sigma,i}^\dagger c_{\sigma,i}^\dagger c_{\sigma,i} c_{\sigma,i} | \varphi \rangle_i \qquad (\sigma = a, b),
\end{aligned} \tag{17.3.1}$$

where, as before, the upper and lower terms correspond to $\sigma = a$ or b, respectively, and the $\pm$ sign refers to the statistics of the particles: a plus sign for bosons and a minus sign for fermions. We note that the last, statistics-dependent term of Eq. (17.3.1) is the sum of nonnegative matrix elements and so itself is nonnegative or nonpositive, according to the plus sign or negative sign, respectively. A quantitative analysis of the contribution of this statistics-dependent term requires a specific model of the coherences between atoms in a dense beam. However, one can qualitatively state that for sufficiently high densities the use of fermionic atoms will tend to *lower* the quantum noise limit. This is because the last term will be negative. Bosons will have the opposite effect. In many experiments of interest, the beam intensity is so low that there is only one atom at a time within a single coherence length. In this case, the statistics-dependent term in the last line of Eq. (17.3.1) is zero, and we are left with the result

$$\langle \Delta N_\sigma \rangle = \frac{\sqrt{N}}{2} \sin \varphi_{\alpha\beta}. \tag{17.3.2}$$

We notice that this result depends on the total number of atoms N. Now, the signal in either branch N_σ is given by Eq. (17.2.9).

The quantum fluctuations in phase $\Delta\varphi_{\alpha\beta}$ in the measured phase difference $\varphi_{\alpha\beta}$ may be determined by (see Eq. (4.4.51))

$$|\Delta\varphi_{\alpha\beta}| \equiv \frac{\langle \Delta N_\sigma \rangle}{|\partial \langle N_\sigma \rangle / \partial \varphi_{\alpha\beta}|} = \frac{1}{\sqrt{N}}, \tag{17.3.3}$$

a result that is *independent* of $\varphi_{\alpha\beta}$. This independence might appear surprising at first, but it is a direct result of the fact that the quantum number state noise $\langle \Delta N_\sigma \rangle$ is proportional to the slope of the signal $\langle N_\sigma \rangle$ for the upper and lower number states considered here.

We conclude by applying this result to the gyroscope problem. Let us note that the atom number N is given by jt_m, where j is the atom flux (in atoms per second) hitting the detector. We have from Eq. (17.3.3) the minimum detectable phase shift, $\varphi_{\min} = 1/\sqrt{jt_m}$, and equating this to the signal derived earlier, $\varphi^{\text{signal}} = 4Am\Omega_r/\hbar$, we find the minimum detectable rotation rate $\Omega_r^{\min}$ is given by

$$\Omega_r^{\min} \cong \frac{\hbar}{2Am} \frac{1}{\sqrt{jt_m}} \quad \text{(matter).} \tag{17.3.4}$$

This should be compared to the same result obtained using an optical interferometer in which the flux j is given by the power P divided by the photon energy $\hbar\nu$, in other words

$$\Omega_r^{\min} \cong \frac{\hbar}{Am_\gamma} \sqrt{\frac{\hbar\nu}{Pt_m}} \quad \text{(light),} \tag{17.3.5}$$

where m_γ is the effective photon mass, defined by $m_\gamma \equiv \hbar\nu/c^2$. As mentioned before, we note that the typical photon effective mass gives an increase in sensitivity of 10^{10}. This mass factor, however, is offset by the low particle flux available for atoms. This fact increases the laser gyroscope sensitivity over that of matter–wave devices by a factor of around 10^2. In addition, the atoms make about one 'round trip' through an interferometer, whereas in a ring laser gyroscope the photons make many ($\approx 10^4$) circuits around the ring and yield an additional sensitivity factor of 10^4 in favor of the laser system. This still leaves the matter–wave device 10^4 times more sensitive.

17.4 Limits to laser cooling

17.4.1 Recoil limit

We turn now to the question of velocity spread i.e., fluctuations in the momentum distribution associated with laser cooling. The z component of the atomic momentum after absorption of n photons is given by

$$p = p_0 + n\hbar k + \sum_{j=1}^{n} \hbar k \cos \theta_j \tag{17.4.1}$$

where $\hbar k \cos \theta_j$ is the projection, onto the z axis, of the jth spontaneously emitted photon.

Note that both the number of absorption events, n, and the emission directions of the emitted photons are random variables. We next give a heuristic derivation of the fluctuations in radiation pressure due to these two random processes.

Considering the fluctuations due to spontaneous emission (SE), we note that the number of SE events per second is given by

$$\frac{dN}{dt} = \Gamma P_{\text{ex}} \tag{17.4.2}$$

where Γ is the spontaneous emission rate and P_{ex} is the probability that the laser-driven atom is in the excited state. The diffusion in momentum in a time Δt is then characterized by

$$\Delta p^2 = \langle p^2 \rangle - \langle p \rangle^2 \tag{17.4.3}$$

and since $\langle p \rangle$ vanishes due to a large number of SE events we have

$$\begin{aligned} \Delta p^2 &= (\hbar k)^2 \frac{dN}{dt} \Delta t \\ &= \hbar^2 k^2 \frac{\Gamma}{2} \Delta t \end{aligned} \tag{17.4.4}$$

where in the last line we have noted that $dN/dt = \Gamma/2$ since $P_{\text{ex}} \Rightarrow 1/2$ for a damped, driven, two-level atom.

Recalling that the diffusion coefficient is $2D_{\text{SE}} = \Delta p^2/\Delta t$ we have

$$D_{\text{SE}} = \frac{1}{4} \hbar^2 k^2 \Gamma, \tag{17.4.5}$$

which is the momentum diffusion coefficient due to spontaneous emission.

Likewise, there are fluctuations in radiation pressure due to fluctuations in the number of absorption events leading to atomic excitation. It turns out that this source of noise is essentially equal to that due to SE fluctuations. Hence the fluctuations due to emission and absorption are characterized by a diffusion in velocity given by

$$D = \hbar^2 k^2 \Gamma / m^2 \,. \tag{17.4.6}$$

Thus may we write the Fokker–Planck equation for the velocity space probability density $P(v,t)$:

$$\frac{\partial P}{\partial t} = \frac{\partial}{\partial v}\left[d(v) + D(v)\frac{\partial}{\partial v}\right] P \,, \tag{17.4.7}$$

where the drift (damping) coefficient d governs the cooling process and the diffusion (fluctuation) coefficient D governs the effective temperature of the cold gas.

From Eq (17.4.7) we see that, at steady state,

$$P(v) = P(0)\exp\left[-\int_0^v d(v')/D(v')dv'\right] \,, \tag{17.4.8}$$

and using the fact that maximum damping occurs when Δ and Ω_R in Eq. (17.1.11) are of order Γ, we have $d \approx \hbar k^2 v/mg$, and taking $D(v)$ from Eq. (17.4.6), the velocity distribution (17.4.8) becomes

$$P(v) = P(0)\exp(-mv^2/\hbar\Gamma) \,. \tag{17.4.9}$$

Comparing (17.4.9) to the usual Boltzmann distribution

$$f(v) = f(0)\exp(-mv^2/2k_B T) \,, \tag{17.4.10}$$

we have the effective temperature for a laser cooled gas

$$T_{\text{eff}} \cong \hbar\Gamma/2k_B. \tag{17.4.11}$$

Eq. (17.4.11) is called the multi-photon recoil ("Doppler") limit to laser cooling, and is in the ballpark of a few hundred micro-Kelvin. Such low temperatures are interesting in many applications, but things get even better when we add the physics of atomic coherence, i.e., population trapping, to the laser cooling problem, as we see in the next section.

*17.4.2 Velocity selective coherent population trapping**

In the previous section we found that for two-level atoms the recoil limit to laser cooling is given by Eq. 17.4.11. But if we extend our considerations to multi-level atoms, and in particular atoms having two lower levels, then it is possible to cool beyond the single photon recoil limit, $k_B T \cong \hbar^2 k^2/2m$.

In order to see this, consider the situation as depicted in Fig. 17.5. There we see an atom of momentum $p_z \equiv p$ driven by two fields of polarization σ^+ and σ^-. Now velocity selective coherent population

* For a good account of this ingenious idea see A. Aspect, E. Arimondo, R. Kaiser, N. Vantseenkiste, and C. Cohen-Tannoudji, *J. Opt. Soc. Am.* **B6**, 2112 (1989); for another clever 'subrecoil cooling' scheme see M. Kasevich and S. Chu, *Phys. Rev. Lett.* **69**, 1741 (1992). Please note that by "subrecoil" we mean below the single photon limit; which is, naturally, below the multi-photon recoil limit.

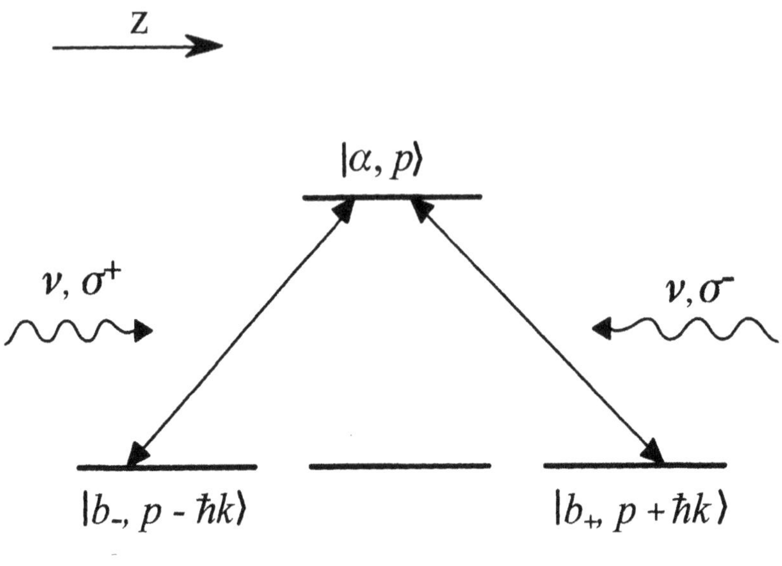

Fig. 17.5
Atom moving in z-direction with momentum p while driven from lower level ($J = 1$, $m = -1$) state $|b_-\rangle$ to upper level ($J = 1$, $m = 0$) state $|a\rangle$ via σ^+ radiation, and ($J = 1$, $m = +1$) state $|b\rangle$ to $|a\rangle$ state via σ^- radiation. Momentum of atom in states $b_\pm$ coupling to excited state having momentum p, is $p \pm \hbar k$.

trapping (VSCPT) occurs when the atom, after being excited, falls into the state

$$|\Psi_{NC}\rangle(p) = \frac{1}{\sqrt{2}}\Big[|b_+, p + \hbar k\rangle + |b_-, p - \hbar k\rangle\Big] . \qquad (17.4.12)$$

We take the interaction with the laser field to be

$$V = \frac{\Omega_R}{2}\left[-|a\rangle\langle b_-|e^{-i(\nu t - kz + \phi)} + |a\rangle\langle b_+|e^{-i(\nu t + kz + \phi)}\right] + \text{h.c.} \qquad (17.4.13)$$

where we have chosen the matrix elements such that $\wp_{a,b_-}$ and $\wp_{a,b_+}$ differ by a sign. From (17.4.12) and (17.4.13) we see that

$$\langle a, p|V|\Psi_{NC}(P)\rangle = 0 . \qquad (17.4.14)$$

When an atom is excited to state $|a, p\rangle$, it can spontaneously emit a photon in any direction, and the atomic momentum can change by any amount, q, from $\hbar k$ to $-\hbar k$. But when it happens to fall into a noncoupled state it will cease to absorb laser photons.

In order to see clearly what is happening in VSCPT consider the situations wherein a spontaneously emitted photon has left the atom in a superposition of the two states

$$|\Psi_{NC}(q)\rangle = \frac{1}{\sqrt{2}}\Big(|b_+, q + \hbar k\rangle + |b_-, q - \hbar k\rangle\Big) \qquad (17.4.15a)$$

$$|\Psi_{C}(q)\rangle = \frac{1}{\sqrt{2}}\Big(|b_+, q + \hbar k\rangle - |b_-, q - \hbar k\rangle\Big) \qquad (17.4.15b)$$

However, these states are, in general, coupled by the kinetic energy operator, that is

$$\langle \Psi_C(q)|\frac{p_z^2}{2m}|\Psi_{NC}(q)\rangle = \frac{\hbar k q}{m} . \tag{17.4.16}$$

Hence an atom which falls into the state $|\Psi_{NC}(q)\rangle$ will evolve into a superposition of $|\Psi_{NC}(q)\rangle$ and $|\Psi_C(q)\rangle$ due to the nonvanishing matrix element (17.4.16) of the kinetic energy operator coupling these states. However, in the case of q = 0, these states are uncoupled, and $|\Psi_{NC}(0)\rangle$ is seen to be *a* 'perfect' trapping state.

To summarize: a multi-level atom, of the type shown in Fig. 17.5, can be cooled below the single photon limit of some micro-Kelvin via VSCPT. Atoms are trapped in the zero-velocity non-coupled state $|\Psi_{NC}(0)\rangle$. At present, VSCPT has been used to yield temperatures in the nano-Kelvin domain.

Problems

17.1 As seen in Eq. (17.1.15), an atom experiences a diffractive scattering in a standing wave with momentum components spaced by the photon momenta $\hbar k$. Show that the momentum width is given by

$$\langle \Delta p^2 \rangle = \frac{1}{2}\hbar^2 k^2 \Omega_R^2 t^2.$$

References and bibliography

Review articles and books on mechanical effects of light

P. E. Toschek, in *New Trends in Atomic Physics*, Vol. 1, Les Houches 1982, Session XXXVIII, ed. G. Grynberg and R. Stora (North-Holland, Amsterdam 1984).
S. Stenholm, *Rev. Mod. Phys.* **58**, 699 (1986).
V. G. Minogin and V. S. Letokhov, *Laser Light Pressure on Atoms*, (Gordon and Breach, New York 1987).
A. P. Kazantsev, G. J. Surdutovich, and V. P. Yakovlev, *Mechanical Action of Light on Atoms*, (World Scientific, Singapore 1990).
Articles in *Laser Manipulation of Atoms and Ions*, Proceedings of the International School of Physics Enrico Fermi, Course CXIII, ed. E. Arimondo, W. D. Phillips, and F. Strumia (North-Holland, Amsterdam 1992).

Laser cooling and radiation pressure

V. S. Letokhov, *JETP Lett.* **7**, 272 (1968).
T. W. Hänsch and A. L. Schawlow, *Opt. Commun.* **13**, 68 (1975).
D. Wineland and H. Dehmelt, *Bull. Am. Phys. Soc.* **20**, 637 (1975).
A. Ashkin, *Phys. Rev. Lett.* **40**, 729 (1978).
J. Javanainen and S. Stenholm, *Appl. Phys.* **21**, 283 (1980); *ibid.* **24**, 71 (1981); *ibid.* **24**, 151 (1981).
S. Chu, L. Hallberg, J. E. Bjorkholm, A. Cable, and A. Ashkin, *Phys. Rev. Lett.* **55**, 48 (1985).
S. Chu, J. E. Bjorkholm, A. Ashkin, and A. Cable, *Phys. Rev. Lett.* **57**, 314 (1986).
P. D. Lett, R. N. Watts, C. I. Westbrook, W. D. Phillips, P. L. Gould, and H. J. Metcalf, *Phys. Rev. Lett.* **62**, 1118 (1988).
F. Diedrich, J. C. Bergquist, W. M. Itano, and D. J. Wineland, *Phys. Rev. Lett.* **62**, 403 (1989).
J. Dalibard and C. Cohen-Tannoudji, *J. Opt. Soc. Am. B* **6**, 2023 (1989).
Y. Castin, H. Wallis, and J. Dalibard, *J. Opt. Soc. Am. B* **6**, 2046 (1989).
P. J. Ungar, D. S. Weiss, E. Riis, and S. Chu, *J. Opt. Am. Soc. B* **6**, 2058 (1989).
Y. Shevy, *Phys. Rev. Lett.* **64**, 2905 (1990).
C. Salomon, J. Dalibard, A. Aspect, A. Metcalf, and C. Cohen-Tannoudji, *Phys. Rev. Lett.* **65**, 559 (1990).
E. P. Storey, M. J. Collett, and D. F. Walls, *Phys. Rev. Lett.* **68**, 472 (1992).

Atomic diffraction and atomic beam-splitter

P. E. Moskowitz, P. L. Gould, S. R. Atlas, and D. E. Pritchard, *Phys. Rev. Lett.* **51**, 370 (1983).
C. Tangvy, S. Reynaud, and C. Cohen-Tannoudji, *J. Phys. B* **17**, 4623 (1984).
P. L. Gould, G. A. Ruff, and D. E. Pritchard, *Phys. Rev. Lett.* **56**, 827 (1986).

P. J. Martin, P. L. Gould, B. G. Oldaker, A. H. Miklich, and D. E. Pritchard, *Phys. Rev. A* **36**, 2495 (1987).
P. J. Martin, B. G. Oldaker, A. H. Miklich, and D. E. Pritchard, *Phys. Rev. A* **60**, 515 (1988).
P. Meystre, E. Schumacher, and S. Stenholm, *Opt. Commun.* **73**, 443 (1989).
E. M. Wright and P. Meystre, *Opt. Commun.* **75**, 388 (1990).
P. L. Gould, P. G. Martin, G. A. Ruff, R. E. Stoner, J.-L. Picque, and D. E. Pritchard, *Phys. Rev. A* **43**, 585 (1991).
V. M. Akulin, F. L. Kien, and W. P. Schleich, *Phys. Rev. A* **44**, 1462 (1991).
B. W. Shore, P. Meystre, and S. Stenholm, *J. Opt. Soc. Am. B* **8**, 903 (1991).
S. M. Tan and D. F. Walls, *Phys. Rev. A* **44**, R2779 (1991).
M. Wilkens, E. Schumacher, and P. Meystre, *Opt. Comun.* **86**, 34 (1991).

Atomic mirror and lens – gradient force

D. B. Pearson, *Phys. Rev. Lett.* **41**, 1361 (1978); *Opt. Lett.* **5**, 111 (1980).
R. Cook and R. Hill, *Opt. Commun.* **43**, 258 (1982).
V. I. Balykin, V. S. Letokhov, V. G. Minogin, and T. Z. Zueva, *Appl. Phys. B* **35**, 149 (1984).
V. I. Balykin, V. S. Letokhov, Yu. B. Ovchinnikov, and A. I. Sidorov, *Phys. Rev. Lett.* **60**, 2137 (1988).

Neutron interferometry

R. Collela, A. W. Overhauser, and S. A. Werner, *Phys. Rev. Lett.* **34**, 1472 (1975).
A. Zeilinger, *Z. Phys. B* **25**, 97 (1976).
H. Rauch, W. Treimer, and U. Bonse, *Phys. Lett. A* **47**, 369 (1977).
S. A. Werner, J.-L. Staudemann, and R. Collela, *Phys. Rev. Lett.* **42**, 1103 (1979).
G. Badurek, H. Rauch, J. Summhammer, U. Kischko, and A. Zeilinger, *Physics B* **151**, 82 (1988).

Noise in atom interferometry

B. Yurke, *Phys. Rev. Lett.* **56**, 1515 (1986).
O. Carnal and J. Mlynek, *Phys. Rev. Lett.* **66**, 2689 (1991).
D. W. Keith, C. R. Ekstrom, Q. A. Turchette, and D. E. Pritchard, *Phys. Rev. Lett.* **66**, 2693 (1991).
F. Riehle, T. Kisters, A. Witte, J. Helmeche, and D. E. Bordé, *Phys. Rev. Lett.* **67**, 177 (1991).
M. Kasevich and S. Chu, *Phys. Rev. Lett.* **67**, 181 (1991).
M. O. Scully and J. P. Dowling, *Phys. Rev. A* **48**, 3186 (1993).

Limits to laser cooling

A. P. Kazantsev, G. I. Surdovich, and V. P. Yakovlev, *Mechanical Action of Light on Atoms* (World Scientific, Singapore 1990).

S. Stenholm, in *Laser Manipulation of Atoms and Ions (Proc. International School of Physics)* ed. E. Arimondo, W. D. Phillips, and F Strumina (North-Holland, Amsterdam 1992), p. 53.

A. Aspect, O. Emile, C. Gerz, R. Kaiser, N. Vansteenkiste, H. Wallis, and C. Cohen-Tannoudji, *ibid.*, p. 401.

CHAPTER 18

The EPR paradox, hidden variables, and Bell's theorem

Quantum mechanics is an immensely successful theory, occupying a unique position in the history of science. It has solved mysteries ranging from macroscopic superconductivity to the microscopic theory of elementary particles and has provided deep insights into the nature of vacuum on the one hand and the description of the nucleon on the other. Whole new fields such as quantum optics and quantum electronics owe their very existence to this body of knowledge.

However, despite the stunning successes of quantum mechanics, there is no general agreement on the conceptual foundations and interpretation of the subject. The theory provides unambiguous information about the outcome of a measurement of a physical object. However, many feel that it does not provide a satisfactory answer to the nature of the "reality" we should attribute to the physical objects between the acts of measurement.

The conceptual difficulty comes about because the wave function $|\psi\rangle$ is usually given by a coherent superposition of various distinguishable experimental outcomes. If we denote the collection of states that represent the possible outcomes of an experiment by $|\psi_j\rangle$, then $|\psi\rangle = \sum_j c_j|\psi_j\rangle$ where $c_j = \langle\psi_j|\psi\rangle$. The probability of the outcome $|\psi_j\rangle$ is $p_j = |c_j|^2$. In the process of measurement, the so called *collapse of the wave function* takes place and a single, definite state $|\psi_i\rangle$ of the physical object is chosen. The difficulty comes about in the interpretation of the mechanism by which this definite state is chosen from amongst all the possible outcomes.

An important consequence of the quantum mechanical formalism is that it does not seem to allow a *local* description of events in the sense discussed below. Alternatively, a *local* theory can be achieved but with the additional difficulties of *negative* probabilities.

This counter-intuitive nonlocal aspect of quantum mechanics has been a subject of debate since the early days. In particular, Einstein, Podolsky, and Rosen (EPR) conjectured, on the basis of a gedanken experiment, that quantum mechanics is an incomplete theory. In the absence of a concrete experimental situation to test the *reality* and *locality* aspects of quantum mechanics, the debate concerning the foundations of quantum mechanics continued to be essentially philosophical in nature for many years.

The situation however changed dramatically when, in 1964, J. S. Bell formulated certain inequalities, known as Bell's inequalities,* which should always be true for any theory that satisfies the *intuitively reasonable* notions of reality and locality. One of the most interesting results of modern physics is that quantum mechanics violates Bell's inequalities in certain situations, and that experimental results agree with the quantum mechanical predictions.

In this chapter, we present the EPR arguments concerning the *incompleteness* of quantum mechanics. We then discuss Bell's inequality and the quantum mechanical results violating it. The disagreement between Bell's inequality and the quantum mechanical predictions is further sharpened by the study of various alternative theories to quantum mechanics, hidden variable (HV) theories being prominent among these. In order to better understand the problem, we show that a 'nonlocal' hidden variable theory can be developed which is in agreement with quantum theory. Finally, we show that a new kind of equality, the so-called Greenberger–Horne–Zeilinger (GHZ) equality, is violated by quantum mechanics.

The present chapter, and the next two chapters as well, deal with interpretational problems of quantum mechanics. In all such studies, we follow the lead of Lamb [1969]; namely, develop the analysis around the theory for an apparatus which is designed to make the appropriate measurements. This sharpens the arguments and keeps the goal in focus.

18.1 The EPR 'paradox'

In 1935, Einstein, Podolsky, and Rosen (EPR) presented an argument to show that there are situations in which the general probabilistic scheme of quantum theory seems to be incomplete. Here we present a variation of this argument due to Bohm.

* For a beautiful account of the subject, see Mermin [1990a,b].

Let us consider a two-component system consisting of two spin-1/2 particles (e.g., the Hg_2 molecule). Up to some time $t = 0$, these particles are taken to be in a bound state of zero angular momentum. We designate the corresponding state vector as $|\Psi_{1,2}\rangle$. At time $t = 0$ we 'turn off' the binding potential (e.g., we photo-disintegrate the molecule) but introduce no angular momentum into the system and do not disturb the spins in any way. The separate parts of the system are now free to move off to opposite sides of the laboratory (or the universe for that matter). We now consider two kinds of experimental arrangements as shown in Figs. 18.1(a) and 18.1(b).

First we consider the case where we measure the z-component of spin 1 as indicated in Fig. 18.1(a). Before making the measurement on spin 1 the state vector for the system is

$$|\Psi_{1,2}\rangle = \frac{1}{\sqrt{2}}(|\uparrow_1, \downarrow_2\rangle - |\downarrow_1, \uparrow_2\rangle), \tag{18.1.1}$$

where $|\uparrow_1, \downarrow_2\rangle$ labels the state of particle 1 with spin projection $+1/2$, and the state of the second particle with spin projection $-1/2$ with respect to the z-axis, etc.

Now one version of the EPR argument runs as follows:

(1) Pick an arbitrary direction, which we can take to be the z-axis, and pass one Hg atom (say atom 1) through a Stern–Gerlach apparatus (SGA) oriented along the z-axis. The particle will now be deflected in either the $+$ or $-z$ direction, say $+z$. Thus we know the value of σ_z of that state is $+1$.

(2) Knowing that the spin of particle 1 is up, we now know the spin of particle 2 is down. But if we then pass atom 2 through a SGA oriented along the x-axis we will find that particle 2 has a definite spin along the x-direction (either $+x$ or $-x$), i.e., we now know the value of σ_x.

(3) Therefore, as the argument goes, we know both the z and x components of spin 2, which is a violation of complementarity.

It is worthwhile to restate the above version of the EPR paradox, which focused on an apparent violation of complementarity (we have "found" both σ_x and σ_z), in terms of a state vector picture.

Let us consider first the case where we measure the z-component of spin 1 as indicated in Fig. 18.1(a). Before making the measurement (at some time t_0) on spin 1 the state vector for the system is

$$|\psi_{1,2}^{<}\rangle = \frac{1}{\sqrt{2}}(|\uparrow_1, \downarrow_2\rangle - |\downarrow_1, \uparrow_2\rangle), \tag{18.1.2}$$

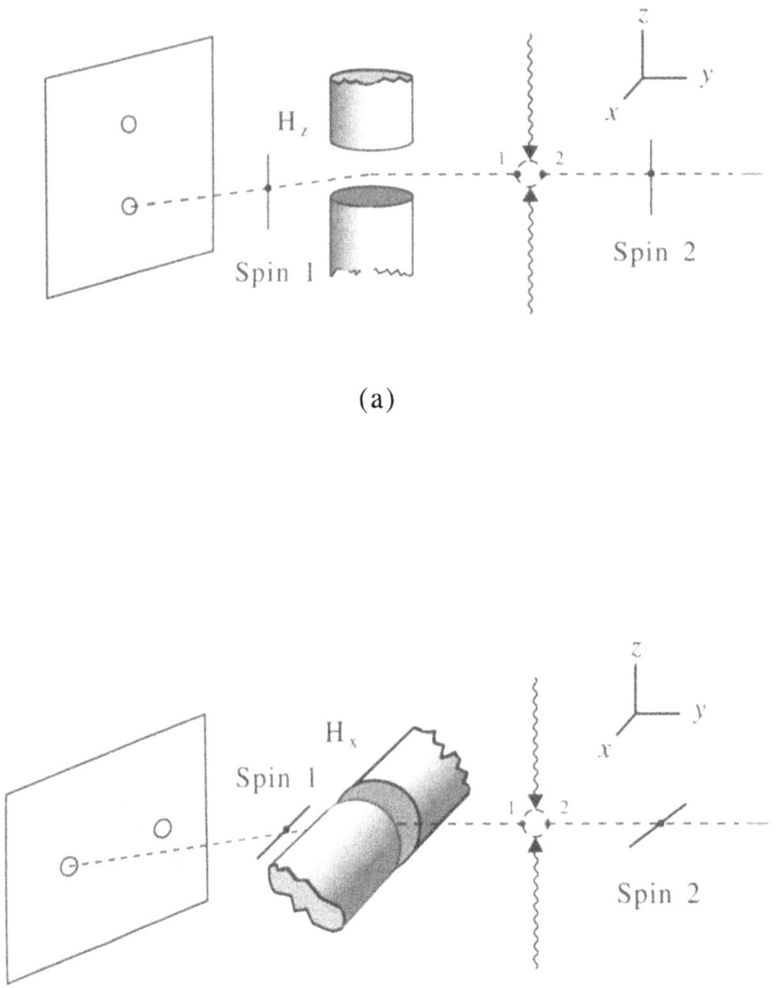

Fig. 18.1
Schematic of the EPR gedanken experiment. A spin-zero system such as orthohydrogen is split by an external field. The two spin-1/2 particles (protons) proceed in opposite directions. Particle 1 passes through a Stern-Gerlach apparatus
(a) oriented along the z-axis and
(b) oriented along the x-axis.

where $\psi^{<}$ denotes the state *before* ($t < t_0$) 'looking' at particle 1. Now if after a measurement on particle 1 we find it to be, say, in the spin down state $|\downarrow_1\rangle$, then the state of particle 2 is given by

$$|\psi_2^{>}\rangle = |\uparrow_2\rangle. \tag{18.1.3}$$

Here $\psi_2^{>}$ denotes the state of system 2 *after* ($t > t_0$) measuring particle 1.

At this point EPR argue as follows: since at the time of measurement the two systems no longer interact, no real change has taken place in the second system as a consequence of anything that may happen to the first system. That is, there exists no interaction between the two systems. Furthermore, EPR argue, since we have not affected particle 2 by looking at 1, the state of particle 2 must be the same before and after the measurement. That is,

$$|\psi_2^<\rangle = |\psi_2^>\rangle = |\uparrow_2\rangle, \tag{18.1.4}$$

where $|\psi_2^<\rangle$ and $|\psi_2^>\rangle$ denote the 'before' and 'after' states.

But we could have just as well decided to measure the x-component of particle 1 as in Fig. 18.1(b). Therefore, we naturally describe our spins in terms of $|\pm x\rangle$ states

$$|\pm x\rangle \equiv |\pm\rangle = \frac{1}{\sqrt{2}}(|\uparrow\rangle \pm |\downarrow\rangle), \tag{18.1.5}$$

and our spin singlet state before the measurment is

$$|\phi_{1,2}^<\rangle = \frac{1}{\sqrt{2}}(|+_1 -_2\rangle - |-_1 +_2\rangle). \tag{18.1.6}$$

Now after finding spin 1 to be in, say, the state $|-_1\rangle$, we have

$$|\phi_2^>\rangle = |+_2\rangle, \tag{18.1.7}$$

which, following the EPR argument as before, implies

$$|\phi_2^<\rangle = |+_2\rangle, \tag{18.1.8}$$

a very unsatisfactory state of affairs! For in the words of EPR: 'Thus, *it is possible to assign two different state vectors* [in our notation $|\uparrow_2\rangle$ and $|+_2\rangle$] *to the same reality.*'

One way out of this problem is to argue that when we are looking at a subsystem (e.g., particle 2 only), then we should be using a density matrix formulation. In general, when we are considering a composite system consisting of two subsystems, A and B, and if we are only interested in expectation values of operators $\hat{Q}_A$ which refer to system A alone, i.e.,

$$\hat{Q} = \hat{Q}_A \otimes 1_B, \tag{18.1.9}$$

then we are led to introduce the reduced density matrix ρ_A. That is, expressed in terms of the total density matrix $\rho_{A,B}$ we have

$$\begin{aligned}\langle Q\rangle &= \mathrm{Tr}_{A,B}(\rho_{AB}\hat{Q}) = \sum_{a,b}\langle a,b|\rho_{AB}\hat{Q}|a,b\rangle \\ &= \sum_a \langle a|\sum_b \langle b|\rho_{AB}1_B|b\rangle\hat{Q}_A|a\rangle = \mathrm{Tr}_A(\rho_A\hat{Q}_A),\end{aligned} \tag{18.1.10}$$

where reduced density matrix for system A is

$$\rho_A^{(r)} = \sum_b \langle b|\rho_{AB} 1_B|b\rangle = \mathrm{Tr}_B(\rho_{AB}). \tag{18.1.11}$$

Hence we should properly be considering the density matrix for system 2 (before looking at 1). In the first experiment, this is given by

$$\begin{aligned}\rho_2^<(I) = \mathrm{Tr}_1[\rho_{12}^<(I)] &= \langle\uparrow_1|\psi_{12}^<\rangle\langle\psi_{12}^<|\uparrow_1\rangle + \langle\downarrow_1|\psi_{12}^<\rangle\langle\psi_{12}^<|\downarrow_1\rangle \\ &= \frac{1}{2}(|\uparrow_2\rangle\langle\uparrow_2| + |\downarrow_2\rangle\langle\downarrow_2|) \\ &= \frac{1}{2}\begin{pmatrix}1 & 0\\ 0 & 1\end{pmatrix}_2.\end{aligned} \tag{18.1.12}$$

Likewise, the density matrix corresponding to the second experiment is

$$\begin{aligned}\rho_2^<(II) = \mathrm{Tr}_1[\rho_{12}^<(II)] &= \langle +_1|\phi_{12}^<\rangle\langle\phi_{12}^<|+_1\rangle + \langle -_1|\phi_{12}^<\rangle\langle\phi_{12}^<|-_1\rangle \\ &= \frac{1}{2}(|-_2\rangle\langle -_2| + |+_2\rangle\langle +_2|),\end{aligned} \tag{18.1.13}$$

which in terms of $|\uparrow_2\rangle$ and $|\downarrow_1\rangle$ spinors becomes

$$\begin{aligned}\rho_2^<(II) &= \frac{1}{4}[(|\uparrow_2\rangle - |\downarrow_2\rangle)(\langle\uparrow_2| - \langle\downarrow_2|) + (|\uparrow_2\rangle + |\downarrow_2\rangle) \\ &\quad (\langle\uparrow_2| + \langle\downarrow_2|)] \\ &= \frac{1}{2}(|\uparrow_2\rangle\langle\uparrow_2| + |\downarrow_2\rangle\langle\downarrow_2|) = \frac{1}{2}\begin{pmatrix}1 & 0\\ 0 & 1\end{pmatrix}_2.\end{aligned} \tag{18.1.14}$$

Hence we now have

$$\rho_2^<(I) = \rho_2^<(II), \tag{18.1.15}$$

which equality demonstrates the internal consistency of quantum theory. That is, we do not have two different descriptions of the same particle if we use the proper density matrix approach.

But even though the internal workings of quantum mechanics can be made self-consistent,* quantum mechanics seems strange. Furthermore, it is just this clever use of 'entangled' (e.g., spin singlet) states as introduced by EPR and expanded by Bell that teaches us just how strange the quantum world really is.

The inability of quantum mechanics to make definite predictions for the outcome of certain measurements led EPR to postulate the existence of 'hidden' variables which are not known and perhaps not measurable. It was hoped that an inclusion of these hidden variables would restore the completeness and determinism to the quantum theory. Bell's inequalities, to which we turn next, provide a basis for a

* See also the treatment of Griffiths and of Gell-Mann and Hartle on a consistent interpretation of quantum mechanics via quantum trajectories as discussed in *Phys. Rev. Lett.* **70**, 2201 (1993) and references therein.

quantitative test of the hidden variable approach. It is shown that, by performing correlation experiments of the type considered in EPR's argument, one can distinguish between the predictions of certain hidden variable theories and quantum mechanics.

18.2 Bell's inequality

We consider the EPR gedanken experiment illustrated in Fig. 18.2. A spin-zero system 'splits' into two spin-1/2 particles which then have anticorrelated values of spin projection along any given axis. For the purpose of proving Bell's theorem we are interested in the probability that particle 1 will pass through a Stern–Gerlach apparatus (SGA_1) in Fig. 18.2 which is oriented at an angle θ_a with the vertical ($+z$) direction and that particle 2 will pass through a Stern–Gerlach apparatus (SGA_2) which is oriented at an angle θ_b to the vertical. We denote this joint passage probability by $P(\theta_a, \theta_b) \equiv P_{ab}$. To proceed with the proof, we first establish our notation by considering the expression,

$$P_{ab} = P\,(\overset{\text{particle 1}}{\underset{a}{+}\;\;\underset{b}{-}\;\;\underset{c}{\bigcirc}} \;\big|\; \overset{\text{particle 2}}{\underset{a}{-}\;\;\underset{b}{+}\;\;\underset{c}{\bigcirc}})\,. \tag{18.2.1}$$

Here, the left side of the partition in the expanded notation refers to particle 1 and the right side to particle 2. As shown in Eq. (18.2.1) there are three 'slots' on each side of the partition in which we have put either a plus sign, a minus sign, or a circle. The first, second, and third slots are reserved for information concerning passage through an SGA oriented at the angles θ_a, θ_b, and θ_c, respectively. A plus sign refers to passage and a minus sign to blockage. A circle means that the particular joint probability in question does not contain information about passage at that angle. So for example in Eq. (18.2.1) the first $+$ means that particle 1 passes the SGA oriented at θ_a but then particle 2 would not pass through a SGA oriented at θ_a and this we denote by a $-$. Likewise if particle 2 passes through a SGA at θ_b we put a $+$ in the record slot to the right of the vertical bar and therefore a $-$ in the record slot associated with particle 1.

Now that we have explained the notation in general, let us return to Eq. (18.2.1). Recall that P_{ab} denotes the probability that particle 1 passes SGA_1 oriented at the angle θ_a to the z-axis and particle 2 passes SGA_2 oriented at the angle θ_b to the vertical. Likewise we write,

$$P_{bc} = P(\bigcirc\;+\;-\;|\;\bigcirc\;-\;+), \tag{18.2.2}$$

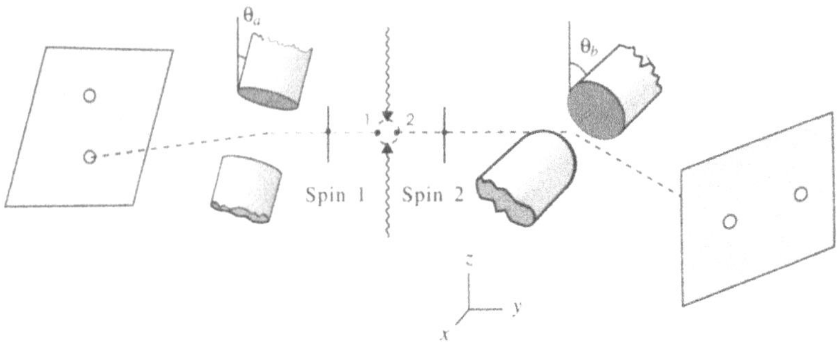

Fig. 18.2
Schematic of the EPR gedanken experiment. A spin-zero system such as orthohydrogen is split by an external field. The two spin-1/2 particles (protons) proceed in the opposite directions where they pass through the Stern-Gerlach apparati oriented at an angle θ_a with the vertical (+z) direction in the case of particle 1 and at an angle θ_b in the case of particle 2. For example the spin-up particle as indicated will 'pass' through the lower hole.

and

$$P_{ac} = P(+\ \bigcirc\ -\mid -\ \bigcirc\ +). \tag{18.2.3}$$

The usefulness of this notation becomes apparent when we take the next step. Although the joint probability P_{ab} says nothing about passage at θ_c we do know that for any given particle the probability that it will pass an SGA oriented at θ_c to the vertical plus the probability that it will not pass such an apparatus must be equal to unity. Using this fact and the anticorrelation of the spin projections we write,

$$\begin{aligned} P_{ab} &= P(+\ -\ \bigcirc \mid -\ +\ \bigcirc) \\ &= P(+\ -\ +\mid -\ +\ -) + P(+\ -\ -\mid -\ +\ +). \end{aligned} \tag{18.2.4}$$

Similarly,

$$\begin{aligned} P_{bc} &= P(\bigcirc\ +\ -\mid \bigcirc\ -\ +) \\ &= P(+\ +\ -\mid -\ -\ +) + P(-\ +\ -\mid +\ -\ +), \end{aligned} \tag{18.2.5}$$

$$\begin{aligned} P_{ac} &= P(+\ \bigcirc\ -\mid -\ \bigcirc\ +) \\ &= P(+\ +\ -\mid -\ -\ +) + P(+\ -\ -\mid -\ +\ +). \end{aligned} \tag{18.2.6}$$

Given Eqs. (18.2.4)–(18.2.6), the proof of Bell's theorem easily follows. We add P_{ab} and P_{bc} to get,

$$\begin{aligned} P_{ab} + P_{bc} = {} & P(+\ -\ +\mid -\ +\ -) + P(+\ -\ -\mid -\ +\ +) \\ & + P(+\ +\ -\mid -\ -\ +) + P(-\ +\ -\mid +\ -\ +). \end{aligned} \tag{18.2.7}$$

We note that, using Eq. (18.2.6), Eq. (18.2.7) can be written as,

$$\begin{aligned} P_{ab} + P_{bc} = {} & P_{ac} + P(+\ -\ +\mid -\ +\ -) \\ & + P(-\ +\ -\mid +\ -\ +). \end{aligned} \tag{18.2.8}$$

Classically, probabilities must be positive so that this implies,

$$P_{ab} + P_{bc} \geq P_{ac}. \tag{18.2.9}$$

This completes our proof of Bell's theorem.

18.3 Quantum calculation of the correlations in Bell's theorem

The quantum calculation for the probability of a spin-1/2 particle described by a state vector $|\Psi\rangle$ passing through a SGA oriented at angle θ is given by

$$P_\Psi(\theta) = |\langle\theta|\Psi\rangle|^2, \tag{18.3.1}$$

where the state $|\theta\rangle$ is formed by rotating a 'spin up' state about the y-axis

$$|\theta\rangle = e^{-i\theta\sigma_y/2}|\uparrow\rangle. \tag{18.3.2}$$

Here we recall that

$$\sigma_x = \begin{pmatrix} 0 & 1 \\ 1 & 0 \end{pmatrix}, \quad \sigma_y = \begin{pmatrix} 0 & -i \\ i & 0 \end{pmatrix}, \quad \sigma_z = \begin{pmatrix} 1 & 0 \\ 0 & -1 \end{pmatrix}, \tag{18.3.3}$$

and

$$|\uparrow\rangle = \begin{pmatrix} 1 \\ 0 \end{pmatrix}, \quad |\downarrow\rangle = \begin{pmatrix} 0 \\ 1 \end{pmatrix}. \tag{18.3.4}$$

We may rewrite (18.3.1) as

$$P_\Psi(\theta) = \langle\Psi|\theta\rangle\langle\theta|\Psi\rangle. \tag{18.3.5}$$

Now the projection operator $|\theta\rangle\langle\theta|$ is a useful quantity which we define as

$$\pi_\theta = |\theta\rangle\langle\theta|. \tag{18.3.6}$$

From Eq. (18.3.2) this may be written as

$$\pi_\theta = e^{-i\theta\sigma_y/2}|\uparrow\rangle\langle\uparrow|e^{i\theta\sigma_y/2}, \tag{18.3.7}$$

and using the fact that

$$e^{-i\theta\sigma_y/2}|\uparrow\rangle = \cos\frac{\theta}{2}|\uparrow\rangle + \sin\frac{\theta}{2}|\downarrow\rangle, \tag{18.3.8}$$

we find that Eq. (18.3.7) becomes

$$\pi_\theta = \frac{1}{2}(1 + \sigma_z \cos\theta + \sigma_x \sin\theta). \tag{18.3.9}$$

Now from the previous discussion we see that the probability of simultaneous passage through SGAs at θ_a and θ_b by the particles described by the spin singlet state Eq. (18.1.1) is

$$P_{ab} = \langle \Psi_{1,2} | \pi_{\theta_a}^{(1)} \pi_{\theta_b}^{(2)} | \Psi_{1,2} \rangle, \tag{18.3.10}$$

where the projection operators $\pi_{\theta_a}^{(1)}$ and $\pi_{\theta_b}^{(2)}$ correspond to particles 1 and 2. After a little algebra, see Problem 18.3, we find

$$P_{ab} = \frac{1}{4}[1 - \cos(\theta_a - \theta_b)] = \frac{1}{2}\sin^2\left(\frac{\theta_a - \theta_b}{2}\right). \tag{18.3.11}$$

Now in our derivation of Bell's inequality (18.2.9) we considered only three angles. Hence, we may use our quantum mechanical result

$$P_{ab} = \frac{1}{2}\sin^2\left(\frac{\theta_a - \theta_b}{2}\right), \tag{18.3.12}$$

to check whether Bell's theorem is 'obeyed' by quantum mechanics. That is, is the 'quantum version' of Bell's inequality obeyed?

$$\frac{1}{2}\sin^2\left(\frac{\theta_a - \theta_b}{2}\right) + \frac{1}{2}\sin^2\left(\frac{\theta_b - \theta_c}{2}\right) \geq \frac{1}{2}\sin^2\left(\frac{\theta_a - \theta_c}{2}\right). \tag{18.3.13}$$

To answer this we need only consider the angles $\theta_a = 0$, $\theta_b = \pi/4$, and $\theta_c = \pi/2$, so that Eq. (18.3.13) implies

$$2\sin^2\frac{\pi}{8} \geq \sin^2\frac{\pi}{4},$$

or

$$0.15 \geq 0.25, \tag{18.3.14}$$

which is false and therefore quantum mechanics violates the Bell's inequality!

Thus we have a clear situation in which the predictions of quantum mechanics and hidden variable theory are at variance. Many experiments have been, and continue to be, carried out and all of the experiments to date favor quantum mechanics, as seen below. There are still a few 'loopholes' which leave the question open but most workers now believe that the ultimate experiment* will support quantum mechanics.

What is it that went wrong in our derivation of Bell's theorem? How could such simple arguments be wrong? Perhaps the best way to answer such question is through the study of the simple examples that are considered in the next section.

* Kwiat, Eberhard, Steinberg, and Chiao [1994] and Fry, Walther, and Li [1995].

It may be noted that there are a number of other Bell inequalities. One useful form of the Bell inequalities (which is usually tested in experiments) is due to Clauser and Horne, and is given by (see Problem 18.1)

$$S \leq 1, \tag{18.3.15}$$

where

$$S = \frac{P_{12}(\theta_a, \theta_b) - P_{12}(\theta_a, \theta'_b) + P_{12}(\theta'_a, \theta_b) + P_{12}(\theta'_a, \theta'_b)}{P_1(\theta'_a) + P_2(\theta_b)}, \tag{18.3.16}$$

with $P_1(\theta'_a)$ and $P_2(\theta_b)$ being the passage probabilities for particles 1 and 2 to pass through the respective Stern–Gerlach apparati at angles θ'_a and θ_b, respectively.

In many experiments to test the Bell's inequalities, certain symmetries help to simplify the inequality (18.3.15). In these experiments $P_1(\theta'_a)$ and $P_2(\theta_b)$ are independent of the angles θ'_a and θ_b respectively, i.e., $P_1(\theta'_a) \equiv P_1$ and $P_2(\theta_b) \equiv P_2$. In addition the joint probabilities $P_{12}(\theta_a, \theta_b)$ depend only on the magnitude of the difference of the angles θ_a and θ_b, i.e., $P_{12}(\theta_a, \theta_b) \equiv P_{12}(|\theta_a - \theta_b|)$. Suppose that we chose θ_a, θ_b, θ'_b, and θ'_a in (18.3.16) so that

$$|\theta_a - \theta_b| = |\theta'_a - \theta_b| = |\theta'_a - \theta'_b| = \frac{1}{3}|\theta_a - \theta'_b| = \alpha. \tag{18.3.17}$$

We then have

$$S(\alpha) = \frac{3P_{12}(\alpha) - P_{12}(3\alpha)}{P_1 + P_2}. \tag{18.3.18}$$

Most experiments have been a variation of an experiment in which one measures the polarization correlations of the photons emitted successively in an atomic cascade. In such experiments, a three-level atom proceeds from, for example, a $J = 0$ level to a $J = 1$ level, and terminates in a $J = 0$ level which is the atomic ground state. Typically the atomic level scheme in calcium is employed where the $4p^2\ {}^1S_0$ level is populated by laser radiation via two-photon excitation. It then decays to the $4s^2\ {}^1S_0$ state via the $4p4s^1\ P_1$ level emitting two photons of wavelengths 5513 Å and 4227 Å (see Fig. 18.3). Due to parity and angular momentum conservation, there is a strong correlation in the polarization of the emitted photons.

The schematics of the experiment are shown in Fig. 18.4. The pair of correlated visible photons are emitted in the atomic cascade in a well-stabilized high-efficiency source S. These photons pass through the switching devices C_1 and C_2, followed by two polarizers in two different orientations: θ_a and θ'_a on side 1, and θ_b and θ'_b on side 2. The

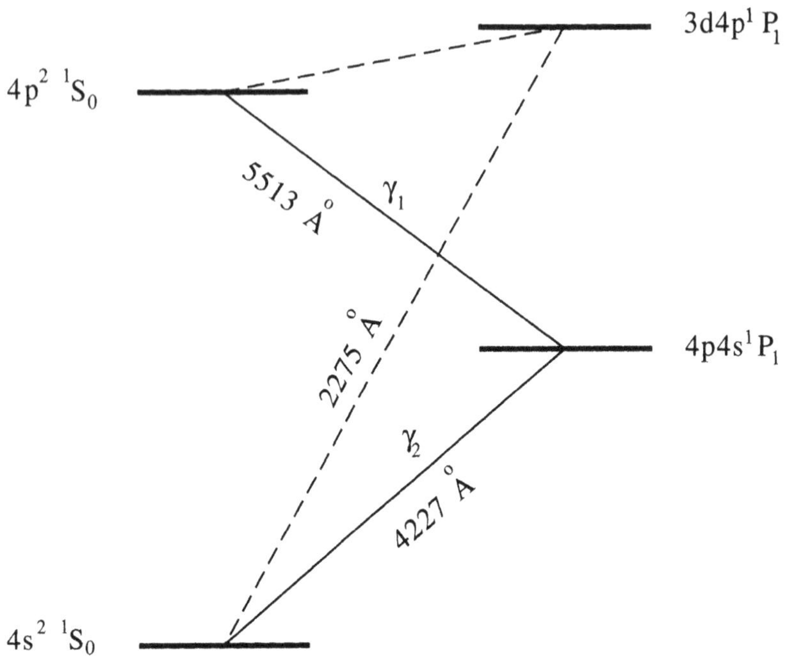

Fig. 18.3
Level scheme of calcium. The two-photon route for excitation to the upper level $4p^2\,{}^1S_0$ is shown by dashed lines.

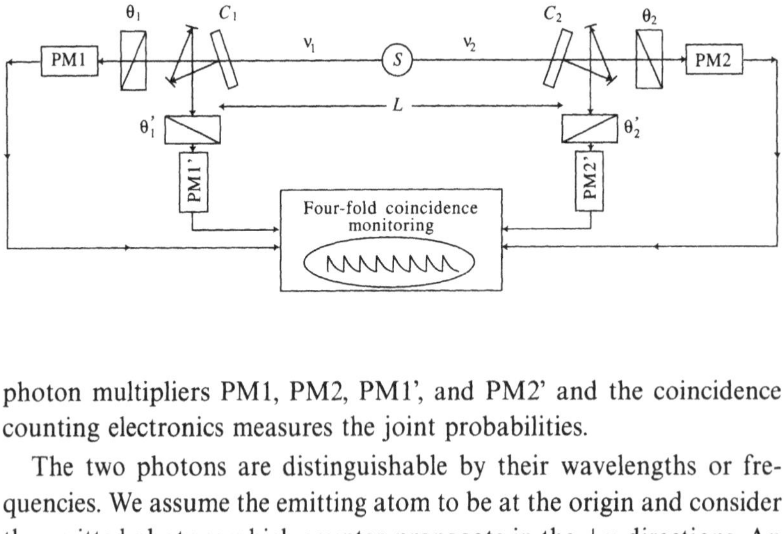

Fig. 18.4
Schematic of the photon correlation experiment in two-photon cascade emission to test Bell's inequality. (From A. Aspect, J. Dalibard, and G. Roger, *Phys. Rev. Lett.* **49**, 1804 (1982).)

photon multipliers PM1, PM2, PM1', and PM2' and the coincidence counting electronics measures the joint probabilities.

The two photons are distinguishable by their wavelengths or frequencies. We assume the emitting atom to be at the origin and consider the emitted photons which counter-propagate in the $\pm y$-directions. An optical filter in the $+y$-direction transmits only photons of frequency ν_1 and a filter in the $-y$-direction transmits only photons of frequency ν_2. As the transition is $J = 0 \to J = 1 \to J = 0$, the initial and final states of the atom have zero angular momentum and the same parity. Similarly, the two-photon state must have zero angular momentum and even parity. The state of the polarization of the two photons, after the passage through the filters, is of the form

$$|\Psi_{1,2}\rangle = \frac{1}{\sqrt{2}}(|R_1\rangle|R_2\rangle + |L_1\rangle|L_2\rangle), \tag{18.3.19}$$

where R and L refer to the photon polarizations being right and left circular and the subscripts 1 and 2 refer to the photons having frequencies ν_1 and ν_2, respectively. A change of basis to linear polarization states $|x\rangle$, $|z\rangle$ allows the state vector (18.3.19) to be rewritten as

$$|\Psi_{1,2}\rangle = \frac{1}{\sqrt{2}}(|z_1\rangle|z_2\rangle + |x_1\rangle|x_2\rangle). \tag{18.3.20}$$

The joint linear polarization measurement made by polarizers at angles θ_a and θ_b to the z-axis projects the state of Eq. (18.3.20) onto the two polarization states

$$|\theta_a\rangle = \cos\theta_a|z_1\rangle + \sin\theta_a|x_1\rangle, \tag{18.3.21}$$

$$|\theta_b\rangle = \cos\theta_b|z_2\rangle + \sin\theta_b|x_2\rangle. \tag{18.3.22}$$

The quantum mechanical probability for passage through the two polarization analyzers is therefore given by

$$\begin{aligned} P_{12}(\theta_a, \theta_b) &= |\langle\theta_a|\langle\theta_b|\Psi_{1,2}\rangle|^2, \\ &= \frac{1}{2}\cos^2(\theta_a - \theta_b). \end{aligned} \tag{18.3.23}$$

Next we calculate $P_1(\theta)$ and $P_2(\theta)$. If the incident photon of frequency ν_1 is polarized along the x-axis, then the probability of passing the polarizer oriented at an angle θ with the x-axis, with

$$|\Psi_1\rangle = \cos\theta|x\rangle + \sin\theta|z\rangle, \tag{18.3.24}$$

is $\cos^2\theta$. However, as the incident beam is unpolarized, we average over all values of θ, i.e.,

$$\begin{aligned} P_1(\theta) &= \frac{1}{2\pi}\int_0^{2\pi} \cos^2\theta d\theta \\ &= \frac{1}{2}. \end{aligned} \tag{18.3.25}$$

Similarly

$$P_2(\theta) = \frac{1}{2}. \tag{18.3.26}$$

We now substitute the values of $P_{12}(\theta_a, \theta_b)$, $P_1(\theta)$, and $P_2(\theta)$ from Eqs. (18.3.23), (18.3.25), and (18.3.26), respectively, in Eq. (18.3.18),

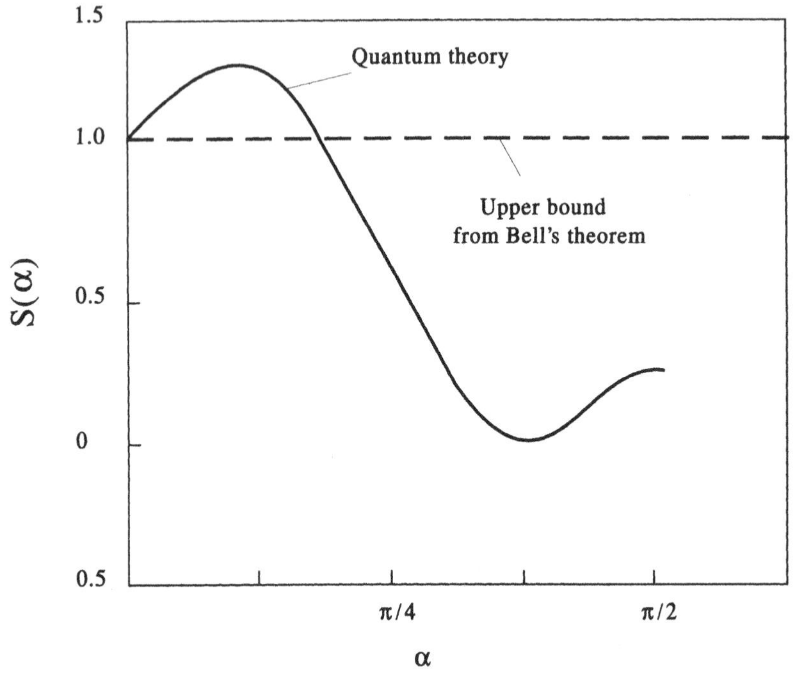

Fig. 18.5
$S(\alpha)$ versus α as given by Eq. (18.3.27). The dashed line shows the upper bound from Bell's inequality. (From J. F. Clauser and A. Shimony, *Rep. Prog. Phys.* **41**, 1881 (1978).)

and obtain

$$\begin{aligned} S(\alpha) &= \frac{3P_{12}(\alpha) - P_{12}(3\alpha)}{P_1 + P_2} \\ &= \frac{1}{2}(3\cos^2\alpha - \cos^2 3\alpha). \end{aligned} \tag{18.3.27}$$

For $\alpha = 22.5°$ this reduces to

$$S(22.5°) = 1.207, \tag{18.3.28}$$

in clear violation of the Bell's inequality (18.3.15). In Fig. 18.5, $S(\alpha)$ is plotted against α. It is seen that Bell's inequality (18.3.15) is violated for $0 < \alpha < 3\pi/16$.

18.4 Hidden variables from a quantum optical perspective*

Belinfante in his scholarly book on hidden variable (HV) theory shows that HV theories are not so far from quantum mechanics (QM) as might be thought. Stimulated by Belinfante's treatment and observations, one is led to apply quantum distribution theory to the

* This section follows Scully [1983]; for further reading see Mermin [1993].

spin-1/2 system, much as we want to do in quantum optics. In so doing a hidden variable theory is suggested which is in agreement with quantum theory insofar as the two-particle correlation experiments are concerned, but is clearly nonlocal.

Let us begin by recalling the joint probability that particle 1 is passed through a SGA oriented at an angle θ_a to the vertical $(+z)$ direction and that particle 2 is passed through a SGA oriented at an angle θ_b to the vertical, as given by the correlation function

$$P_{12}(\theta_a, \theta_b) = \langle\Psi|\pi^{(1)}_{\theta_a}\pi^{(2)}_{\theta_b}|\Psi\rangle, \tag{18.4.1}$$

and for the spin singlet we found (see Eq. (18.3.11))

$$P_{12}(\theta_a, \theta_b) = \frac{1}{4}[1 - \cos(\theta_a - \theta_b)]. \quad \text{(QM)} \tag{18.4.2}$$

Next we consider the same problem following Belinfante; we require that in order to give hidden variable theories an air of possibility we want them to yield the same results as quantum mechanics, at least in the simplest cases. For example, in an unpolarized beam only 1/2 the particles should pass through a given SGA. Further, the probability of passing through a second SGA placed behind (and at an angle θ) relative to the previous (vertical) SGA should be given by $\langle\uparrow|\pi_\theta|\uparrow\rangle$ which is $(1 + \cos\theta)/2$. Or, more generally, if a spin emerges from a SGA oriented at an angle α and then passes into a SGA tipped through an angle θ relative to the vertical, then the likelihood that the particle will emerge from the second SGA is given by

$$\frac{1}{2}[1 + \cos(\theta - \alpha)]. \tag{18.4.3}$$

Thus we might say that a 'hidden variable' α determined whether the spin passed through the apparatus whose angle θ is determined by the experimenter.

With this in mind we define the hidden variable probability function

$$\tilde{\pi}_\theta(\alpha) \equiv \frac{1}{2}[1 + \cos(\theta - \alpha)], \tag{18.4.4}$$

as giving the probability of 'simultaneous passage' through the SGAs oriented at θ and α.

We proceed now to consider the case where the two spins of our singlet system of Fig. 18.2 (having polarization angles α and β for spins 1 and 2) are correlated such that

$$I(\alpha, \beta)d\alpha d\beta \tag{18.4.5}$$

is the probability that the spins carry polarizations α and β, while all other hidden variables can be randomly distributed. For maximum polarization correlation, Belinfante then makes the reasonable ansatz

$$I(\alpha, \beta) = \delta(\alpha - \beta - \pi)\frac{1}{2\pi}. \tag{18.4.6}$$

Let us next ask: what is the probability of simultaneous passage of spins 1 and 2 through the double SGA system of Fig. 18.2? That is, what is $P_{12}(\theta_a, \theta_b)$ in the hidden variable theory? Following the above discussion and in view of Eqs. (18.4.4) and (18.4.5) we reasonably answer

$$\tilde{P}_{12}(\theta_a, \theta_b) = \iint d\alpha d\beta I(\alpha, \beta)\tilde{\pi}^{(2)}_{\theta_b}(\beta)\tilde{\pi}^{(1)}_{\theta_a}(\alpha). \tag{18.4.7}$$

And from Eqs. (18.4.6) and (18.4.4) this implies

$$\tilde{P}_{12}(\theta_a, \theta_b) = \iint d\alpha d\beta \delta(\alpha - \beta - \pi)$$
$$\frac{1}{2\pi}\left\{\frac{1}{2}[1+\cos(\theta_b-\beta)]\frac{1}{2}[1+\cos(\theta_a-\alpha)]\right\}. \tag{18.4.8}$$

Carrying out the simple integrations in Eq. (18.4.8) we find

$$\tilde{P}_{12}(\theta_a, \theta_b) = \frac{1}{4}\left[1 - \frac{1}{2}\cos(\theta_a - \theta_b)\right].\text{(HV)} \tag{18.4.9}$$

It is precisely the difference between the quantum correlation (18.4.2) and the hidden variable result (18.4.9) which concerns us here.

What then should we think of the hidden variable prediction Eq. (18.4.9)? It is not all that different from the quantum prediction. Might there not be a germ of 'truth' hidden in Eq. (18.4.9) and, more to the point, the arguments leading to it? In this context we quote Belinfante [1973]:

"The polarization (spin) hidden-variable here introduced is, of course, a quantity which does not exist in quantum theory ... in quantum theory no such thing as α even exists."

In the following we shall argue that a rigorous quantum mechanical treatment of the present problem can in fact be couched in terms of the angular variable α. In so doing we shall be led to reconsider the correlation function (Eq. (18.4.7)) and by a simple extension of Eq. (18.4.6) regain the quantum result (Eq. (18.4.2)) via a 'hidden variable' theory made to be a 'look-alike' to the quantum theory.

Proceeding toward a quantum mechanical description of spin-1/2 correlation we first calculate the two-dimensional spin distribution function

Fig. 18.6
Description of a spin-up particle in quantum distribution theory.

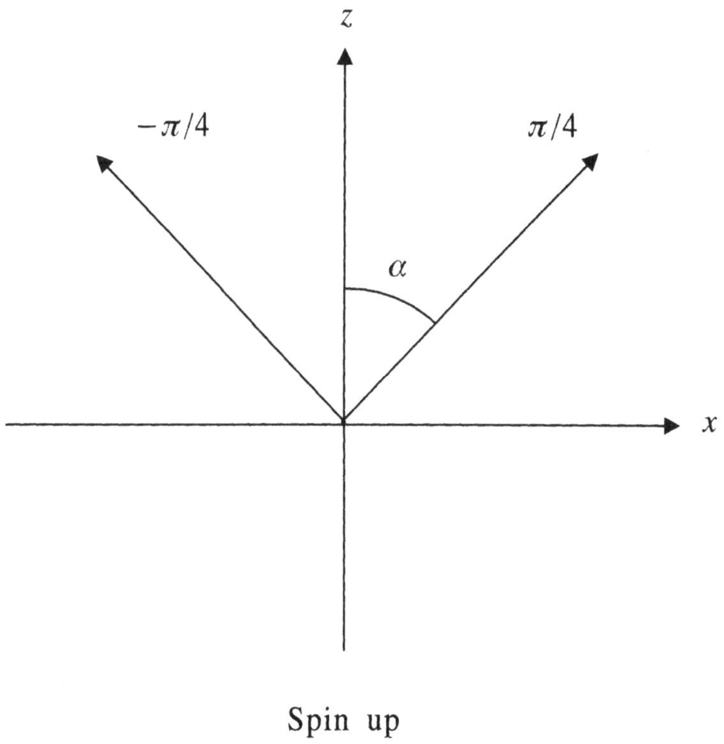

$$P_{\uparrow}(m_z, m_x) = \mathrm{Tr}[\rho\delta(m_x - \sigma_x)\delta(m_z - \sigma_z)], \tag{18.4.10}$$

for the case in which $\rho = |\uparrow\rangle\langle\uparrow|$, i.e., a spin-up particle. As shown in Appendix 18.A, this is given by

$$P_{\uparrow}(m_z, m_x) = \frac{1}{2}[\delta(m_x+1)\delta(m_z-1)+\delta(m_x-1)\delta(m_z-1)]. \tag{18.4.11}$$

Consulting Fig. 18.6 we see that the quantum distribution function for the state $|\uparrow\rangle$ corresponds to equal admixtures of 'probability' at $\pm\pi/4$.

We may rewrite (18.4.11) in terms of an angle α as in Fig. 18.6 such that $m_x = m\sin\alpha$ and $m_z = m\cos\alpha$, we find

$$P_{\uparrow}(\alpha, m) = \frac{1}{2}\left[\delta\left(\alpha - \frac{\pi}{4}\right) + \delta\left(\alpha + \frac{\pi}{4}\right)\right]\delta(m - \sqrt{2}). \tag{18.4.12a}$$

Similarly we obtain, for spin-down,

$$P_{\downarrow}(\alpha, m) = \frac{1}{2}\left[\delta\left(\alpha - \frac{3\pi}{4}\right) + \delta\left(\alpha + \frac{3\pi}{4}\right)\right]\delta(m - \sqrt{2}). \tag{18.4.12b}$$

Next we rewrite the operator π_θ of Eq. (18.3.7) in its associated c-number representation, i.e.,

$$\pi_\theta = \frac{1}{2}(1 + m_z \cos\theta + m_x \sin\theta), \tag{18.4.13}$$

which may be written in terms of m and α as becomes

$$\pi_\theta = \frac{1}{2}[1 + m\cos(\theta - \alpha)]. \tag{18.4.14}$$

The vector length m in (18.4.14) can be determined by requiring that $\langle\uparrow|\hat{\pi}_0|\uparrow\rangle$ be unity, i.e.,

$$\begin{aligned} 1 &= \int d\alpha \frac{1}{2}\left[\delta(\alpha - \frac{\pi}{4}) + \delta(\alpha + \frac{\pi}{4})\right]\frac{1}{2}(1 + m\cos\alpha) \\ &= \frac{1}{2}\left[1 + m\cos\frac{\pi}{4}\right]. \end{aligned} \tag{18.4.15}$$

This fixes the value of m to be $\sqrt{2}$.

In conclusion we note that the probability density calculated for the state corresponding to an arbitrary angle ϕ is given by

$$P_{+\phi}(\alpha) = \frac{1}{2}\left[\delta\left(\alpha - \phi - \frac{\pi}{4}\right) + \delta\left(\alpha - \phi + \frac{\pi}{4}\right)\right]. \tag{18.4.16}$$

To make a connection with the hidden variable theory of Belinfante we note that the probability/projection functions in hidden variable and quantum theory, i.e., $\tilde{\pi}_\theta(\alpha)$ from Eq. (18.4.4) and $\pi_\theta(\alpha)$ as given by Eq. (18.4.14) are quite similar. Indeed the quantum probability densities given by Eq. (18.4.16) suggest that we introduce similar probability densities into our hidden variable considerations as

$$\tilde{P}_{+\phi}(\alpha) = \delta(\alpha - \phi). \tag{18.4.17}$$

This is summarized in Table 18.1.

Now we want to extend the results of Table 18.1 to a two-particle spin-1/2 singlet with an eye toward the corresponding HV theory which such studies suggest. Corresponding to the anticorrelated (singlet) state for a pair of spin-1/2 particles

$$|\Psi\rangle = \frac{1}{\sqrt{2}}(|+_1 -_2\rangle - |-_1 +_2\rangle) \tag{18.4.18}$$

we have the associated probability distribution

$$\begin{aligned} &P(\mathbf{m}^{(1)}, \mathbf{m}^{(2)}) \\ &= \mathrm{Tr}\{\rho\delta(m_x^{(1)} - \sigma_x^{(1)})\delta(m_z^{(1)} - \sigma_z^{(1)})\delta(m_x^{(2)} - \sigma_x^{(2)}) \\ &\quad \delta(m_z^{(2)} - \sigma_z^{(2)})\}. \end{aligned} \tag{18.4.19}$$

In Eq. (18.4.19) $\rho = |\Psi\rangle\langle\Psi|$ with $|\Psi\rangle$ given by Eq. (18.4.18), $\sigma_x^{(i)}$ and $\sigma_z^{(i)}$ are the usual Pauli operators for the ith particle ($i = 1, 2$), and $m_x^{(i)}$ and $m_z^{(i)}$ are the associated quasiclassical variables.

Table 18.1. *Tabulation emphasizing the close similarity between the single-particle description in quantum mechanics and in the present HV theory.*

	Quantum mechanics	Present HV
Passage 'probability'	$\pi_\theta(\alpha) = \frac{1}{2}[1 + m\cos(\theta - \alpha)]$ $m = \sqrt{2}$	$\tilde{\pi}_\theta(\alpha) = \frac{1}{2}[1 + \cos(\theta - \alpha)]$
Distribution for spin-up	$P_\uparrow(\alpha) = \frac{1}{2}\times$ $\times\left[\delta\left(\alpha - \frac{\pi}{4}\right) + \delta\left(\alpha + \frac{\pi}{4}\right)\right]$	$\tilde{P}_\uparrow(\alpha) = \delta(\alpha)$
Expectation of passing	$\mathscr{P}_\uparrow = \int P_\uparrow(\alpha)\pi_0(\alpha)d\alpha$ $= 1$	$\tilde{\mathscr{P}}_\uparrow = \int \tilde{P}_\uparrow(\alpha)\tilde{\pi}_0(\alpha)d\alpha$ $= 1$

We must now carry out a calculation for $P(\mathbf{m}^{(1)}, \mathbf{m}^{(2)})$ similar to the calculation of $P(\mathbf{m})$ given above. After some algebra, we find

$$\begin{aligned} P_\phi(\alpha, \beta) = {} & \delta(\alpha - \beta - \pi) \\ & \frac{1}{2}\left\{\frac{1}{2}\left[\delta\left(\alpha - \phi - \frac{\pi}{4}\right) + \delta\left(\alpha - \phi + \frac{\pi}{4}\right)\right]\right. \\ & \left. + \frac{1}{2}\left[\delta\left(\alpha - \phi - \frac{3\pi}{4}\right) + \delta\left(\alpha - \phi + \frac{3\pi}{4}\right)\right]\right\} \end{aligned} \tag{18.4.20}$$

where α and β are the angular variables corresponding to particles 1 and 2. The physical meaning of the various terms in Eq. (18.4.20) is indicated in Fig. 18.7 for the case $\phi = 0$. We note immediately that the 'spin anticorrelation' factor occurring in Eq. (18.4.20), i.e., the $\delta(\alpha - \beta - \pi)$ term, is identical with that contained in Belinfante's $I(\alpha, \beta)$ given by Eq. (18.4.5). The difference, of course, is that the present result follows directly from quantum theory. Furthermore the curly bracketed factor corresponds to the sum of single-particle probability density functions for spin along $+\phi$ and $-\phi$.

In view of Eqs. (18.4.6), (18.4.16), (18.4.17), and (18.4.20), we are motivated to consider the 'hidden variable' probability function

$$\tilde{P}_\phi(\alpha, \beta) = \delta(\alpha - \beta - \pi)\frac{1}{2}[\delta(\alpha - \phi) + \delta(\alpha - \phi - \pi)]. \tag{18.4.21}$$

The joint count probabilities such as Eq. (18.4.9) can now be easily calculated via the general expression

$$P_0(\alpha,\beta)=\underbrace{\tfrac{1}{2}\,\delta(\alpha-\beta-\pi)}\{\underbrace{\tfrac{1}{2}[\delta(\alpha-\tfrac{\pi}{4})+\delta(\alpha+\tfrac{\pi}{4})]}+\underbrace{\tfrac{1}{2}[\delta(\alpha-\tfrac{3\pi}{4})+\delta(\alpha+\tfrac{3\pi}{4})]}\}$$

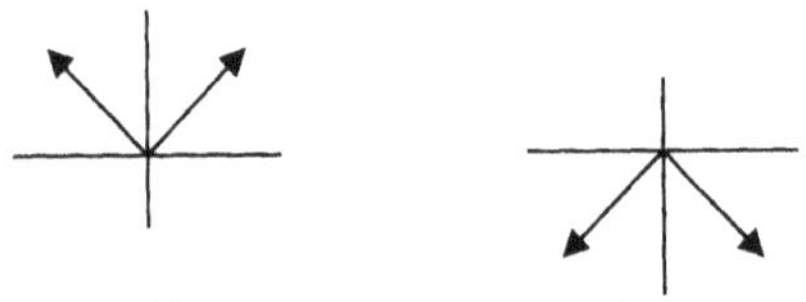

Fig. 18.7 Physical interpretation of the joint probability density for the spin–spin correlation function.

$$\tilde{P}_{12}(\theta_a,\theta_b)=\int\int \tilde{P}_\phi(\alpha,\beta)\tilde{\pi}_{\theta_b}(\beta)\tilde{\pi}_{\theta_a}(\alpha)d\alpha d\beta. \tag{18.4.22}$$

The present hidden variable 'theory' involves inserting Eqs. (18.4.4) and (18.4.21) into Eq. (18.4.22) to obtain

$$\tilde{P}_{12}(\theta_a,\theta_b)=\int\int \delta(\alpha-\beta-\pi)\frac{1}{2}[\delta(\alpha-\phi)+\delta(\alpha-\phi-\pi)] \times\frac{1}{2}[1+\cos(\theta_b-\beta)]\frac{1}{2}[1+\cos(\theta_a-\alpha)]d\alpha d\beta \tag{18.4.23}$$

and, upon carrying out a couple of simple integrations, this yields

$$\tilde{P}_{12}(\theta_a,\theta_b)=\frac{1}{4}[1-\cos(\theta_a-\theta_b)] \tag{18.4.24}$$

for $\phi=\theta_a$. The present HV result (18.4.24) agrees with the quantum mechanical prediction (Eq. (18.4.2)). The results of the various theories are summarized in Table 18.2.

In conclusion we must address the question: what is the difference between the present HV theory, which agrees with QM, and the Bell treatment (as embodied in the Bell inequality), which differs from QM? The answer is: the present HV theory involves negative probability in the first instance and furthermore the theory is nonlocal.

Concerning the negative probability aspect of the quantum mechanical description, we recall the general form of the classical description of $\tilde{P}_{12}(\theta_a,\theta_b)$ as given by Eq. (18.4.22).

In Eq. (18.4.22) $\tilde{\pi}_{\theta_a}(\alpha)$ and $\tilde{\pi}_{\theta_b}(\beta)$ represent the probabilities that particle 2 with the 'hidden' variable α would pass through a SGA_1 oriented at an angle θ_a to the vertical and particle 1 with the 'hidden' variable β would pass through SGA_1 oriented at an angle θ_b to the vertical. Classically these probabilities are, of course, positive. The function $\tilde{P}_\phi(\alpha,\beta)$ represents the correlation between the variables α and β and so integration over these leads to the joint probability of passage, $\tilde{P}_{12}(\theta_a,\theta_b)$. The general form for $\tilde{P}_{12}(\theta_a,\theta_b)$ from quantum distribution theory closely resembles the classical form (Eq. (18.4.22)) and is given by

Table 18.2. *Table summarizing the correspondence between the two-particle spin-correlation functions in Belinfante's HV, quantum mechanics, and the present HV.*

	$P(\alpha, \beta)$ Two-particle distribution	$P(\theta_a, \theta_b)$ Joint probability
Belinfante HV	$\frac{1}{2\pi}\delta(\alpha - \beta\pi)$	$\frac{1}{4}\left[1 - \frac{1}{2}\cos(\theta_a - \theta_b)\right]$
QM	$\frac{1}{2}\delta(\alpha - \beta - \pi)$ $\times\left\{\frac{1}{2}\left[\delta\left(\alpha - \phi - \frac{\pi}{4}\right) + \delta\left(\alpha - \phi + \frac{\pi}{4}\right)\right] + \frac{1}{2}\left[\delta\left(\alpha - \phi - \frac{3\pi}{4}\right) + \delta\left(\alpha - \psi + \frac{3\pi}{4}\right)\right]\right\}$	$\frac{1}{4}[1 - \cos(\theta_a - \theta_b)]$
Present HV	$\delta(\alpha - \beta - \pi)\frac{1}{2}[\delta(\alpha - \phi) + \delta(\alpha - \phi - \pi)]$	$\frac{1}{4}[1 - \cos(\theta_a - \theta_b)]$

$$\tilde{P}_{12}(\theta_a, \theta_b) = \int\int P_\phi(\alpha, \beta)\pi_{\theta_b}(\beta)\pi_{\theta_a}(\alpha)d\alpha d\beta. \tag{18.4.25}$$

As in the classical case $P_\phi(\alpha, \beta)$ plays the role of a correlation function between the variables α and β. However the projection operators $\pi_{\theta_b}(\beta)$ and $\pi_{\theta_a}(\alpha)$ given by

$$\pi_{\theta_a}(\alpha) = \frac{1}{2}\left[1 + \sqrt{2}\cos(\theta_a - \alpha)\right] \tag{18.4.26a}$$

and

$$\pi_{\theta_b}(\beta) = \frac{1}{2}\left[1 + \sqrt{2}\cos(\theta_b - \beta)\right] \tag{18.4.26b}$$

can now be negative. Herein lies an essential difference between the classical and quantum descriptions of the EPR spin-singlet problem. The functions $\pi_{\theta_b}(\beta)$ and $\pi_{\theta_a}(\alpha)$, which are the closest quantum mechanical analogs to the classical probabilities $\tilde{\pi}_{\theta_b}(\beta)$ and $\tilde{\pi}_{\theta_a}(\alpha)$ can be negative. We recall that the proof of Bell's theorem does not hold if one allows negative probabilities.

We have shown that an attempt to make quantum mechanics look classical requires that we make correspondences between classical probabilities and quantum distribution theory functions which are not positive semidefinite. Thus we see (in this example) how it is that we have a quantum mechanical violation of Bell's theorem.

We now turn to a somewhat subtle but important point. Upon calculating $P_{12}(\theta_a, \theta_b)$ via $P_0(\alpha, \beta)$ from Eq. (18.4.20) with $\phi = 0$ we find

$$\begin{aligned} P_{12}(\theta_a, \theta_b) = \iint \delta(\alpha - \beta - \pi)\frac{1}{2}\left\{\frac{1}{2}\left[\delta\left(\alpha - \frac{\pi}{4}\right) + \delta\left(\alpha + \frac{\pi}{4}\right)\right]\right. \\ \left. + \frac{1}{2}\left[\delta\left(\alpha - \frac{3\pi}{4}\right) + \delta\left(\alpha + \frac{3\pi}{4}\right)\right]\right\} \\ \times\frac{1}{2}[1 + m\cos(\theta_b - \beta)]\frac{1}{2}[1 + m\cos(\theta_a - \alpha)]d\alpha d\beta \\ = \frac{1}{4}[1 - \cos(\theta_a - \theta_b)], \end{aligned} \tag{18.4.27}$$

which is the same result as that obtained when $P_\phi(\alpha, \beta)$ from Eq. (18.4.20) is used. This is as would be expected since use of either spherically symmetric state

$$|\Psi\rangle = \frac{1}{\sqrt{2}}[|\uparrow_1\downarrow_2\rangle - |\downarrow_1\uparrow_2\rangle] \rightarrow P_0(\alpha, \beta) \tag{18.4.28a}$$

or

$$|\Psi\rangle = \frac{1}{\sqrt{2}}[|+_1 -_2\rangle - |-_1 +_2\rangle] \rightarrow P_\phi(\alpha, \beta) \tag{18.4.28b}$$

should give the same results.

However, if, within the present HV theory, we carry out the calculation of $\tilde{P}_{12}(\theta_a, \theta_b)$, using $\tilde{P}_0(\alpha, \beta)$ as given in Fig. 18.7, we would have

$$\begin{aligned} \tilde{P}_{12}(\theta_a, \theta_b) = \int\int \delta(\alpha - \beta - \pi)\frac{1}{2}[\delta(\alpha) + \delta(\alpha - \pi)]\frac{1}{2} \\ \times[1 + \cos(\theta_b - \beta)]\frac{1}{2}[1 + \cos(\theta_a - \alpha)]d\alpha d\beta \end{aligned} \tag{18.4.29}$$

which yields

$$\tilde{P}_{12}(\theta_a, \theta_b) = \frac{1}{4}\left[1 - \frac{1}{2}\cos(\theta_a + \theta_b) - \frac{1}{2}\cos(\theta_a - \theta_b)\right]. \tag{18.4.30}$$

This result is in clear agreement with the quantum prediction only if $\theta_b = 0$.

The physical content and interpretation of passage (projection) functions, e.g., $\pi_\theta(\alpha)$ and the conditional probability distribution $P_\phi(\alpha, \beta)$ in quantum theory and the present HV theory differ, and the roles of state preparation and experimental parameterization enter into the theories in different ways.

In the quantum theory, state preparation involves choosing $|\psi\rangle$, i.e., $P_\phi(\alpha,\beta)$ and then the experiment in question tells us what operator, e.g., π_0, we should calculate. However, if we change from $P_\phi(\alpha,\beta)$ to $P_0(\alpha,\beta)$ the calculation should remain unchanged since we are considering a spherically symmetric state.

In the present HV theory, on the other hand, the choice of $\tilde{P}_0(\alpha,\beta)$, for the present spherically symmetric problem, must 'match' the choice of $\pi_\theta(\alpha)$ as the calculations of Eqs. (18.4.29), (18.4.30), and (18.4.26a,18.4.26b) indicate. Both $\tilde{P}_\phi(\alpha,\beta)$ and $\tilde{\pi}_\theta(\alpha)$ involve state preparation and experimental specification. Thus while the present HV theory reproduces the quantum spin–spin correlation function, it differs from quantum theory in this important aspect. Thus we see that the present HV theory is nonlocal.

18.5 Bell's theorem without inequalities: Greenberger–Horne–Zeilinger (GHZ) equality

The results of Section 18.3 tell us that quantum theory, and not the hidden variable theories, are supported by experiments, but the hidden variable results are only ruled out in some regions of angle α. Furthermore, as in Section 18.4, we see that (nonlocal) hidden variable theories can even be derived that agree with quantum mechanics. It would therefore be very interesting to find a situation in which hidden variables and quantum mechanical correlations are completely at variance. It is to this end that we now turn.

Greenberger, Horne, and Zeilinger (GHZ) have shown, by considering certain three-particle correlation experiments, that the results of the hidden variable theories and quantum mechanics are in complete contradiction. There are no inequalities involved in the GHZ work and the predictions of the hidden variable theories and quantum mechanics are simply different. This provides a stronger refutation of the hidden variable theories.

We consider the three-particle state

$$|\Psi\rangle_3 = \frac{1}{\sqrt{2}}(|\uparrow_1\uparrow_2\uparrow_3\rangle - |\downarrow_1\downarrow_2\downarrow_3\rangle), \tag{18.5.1}$$

instead of the earlier two-particle state

$$|\Psi\rangle_2 = \frac{1}{\sqrt{2}}(|\uparrow_1\downarrow_2\rangle - |\downarrow_1\uparrow_2\rangle). \tag{18.5.2}$$

The state $|\Psi\rangle_3$ is an eigenstate of the operators $\sigma_x^{(1)}\sigma_y^{(2)}\sigma_y^{(3)}$, $\sigma_y^{(1)}\sigma_x^{(2)}\sigma_y^{(3)}$, and $\sigma_y^{(1)}\sigma_y^{(2)}\sigma_x^{(3)}$. This can be checked by recalling that

$$\sigma_x|\uparrow\rangle=|\downarrow\rangle, \quad \sigma_x|\downarrow\rangle=|\uparrow\rangle, \quad \sigma_y|\uparrow\rangle=i|\downarrow\rangle, \quad \sigma_y|\downarrow\rangle=-i|\uparrow\rangle, \tag{18.5.3}$$

so that, for example,

$$\begin{aligned}\sigma_x^{(1)}\sigma_y^{(2)}\sigma_y^{(3)}|\Psi\rangle_3 &= \sigma_x^{(1)}\sigma_y^{(2)}\sigma_y^{(3)}\frac{1}{\sqrt{2}}(|\uparrow_1\uparrow_2\uparrow_3\rangle - |\downarrow_1\downarrow_2\downarrow_3\rangle)\\ &= \frac{1}{\sqrt{2}}[i^2|\downarrow_1\downarrow_2\downarrow_3\rangle - (-i)^2|\uparrow_1\uparrow_2\uparrow_3\rangle]\\ &= |\Psi\rangle_3.\end{aligned} \tag{18.5.4}$$

So $|\Psi\rangle_3$ is an eigenstate of the operators $\sigma_x^{(1)}\sigma_y^{(2)}\sigma_y^{(3)}$ with eigenvalue $+1$. In a similar manner, we can show that $|\Psi\rangle_3$ is also an eigenstate of the operators $\sigma_y^{(1)}\sigma_x^{(2)}\sigma_y^{(3)}$ and $\sigma_y^{(1)}\sigma_y^{(2)}\sigma_x^{(3)}$ with eigenvalue $+1$.

Therefore we have a new kind of EPR correlation such that if we find the measured values of $\sigma_y^{(1)}$ and $\sigma_y^{(2)}$ to both be $+1$ then $\sigma_x^{(3)}$ must also be $+1$. However, if $\sigma_y^{(1)}$ and $\sigma_y^{(2)}$ are found to be $+1$ and -1, respectively, then $\sigma_x^{(3)}$ will be -1. So we now have a new kind of EPR 'action at a distance'.

Next we proceed to note that

$$\sigma_x^{(1)}\sigma_x^{(2)}\sigma_x^{(3)}|\Psi\rangle_3 = -|\Psi\rangle_3 \tag{18.5.5}$$

since

$$\sigma_x^{(1)}\sigma_x^{(2)}\sigma_x^{(3)} = -[\sigma_x^{(1)}\sigma_y^{(2)}\sigma_y^{(3)}][\sigma_y^{(1)}\sigma_x^{(2)}\sigma_y^{(3)}][\sigma_y^{(1)}\sigma_y^{(2)}\sigma_x^{(3)}] \tag{18.5.6}$$

(recall that $\sigma_x^{(i)}\sigma_y^{(i)} = -\sigma_y^{(i)}\sigma_x^{(i)}$ $(i = 1,2,3)$) and the operators of each of the three bracketed expressions on $|\Psi\rangle_3$ yields $+|\Psi\rangle_3$. Hence the measurement of $\sigma_x^{(1)}$, $\sigma_x^{(2)}$, and $\sigma_x^{(3)}$ will *always* result in a situation where the product of the three outcomes will be -1.

Next we consider the predictions of the outcome in the hidden variable theories. To each operator $\sigma_x^{(i)}$ and $\sigma_y^{(i)}$, we assign the corresponding classical quantities $m_x^{(i)}$ and $m_y^{(i)}$ where $m_x^{(i)}$ and $m_y^{(i)}$ can be $+1$ or -1. It follows from Eq. (18.5.4) that

$$m_x^{(1)}m_y^{(2)}m_y^{(3)} = +1, \tag{18.5.7}$$

and also

$$m_y^{(1)}m_x^{(2)}m_y^{(3)} = +1, \tag{18.5.8}$$

$$m_y^{(1)}m_y^{(2)}m_x^{(3)} = +1. \tag{18.5.9}$$

Therefore

$$[m_x^{(1)}m_y^{(2)}m_y^{(3)}][m_y^{(1)}m_x^{(2)}m_y^{(3)}][m_y^{(1)}m_y^{(2)}m_x^{(3)}] = +1, \tag{18.5.10}$$

or

$$m_x^{(1)}m_x^{(2)}m_x^{(3)}[m_y^{(1)}]^2[m_y^{(2)}]^2[m_y^{(3)}]^2 = +1. \tag{18.5.11}$$

Since $[m_y^{(1)}]^2 = +1$ etc., we see that

$$m_x^{(1)}m_x^{(2)}m_x^{(3)} = +1, \tag{18.5.12}$$

in contradiction with the quantum prediction, Eq. (18.5.5), according to which this quantity should be equal to -1.

18.6 Quantum cryptography

An interesting application of the quantum correlations between the single photons in an EPR type setup discussed in Section 18.1 is in the field of quantum cryptography. As we have discussed before, the measurement of a quantum system in general causes a disturbance and provides only incomplete information of the system before the measurement. In quantum cryptography, this aspect of quantum mechanics is used to allow two parties, the sender and the receiver, to communicate in absolute secrecy, even in the presence of an eavesdropper.

In any exchange of secret information, the data (usually a sequence of bits 1 and 0) is combined with a random sequence of bits, called the key, and is sent through a communication channel. The key is known only to the sender and the receiver. The randomness of the key ensures that the transmitted data is also random and is inaccessible to a potential eavesdropper who does not have the key. The safety of the channel therefore depends on the secrecy of the key. A problem with a classical channel is that, in principle, eavesdropping can take place without the sender or the receiver knowing. This is not true of quantum cryptography, in which (as seen below) eavesdropping will disturb the measured sequence in a detectable way. In the following, we present different quantum cryptographic systems.

18.6.1 Bennett–Brassard protocol

In this scheme, the quantum cryptography apparatus consists of a transmitter and a receiver. At the transmitter end, the sender transmits photons in one of the four polarization states, at angles 0°, 45°,

90°, and 135° with the vertical direction. The two polarization states, say along 0° and 45°, stand for bit 0, while the other two, along 90° and 135°, stand for 1. The stream of photons appropriately polarized thus stands for a sequence of 1's and 0's. At the receiving end, the receiver measures the polarization of the arriving photons by randomly choosing the basis ⊕ (polarizer oriented in the vertical direction) or ⊗ (polarizer oriented at an angle of 45°). According to the laws of quantum mechanics, the receiver can distinguish between the rectilinear polarization (i.e., at angles 0° and 90°) or the diagonal polarization (i.e., at angles 45° and 135°), but not both.

When the basis chosen by the receiver is the same as that of the sender, the polarization of the received photon is perfectly correlated with the sender's. Thus a photon polarized along 90° received through a polarizer oriented along the vertical direction will be found polarized along 90° and so on. However no such correlation exists when the basis chosen by the receiver is conjugate to that of the sender, i.e., a photon polarized along 90° will be found polarized either along 45° or 135° with equal probability if received through a polarizer oriented at an angle of 45°. Thus a sequence of polarizations along 0°, 90°, 135°, 0°, 45°, 135°, 45°, 45°, 90° received through a sequence of basis ⊗, ⊕, ⊕, ⊗, ⊗, ⊗, ⊕, ⊗, ⊕ may yield the outcome sequence along 135°, 90°, 0°, 45°, 45°, 135°, 90°, 45°, 90°. The receiver records the outcome of his measurements in secrecy, but the receiver and the sender compare their sequences of basis through a public channel. They retain the instances where they use the same basis and discard the rest. Thus, in the above example, outcomes at 2, 5, 6, 8, and 9 are retained and the outcomes 1, 3, 4, and 7 are discarded. When translated into bits 0 or 1 (1 0 1 0 1 in the above example), the key is obtained.

In the Bennett–Brissard protocol the choice of basis is completely hidden from the eavesdropper. A passive eavesdropping in this protocol is not possible as any attempt at eavesdropping would lead to discrepancies between the sequences. The sender and the receiver can try to infer the presence of an eavesdropper by comparing part of their data . If the discrepancies are found, they can then reject their data and start over.

18.6.2 Quantum cryptography based on Bell's theorem

This potential scheme for quantum cryptography is based on the EPR gedanken experiment (Section 18.2) and Bell's theorem is used to test for eavesdropping. The set up is similar to the one to test Bell's inequality (Fig. 18.4). The channel consists of a source that emits pairs of spin-1/2 particles in a singlet state. The particles move off in opposite directions, toward the sender and the receiver, who then make measurements on the polarization components along different directions as in the experiment to test Bell's inequality.

After the measurements on the polarization components are made, the sender and receiver compare on a public channel the sequence of orientations of their analyzers. They divide their measurements in two group. In the first group, they include those measurements when their analyzers were oriented in the same direction and the second group these in which they were oriented in different directions. They discard all those measurements in which either of them failed to register a particle.

The outcome of the first group of measurements is not revealed on the public channel and is used to form the key as the measurements are perfectly correlated. The outcome of the second group of measurements is communicated between the sender and receiver on the public channel and the resulting data is used to evaluate the quantity S as given by Eq. (18.3.16). If no eavesdropping has taken place then the result should satisfy the predictions of quantum mechanics. Thus a test of eavesdropping becomes available.

18.A Quantum distribution function for a single spin-up particle

Proceeding toward an 'α' description of our spin-singlet problem, consider first the expectation value of an operator $Q(\sigma_x, \sigma_y, \sigma_z)$ given in terms of the density matrix ρ:

$$\langle Q \rangle = \mathrm{Tr}[\rho(t) Q(\sigma_x, \sigma_y, \sigma_z)]. \tag{18.A.1}$$

Introducing the operator δ-function

$$\delta(\beta - b) \equiv \int \frac{d\xi}{2\pi} \exp[-i\xi(\beta - b)], \tag{18.A.2}$$

where b is an operator (e.g., σ_z) and β is the associated classical variable (e.g., m_z); we may the write $Q(\sigma_x, \sigma_y, \sigma_z)$ as

$$Q(\sigma_x, \sigma_y, \sigma_z) = \int d^3m Q(m_x, m_y, m_z) \delta(m_x - \sigma_x) \delta(m_y - \sigma_y) \delta(m_z - \sigma_z). \tag{18.A.3}$$

Inserting Eq. (18.A.3) into Eq. (18.A.1) we then find the expectation value for Q to be given by

$$\langle Q \rangle = \int d^3 m P(m_x, m_y, m_z, t) Q(m_x, m_y, m_z), \tag{18.A.4}$$

where

$$P(m_x, m_y, m_z, t) \equiv \mathrm{Tr}[\rho(t)\delta(m_x - \sigma_x)\delta(m_y - \sigma_y)\delta(m_z - \sigma_z)]. \tag{18.A.5}$$

Recalling, however, from the discussion of Section 18.4 that we shall only be interested in operators Q involving σ_x and σ_z, e.g., $\hat{\pi}_\theta$, we may rigorously restrict our treatment to operator expansions of the form

$$Q(\sigma_x, \sigma_z) = \int\int Q(m_x, m_z)\delta(m_x - \sigma_x)\delta(m_z - \sigma_z) dm_x dm_z,$$

and the associated quantum distribution function

$$P(m_x, m_z, t) = \mathrm{Tr}[\rho(t)\delta(m_x - \sigma_x)\delta(m_z - \sigma_z)]. \tag{18.A.6}$$

The distribution function Eq. (18.A.6) is the vehicle by which we shall realize a quantum treatment of the present spin-1/2 problem in terms of the angle α. In the following we treat the simple problem of a single 'spin-up' particle.

For the 'spin-up' case the density matrix is, of course,

$$\rho = |\uparrow\rangle\langle\uparrow| \tag{18.A.7}$$

and the associated distribution function is

$$\begin{aligned} P_\uparrow(m_x, m_z) &= \mathrm{Tr}[|\uparrow\rangle\langle\uparrow|\delta(m_x - \sigma_x)\delta(m_z - \sigma_z)] \\ &= \langle\uparrow|\delta(m_x - \sigma_x)\delta(m_z - \sigma_z)|\uparrow\rangle, \end{aligned} \tag{18.A.8}$$

which by Eq. (18.A.2) becomes

$$\begin{aligned} P_\uparrow(m_x, m_z) &= \int \frac{d\xi}{2\pi} \int \frac{d\eta}{2\pi} \langle\uparrow| e^{i\xi\sigma_x} e^{i\eta\sigma_z} |\uparrow\rangle e^{-im_x\xi} e^{-im_z\eta} \\ &= \int \frac{d\xi}{2\pi} \int \frac{d\eta}{2\pi} \cos(\xi) e^{i\eta} e^{-im_x\xi} e^{-im_z\eta} \\ &= \frac{1}{2}[\delta(m_x + 1)\delta(m_z - 1) + \delta(m_x - 1)\delta(m_z - 1)]. \end{aligned} \tag{18.A.9}$$

18.B Quantum distribution function for two particles

Here, we fill in the gap between Eqs. (18.4.19) and (18.4.20). First, note that (18.4.19) may be written

$$P(\mathbf{m}^{(1)}, \mathbf{m}^{(2)})$$
$$= \langle \Psi | \delta(m_x^{(1)} - \hat{\sigma}_x^{(1)}) \delta(m_z^{(1)} - \hat{\sigma}_z^{(1)}) \delta(m_x^{(2)} - \hat{\sigma}_x^{(2)}) \delta(m_z^{(2)} - \hat{\sigma}_z^{(2)}) | \Psi \rangle \tag{18.B.1}$$

and

$$|\Psi\rangle = \frac{1}{\sqrt{2}} \left(| \uparrow_1 \downarrow_2 \rangle - | \downarrow_1 \uparrow_2 \rangle \right) \tag{18.B.2}$$
$$= -\frac{1}{\sqrt{2}} \left(| +_1 -_2 \rangle - | -_1 +_2 \rangle \right) , \tag{18.B.3}$$

where, for the purposes of this appendix, $| \uparrow\rangle$ is the positive eigenstate of $\hat{\sigma}_z$, and $|+\rangle$ is the positive eigenstate of $\hat{\sigma}_x$; hence $\langle \pm | \uparrow \rangle = \frac{1}{\sqrt{2}}$, $\langle \pm | \downarrow \rangle = \pm \frac{1}{\sqrt{2}}$, so, for example,

$$\langle +_1 -_2 | \uparrow_1 \downarrow_2 \rangle = \langle +_1 | \uparrow_1 \rangle \langle -_2 | \downarrow_2 \rangle = \frac{1}{\sqrt{2}} \left(-\frac{1}{\sqrt{2}} \right) = -\frac{1}{2} .$$

We can handle a function of an operator by breaking its operand into eigenstates, upon which the function may be evaluated at the eigenvalue. For this case in particular

$$\delta(m_z^{(2)} = \hat{\sigma}_z^{(2)}) | \Psi \rangle$$
$$= \frac{1}{\sqrt{2}} \delta(m_z^{(2)} - \hat{\sigma}_z^{(2)}) \left[| \uparrow_1 \downarrow_2 \rangle - | \downarrow_1 \uparrow_2 \rangle \right]$$
$$= \frac{1}{\sqrt{2}} \left[\delta(m_z^{(2)} + 1) | \uparrow_1 \downarrow_2 \rangle - \delta(m_z^{(2)} - 1) | \downarrow_1 \uparrow_2 \rangle \right] , \tag{18.B.4}$$

$$\langle \Psi | \delta(m_x^{(1)} - \hat{\sigma}_x^{(1)}$$
$$= -\frac{1}{\sqrt{2}} \left[\langle +_1 -_2 | \right] \delta(m_x^{(1)} - \hat{\sigma}_x^{(1)})$$
$$= -\frac{1}{\sqrt{2}} \left[\langle +_1 -_2 | \delta(m_x^{(1)} - 1) - \langle -_1 +_2 | \delta(m_x^{(1)} + 1) \right] . \tag{18.B.5}$$

Picking up with (18.B.1), then

$$P(\mathbf{m}^{(1)}, \mathbf{m}^{(2)})$$
$$= \frac{1}{4} \Big[\delta(m_x^{(1)} - 1) \delta(m_z^{(1)} + 1) \delta(m_x^{(2)} + 1) \delta(m_z^{(2)} - 1)$$
$$+ \delta(m_x^{(1)} - 1) \delta(m_z^{(1)} - 1) \delta(m_x^{(2)} + 1) \delta(m_z^{(2)} + 1)$$
$$+ \delta(m_x^{(1)} + 1) \delta(m_z^{(1)} + 1) \delta(m_x^{(2)} - 1) \delta(m_z^{(2)} - 1)$$
$$+ \delta(m_x^{(1)} + 1) \delta(m_z^{(1)} - 1) \delta(m_x^{(2)} - 1) \delta(m_z^{(2)} + 1) \Big] . \tag{18.B.6}$$

Now, following the same approach as before, and letting α represent the orientation of $\mathbf{m}^{(1)}$, and β that of $\mathbf{m}^{(2)}$, with ϕ the arbitrary orientation of the z-axis, this may be rewritten as

$$\begin{aligned}
P(\mathbf{m}^{(1)}, \mathbf{m}^{(2)}) & \\
= \frac{1}{4}\Bigg[& \delta\left(\alpha - \phi - \frac{3\pi}{4}\right)\delta\left(\beta - \phi + \frac{\pi}{4}\right) \\
& + \delta\left(\alpha - \phi - \frac{\pi}{4}\right)\delta\left(\beta - \phi + \frac{3\pi}{4}\right) \\
& + \delta\left(\alpha - \phi + \frac{3\pi}{4}\right)\delta\left(\beta - \phi - \frac{\pi}{4}\right) \\
& + \delta\left(\alpha - \phi + \frac{\pi}{4}\right)\delta\left(\beta - \phi - \frac{3\pi}{4}\right)\Bigg] \\
= \frac{1}{4}\delta(\alpha - \beta - \pi)\Bigg[& \delta\left(\alpha - \phi - \frac{3\pi}{4}\right) + \delta\left(\alpha - \phi - \frac{\pi}{4}\right) \\
& + \delta\left(\alpha - \phi + \frac{3\pi}{4}\right) + \delta\left(\alpha - \phi + \frac{\pi}{4}\right)\Bigg],
\end{aligned} \tag{18.B.7}$$

which is Eq. (18.4.20).

Problems

18.1 (a) Consider four numbers x_1, x_2, x_3, and x_4 such that $0 \leq x_i < 1$, $(i = 1, 2, 3, 4)$. Show that the function

$$X = x_1x_2 - x_1x_4 + x_2x_3 + x_3x_4 - x_2 - x_3,$$

is constrained by the inequality

$$-1 \leq X \leq 0.$$

(b) If we choose $x_1 = P_1(\mu, \theta_a)$, $x_2 = P_2(\mu, \theta_b)$, $x_3 = P_1(\mu, \theta'_a)$, and $x_4 = P_2(\mu, \theta'_b)$ in the above inequality, then prove the following Bell's inequality:

$$S \leq 1,$$

where

$$S = \frac{P_{12}(\theta_a, \theta_b) - P_{12}(\theta_a, \theta'_b) + P_{12}(\theta'_a, \theta_b) + P_{12}(\theta'_a, \theta'_b)}{P_1(\theta'_a) + P_2(\theta_b)}$$

with

$$P_1(\theta_a) = \int d\Lambda P_1(\mu, \theta_a),$$

$$P_2(\theta_b) = \int d\Lambda P_2(\mu, \theta_b),$$

$$P_{12}(\theta_a, \theta_b) = \int d\Lambda P_1(\mu, \theta_a) P_2(\mu, \theta_b).$$

Here $P_1(\mu, \theta_a)$ and $P_2(\mu, \theta_b)$ are the probabilities of detecting particles 1 and 2 with the orientation of Stern–Gerlach apparati in Fig. 18.2 at angles θ_a and θ_b, respectively, where μ are the hidden variables that describe 'completely' the emission process in the source, and $d\Lambda$ is a measure of the variables μ. Now $P_1(\theta'_a)$ and $P_2(\theta_b)$ are the passage probabilities for particles 1 and 2 to pass through the respective Stern–Gerlach apparati oriented at angles θ'_a and θ_b, respectively, and $P_{12}(\theta_a, \theta_b)$ is the joint probability that particles 1 and 2 will pass through their respective Stern–Gerlach apparati oriented at angles θ_a and θ_b, respectively. (Hint: See J. F. Clauser and M. A. Horne, *Phys. Rev. D* **10**, 526 (1974).)

18.2 (a) Show that

$$e^{-i\theta\sigma_y/2} = \cos\left(\frac{\theta}{2}\right) - i\sigma_y \sin\left(\frac{\theta}{2}\right).$$

(b) Use this result to show that

$$e^{-i\theta\sigma_y/2} |\uparrow\rangle\langle\uparrow| e^{i\theta\sigma_y/2}$$
$$= \frac{1}{2}(1 + \sigma_z \cos\theta + \sigma_x \sin\theta).$$

18.3 For a spin singlet state

$$|\Psi_{1,2}\rangle = \frac{1}{\sqrt{2}}[|\uparrow_1, \downarrow_2\rangle - |\downarrow_1, \uparrow_2\rangle],$$

show that

$$\langle\Psi_{1,2}|\pi_{\theta_a}\pi_{\theta_b}|\Psi_{1,2}\rangle = \frac{1}{4}[1 - \cos(\theta^1_a - \theta^2_b)],$$

where

$$\pi_{\theta_a}^{(1)} = \frac{1}{2}[1 + \sigma_z^{(1)} \cos\theta_a + \sigma_x^{(1)} \sin\theta_a],$$

$$\pi_{\theta_b}^{(2)} = \frac{1}{2}[1 + \sigma_z^{(2)} \cos\theta_b + \sigma_x^{(2)} \sin\theta_b].$$

Here $\sigma_x^{(i)}$ and $\sigma_z^{(i)}$ are the Pauli matrices for the ith spin.

18.4 Show that

$$|\Psi\rangle_3 = \frac{1}{\sqrt{2}}(|\uparrow_1\uparrow_2\uparrow_3\rangle - |\downarrow_1\downarrow_2\downarrow_3\rangle)$$

is an eigenstate of the operators $\sigma_y^{(1)}\sigma_x^{(2)}\sigma_y^{(3)}$ and $\sigma_y^{(1)}\sigma_y^{(2)}\sigma_x^{(3)}$.

References and bibliography

The theory of measurement

W. E. Lamb, Jr., *Physics Today*, **22**, 23 (1969).
R. Griffiths, *J. Stat. Phys.* **36**, 219 (1984).
R. Griffiths, *Phys. Rev. Lett.* **70**, 2201 (1993).
M. Gell-Mann and J. B. Hartle, in *Complexity, Entropy, and the Physics of Information*, edited by W. Zurek (Addison-Wesley, Reading, 1990).

EPR 'paradox'

A. Einstein, B. Podolsky, and N. Rosen, *Phys. Rev.* **47**, 777 (1935).
N. Bohr, *Phys. Rev.* **48**, 696 (1935). Bohr's reply to EPR has come down to us as the so-called *Copenhagen interpretation* of quantum mechanics.
D. Bohm, *Quantum Theory*, (Prentice Hall, Englewood Cliffs, NJ 1951).
F. Belinfante, *A Survey of Hidden Variable Theories*, (Pergamon, New York 1973).
C. D. Cantrell and M. O. Scully, *Phys. Rep. C* **43**, 499 (1978).

Bell's inequality

J. S. Bell, *Physics* **1**, 195 (1965); *Rev. Mod. Phys.* **38**, 447 (1966).
E. P. Wigner, *Am. J. Phys.* **38**, 1005 (1970).
J. F. Clauser and M. A. Horne, *Phys. Rev. D* **10**, 526 (1974).
J. F. Clauser and A. Shimony, *Rep. Prog. Phys.* **41**, 1881 (1978). This is an excellent review of the subject.
F. M. Pipkin, in *Advances in Atomic and Molecular Physics*, Vol. 14, ed. D. R. Bates and B. Bederson (Academic, New York 1978), p. 281.
B. de' Espagnat, *Sci. Am.* **241**, 158 (Nov. 1979). This paper gives a popular account of Bell's inequality.
L. M. Pedrotti and M. O. Scully, in *Quantum Electrodynamics and Quantum Optics*, ed. A. O. Barut (Plenum, New York 1984).
N. D. Mermin, *Rev. Mod. Phys.* **65**, 803 (1993).

Optical tests of Bell's inequality

S. J. Freedman and J. F. Clauser, *Phys. Rev. Lett.* **28**, 938 (1972).
E. S. Fry and R. C. Thompson, *Phys. Rev. Lett.* **37**, 465 (1976).
A. Aspect, P. Grangier, and G. Roger, *Phys. Rev. Lett.* **49**, 91 (1982).
A. Aspect, J. Dalibard, and G. Roger, *Phys. Rev. Lett.* **49**, 1804 (1982).
Z. Y. Ou and L. Mandel, *Phys. Rev. Lett.* **61**, 50 (1988).
Y. H. Shih and C. O. Alley, *Phys. Rev. Lett.* **61**, 2921 (1988).
P. G. Kwiat, P. H. Eberhard, A. M. Steinberg, and R. Y. Chiao, *Phys. Rev. A* **49**, 3209 (1994).
E. S. Fry, T. Walther, and S. Li, *Phys. Rev. A* **52**, 4381 (1995).

Greenberger, Horne, and Zeilinger (GHZ) equality

D. M. Greenberger, M. Horne, and A. Zeilinger, in *Bell's Theorem, Quantum Theory, and Conceptions of the Universe*, ed. M. Kafatos (Kluwer, Dordrecht 1989), p. 73.
N. D. Mermin, *Am. J. Phys.* **58**, 731 (1990a); *Physics Today* **43** (6), 9 (1990b).
D. M. Greenberger, M. A. Horne, A. Shimony, and A. Zeilinger, *Am. J. Phys.* **58**, 1131 (1990).

Distribution functions for spin-1/2

R. L. Stratnovich, *Sov. Phys. JETP* **4**, 891 (1957).
Arecchi, E. Courtens, R. Gilmore, and H. Thomas, *Phys. Rev. A* **6**, 2211 (1972).
M. O. Scully, *Phys. Rev. D* **28**, 2477 (1983).
L. Cohen and M. O. Scully, *Found. Phys.* **16**, 295 (1986).
C. Chandler, L. Cohen, C. Lee, M. O. Scully, and K. Wódkiewicz, *Found. Phys.* **22**, 867 (1992).
J. P. Dowling, G. S. Agarwal, and W. P. Schleich, *Phys. Rev. A* **49**, 4101 (1994).
M. O. Scully and K. Wódkiewicz, *Found. Phys.* **24**, 85 (1994).

Quantum cryptography

C. H. Bennett and G. Brassard, in *Proceedings of IEEE International Conference on Computers, Systems, and Signal Processing, Banglore, India* (IEEE, New York, 1984), p. 175. This paper presents the Bennett–Brassard protocol.
A. K. Ekert, *Phys. Rev. Lett.* **67**, 661 (1991). This paper presents quantum cryptography based on Bell's theorem.
C. H. Bennett, *Phys. Rev. Lett.* **68**, 3121 (1992).
C. H. Bennett, G. Brassard, and N. D. Mermin, *Phys. Rev. Lett.* **68**, 557 (1992).
P. D. Townsend, J. G. Rarity, and P. R. Tapster, *Electron. Lett.* **29**, 634 (1993).
B. Huttner, N. Imoto, N. Gisin, and T. Mor, *Phys. Rev. A* **51**, 1863 (1995).

CHAPTER 19

Quantum nondemolition measurements*

In many quantum systems, the process of measurement of an observable introduces noise so that successive measurements of the same observable yield different results. The simplest example of a quantum system is a free particle described by the Hamiltonian

$$\mathscr{H} = \frac{p^2}{2m}, \tag{19.1}$$

where p is the momentum operator and m is the mass of the particle. The position operator x can be measured with arbitrary accuracy in an instantaneous measurement. However, according to the Heisenberg uncertainty relation between x and p, an initial precise measurement of x perturbs p sharply so that

$$\Delta p \geq \frac{\hbar}{2\Delta x}. \tag{19.2}$$

During subsequent free evolution, p introduces changes in x, according to

$$\dot{x} = \frac{1}{i\hbar}[x, \mathscr{H}] = \frac{p}{m}, \tag{19.3}$$

or

$$x(t) = x(0) + \frac{p(0)t}{m}. \tag{19.4}$$

As a result

$$\begin{aligned}[\Delta x(t)]^2 &= [\Delta x(0)]^2 + \left[\frac{\Delta p(0)}{m}\right]^2 t^2 \\ &\geq [\Delta x(0)]^2 + \left[\frac{\hbar}{2m\Delta x(0)}\right]^2 t^2,\end{aligned} \tag{19.5}$$

i.e., the accuracy of a second measurement is spoiled.

* The standard reference on this subject is the review article by Caves, Thorne, Drever, Sandberg, and Zimmermann [1980]. More recently, the textbook by pioneers Braginsky and Khalili [1992] provides many useful insights into the subject.

In quantum optical systems, the detection of the field variables is usually done by photocounting techniques which are field destructive. As a consequence successive measurements of the field variables yield different results. It is desirable to consider schemes to avoid *back action* of the measurement on the detected observable due to the measuring process. Such a measurement in which one monitors an observable that can be measured repeatedly with the result of each measurement being completely determined by the result of an initial, precise measurement is called a *quantum nondemolition* (QND) *measurement.*

In this chapter we first give conditions for QND measurements in a general quantum system and then discuss QND schemes for the measurement of the number of photons in certain quantum optical systems

19.1 Conditions for QND measurements

In a general QND measurement, a signal observable A_S of a quantum system S is measured by detecting a change in an observable A_P of the probe system P coupled to S during the measurement time T, without perturbing the subsequent evolution of A_S. We can therefore have a sequence of precise measurements of A_S such that the result of each measurement is completely predictable from the result of the preceding measurement. Such an observable is called QND observable.

The total Hamiltonian for the S–P system is expressed as

$$\mathscr{H} = \mathscr{H}_S + \mathscr{H}_P + \mathscr{H}_I, \tag{19.1.1}$$

where $\mathscr{H}_S$, $\mathscr{H}_P$, and $\mathscr{H}_I$ are the Hamiltonians of system, probe, and their interaction, respectively. The equations of motion for the operators A_S and A_P are

$$i\hbar\dot{A}_S = [A_S, \mathscr{H}_S + \mathscr{H}_I] \tag{19.1.2a}$$

$$i\hbar\dot{A}_P = [A_P, \mathscr{H}_P + \mathscr{H}_I]. \tag{19.1.2b}$$

We now consider the conditions that define a QND measurement process. A measurement is the QND type if the observable A_S to be measured, the probe or readout observable A_P, and the interaction Hamiltonian $\mathscr{H}_I$ satisfy the following conditions:

(a) Since A_S is to be measured, $\mathscr{H}_I$ must be a function of A_S, i.e.,

$$\frac{\partial \mathscr{H}_I}{\partial A_S} \neq 0. \tag{19.1.3}$$

(b) As the probe P is being used to measure A_S, it cannot be a constant of motion $\dot{A}_P \neq 0$. The commutator of A_P and $\mathscr{H}_I$ must therefore be different from zero, i.e.,

$$[A_P, \mathscr{H}_I] \neq 0. \tag{19.1.4}$$

(c) The observable A_S should not be affected by its coupling to A_P during measurement, so that

$$[A_S, \mathscr{H}_I] = 0. \tag{19.1.5}$$

(d) The unperturbed Hamiltonian $\mathscr{H}_S$ is not a function of A_S^c, the conjugate observable of A_S, i.e.,

$$\frac{\partial \mathscr{H}_S}{\partial A_S^c} = 0. \tag{19.1.6}$$

This requirement ensures that the motion of A_S does not become unpredictable due to the uncertainty imposed on the conjugate variable A_S^c by the measurement of A_S, and the measurement of A_S does not affect A_S itself.

Conditions (19.1.3)–(19.1.6) define a QND measurement process. We now consider QND measurement schemes in which the photon number of the signal photons are measured.

19.2 QND measurement of the photon number via the optical Kerr effect

The refractive index of many materials can be described by the relation

$$n = n_0 + n_2 E^2, \tag{19.2.1}$$

where n_0 represents the usual, weak field refractive index and n_2 is a constant which gives the rate of change of the refractive index with changing optical intensity. If a strong signal wave is incident on such a nonlinear material, it causes a change in the refractive index and a weak probe wave therefore experiences a phase shift in propagating through the material which is proportional to the intensity of the signal wave. The change in refractive index described by Eq. (19.2.1) is called the optical Kerr effect. In this section we describe a QND measurement scheme in which the photon number of the signal wave is measured via the phase of the probe wave using the optical Kerr effect.

The configuration for the QND measurement of the signal photon number is given in Fig. 19.1. The probe wave undergoes a phase shift

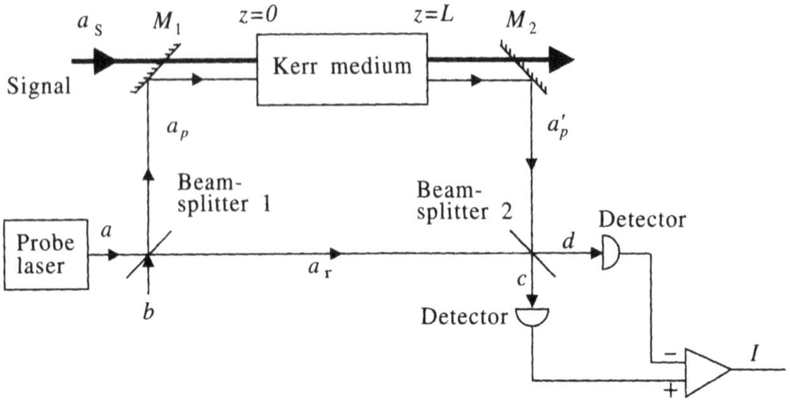

Fig. 19.1 Schematic of the QND measurement of the signal photon number. Probe, reference, and signal waves are denoted by the annihilation operators a_P, a_r, and a_S, respectively. The vacuum fluctuation, b is mixed at beam-splitter 1. Transmissions of mirrors M_1 and M_2 are unity for the signal wave. (From N. Imoto, H. A. Haus, and Y. Yamamoto, *Phys. Rev. A* **32**, 2287 (1985).)

due to the refractive index change, which is proportional to the signal intensity in the Kerr medium. The signal wave passes through the Kerr medium without changing its photon number and the phase of the probe field is modulated by the signal photon number. The phase shift for the probe wave is measured in terms of the photocurrent of the balanced-mixer detector. In Fig. 19.1, the reflectivity of mirrors M_1 and M_2 is zero for the signal frequency ν_S and unity for the probe frequency ν_P so that the interferometer is formed only for the probe field.

The Hamiltonian for the interaction of the signal and pump waves in the Kerr medium is given by

$$\mathscr{H} = \mathscr{H}_S + \mathscr{H}_P + \mathscr{H}_I, \tag{19.2.2}$$

where

$$\mathscr{H}_S = \hbar\nu_S\left(a_S^\dagger a_S + \frac{1}{2}\right), \tag{19.2.3}$$

$$\mathscr{H}_P = \hbar\nu_P\left(a_P^\dagger a_P + \frac{1}{2}\right), \tag{19.2.4}$$

$$\mathscr{H}_I = \hbar\kappa a_S^\dagger a_S a_P^\dagger a_P. \tag{19.2.5}$$

Here κ is a coupling constant which depends upon the third-order nonlinear susceptibility for the optical Kerr effect. The QND observable to be measured is the photon number operator

$$A_S = a_S^\dagger a_S, \tag{19.2.6}$$

and the probe or readout observable is a suitable phase operator. A phase operator ϕ_P may be defined by the relations:

$$a_P = \sqrt{a_P^\dagger a_P + 1}\,\exp(i\phi_P), \tag{19.2.7}$$

$$a_P^\dagger = \exp(-i\phi_P)\sqrt{a_P^\dagger a_P + 1}. \tag{19.2.8}$$

It can be verified that $\exp(i\phi_P)$ and $\exp(-i\phi_P)$ are not Hermitian operators, and they cannot therefore represent observable properties of the field. They can, however, be combined to produce

$$\cos\phi_P = \frac{1}{2}[\exp(i\phi_P) + \exp(-i\phi_P)], \tag{19.2.9a}$$

$$\sin\phi_P = \frac{1}{2i}[\exp(i\phi_P) - \exp(-i\phi_P)], \tag{19.2.9b}$$

which are Hermitian operators. These operators can be regarded as the quantum mechanical operators that represent the observable phase properties of the electromagnetic field. We can thus choose

$$A_P = \sin\phi_P. \tag{19.2.10}$$

It follows from Eqs. (19.2.7), (19.2.8), and (19.2.10) that, in terms of a_P and $a_P^\dagger$

$$A_P = \frac{1}{2i}\left(\frac{1}{\sqrt{a_P^\dagger a_P + 1}}a_P - a_P^\dagger\frac{1}{\sqrt{a_P^\dagger a_P + 1}}\right). \tag{19.2.11}$$

The conditions (19.1.3)–(19.1.6) can be checked for the operators A_S and A_P (see Problem 19.2), which indicates that the present scheme provides a QND measurement of the signal photon number $A_S = a_S^\dagger a_S$. In particular it is seen that A_S is a constant of motion.

The Heisenberg equation of motion for the probe operator a_P inside the Kerr medium is

$$\dot{a}_P = -i\kappa A_S a_P. \tag{19.2.12}$$

This time-evolution equation may be rewritten as a spatial-evolution equation for the present travelling-wave problem by replacing t by $-z/v$ for propagation toward $+z$-direction with a velocity v, i.e.,

$$\frac{d}{dz}a_P(z) = \frac{i\kappa}{v}A_S a_P(z). \tag{19.2.13}$$

Integrating Eq. (19.2.13) from $z = 0$ to L, we obtain

$$a_P(L) = \exp\left(\frac{i\kappa}{v}A_S L\right)a_P(0), \tag{19.2.14}$$

where we use the fact that A_S is a constant of motion. The operator $\kappa A_S L/v$ in Eq. (19.2.14) corresponds to the phase shift in a_P. The operator A_P for the probe field at $z = L$ is then given by

$$A_P(L) = \frac{1}{2i}\left[\frac{1}{\sqrt{a_P^\dagger(0)a_P(0)+1}}a_P(0)\exp\left(\frac{i\kappa}{v}A_S L\right)\right.$$

$$\left. -\exp\left(\frac{-i\kappa}{v}A_S L\right)a_P^\dagger(0)\frac{1}{\sqrt{a_P^\dagger(0)a_P(0)+1}}\right]. \quad (19.2.15)$$

This phase shift in the probe field, and hence the signal photon number, can be measured in the balanced-mixer detector.

We consider the output of the balanced-mixer detector in Fig. 19.1 where the annihilation operator for each part of the interferometer is specified. As discussed in Section 4.4, the output signal in a two-port homodyne detection scheme is determined by the operator

$$n_{cd} = d^\dagger d - c^\dagger c. \quad (19.2.16)$$

We relate the field operators c and d to the input field operators a and b for the probe laser output and the vacuum fluctuations, respectively. The probe field a_P and the reference field a_r are related to a and b via

$$a_P = i\sqrt{1-T_1}a + \sqrt{T_1}b, \quad (19.2.17)$$

$$a_r = \sqrt{T_1}a + i\sqrt{1-T_1}b, \quad (19.2.18)$$

where T_1 is the transmission coefficient of beam-splitter 1, and we have assumed a $\pi/2$ phase shift in the reflected mode. The Kerr medium shifts the phase of the probe wave according to Eq. (19.2.14):

$$a_P' = \exp\left[i\left(\frac{\kappa}{v}A_S L + \frac{\pi}{2}\right)\right]a_P(0), \quad (19.2.19)$$

where the phase shift $\pi/2$ is added by adjusting the interferometer configuration. With the value T_2 for the transmission coefficient of beam-splitter 2, the fields c and d are written as

$$c = \sqrt{T_2}a_P' + i\sqrt{1-T_2}a_r, \quad (19.2.20)$$

$$d = i\sqrt{1-T_2}a_P' + \sqrt{T_2}a_r. \quad (19.2.21)$$

It follows from Eqs. (19.2.16)–(19.2.21) that, for $T_2 = 0.5$,

$$n_{cd} = 2\sqrt{T_1(1-T_1)}(a^\dagger a - b^\dagger b)\sin\left(\frac{\kappa L}{v}A_S\right)$$

$$-2iT_1(a^\dagger b - b^\dagger a)\sin\left(\frac{\kappa L}{v}A_S\right)$$

$$-\left[a^\dagger b\exp\left(-i\frac{\kappa L}{v}A_S\right) + b^\dagger a\exp\left(i\frac{\kappa L}{v}A_S\right)\right]. \quad (19.2.22)$$

As the field b is in the vacuum mode,

$$\langle n_{cd} \rangle = 2\sqrt{T_1(1-T_1)}\langle a^\dagger a \rangle \frac{\kappa L}{v} \langle A_S \rangle, \tag{19.2.23}$$

where we use the approximation $\sin(\kappa L A_S/v) \simeq \kappa L A_S/v$, which is valid if $\kappa L A_S/\ v \ll 1$. The measured photocurrent in the balanced detector is therefore proportional to the mean photon number in the signal field.

19.3 QND measurement of the photon number by dispersive atom–field coupling*

In Chapter 13, we discussed possible schemes to generate number states in a micromaser. We now consider a QND method to measure the number of photons in the number state $|n\rangle$ prepared in a high-Q cavity. This method is based on the detection of the dispersive phase shift produced by the cavity field on the wave function of nonresonant atoms crossing the cavity. The shift can be measured by atomic interferometry, using the Ramsey method described below. In this QND method, the probe is no longer a field but a beam of atoms which interact nonlinearly and nonresonantly with the signal field.

We consider a three-level atomic system in a cascade configuration as shown in Fig. 19.2. The transitions $|a\rangle \rightarrow |b\rangle$ and $|b\rangle \rightarrow |c\rangle$ are allowed and the transition $|a\rangle \rightarrow |c\rangle$ is forbidden. The cavity mode of frequency ν is detuned from the $|a\rangle \rightarrow |b\rangle$ transition by an amount $\Delta = \nu - \omega_{ab}$, which is assumed to be small as compared to ω_{ab} and all the other transitions in the atomic spectrum, particularly the $|b\rangle \rightarrow |c\rangle$ transition. As a consequence, only levels $|a\rangle$ and $|b\rangle$ are affected by the nonresonant atom–field coupling and level $|c\rangle$ remains unperturbed.

The basic physics involved in the QND measurement of the number of photons inside the cavity is that the interaction of the nonresonant field with the atom introduces a dispersive energy shift (no absorption) in the state $|b, n\rangle$ (atom in level $|b\rangle$ and field in the number state $|n\rangle$) which is proportional to the number of photons in the field, n. This energy shift is then detected by measuring the dephasing accumulated between the levels $|b\rangle$ and $|c\rangle$ in a Ramsey method of separated oscillating fields. The fact that the energy shift depends on n is due to the dispersive effect, which does not affect the number of photons, is responsible for the nondemolition character of the method.

Before discussing the Ramsey setup for the QND measurement, we first derive the dispersive energy shift δ_b of the state $|b, n\rangle$.

* This section is based on Brune, Haroche, Lefevre, Raimond, and Zagury [1990].

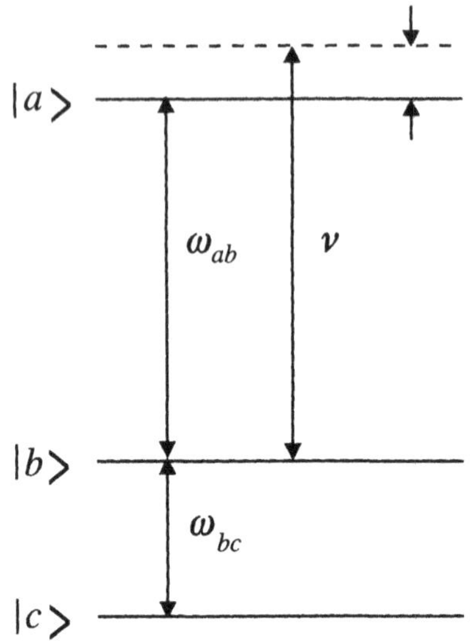

Fig. 19.2
Three-level atomic system for the QND measurement of the photon number by dispersive atom–field coupling.

The interaction of the atom with the field is described by the Hamiltonian (Eqs. (6.2.1)–(6.2.3))

$$\mathscr{H} = \frac{\hbar\omega_{ab}}{2}(|a\rangle\langle a| - |b\rangle\langle b|) + \hbar\nu a^\dagger a + \hbar g(|a\rangle\langle b|a + a^\dagger|b\rangle\langle a|), \tag{19.3.1}$$

where we have considered the interaction of the field of frequency ν with the atomic levels $|a\rangle$ and $|b\rangle$ only and neglected the far detuned level $|c\rangle$. The Hamiltonian $\mathscr{H}$ can be diagonalized to obtain atom–field dressed states

$$|+\rangle = \cos\theta_n|a, n-1\rangle - \sin\theta_n|b, n\rangle \tag{19.3.2a}$$

$$|-\rangle = \sin\theta_n|a, n-1\rangle + \cos\theta_n|b, n\rangle \tag{19.3.2b}$$

with the corresponding eigenvalues

$$E_{+n} = \hbar\left[(n-1)\nu + \frac{1}{2}\omega_{ab}\right] - \frac{\hbar}{2}(\Omega_n - \Delta), \tag{19.3.3a}$$

$$E_{-n} = \hbar\left(n\nu - \frac{1}{2}\omega_{ab}\right) + \frac{\hbar}{2}(\Omega_n - \Delta), \tag{19.3.3b}$$

where

$$\sin\theta_n = \frac{\Omega_n - \Delta}{\sqrt{(\Omega_n - \Delta)^2 + 4g^2 n}}, \tag{19.3.4}$$

$$\cos\theta_n = \frac{2g\sqrt{n}}{\sqrt{(\Omega_n - \Delta)^2 + 4g^2 n}}, \tag{19.3.5}$$

$$\Omega_n = \sqrt{\Delta^2 + 4g^2 n}. \tag{19.3.6}$$

(19.3.7)

Let us assume that Δ is large compared to $2g\sqrt{n}$ so that

$$\frac{4g^2 n}{\Delta^2} \ll 1. \tag{19.3.8}$$

A simple analysis shows that, if condition (19.3.8) is valid, $\cos\theta_n \simeq 1$ and $\sin\theta_n \simeq 0$ so that

$$|+\rangle \simeq |a, n-1\rangle, \tag{19.3.9a}$$

$$|-\rangle \simeq |b, n\rangle, \tag{19.3.9b}$$

with the corresponding eigenvalues

$$E_{+n} = \hbar\left[(n-1)v + \frac{1}{2}\omega_{ab}\right] - \frac{\hbar g^2 n}{\Delta}, \tag{19.3.10a}$$

$$E_{-n} = \hbar\left(nv - \frac{1}{2}\omega_{ab}\right) + \frac{\hbar g^2 n}{\Delta}. \tag{19.3.10b}$$

The net effect is that there is no change in the photon numbers but the state $|b, n\rangle$, whose unperturbed energy is $\hbar(nv - \omega_{ab}/2)$, experiences the energy shift

$$\hbar\delta_b(n) = \frac{\hbar g^2 n}{\Delta}. \tag{19.3.11}$$

The energy shift per photon, $\hbar g^2/\Delta$, can be large in Rydberg atom systems. In the Rydberg levels of alkali atoms, a shift per photon of 10^5 sec^{-1} is possible for appropriate detunings. The important point of this analysis is that the energy shift is proportional to the number of signal photons and it is dispersive, i.e., the shift is not accompanied by the absorption of photons.

We now restrict our discussion to the atomic Hilbert space spanned by the states $|b\rangle$ and $|c\rangle$ only. The effective Hamiltonian for the atom–field interaction is

$$\mathscr{H} = \mathscr{H}_a + \mathscr{H}_f + \mathscr{H}_I, \tag{19.3.12}$$

where

$$\mathscr{H}_a = \frac{\hbar\omega_{bc}}{2}\sigma_z, \tag{19.3.13}$$

$$\mathscr{H}_f = \hbar v a_S^\dagger a_S, \tag{19.3.14}$$

$$\mathscr{H}_I = \frac{\hbar g^2}{\Delta} a_S^\dagger a_S \sigma_+ \sigma_-, \tag{19.3.15}$$

with $\sigma_z = (|b\rangle\langle b| - |c\rangle\langle c|)$, $\sigma_+ = |b\rangle\langle c|$, and $\sigma_- = |c\rangle\langle b|$. This effective Hamiltonian can be seen to introduce a phase shift proportional to $\delta_b(n)$ in level $|b\rangle$ and leave the level $|c\rangle$ unpertubed. Notice that n is the eigenvalue of the signal operator $A_S = a_S^\dagger a_S$. The probe observable A_P is chosen to be the atomic dipole operator

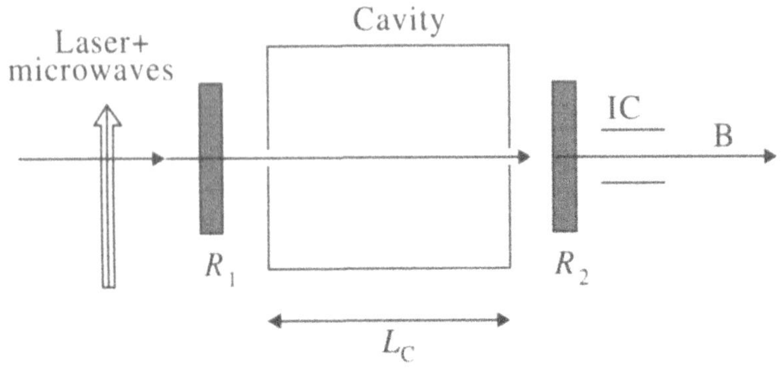

Fig. 19.3
Schematic of the QND measurement of the photon number by dispersive atomfield coupling. The atoms prepared in level $|b\rangle$ by laser excitation cross successively the microwave field zone R_1, the cavity, and the microwave field zone R_2 before their state ($|b\rangle$ or $|c\rangle$) is determined by the field ionization counter IC. (From M. Brune, S. Haroche, V. Lefevre, J. M. Raimond, and N. Zagury, *Phys. Rev. Lett.* **65**, 976 (1990).)

$$A_P = \frac{1}{2i}(\sigma_+ - \sigma_-). \tag{19.3.16}$$

The operators A_S and A_P satisfy all the QND criteria. This system is analogous to the optical Kerr effect discussed in the previous section, with the atomic operators σ_+ and σ_- merely replacing the probe field operators $a_P^\dagger$ and a_P, respectively.

We now consider the Ramsey method of separated oscillatory fields to detect the dephasing accumulated between the states $|b\rangle$ and $|c\rangle$. The experimental setup for this method is shown in Fig. 19.3. The cavity containing n photons is placed between two field zones R_1 and R_2 driving the $|b\rangle \rightarrow |c\rangle$ transition. Each atom is initially prepared in the level $|b\rangle$ by laser excitation. It then interacts in the first zone R_1 with a microwave field tuned at frequency ν_r, resonant with the $|b\rangle \rightarrow |c\rangle$ transition, and is thus prepared in a coherent superposition of levels $|b\rangle$ and $|c\rangle$. It then crosses the cavity before interacting with a second microwave field in the zone R_2 tuned at frequency ν_r. During the passage through the cavity, a phase shift proportional to $\delta_b(n)$ is introduced in the amplitude of the state $|b\rangle$. Finally after passing zone R_2 it is detected in the state $|b\rangle$ or $|c\rangle$ by a field ionization counter. The probability of detecting an atom in level $|b\rangle$ or $|c\rangle$ is a periodic function of $\delta_b(n)$, and thus gives a characteristic pattern of fringes whose spacing depends on the number of photons contained in the cavity. These *Ramsey fringes* therefore can be used to monitor the number of photons in a QND measurement.

If an atom with velocity v crosses the cavity containing n photons, the atom–field system ends up in the superposition state:

$$|\psi_{\text{final}}^{\text{atom-field}}\rangle = c_{b,n}|b,n\rangle + c_{c,n}|c,n\rangle. \tag{19.3.17}$$

We now determine the probability amplitudes $c_{b,n}$ and $c_{c,n}$ for finding the atom in levels $|b\rangle$ and $|c\rangle$, respectively, with n photons inside the cavity after passage through the microwave and the cavity fields. We assume, for the sake of simplicity, that the microwave fields in the zones R_1 and R_2 are resonant with the $|b\rangle \to |c\rangle$ transition and they are uniform in R_1 and R_2 with a Rabi frequency Ω_R, which is taken to be real. The initial conditions for the probability amplitudes, before the atom enters zone R_1, are

$$c_{b,n} = 1, \quad c_{c,n} = 0. \tag{19.3.18}$$

The evolution of these amplitudes during the interaction of the atom with the microwave field in the region R_1 for a time $\tau = L/v$ (where L is the length of the zones R_1 and R_2 and v is the velocity of the atom) is given by (see Eqs. (5.2.12) and (5.2.13))

$$\dot{c}_{b,n} = i\frac{\Omega_R}{2}c_{c,n}, \tag{19.3.19a}$$

$$\dot{c}_{c,n} = i\frac{\Omega_R}{2}c_{b,n}.. \tag{19.3.19b}$$

The solution of these equations, subject to the boundary conditions (19.3.18), at time τ is (see Eqs. (5.2.12) and (5.2.13))

$$c_{b,n} = \cos\left(\frac{\Omega_R\tau}{2}\right), \tag{19.3.20a}$$

$$c_{c,n} = i\sin\left(\frac{\Omega_R\tau}{2}\right). \tag{19.3.20b}$$

After passage through the cavity containing n photons, $c_{b,n}$ is phase shifted by an amount $\delta_b(n)\tau_c = g^2 n\tau_c/\Delta$ (where $\tau_c = L_c/v$ is the time spent by the atom inside the cavity) while $c_{c,n}$ remains unchanged, so that the atomic amplitudes before the atom enters zone R_2 are

$$c_{b,n} = \cos\left(\frac{\Omega_R\tau}{2}\right)\exp\left(-i\frac{g^2\tau_c n}{\Delta}\right), \tag{19.3.21}$$

$$c_{c,n} = i\sin\left(\frac{\Omega_R\tau}{2}\right). \tag{19.3.22}$$

In zone R_2, the amplitudes evolve again according to Eqs. (19.3.19a) and (19.3.19b) which can be solved subject to the initial conditions (19.3.21) and (19.3.22). The atomic amplitudes after passing through zone R_2 are then given by (see Eqs. (5.2.21) and (5.2.22))

$$c_{b,n} = \cos^2\left(\frac{\Omega_R\tau}{2}\right)\exp\left(-i\frac{g^2\tau_c n}{\Delta}\right) - \sin^2\left(\frac{\Omega_R\tau}{2}\right), \tag{19.3.23}$$

$$c_{c,n} = \frac{i}{2}\sin(\Omega_R\tau)\left[\exp\left(-i\frac{g^2\tau_c n}{\Delta}\right) + 1\right]. \tag{19.3.24}$$

These expressions simplify considerably if the Rabi frequency Ω_R of the microwave fields and the interaction time τ are chosen such that

$\Omega_R\tau = \pi/2$, and we obtain the following probabilities for finding the atom in levels $|b\rangle$ and $|c\rangle$

$$P_b = |c_{b,n}|^2 = \frac{1}{2}\left[1 - \cos\left(\frac{g^2\tau_c n}{\Delta}\right)\right], \tag{19.3.25a}$$

$$P_c = |c_{c,n}|^2 = \frac{1}{2}\left[1 + \cos\left(\frac{g^2\tau_c n}{\Delta}\right)\right]. \tag{19.3.25b}$$

These are periodic functions of n exhibiting a characteristic pattern of fringes whose spacing depends on n. Thus, by monitoring the atoms in level $|b\rangle$ or $|c\rangle$, the number of photons in the cavity can be measured.

Physically we can understand the oscillatory nature of the probabilities P_b and P_c in terms of an atomic interference process. The probability amplitude $c_{c,n}$ of finding the atom in level $|c\rangle$ when it is initially prepared in level $|b\rangle$ consists of two terms, as given by Eq. (19.3.24). The term unity in the bracket corresponds to the process in which the atom is transferred from $|b\rangle$ to $|c\rangle$ in zone R_1 so that it does not experience a phase shift due to the field inside the cavity and, in zone R_2 it stays in level $|c\rangle$. The term $\exp(-ig^2\tau_c n/\Delta)$ corresponds to the process in which the atom does not make a transition to level $|c\rangle$ and experience a phase shift inside the cavity. The probability P_c is the squared sum of partial amplitudes corresponding to these processes, leading to the fringe pattern exhibited in Eq. (19.3.25b). In other words, the atom is prepared in zone R_1 in a coherent superposition of two states, one that is affected by the cavity field and another that is not. When the atom is detected in level $|b\rangle$ or $|c\rangle$, it is not possible to tell through *which path* the system has evolved and the resulting detection probability reveals the corresponding quantum coherence in the form of a Ramsey fringe pattern. We shall examine, in detail, the role of *which-path* information in quantum interference experiments in the next chapter.

We now consider a more complicated but more realistic situation where the initial field inside the cavity is not in a number state, but rather in a superposition of number states, i.e., the field is described by

$$|\psi\rangle = \sum_n b_n|n\rangle. \tag{19.3.26}$$

An example of such a state of the field is a coherent state $|\alpha\rangle$, for which (see Eq. (2.2.2))

$$b_n = \frac{e^{-|\alpha|^2/2}\alpha^n}{\sqrt{n!}}. \tag{19.3.27}$$

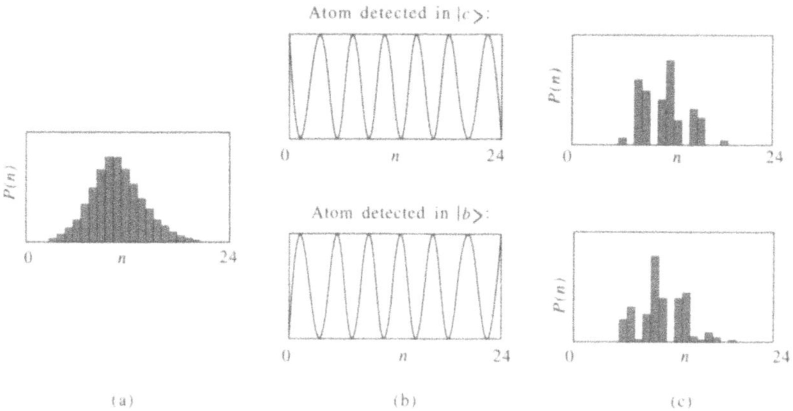

Fig. 19.4
The transformation of an initial Poisson distribution with $\langle n \rangle = |\alpha|^2 = 10$ by a single atom detection event. The initial distribution $|b_n|^2$ (displayed in (a)) is multiplied by the oscillating fringe function $|c_i(n)|^2$ ($i = b$ or c) (displayed in (b)). In the resulting distributions (c), photon numbers closest to *dark fringes* are decimated. The patterns obtained after detection of the atom in levels $|b\rangle$ and $|c\rangle$ are complementary. (From M. Brune, S. Haroche, J. M. Raimond, L. Davidovich, and N. Zagury, *Phys. Rev. A* **45**, 5193 (1992).)

After interaction with an atom with velocity v, the state of the atom–field system is obtained by a superposition state of the type:

$$|\psi_{\text{final}}^{\text{atom-field}}\rangle = \sum_n b_n[c_{b,n}|b, n\rangle + c_{c,n}|c, n\rangle]. \tag{19.3.28}$$

The atom–field system is obviously in an *entangled* state which cannot be expressed as a product of atom and field contributions, thus presenting strong correlations between the atom and field parts.

A detection of the atom in, say, level $|b\rangle$ projects the state (19.3.28) into

$$\sum_n b_n c_{b,n}|b, n\rangle \Bigg/ \left[\sum_n |b_n|^2 |c_{b,n}|^2\right]^{1/2}, \tag{19.3.29}$$

resulting in system *disentanglement*. The field state is now a superposition of photon number states with probability amplitudes given within a normalization factor by $b_n c_{b,n}$. The photon number distribution is essentially multiplied by an oscillating function of n, $|c_{b,n}|^2$. In Fig. 19.4, we see how the photon number distribution

$$p(n) = \frac{|b_n|^2 |c_{i,n}|^2}{\sum_n |b_n|^2 |c_{i,n}|^2} \quad (i = b \ or \ c), \tag{19.3.30}$$

initially in a coherent state (b_n given by Eq. (19.3.27)), is transformed after detection of the atom in level $|b\rangle$ or $|c\rangle$. The photon numbers, for which the fringe function $|c_{i,n}|^2$ is zero are efficiently decimated.

If a large number of atoms with random velocities v (random τ_c in Eqs. (19.3.23) and (19.3.24)) pass through the Ramsey setup and are found to be in the levels $|b\rangle$ or $|c\rangle$, other photon numbers are suppressed. The decimation goes on until we are left with only one photon number. A photon number state is finally obtained inside the cavity, as shown in Fig. 19.5, even though no energy has been exchanged between the atoms and the field. This dissipation-free process represents a QND measurement of the photon number inside the cavity.

19.4 QND measurements in optical parametric processes

In this section we consider QND measurements in optical parametric processes which are characterized by a generalized relation between the incoming and outgoing modes of the form

$$a_{out} = e^{i\theta_1} a_{in} + \mathcal{G} e^{i(\theta_1+\psi)}(e^{-i\phi} b_{in} + e^{i\phi} b_{in}^\dagger), \tag{19.4.1a}$$

$$b_{out} = e^{i\theta_2} b_{in} + \mathcal{G} e^{i(\theta_2+\phi)}(-e^{-i\psi} a_{in} + e^{i\psi} a_{in}^\dagger), \tag{19.4.1b}$$

where a_{in}, a_{out}, b_{in}, and b_{out} are the annihilation operators for the modes entering and leaving the *black box* representing the QND measurement or back-action-evading amplifier and $\mathcal{G}$ is a gain parameter. The input–output transformations of the type given by Eqs. (19.4.1) may not result from a single device but, as we shall see in the example discussed in this section, from an appropriate combination of parametric devices

In order to show that a black box with the input–output transformations described by Eqs. (19.4.1) performs QND measurements, we define the quadrature component operators for the incoming field modes

$$X_a^{in}(\psi) = \frac{1}{\sqrt{2}}(e^{-i\psi} a_{in} + e^{i\psi} a_{in}^\dagger), \tag{19.4.2}$$

$$Y_a^{in}(\psi) = \frac{i}{\sqrt{2}}(e^{-i\psi} a_{in} - e^{i\psi} a_{in}^\dagger), \tag{19.4.3}$$

$$X_b^{in}(\phi) = \frac{1}{\sqrt{2}}(e^{-i\phi} b_{in} + e^{i\phi} b_{in}^\dagger), \tag{19.4.4}$$

$$Y_b^{in}(\phi) = \frac{i}{\sqrt{2}}(e^{-i\phi} b_{in} - e^{i\phi} b_{in}^\dagger), \tag{19.4.5}$$

and for the outgoing field modes

$$X_a^{out}(\theta_1 + \psi) = \frac{1}{\sqrt{2}}[e^{-i(\theta_1+\psi)} a_{out} + e^{i(\theta_1+\psi)} a_{out}^\dagger], \tag{19.4.6}$$

$$Y_a^{out}(\theta_1 + \psi) = \frac{i}{\sqrt{2}}[e^{-i(\theta_1+\psi)} a_{out} - e^{i(\theta_1+\psi)} a_{out}^\dagger], \tag{19.4.7}$$

$$X_b^{out}(\theta_2 + \phi) = \frac{1}{\sqrt{2}}[e^{-i(\theta_2+\phi)} b_{out} + e^{i(\theta_2+\phi)} b_{out}^\dagger], \tag{19.4.8}$$

$$Y_b^{out}(\theta_2 + \phi) = \frac{i}{\sqrt{2}}[e^{-i(\theta_2+\phi)} b_{out} - e^{i(\theta_2+\phi)} b_{out}^\dagger]. \tag{19.4.9}$$

Fig. 19.5
Evolution of the photon number distribution in a QND sequence. The initial photon distribution is (a) Poisson with $\langle n \rangle = |\alpha|^2 = 5$. Traces (b)–(f) correspond to the detection of 1, 3, 6, 10, and 15 atoms, respectively. The field finally collapses to $n = 3$ Fock state. (From M. Brune, S. Haroche, J. M. Raimond, L. Davidovich, and N. Zagury, *Phys. Rev. A* **45**, 5193 (1992).)

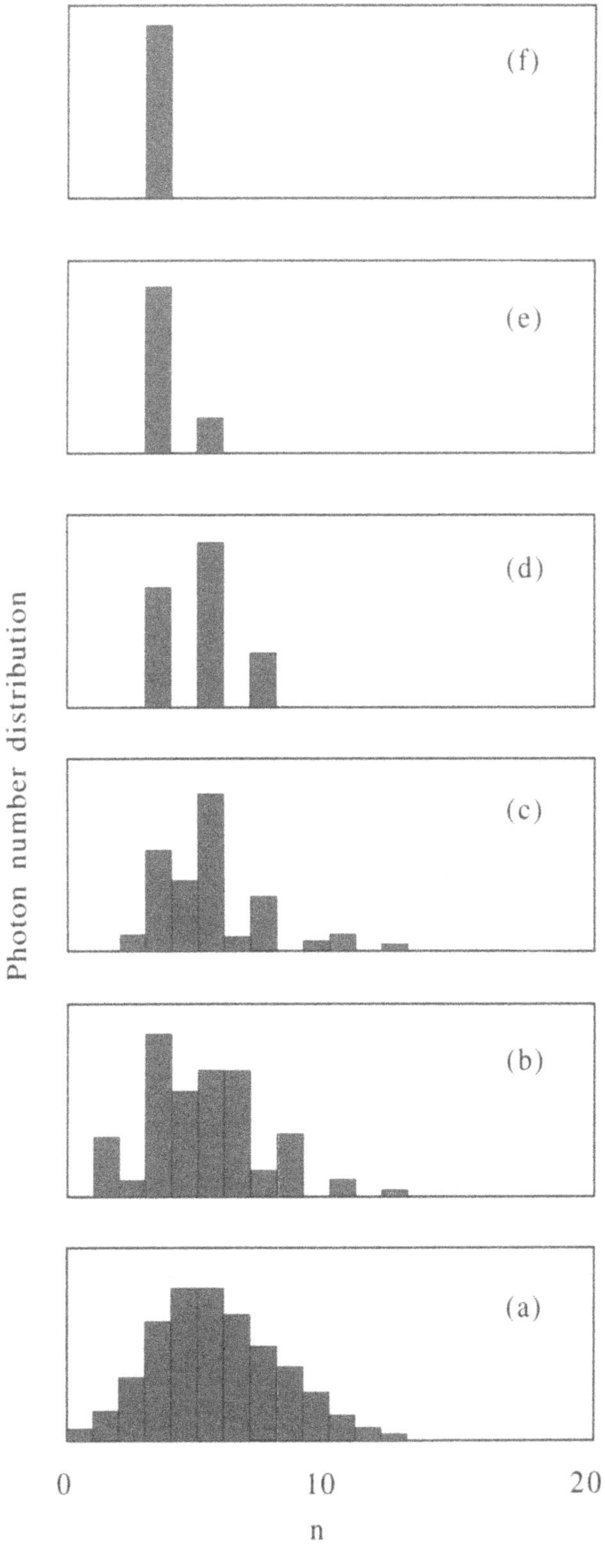

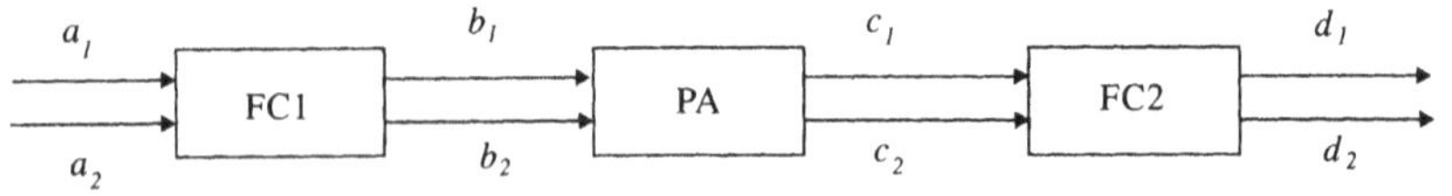

Fig. 19.6 Schematic of a QND amplifier constructed out of a combination of frequency converters, FC1 and FC2, and a parametric amplifier PA. The light propagating along a_1, b_1, c_1, and d_1 oscillates at the signal frequency ν_s, and the light propagating along a_2, b_2, c_2, and d_2 oscillates at the probe frequency ν_r. FC1 and FC2 are pumped at the difference frequency $|\nu_s - \nu_r|$, while PA is pumped at the sum frequency $\nu_s + \nu_r$. The pump beams are now shown. (From B. Yurke, *J. Opt. Soc. Am. B* **2**, 732 (1985).)

It then follows readily from Eqs. (19.4.1) that

$$X_a^{\text{out}}(\theta_1 + \psi) = X_a^{\text{in}}(\psi) + 2\mathcal{G}X_b^{\text{in}}(\phi), \tag{19.4.10}$$

$$Y_a^{\text{out}}(\theta_1 + \psi) = Y_a^{\text{in}}(\psi), \tag{19.4.11}$$

$$X_b^{\text{out}}(\theta_2 + \phi) = X_b^{\text{in}}(\phi), \tag{19.4.12}$$

$$Y_b^{\text{out}}(\theta_2 + \phi) = Y_b^{\text{in}}(\phi) - 2\mathcal{G}Y_a^{\text{in}}(\psi). \tag{19.4.13}$$

From these relations, back-action evasion is readily apparent. From Eq. (19.4.13) we see that $Y_b^{\text{out}}(\theta_2+\phi)$ carries information about $Y_a^{\text{in}}(\psi)$ so that the component $Y_a(\psi)$ of the incoming signal is measured. From Eq. (19.4.11), we however see that the component $Y_a^{\text{in}}(\psi)$ is kept clean whereas, according to Eq. (19.4.10), all the back-action noise $2\mathcal{G}X_b^{\text{in}}(\phi)$, is dumped in the conjugate variable X_a.

We now consider an example of the QND input–output relations (19.4.1) in a system consisting of parametric devices as shown in Fig. 19.6. Here the light propagating along a_1, b_1, c_1, and d_1 oscillates at the signal frequency ν_s and the light propagating along a_2, b_2, c_2, and d_2 oscillates at the probe or readout frequency ν_r.

The frequency converters FC1 and FC2 are pumped at the difference frequency $\nu_s - \nu_r$ and are described, in the parametric approximation (see Section 16.1), by the interaction picture Hamiltonian

$$\mathscr{H} = \hbar\kappa_1\beta_1(a_s^\dagger a_r e^{-i\phi_1} + a_r^\dagger a_s e^{i\phi_1}). \tag{19.4.14}$$

Here κ_1 is a coupling constant which depends on the nonlinear susceptibility of the medium, β_1 and ϕ_1 are the amplitude and the phase of the classical pump field, and a_s and a_r are the annihilation operators for the signal and the readout fields. The Heisenberg equations of motion for the operators a_s and a_r are

$$\dot{a}_s = -i\kappa_1\beta_1 a_r e^{-i\phi_1}, \tag{19.4.15}$$

$$\dot{a}_r = -i\kappa_1\beta_1 a_s e^{i\phi_1}. \tag{19.4.16}$$

A solution of these equations is given by

$$a_s(L) = \cos(\kappa_1\beta_1 L/c)a_s(0) - i\sin(\kappa_1\beta_1 L/c)a_r(0)e^{-i\phi_1}, \tag{19.4.17a}$$

$$a_r(L) = \cos(\kappa_1\beta_1 L/c)a_r(0) - i\sin(\kappa_1\beta_1 L/c)a_s(0)e^{i\phi_1}, \tag{19.4.17b}$$

where L is the length of the nonlinear medium and we replace t by L/c where c is the speed of light. The input–output relationships for the frequency converters FC1 and FC2 in Fig. 19.6 are therefore obtained by replacing $a_s(0)$, $a_r(0)$, $a_s(L)$, and $a_r(L)$ by a_1 a_2, b_1, and b_2, respectively, for FC1 and by c_1, c_2, d_1, and d_2, respectively, for FC2 in Eqs. (19.4.17a) and (19.4.17b). The resulting equations are

$$b_1 = K_1 a_1 - i(1 - K_1^2)^{1/2} e^{-i\phi_1} a_2, \tag{19.4.18}$$

$$b_2 = -i(1 - K_1^2)^{1/2} e^{i\phi_1} a_1 + K_1 a_2, \tag{19.4.19}$$

$$d_1 = K_1 c_1 - i(1 - K_1^2)^{1/2} e^{-i\phi_1} c_2, \tag{19.4.20}$$

$$d_2 = -i(1 - K_1^2)^{1/2} e^{i\phi_1} c_1 + K_1 c_2, \tag{19.4.21}$$

where $K_1 = \cos(\kappa_1 \beta_1 L/c)$.

The parametric amplifier PA in Fig. 19.6 is pumped at the sum frequency $\nu_s + \nu_r$ and is described, in the parametric approximation, by the interaction picture Hamiltonian (see Eq. (16.1.2))

$$\mathscr{H} = \hbar\kappa_2\beta_2(a_s^\dagger a_r^\dagger e^{-i\phi_2} + a_s a_r e^{i\phi_2}). \tag{19.4.22}$$

The corresponding Heisenberg equations of motion for the operators a_s and $a_r^\dagger$

$$\dot{a}_s = -i\kappa_2\beta_2 a_r^\dagger e^{-i\phi_2}, \tag{19.4.23a}$$

$$\dot{a}_r^\dagger = i\kappa_2\beta_2 a_s e^{i\phi_2}, \tag{19.4.23b}$$

can be solved to yield

$$a_s(L) = a_s(0)\cosh(\kappa_2\beta_2 L/c) - i a_r^\dagger(0)\sinh(\kappa_2\beta_2 L/c) e^{-i\phi_2}, \tag{19.4.24}$$

$$a_r(L) = a_r(0)\cosh(\kappa_2\beta_2 L/c) - i a_s^\dagger(0)\sinh(\kappa_2\beta_2 L/c) e^{-i\phi_2}. \tag{19.4.25}$$

The relationship between the field operators b_1, b_2, c_1, and c_2 in Fig. 19.6 is then obtained by replacing $a_s(0)$, $a_r(0)$, $a_s(L)$, and $a_r(L)$ in Eqs. (19.4.24) and (19.4.25) by b_1, b_2, c_1, and c_2, respectively. We obtain

$$c_1 = K_2 b_1 - i(K_2^2 - 1)^{1/2} e^{-i\phi_2} b_2^\dagger, \tag{19.4.26}$$

$$c_2 = -i(K_2^2 - 1)^{1/2} e^{-i\phi_2} b_1^\dagger + K_2 b_2, \tag{19.4.27}$$

where $K_2 = \cosh(\kappa_2\beta_2 L/c)$.

We are interested in relating the output operators d_1 and d_2 in terms of the input operators a_1 and a_2. Equations (19.4.18)–(19.4.21),

(19.4.26), and (19.4.27) can be solved to yield

$$d_1 = K_2(2K_1^2 - 1)a_1 - 2iK_1K_2(1 - K_1^2)^{1/2}e^{-i\phi_1}a_2 \\ -i(K_2^2 - 1)^{1/2}e^{-i\phi_2}a_2^\dagger, \tag{19.4.28}$$

$$d_2 = K_2(2K_1^2 - 1)a_2 - 2iK_1K_2(1 - K_1^2)^{1/2}e^{i\phi_1}a_1 \\ -i(K_2^2 - 1)^{1/2}e^{-i\phi_2}a_1^\dagger. \tag{19.4.29}$$

If K_1 and K_2 are adjusted (by adjusting the pump amplitudes) such that

$$K_1 = \left(\frac{K_2 + 1}{2K_2}\right)^{1/2}, \tag{19.4.30}$$

then Eqs. (19.4.28) and (19.4.29) reduce to

$$d_1 = a_1 - i(K_2^2 - 1)^{1/2}(e^{-i\phi_1}a_2 + e^{-i\phi_2}a_2^\dagger), \tag{19.4.31}$$

$$d_2 = a_2 - i(K_2^2 - 1)^{1/2}(e^{i\phi_1}a_1 + e^{-i\phi_2}a_1^\dagger). \tag{19.4.32}$$

These equations are of the same form as Eqs. (19.4.1) for

$$\phi_1 = \phi - \psi - \pi/2, \tag{19.4.33}$$

$$\phi_2 = -(\phi + \psi) - \pi/2. \tag{19.4.34}$$

This shows that the chain of the parametric devices in Fig. 19.6 is capable of making QND measurements on the field variables provided that the pump amplitudes and phases for the frequency converters and the parametric amplifier are adjusted properly.

Problems

19.1 Prove the following properties of the phase operators $\cos\phi$ and $\sin\phi$ as introduced in Eqs. (19.2.9a) and (19.2.9b).

$$\langle n-1|\cos\phi|n\rangle = \langle n|\cos\phi|n-1\rangle = \frac{1}{2},$$

$$\langle n-1|\sin\phi|n\rangle = -\langle n|\sin\phi|n-1\rangle = \frac{1}{2i},$$

$$[\cos\phi, \sin\phi] = \frac{1}{2i}\left[a^\dagger\frac{1}{(a^\dagger a + 1)}a - 1\right],$$

$$[a^\dagger a, \cos\phi] = -i\sin\phi,$$

$$[a^\dagger a, \sin\phi] = i\cos\phi.$$

19.2 For the Hamiltonian (19.2.2)–(19.2.5), show that A_S and A_P as given by Eqs. (19.2.6) and (19.2.11) satisfy all the conditions for QND measurement.

19.3 Show that Eq. (19.4.1b) is uniquely determined from Eq. (19.4.1a) by the requirement that b_{out} depends linearly on the operators b_{in}, a_{in}, and $a_{\text{in}}^{\dagger}$, and that the creation and annihilation operators satisfy the commutation relations

$$[a_{\text{in}}, a_{\text{in}}^{\dagger}] = 1,$$
$$[b_{\text{in}}, b_{\text{in}}^{\dagger}] = 1,$$

with all other independent commutation relations among incoming creation and annihilation operators being zero, and

$$[a_{\text{out}}, a_{\text{out}}^{\dagger}] = 1,$$
$$[b_{\text{out}}, b_{\text{out}}^{\dagger}] = 1,$$

with all other independent commutation relations among outgoing creation and annihilation operators being zero.

References and bibliography

Quantum nondemolition measurement (QND) concept

V. B. Braginsky, Y. I. Vorontsov, and F. I. Khalili, *Zh. Eksp. Teor. Fiz.* **73**, 1340 (1977) (*Sov. Phys. JETP* **46**, 705 (1977)).

V. B. Braginsky, Y. I. Vorontsov, and F. Y. Khalili, *Zh. Exsp. Teor. Fiz.* **73**, 1340 (1977) (*Sov. Phys. JETP* **46**, 705 (1977)).

W. G. Unruh, *Phys. Rev. D* **18**, 1764 (1978).

C. M. Caves, K. S. Thorne, R. W. D. Drever, V. D. Sandberg, and M. Zimmerman, *Rev. Mod. Phys.* **52**, 341 (1980).

V. B. Braginsky and F. Y. Khalili, *Quantum Measurement*, (Cambridge Univ. Press, 1992).

QND schemes

G. J. Milburn and D. F. Walls, *Phys. Rev. A* **28**, 2065 (1983).

M. Hillery and M. O. Scully in *Quantum Optics, Experimental Gravity and Measurement Theory*, ed. P. Meystre and M. O. Scully (Plenum, New York 1983), p. 661.

B. Yurke and L. R. Corruccini, *Phys. Rev. A* **30**, 895 (1984).

N. Imoto, H. A. Haus, and Y. Yamamoto, *Phys. Rev. A* **32**, 2287 (1985).

B. Yurke, *J. Opt. Soc. Am. B* **2**, 732 (1985).

M. Brune, S. Haroche, V. Lefevre, J. M. Raimond, and N. Zagury, *Phys. Rev. Lett.* **65**, 976 (1990).

M. D. Levenson, *Phys. Rev. A* **42**, 2935 (1990).

M. J. Holland, D. F. Walls, and P. Zoller, *Phys. Rev. Lett.* **67**, 1716 (1991).

A. Shimizu, *Phys. Rev. A* **43**, 3819 (1991).

M. Brune, S. Haroche, J. M. Raimond, L. Davidovich, and N. Zagury, *Phys. Rev. A* **45**, 5193 (1992).

S. Haroche, M. Brune, and J. M. Raimond, *Appl. Phys. B* **54**, 355 (1992); *J. Phys. II France* **2**, 659 (1992).

Experimental demonstration of the QND schemes

M. D. Levenson, R. M. Shelby, M. Reid, and D. F. Walls, *Phys. Rev. Lett.* **57**, 2473 (1986).

N. Imoto, S. Watkins, and Y. Sasaki, *Opt. Commun.* **61**, 159 (1987).

A. LaPorta, R. E. Slusher, and B. Yurke, *Phys. Rev. Lett.* **62**, 28 (1989).

P. Grangier, J. F. Roch, and G. Roger, *Phys. Rev. Lett.* **66**, 1418 (1991).

Ramsey separated-oscillatory-field method

N. F. Ramsey, *Molecular Beams*, (Oxford University Press, New York 1985).

CHAPTER 20

Quantum optical tests of complementarity*

Complementarity, e.g., the wave–particle duality of nature, lies at the heart of quantum mechanics. It distinguishes the world of quantum phenomena from the reality of classical physics. In the 1920s, quantum theory as we know it today was still new, and examples used to illustrate complementarity emphasized the position (particle-like) and momentum (wave-like) attributes of a quantum mechanical object, be it a photon or a massive particle. This is the historical reason why complementarity is often superficially identified with the so-called *wave–particle duality of matter.*

Complementarity, however, is a more general concept. We say that two observables are *complementary* if precise knowledge of one of them implies that all possible outcomes of measuring the other one are equally probable. We may illustrate this by two extreme examples. The first example consists of the position and momentum (along one direction) of a particle: if, say, the position is predetermined then the result of a momentum measurement cannot be predicted, all momentum values are equally probable (in a large range). The second extreme involves two orthogonal spin components of a spin-1/2 particle: if, say, the vertical spin component has a definite value (*up* or *down*) then upon measuring a horizontal component both values (*left* or *right*, for instance) are found, each with a probability of 50%. Thus, in the microcosmos complete knowledge in the sense of classical physics is not available. The classic example of the merger of wave and particle behavior is provided by Young's double-slit experiment. There we find that it is impossible to tell which slit light went through and still observe an interference pattern. In other words, any attempt

* The material in this chapter draws from Scully, Englert, and Schwinger [1989]; and Scully, Englert, and Walther [1991].

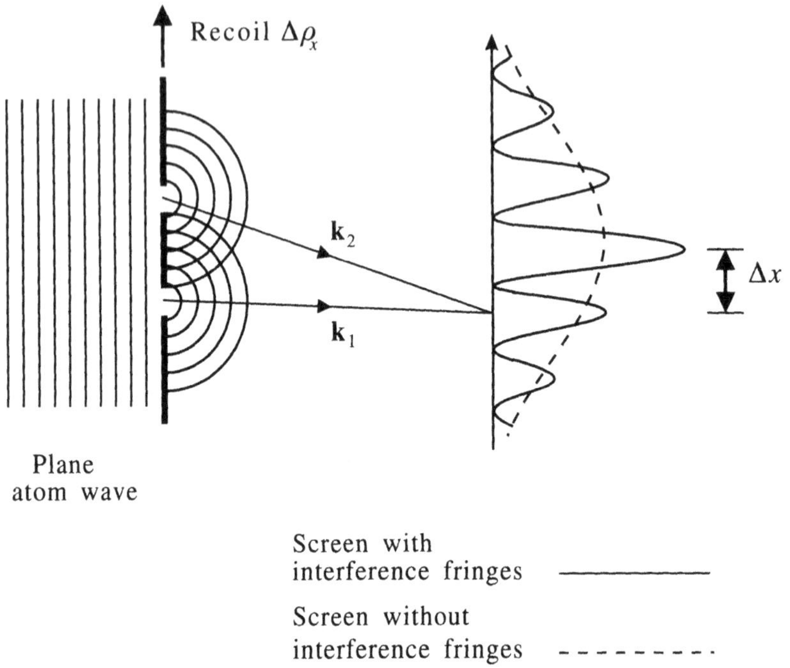

Fig. 20.1
Schematic diagram of the double-slit experiment. In this gedanken experiment the slits can recoil and reveal through which slit the photon reached the screen.

to gain which-path information will disturb the light so as to wash out the interference fringes. This point is made especially clear in the Einstein–Bohr dialogues, whose arguments we recall in the following paragraphs.

Einstein invites us to consider a Young's double-slit experiment in which the slits can recoil, as indicated in Fig. 20.1. The interference pattern is constructed by, for example, measuring the output of a photodetector array due to light passing through the slits. Now if the mass of the optical baffle (double-slit assembly) is small enough, it will recoil when the light is *emitted* by a given slit, then by conservation of momentum, we could tell which wave vector $\mathbf{k}_1$ or $\mathbf{k}_2$, the *photon* has (see Fig. 20.1). That is, we would then have which-path information.

However, Bohr points out that we must also treat the recoiling plate by the rules of quantum mechanics. Specifically, Bohr argues that the physical position of the recoiling plate is only known to within Δx due to the uncertainty principle. This error will contribute a random phase shift $\Delta\phi$ to our light beams which will destroy the interference pattern.

Such random-phase arguments, showing how which-path information destroys the coherent wave-like interference aspects of a given experimental setup, are appealing. This is in the spirit of Heisenberg's *γ-ray microscope.* In all such arguments, one notes that the act of

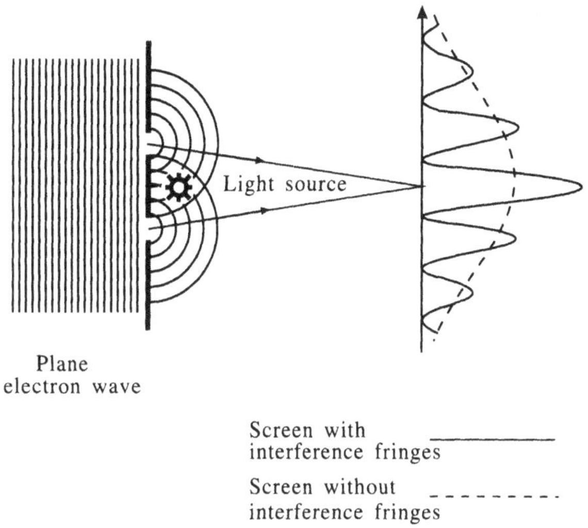

Fig. 20.2
Feynman's version of a double-slit experiment. Here, electrons interfere and the scattering of photons is used to detect their position just behind the slits, revealing through which slit the electron reached the screen.

measuring invariably disturbs the system being measured and the loss of coherence is the inevitable result of such disturbance.*

In another example, much in the spirit of the present chapter, Feynman [1965] replaces the photons by electrons. As the wave nature of matter is well known, interference between the electrons passing through slits, as in Fig. 20.2, would be expected to lead to the usual fringe patterns on the screen. In this scheme we now have an extra 'handle' on the interfering particles as electrons can be observed by, for example, light scattering. This is depicted in Fig. 20.2 where we see a light source which would scatter light from the vicinity of either slit depending on which slit the electron comes through. Feynman then goes on to explain that this observation procedure destroys the interference patterns as seen on the screen. He concludes his analysis of this interesting example with the following statement:

> If an apparatus is capable of determining which hole the electron goes through, it cannot be so delicate that it does not disturb the pattern in an essential way. No one has ever found (or even thought of) a way around the uncertainty principle.

In the experimental situations discussed so far, as in all standard examples, including Heisenberg's famous microscope, complementarity is enforced with the aid of Heisenberg's position–momentum uncertainty relation. Is this mechanism always at work? No! As we shall see, it is possible in principle and in practice to design experiments which

* See however Wooters and Zurek [1979].

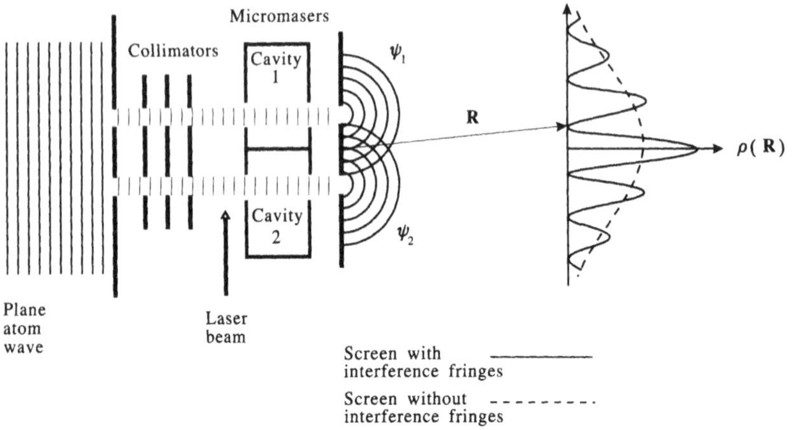

Fig. 20.3 Schematic diagram of the double-slit experiment with atoms. (From M. O. Scully, B.-G. Englert, and H. Walther, *Nature* **351**, 111 (1991).)

provide which-path information via detectors which *do not disturb the system* in any noticeable way.

In this chapter we consider some optical systems in which the loss of coherence occasioned by which-path information, i.e., by the presence of a *which-path* detector, is due simply to the establishing of quantum correlations* and is in no way associated with large random-phase factors as in Einstein's recoiling slits or Heisenberg's microscope. The principle of complementarity is therefore manifest although the position–momentum uncertainty relation plays no role.†

20.1 A micromaser which-path detector

In this section, we consider a two-slit experiment with atoms as shown in Fig. 20.3. A set of wider slits collimates two atom beams which illuminate the narrow slits where the interference pattern originates. This setup is supplemented by two high-Q micromaser cavities and a laser beam to excite all the atoms from the lower level $|b\rangle$ to the excited level $|a\rangle$. The cavity length is chosen such that the Rydberg atom makes the transition $|a\rangle \rightarrow |b\rangle$ with unit probability when passing through the cavity, through spontaneous emission of a microwave photon, even when the cavity does not contain photons initially.

In the absence of the laser–cavity system, we now describe the atomic beam, after passing through the double slits, by the state vector

* For another earlier discussion of this point, see Scully, Shea, and McCullen [1978].

† We note that it is possible to rearrange the analysis such that it "looks like" measurement back action random phasing has destroyed the interference pattern; see Stern, Aharonov, and Imry [1990]. We will comment further on this in Section 20.3.

$$\psi(\mathbf{r}) = \frac{1}{\sqrt{2}} [\psi_1(\mathbf{r}) + \psi_2(\mathbf{r})] \, |i\rangle, \tag{20.1.1}$$

where $\mathbf{r}$ is the center-of-mass coordinate and i denotes the internal state of the atom. Hence the probability density for particles on the screen at $\mathbf{r} = \mathbf{R}$ is given by the squared modulus of $\Psi(\mathbf{R})$,

$$P(\mathbf{R}) = \frac{1}{2}[(|\psi_1|^2 + |\psi_2|^2) + (\psi_1^* \psi_2 + \psi_2^* \psi_1)]\langle i|i\rangle, \tag{20.1.2}$$

We note that the usual interference behavior is represented by the cross-term $\psi_1^* \psi_2 + \psi_2^* \psi_1$.

Next we consider the situation with the laser turned on and the ultracold (vacuum) micromaser cavities put into the two paths, as in Fig. 20.3. Before entering the cavities, the laser beam excites the atoms to the long-lived Rydberg state $|a\rangle$. After passing through the cavities and making the transition $|a\rangle \to |b\rangle$, say, by spontaneous emission of a photon, the state of the correlated atomic beam and maser cavity system is given by

$$\psi(\mathbf{r}) = \frac{1}{\sqrt{2}}[\psi_1(\mathbf{r})|1_1 0_2\rangle + \psi_2(\mathbf{r})|0_1 1_2\rangle]|b\rangle, \tag{20.1.3}$$

where, for example, $|1_1 0_2\rangle$ denotes the state in which there is one photon in cavity 1 and none in cavity 2. Note that unlike Eq. (20.1.1) this $\psi(\mathbf{r})$ is not a product of two factors, one referring to the atomic and the other to the photonic degrees of freedom. The system and the detector have become entangled by their interaction. In contrast to Eq. (20.1.2), the probability density at the screen is now given by

$$\begin{aligned} P(\mathbf{R}) = \frac{1}{2}[(|\psi_1|^2 + |\psi_2|^2) + (\psi_1^* \psi_2 \langle 1_1 0_2 | 0_1 1_2\rangle \\ + \psi_2^* \psi_1 \langle 0_1 1_2 | 1_1 0_2\rangle)]\langle b|b\rangle. \end{aligned} \tag{20.1.4}$$

But because $\langle 1_1 0_2 | 0_1 1_2\rangle$ vanishes, the interference terms disappear here, so that

$$P(\mathbf{R}) = \frac{1}{2}(|\psi_1|^2 + |\psi_2|^2), \tag{20.1.5}$$

does not show fringes.

The micromasers will serve as which-path detectors only if the one extra photon left by the atom changes the photon field in a detectable manner. Thus whether which-path information is available or not depends on the photon states initially prepared in the cavities. One extreme situation has just been discussed: no photons initially, one photon in one of the detectors finally. Clearly, here one can tell through which cavity, and therefore through which slit, the atom came to the screen. The situation is quite different when the cavities contain classical microwave radiation with large (average) numbers of photons, n_1 and n_2, which have spreads given by their square roots. For instance, the change in photon number in cavity 1 is now from $n_1 \pm \sqrt{n_1}$ to $n_1 + 1 \pm \sqrt{n_1}$. This change cannot be detected, because $\sqrt{n_1} \gg 1$, so that there is no which-path information available.

In the latter situation (classical radiation in the micromaser cavities), we cannot tell through which slit the atom reached the screen and the interference pattern is just the same as in the absence of the micromaser cavities. In contrast, cavities containing no photons initially store which-path information and therefore the interference pattern is lost. It is changed to the incoherent superposition (20.1.5) of one-slit patterns.

Thus we have shown that the interference cross term, which goes as

$$\psi_1^*(\mathbf{r})\psi_2(\mathbf{r})\langle 1_1 0_2 | 0_1 1_2 \rangle \tag{20.1.6}$$

vanishes because $\langle 1_1 0_2 | 0_1 1_2 \rangle = 0$. But how do we know that the 'uncertainty principle' does not cause the $\psi_2^*\psi_1$ term to vanish as well? In such a case we could argue that it is the introduction of random phase factors, due to the interaction with the field, which are responsible for the loss of interference. We next show that that is not the case.

20.2 The resonant interaction of atoms with a microwave field and its effect on atomic center-of-mass motion*

The resonant atom–field Hamiltonian for our problem is

$$\mathscr{H} = \frac{p^2}{2m} + \frac{\hbar\omega}{2}\sigma_z + \hbar\omega a^\dagger a + \hbar g U(z)\mathscr{N} \tag{20.2.1}$$

with

$$\mathscr{N} = \sigma_+ a + a^\dagger \sigma_-. \tag{20.2.2}$$

Here we have added the term $p^2/2m$ associated with the center-of-mass motion of the atom to the usual interaction Hamiltonian for a two-level atom interacting with a single-mode field (Eq. (6.2.8)). In Eq. (20.2.1), p is the operator of the center-of-mass momentum of the atom with mass m, and the function $U(z)$ describes the spatial dependence of the coupling between the atomic transition and the maser photons – ideally it is a mesa function as sketched in Fig. 20.4.

* For further reading on this subject, see Englert, Schwinger, and Scully [1990].

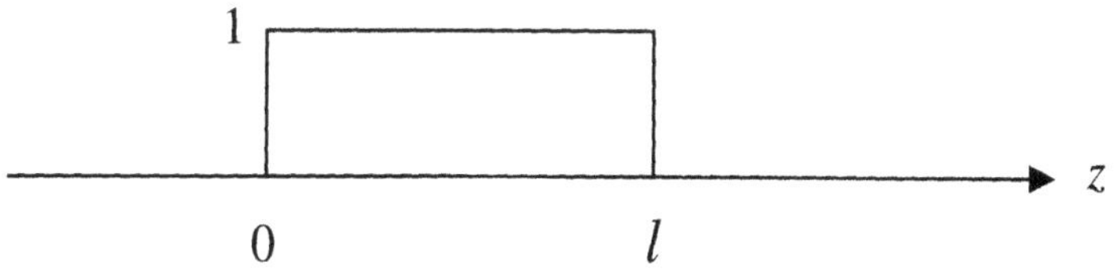

Fig. 20.4
The mesa function $U(z)$.

Now the eigenstates of the operator $\mathcal{N}$ are given by

$$|\pm, n\rangle = \frac{1}{\sqrt{2}}(|a, n\rangle \pm |b, n+1\rangle), \tag{20.2.3}$$

corresponding to the eigenvalues $\pm\sqrt{n+1}$ (see Section 10.2.2). The (constant) energy eigenvalues of the system take the form

$$z < 0 \qquad E = \frac{p^2}{2m}, \tag{20.2.4a}$$

$$0 < z < l \qquad E = \frac{p_\pm^2}{2m} \pm \hbar g\sqrt{n+1}, \tag{20.2.4b}$$

$$z > l \qquad E = \frac{p^2}{2m}, \tag{20.2.4c}$$

and therefore

$$\frac{p^2}{2m} = \frac{p_\pm^2}{2m} \pm \hbar g\sqrt{n+1}, \tag{20.2.5}$$

which implies

$$p_\pm = (p^2 \mp 2m\hbar g\sqrt{n+1})^{1/2}. \tag{20.2.6}$$

As the center-of-mass energy is much larger than the interaction energy

$$p_\pm \cong p \mp \frac{m\hbar g}{p}\sqrt{n+1}. \tag{20.2.7}$$

Consider now the case in which the atom enters into the cavity having n photons in its upper state $|a\rangle$, that is

$$\psi(z=0) = |a, n\rangle = \frac{1}{\sqrt{2}}(|+, n\rangle + |-, n\rangle), \tag{20.2.8}$$

and at the cavity exit $z = l$ we have

$$\begin{aligned}\psi(z=l) &= \frac{1}{\sqrt{2}}\left(|+, n\rangle e^{-ip_+l/\hbar} + |-, n\rangle e^{-ip_-l/\hbar}\right) \\ &= \left[|a, n\rangle \cos\left(g\sqrt{n+1}\frac{l}{v_0}\right)\right. \\ &\qquad \left. + i|b, n+1\rangle \sin\left(g\sqrt{n+1}\frac{l}{v_0}\right)\right] e^{-ipl/\hbar},\end{aligned} \tag{20.2.9}$$

where $v_0 = p/m$.

Thus we see that the phase factors $p_{\pm}l/\hbar$ which occur due to interaction with the field are not random but just the Rabi factor $g\tau\sqrt{n+1}$ where $\tau = l/v_0$ is the time of flight.

In Appendix 20.A we carry out the explicit calculation dealing with the complete problem including the (tiny) photon recoil and show that here again no random phases are present.

The point is that the micromaser which-path detectors do not lead to random phases due to the atom–field interaction. There is no significant change in the spatial wave function of the atoms due to the atom–field interaction. It is the correlation of the center-of-mass wave function to the photon degrees of freedom in the cavities that is responsible for the loss of interference.

20.3 Quantum eraser

In the preceding sections we have seen that it is the system detector correlations which account for the dramatic effects of the measuring apparatus on the system of interest. It is no surprise that coherence is destroyed as soon as one has which-path information, but in Section 20.2 no uncontrollable scattering events were involved in destroying the interference (wave-like) behavior.

One then wonders whether it might not be possible to retrieve the coherent interference cross terms by removing ('erasing') the which-path information contained in the detectors. In this sense, we are here considering the quantum eraser problem. Notice that if we considered the coherence to be lost because of random scattering or other stochastic perturbations, as discussed at the beginning of this chapter, for example, this question would never come up. In fact, we shall see that interference effects can be restored by manipulating the which-path detectors long after the atoms have passed.

Consider now the arrangement of the atomic beam/micromaser system as indicated in Fig. 20.5. There we see that the atoms pass through the two maser cavity detectors, but now we will imagine that the which-path detectors are separated by a shutter–detector combination. So we now have a configuration in which the quantum eraser becomes possible. In particular, consider the cavity system in Fig. 20.5(a). There we see two shutters arranged such that radiation will be constrained to remain either in the upper or the lower cavity, when the shutters are closed. We further imagine that on opening the shutters, light will be allowed to interact with the photodetector wall. In this way the radiation, which is left either in the upper or in the

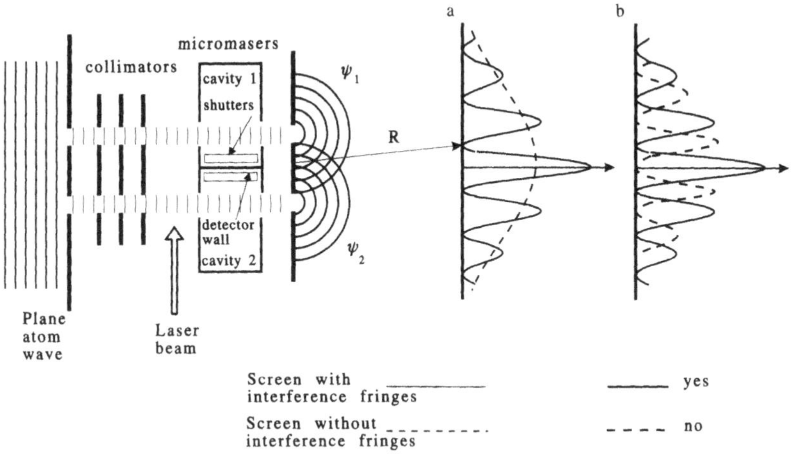

Fig. 20.5
(a) Quantum eraser configuration in which electrooptic shutters separate microwave photons in two cavities from the detector wall which absorbs microwave photons and acts as a photodetector. (b) Density of the particles on the screen depending upon whether a photocount is observed in the detector wall ('yes') or not ('no') demonstrating that correlations between the event on the screen and the eraser photocount are necessary to retrieve the interference pattern. (From M. O. Scully, B.-G. Englert, and H. Walther, *Nature* **351**, 111 (1991).)

lower cavity, depending upon whether the atom travelled along the upper or lower path, will now be absorbed and the *memory of passage* (the which-path information) could be said to be erased.

Do we now (after erasure) regain interference fringes? The answer is yes, but how can that be? The atom is now far removed from the micromaser cavities and so there can be no thought of any physical influence on the atom's center-of-mass wave function. The fact that this situation is seemingly paradoxical has been noted by Jaynes* (whom we paraphrase):

> We have, then, the full EPR [Einstein-Podolsky-Rosen] paradox – and more. By applying or not applying the eraser mechanism before measuring the state of the microwave cavities we can, at will, for the atomic beam into either : (1) a state with a known path, and no possibility of interference effects in any subsequent measurement; (2) a state with both ψ_1 and ψ_2 present with a known relative phase. Interference effects are then not only observable, but predictable. And we can decide which to do after the interaction is over and the atom is far from the cavities, so there can be no thought of any physical influence on the atom's centre-of-mass wavefunction!
>
> From this, it is pretty clear that present quantum theory not only does not use – it does not even dare to mention – the notion of a 'real physical situation'. Defenders of the theory say that this notion is philosophically naive, a throwback to outmoded ways of thinking, and that recognition of this constitutes deep new wisdom about the nature of human

* Jaynes [1980]. The quantum eraser was first proposed independently by Scully and Drühl [1982], and the connection between that work and the Jaynes paradox was first made by Scully, Englert, and Walther [1992].

> knowledge. I say that it constitutes a violent irrationality, that somewhere in this theory the distinction between reality and our knowledge of reality has become lost, and the result has more the character of medieval necromancy than of science. It has been my hope that quantum optics, with its vast new technological capability, might be able to provide the experimental clue that will show us to resolve these contradictions.

We proceed to resolve the "Jaynes paradox" by extending the mathematical description to include the detector, which is initially in its ground state $|d\rangle$, we have

$$\psi(\mathbf{r}) = \frac{1}{\sqrt{2}}[\psi_1(\mathbf{r})|1_1 0_2\rangle + \psi_2(\mathbf{r})|0_1 1_2\rangle]|b\rangle|d\rangle, \tag{20.3.1}$$

which replaces Eq. (20.1.3). After absorbing a photon, the detector would be found in the excited state $|e\rangle$.

It is now convenient to introduce symmetric, ψ_+, and antisymmetric, ψ_-, atomic states defined as

$$\psi_\pm(\mathbf{r}) = \frac{1}{\sqrt{2}}[\psi_1(\mathbf{r}) \pm \psi_2(\mathbf{r})]. \tag{20.3.2}$$

Likewise, we introduce symmetric, $|+\rangle$, and antisymmetric, $|-\rangle$, states of the radiation fields contained in the which-path cavities,

$$|\pm\rangle = \frac{1}{\sqrt{2}}(|1_1 0_2\rangle \pm |0_1 1_2\rangle). \tag{20.3.3}$$

In terms of Eqs. (20.3.2) and (20.3.3), the state (20.3.1) of the atom–beam/micromaser cavity/detector system appears as

$$\psi(\mathbf{r}) = \frac{1}{\sqrt{2}}[\psi_+(\mathbf{r})|+\rangle + \psi_-(\mathbf{r})|-\rangle]|b\rangle|d\rangle. \tag{20.3.4}$$

We now consider the interaction between the radiation field existing in the cavity and the detector. As mentioned earlier, we envisage the detector to consist of an atom with a lower state $|d\rangle$ and an excited state $|e\rangle$. The interaction Hamiltonian between field and detector depends on symmetric combinations of the field variables, so that only the symmetric state $|+\rangle$ will couple to the fields.

We then find that the action of the detector (eraser) system produces the state

$$\psi(\mathbf{r}) = \frac{1}{\sqrt{2}}[\psi_+(\mathbf{r})|0_1 0_2\rangle|e\rangle + \psi_-(\mathbf{r})|-\rangle|d\rangle]|b\rangle, \tag{20.3.5}$$

that is, the symmetric interaction couples only to the symmetric radiation state $|+\rangle$; the antisymmetric state $|-\rangle$ remains unchanged.

Now, the atomic probability density for (20.3.4) at the screen goes as

$$\begin{aligned} P(\mathbf{R}) &= \frac{1}{2}[\psi_+^*(\mathbf{R})\psi_+(\mathbf{R}) + \psi_-^*(\mathbf{R})\psi_-(\mathbf{R})] \\ &= \frac{1}{2}[\psi_1^*(\mathbf{R})\psi_1(\mathbf{R}) + \psi_2^*(\mathbf{R})\psi_2(\mathbf{R})] \end{aligned} \tag{20.3.6}$$

and does not show any interference fringes as long as the final state of the detector is unknown. But if one asks what is the probability density $P_e(\mathbf{R})$ for finding both the detector excited and the atom at $\mathbf{R}$ on the screen, the answer is

$$\begin{aligned} P_e(\mathbf{R}) &= |\psi_+(\mathbf{R})|^2 \\ &= \frac{1}{2}[|\psi_1(\mathbf{R})|^2 + |\psi_2(\mathbf{R})|^2] + \mathrm{Re}[\psi_1^*(\mathbf{R})\psi_2(\mathbf{R})], \end{aligned} \tag{20.3.7}$$

which exhibits the same fringes as Eq. (20.1.2), indicated as a solid line in Fig. 20.5(b). In contrast, the probability density $P_d(\mathbf{R})$ for finding both the detector deexcited and the atom at $\mathbf{R}$ on the screen is

$$\begin{aligned} P_d(\mathbf{R}) &= |\psi_-(\mathbf{R})|^2 \\ &= \frac{1}{2}[|\psi_1(\mathbf{R})|^2 + |\psi_2(\mathbf{R})|^2 - \mathrm{Re}[\psi_1^*(\mathbf{R})\psi_2(\mathbf{R})], \end{aligned} \tag{20.3.8}$$

giving rise to the antifringes indicated by the broken line in Fig. 20.5(b). If the eraser photon signal is disregarded, we obtain the superposition (Eq. (20.3.6)), equal to half the sum of P_e and P_d, which is fringeless, and, of course, identical with Eq. (20.1.5), see Fig. 20.3.

Here is the physical interpretation of this calculation. After an atom has travelled from the oven to the screen, passing through the micromasers and leaving its tell-tale photon, we record an event somewhere on the screen. Then we return to the which-path micromasers, open the shutters and allow the absorption of the microwave photon. When we observe a photocount in the detector we know that erasure has been completed. In this event the atom is counted as a *yes*-atom.

Then we wait for another atom to pass through the system from oven to screen. Again we record an event on the screen and then turn to the micromaser cavities. This time suppose that, upon opening the shutter, we observe no photocount in the quantum eraser detector. This will be the case half of the time, as explained above. Now we count the atom as a *no*-atom.

We repeat the above sequence many times. Eventually, the *yes*-atoms will build up the solid-line fringes in Fig. 20.5, and the *no*-atoms

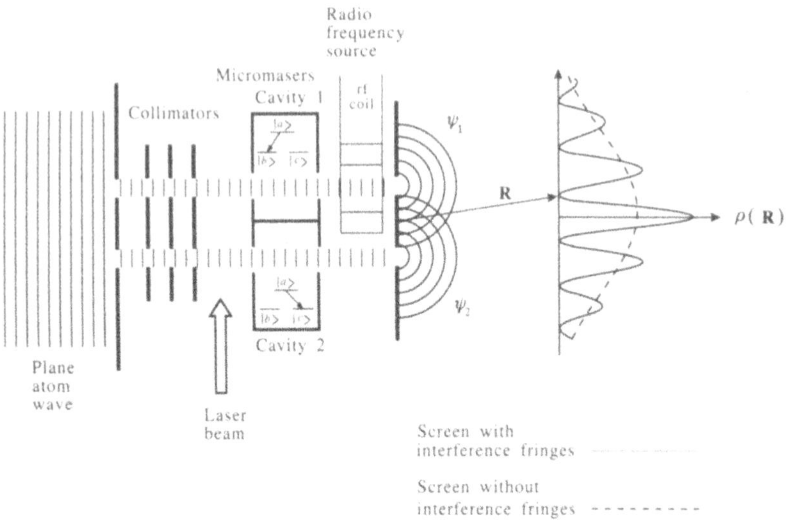

Fig. 20.6
Asymmetric setup in which cavity 1 induces the transition $|a\rangle \rightarrow |b\rangle$ and cavity 2 induces $|a\rangle \rightarrow |c\rangle$. Which-path information is erased by the radio frequency in the coil where $|b\rangle \rightarrow |c\rangle$ happens. (From M. O. Scully, B.-G. Englert, and H. Walther, *Nature* **351** 111 (1991).)

produce the broken-line antifringes. Finally we note that the fringes and antifringes will cancel if we do not correlate them to the state of the eraser detector.

Having presented the physics of quantum erasure we now turn to an experimentally more realizable scheme which has much in common with the quantum eraser idea. We consider as in Fig. 20.6, the asymmetric situation in which cavity 1 is tuned to the transition $|a\rangle \rightarrow |b\rangle$, and cavity 2 is tuned to the transition $|a\rangle \rightarrow |c\rangle$.

Even if the cavities contain classical microwave radiation, as we shall assume in the sequel, and therefore do not store which-path information, the screen will not show interference fringes because the internal atomic states $|b\rangle$ and $|c\rangle$ are orthogonal. This is analogous to the disappearance of the interference terms in Eq. (20.1.4), except that now the atoms themselves carry the which-path information.

The latter circumstance again invites the question: could one induce the transitions $|b\rangle \rightarrow |c\rangle$ in the atoms that traversed cavity 1, so that the which-path information is erased, and thereby make the interference pattern reappear? The answer is affirmative. The actual experimental realization, however, is a delicate matter, because one must exert careful control on the phases of the various classical radiation fields. To appreciate what is involved, suppose that between cavity 1 and the slit plate there is a coil that can be fed with radio frequency of $\sim$ 50 MHz with the right strength to ensure the transition $|b\rangle \rightarrow |c\rangle$, as depicted in Fig. 20.6. In the interference region the state of the atom is essentially

$$\psi(\mathbf{r}) = \frac{1}{\sqrt{2}}[\psi_1(\mathbf{r}) + e^{i\beta}\psi_2(\mathbf{r})]|c\rangle \tag{20.3.9}$$

where the relative phase angle β is determined by the phases of the microwave fields in the two maser cavities and the radio-wave field in the coil. As these fields have different frequencies, β really refers to a certain instant, the moment, say, when the atom is excited to state $|a\rangle$ by the laser beam. The probability density at the screen

$$P(\mathbf{R}) = |\psi(\mathbf{R})|^2 = \frac{1}{2}(|\psi_1|^2 + |\psi_2|^2) + \text{Re}(\psi_1^* e^{i\beta}\psi_2), \tag{20.3.10}$$

now exhibits an interference term that depends on β very sensitively. If, therefore, the value of β varies from atom to atom, the interference pattern will not build up. This illustrates quite well the omnipresent phenomenon of coherence loss caused by random phases. Consequently, we must ensure that the phase angle β is the same for all atoms to make the interference fringes reappear. In the setup of Fig. 20.6 this can be achieved by adjusting the phase of the radio frequency radiation in the coil to the phases that the microwave fields in the cavities have at the moment when the laser excites the atom. An additional bonus is the possibility of varying the chosen value of β, which enables us to shift the interference pattern on the screen. In summary, the control* over the phase angle β represents a switch with which the experimenter can turn the interference fringes on and off, or relocate them.

20.4 Quantum optical Ramsey fringes

Another interesting example allowing us to probe the way in which a measurement process (the presence of a detector) influences the investigated system is provided by a possible quantum optical Ramsey fringes experiment. Such a possibility exists due to the fact that number states of the radiation field can be generated in a micromaser as discussed in Section 13.3. The present scheme is close to being realized as opposed to the double-slit experiment supplemented by micromaser cavities discussed in the previous section.

We consider an interferometer in which a two-level atom traverses two identical high-Q micromaser cavities in succession as shown in Fig. 20.7. Before entering the first cavity, the atoms are prepared in either the upper state $|a\rangle$ or the lower state $|b\rangle$. After leaving the

* Here the 'system-detector-correlation destroys interference' point of view (as opposed to random phasing) receives strong support. If we focus on the "back-action-random-phase" point of view, we are apt to miss this point.

a

b

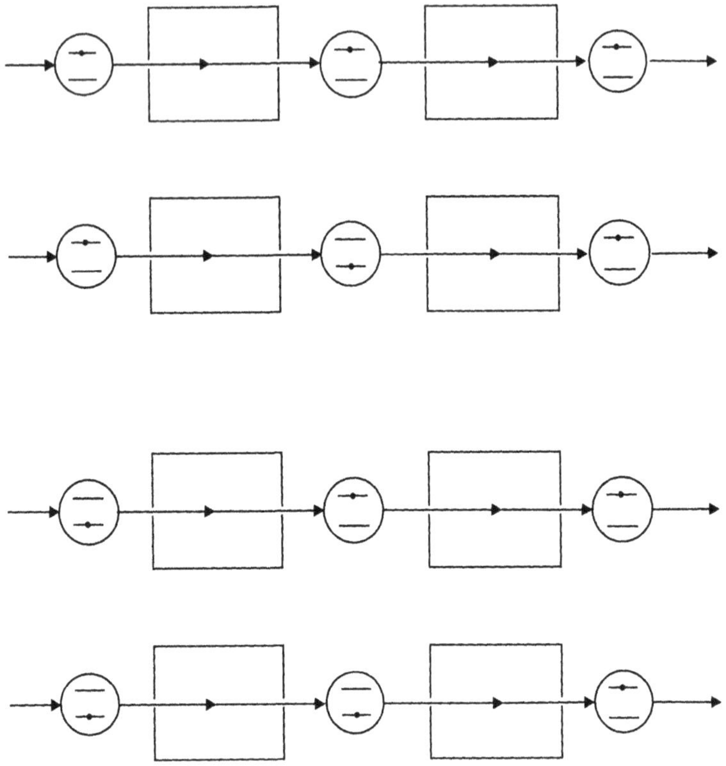

Fig. 20.7
Alternative routes for reaching the final upper state. If the atom is initially in the upper state (a) then the photon number is changed either in none of the cavities or in both. If the atom is initially in the lower state (b) then the photon number is changed either only in the first cavity or only in the second. (From B.-G. Englert, H. Walther, and M. O. Scully, *Appl. Phys. B* **54**, 366 (1991).)

second cavity, the internal state of the atom is tested by state-selective field ionization, so that the probability for ending up in the upper state is measured. For each initial state there are two different *paths* along which the final upper state can be reached, as seen in Fig. 20.7. As long as no information is available about which alternative has been realized, interference may be observed. On the other hand, if which-path information is stored in the cavities, then complementarity does not allow for interference.

It can be shown, using the solutions (6.2.13) and (6.2.14) for the probability amplitudes for the interaction of the two-level atom with a single-mode quantized radiation field, that the probabilities of finding the atom in the upper state $|a\rangle$ after passing through the two micromaser cavities are (see Problem 20.1)

$$\begin{aligned} P_a^{(+)} &= \langle C_1^\dagger C_1\rangle\langle C_2^\dagger C_2\rangle + \langle S_1^2 a_1 a_1^\dagger\rangle\langle \tilde{S}_2^2 a_2^\dagger a_2\rangle \\ &\quad -2\mathrm{Re}(\langle C_1^\dagger a_1^\dagger S_1\rangle\langle C_2^\dagger a_2 \tilde{S}_2\rangle e^{i\Delta T}), \end{aligned} \tag{20.4.1}$$

$$\begin{aligned} P_a^{(-)} &= \langle \tilde{C}_1^\dagger \tilde{C}_1\rangle\langle \tilde{S}_2^2 a_2^\dagger a_2\rangle + \langle \tilde{S}_1^2 a_1^\dagger a_1\rangle\langle C_2^\dagger C_2\rangle \\ &\quad +2\mathrm{Re}(\langle \tilde{S}_1 a_1^\dagger \tilde{C}_1\rangle\langle C_2^\dagger a_2 \tilde{S}_2\rangle e^{i\Delta T}). \end{aligned} \tag{20.4.2}$$

Here $P_a^{(+)}$ and $P_a^{(-)}$ represent the final probability for finding the atom in the upper state if the atom is initially in the upper and lower states, respectively, $\Delta = \omega_{ab} - \nu$ is the atom–field detuning, T is the time the atom needs to cover the distance between the two cavities, and

$$S_j = \frac{2g \sin(\sqrt{\Delta^2 + 4g^2 a_j a_j^\dagger}\ \tau/2)}{\sqrt{\Delta^2 + 4g^2 a_j a_j^\dagger}} \tag{20.4.3}$$

$$C_j = \cos(\sqrt{\Delta^2 + 4g^2 a_j a_j^\dagger}\ \tau/2) + i\frac{\Delta}{2g} S_j, \tag{20.4.4}$$

where a_j and $a_j^\dagger$ $(j = 1, 2)$ are the destruction and creation operators of the field in the two cavities, g is the atom–field coupling constant, τ is the duration of the interaction between the atom and either one of the cavity fields, and $\tilde{S}_j$ and $\tilde{C}_j$ are obtained upon replacing $a_j a_j^\dagger$ by $a_j^\dagger a_j$. The expectation values in Eqs. (20.4.1) and (20.4.2) refer to an instant prior to the atom entering the first cavity, so they are expectation values of the photon fields as they are initially prepared. In deriving Eqs. (20.4.1) and (20.4.2) the cavity damping and spontaneous decay of the atom are neglected.

We now consider two different situations. In the first case we assume that the initial field states inside the micromaser cavities are classical states (coherent states with a large amplitude). We further assume, for simplicity, that the mean photon numbers in these field states are equal $(\langle a_1^\dagger a_1 \rangle = \langle a_2^\dagger a_2 \rangle = n)$ and large $(n \gg 1, (\Delta/2g)^2)$, and are such that

$$\cos(g\sqrt{n}\tau) = \sin(g\sqrt{n}\tau) = \frac{1}{\sqrt{2}}. \tag{20.4.5}$$

Under these conditions, Eqs. (20.4.1) and (20.4.2) simplify considerably. If ϕ is the relative phase between the fields in the two cavities, then the probability of finding the atom in the upper state finally is

$$P_a^{(\pm)} = \frac{1}{2}[1 \mp \cos(\Delta T + \phi)]. \tag{20.4.6}$$

It is evident that, upon varying T – for example, by changing the distance between the cavities – the *Ramsey fringes* are observed. These Ramsey fringes signify the presence of a coherent superposition of the two atomic states in the region between the cavities. The *which-path* information is not stored and, therefore, interferences are permitted.

In the other extreme situation, number states are prepared in both cavities. The alternatives of Fig. 20.7 then correspond to these changes in the photon states:

$$\text{(a)} \qquad |n, n\rangle \to |n, n\rangle \text{ or } |n+1, n-1\rangle, \tag{20.4.7a}$$

$$\text{(b)} \qquad |n, n\rangle \to |n-1, n\rangle \text{ or } |n, n-1\rangle. \tag{20.4.7b}$$

Consequently, the final numbers of photons in the cavities indicate which path has actually been realized, so that no interferences are possible. This is readily seen from Eqs. (20.4.1) and (20.4.2) where the interference terms vanish for initial number states. When the conditions discussed above are satisfied

$$P_a^{(\pm)} = \frac{1}{2}, \tag{20.4.8}$$

i.e., the probabilities for finding the atom finally in the upper state do not display the Ramsey fringes.

Thus, by preparing number states or classical states (equal to coherent states with a large amplitude) in the micromaser cavities, the experimenter can either obtain which-path information or Ramsey fringes. Both at the same time are not possible. It is interesting to note that the loss of the capability of interfering brought about by the number states has nothing to do with random phases, nor with a variant of Heisenberg's uncertainty relations. It is the orthogonality, or in physical terms the distinguishability of the two final states in Eqs. (20.4.7), that causes this loss.

In summary, the atom interferometer can be employed to demonstrate how complementarity is enforced by a mechanism different from random phases or bounds on uncertainty products. Ramsey fringes must always disappear as soon as the micromaser cavities contain information about which alternative has been realized. Although it may be rather difficult to retrieve this information, it is there whether we *look at it* or not. The mere fact that we could in principle have which-path information is enough to rub out the fringes.

20.A Effect of recoil in a micromaser which-path detector*

We write the wave function for the atoms at the slits, $z = 0$, in the form of an entagled state

$$\begin{aligned} |\psi(x, y, 0)\rangle = {} & A(x - d/2, y)|1_1, 0_2\rangle \psi_i(x) g_1(x, y) \\ & + A(x + d/2, y)|0_1, 1_2\rangle \psi_i(x) g_2(x, y), \end{aligned} \tag{20.A.1}$$

where $A(x \pm d/2, y)$ are the aperture functions which are nonzero only in the vicinity of $x = \mp d/2$, $y = 0$, $\psi_i(x)$ is the probability

* This section follows from Englert et al. [1994] and Yelin et al. [1996].

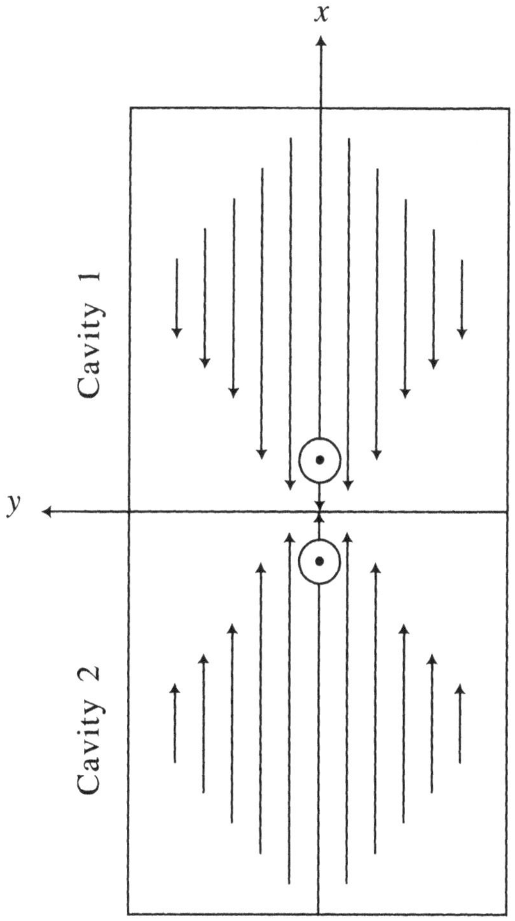

Fig. 20.8
The cavity orientation and the direction of the electric vector of the microwave fields. The cavities are shown as seen by someone standing at the screen. The circles indicate the holes through which the atoms approach this observer.

amplitude before the interaction, and the functions $g_1(x, y)$ and $g_2(x, y)$ summarize the effect of the interaction and determine the probabilities that the atom emits a photon in cavities 1 and 2, respectively.

For the mode function $U(x, y) = \cos(ky)$ corresponding to the geometry in Fig. 20.8, the functions g_1 and g_2 are given by

$$g_1(x, y) = \sin[g \cos(ky)\tau] \quad \text{for } x > 0, \tag{20.A.2a}$$

$$g_2(x, y) = \sin[g \cos(ky)\tau] \quad \text{for } x < 0. \tag{20.A.2b}$$

In order to calculate the fringe pattern we must first find the wave function on the screen at $z = D$ in terms of what it is in the aperture at $z = 0$. This relation is given by

$$\psi(x, y, D) = \psi_1(x, y, D) + \psi_2(x, y, D) \tag{20.A.3}$$

with

$$\psi_1(x, y, D) = \int\int dx'\, dy'\, G(x, y, D; x', y', 0) A(x' - d/2, y') \times |1_1, 0_2\rangle \psi_i(x') g_1(x', y'), \tag{20.A.4}$$

$$\psi_2(x, y, D) = \int\int dx'\, dy'\, G(x, y, D; x', y', 0) A(x' + d/2, y') \times |0_1, 1_2\rangle \psi_i(x') g_2(x', y'), \tag{20.A.5}$$

where $G(x, y, D; x', y', 0)$ is the Green's function for a wave with de Broglie wavelength λ_{dB}:

$$G(x, y, D; x', y', 0) = \mathcal{N} \exp\left\{ \frac{i\pi}{D\lambda_{dB}} \left[(x - x')^2 + (y - y')^2\right] \right\}. \tag{20.A.6}$$

The factor $\mathcal{N}$ contains all the constants that are irrelevant here. This Green's function applies in the Fraunhofer regime, when the screen is far away from the slits and the scattering is at small angles.

To ensure that the probability of no-emission in either microwave cavity is small, $g\tau = \pi/2$. We need $g_{1,2}(x, y)$ only in the aperture at $x = \pm d/2$, $y = 0$. Under these circumstances

$$g_{1,2}(x, y) \cong 1 - \frac{\pi^2}{32} k^4 y^4, \tag{20.A.7}$$

where y ranges across the apertures, so $g_{1,2}(x, y)$ are measured on the scale set by the slit width a ($\sim$ 5 nm). Therefore, the deviation from unity in Eq. (20.A.7) is roughly given by $(ka)^4 \sim [10^2 \mathrm{m}^{-1}(5 \times 10^{-9}\mathrm{m})]^4 \sim 10^{-25}$ which is quite negligible.

With the ingredients of Eq. (20.A.3) now at hand, we proceed to calculate the branches of the center-of-mass wave function at the screen and so demonstrate explicitly that the capability for interfering has not been lost as a result of the interaction. For the sake of simplicity, we choose a Gaussian aperture function

$$A(x \mp d/2, y) = \exp\left\{ -\frac{1}{2a^2} \left[(x \mp d/2)^2 + y^2\right] \right\}. \tag{20.A.8}$$

The upper slit produces

$$\begin{aligned} &\psi_1(x, y, D) \\ &= \mathcal{N} \int\int dx' dy' \exp\left\{ \frac{i\pi}{D\lambda_{dB}} \left[(x - x')^2 + (y - y')^2\right] \right\} \\ &\quad \times \exp\left\{ -\frac{1}{2a^2} \left[(x' - d/2)^2 + y'^2\right] \right\} \left[1 - \frac{\pi^2}{32}(ky')^4\right]. \end{aligned} \tag{20.A.9}$$

A similar expression is obtained for $\psi_2(x, y, D)$.

The two-slit interference pattern is produced by

$$\mathrm{Re}\left[\psi_2^*(x,y,D)\psi_1(x,y,D)\right] = \mathrm{Re}\left[\psi_1^*(-x,y,D)\psi_1(x,y,D)\right]. \tag{20.A.10}$$

Without the atom–photon interaction it is given by

$$\left(2\pi a^2|\mathcal{N}|\right)^2 \exp\left[-\frac{x^2+y^2+(d/2)^2}{b^2}\right]\cos\left(\frac{xd}{ab}\right), \tag{20.A.11}$$

where $b = \lambda_{dB}D/2\pi a$. This expression identifies the fringe spacing $2\pi ab/d = \lambda_{dB}D/d$. Thus the interference pattern is not rubbed out by transverse recoil effects.

Problems

20.1 Derive Eqs. (20.4.1) and (20.4.2).

20.2 Discuss whether or not the which-path information will be stored in the atom interferometer discussed in Section 20.4 if the radiation field inside the micromaser cavities is (a) a single mode thermal field at temperature T, (b) a squeezed vacuum state $|\xi, 0\rangle$, or (c) a state described by the density operator

$$\rho = e^{-\langle n\rangle/2}\sum_{n=0}^{\infty}\frac{\langle n\rangle^n}{2^n n!}|2n\rangle\langle 2n|.$$

References and bibliography

General references

R. Feynman, R. Leighton, and M. Sands, *The Feynman Lectures on Physics*, Vol. III, (Addison Wesley, Reading, MA 1965).
M. O. Scully, R. Shea, and J. D. McCullen, *Phys. Rep.* **43**, 485 (1978).
J. A. Wheeler, in *Problems in the Foundations of Physics*, ed. G. T. di Francia (North-Holland, Amsterdam 1979).
H. W. K. Wootters and W. H. Zurek, *Phys. Rev. D* **19**, 473 (1979).
A. Barut, P. Meystre, and M. O. Scully, *Laser Focus* (Oct. 1982), pp. 49–56.
A. Wheeler and W. H. Zurek, *Quantum Theory and Measurement* (Princeton Univ. Press, 1983)
B. Englert, J. Schwinger, and M. O. Scully, *Found. Phys.* **18**, 1045 (1988).
J. Schwinger, M. O. Scully, and B.-G. Englert, *Z. Phys. D* **10**, 135 (1988).
M. O. Scully, B.-G. Englert, and J. Schwinger, *Phys. Rev. A* **40**, 1775 (1989).
B.-G. Englert, M. O. Scully, and H. Walther, *Sci. Am.* **271**, 56 (1994).

Micromaser which-path detector

M. O. Scully and H. Walther, *Phys. Rev. A* **39**, 5229 (1989).
A. Stern, Y. Aharonov, and Y. Imry, *Phys. Rev. A* **41**, 3436 (1990).
M. O. Scully, B.-G. Englert, and H. Walther, *Nature* **351**, 111 (1991).
S. M. Tan and D. F. Walls, *Phys. Rev. A* **47**, 4663 (1993).

Center-of-mass motion of masing atoms

B.-G. Englert, J. Schwinger, A. O. Barut, and M. O. Scully, *Europhys. Lett.* **14**, 25 (1991).
S. Haroche, M. Brune, and J. M. Raimond, *Europhys. Lett.* **14**, 19 (1991).
B.-G. Englert, J. Schwinger, and M. O. Scully, in *New Frontiers in Quantum Electrodynamics and Quantum Optics*, ed. A. O. Barut (Plenum, New York, 1990).
B.-G. Englert, H. Fearn, M. O. Scully, and H. Walther, in *Quantum Interferometry*, (World Scientific, Singapore, 1994), p. 103.
S. F. Yelin, C. J. Bednar, and C.-R. Hu, *Opt. Commun.* **136**, 171 (1997).

Quantum optical Ramsey fringes

B.-G. Englert, H. Walther, and M. O. Scully, *Appl. Phys. B* **54**, 366 (1991).

Quantum eraser

E. Jaynes, in *Foundation of Radiation Theory and Quantum Electronics*, ed. A. O. Barut (Plenum, New York 1980), p. 37.
M. O. Scully and K. Drühl, *Phys. Rev. A* **25**, 2208 (1982).

M. Hillery and M. O. Scully in *Quantum Optics, Experimental Gravity and Measurement Theory*, ed. P. Meystre and M. O. Scully (Plenum, New York 1983), p. 65.

Z. Y. Ou, L. J. Wang, X. Y. Zou, and L. Mandel, *Phys. Rev. A* **41**, 566 (1990).

A. G. Zajonc, L. J. Wang, X. Y. Zou, and L. Mandel, *Nature* **353**, 507 (1991).

X. Y. Zou, L. J. Wang, and L. Mandel, *Phys. Rev. Lett.* **67**, 318 (1991).

P. G. Kwiat, A. M. Steinberg, and R. Y. Chiao, *Phys. Rev. A* **45**, 7729 (1992); *ibid.* **49**, 61 (1994).

T. J. Herzog, P. G. Kwiat, H. Weinfurter, and A. Zeilinger, *Phys. Rev. Lett.* **75**, 3034 (1995).

CHAPTER 21

Two-photon interferometry, the quantum measurement problem, and more

As was demonstrated in the previous chapter, the process of observation and acquisition of information or at least the possibility of 'knowing' (whether or not we bother to 'look') can profoundly change the outcome of an experiment. For example, in the case of the micromaser *which-path* detector, we do not need to 'look at' or 'interrogate' the masers in order to lose the interference cross term; it is enough that we could have known. Experiments along these lines provide a dramatic example of the importance of which-path, or 'Welcher-Weg', information.

The present chapter treats the Welcher-Weg quantum eraser problem from a different vantage. We first consider the interference of light as it is scattered from simple atomic systems* consisting of single atoms located at two neighboring sites. From this simple model, we can gain a wealth of insight into such problems as complementarity, delayed choice, and the quantum eraser via field–field and photon–photon correlation functions, i.e., via $G^{(1)}(\mathbf{r}, t)$ and $G^{(2)}(\mathbf{r}, \mathbf{r}'; t, t')$. The chapter concludes with a demonstration that such considerations can, in principle, even lead to new kinds of high-resolution spectroscopy.

21.1 The field–field correlation function of light scattered from two atoms

In order to set the stage for our problem, consider a 'two-slit' experiment in which the slits are replaced by two two-level atoms resonant

* For further reading on atomic state Welcher-Weg detectors see Scully and Drühl [1982] and Eichmann, Bergquist, Bollinger, Gilligan, Itano, Wineland, and Raizen [1993].

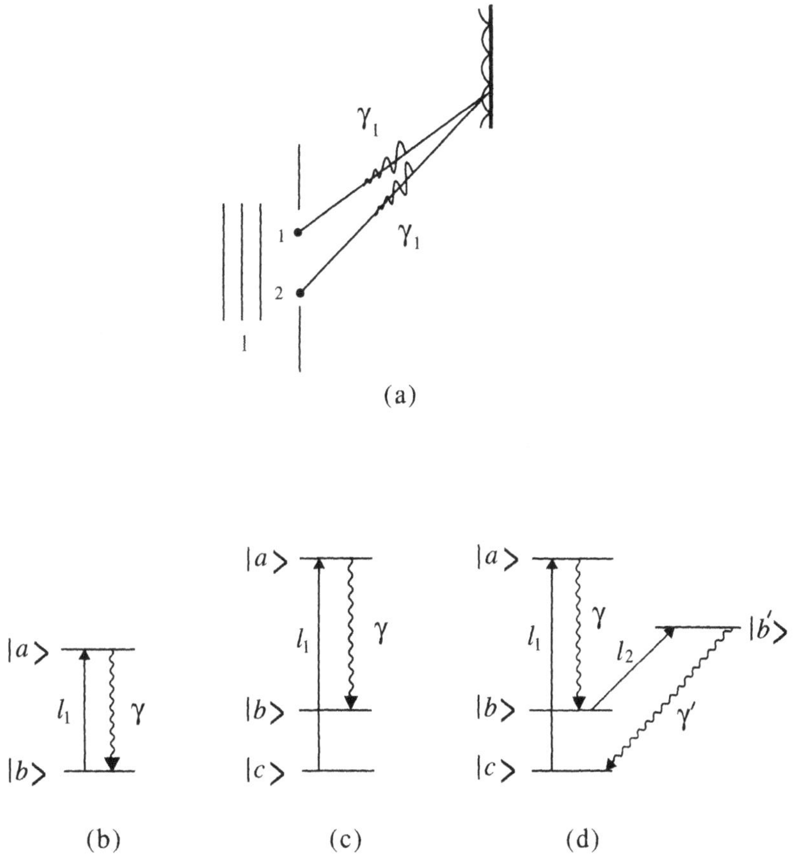

Fig. 21.1
(a) Figure depicting light impinging from the left on atoms at sites 1 and 2. Scattered photons γ_1 and γ_2 produce an interference pattern on the screen.
(b) Two-level atoms are excited by a laser pulse l_1, and emit γ-photons in the $|a\rangle \rightarrow |b\rangle$ transition.
(c) Three-level atoms excited by a pulse l_1 from $|c\rangle \rightarrow |a\rangle$ followed by emission of γ-photons in the $|a\rangle \rightarrow |b\rangle$ transition.
(d) As (c) but a second pulse l_2 takes atoms from $|b\rangle \rightarrow |b'\rangle$. Decay from $|b'\rangle \rightarrow |c\rangle$ results in emission of ϕ-photons. (From M. O. Scully and K. Drühl, *Phys. A* **25**, 2208 (1982).)

with the incident pulse l_1, as in Fig. 21.1(a). The field correlation function

$$G^{(1)}(\mathbf{r}, t) = \langle\psi|E^{(-)}(\mathbf{r}, t)E^{(+)}(\mathbf{r}, t)|\psi\rangle \tag{21.1.1}$$

describes the interference pattern associated with the scattered light.

Here the positive frequency part of the field $E^{(+)}(\mathbf{r}, t)$ is given by the usual Fourier sum involving the annihilation operators $a_{\mathbf{k}}$ as (Eq. (1.1.30))

$$E^{(+)}(\mathbf{r}, t) = \sum_{\mathbf{k}} \mathscr{E}_{\mathbf{k}} a_{\mathbf{k}} e^{i(\mathbf{k}\cdot\mathbf{r} - \nu_{\mathbf{k}} t)}, \tag{21.1.2}$$

with a corresponding expression for the negative frequency part $E^{(-)}(\mathbf{r}, t)$. The relevant portion of our atom–scattered field system is now described by a state vector of the form

$$|\psi\rangle = \frac{1}{\sqrt{2}}|b_1 b_2\rangle(|\gamma_1\rangle + |\gamma_2\rangle). \tag{21.1.3}$$

The state vector for the photon scattered from the ith atom is given by Eq. (6.3.18) as

$$|\gamma_i\rangle = \sum_{\mathbf{k}} \frac{g_{\mathbf{k}} e^{-i\mathbf{k}\cdot\mathbf{r}_i}}{(\nu_k - \omega) + i\Gamma/2} |1_{\mathbf{k}}\rangle, \tag{21.1.4}$$

where $g_{\mathbf{k}}$ is a constant depending on the strength of the atom–field coupling, ω is the atomic frequency between levels $|a\rangle$ and $|b\rangle$, Γ is the decay rate for the $|a\rangle \to |b\rangle$ transition, and $\mathbf{r}_i$ is the atomic position of the ith atom. From Eqs. (21.1.1)–(21.1.4) the correlation function for the scattered field is found to be

$$\begin{aligned} G^{(1)}(\mathbf{r},t) &= \frac{1}{2}(\langle\gamma_1| + \langle\gamma_2|)E^{(-)}(\mathbf{r},t)|0\rangle\langle 0|E^{(+)}(\mathbf{r},t)(|\gamma_1\rangle + |\gamma_2\rangle) \\ &= \frac{1}{2}|\langle 0|E^{(+)}(\mathbf{r},t)|\gamma_1\rangle + \langle 0|E^{(+)}(\mathbf{r},t)|\gamma_2\rangle|^2 \\ &= \frac{1}{2}|\Psi_1(\mathbf{r},t) + \Psi_2(\mathbf{r},t)|^2, \end{aligned} \tag{21.1.5}$$

and from Eq. (6.3.24) we have

$$\begin{aligned} \Psi_i(\mathbf{r},t) &= \langle 0|E^{(+)}(\mathbf{r},t)|\gamma_i\rangle \\ &= \frac{\mathscr{E}_0}{\Delta r_i}\Theta(t - \Delta r_i/c)e^{-(i\omega+\Gamma/2)(t-\Delta r_i/c)}, \end{aligned} \tag{21.1.6}$$

where $\Delta r_i = |\mathbf{r} - \mathbf{r}_i|$ is the distance from the ith atom to the detector. Thus we have the interference cross term

$$\begin{aligned} \Psi_1^*\Psi_2 + \text{c.c.} = &\frac{2|\mathscr{E}_0|^2}{\Delta r_1 \Delta r_2}\Theta(t - \Delta r_1/c)\Theta(t - \Delta r_2/c) \\ &\times e^{-(t-\Delta r_1/c)\Gamma/2}e^{-(t-\Delta r_2/c)\Gamma/2}\cos[k_0(\Delta r_1 - \Delta r_2)], \end{aligned} \tag{21.1.7}$$

with $k_0 = \omega/c$.

Equation (21.1.7) is just the interference pattern associated with a Young's double-slit experiment generalized to the present scattering problem. Note that when the γ_1 and γ_2 photons arrive at the detector at the 'same time', interference fringes are present. If there is no 'overlap', i.e., if the coherence time of the scattered light is too short, then there will be no interference. This is the temporal version of the frequency domain statement to the effect that incoherent light does not produce an interference pattern. This is clearly not the easiest way to get this elementary result but it is amusing none the less. Finally, we note that an operator approach to the present problem is given in the Appendix 21.A. The operator approach is useful when dealing with more complicated problems.

21.1.1 Correlation function $G^{(1)}(\mathbf{r}, t)$ generated by scattering from two excited atoms

We now use our operator formalism to find out whether two excited atoms, again at $\mathbf{r}_1$ and $\mathbf{r}_2$, will produce an interference pattern, see Fig. 21.1(a). We assume as before that the atoms are far enough apart that cooperative effects can be ignored. We start at $t = 0$ with the atoms in the state $|a_1 a_2\rangle|0\rangle$. For times $t \gg 1/\Gamma$ the state of the system is

$$|\psi\rangle = |b_1 b_2\rangle|\gamma_1 \gamma_2\rangle, \tag{21.1.8}$$

and the correlation function is now

$$\begin{aligned}
&\langle\psi|E^{(-)}(\mathbf{r}, t)E^{(+)}(\mathbf{r}, t)|\psi\rangle \\
&= \langle\gamma_1\gamma_2|E^{(-)}(\mathbf{r}, t)E^{(+)}(\mathbf{r}, t)|\gamma_1\gamma_2\rangle \\
&= |\Psi_1(\mathbf{r}, t)|^2 + |\Psi_2(\mathbf{r}, t)|^2 + [\Psi_1^*(\mathbf{r}, t)\Psi_2(\mathbf{r}, t)\langle\gamma_2|\gamma_1\rangle + \text{c.c.}].
\end{aligned} \tag{21.1.9}$$

We see, then, that the interference pattern, which comes from the last term in Eq. (21.1.9), is proportional to $\langle\gamma_2|\gamma_1\rangle$. It follows from the definition of $|\gamma_i\rangle$ in Eq. (21.1.4), that (Problem 21.2)

$$\langle\gamma_2|\gamma_1\rangle = \frac{\sin(k_0|\mathbf{r}_2 - \mathbf{r}_1|)}{k_0|\mathbf{r}_2 - \mathbf{r}_1|} e^{-|\mathbf{r}_2 - \mathbf{r}_1|\Gamma/(2c)}. \tag{21.1.10}$$

Therefore, if $|\mathbf{r}_1 - \mathbf{r}_2| \gg \lambda$, then $|\langle\gamma_2|\gamma_1\rangle| \ll 1$ and the interference pattern will be very weak.

21.1.2 Excitation by laser light

Now we look at what would happen if we do a two-atom interference experiment using a laser pulse. We will assume that the pulse is strong enough that we can treat it classically and that its duration is much shorter than the decay time of the atoms. This means that we can assume that the pulse, essentially instantaneously, puts the atoms in some superposition of ground and excited states. To be more specific, before the pulse hits one of the atoms, it is in the state $|b\rangle$. Immediately after the pulse has passed, it is in the state $c_a|a\rangle + c_b|b\rangle$ where $|c_a|^2 + |c_b|^2 = 1$. Therefore, at $t = 0$ (just after the pulse has passed) the two-atom system (we assume that the pulse hits both atoms) is in the state

$$\begin{aligned}
|\psi\rangle &= (c_a|a_1\rangle + c_b|b_1\rangle)(c_a|a_2\rangle + c_b|b_2\rangle)|0\rangle \\
&= (c_a^2|a_1 a_2\rangle + c_a c_b|a_1 b_2\rangle + c_b c_a|b_1 a_2\rangle + c_b^2|b_1 b_2\rangle)|0\rangle.
\end{aligned} \tag{21.1.11}$$

After a time $t \gg 1/\Gamma$, the state (21.1.11) becomes

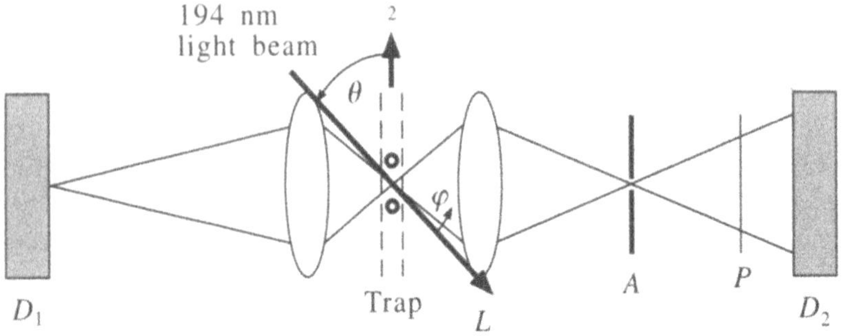

Fig. 21.2 Schematic diagram of the experiment. Scattered light from two ions (represented by small circles near the center of the trap) is imaged onto detector D_2 via the collecting lens L, aperture A, and an optional polarizer P. Detector D_1 serves as a monitor of ion number. (From U. Eichmann, J. C. Bergquist, J. J. Bollinger, J. M. Gilligan, W. M. Itano, D. J. Wineland, and M. G. Raizen, *Phys. Rev. Lett.* **70**, 2359 (1993).)

$$|\psi\rangle = (c_a^2|\gamma_1\gamma_2\rangle + c_a c_b|\gamma_1\rangle + c_b c_a|\gamma_2\rangle + c_b^2|0\rangle) . \tag{21.1.12}$$

We now want to compute the correlation function. If we assume that the atoms are far enough apart that $\langle\gamma_2|\gamma_1\rangle \cong 0$, we have

$$\begin{aligned} &\langle\psi|E^{(-)}(\mathbf{r},t)E^{(+)}(\mathbf{r},t)|\psi\rangle \\ &= \left[|\Psi_1(\mathbf{r},t)|^2 + |\Psi_2(\mathbf{r},t)|^2\right](|c_a|^4 + |c_a c_b|^2) \\ &\quad + 2|c_a c_b|^2 \mathrm{Re}\left[\Psi_1^*(\mathbf{r},t)\Psi_2(\mathbf{r},t)\right] . \end{aligned} \tag{21.1.13}$$

The last term in Eq. (21.1.13) is the interference term and is a result of the terms in Eq. (21.1.11) in which only one of the two atoms is excited. If we want to maximize the size of the interference pattern how should we choose the incoming pulse? The interference term is proportional to $|c_a c_b|^2 = |c_a|^2(1 - |c_a|^2)$. This expression is maximal when $|c_a|^2 = 1/2$ so that we should choose a pulse which will produce an atomic state in which $|c_a| = |c_b|$. A pulse which will do this is known as a $(\pi/2)$-pulse. Note that a π-pulse, which would invert both atoms, would lead to no interference pattern at all. In recent experiments such two-atom interference has been observed. In these experiments, two $^{198}Hg^+$ ions were localized in a Paul trap and irradiated by resonant laser light at 194 nm tuned to the $6s^2S_{1/2} \rightarrow 6p^2P_{1/2}$ transition, see Fig. 21.2. The scattered light showed the interference pattern predicted by Eq. (21.1.13).

21.1.3 Using three atomic levels as a which-path flag

Next let us alter our 'experiment' so as to replace the two-level atoms in Fig. 21.1(b) by atoms having three levels as in Fig. 21.1(c). Our atoms are now excited to $|a\rangle$ by the incident laser pulse l_1, and then decay to $|b\rangle$ or $|c\rangle$ via γ-photon emission. Let us now arrange our detection system so that it is sensitive only to radiation emitted in the $|a\rangle \rightarrow |b\rangle$ transition, i.e., we ignore radiation from the $|a\rangle \rightarrow |c\rangle$ transition. We wish to again consider the scattered field correlation function just as in the previous experiment. At first glance, one might think that fringes would again be observed since the atomic configuration of Fig. 21.1(c) is not that different from that of Fig. 21.1(b). However, a little reflection will suffice to convince oneself that this is not true. We need only look to see which atom (1 or 2) is in the $|b\rangle$ state in order to determine which atom did the scattering. Now, if we know (or could know) which source (slit or atom) the light came from we could expect the interference fringes to disappear. Detailed calculation bears out this expectation as shown below. The state of the system describing the coupled atom–field system of Fig. 21.1(c) is now

$$|\psi_1\rangle = \frac{1}{\sqrt{2}}\left(|b_1c_2\rangle|\gamma_1\rangle + |c_1b_2\rangle|\gamma_2\rangle\right) , \tag{21.1.14}$$

and the field correlation function implied by this state vector is given by

$$\begin{aligned} G^{(1)}(\mathbf{r},t) &= \frac{1}{2}\Big(|\Psi_1(\mathbf{r},t)|^2 + |\Psi_2(\mathbf{r},t)|^2 \\ &\qquad + \left[\Psi_1^*(\mathbf{r},t)\Psi_2(\mathbf{r},t)\langle b_1c_2|c_1b_2\rangle + \text{c.c.}\right]\Big) \\ &= \frac{1}{2}\left(|\Psi_1(\mathbf{r},t)|^2 + |\Psi_2(\mathbf{r},t)|^2\right) . \end{aligned} \tag{21.1.15}$$

From Eq. (21.1.15) we see that the interference terms have disappeared, since the states $|b\rangle$ and $|c\rangle$ are orthogonal, in accord with our intuitive notion as discussed earlier. The utility of such a simple *which-path* setup has also been confirmed by Eichmann et al.* When they set up their experiment so that $|a\rangle \to |b\rangle$ decay, see Fig. 21.1(c), took place with σ-polarized light they saw no interference. However, when they looked at the light corresponding to the $|a\rangle \to |c\rangle$ transition of Fig. 21.1(c), which involves π-polarized light, they did see interference. This is summarized in Fig. 21.3.

21.2 The field–field and photon–photon correlations of light scattered from two multi-level atoms: quantum eraser

As we saw in Section 21.1.3, the three-level atoms of Fig. 21.1(c) provide possible which-path information, which leads to the loss of interference. Specifically, the interference terms in the field–field correlation function of Eq. (21.1.15) are multiplied by the atomic innerproduct

* Phys. Rev. Lett. **70**, 2359 (1993), this classic paper is recommended reading.

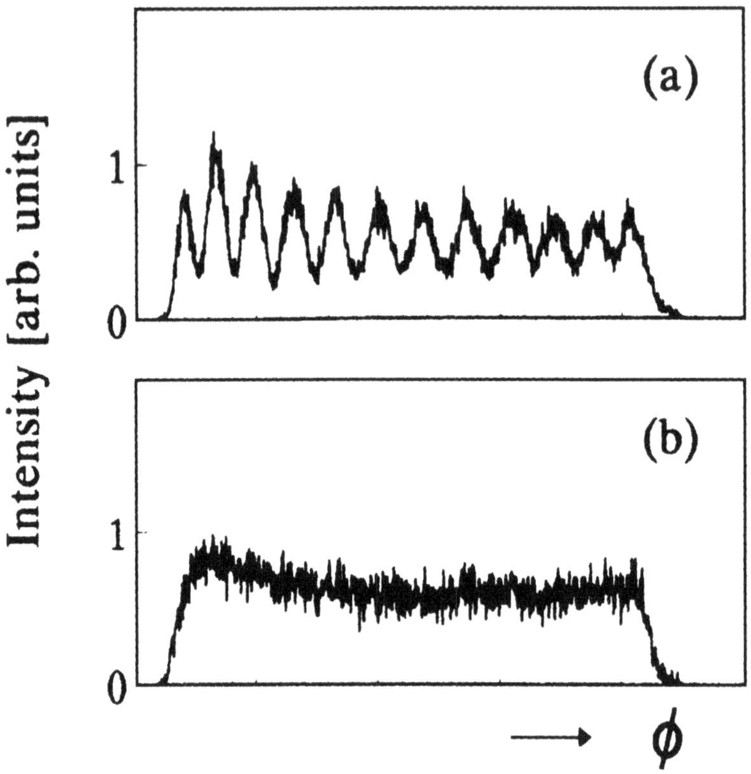

Fig. 21.3 Polarization-sensitive detection of the scattered light (unnormalized): (a) π-polarized scattered light, showing interference; (b) σ-polarized scattered light, showing no interference pattern. (From U. Eichmann, J. C. Bergquist, J. J. Bollinger, J. M. Gilligan, W. M. Itano, D. J. Wineland, and M. G. Raizen, *Phys. Rev. Lett.* **70**, 2359 (1993).)

$\langle b_1c_2|c_1b_2\rangle$ and therefore vanish. We are therefore led to ask:* what if we could 'force' the atoms from their $|b\rangle$ state to the $|c\rangle$ state so that they are no longer orthogonal? If we could do this, then we would have an interesting situation. That is, the γ-photons could be well on their way to the detector (i.e., far removed from atoms 1 and 2) and fringes made to appear or not depending on what we do with the atoms long after the γ-emission has taken place.

Suppose we couple the $|b\rangle$ state to the $|c\rangle$ state via an appropriate laser or microwave field. Then the unitary time evolution of the $|b\rangle \rightarrow |c\rangle$ transition on atoms 1 and 2 implies

$$\langle b_1c_2|c_1b_2\rangle \rightarrow \langle b_1c_2|U^\dagger U|c_1b_2\rangle. \tag{21.2.1}$$

Thus if our time-evolution matrix U is unitary, $U^\dagger U = 1$, and we see that we have *not* succeeded in producing fringes by applying the second pulse. Yet one wonders if some other scheme designed to retrieve the interference fringes might work. After all, the presence of the information contained in our three-level atoms is analogous

* We note that this presentation of the quantum erasure concept preceeded that presented in Chapter 20, see Scully and Drühl [1982], and is here presented in a "stand alone" fashion.

to having information stored in the form of an observation, and we know that the process of observation changes the state vector in a nonunitary fashion. More pictorially, the question may well be asked: can we *erase* the information (memory) locked in our atoms and thus recover fringes? Motivated by these considerations, let us consider the following information eraser: allow our atoms 1 and 2 to take on a slightly more involved level structure involving four relevant levels as depicted in Fig. 21.1(d). The second laser pulse l_2 is tuned so as to be resonant with the $|b\rangle \rightarrow |b'\rangle$ transition and tailored such that it transfers 100 percent of the population from $|b\rangle$ to $|b'\rangle$. That is, we let the second laser pulse be a π-pulse. Such a pulse is defined by the requirement that the integrated amplitude of the laser pulse envelope be such that

$$\frac{\wp_{bb'}}{\hbar}\int_{-\infty}^{\infty} dt'\mathscr{E}(t') = \pi \tag{21.2.2}$$

where $\wp_{bb'}$ is the dipole matrix element connecting the $|b\rangle$ and $|b'\rangle$ states. Such a π-pulse will take every atom it encounters in $|b\rangle$ to $|b'\rangle$. Hence, the state of the system after interacting with the l_2 pulse is

$$|\psi_2\rangle = \frac{1}{\sqrt{2}}\left(|b_1'c_2\rangle|\gamma_1\rangle + |c_1b_2'\rangle|\gamma_2\rangle\right) . \tag{21.2.3}$$

However, as indicated in Fig. 21.1(d), $|b'\rangle$ is strongly coupled to $|c\rangle$, so that after a short time we may be sure that the ith atom has decayed to the $|c\rangle$ state via the emission of a photon which we designate as $|\phi_i\rangle$. The state $|\phi\rangle$ is the same as that of the $|\gamma\rangle$-photon state, with the obvious changes in wave vector, decay rate, etc. The state vector after ϕ-emission now reads

$$|\psi_3\rangle = \frac{1}{\sqrt{2}}\left(|c_1c_2\rangle(|\phi_1\gamma_1\rangle + |\phi_2\gamma_2\rangle)\right) . \tag{21.2.4}$$

Consider next an experimental arrangement which, in effect, allows us to 'reduce' the photon states $|\phi_1\rangle$ and $|\phi_2\rangle$ to the vacuum with the excitation of a common photodetector. This is shown in Fig. 21.4. The photodetection of a ϕ-photon (at $\mathbf{r}', t'$) followed by detection of γ-radiation (at $\mathbf{r}, t$) is described by the intensity correlation function

$$G^{(2)}(\mathbf{r},\mathbf{r}';t,t') = \langle\psi|E^{(-)}(\mathbf{r},t)E^{(-)}(\mathbf{r}',t')E^{(+)}(\mathbf{r}',t')E^{(+)}(\mathbf{r},t)|\psi\rangle, \tag{21.2.5}$$

where $|\psi\rangle$ is given by Eq. (21.2.4). Let us simplify matters at this point by considering our atoms to be of the type given in Fig. 21.1(d) and we assume that a cascade of $|a\rangle \rightarrow |b\rangle$ takes place with the emission of a γ-photon and $|b\rangle \rightarrow |c\rangle$ with the emission of a ϕ-photon.

That is, we can think of the coupling to level $|b'\rangle$ as being always 'on' and the state we are calling $|b\rangle$ is really $|b'\rangle$ or we can think of the initial $|c\rangle \to |a\rangle$ excitation as being driven by a two-photon excitation in which case both $|a\rangle \to |b\rangle$ and $|b\rangle \to |c\rangle$ are allowed transitions. We proceed to calculate $G^{(2)}(\mathbf{r}, \mathbf{r}'; t, t')$ by defining the ϕ-photon states

$$|\phi_i\rangle = \sum_{\mathbf{k}} \frac{g_{\mathbf{k}} e^{-i\mathbf{k}\cdot\mathbf{r}_i}}{(\nu_k - \omega_{bc}) + i\Gamma_{bc}/2} |1_{\mathbf{k}}\rangle \qquad (i = 1, 2), \tag{21.2.6}$$

as in the case of the γ-photons. We now have the ϕ-photon probability amplitude

$$\langle 0|E^{(+)}(\mathbf{r}, t)|\phi_i\rangle = \Psi_{\phi,i}(\mathbf{r}, t), \tag{21.2.7}$$

where here, and hence forth, we attach the subscript ϕ or γ to denote which photon and the subscript i to tell us which atom. The γ-photon counterpart to (21.2.7) reads

$$\langle 0|E^{(+)}(\mathbf{r}, t)|\gamma_i\rangle = \Psi_{\gamma,i}(\mathbf{r}, t). \tag{21.2.8}$$

We therefore find the two-photon correlation function (21.2.5) to be

$$\frac{1}{2}\left(\langle\gamma_1\phi_1| + \langle\gamma_2\phi_2|\right) E^{(-)}E'^{(-)}E'^{(+)}E^{(+)} \left(|\gamma_1\phi_1\rangle + |\gamma_2\phi_2\rangle\right)$$
$$= |\Psi_{\gamma,1}(\mathbf{r}, t)\Psi_{\phi,1}(\mathbf{r}', t') + \Psi_{\gamma,2}(\mathbf{r}, t)\Psi_{\phi,2}(\mathbf{r}', t')|^2. \tag{21.2.9}$$

Here again the cross terms produce an interference pattern. We would not have the interference effect if we only measured the γ-photon. The measurement of the ϕ-photon is necessary to erase the information about the path of the γ-photon (i.e., whether it is scattered from atom 1 or atom 2).

21.2.1 *Alternative photon basis*

A significant simplification can be achieved by going to a different basis. To this end, we rewrite our state (21.2.3) in terms of symmetric and antisymmetric combinations. That is, if we define the photon states

$$|\chi_\pm\rangle = \frac{1}{\sqrt{2}}(|\chi_1\rangle \pm |\chi_2\rangle), \tag{21.2.10}$$

where χ is either ϕ or γ, then Eq. (21.2.4) may be written as

$$|\psi\rangle = \frac{1}{\sqrt{2}}\left[|c_1c_2\rangle(|\phi_+\gamma_+\rangle + |\phi_-\gamma_-\rangle)\right]. \tag{21.2.11}$$

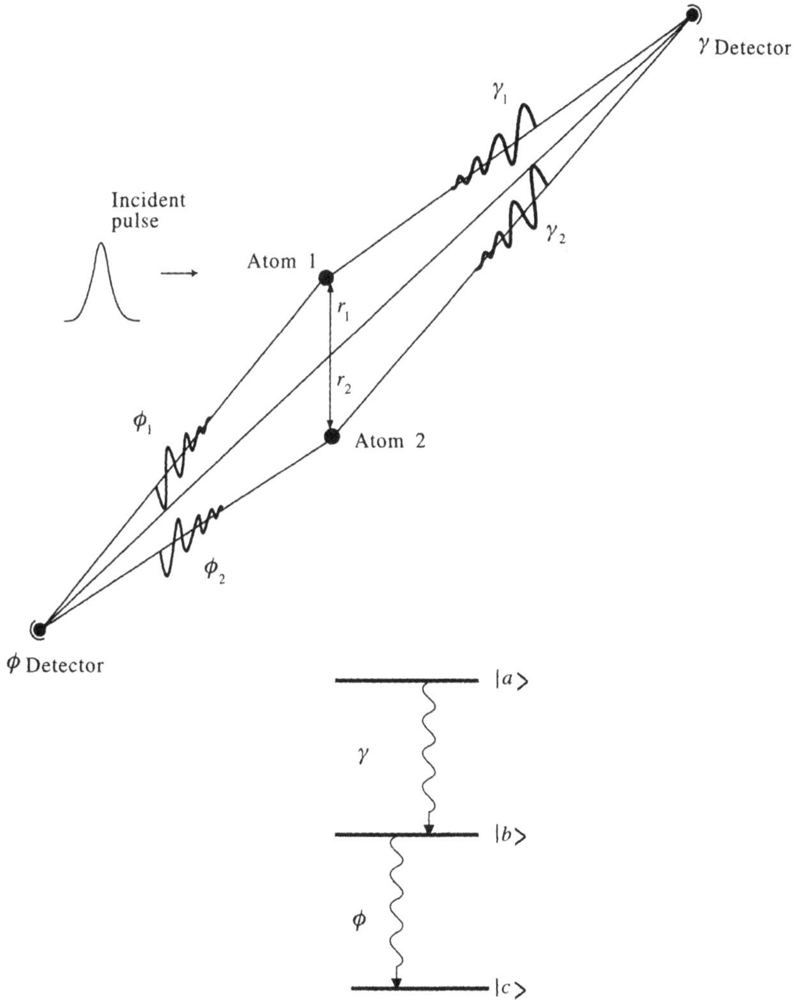

Fig. 21.4
Schematic of a two-atom correlation experiment.

However, if we take $|\mathbf{r}-\mathbf{r}_1| = |\mathbf{r}-\mathbf{r}_2|$, then we find

$$\begin{aligned}\langle 0|E^{(+)}(\mathbf{r},t)|\phi_-\rangle &= \frac{1}{\sqrt{2}}\left(\Psi_{\phi,1}(\mathbf{r},t)-\Psi_{\phi,2}(\mathbf{r},t)\right)\\ &= 0,\end{aligned}\tag{21.2.12}$$

since from Eq. (21.1.6) we see that $\Psi_{\phi,i}(\mathbf{r},t)$ depends only on $\Delta r_i = |\mathbf{r}-\mathbf{r}_i|$ and for the symmetric location $\Psi_{\phi,1} = \Psi_{\phi,2}$. Since all terms like $\langle 0|E^{(+)}(\mathbf{r},t)|\phi_-\rangle$ vanish, we can write the correlation function in the $\pm$ basis as

$$
\begin{aligned}
&G^{(2)}(\mathbf{r},\mathbf{r}';t,t') \\
&= \frac{1}{2}(\langle\gamma_1\phi_1| + \langle\gamma_2\phi_2|)E^{(-)}(\mathbf{r},t)E^{(-)}(\mathbf{r}',t') \\
&\qquad\qquad \times E^{(+)}(\mathbf{r}',t')E^{(+)}(\mathbf{r},t)(|\gamma_1\phi_1\rangle + |\gamma_2\phi_2\rangle) \\
&= \frac{1}{2}(\langle\gamma_+\phi_+| + \langle\gamma_-\phi_-|)E^{(-)}(\mathbf{r},t)E^{(-)}(\mathbf{r}',t') \\
&\qquad\qquad \times E^{(+)}(\mathbf{r}',t')E^{(+)}(\mathbf{r},t)(|\gamma_+\phi_+\rangle + |\gamma_-\phi_-\rangle) \\
&= \frac{1}{2}\langle\gamma_+\phi_+|E^{(-)}(\mathbf{r},t)E^{(-)}(\mathbf{r}',t')E^{(+)}(\mathbf{r}',t')E^{(+)}(\mathbf{r},t)|\gamma_+\phi_+\rangle.
\end{aligned}
\tag{21.2.13}
$$

We see that only the $|\phi_+\rangle$ states contribute. This means that, in a given ensemble of scattering events, the $|\phi_+\rangle$-half of the scattering events are expected to lead to a count, while the $|\phi_-\rangle$-half will lead to no count. By keeping only those events which lead to a ϕ-photon count, interference fringes are found in the statistical distribution of γ-photon counts on the observation screen. If on the other hand, we choose not to read our ϕ-photon counter and keep all scattering events, no interference pattern will be found in the complete ensemble of all γ-photon counts. Hence in our experiment the total ensemble of scattering events is decomposed into two subensembles showing interference fringes and 'antifringes'.

21.3 Bell's inequality experiments via two-photon correlations

A two-atom correlation experiment of the type shown in Fig. 21.4 has the basic ingredients necessary for testing Bell's inequality. We recall that the correlated state resulting from two-photon emission by atoms at the sites 1 and 2 is

$$
|\psi\rangle = \frac{1}{\sqrt{2}}\left(|\gamma_1\phi_1\rangle + |\gamma_2\phi_2\rangle\right) , \tag{21.3.1}
$$

and the resulting joint count probability $P_{\gamma\phi}$ for detection of a photon in both the γ- and ϕ-counters can be calculated via the second-order correlation function.

That the present configuration could provide an experimental test of Bell's theorem* is most easily established by making contact with

* A simple discussion of Bell's theorem and two-photon correlations in this context is given by Scully [1981].

the polarization correlation experiments discussed in Section 18.3. In these photon cascade experiments, a 'polarization correlated photon state' such as

$$|\psi\rangle = \frac{1}{\sqrt{2}}\left(|\gamma_x\phi_x\rangle + |\gamma_y\phi_y\rangle\right) , \tag{21.3.2}$$

is generated, see Fig. 18.4. The joint count distribution $P_{\gamma\phi}$ as a function of the angle between the analyzers 1 and 2 is then measured. It is clear that there is a direct correspondence between the two problems and that the analysis of the atomic cascade experiment would apply, with obvious changes, to the present problem involving two atoms. Thus, correlations involving the two-photon state given by Eq. (21.3.1) could provide a new test of hidden variable theories via Bell's theorem.

The two-photon states are generated in a parametric down-conversion process. We recall from the discussion in Chapter 16 that, in a nondegenerate parametric amplification process, the interaction of a pump wave of frequency ν_p with a nonlinear $\chi^{(2)}$ crystal generates signal and idler waves of frequencies ν_s and ν_i, respectively, such that $\nu_p = \nu_s + \nu_i$. In the experiments to test Bell's inequality, the correlation measurements of mixed signal and idler photons are performed as a function of the two linear polarizer settings.

Shih and Alley* carried out an experiment based on the production of a correlated two-photon state produced by down-conversion in a nonlinear crystal. In this experiment, 266 nm radiation was sent through a deutrated potassium dihydrogen phosphate (KD*P) nonlinear optical crystal to produce correlated photon pairs as in Fig. 21.5. Each photon was converted into a definite polarization eigenstate by inserting quarter wave ($\lambda/4$) plates into the paths A and B to transform the linear-polarization states into circular-polarization states. The photons were superposed when they met at the beam-splitter (BS). In this way, using coincidence detection, polarizations of the EPR type were observed and Bell's inequality was shown to be violated.

In a similar experiment, Ou and Mandel† showed a clear violation of Bell's inequality, and showed that the classical probability relations for waves are violated as well. As shown in Fig. 21.6, uv pump light falls on a nonlinear crystal of potassium dihydrogen phosphate (KDP) and the signal and the idler photons are produced in the down-conversion process. When the phase matching condition for the degenerate parametric process ($\nu_s = \nu_i$) is satisfied, the signal and

* Shih and Alley [1988]; see also Franson [1989].
† Ou and Mandel [1988].

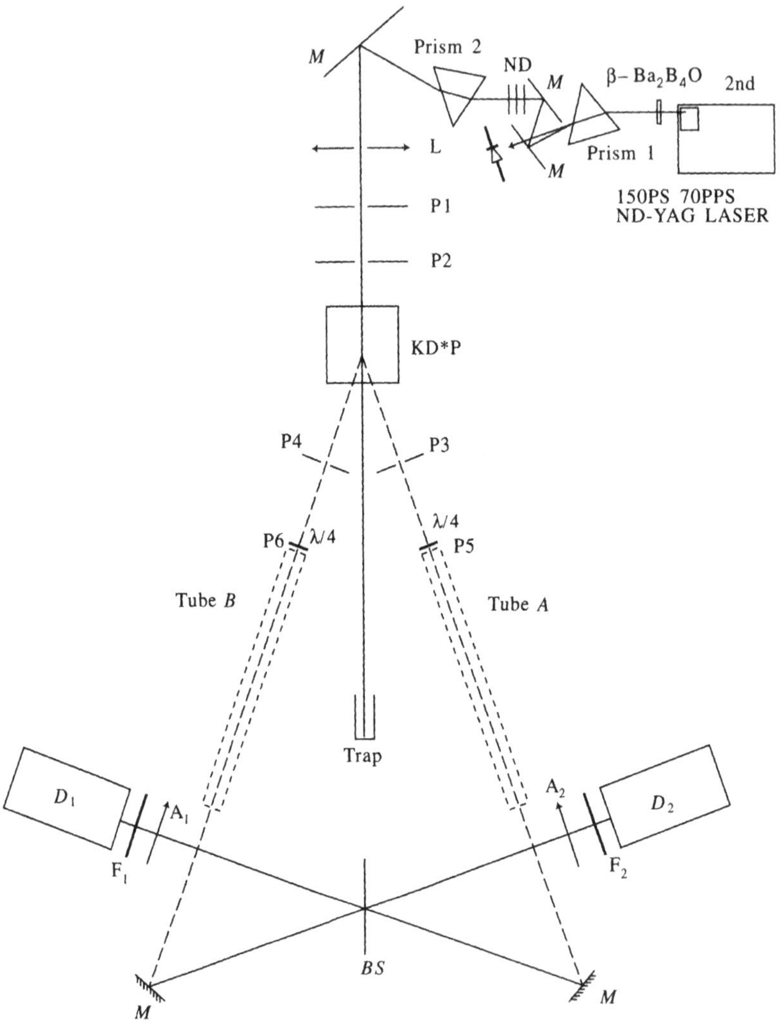

Fig. 21.5
Outline of the experiment to test Bell's inequality. (From Y. H. Shih and C. O. Alley, *Phys. Rev. Lett.* **61**, 2921 (1988).)

idler photons emerge at small angles relative to the pump beam. The signal and idler photons are linearly polarized with their electric field vectors in the plane of the diagram. The idler photons pass through a 90° polarization rotator whereas the signal photons pass through a compensating glass plate C_1 to produce equal time delay. The two beams are then incident on a beam-splitter (BS) from opposite directions which results in the two beams consisting of mixed signal and idler photons. These beams pass through linear polarizers set at adjustable angles θ_1 and θ_2 and finally fall on two photodetectors which are connected to coincidence counting electronics. As before, the coincidence counting provides a measure of the joint probability

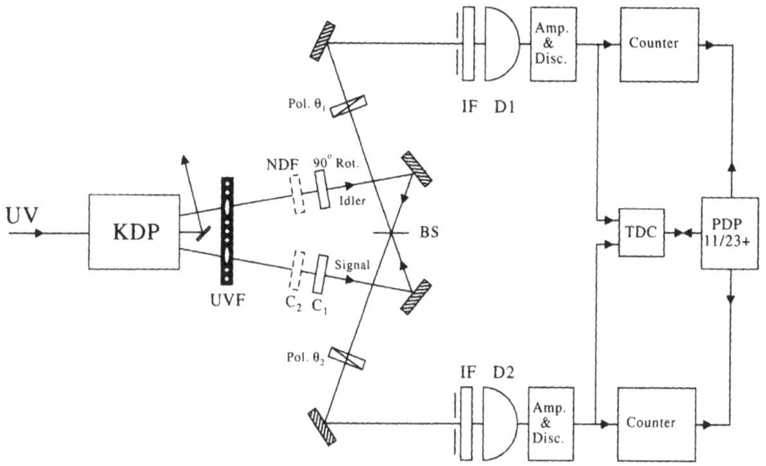

Fig. 21.6
Outline of the experiment to test Bell's inequality. (From Z. Y. Ou and L. Mandel, *Phys. Rev. Lett.* **61**, 50 (1988).)

$P_{12}(\theta_1, \theta_2)$ of detecting two photons for various settings θ_1 and θ_2 of the two linear polarizers.

21.4 Two-photon cascade interferometry

In Section 21.2, we considered two-photon emission from two separated atoms in the context of a quantum eraser gedanken experiment. In so doing, we were naturally led to a field state corresponding to interrupted emission and the first and second photons were delayed relative to one another. As will be shown in this section, this has much in common with the problem of two-photon cascade emission of Section 6.4. In particular, the problem of interrupted two-photon cascade emission and the general relation of two-photon cascade emission, as determined by Eq. (6.4.13), yield the same $\Psi^{(2)}$ function when the intermediate level is much longer lived than the upper level.

In the present section, we consider the two-photon correlation function associated with two-photon cascade emission and apply the results to the Franson–Chiao interferometer.

21.4.1 Two-photon correlations produced by atomic cascade emission

The radiation field produced by interrupted cascade emission from a single atom, as in Section 21.2, takes the form

$$|\psi\rangle = |\gamma\phi\rangle \,, \tag{21.4.1}$$

where $|\gamma\rangle$ and $|\phi\rangle$ are given by expressions of the form (21.1.4). Then from Eq. (21.2.9), we see that the photon–photon correlation function may be written as

$$G^{(2)}(\mathbf{r},\mathbf{r}';t,t') = \Psi^{(2)*}(\mathbf{r},t;\mathbf{r}',t')\Psi^{(2)}(\mathbf{r},t;\mathbf{r}',t') \tag{21.4.2}$$

with

$$\Psi^{(2)}(\mathbf{r},t;\mathbf{r}',t') = \Psi_\gamma(\mathbf{r},t)\Psi_\phi(\mathbf{r}',t') + \Psi_\gamma(\mathbf{r}',t')\Psi_\phi(\mathbf{r},t), \tag{21.4.3}$$

in which Ψ_γ and Ψ_ϕ are the amplitudes for γ- and ϕ-photons exciting the detectors at $\mathbf{r}$ at time t and at $\mathbf{r}'$ at time t'.

We now consider the photon correlation problem in which the three-level atoms have a sharp upper level and a rapidly decaying intermediate state (see Fig. 21.7). As we shall see, this atomic configuration is in contrast to that of the quantum eraser problem, in which the upper state is broad and the intermediate state is sharp. In the case of Fig. 21.7 (a long lived $|a\rangle$ level with a rapidly decaying $|b\rangle$ level), the state of the radiation field can no longer be written in the simple factorized form of the quantum eraser problem but takes the form

$$|\psi\rangle = \sum_{\mathbf{k},\mathbf{q}} c_{\mathbf{k},\mathbf{q}}|1_\mathbf{k}1_\mathbf{q}\rangle, \tag{21.4.4}$$

where $c_{\mathbf{k},\mathbf{q}}$ is given by Eq. (6.4.12). The key point is that the energy $\epsilon_a - \epsilon_c = \hbar\omega_\mathbf{k} + \hbar\omega_\mathbf{q}$ so that the $\mathbf{k}$ and $\mathbf{q}$ photons are correlated. Hence a factorized state vector of the form (21.4.1) is not expected. As before, we seek the two-photon correlation function

$$\begin{aligned} &G^{(2)}(\mathbf{r}_1,\mathbf{r}_2;t_1,t_2) \\ &= \langle\psi|E^{(-)}(\mathbf{r}_1,t_1)E^{(-)}(\mathbf{r}_2,t_2)E^{(+)}(\mathbf{r}_1,t_1)E^{(+)}(\mathbf{r}_2,t_2)|\psi\rangle \end{aligned}$$

corresponding to the two detectors at the points $\mathbf{r}_1$ and $\mathbf{r}_2$. The interaction with the photon field, described by $|\psi\rangle$, is switched on at times t_1 and t_2, respectively. We show in the Appendix 21.B that $G^{(2)}$ is now governed by the double photoexciting probability amplitude

$$\begin{aligned} &\Psi^{(2)}(\mathbf{r}_1,t_1;\mathbf{r}_2,t_2) \\ &= \frac{-\kappa}{\Delta r_1 \Delta r_2} \exp\left[-(i\omega_{ac} + \Gamma_a/2)\left(t_1 - \frac{\Delta r_1}{c}\right)\right] \Theta\left(t_1 - \frac{\Delta r_1}{c}\right) \\ &\times \exp\left\{-(i\omega_{bc} + \Gamma_b/2)\left[\left(t_2 - \frac{\Delta r_2}{c}\right) - \left(t_1 - \frac{\Delta r_1}{c}\right)\right]\right\} \\ &\times\Theta\left[\left(t_2 - \frac{\Delta r_2}{c}\right) - \left(t_1 - \frac{\Delta r_1}{c}\right)\right] + (1 \leftrightarrow 2), \end{aligned} \tag{21.4.5}$$

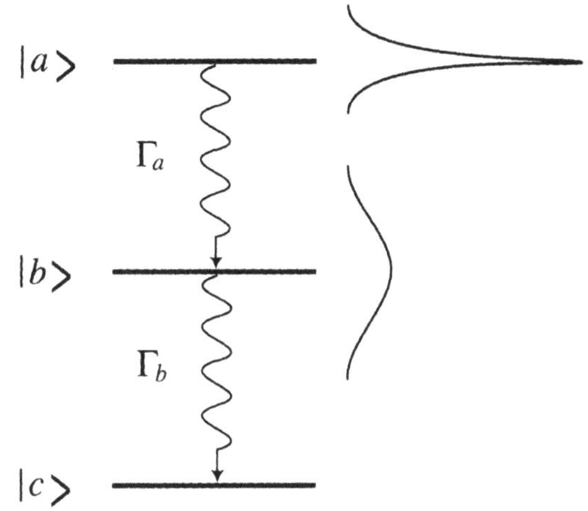

Fig. 21.7
The radiative broadening of the upper and intermediate levels of a three-level atom in the cascade configuration, depicted schematically for the case $\Gamma_b \gg \Gamma_a$.

where κ is an uninteresting constant at $\Delta r_i = |\mathbf{r}_i - \mathbf{r}|$. We emphasize again that $\Psi^{(2)}$ is not of the simple factorized form of Eq. (21.4.3). We also note that Eq. (21.4.5) is generally valid for all values of Γ_a and Γ_b. Since $\Psi^{(2)}(\mathbf{r}_1, \mathbf{r}_2; t_1, t_2)$ is nonzero only for times $t_1 = \Delta r_1/c + O(\Gamma_a^{-1})$, we can write Eq. (21.4.5) as (with $\Gamma_a \gg \Gamma_b$)

$$
\begin{aligned}
&\Psi^{(2)}(\mathbf{r}_1, t_1; \mathbf{r}_2, t_2)\\
&= \frac{-\kappa}{\Delta r_1 \Delta r_2} e^{-(i\omega_{ab}+\Gamma_a/2-\Gamma_b/2)\left(t_1-\frac{\Delta r_1}{c}\right)} \Theta\left(t_1 - \frac{\Delta r_1}{c}\right)\\
&\quad \times e^{-(i\omega_{bc}+\Gamma_b/2)\left(t_2-\frac{\Delta r_2}{c}\right)} \Theta\left[t_2 - \frac{\Delta r_2}{c} + O(\Gamma_a^{-1})\right]\\
&\quad +(1 \leftrightarrow 2)\\
&\cong \frac{-\kappa}{\Delta r_1 \Delta r_2} e^{-(i\omega_{ab}+\Gamma_a/2)(t_1-\frac{\Delta r_1}{c})} \Theta\left(t_1 - \frac{\Delta r_1}{c}\right)\\
&\quad \times e^{-(i\omega_{bc}+\Gamma_b/2)\left(t_2-\frac{\Delta r_2}{c}\right)} \Theta\left(t_2 - \frac{\Delta r_2}{c}\right) + (1 \leftrightarrow 2), \qquad (21.4.6)
\end{aligned}
$$

i.e., in the limit $\Gamma_a \gg \Gamma_b$, Eq. (21.4.5) reduces to a factorized form. This is intuitively reasonable since the interrupted cascade problem is essentially one in which the $|b\rangle$ state is 'long lived' and therefore we are pleased to see that $\Psi^{(2)}$ as given by Eq. (21.4.6) is of the form of Eq. (21.4.3).

21.4.2 *Franson–Chiao interferometry*

We now turn to an intriguing example of two-photon correlation interferometry first suggested by Franson* and realized by Chiao†

* Franson [1989].

† Kwiat, Steinberg, and Chiao [1992].

and co-workers. The setup is depicted in Fig. 21.8. There we see a two-photon interferometer with a single excited three-level atom as the light source. Upon decaying from the upper state $|a\rangle$ to an intermediate level $|b\rangle$ with a rate Γ_a and from $|b\rangle$ to the ground state $|c\rangle$ with a rate Γ_b, the atom emits two photons γ and ϕ, respectively. The radiation travels from the atom to the detectors D_1 and D_2 via short and long optical paths of lengths S and L as determined by the beamsplitters BS and mirrors M. Now the coherence length of the emitted light is much less than the difference in path lengths in the two arms of the Mach–Zehnder interferometer (BS–M–M–BS and BS–BS in Fig. 21.8). One might therefore expect that there would be no interference fringes observed, which is indeed the case if we look at either detector alone. That is, there is no $G^{(1)}$ coherence. However, if we look at the correlation of photon counts in detectors D_1 and D_2, i.e., if we look at $G^{(2)}$, then we do find interference. We proceed to analyze this problem by modeling the action of the beamsplitters and mirrors via a Heisenberg treatment of the operators

$$E^{(+)}(\mathbf{r}_i, t_i) = \frac{1}{2}[E^{(+)}(S_i, t_i) + E^{(+)}(L_i, t_i)], \tag{21.4.7}$$

where the short and long paths are described by S_i and L_i, respectively. Using the two-photon state $|\Psi\rangle$ given by Eq. (6.4.13), we may write $G^{(2)}$ in terms of the pair probability amplitude

$$G^{(2)}(1,2) = \Psi^*(1,2)\Psi(1,2), \tag{21.4.8}$$

where

$$\Psi(1,2) = \frac{1}{4}\langle 0|[E^{(+)}(S_2, t_2) + E^{(+)}(L_2, t_2)] \\ [E^{(+)}(S_1, t_1) + E^{(+)}(L_1, t_1)]|\psi\rangle. \tag{21.4.9}$$

For simplicity, we focus on the situation in which the detectors are located at $\mathbf{r}_1 = -\mathbf{r}_2$. Then we have

$$\Psi(1,2) = \Psi(S_1, t_1; S_2, t_2) + \Psi(S_1, t_1; L_2, t_2) \\ +\Psi(L_1, t_1; S_2, t_2) + \Psi(L_1, t_1; L_2, t_2), \tag{21.4.10}$$

with

$$\Psi(\mathbf{r}_1, t_1; \mathbf{r}_2, t_2) \\ = \frac{\kappa_a \kappa_b}{4 r_1 r_2} e^{(i\omega_{ac}+\Gamma_a/2)(t_1 - r_1/c)}\Theta(t_1 - r_1/c) \\ \times e^{(i\omega_{bc}+\Gamma_b/2)[(t_2 - r_2/c)-(t_1 - r_1/c)]}\Theta[(t_2 - r_2/c) - (t_1 - r_1/c)] \\ +(1 \leftrightarrow 2), \tag{21.4.11}$$

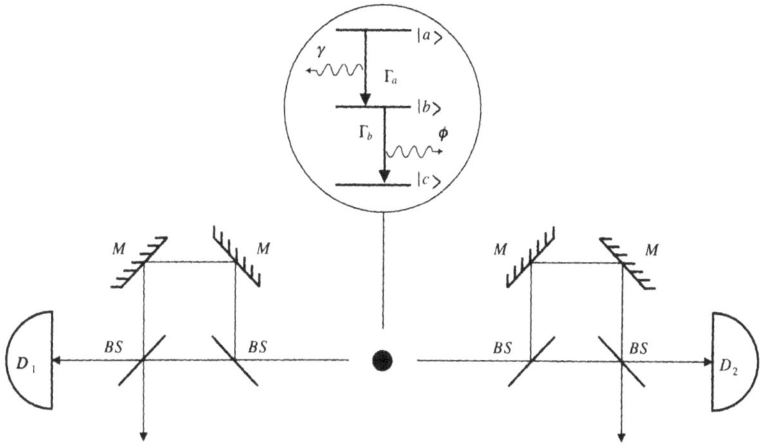

Fig. 21.8
Two-photon interferometer with a single excited three-level atom as the light source. Decaying from the upper state $|a\rangle$ to an intermediate state $|b\rangle$ with a rate Γ_a and from $|b\rangle$ to the ground state $|c\rangle$ with a rate Γ_b, the atom emits two photons γ and ϕ, respectively. The radiation travels from the atom to the detectors D_1 and D_2 via short and long optical paths of lengths S and L as determined by the beam-splitters BS and mirrors M.

where $r_i = S_i$ or L_i $(i = 1, 2), \kappa_a = \pi g_{ak_0}\mathscr{E}_{k_0}D(\omega_{ab})/k_0$, $\kappa_b = \pi g_{bq_0}\mathscr{E}_{q_0} D(\omega_{bc})/q_0$, $k_0 = \omega_{ab}/c$, $q_0 = \omega_{bc}/c$, and $D(v) = Vv^2/\pi^2c^3$ is the mode density of the free space.

It is useful to catalog the various terms in Eq. (21.4.10) as in Fig. 21.9. Each of the four combinations of paths is symmetric and is depicted by the diagrams as indicated. For example, in the case of $\Psi(S_1, t_1; L_2, t_2)$, the first diagram (associated with B_{12}) denotes photon ϕ taking the short path to detector D_1 and photon γ the long path to detector D_2.

Now we can understand the physics of the problem and also shorten much of the algebra by assuming that $\Gamma_a \ll \Gamma_b$, and $L - S \gg c\Gamma_b^{-1}$. That is, we assume that the atom decays from $|b\rangle \rightarrow |c\rangle$ very rapidly so that the emission of the γ-photon ($|a\rangle \rightarrow |b\rangle$) is accompanied by a ϕ-photon ($|b\rangle \rightarrow |c\rangle$). Then only the first and last lines of Fig. 21.9, i.e., $\Psi(S_1, t_1; S_2, t_2)$ and $\Psi(L_1, t_1; L_2, t_2)$, will contribute to $G^{(2)}$. The other terms $\Psi(S_1, t_1; L_2, t_2)$ and $\Psi(L_1, t_1; S_2, t_2)$ will vanish because the γ- and ϕ-photons will never 'overlap' at the detectors.

This makes the calculations much shorter and we find the joint count probability to be given by

$$P(1, 2) = \int_0^\infty dt_1 \int_0^\infty dt_2 |\Psi(S_1, t_1; S_2, t_2) + \Psi(L_1, t_1; L_2, t_2)|^2. \tag{21.4.12}$$

After carrying out the necessary integrations, Eq. (21.4.12) yields the simple answer

$$P(1, 2) = \frac{\mathcal{N}}{\Gamma_a\Gamma_b}(2 + e^{-\Gamma_a\tau/2}\cos\omega_{ac}\tau), \tag{21.4.13}$$

$$\Psi(S_1, t_1; S_2, t_2) = A_{12} + A_{21} =$$

$$\Psi(S_1, t_1; L_2, t_2) = B_{12} + B_{21} =$$

$$\Psi(L_1, t_1; S_2, t_2) = C_{12} + C_{21} =$$

$$\Psi(L_1, t_1; L_2, t_2) = D_{12} + D_{21} =$$

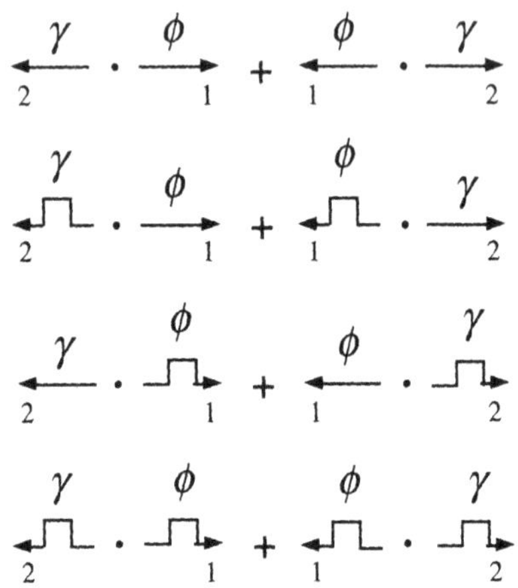

Fig. 21.9
Illustration of the various terms in Eq. (21.4.10), e.g., B_{12} can be interpreted as the contribution arising from photon γ taking the long path to detector D_2 and photon ϕ taking the short path to D_1.

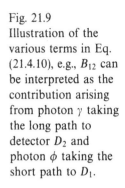

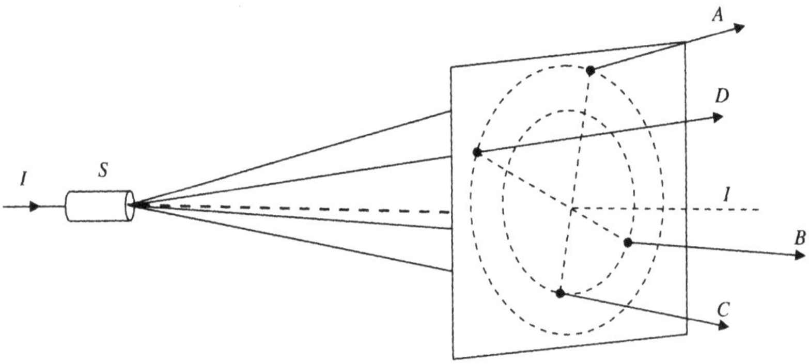

Fig. 21.10
A three-dimensional arrangement of four beams selected from the output of a down-converting crystal S. The diaphragm downstream from S is normal to the incident beam direction I. The pinholes from which the beams A and D emerge lie on a circle centered about I, and the pinholes from which B and C emerge lie on another circle centered about I. The plane AC intersects the plane BD along I. Beams A and D and beams B and C are to be recombined. (From M. A. Horne, A. Shimony, and A. Zeilinger, *Phys. Rev. Lett.* **62**, 2209 (1989).)

where $\mathcal{N} = (\kappa_a \kappa_b / 2SL)^2$, $\omega_{ac} = \omega_{ab} + \omega_{bc}$, and $\tau = (L - S)/c$, and we have taken $S_1 = S_2 = S$ and $L_1 = L_2 = L$.

21.5 Two-particle interferometry via nonlinear down-conversion and momentum selected photon pairs

In a pioneering paper Horne, Shimony, and Zeilinger (HSZ)* studied the problem of two-particle interferometry and suggested an arrangement in which four beams are selected from the output of a nonlinear crystal as in Fig. 21.10. Here they are taking specific advantage of the fact that phase matching requires that

$$\mathbf{k}_A + \mathbf{k}_C = \mathbf{k}_B + \mathbf{k}_D = \mathbf{k},$$

where $\mathbf{k}$ is the wave vector of the driving laser. If a photon is found to have a wave vector, say $\mathbf{k}_A$, then its correlated down-conversion

* Horne, Shimony, and Zeilinger [1989].

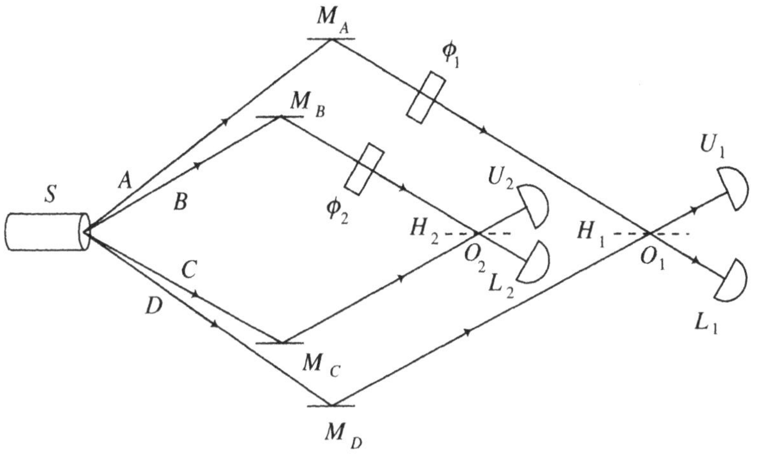

Fig. 21.11 An arrangement for two-particle interferometry with variable phase shifters. The source S emits two particles, 1 and 2, into four beams A, B, C, and D. Index i ($i = 1, 2$) labels the particle that is registered in detectors U_i or L_i. The state of the pair is assumed to be given by Eq. (21.4.1), which is a superposition of two amplitudes: (I) particle 1 in beam A and particle 2 in beam C, and (II) particle 1 in beam D adn particle 2 in beam B. The two beams A and D of particle 1 are given a variable relative phase shift ϕ_1 before recombination near the point O_1 on the half-silvered mirror H_1 (Mach-Zehnder interferometry). Likewise, the two beams B and C of particle 2 are given a variable phase shift ϕ_2 before recombination near O_2 on the half-silvered mirror H_2. The observed quantities of interest are the two-particle coincident count rates as functions of ϕ_1 and ϕ_2. (From M. A. Horne, A. Shimony, and A. Zeilinger, *Phys. Rev. Lett.* **62**, 2209 (1989).)

partner will be found to have wave vector $\mathbf{k}_C$. They showed that if two beams are combined on one detector, and two on another, then the coincident rates $G^{(2)}$ will show interference but the single-detector count rates $G^{(1)}$ will not. Their suggested experimental arrangement is shown in Fig. 21.11. The nonlinear conversion produces a correlated state

$$|\psi\rangle = |1_{\mathbf{k}_A} 1_{\mathbf{k}_C}\rangle + |1_{\mathbf{k}_D} 1_{\mathbf{k}_B}\rangle. \tag{21.5.1}$$

21.5.1 Two-site down-conversion interferometry

Stimulated by the work of HSZ, Alley and Shih and Mandel and co-workers proceeded to carry out several beautiful two-photon interference experiments using a double down-conversion scheme. In one experiment, Ou, Wang, Zou, and Mandel (OWZM) used two nonlinear crystals to demonstrate the effect. In this experiment, which is conceptually sketched in Fig. 21.12, there are two nonlinear $LiIO_3$ crystals, designated as NL 1 and NL 2, each of which produces down-converted photon pairs. The radiation from each is incident on the beam-splitters BS_A and BS_B and the combined signals fall on the detectors D_A and D_B. In order to analyze the problem in a simple but complete form, let us revert to the scheme of Section 21.2 and consider the case in which NL 1 and NL 2 are nature's simplest down-

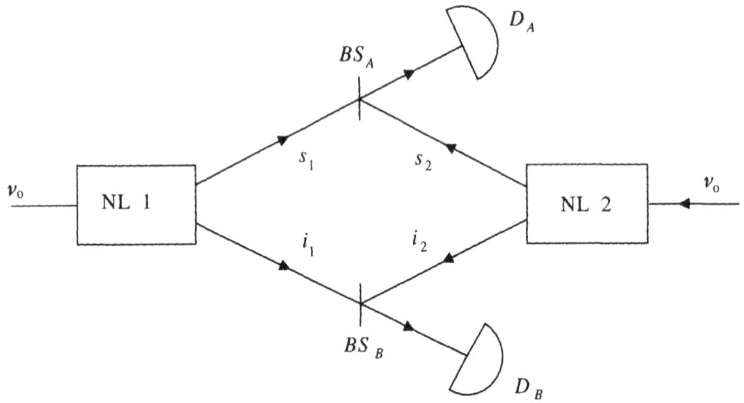

Fig. 21.12 Schematic of the two-photon interference experiment using a double down-converter scheme. (From Z. Y. Ou, L. J. Wang, X. Y. Zou, and L. Mandel, *Phys. Rev. A* **41**, 566 (1990).)

converters, namely, the simple three-level atoms of Fig. 21.1. Naturally, the utilization of our simple atomic down-converter is an idealization. In the setup of Fig. 21.13 we can arrange, following the suggestion of HSZ, the experiment such that a γ-photon in some direction is always accompanied by a ϕ-photon due to phase matching. However, in the case of atomic down-converters, this is not the case. Nevertheless, correlations will be observed, and the general principles remain as before.

The essential ingredients of the physics are the same in the actual experimental arrangement of OWZM and the setup of Fig. 21.13, and we proceed with the analysis of Fig. 21.13. In this case we may apply the results of Section 21.2 directly to the present problem (see Eq. (21.2.9))

$$\begin{aligned} &G^{(2)}(\mathbf{r},\mathbf{r}';t,t') \\ &= |\Psi_\gamma(|\mathbf{r}-\mathbf{r}_1|,t)|^2|\Psi_\phi(|\mathbf{r}'-\mathbf{r}_1|,t')|^2 \\ &\quad +|\Psi_\gamma(|\mathbf{r}-\mathbf{r}_2|,t)|^2|\Psi_\phi(|\mathbf{r}'-\mathbf{r}_2|,t')|^2 \\ &\quad +\Big[\Psi_\gamma^*(|\mathbf{r}-\mathbf{r}_1|,t)\Psi_\gamma(|\mathbf{r}-\mathbf{r}_2|,t)\Psi_\phi^*(|\mathbf{r}'-\mathbf{r}_1|,t')\Psi_\phi(|\mathbf{r}'-\mathbf{r}_2|,t') \\ &\qquad +\text{c.c.}\Big], \end{aligned} \tag{21.5.2}$$

where the notation is the same as in Fig. 21.4, except that the path lengths $|\mathbf{r}-\mathbf{r}_i|$ and $|\mathbf{r}'-\mathbf{r}_i|$, $i=\{1,2\}$, are now the folded optical path lengths taking the beam-splitters into account. On substituting Ψ_γ and Ψ_ϕ from Eq. (21.1.6) into Eq. (21.5.2), we obtain

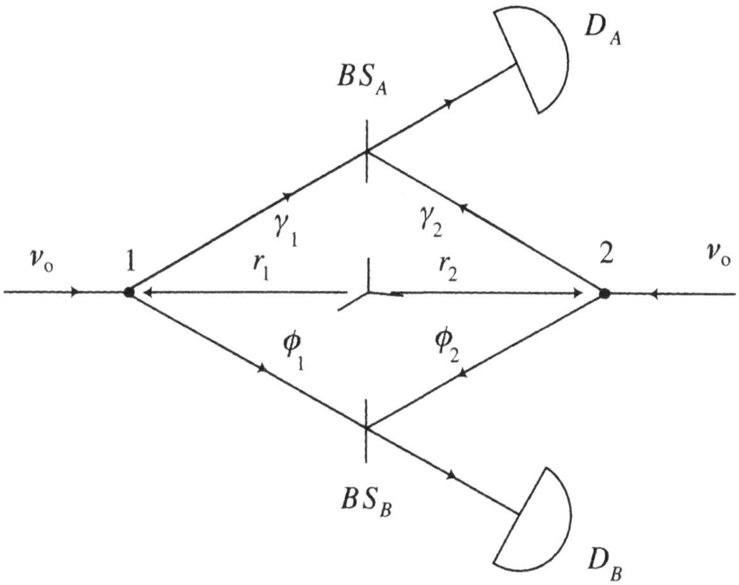

Fig. 21.13 Setup for the single-atom down-converters to analyze the OWZM interference experiment of Fig. 21.12.

$$
\begin{aligned}
&G^{(2)}(\mathbf{r},\mathbf{r}';t,t')\\
&=\frac{\mathscr{E}_{0\gamma}^2}{\Delta r_1^2}\Theta\left(t-\frac{\Delta r_1}{c}\right)e^{-\left(t-\frac{\Delta r_1}{c}\right)\Gamma}\frac{\mathscr{E}_{0\phi}^2}{\Delta r_1'^2}\Theta\left(t'-\frac{\Delta r_1'}{c}\right)e^{-\left(t-\frac{\Delta r_1'}{c}\right)\Gamma}\\
&+\frac{\mathscr{E}_{0\gamma}^2}{\Delta r_2^2}\Theta\left(t-\frac{\Delta r_2}{c}\right)e^{-\left(t-\frac{\Delta r_2}{c}\right)\Gamma}\frac{\mathscr{E}_{0\phi}^2}{\Delta r_2'^2}\Theta\left(t'-\frac{\Delta r_2'}{c}\right)e^{-\left(t-\frac{\Delta r_2'}{c}\right)\Gamma}\\
&+2\frac{\mathscr{E}_{0\gamma}^2}{\Delta r_1\Delta r_2}\Theta\left(t-\frac{\Delta r_1}{c}\right)e^{-\left(t-\frac{\Delta r_1}{c}\right)\Gamma/2}\Theta\left(t-\frac{\Delta r_2}{c}\right)e^{-\left(t-\frac{\Delta r_2}{c}\right)\Gamma/2}\\
&\times\frac{\mathscr{E}_{0\phi}^2}{\Delta r_1'\Delta r_2'}\Theta\left(t'-\frac{\Delta r_1'}{c}\right)e^{-\left(t-\frac{\Delta r_1'}{c}\right)\Gamma/2}\Theta\left(t'-\frac{\Delta r_2'}{c}\right)e^{-\left(t-\frac{\Delta r_2'}{c}\right)\Gamma/2}\\
&\times\cos[k_\gamma(\Delta r_1-\Delta r_2)+k_\phi(\Delta r_1'-\Delta r_2')],
\end{aligned}
\tag{21.5.3}
$$

where $\Delta r_i = |\mathbf{r}-\mathbf{r}_i|$ and $\Delta r_i' = |\mathbf{r}'-\mathbf{r}_i|$. Thus, by varying the relative phases by, e.g., varying the atomic position $\mathbf{r}_i$, interference should show up as a function of path length difference, and this is confirmed in the OWZM experiment.

In another ingenious double down-conversion experiment (see Fig. 21.14), Zou, Wang, and Mandel (ZWM)* show that it is possible to change the configuration of Fig. 21.12 in such a way that mixing the idlers restores interference without the need for coincidence detection.

* Zou, Wang, and Mandel [1991].

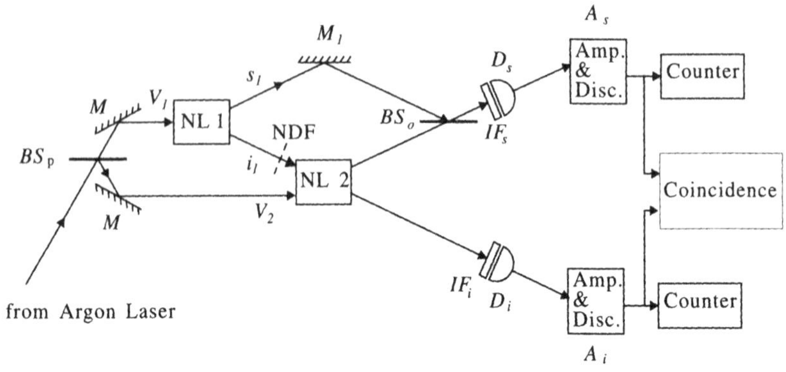

Fig. 21.14 Schematic diagram of the interference experiment. (From X. Y. Zou, L. J. Wang, and L. Mandel, *Phys. Rev. Lett.* **67**, 318 (1991).)

In Fig. 21.14, we see two nonlinear crystals NL 1 and NL 2 pumped coherently to produce two pairs of signal and idler beams via spontaneous parametric down-conversion. The two idlers are then aligned and detected with one common photodetector D_i, thereby making them indistinguishable. The two signals are joined at detector D_s and the question is asked whether these two beams would show interference. It is to be noted that the intensity of the idler beam coming from NL 1 is too weak to induce any stimulated down-conversion in NL 2, so a detected coherence is certainly not due to an induced emission process that would obviously result in coherent signal beams.

The result is coherence, and therefore interference in the signal beams, if the idlers are aligned and loss of coherence if the idlers are distinguishable. (The idlers can be made distinguishable by placing a beamstop between NL 1 and NL 2.) This can be understood in principle from the point of view of which-path information. Let us first consider the case with a beamstop inserted between NL 1 and NL 2. If a photon is measured in D_s, then, by looking at D_i, one can tell whether the photon pair originated in NL 1 or NL 2: if D_i also measured a photon, then the photon pair must have originated in NL 2, otherwise it must have come from NL 1. In other words, one has which-path information with the beamstop inserted and therefore the signal beams do not show interference. Without the beamstop, however, there is no way of telling which crystal sent out a signal photon measured at D_s, because D_i will record a photon in any case. So without the beamstop we do not have which-path information and therefore have to add probability amplitudes instead of probabilities, hence giving rise to interference.

Note, however, that the interference Mandel and co-workers reported is due to *first-order* coherence of the signal beams. In other words, the detector D_i could as well have been absent in the experi-

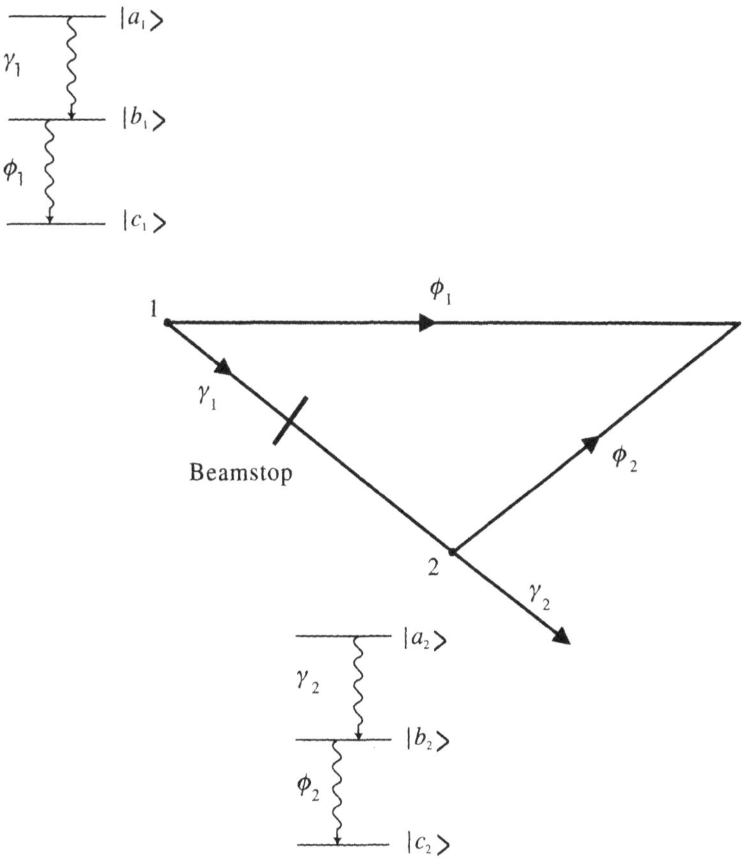

Fig. 21.15 Setup for the single-atom down-converters to analyze the ZWM experiment of Fig. 21.14.

ment. It is only the *potential* for which-path information that impacts on the question of interference.

We proceed to analyze the problem by once again using single-atom down-converters, see Fig. 21.15. The first-order coherence function for the setup of Fig. 21.15 is

$$\begin{aligned} G^{(1)}(\mathbf{r}, t) = {} & \langle \phi_1 | E^{(-)}(\mathbf{r}, t) E^{(+)}(\mathbf{r}, t) | \phi_1 \rangle \langle \gamma_1 | \gamma_1 \rangle \\ & + \langle \phi_2 | E^{(-)}(\mathbf{r}, t) E^{(+)}(\mathbf{r}, t) | \phi_2 \rangle \langle \gamma_2 | \gamma_2 \rangle \\ & + [\langle \phi_1 | E^{(-)}(\mathbf{r}, t) E^{(+)}(\mathbf{r}, t) | \phi_2 \rangle \langle \gamma_1 | \gamma_2 \rangle + \text{c.c.}]. \end{aligned} \quad (21.5.4)$$

For interatomic separations larger than the optical wavelength, it can be shown that the states $|\gamma_i\rangle$ are orthogonal. Therefore the interference term in Eq. (21.5.4) vanishes and the photons $|\phi_i\rangle$ do not interfere. This result is not too surprising since the γ photon carries which-path information concerning the emission process.

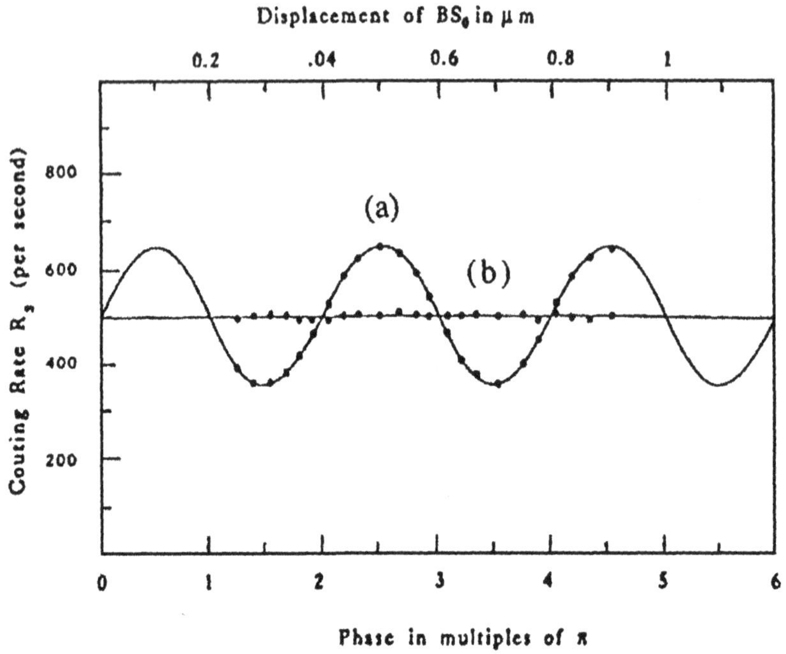

Fig. 21.16 Measured photon counting rate R_s as a function of beam-splitter BS_0 displacement: (a) Neutral density filter with transmittivity $T = 0.91$, and (b) beamstop with $T = 0$ inserted between NL 1 and NL 2. The solid curves are the best fitting sinusoidal functions of period 394 nm. (From X. Y. Zou, L. J. Wang, and L. Mandel, *Phys. Rev. Lett.* **67**, 318 (1991).)

In order to regain interference we proceed as in Section 21.2: a γ-photodetector is placed at $\mathbf{r}$, and the relevant expression is now the second-order correlation function

$$\begin{aligned} &G^{(2)}(\mathbf{r},\mathbf{r}';t,t') \\ &= \langle\phi_1|E^{(-)}(\mathbf{r}',t')E^{(+)}(\mathbf{r}',t')|\phi_1\rangle\langle\gamma_1|E^{(-)}(\mathbf{r},t)E^{(+)}(\mathbf{r},t)|\gamma_1\rangle \\ &\quad+\langle\phi_2|E^{(-)}(\mathbf{r}',t')E^{(+)}(\mathbf{r}',t')|\phi_2\rangle\langle\gamma_2|E^{(-)}(\mathbf{r},t)E^{(+)}(\mathbf{r},t)|\gamma_2\rangle \\ &\quad+[\langle\phi_1|E^{(-)}(\mathbf{r}',t')E^{(+)}(\mathbf{r}',t')|\phi_2\rangle\langle\gamma_1|E^{(-)}(\mathbf{r},t)E^{(+)}(\mathbf{r},t)|\gamma_2\rangle \\ &\qquad+\text{c.c.}], \end{aligned} \tag{21.5.5}$$

and interference is restored.

But now, and this is the intriguing point made by ZWM, if we align the down-converters 1 and 2 properly, then we can insure that $|\gamma_1\rangle = |\gamma_2\rangle$ and the first-order correlation function (21.5.4) shows coherence! This is 'reasonable' from the point of view of orthodox quantum mechanics which says that we no longer have which-path information and so interference returns.

Furthermore, when we put a beamstop, e.g., a mirror between the down-converters 1 and 2, the interference disappears. This is because the innerproduct $\langle\gamma_1|\gamma_2\rangle$ now vanishes since $|\gamma_1\rangle$ and $|\gamma_2\rangle$ are orthogonal. Physically, we now have which-path information. These results were confirmed by ZWM and are summarized in Fig. 21.16.

21.6 A vacuum–fluctuation picture of the ZWM experiment

The common feature of the quantum eraser, the micromaser 'Welcher-Weg' detector, and Mandel's experiments is the fact that apparently we have to invoke QED to fully understand them. This motivates the question whether there may be a simple mechanistic picture that can help us to better understand the physics.* In this regard, we view the situation much as we do the Lamb shift in QED. There can be no doubt that QED gives an excellent account of the effect. However, one is gratified to also have the Welton vacuum fluctuation heuristic arguments which provide a simple insight into the physics. To be sure, the Welton treatment is no replacement for QED but it is nice to have. Thus motivated, we here sketch such a simple pictorial explanation for the ZWM experiment. To this end, we invoke the notion of stochastically fluctuating electromagnetic fields.

For our 'explanation' of the ZWM experiment based on classical fluctuating fields, we replace the two nonlinear crystals by two atoms 1 and 2, each with three levels $|a\rangle$, $|b\rangle$, and $|c\rangle$. Note that, as mentioned earlier, we enforce a directionality onto the idler and signal fields in order to account for the missing directionality of spontaneous emission. This could be accomplished, for example, by placing the two atoms into optical waveguides for the respective frequencies. These two atoms are pumped by a weak pump P, and, once excited, they emit two classical fields γ_i and ϕ_i where $i = 1, 2$ (see Fig. 21.15). The state of each atom after the excitation is $|\psi\rangle = c_a|a\rangle + c_c|c\rangle$. Such an atom is subject to perturbations by the vacuum fluctuations that induce some population transfer from $|a\rangle$ to $|b\rangle$ yielding

$$|\psi\rangle = c_a|a\rangle + \delta c_b|b\rangle + c_c|c\rangle. \tag{21.6.1}$$

The amplitude δc_b has a random phase $\phi_{\gamma,i}$ governed by the inducing field, so that the dipole formed by the levels $|a\rangle$ and $|b\rangle$ radiates with the phase $\phi_{a,i} - \phi_{\gamma,i}$. As soon as there is some population in level $|b\rangle$, the dipole $|b\rangle - |c\rangle$ starts radiating as well. The phase of this radiation is $\phi_{\gamma,i} - \phi_{c,i}$, where $\phi_{a,i} - \phi_{c,i}$ is the relative phase between levels $|a\rangle$ and $|c\rangle$ (of the ith atom) as determined by the incident weak field inducing the $|c\rangle \to |a\rangle$ transition. Note that we do not have to take into account the vacuum fluctuations interacting with the transition $|b\rangle \to |c\rangle$, as

* Scully and Rathe [1994].

they are second order in the field. The total dipole moments are thus

$$P_i = \wp_{ab}|c_a||\delta c_b| \exp\left[-i\omega_{ab}t + i(\phi_{a,i} - \phi_{\gamma,i})\right]$$
$$+ \wp_{bc}|\delta c_b||c_c| \exp\left[-i\omega_{bc}t + i(\phi_{\gamma,i} - \phi_{c,i})\right]$$
$$+ \text{higher orders}\,. \tag{21.6.2}$$

These dipole moments radiate fields

$$E_i(\mathbf{r}, t) = E_i^{\gamma}(\mathbf{r}, t) + E_i^{\phi}(\mathbf{r}, t), \tag{21.6.3}$$

where the two contributions with different frequencies are given by

$$E_i^{\gamma}(\mathbf{r}, t) = |\mathscr{E}_i^{\gamma}(\mathbf{r})|\ \exp\left[-i\omega_{ab} + i(\phi_{a,i} - \phi_{\gamma,i}) + i\mathbf{k}_{\gamma,i}\cdot(\mathbf{r} - \mathbf{r}_i)\right], \tag{21.6.4}$$

$$E_i^{\phi}(\mathbf{r}, t) = |\mathscr{E}_i^{\phi}(\mathbf{r})|\ \exp\left[-i\omega_{bc} + i(\phi_{\gamma,i} - \phi_{c,i}) + i\mathbf{k}_{\phi,i}\cdot(\mathbf{r} - \mathbf{r}_i)\right]. \tag{21.6.5}$$

To obtain the degree of coherence these fields exhibit, we evaluate the classical counterpart of the Glauber coherence function

$$G^{(1)}(\mathbf{r}, t) = \langle E^*(\mathbf{r}, t)E(\mathbf{r}, t)\rangle. \tag{21.6.6}$$

The key term in the expansion of this expression according to Eq. (21.6.3) determining the degree of coherence is the two-atom cross term $E_1^{\phi *}(\mathbf{r}, t)E_2^{\phi}(\mathbf{r}, t)$

$$G_{\phi}^{(1)}(\mathbf{r}, t) = |\mathscr{E}_1^{\phi}(\mathbf{r})||\mathscr{E}_2^{\phi}(\mathbf{r})|\langle\exp\left[-i(\phi_{\gamma,1} - \phi_{\gamma,2})\right]\rangle$$
$$\times \exp\left[i(\phi_{c,1} - \phi_{c,2}) - i\mathbf{k}_{\phi,1}\cdot(\mathbf{r} - \mathbf{r}_1) + i\mathbf{k}_{\phi,2}\cdot(\mathbf{r} - \mathbf{r}_2)\right]$$
$$+ \ldots \tag{21.6.7}$$

In the present case, the average over the statiscally independent vacuum phases $\phi_{\gamma,i}$ yields zero. Vacuum fluctuation physics therefore predicts the absence of interference, as does QED.

We next recall the relevant arguments for the treatment of the ZWM problem from the point of view of QED. The interaction of the atoms with the pump and subsequent decay leads to interference as discussed in Section 21.5. The key term in the first-order coherence function expression, determining whether interference fringes will be detected, is

$$G^{(1)}(\mathbf{r}, t) = \langle\phi_1|E^{(-)}(\mathbf{r}, t)E^{(+)}(\mathbf{r}, t)|\phi_2\rangle\langle\gamma_1|\gamma_2\rangle + \ldots \tag{21.6.8}$$

If the beamstop keeps the idler photon $|\gamma_1\rangle$ from overlapping with photon $|\gamma_2\rangle$, then $|\gamma_1\rangle$ and $|\gamma_2\rangle$ are states of two different modes and are orthogonal, that is

$$\text{QED argument, beamstop in}: \langle\gamma_1|\gamma_2\rangle = 0 \Longrightarrow \text{no interference.} \tag{21.6.9}$$

If, however, the beamstop is absent, then the two photons correspond to states of the same mode

$$\text{QED argument, beamstop out}: \quad \langle\gamma|\gamma\rangle = 1 \Longrightarrow \text{interference.} \tag{21.6.10}$$

Turning now to the vacuum fluctuation (VF) logic, we note that the relevant first-order coherence function, as given by Eq. (21.6.7), is

$$G^{(1)}(\mathbf{r}, t) = \langle E^*(\mathbf{r}, t)E(\mathbf{r}, t)\rangle = \kappa\langle\exp\left[-i(\phi_{\gamma,1} - \phi_{\gamma,2})\right]\rangle + \ldots, \tag{21.6.11}$$

where we have written only the key term. With the beamstop in, the two vacuum phases are statistically independent and the averaging process will lead to a zero,

$$\text{VF argument, beamstop in}:$$
$$\langle\exp\left[-i(\phi_{\gamma,1} - \phi_{\gamma,2})\right]\rangle = 0 \Longrightarrow \text{no interference.} \tag{21.6.12}$$

If the beamstop is removed, then the vacuum field stimulating the emission in atom 1 will travel to atom 2 and therefore impart the same phase to both atoms. Hence the average in this case will produce unity (since $\phi_{\gamma,1} = \phi_{\gamma,2}$)

$$\text{VF argument, beamstop out}:$$
$$\langle\exp\left[-i(\phi_{\gamma,1} - \phi_{\gamma,2})\right]\rangle = 1 \Longrightarrow \text{interference.} \tag{21.6.13}$$

In summary, our vacuum fluctuation logic yields the same result as QED, in accord with the 'pseudo-ZWM' experiment.

In general, we often find stochastic electrodynamics useful in providing insight. We stress, however, that the present heuristic approach is no substitute for a full QED analysis. In fact, stochastic electrodynamics provides answers contradicting QED for certain problems. A case in point is the analysis of quantum beats in Λ type systems presented in Chapter 1. We also run into trouble here if we push the vacuum fluctuation logic too far.

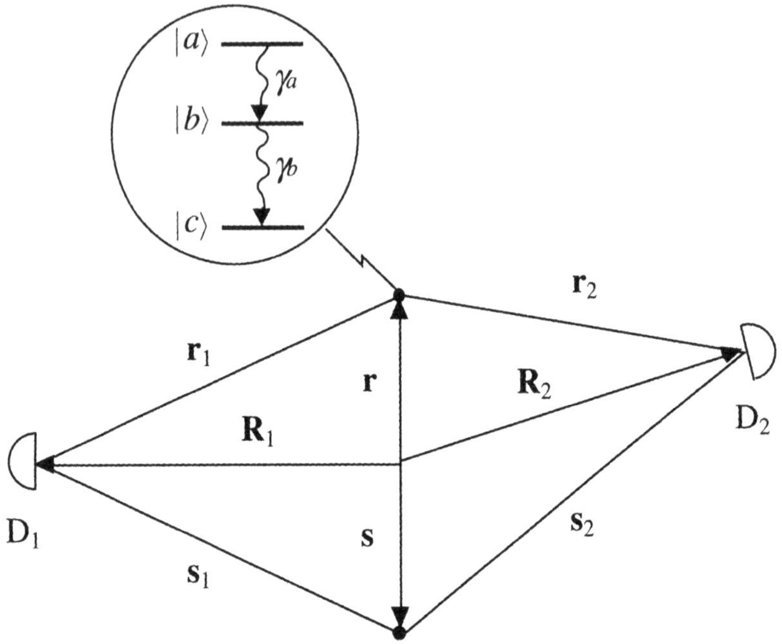

Fig. 21.17
Two three-level atoms are stimulated weakly such that either the atom at **r** or the atom at **s** is excited.

21.7 High-resolution spectroscopy via two-photon cascade interferometry*

As a capstone to this chapter, we next demonstrate that it is possible to achieve a new kind of line narrowing in spontaneously emitted radiation via two-photon correlation measurements on atomic cascade radiation. The width of an optical transition between two atomic states $|a\rangle$ and $|b\rangle$, as measured by conventional spectroscopic means, is governed by the sum of the radiative decay rates, $\Gamma_a + \Gamma_b$, out of the two levels in question. An exception to this is provided by the 'time delay spectroscopy' of Section 5.7, where the resolution is shown to go as $\Gamma_a - \Gamma_b$.

It is to be noted that in such measurement schemes, we are dealing with ordinary single-photon events, i.e., single-photon emission and detection. However, as we show in the following, the situation is radically changed when we instead consider two-photon correlated spontaneous emission radiation together with a two-photon detection scheme as in Fig. 21.17. In such a case, we find that the second-order correlation function displays spectral features of width Γ_a *independent* of Γ_b, where Γ_a is the decay rate out of state $|a\rangle$ into $|b\rangle$ and Γ_b is the decay rate out of $|b\rangle$ into $|c\rangle$.

* For further reading see Rathe and Scully [1994].

We note that the (single-photon) time delay spectroscopy of Chapter 5, is most useful when $\Gamma_a \cong \Gamma_b$ since the line narrowing there is governed by $\Gamma_a - \Gamma_b$. However, in the present case, the (two-photon) cascade correlation interferometry yields, in principle, an improved resolution when $\Gamma_b \gg \Gamma_a$.

We proceed by considering the experimental setup depicted in Fig. 21.17, with two atoms located at $\mathbf{r}$ and $\mathbf{s}$, and two detectors at $\mathbf{r}_1$ and $\mathbf{r}_2$. As a starting point, we prepare an atomic state such that one of the atoms is excited to state $|a\rangle$ and the other one is in the ground state $|c\rangle$ with the field in the vacuum state $|0\rangle$. This initial atom–field state may be written as

$$|\psi\rangle = \frac{1}{\sqrt{2}}(|a,c\rangle + |c,a\rangle)|0\rangle. \tag{21.7.1}$$

For times $t \gg \Gamma_a^{-1}, \Gamma_b^{-1}$ this state evolves to

$$|\psi\rangle = \frac{1}{\sqrt{2}}|c,c\rangle(|\psi_\mathbf{r}, \phi_\mathbf{r}\rangle + |\psi_\mathbf{s}, \phi_\mathbf{s}\rangle), \tag{21.7.2}$$

where $|\psi_\mathbf{r}, \phi_\mathbf{r}\rangle$ and $|\psi_\mathbf{s}, \phi_\mathbf{s}\rangle$ are the field states generated by the atoms at $\mathbf{r}$ and $\mathbf{s}$, respectively, and $\psi(\phi)$ is associated with the $|a\rangle \to |b\rangle$ ($|b\rangle \to |c\rangle$) transition. Such states are given by Eq. (6.4.13), which in the present notation reads

$$|\psi_\mathbf{d}, \phi_\mathbf{d}\rangle = \sum_{\mathbf{k},\mathbf{q}} \frac{-g_{a,\mathbf{k}} g_{b,\mathbf{q}} e^{-i\mathbf{k}\cdot\mathbf{d} - i\mathbf{q}\cdot\mathbf{d}}}{[i(\nu_k + \nu_q - \omega_{ac}) - \Gamma_a/2][i(\nu_q - \omega_{bc}) - \Gamma_b/2]} |1_\mathbf{k}, 1_\mathbf{q}\rangle, \tag{21.7.3}$$

where $\mathbf{d} = \mathbf{r}$ or $\mathbf{s}$.

The corresponding two-photon probability amplitude

$$\Psi^{(2)}(\mathbf{r}_1, t_1; \mathbf{r}_2, t_2) = \langle 0|E^{(+)}(\mathbf{r}_2, t_2) E^{(+)}(\mathbf{r}_1, t_1)|\psi\rangle \tag{21.7.4}$$

allows us to calculate the joint count probability that both detectors at $\mathbf{r}_1$ and $\mathbf{r}_2$ register a count. We obtain

$$P^{(2)} = \int_0^\infty dt_1 \int_0^\infty dt_2 G^{(2)}(\mathbf{r}_1, \mathbf{r}_2; t_1, t_2), \tag{21.7.5}$$

where the intensity correlation function $G^{(2)}(\mathbf{r}_1, \mathbf{r}_2; t_1, t_2)$ may be written as (see Section 21.4.2)

$$G^{(2)}(\mathbf{r}_1, \mathbf{r}_2; t_1, t_2) = \Psi^{(2)*}(\mathbf{r}_1, t_1; \mathbf{r}_2, t_2)\Psi^{(2)}(\mathbf{r}_1, t_1; \mathbf{r}_2, t_2). \tag{21.7.6}$$

Inserting the state $|\psi\rangle$ given by (21.7.2) into (21.7.4) yields

$$\Psi^{(2)}(\mathbf{r}_1,t_1;\mathbf{r}_2,t_2) = \frac{1}{\sqrt{2}}\Big[\langle 0|E^{(+)}(\mathbf{r}_1,t_1)E^{(+)}(\mathbf{r}_2,t_2)|\psi_\mathbf{r},\phi_\mathbf{r}\rangle$$
$$+\langle 0|E^{(+)}(\mathbf{r}_1,t_1)E^{(+)}(\mathbf{r}_2,t_2)|\psi_\mathbf{s},\phi_\mathbf{s}\rangle\Big]$$
$$= \frac{1}{\sqrt{2}}\Big[\Psi^{(2)}_\mathbf{r}(\mathbf{r}_1,t_1;\mathbf{r}_2,t_2) + \Psi^{(2)}_\mathbf{s}(\mathbf{r}_1,t_1;\mathbf{r}_2,t_2)\Big], \tag{21.7.7}$$

with $\Psi^{(2)}_\mathbf{d}(\mathbf{r}_1,\mathbf{r}_2;t_1,t_2) = \langle 0|E^{(+)}(\mathbf{r}_1,t_1)E^{(+)}(\mathbf{r}_2,t_2)|\psi_\mathbf{d},\phi_\mathbf{d}\rangle$ ($\mathbf{d} = \mathbf{r}$ or $\mathbf{s}$). Using Eq. (21.7.3) we find, as in Appendix 21.B

$$\Psi^{(2)}_\mathbf{r}(\mathbf{r}_1,\mathbf{r}_2;t_1,t_2)$$
$$= \kappa \exp\left[-(i\omega_{ab}+\Gamma_a/2)\left(t_1-\frac{r_1}{c}\right)\right]\Theta\left(t_1-\frac{r_1}{c}\right)$$
$$\times \exp\left\{-(i\omega_{bc}+\Gamma_b/2)\left[\left(t_1-\frac{r_1}{c}\right)-\left(t_2-\frac{r_2}{c}\right)\right]\right\}$$
$$\times\Theta\left[\left(t_1-\frac{r_1}{c}\right)-\left(t_2-\frac{r_2}{c}\right)\right]$$
$$+(1\leftrightarrow 2), \tag{21.7.8}$$

$$\Psi^{(2)}_\mathbf{s}(\mathbf{r}_1,t_1;\mathbf{r}_2,t_2) = \Psi^{(2)}_\mathbf{r}(\mathbf{r}_1,t_1;\mathbf{r}_2,t_2)\Big|_{r_i\to s_i}. \tag{21.7.9}$$

In Eqs. (21.7.8) and (21.7.9), $r_i = |\mathbf{r}_i - \mathbf{r}|$ and $s_i = |\mathbf{r}_i - \mathbf{s}|$ ($i = 1, 2$), κ is an uninteresting constant, and $\Theta(x)$ is the usual step function. The notation $(1 \leftrightarrow 2)$ in Eq. (21.7.8) indicates that the ω_{ab} photon now goes to detector D_2 and the ω_{bc} photon to detector D_1. Finally, we note that identical physics applies to the atom located at $\mathbf{s}$, and the corresponding two-photon probability amplitude $\Psi^{(2)}_\mathbf{s}(\mathbf{r}_1,t_1;\mathbf{r}_2,t_2)$ is defined by Eq. (21.7.9).

In order to simplify the calculation, we consider the geometry of Fig. 21.17, in which $s_2 > r_2$ and $s_1 = r_1$. With this case in mind, we insert Eqs. (21.7.8) and (21.7.9) into Eq. (21.7.6) and carry out the integrations of Eq. (21.7.5). After a somewhat lengthy calculation, we find

$$P^{(2)}(\mathbf{r}_1,\mathbf{r}_2;t_1,t_2) = \kappa^2\left\{\frac{1}{\Gamma_a\Gamma_b} + e^{-\Gamma_b\tau}\left[\frac{f(\tau,\Delta)}{2\Gamma_a\Gamma_b} + \frac{g(\tau,\Delta)}{\Delta^2+\Gamma_a^2}\right]\right\}, \tag{21.7.10}$$

with

$$f(\tau,\Delta) = \cos\omega_{bc}\tau + e^{-\Gamma_a\tau}\cos\omega_{ab}\tau, \tag{21.7.11}$$

$$g(\tau,\Delta) = \cos\omega_{bc}\tau - \frac{\Delta}{\Gamma_a}\sin\omega_{bc}\tau$$
$$- e^{-\Gamma_a\tau}\left(\cos\omega_{ab}\tau - \frac{\Delta}{\Gamma_a}\sin\omega_{ab}\tau\right). \tag{21.7.12}$$

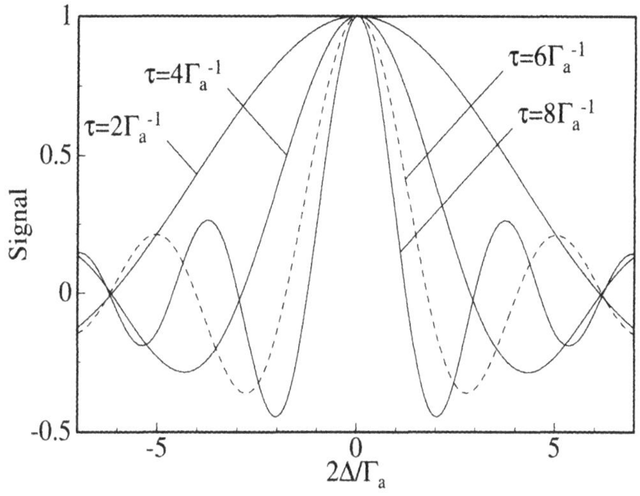

Fig. 21.18 Correlated emission spectroscopy signal part $g(\tau, \Delta)/(\Delta^2 + (\Gamma_a/2)^2)$ in the joint count probability for different τ. The different curves have been normalized for simplicity.

In Eq. (21.7.10), $\Delta \equiv \omega_{ab} - \omega_{bc}$, $\tau = (s_2 - r_2)/c$, and c is the speed of light. It is the presence of the sharp Lorentzian, of width Γ_a, which interests us here since one normally (i.e., in uncorrelated spontaneous emission) expects a width $\Gamma_a + \Gamma_b$. In Fig. 21.18, we plot the signal $g(\tau, \Delta)/(\Delta^2 + \Gamma_a^2)$ for different τ to demonstrate the line narrowing. For example, the dashed curve corresponding to $\tau = 6\Gamma_a^{-1}$ shows a width of approximately Γ_a, independent of Γ_b.

The term proportional to $(\Delta^2 + \Gamma_a^2)^{-1}$ in Eq. (21.7.10) enables us to envision a high-resolution measurement of the atomic transition frequencies ω_{ab} and ω_{bc} in the following way. As indicated in Fig. 21.19, the intermediate level $|b\rangle$ of the atomic cascade may be taken to be a magnetic sublevel with $m = -1$ so that we can vary $\Delta = \omega_{ab} - \omega_{bc}$ around $\Delta = 0$ by applying a magnetic field. In this way we could map out the sharp Lorentzian $(\Delta^2 + \Gamma_a^2)^{-1}$ and thus provide a good measurement of the magnetic field strength B_0 for which $\Delta = 0$. With this knowledge of B_0, we are able to determine the difference of the energy spacing of the unshifted transition frequencies between $|a\rangle$ and $|b_0\rangle$, $|b_0\rangle$ being the intermediate state with $m = 0$, and between $|b_0\rangle$ and $|c\rangle$. This procedure thus enables us, in principle, to measure $\omega_{ab_0} - \omega_{b_0c}$ limited only by the linewidth of the atomic level $|a\rangle$. An additional measurement for $\omega_{ab_0} + \omega_{b_0c} = \omega_{ac}$ could be performed, also limited by Γ_a. So that we are finally in a position to determine ω_{ab_0} and ω_{b_0c} to a precision governed only by Γ_a.

Note at this point, however, that $g(\tau, \Delta)$ varies with Δ as well as the Lorentzian denominator. This functional dependence is such that

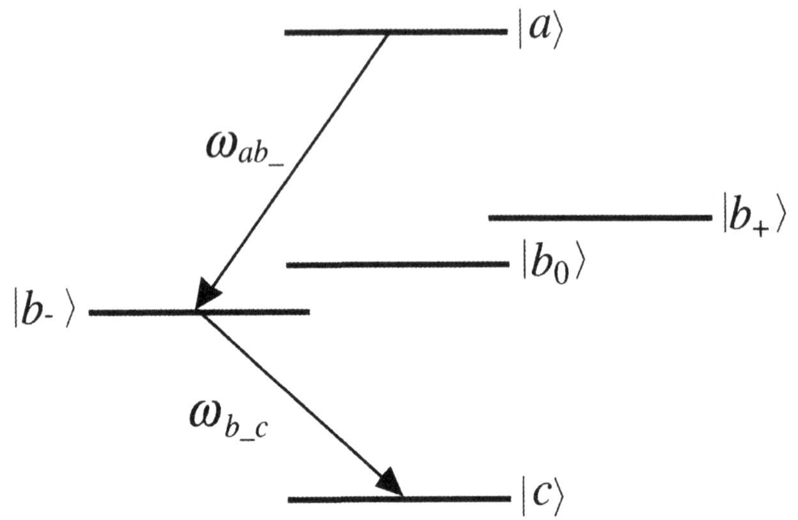

Fig. 21.19
Cascade radiation is 'tuned' by an external magnetic field such that $\omega_{ab\_} = \omega_{b\_c} \cong \omega_0$. Then $\Delta = \omega_{ab\_} - \omega_{b\_}c$ can be varied around zero. (The level $|b_-\rangle$ plays the role of $|b\rangle$ in the calculation.)

for small $\tau \ll \Gamma_a^{-1}$ the signal $g(\tau, \Delta)/(\Delta^2 + \Gamma_a^2)$ goes to a constant independent of Δ. Therefore, the time delay τ has to be of order Γ_a^{-1}. This, however, leads to an exponential decrease of the signal amplitude by means of the prefactor $\exp(-\Gamma_b\tau)$ in Eq. (21.7.10), and eventually to an oscillatory behavior of the signal by means of the sine and cosine functions in Eq. (21.7.10). Both effects are well known from time delay spectroscopy, see Section 5.7.

The devoted reader may well ask: what is the physics behind this 'two-photon line narrowing'? For the answer to this well taken question, we refer the reader to the literature.* It is perhaps appropriate that we leave some questions open; the blossoming field of Quantum Optics is itself very open ended.

21.A Scattering from two atoms via an operator approach

It is useful to introduce an operator formalism for treating the problem of Section 21.1 and also for the more complicated problems in Chapter 21. To that end, we introduce the operator $\gamma_1^\dagger$ which we define as

$$\gamma_i^\dagger = \sum_{\mathbf{k}} \frac{g_{\mathbf{k}} e^{-i\mathbf{k}\cdot\mathbf{r}_i}}{(\nu_k - \omega) + i\Gamma/2} a_{\mathbf{k}}^\dagger \qquad (i = 1, 2), \tag{21.A.1}$$

which, in view of Eq. (21.1.4), implies

* See, e.g., Rathe and Scully [1994].

$$|\gamma_1\rangle = \gamma_1^\dagger|0\rangle. \tag{21.A.2}$$

This can be used to rewrite the correlation function as

$$\langle\gamma_1|E^{(-)}(\mathbf{r},t)E^{(+)}(\mathbf{r},t)|\gamma_1\rangle = \langle 0|\gamma_1 E^{(-)}(\mathbf{r},t)E^{(+)}(\mathbf{r},t)\gamma_1^\dagger|0\rangle. \tag{21.A.3}$$

Now the operator product $E^{(+)}(\mathbf{r},t)\gamma_1^\dagger$ may be written as

$$E^{(+)}(\mathbf{r},t)\gamma_1^\dagger = \gamma_1^\dagger E^{(+)}(\mathbf{r},t) + [E^{(+)}(\mathbf{r},t),\gamma_1^\dagger], \tag{21.A.4}$$

where $[E^{(+)}(\mathbf{r},t),\gamma_1^\dagger]$ is the usual commutator expression.

This is useful because $\gamma_1^\dagger E^{(+)}(\mathbf{r},t)|0\rangle = 0$ and therefore from Eqs. (21.A.3) and (21.A.4) we have

$$\langle\gamma_1|E^{(-)}(\mathbf{r},t)E^{(+)}(\mathbf{r},t)|\gamma_1\rangle = \left|\langle 0|[E^{(+)}(\mathbf{r},t),\gamma_1^\dagger]|0\rangle\right|^2, \tag{21.A.5}$$

and this can be simply evaluated since (Problem 21.1)

$$\begin{aligned}\left|\langle 0|[E^{(+)}(\mathbf{r},t),\gamma_1^\dagger]|0\rangle\right| &= \frac{\mathscr{E}_0}{\Delta r_1}\Theta(t-\Delta r_1/c)e^{-(t-\Delta r_1/c)(i\omega+\Gamma/2)} \\ &\equiv \Psi_\gamma(\Delta r_1,t).\end{aligned} \tag{21.A.6}$$

We now apply this approach to the previous two-atom problem. As before, suppose that we have two atoms at $\mathbf{r}_1$ and $\mathbf{r}_2$ and that at $t=0$ they are in the state $(1/\sqrt{2})(|a_1b_2\rangle + |b_1a_2\rangle)|0\rangle$, i.e., one atom excited and the other not. For times $t \gg 1/\gamma$ the state of the system will be (if the atoms are far enough apart so that the probability that one atom absorbs the photon emitted by the other is negligible, i.e., $|\mathbf{r}_1-\mathbf{r}_2| \gg \lambda$)

$$|\psi\rangle = \frac{1}{\sqrt{2}}(|\gamma_1\rangle + |\gamma_2\rangle)|b_1b_2\rangle = \frac{1}{\sqrt{2}}(\gamma_1^\dagger + \gamma_2^\dagger)|0\rangle|b_1b_2\rangle. \tag{21.A.7}$$

The correlation function is now

$$\begin{aligned}&\langle\psi|E^{(-)}(\mathbf{r},t)E^{(+)}(\mathbf{r},t)|\psi\rangle \\ &= \frac{1}{2}\langle 0|(\gamma_1+\gamma_2)E^{(-)}(\mathbf{r},t)E^{(+)}(\mathbf{r},t)(\gamma_1^\dagger+\gamma_2^\dagger)|0\rangle \\ &= \frac{1}{2}\left|\langle 0|[E^{(+)}(\mathbf{r},t),\gamma_1^\dagger] + [E^{(+)}(\mathbf{r},t),\gamma_2^\dagger]|0\rangle\right|^2 \\ &= \frac{1}{2}\left|\Psi_\gamma(\Delta r_1,t) + \Psi_\gamma(\Delta r_2,t)\right|^2.\end{aligned} \tag{21.A.8}$$

The cross term in Eq. (21.A.8) is responsible for interference effects and describes the fringes that would be produced on a screen which caught the light emitted by the atoms. We can see this by actually calculating the cross term and finding that if $|\mathbf{r}| \gg |\mathbf{r}_1|$ and $|\mathbf{r}_2|$, then

$$\begin{aligned}&2\mathrm{Re}[\Psi_\gamma^*(\Delta r_1,t)\Psi_\gamma(\Delta r_2,t)] \\ &= \frac{2\mathscr{E}_0^2}{r^2}\Theta(t-r/c)e^{-(t-r/c)\Gamma}\cos[k_0(\mathbf{r}_2-\mathbf{r}_1)\cdot\mathbf{r}/|\mathbf{r}|].\end{aligned} \tag{21.A.9}$$

The cosine factor describes the interference pattern. For example, if $\mathbf{r}_1 = (a/2)\hat{z}$ and $\mathbf{r}_2 = -(a/2)\hat{z}$ and we have a screen perpendicular to the x-axis at a distance R from the origin then the spacing between interference fringes will be just $\lambda R/a$ where λ is the wavelength of the light.

21.B Calculation of the two-photon correlation function in atomic cascade emission

We note that for the radiation from a single atom, only two photons are involved so that

$$\begin{aligned}\langle\psi|E_1^{(-)}E_2^{(-)}E_2^{(+)}E_1^{(+)}|\psi\rangle &= \sum_{\{n\}}\langle\psi|E_1^{(-)}E_2^{(-)}|\{n\}\rangle\langle\{n\}|E_2^{(+)}E_1^{(+)}|\psi\rangle\\ &= \langle\psi|E_1^{(-)}E_2^{(-)}|0\rangle\langle 0|E_2^{(+)}E_1^{(+)}|\psi\rangle,\end{aligned}$$

and therefore it is the two-photon 'wave function' which is of interest

$$\Psi^{(2)}(\mathbf{r}_1,t_1;\mathbf{r}_2,t_2) \equiv \langle 0|E^{(+)}(\mathbf{r}_2,t_2)E^{(+)}(\mathbf{r}_1,t_1)|\psi\rangle. \tag{21.B.1}$$

Using the fact that

$$E(\mathbf{r}_i,t_i) = \sum_{\mathbf{k}} a_{\mathbf{k}}e^{-i\nu_k t_i + i\mathbf{k}\cdot\mathbf{r}_i} \qquad (i = 1,2), \tag{21.B.2}$$

and taking $|\psi\rangle$ from Eq. (6.4.13) we have

$$\begin{aligned}&\Psi^{(2)}(\mathbf{r}_1,t_1;\mathbf{r}_2,t_2)\\ &= \sum_{\mathbf{k},\mathbf{q}}\sum_{\mathbf{p},\mathbf{s}}\langle 0|a_{\mathbf{p}}a_{\mathbf{s}}e^{-i\nu_p t_1 + i\mathbf{p}\cdot\mathbf{r}_1}e^{-i\nu_s t_2 + i\mathbf{s}\cdot\mathbf{r}_2}\\ &\quad\times\frac{-e^{-i\mathbf{k}\cdot\mathbf{r} - i\mathbf{q}\cdot\mathbf{r}}g_{a,\mathbf{k}}g_{b,\mathbf{q}}}{\left[i(\nu_k + \nu_q - \omega_{ac}) - \Gamma_a/2\right]\left[i(\nu_q - \omega_{bc}) - \Gamma_b/2\right]}|1_{\mathbf{k}}1_{\mathbf{q}}\rangle\\ &= \sum_{\mathbf{k},\mathbf{q}}\left(e^{-i\nu_k t_1 + i\mathbf{k}\cdot\mathbf{r}_1}e^{-i\nu_q t_2 + i\mathbf{q}\cdot\mathbf{r}_2} + e^{-i\nu_k t_2 + i\mathbf{k}\cdot\mathbf{r}_2}e^{-i\nu_q t_1 + i\mathbf{q}\cdot\mathbf{r}_1}\right)\\ &\quad\times\frac{-e^{-i\mathbf{k}\cdot\mathbf{r} - i\mathbf{q}\cdot\mathbf{r}}g_{a,\mathbf{k}}g_{b,\mathbf{q}}}{\left[i(\nu_k + \nu_q - \omega_{ac}) - \Gamma_a/2\right]\left[i(\nu_q - \omega_{bc}) - \Gamma_b/2\right]}.\end{aligned} \tag{21.B.3}$$

In order to evaluate $\Psi^{(2)}$, we change the sums on $\mathbf{k}$ and $\mathbf{q}$ into integrals, evaluate all coupling constants and density of state factors at the atomic resonances $ck_0 = \omega_{ab}$ and $cq_0 = \omega_{bc}$, and choose the z-axis for the $\mathbf{k}$ and $\mathbf{q}$ integrations as $\Delta\mathbf{r}_1 = \mathbf{r}_1 - \mathbf{r}$ and $\Delta\mathbf{r}_2 = \mathbf{r}_2 - \mathbf{r}$

for the first term in the square brackets and $\mathbf{r}_2 - \mathbf{r}$ and $\mathbf{r}_1 - \mathbf{r}$ for the second. On carrying out these steps we can write Eq. (21.B.3) as

$$\begin{aligned}
&\Psi^{(2)}(\mathbf{r}_1, t_1; \mathbf{r}_2, t_2) \\
&= -\int_0^\infty k^2 dk \int_0^\pi \sin\theta_k d\theta_k \int_0^{2\pi} d\phi_k \int_0^\infty q^2 dq \int_0^\pi \sin\theta_q d\theta_q \int_0^{2\pi} d\phi_q \\
&\quad \times g_{a,\mathbf{k}_0} g_{b,\mathbf{q}_0} \sigma(k_0)\sigma(q_0) \\
&\quad \times \left[e^{-ickt_1 + ik\Delta r_1 \cos\theta_k} e^{-icqt_2 + iq\Delta r_2 \cos\theta_q} + (1 \leftrightarrow 2)\right] \\
&\quad \times \frac{1}{\left[i(ck + cq - \omega_{ac}) - \Gamma_a/2\right]\left[i(cq - \omega_{bc}) - \Gamma_b/2\right]}. \qquad (21.B.4)
\end{aligned}$$

The angular integrations yield

$$\begin{aligned}
&\Psi^{(2)}(\mathbf{r}_1, t_1; \mathbf{r}_2, t_2) \\
&= -\frac{(2\pi)^2}{i^2 \Delta r_1 \Delta r_2} k_0 g_{a,k_0} \sigma(k_0) q_0 g_{b,q_0} \sigma(q_0) \\
&\quad \times \int_0^\infty \int_0^\infty dk dq \frac{e^{-ickt_1}\left(e^{-ik\Delta r_1} - e^{ik\Delta r_1}\right) e^{-icqt_2}\left(e^{-iq\Delta r_2} - e^{iq\Delta r_2}\right)}{\left[i(ck + cq - \omega_{ac}) - \Gamma_a/2\right]\left[i(cq - \omega_{bc}) - \Gamma_b/2\right]} \\
&\quad + (1 \leftrightarrow 2). \qquad (21.B.5)
\end{aligned}$$

We proceed to do the k-integration. The limits of integration on k and q may be extended to $-\infty$ because of the sharply peaked Lorentzians. The k contour is closed in the lower half-plane, as in Fig. 21.20(a), when $ct_1 > \Delta r_1$ and in the upper half-plane when $ct_1 < \Delta r_1$. Hence, neglecting the unphysical incoming waves which go as $e^{-ik\Delta r_1}$, we find

$$\begin{aligned}
&\Psi^{(2)}(\mathbf{r}_1, t_1; \mathbf{r}_2, t_2) \\
&= i\frac{(2\pi)^3 k_0 g_{a,k_0} \sigma(k_0) q_0 g_{b,q_0} \sigma(q_0)}{\Delta r_1 \Delta r_2} e^{-(i\omega_{ac} + \Gamma_a/2)\left(t_1 - \frac{\Delta r_1}{c}\right)} \Theta\left(t_1 - \frac{\Delta r_1}{c}\right) \\
&\quad \times \int_{-\infty}^\infty dq \frac{\exp\left\{-icq\left[\left(t_2 - \frac{\Delta r_2}{c}\right) - \left(t_1 - \frac{\Delta r_1}{c}\right)\right]\right\}}{i(cq - \omega_{bc}) - \Gamma_b/2} + (1 \leftrightarrow 2). \\
&\qquad (21.B.6)
\end{aligned}$$

In a like manner we carry out the q-integration via the contour of Fig. 21.20(b) to find

$$\begin{aligned}
&\Psi^{(2)}(\mathbf{r}_1, t_1; \mathbf{r}_2, t_2) \\
&= \frac{-\kappa}{\Delta r_1 \Delta r_2} \exp\left[-(i\omega_{ac} + \Gamma_a/2)\left(t_1 - \frac{\Delta r_1}{c}\right)\right] \Theta\left(t_1 - \frac{\Delta r_1}{c}\right) \\
&\quad \times \exp\left\{-(i\omega_{bc} + \Gamma_b/2)\left[\left(t_2 - \frac{\Delta r_2}{c}\right) - \left(t_1 - \frac{\Delta r_1}{c}\right)\right]\right\} \\
&\quad \times \Theta\left[\left(t_2 - \frac{\Delta r_2}{c}\right) - \left(t_1 - \frac{\Delta r_1}{c}\right)\right] + (1 \leftrightarrow 2), \qquad (21.B.7)
\end{aligned}$$

where $\kappa = (2\pi)^4 k_0 g_{a,k_0} \sigma(k_0) q_0 g_{b,q_0} \sigma(q_0)$.

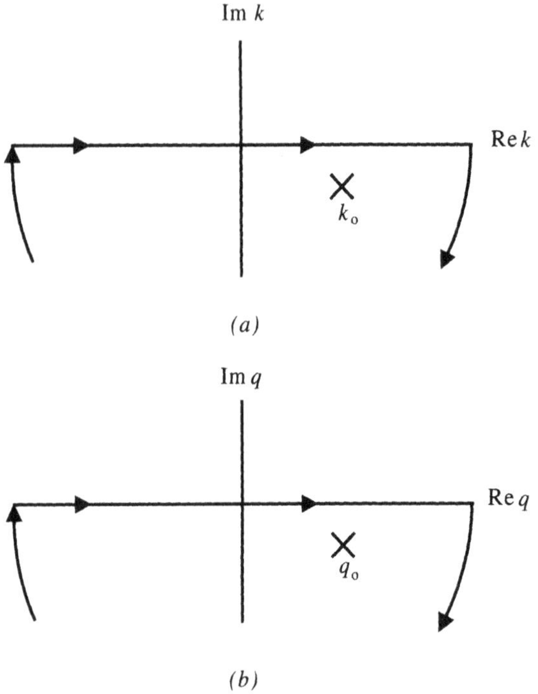

Fig. 21.20
(a) Figure illustrating pole at $k_0 = (\omega_{ac} - i\Gamma_a/2)/c - q$ and contour for $ct_1 > \Delta r_1$. Integral in Eq. (21.B.5) vanishes when $ct_1 < \Delta r_1$. (b) Pole at $q_0 = (\omega_{bc} - i\Gamma_b/2)/c$ and contour for $(t_2 - \Delta r_2/c) > (t_1 - \Delta r_1/c)$.

21.C Calculation of the joint count probability in Franson–Chiao interferometry

According to Eq. (6.4.13), the two-photon state produced by cascade emission from a single excited three-level atom at the origin is given by the entangled state

$$|\Psi\rangle = \sum_{\mathbf{k},\mathbf{q}} \frac{-g_{a\mathbf{k}}\, g_{b\mathbf{q}}}{[\mathrm{i}(\nu_k + \nu_q - \omega_{ac}) - \Gamma_a/2][\mathrm{i}(\nu_q - \omega_{bc}) - \Gamma_b/2]} |1_{\mathbf{k}}, 1_{\mathbf{q}}\rangle \,. \tag{21.C.1}$$

The state $|\Psi\rangle$ allows us to calculate the joint count probability that both detectors in the experimental configuration of Fig. 21.8 are excited (within an infinite coincidence window), which is proportional to the time integral

$$P(1,2) = \int_0^\infty \mathrm{d}t_1 \int_0^\infty \mathrm{d}t_2\, G^{(2)}(1,2) \tag{21.C.2}$$

over the Glauber second-order correlation function

$$G^{(2)}(1,2) = \Psi^*(1,2)\,\Psi(1,2) \tag{21.C.3}$$

with the "probability amplitude for a photo-electron pair" $\Psi(1,2)$ as in Eq. (21.4.10).

Using Eq. (21.4.10) and the notation of Fig. 21.9, we find after a somewhat lengthy calculation for $S_1 = S_2 \equiv S \approx L_1 = L_2 \equiv L$, $T \equiv (L - S)/c$, and $S/c \gg \Gamma_a^{-1} + \Gamma_b^{-1}$

$$\begin{aligned}
&P(1,2)\\
&= \int_0^\infty dt_1 \int_0^\infty dt_2\\
&\quad \left\{ \left[(|A_{12}|^2 + |B_{12}|^2 + |C_{12}|^2 + |D_{12}|^2) + (1 \leftrightarrow 2)\right]_1 \right.\\
&\quad + \left[(A_{12}^* D_{12} + \text{c.c.}) + (1 \leftrightarrow 2)\right]_2\\
&\quad + \left[(A_{12}^* B_{12} + C_{12}^* D_{12} + \text{c.c.}) + (1 \leftrightarrow 2)\right]_3\\
&\quad + \left[(A_{12}^* C_{12} + B_{12}^* D_{12} + \text{c.c.}) + (1 \leftrightarrow 2)\right]_4\\
&\quad + \left[(B_{12}^* C_{12} + \text{c.c.}) + (1 \leftrightarrow 2)\right]_5\\
&\quad \left. + \left[A_{12}^* B_{21} + C_{12}^* A_{21} + C_{12}^* B_{21} + C_{12}^* D_{21} + D_{12}^* B_{21} + \text{c.c.}\right]_6 \right\}
\end{aligned} \tag{21.C.4}$$

$$\begin{aligned}
&= \left[\frac{2\kappa}{\Gamma_a\Gamma_b}\right]_1 + \left[\frac{\kappa e^{-\Gamma_a T/2}}{\Gamma_a\Gamma_b}\cos(\omega_{ac}T)\right]_2 + \left[\frac{2\kappa e^{-\Gamma_b T/2}}{\Gamma_a\Gamma_b}\cos(\omega_{bc}T)\right]_3\\
&\quad + \left[\frac{2\kappa e^{-(\Gamma_a+\Gamma_b)T/2}}{\Gamma_a\Gamma_b}\cos(\omega_{ab}T)\right]_4 + \left[\frac{\kappa e^{-(\Gamma_a/2+\Gamma_b)T}}{\Gamma_a\Gamma_b}\cos(\Delta T)\right]_5\\
&\quad + \left[\frac{4\kappa e^{-\Gamma_b T/2}}{4\Delta^2 + \Gamma_a^2}\left\{\cos(\omega_{bc}T) - \frac{2\Delta}{\Gamma_a}\sin(\omega_{bc}T) + \frac{e^{-\Gamma_b T/2}}{2}\right.\right.\\
&\qquad\qquad - e^{-\Gamma_a T/2}\left[\cos(\omega_{ab}T) - \frac{2\Delta}{\Gamma_a}\sin(\omega_{ab}T)\right]\\
&\qquad\qquad \left.\left. - \frac{e^{-(\Gamma_a+\Gamma_b)T/2}}{2}\left[\cos(\Delta T) - \frac{2\Delta}{\Gamma_a}\sin(\Delta T)\right]\right\}\right]_6,
\end{aligned} \tag{21.C.5}$$

where $\kappa = (\kappa_a\kappa_b/2SL)^2$, $\omega_{ac} = \omega_{ab} + \omega_{bc}$, $\Delta = \omega_{ab} - \omega_{bc}$. The square bracket expressions in (21.C.4), enumerated by subscripts, are the origin of the corresponding terms in (21.C.5).*

* The various terms are discussed by Meyer, Agarwal, Huang, and Scully [1994].

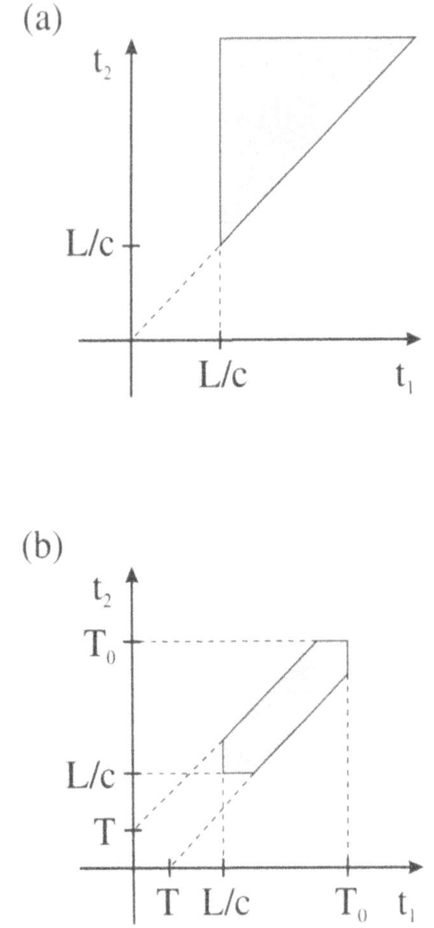

Fig. 21.21
(a) The integration region for Eq. (21.C.6), a term from Eq. (21.C.4) for the joint count probability $P(1,2)$ in Franson–Chiao interferometry.
(b) The integration region for $C^*_{12}B_{21}$ of Eq. (21.C.4).

As an e[illegible]on over t_1 and t_2 for the term

$$A^*_{12}D_{12} + \text{c.c.} = \frac{\kappa}{2}e^{(\Gamma_b-\Gamma_a)t_1-\Gamma_b t_2+\Gamma_a(L+S)/2c}\cos(\omega_{ac}T) \times \Theta(t_1 - L/c)\Theta(t_2 - t_1), \tag{21.C.6}$$

which leads to sum frequency oscillations (with the oscillation frequency ω_{ac}). The integration region is depicted in Fig. 21.21(a). We obtain

$$\begin{aligned}\int_0^\infty \mathrm{d}t_1 \int_0^\infty \mathrm{d}t_2\,(A^*_{12}D_{12} + \text{c.c.}) &= \frac{\kappa}{2}e^{\Gamma_a(L+S)/2c}\int_{L/c}^\infty \mathrm{d}t_1 e^{(\Gamma_b-\Gamma_a)t_1}\int_{t_1}^\infty \mathrm{d}t_2 e^{-\Gamma_b t_2} \\ &= \frac{\kappa e^{-\Gamma_a T/2}}{2\Gamma_a\Gamma_b}\cos(\omega_{ac}T).\end{aligned} \tag{21.C.7}$$

The integration regions for some of the other terms are slightly more complicated, especially if we choose to have a finite coincidence window. For example, Figure 21.21(b) shows the integration region for the term $C_{12}^* B_{21}$ with a coincidence window of time T_0.

Problems

21.1 Using the results of Section 6.3, show that

$$\left| \langle 0| \left[E^{(+)}(\mathbf{r}, t), \gamma_1^\dagger \right] |0\rangle \right|$$
$$= \frac{\mathscr{E}_0}{\Delta r_1} \Theta \left(t - \frac{\Delta r_1}{c} \right) e^{-i(t-\Delta r_1/c)(\omega - i\Gamma/2)},$$

where

$$\gamma_1^\dagger = \sum_{\mathbf{k}} \frac{g_{\mathbf{k}} \exp(-i\mathbf{k} \cdot \mathbf{r}_1)}{(v_k - \omega) + i\Gamma/2} a_{\mathbf{k}}^\dagger.$$

21.2 Show that

$$\langle \gamma_2 | \gamma_1 \rangle = \frac{\sin(k_0 |\mathbf{r}_2 - \mathbf{r}_1|)}{k_0 |\mathbf{r}_2 - \mathbf{r}_1|} e^{-|\mathbf{r}_2 - \mathbf{r}_1|\Gamma/(2c)},$$

where $|\gamma_i\rangle$ $(i = 1, 2)$ are the single-photon states given in Eq. (21.1.4).

References and bibliography

Other which-path detectors

M. O. Scully, R. Shea, and J. McCullen, *Phys. Rep.* **43**, 486 (1978).
M. O. Scully, B.-G. Englert, and J. Schwinger, *Phys. Rev. A* **40**, 1775 (1989).

Scattering of light from atoms – quantum interference, photon correlation, and Bell's theorem

M. O. Scully, *Laser Spectroscopy V*, ed. M. Kellar and B. Stoichoff (1981), p. 41.
M. O. Scully and K. Drühl, *Phys. Rev. A* **25**, 2208 (1982).
M. Hillery and M. O. Scully, in *Quantum Optics, Experimental Gravity, and Measurement Theory*, ed. P. Meystre and M. O. Scully (Plenum, New York 1983), p. 65.
L. Mandel, *Phys. Rev. A* **28**, 929 (1983).
H. Fearn and R. Loudon, *Opt. Commun.* **64**, 485 (1987); *J. Opt. Soc. Am. B* **6**, 917 (1989).
G. M. Meyer, G. S. Agarwal, H. Huang, and M. O. Scully, in *Proceedings of the NATO Advanced Study Institute on Electron Theory and Quantum Electrodynamics*, (Edirne, Turkey, 1994).

Interference experiments

M. A. Horne, A. Shimony, and A. Zeilinger, *Phys. Rev. Lett.* **62**, 2209 (1989).
U. Eichmann, J. C. Bergquist, J. J. Bollinger, J. M. Gilligan, W. M. Itano, D. J. Wineland, and M. G. Raizen, *Phys. Rev. Lett.* **70**, 2359 (1993).

Quantum interference in parametric down-converters

C. K. Hong, Z. Y. Ou, and L. Mandel, *Phys. Rev. Lett.* **59**, 2044 (1987).
Z. Y. Ou and L. Mandel, *Phys. Rev. Lett.* **61**, 50 (1988).
Y. H. Shih and C. O. Alley, *Phys. Rev. Lett.* **61**, 2921 (1988).
J. D. Franson, *Phys. Rev. Lett.* **62**, 2205 (1989).
Z. Y. Ou, L. J. Wang, X. Y. Zou, and L. Mandel, *Phys. Rev. A* **41**, 566 (1990).
Z. Y. Ou, X. Y. Zou, L. J. Wang, and L. Mandel, *Phys. Rev. Lett.* **65**, 321 (1990); *Phys. Rev. A* **42**, 2957 (1990).
G. Rarity and P. R. Tapster, *Phys. Rev. Lett.* **64**, 2495 (1990).
X. Y. Zou, L. J. Wang, and L. Mandel, *Phys. Rev. Lett.* **67**, 318 (1991).
P. G. Kwiat, A. M. Steinberg, and R. Y. Chiao, *Phys. Rev. A* **45**, 7729 (1992).
T. E. Keiss, Y. H. Shih, A. V. Sergienko, and C. O. Alley, *Phys. Rev. Lett.* **71**, 3833 (1993).
M. O. Scully and U. W. Rathe, *Opt. Commun.* **110**, 373 (1994).
Y. H. Shih and A. V. Sergienko, *Phys. Rev.* A **50**, 2564 (1994).
Y. H. Shih, A. V. Sergienko, M. H. Rubin, T. E. Keiss, and C. O. Alley, *Phys. Rev.* A **50**, 23 (1994).

T. J. Herzog, P. G. Kwiat, H. Weinfurter, and A. Zeilinger, *Phys. Rev. Lett.* **75**, 3034 (1995).
P. G. Kwait, K. Mattle, H. Weinfurter, A. Zeilenger, A. V. Sergienko, and Y. Shih, *Phys. Rev. Lett.* **75**, 4337 (1995).
D. V. Strekalov, A. V. Sergienko, D. N. Klyshko, and Y. H. Shih, *Phys. Rev. Lett.* **74**, 3600 (1995).
T. B. Pittman, D. V. Strekalov, A. Migdall, M. H. Rubin, A. V. Sergienko, and Y. H. Shih, *Phys. Rev. Lett.* **77**, 1917 (1996).

Line narrowing via correlated spontaneous emission

H. Huang and J. H. Eberly, *J. Mod. Opt.* **40**, 915 (1993).
M. O. Scully, U. W. Rathe, S. Chang, G. S. Agarwal, *Opt. Commun.* **136**, 39 (1997).
U. W. Rathe and M. O. Scully, *Lett. Math. Phys.* **34**, 297 (1995).
U. W. Rathe and M. O. Scully, *Annals of the New York Academy of Sciences* **755**, 28 (1995).

Index

Lightning Source UK Ltd.
Milton Keynes UK
UKOW04f0557290117
293095UK00008B/297/P

9 780521 435956